NUMERICAL ANALYSIS WITH ALGORITHMS AND PROGRAMMING

NUMERICAL ANALYSIS WITH ALGORITHMS AND PROGRAMMING

Santanu Saha Ray

CRC Press
Taylor & Francis Group
Boca Raton London New York

CRC Press is an imprint of the
Taylor & Francis Group, an **informa** business
A CHAPMAN & HALL BOOK

CRC Press
Taylor & Francis Group
6000 Broken Sound Parkway NW, Suite 300
Boca Raton, FL 33487-2742

Printed and bound by CPI UK on sustainably sourced paper
Version Date: 20160401

International Standard Book Number-13: 978-1-4987-4174-3 (Hardback)

Library of Congress Cataloging-in-Publication Data

Names: Saha Ray, Santanu.
Title: Numerical analysis with algorithms and programming / Santanu Saha Ray.
Description: Boca Raton : Taylor & Francis, 2016. | "A CRC title."
Identifiers: LCCN 2016003230 | ISBN 9781498741743 (alk. paper)
Subjects: LCSH: Numerical analysis--Problems, exercises, etc. | Engineering mathematics--Problems, exercises, etc. | Numerical analysis--Data processing--Problems, exercises, etc. | Algorithms--Problems, exercises, etc.
Classification: LCC QA297 .S24 2016 | DDC 518--dc23
LC record available at http://lccn.loc.gov/2016003230

Visit the Taylor & Francis Web site at
http://www.taylorandfrancis.com

and the CRC Press Web site at
http://www.crcpress.com

Dedication

This work is dedicated to my grandfather the late Sri Chandra Kumar Saha Ray, my parents, my beloved wife Lopamudra, and my son Sayantan.

Contents

Preface

The utmost aim of this book is to provide an extensive study of expedient procedures for obtaining useful acceptable solutions to the desired accuracy of mathematical problems occurring in disciplines of science and engineering and for acquiring useful information from available solutions. The main feature of this book is its multidisciplinary aspect involving science, computer science, engineering, and mathematics. It can be used as a text for various disciplines of science and engineering in which this subject is pertinent to a given curriculum. This book provides a comprehensive foundation of numerical analysis that includes substantial ground work in the algorithms of computation, approximation, numerical solutions of nonlinear equations, interpolation, numerical differentiation and integration, numerical solutions of linear algebraic equation systems, numerical solutions of ordinary differential equations, eigenvalue problems in matrix, approximations of functions, and numerical solutions of partial differential equations (PDEs). In addition, a brief introduction to the finite element method (FEM) is also provided. To gain practical knowledge for applications of the methods, MATHEMATICA® programs are provided at the end of almost each and every method. In very few cases, programs have been intentionally excluded and left for the exercises of the readers. It is a comprehensive textbook in which the subject matter is presented in a well-organized and systematic manner.

This book has 11 chapters. Each chapter presents a thorough analysis of the theory, principles, and methods, followed by many illustrative examples. There are a large number of problems given as exercises for the students to practice, in order to enhance their knowledge and skill involved while solving these problems. In addition, this book may be helpful for numerical computation with high-end digital computer. Nowadays, computational experience is very important and indispensable, and it manifests perception to enter into the deeper sense for most of the theoretical aspects. This experience has prompted the real impetus for preparation of this book.

It is a well-known fact that analytical solutions of many important physical problems are not readily available; hence, numerical approximate solutions are the only alternative. These solutions will contain errors, for which a discussion at the beginning of the book is a must.

Chapter 1 provides fundamental concepts of errors in numerical computations. Some basic ideas about numerical stability, condition number, and convergence are also discussed.

Chapter 2 presents detailed discussions of several methods for solving nonlinear algebraic and transcendental equations. The order of convergence and condition of convergence have also been described in detail. Numerical solutions of systems of nonlinear equations have also been discussed using different numerical methods.

In Chapter 3, different types of interpolation formulas are presented. All the forward, backward, and central interpolation formulas have been included explicitly. Cubic spline has been described exhaustively. A pretty clear idea may be perceived from the pictorial representation of the cubic spline graph. The error analysis of the cubic spline has been presented very elegantly.

Numerical differentiation and numerical integration have been discussed rigorously in Chapters 4 and 5, respectively. Different numerical integration formulas have been derived from Newton–Cotes quadrature formula as well as from the interpolation formula. The numerical integration procedures are also graphically presented. Richardson extrapolation along with Romberg integration and different types of Gauss quadrature formulas are extensively discussed. Different numerical methods for double integration have also been presented. At the end of Chapter 5, theories and properties of Bernoulli polynomials, Bernoulli numbers, and the Euler–Maclaurin formula have been discussed.

Chapter 6 is devoted to the presentation of various direct methods along with their algorithms and operational counts or time complexities. Several important iterative methods have been presented along with their algorithms and convergence analysis. Ill-conditioned systems are also discussed in detail. Moreover, for solving tridiagonal systems, the Thomas algorithm has also been included.

In Chapter 7, several numerical methods including single-step and multistep ones are discussed in detail for numerical solutions of differential equations. Various types of Runge–Kutta (R–K) methods are derived according to their order. Particular attention has been paid to the derivation of the fourth order R–K method. The Runge–Kutta–Fehlberg method is discussed rigorously. The multistep methods, especially the Adams–Bashforth and the Adams–Moulton predictor–corrector methods are also discussed. The numerical solutions of differential equation systems are also taken into consideration. Various methods, such as finite difference method, shooting method, collocation method, and the Galerkin method, have been implemented for solving boundary value problems. Algorithms are also presented with implementations of various numerical techniques for numerical solutions of differential equations. Stability analysis of single-step and multistep methods has also been presented extensively. The fundamental concepts of stiff differential equations, that is, A-stability and L-stability, are also well explored. To get rid of the lack of organized algorithms for the implementation of the methods discussed in this chapter, an emphasis is laid on details regarding the description of algorithms with its applications through computer programs, along with solved examples, which yields the lengthiest chapter in this book.

In Chapter 8, various methods have been included to determine the eigenvalues of a square matrix. The Householder's method and *QR* method are elegantly described with great intent. A meticulous effort has been paid for the comprehensive descriptions of the power method, inverse power method, and other relevant methods for finding eigenvalues of a square matrix, because the matrix is a very good tool for solving engineering problems.

Chapter 9 deals with the approximation of functions. First, Bernstein polynomials and their properties are introduced. Next, least square curve fitting techniques are presented. The Gram–Schmidt orthogonalization process is also discussed to find a set of orthogonal polynomials. Special emphasis is laid on the minimax polynomial approximation and its corresponding theorems with proofs. Function approximation by cubic B-spline has been also introduced. In the end, the Padé approximation has been discussed considerably.

Chapter 10 deals with indispensable salient methods for the numerical solutions of parabolic, hyperbolic, and elliptic PDEs. At the end of the descriptions of each technique, algorithms with MATHEMATICA® programs with some solved problems have been provided for the better perception and comprehension of these different numerical techniques applied to the PDEs. Moreover, for numerical solutions of two-dimensional parabolic PDEs, an alternating direction implicit method has also been described in detail with its algorithm, along with the corresponding computer program. In the end, the stability analysis of the numerical schemes has been explored.

Finally, in Chapter 11, a brief introduction to the FEM is also presented. The FEM constitutes a general tool for the numerical solution of PDEs appearing in applied science and engineering. The notable work of L. Rayleigh (1870) and W. Ritz (1909) on variational methods and the weighted-residual approach adopted by B. G. Galerkin (1915) and others form the theoretical groundwork for the FEM. For this purpose, the Rayleigh–Ritz method is explained intensively with its algorithm and the corresponding computer program. Furthermore, relevant literature has also been referred to for further details regarding the mathematical theory and implementation of the FEM.

This book contains sufficient materials to be adjudged as a text book with respect to the scenario that it covers the numerical analysis course thoroughly. In this book, every concept is illustrated by worked-out examples. In addition, it contains many exercises, covering various application areas. A number of computer programs have been developed by using MATHEMATICA®, with the aid of implementation of the corresponding algorithms related to numerical methods.

The bibliographic material to the relevant literature has been provided to serve as helpful sources for further study and research for interested readers.

Santanu Saha Ray

Acknowledgments

I express my deepest sense of sincere gratitude to Dr. R. K. Bera, former professor and head, Department of Science, National Institute of Technical Teacher's Training and Research, Kolkata, India, and Dr. K. S. Chaudhuri, FNA (India), FIMA (United Kingdom), former professor and emeritus fellow, Department of Mathematics, Jadavpur University, Kolkata, India, for their encouragement in the preparation of this book. I acknowledge, with the deepest thanks, the valuable suggestions provided by Professor Lokenath Debnath, Department of Mathematics, The University of Texas–Pan American, Edinburg, Texas; Professor Abdul-Majid Wazwaz, Department of Mathematics and Computer Science, Saint Xavier University, Chicago, Illinois. I am also highly thankful to Professor Edward J. Allen, Texas Tech University, Lubbock, Texas; Professor Xiao-Jun Yang, China University of Mining and Technology, Xuzhou, Jiangsu, China; Professor Vasily E. Tarasov, Lomonosov Moscow State University, Moscow, Russia; Professor Mehdi Dehghan, Department of Applied Mathematics, Faculty of Mathematics and Computer Science, Amirkabir University of Technology, Tehran, Iran; Professor Santos Bravo Yuste, Departamento de Física, Facultad de Ciencias, Universidad de extremadura, Badajoz, Spain; Professor Dumitru Baleanu, Çankaya University, Ankara, Turkey; Professor Siddhartha Sen, Department of Electrical Engineering, Indian Institute of Technology, Kharagpur, India; Professor J. A. Tenreiro Machado, Institute of Engineering, Polytechnic of Porto, Portugal; and Professor Shantanu Das, senior scientist, Reactor Control Division, Bhabha Atomic Research Centre, Mumbai, India.

It is not out of place to acknowledge the sincere and meticulous efforts of my PhD scholar students who had worked hard to assist me in the preparation of this book.

I also express my sincere gratitude to the director of National Institute of Technology (NIT), Rourkela, India, for his kind cooperation and support. The moral support received from my colleagues at the NIT, Rourkela, India, is also acknowledged.

I express my sincere thanks to everyone involved in the preparation of this book. I take the opportunity to acknowledge the efforts of all those who were directly or indirectly involved in helping me throughout this difficult mission.

Moreover, I am especially grateful to the Taylor & Francis Group/CRC Press for their cooperation in all aspects for the production of this book.

Last, but not the least, special mention should be made of my parents and my beloved wife Lopamudra, for her patience, unequivocal support, and generous encouragement throughout the period of my work. In addition, I acknowledge the allowance of my only son Sayantan for not sparing my pleasant company in his childhood playtime and also in his valuable educational activities.

I look forward to receive comments and suggestions from students, teachers, and researchers.

Santanu Saha Ray

Acknowledgments

Author

Dr. Santanu Saha Ray is an associate professor in the Department of Mathematics, National Institute of Technology, Rourkela, India. Dr. Saha Ray obtained his PhD in 2008 from Jadavpur University, Kolkata, India and MCA (Masters of Computer Applications) degree in 2001 from the Indian Institute of Engineering Science and Technology (IIEST; formerly the Bengal Engineering College) Sibpur, India. He completed a master's degree in applied mathematics at the Calcutta University, Kolkata, India, in 1998 and a bachelor's (honors) degree in mathematics at St. Xavier's College, Kolkata, India, in 1996.

Dr. Saha Ray has 15 years of teaching experience at the undergraduate and postgraduate levels. He has also about 14 years of research experience in various fields of applied mathematics. He has published many peer-reviewed research papers in numerous fields and various international SCI journals of repute, including *Applied Mathematics and Computation*, *Communication in Nonlinear Science and Numerical Simulation*, *Transaction ASME Journal of Applied Mechanics*, *Journal of Computational and Nonlinear Dynamics*, *Computers and Mathematics with Applications*, *Journal of Computational and Applied Mathematics*, *Mathematical Methods in the Applied Sciences*, *Computers & Fluids*, *Physica Scripta*, *Communications in Theoretical Physics*, *Nuclear Engineering and Design*, *International Journal of Nonlinear Science and Numerical Simulation*, *Annals of Nuclear Energy*, and *Journal of Mathematical Chemistry*. For a detail citation overview, the reader may refer to *Scopus*. To date, he has more than 100 research papers published in journals of international repute, including more than 80 SCI journal papers.

He is the author of *Graph Theory with Algorithms and Its Applications: in Applied Science and Technology* (Springer, 2013) and *Fractional Calculus with Applications for Nuclear Reactor Dynamics* (CRC Press, 2015). He is the editor-in-chief for the Springer journal entitled *International Journal of Applied and Computational Mathematics*.

He has contributed papers on several topics, such as fractional calculus, mathematical modeling, mathematical physics, stochastic modeling, integral equations, and wavelet methods. He is a member of the Society for Industrial and Applied Mathematics and the American Mathematical Society.

He was the principal investigator of a Board of Research in Nuclear Sciences project, with grants from the Bhabha Atomic Research Centre, Mumbai, India. He was also the principal investigator of a research project financed by the Department of Science and Technology, Government of India. He is the principal investigator of another two research projects financed by the Board of Research in Nuclear Sciences, Bhabha Atomic Research Centre, Mumbai, India and National Board for Higher Mathematics, Department of Atomic Energy, Government of India, respectively.

A research scholar was awarded with PhD from the National Institute of Technology, Rourkela, India under his supervision. In addition, he is supervising five research scholars, including three senior research fellowship scholars. He had also been the lead guest editor of the international SCI journals of the Hindawi Publishing Corporation, New York.

1 Errors in Numerical Computations

1.1 INTRODUCTION

Numerical analysis is a subject that involves computational methods for studying and solving mathematical problems. It is a branch of mathematics and computer science that creates, analyzes, and implements algorithms for solving mathematical problems numerically. Numerical methods usually emphasize on the implementation of the numerical algorithms. The aim of these methods is, therefore, to provide systematic techniques for solving mathematical problems numerically. Numerical methods are well suited for solving mathematical problems by using modern digital computers, which are very fast and efficient in performing arithmetic operations. The process of solving problems using high precision digital computers generally involves starting from an initial data; the concerned appropriate algorithms are then executed to yield the required results. Inevitably, the numerical data and the methods used are approximate ones. Hence, the error in the computed result may certainly be caused by the errors in the data, or the errors in the method, or both.

We begin this chapter with some preliminary mathematical theorems that are invariably useful concerning the study of numerical analysis. This chapter presents the various kinds of errors that may occur in a problem. The representation of numbers in computers is introduced. The general results on the propagation of errors in numerical computation are also given. Finally, the concepts of stability and conditioning of problems and a brief idea of convergence of numerical methods are also introduced.

1.2 PRELIMINARY MATHEMATICAL THEOREMS

Theorem 1.1: Intermediate Value Theorem

Let $f(x)$ be a real valued continuous function on the finite interval $[a,b]$ and define

$$m = \operatorname*{Inf}_{a \le x \le b} f(x), \quad M = \operatorname*{Sup}_{a \le x \le b} f(x)$$

Then, for any number μ in $[m, M]$, there exists at least one point ξ in $[a,b]$ for which

$$f(\xi) = \mu$$

In particular, there are points $\underline{\xi}$ and $\bar{\xi}$ in $[a,b]$ for which $m = f(\underline{\xi})$ and $M = f(\bar{\xi})$.

Theorem 1.2: Mean Value Theorem

Let $f(x)$ be a real valued continuous function on the finite interval $[a,b]$ and differentiable in (a,b). Then, there exists at least one point ξ in (a,b) for which

$$f(b) - f(a) = (b-a)f'(\xi)$$

Theorem 1.3: Integral Mean Value Theorem

Let $w(x)$ be nonnegative and integrable on $[a,b]$, and let $f(x)$ be continuous on $[a,b]$. Then,

$$\int_a^b w(x)f(x)dx = w(\xi)\int_a^b f(x)dx, \quad \text{for some} \quad \xi \in [a,b]$$

One of the most important and useful tools for approximating functions $f(x)$ by polynomials in numerical analysis is Taylor's theorem and the associated Taylor series. These polynomials expressed in Taylor series are used extensively in numerical analysis.

Theorem 1.4: Taylor's Theorem

Let $f(x)$ be a real valued continuous function on the finite interval $[a,b]$ and have $n+1$ continuous derivatives on $[a,b]$ for some $n \geq 0$, and let $x, x_0 \in [a,b]$. Then

$$f(x) = P_n(x) + R_{n+1}(x)$$

where:

$$P_n(x) = f(x_0) + \frac{(x-x_0)}{1!}f'(x_0) + \cdots + \frac{(x-x_0)^n}{n!}f^{(n)}(x_0)$$

$$R_{n+1}(x) = \frac{(x-x_0)^{n+1}}{(n+1)!}f^{(n+1)}(\xi), \quad a < \xi < b$$

Taylor's Theorem in Two Dimensions

Let $f(x,y)$ be a function of two independent variables x and y and suppose $f(x)$ possesses continuous partial derivatives of order n in some neighborhood N of a point (a,b) in the domain of definition of $f(x)$. Let $(a+h, b+k) \in N$, then there exists a positive number θ $(0 < \theta < 1)$, such that

$$f(a+h,b+k) = f(a,b) + \left(h\frac{\partial}{\partial x} + k\frac{\partial}{\partial y}\right)f(a,b) + \frac{1}{2!}\left(h\frac{\partial}{\partial x} + k\frac{\partial}{\partial y}\right)^2 f(a,b) + \cdots$$
$$+ \frac{1}{(n-1)!}\left(h\frac{\partial}{\partial x} + k\frac{\partial}{\partial y}\right)^{n-1} f(a,b) + R_{n+1}(x)$$

where

$$R_n(x) = \frac{1}{n!}\left(h\frac{\partial}{\partial x} + k\frac{\partial}{\partial y}\right)^n f(a+\theta h, b+\theta k), \quad 0 < \theta < 1$$

$R_n(x)$ is called the remainder after n terms and the theorem is called Taylor's theorem with remainder or Taylor's expansion about the point (a,b).

The above theorems are present in most of the elementary calculus textbooks, and thus their proofs have been omitted.

1.3 APPROXIMATE NUMBERS AND SIGNIFICANT FIGURES

1.3.1 Significant Figures

Significant figures is any one of the digits 1, 2, 3,…, and 0.

In the number 0.00134, the significant figures are 1, 3, and 4. The zeros are used here merely to fix the decimal point and are, therefore, not significant. But in the number 0.1204, the significant figures are 1, 2, 0, and 4. Similarly, 1.00317 has six significant figures.

1.3.1.1 Rules of Significant Figures

Rule 1: All nonzero digits 1, 2, …, 9 are significant.

Rule 2: Zeros between nonzero digits are significant, for example, in reading the measurement 9.04 cm, the zero represents a measured quantity, just as 9 and 4 and is, therefore, a significant number. Similarly, in another example, there are four significant numbers in the number 1005.

Rule 3: Zeros to the left of the first nonzero digit in a number are not significant, for example, 0.0026. Also, in the measurement 0.07 kg, the zeros are used merely to locate the decimal point and are, therefore, not significant.

Rule 4: When a number ends in zeros that are to the right of the decimal point, then the zeros are significant, for example, in the number 0.0200, there are three significant numbers. Another example is that in reading the measurement 11.30 cm, the zero is an estimate and represents a measured quantity. It is therefore significant. Thus, zeros to the right of the decimal point and at the end of the number are significant figures.

Rule 5: When a number ends in zeros that are not to the right of the decimal point, then zeros are not necessarily significant, for example, if a distance is reported as 1200 ft, one may assume two significant figures. However, reporting measurements in scientific notation removes all doubt, since all numbers written in scientific notation are considered significant.

$$1200\,\text{ft} \quad 1.2\times10^3\,\text{ft} \quad \text{Two significant figures}$$

$$1200\,\text{ft} \quad 1.20\times10^3\,\text{ft} \quad \text{Three significant figures}$$

$$1200\,\text{ft} \quad 1.200\times10^3\,\text{ft} \quad \text{Four significant figures}$$

Thus, we may conclude that if a zero represents a measured quantity, it is a significant figure. If it merely locates the decimal point, it is not a significant figure.

1.4 ROUNDING OFF NUMBERS

Numbers are rounded off so as to cause the least possible errors. The general rule for rounding off a number to n significant digits is as follows:

Discard all digits to the right of the nth place. If the discarded number is less than half a unit in the nth place, leave the nth digit unchanged; if the discarded number is greater than half a unit in the nth place, add 1 to the nth digit. If the discarded number is exactly half a unit in the nth place, leave the nth digit unaltered if it is an even number, but increase it by 1 if it is an odd number.

When a number is rounded off according to the above-stated rule, then it is said to be correct to n significant digits.

To illustrate, the following numbers are corrected to four significant figures:

27.1345 becomes 27.13

27.8793 becomes 27.88

27.355 becomes 27.36

27.365 becomes 27.36

We will now proceed to present the classification of the ways by which errors are involved in numerical computation. Let us start with some simple definitions of error.

1.4.1 Absolute Error

Let x_T be the exact value or true value of a number and x_A be its approximate value, then $|x_T - x_A|$ is called the *absolute error*, E_a. Therefore, absolute error can be defined by

$$E_a \equiv |x_T - x_A|$$

1.4.2 Relative and Percentage Errors

Relative error is defined by

$$E_r \equiv \left|\frac{x_T - x_A}{x_T}\right|, \quad \text{provided } x_T \neq 0 \text{ or } x_T \text{ is not close to zero}$$

The percentage relative error is defined by

$$E_p \equiv E_r \times 100 = \left|\frac{x_T - x_A}{x_T}\right| \times 100, \quad \text{provided } x_T \neq 0 \text{ or } x_T \text{ is not close to zero}$$

1.4.2.1 Measuring Significant Digits in x_A

We say x_A has k significant digits with respect to x_T if the error $|x_T - x_A|$ has magnitude less than or equal to 5 in the $(k+1)$th digit of x_T, counting to the right starting from the first nonzero digit in x_T.

Example 1.1

$$x_T = \frac{2}{3} \text{ and } x_A = 0.667, \quad |x_T - x_A| = 0.000333333$$

Since, the error is less than 5 in the fourth digit to the right of the first nonzero digit in x_T, we say that x_A has three significant digits with respect to x_T.

Example 1.2

$$x_T = 21.956 \text{ and } x_A = 21.955, \quad |x_T - x_A| = 0.001$$

In this case, x_A has four significant digits with respect to x_T, since the error is less than 5 in the fifth place to the right of the first nonzero digit in x_T.

Equation 1.1 is sometimes used in measuring number of significant digits in x_A. If

$$\left|\frac{x_T - x_A}{x_T}\right| \leq 5 \times 10^{-k-1} \tag{1.1}$$

then x_A has k significant digits with respect to x_T. In order to verify this, consider $0.1 \leq |x_T| < 1$. Then Equation 1.1 implies that

$$|x_T - x_A| \leq 5 \times 10^{-k-1} |x_T| < 0.5 \times 10^{-k}$$

Thus, x_A has k significant digits with respect to x_T.

For general x_T, taking

$$x_T = \hat{x}_T \times 10^d$$

where d is an integer, with $0.1 \leq |\hat{x}_T| < 1$, the proof is essential the same. Furthermore, Equation 1.1 is a sufficient condition rather than a necessary condition, in order that x_A has k significant digits with respect to x_T.

1.4.3 Inherent Error

Inherent error is that quantity which is already present in the statement of the problem before its solution. This error arises either due to the straight assumptions in the mathematical forms of the problem or due to the physical measurements of the parameters of problem. Inherent error cannot be completely eliminated but can be minimized by selecting better data or by employing high precision computer computations.

1.4.4 Round-Off and Chopping Errors

A number x is said to be rounded correct to a d-decimal digit number $x^{(d)}$ if the error satisfies

$$\left|x - x^{(d)}\right| \leq \frac{1}{2} \times 10^{-d} \tag{1.2}$$

The error arising out of rounding of a number, as defined in Equation 1.2, is known as *round-off error.*

Suppose an arbitrarily given real number x with the following representation:

$$x = .d_1 d_2 \ldots d_k d_{k+1} \ldots \times b^e \tag{1.3}$$

where

b is the base, $d_1 \neq 0$, d_2, …, d_k are integers and satisfies $0 \leq d_i \leq b-1$

the exponent e is such that $e_{\min} \leq e \leq e_{\max}$

The fractional part $.d_1 d_2 \ldots d_k d_{k+1} \ldots$ is called the *mantissa*, and it lies between −1 and 1.

Now, the floating point number $fl(x)$ in k-digit mantissa standard form can be obtained in the following two ways:

1. *Chopping*: In this case, we simply discard the digits $d_{k+1}, d_{k+2}, \ldots$ in Equation 1.3, and obtain

$$fl(x) = .d_1 d_2 \ldots d_k \times b^e \tag{1.4}$$

2. *Rounding*: In this case, $fl(x)$ is chosen as the normalized floating point number nearest to x, together with the rule of symmetric rounding, according to which, if the truncated part be exactly half a unit in the kth position, then if the kth digit be odd, it is rounded up to an even digit and if it is even, then it is left unchanged.

Thus, the relative error for k-digit mantissa standard form representation of x becomes

$$\left|\frac{x - fl(x)}{x}\right| \le \begin{cases} b^{1-k}, & \text{for chopping} \\ \frac{1}{2}b^{1-k}, & \text{for rounding} \end{cases} \tag{1.5}$$

Therefore, the bound on the relative error of a floating point number is reduced by half when rounding is used instead of chopping. For this reason, on the most of the computers rounding is used.

Now, if we write,

$$fl(x) = x(1+\delta)$$

where $\delta = \delta(x)$, depends on x, is called the *machine epsilon*, then from Equation 1.5 we have

$$|\delta(x)| \le \begin{cases} b^{1-k}, & \text{for chopping} \\ \frac{1}{2}b^{1-k}, & \text{for rounding} \end{cases} \tag{1.6}$$

Example 1.3

Approximate values of 1/6 and 1/13, correct to four decimal places are 0.1667 and 0.0769, respectively. Find the possible relative error and absolute error in the sum of 0.1667 and 0.0769.

Solution:

Let $x = 0.1667$ and $y = 0.0769$.

The maximum absolute error in each x and y is given by

$$\frac{1}{2}\times 10^{-4} = 0.00005$$

1. Relative error in $(x+y)_A$

$$E_r\left[(x+y)_A\right] = \left|\frac{(x+y)_T - (x+y)_A}{(x+y)_T}\right| \le \frac{E_a(x)}{(x+y)_T} + \frac{E_a(y)}{(x+y)_T}$$

$$\le \frac{0.00005}{0.1667} + \frac{0.00005}{0.0769} \le 0.000950135$$

2. Absolute error in $(x+y)_A$

$$E_a\left[(x+y)_A\right] = E_r(x+y)\left|(x+y)_T\right| \le 0.000950135 \times 0.2436 = 0.000231452886$$

Example 1.4

Evaluate the sum $S = \sqrt{2} + \sqrt{6} + \sqrt{7}$ correct to four significant digits and find its absolute and relative errors.

Solution:

We have

$$\sqrt{2} = 1.414,\ \sqrt{6} = 2.449,\ \text{and}\ \sqrt{7} = 2.646$$

Therefore, $S = 6.509$ and the maximum absolute error is

$$0.0005 + 0.0005 + 0.0005 = 0.0015$$

Therefore, the total absolute error shows that the sum is correct to three significant figures only. Hence, we take $S = 6.51$, and then the relative error is $0.0015/6.51 = 0.00023$.

Example 1.5

If the number $\pi = 4\tan^{-1}(1)$ is approximated using five significant digits, find the percentage relative error due to

1. Chopping
2. Rounding

Solution:

1. Percentage relative error due to chopping is given by

$$\left|\frac{\pi - 3.1415}{\pi}\right| \times 100 = 0.00294926\%$$

2. Percentage relative error due to rounding is given by

$$\left|\frac{\pi - 3.1416}{\pi}\right| \times 100 = 0.000233843\%$$

From the above errors, it may be easily observed that rounding reduces error.

1.5 TRUNCATION ERRORS

These are the errors due to approximate formulae used in the computations. Truncation errors result from the approximate formulae used, which are generally based on the truncated series. The study of this error is usually associated with the problem of convergence.

For example, let us assume that a function $f(x)$ and all its higher order derivatives with respect to the independent variable x at a point, say $x = x_0$, are known. Now in order to find the function value at a neighboring point, say $x = x_0 + \Delta x$, one can use the Taylor series expansion for the function $f(x)$ about $x = x_0$ as

$$f(x) = f(x_0) + (x - x_0) f'(x_0) + \frac{(x - x_0)^2}{2!} f''(x_0) + \cdots \tag{1.7}$$

The right-hand side of Equation 5.1 is an infinite series, and one has to truncate it after some finite number of terms to calculate $f(x_0 + \Delta x)$ either by using a computer or by manual calculations.

If the series is truncated after n terms, then it is equivalent to approximating $f(x)$ with a polynomial of degree $n-1$. Therefore, we have

$$f(x) \approx P_{n-1}(x) = f(x_0) + (x - x_0) f'(x_0) + \frac{(x - x_0)^2}{2!} f''(x_0) + \cdots + \frac{(x - x_0)^{n-1}}{(n-1)!} f^{(n-1)}(x_0) \tag{1.8}$$

The truncated error is given by

$$E_T(x) = f(x) - P_{n-1}(x) = \frac{(x - x_0)^n}{n!} f^{(n)}(\xi) \tag{1.9}$$

Now, let

$$M_n(x) = \max_{a \le \xi \le x} \left| f^{(n)}(\xi) \right| \tag{1.10}$$

Then the bound of the truncation error is given by

$$\left| E_T(x) \right| \le \frac{M_n(x) \left| x - x_0 \right|^n}{n!} \tag{1.11}$$

If $h = x - a$, then the truncation error $E_T(x)$ is said to be of order $O(h^n)$.

Hence, from Equation 1.8, an approximate value of $f(x_0 + \Delta x)$ can be obtained with the truncation error estimate as given in Equation 1.9.

Example 1.6

Obtain a second degree polynomial approximation to

$$f(x) = (1+x)^{2/3}, \quad x \in [0, 0.1]$$

using Taylor series expansion about $x = 0$. Use the expansion to approximate $f(0.04)$ and find a bound of the truncation error.

Solution:

We have

$$f(x) = (1+x)^{2/3}, \quad f(0) = 1$$

$$f'(x) = \frac{2}{3(1+x)^{1/3}}, \quad f'(0) = \frac{2}{3}$$

$$f''(x) = -\frac{2}{9(1+x)^{4/3}}, \quad f''(0) = -\frac{2}{9}$$

$$f'''(x) = \frac{8}{27(1+x)^{7/3}}$$

Thus, the Taylor series expansion with the remainder term is given by

$$(1+x)^{2/3} = 1 + \frac{2}{3}x - \frac{x^2}{9} + \frac{4}{81}\frac{x^3}{(1+\xi)^{7/3}}, \quad 0 < \xi < 0.1$$

Therefore, the truncation error is

$$E_T(x) = (1+x)^{2/3} - \left(1 + \frac{2}{3}x - \frac{x^2}{9}\right) = \frac{4}{81}\frac{x^3}{(1+\xi)^{7/3}}$$

The approximate value of $f(0.04)$ is

$$f(0.04) \approx 1 + \frac{2}{3}(0.06) - \frac{(0.06)^2}{9} = 1.026488888889, \text{ correct to 12 decimal places}$$

The truncation error bound in $x \in [0, 0.1]$ is given by

$$|E_T| \le \max_{0\le x\le 0.1} \frac{4}{81}\frac{(0.1)^3}{(1+x)^{7/3}}$$

$$\le \frac{4}{81}(0.1)^3 = 0.493827\times 10^{-4}$$

The exact value of $f(0.04)$ correct up to 12 decimal places is 1.026491977549.

Example 1.7

The function $f(x) = \tan^{-1}x$ can be expanded as

$$\tan^{-1}x = x - \frac{x^3}{3} + \frac{x^5}{5} - \cdots + (-1)^{n-1}\frac{x^{2n-1}}{(2n-1)} + \cdots$$

Find n such that the series determines $\tan^{-1}1$ correct to eight significant figures.

Solution:

In an alternating series of positive terms, the truncation error is less than the absolute value of the first term of the truncated infinite series.

This implies that

$$\frac{1}{2n+1} \le \frac{1}{2}\times 10^{-8}$$

which yields

$$n > 10^8$$

Hence, n must be $10^8 + 1$ such that the given series determines $\tan^{-1}1$ correct to eight significant figures.

1.6 FLOATING POINT REPRESENTATION OF NUMBERS

A floating point number is represented in the following form

$$\pm .d_1d_2 \ldots d_k \times b^e$$

where

b is the base, $d_1 \ne 0$, d_2, …, d_k are digits or bits satisfying $0 \le d_i \le b-1$

k is the number of significant digits or bits, which indicates the precision of the number and the exponent e is such that $e_{\min} \le e \le e_{\max}$

The fractional part $d_1d_2 \ldots d_kd_{k+1} \ldots$ is called the *mantissa* or *significand* and it lies between −1 and 1.

1.6.1 Computer Representation of Numbers

Nowadays, usually digital computers are used for numerical computation. Most digital computers use floating point mode for storing real numbers.

The fundamental unit of data stored in a computer memory is called *computer word*. The number of bits a word can hold is called *word length*. The word length is fixed for a computer, although it varies from computer to computer. The typical word lengths are 16, 32, 64 bits, or higher bits.

TABLE 1.1
Effective Floating Point Range

	Effective Floating Point Range	
	Binary Number	**Decimal Number**
Single precision	$\pm(2 - 2^{-23}) \times 2^{127}$	$\sim\pm 10^{38.53}$
Double precision	$\pm(2 - 2^{-52}) \times 2^{1023}$	$\sim\pm 10^{308.25}$

The largest number that can be stored in a computer depends on word length. To store a number in floating point representation, a computer word is divided into three fields. The first part consists of one bit, called the *sign bit*. The next set of bits represent the exponent, and the final set of bits represent the mantissa. For example, in the single-precision floating point format, a 32-bit word is divided into three fields as follows: 1 bit for the sign, 8 bits for the exponent, and 23 bits for the mantissa. The exponent is an 8-bit signed integer from −128 to 127. On the other hand, in the double-precision floating point format, a 64-bit word is divided into three fields as follows: 1 bit for the sign, 11 bits for the exponent, and 52 bits for the mantissa.

In the normalized floating point representation, the exponent is so adjusted that the bit d_1 immediately after the binary point is always 1. Formally, a nonzero floating point number is in normalized floating point form if

$$\frac{1}{b} \leq \text{mantissa} < 1$$

The range of exponents that a typical computer can handle is very large. Table 1.1 shows the effective range of Institute of Electrical and Electronics Engineers (IEEE) floating point numbers.

If in a numerical computation a number lies outside the range, then the following cases arise:

1. *Overflow*: It occurs when the number is larger than the range specified in Table 1.1.
2. *Underflow*: It occurs when the number is smaller than the range specified in Table 1.1.

In case of underflow, the number is usually set to zero and computation continues. But in cases of overflow, the computer execution halts.

1.7 PROPAGATION OF ERRORS

In this section, we consider the effect of arithmetic operations that involve errors. Let, x_A and y_A be the approximate numbers used in the calculations. Suppose they are in error with the true values x_T and y_T, respectively.

Thus, we can write $x_T = x_A + \varepsilon_x$ and $y_T = y_A + \varepsilon_y$. Now, we examine the propagated error in some particular cases:

- *Case 1*: Multiplication
  In multiplication, for the error in $x_A y_A$, we have

$$\begin{aligned} x_T y_T - x_A y_A &= x_T y_T - (x_T - \varepsilon_x)(y_T - \varepsilon_y) \\ &= x_T \varepsilon_y + y_T \varepsilon_x - \varepsilon_x \varepsilon_y \end{aligned}$$

Thus, the relative error in $x_A y_A$ is

$$E_r(x_A y_A) = \left| \frac{x_T y_T - x_A y_A}{x_T y_T} \right|$$

$$= \left| \frac{\varepsilon_x}{x_T} + \frac{\varepsilon_y}{y_T} - \frac{\varepsilon_x}{x_T} \cdot \frac{\varepsilon_y}{y_T} \right|$$

$$\leq E_r(x_A) + E_r(y_A), \text{ provided } E_r(x_A), E_r(y_A) << 1$$

- *Case 2*: Division
  Proceeding with same argument as in multiplication, we get

$$E_r(x_A / y_A) \leq E_r(x_A) + E_r(y_A), \text{ provided } E_r(y_A) << 1$$

- *Case 3*: Addition and subtraction
  In cases of addition and subtraction, we have

$$(x_T \pm y_T) - (x_A \pm y_A) = (x_T - x_A) \pm (y_T - y_A) = \varepsilon_x \pm \varepsilon_y$$

Thus, the absolute error in $(x_A \pm y_A)$ is given by

$$E_a(x_A \pm y_A) \leq E_a(x_A) + E_a(y_A)$$

Notes:

1. The relative error in a product is bounded by the sum of the relative errors in the multiplicands; the relative error in a quotient is bounded by the sum of the relative errors in the dividend and divisor. The relative errors in multiplication or division do not propagate very rapidly.
2. The absolute error in the sum or difference of two numbers is bounded by the sum of the absolute values of the errors in the corresponding numbers. The relative error in $(x_A \pm y_A)$ can be quite poor in comparison with $E_r(x_A)$ and $E_r(y_A)$.

1.8 GENERAL FORMULA FOR ERRORS

Let us consider the differentiable function

$$u = f(x_1, x_2, \ldots, x_n) \tag{1.12}$$

of several independent variables $x_1, x_2, \ldots, x_n$.

Suppose that Δx_i represents error in each x_i, so that the error in u is given by

$$u + \Delta u = f(x_1 + \Delta x_1, x_2 + \Delta x_2, \ldots, x_n + \Delta x_n) \tag{1.13}$$

Taylor series expansion of the right-hand side of Equation 1.13 gives

$$u + \Delta u = f(x_1, x_2, \ldots, x_n) + \sum_{i=1}^{n} \frac{\partial f}{\partial x_i} \Delta x_i + O(\Delta x_i^2) \tag{1.14}$$

If we assume that the errors $\Delta x_1, \Delta x_2 \ldots \Delta x_n$ are relatively very small, we can neglect the second and higher powers of Δx_i. Thus, from Equation 1.14, we get

$$\Delta u \cong \sum_{i=1}^{n} \frac{\partial f}{\partial x_i}\Delta x_i = \frac{\partial f}{\partial x_1}\Delta x_1 + \frac{\partial f}{\partial x_2}\Delta x_2 + \cdots + \frac{\partial f}{\partial x_n}\Delta x_n \tag{1.15}$$

This is the general formula for computing the error of a function $u = f(x_1, x_2, \ldots, x_n)$.

The relative error E_r is then given by

$$E_r = \left|\frac{\Delta u}{u}\right| \cong \left|\frac{\partial f}{\partial x_1}\frac{\Delta x_1}{f} + \frac{\partial f}{\partial x_2}\frac{\Delta x_2}{f} + \cdots + \frac{\partial f}{\partial x_n}\frac{\Delta x_n}{f}\right|$$

Example 1.8

If $u = xyz^3 + 3/2x^2y^3$ and errors in x,y,z are 0.005, 0.001, 0.005, respectively, at $x = 2$, $y = 1$, and $z = 1$, compute the maximum absolute and relative errors in evaluating u.

Solution:

Let

$$u = f(x,y,z) = xyz^3 + \frac{3}{2}x^2y^3$$

We have

$$\frac{\partial f}{\partial x} = yz^3 + 3xy^3, \quad \frac{\partial f}{\partial y} = xz^3 + \frac{9}{2}x^2y^2, \text{ and } \frac{\partial f}{\partial z} = 3xyz^2$$

From Equation 1.15, we get

$$\Delta u \cong \frac{\partial f}{\partial x}\Delta x + \frac{\partial f}{\partial y}\Delta y + \frac{\partial f}{\partial z}\Delta z = (yz^3 + 3xy^3)\Delta x + (xz^3 + \frac{9}{2}x^2y^2)\Delta y + 3xyz^2\Delta z$$

Given that $x = 2$, $y = 1$, $z = 1$, $\Delta x = 0.005$, $\Delta y = 0.001$, and $\Delta z = 0.005$ and, therefore, we obtain

$$|\Delta u| \le \left|\left(yz^3 + 3xy^3\right)\Delta x\right| + \left|\left(xz^3 + \frac{9}{2}x^2y^2\right)\Delta y\right| + \left|3xyz^2\Delta z\right|$$

$$= 7 \times 0.005 + 20 \times 0.001 + 6 \times 0.005 = 0.085$$

Hence, the maximum absolute error in u is 0.085.

The maximum relative error in u is given by

$$(E_r)_{\max} = \max\left|\frac{\Delta u}{u}\right| \approx \frac{0.085}{8} = 0.010625$$

1.9 LOSS OF SIGNIFICANCE ERRORS

Loss of significance is an undesirable effect in calculations using floating point arithmetic. It occurs when an operation on two numbers increases relative error substantially more than it increases absolute error, for example, in subtracting two nearly equal numbers. The effect is that the number of significant digits in the result is reduced unacceptably.

Example 1.9

Consider the evaluation of

$$f(x) = x\left[\sqrt{x+1} - \sqrt{x}\right]$$

Let us tabulate the values of $f(x)$ for an increasing sequence of values of x (Table 1.2).

As x increases, there are fewer digits of accuracy in the computed value $f(x)$.

Let us see for $x = 100$,

$$\sqrt{100} = \underbrace{10.0000}_{\text{exact}}, \quad \sqrt{101} = \underbrace{10.0499}_{\text{rounded}}$$

where $\sqrt{101}$ is correctly rounded to six significant digits of accuracy.

Now,

$$\sqrt{x+1} - \sqrt{x} = \sqrt{101} - \sqrt{100} = 0.0499000$$

while the true value should be 0.0498756.

Thus, the calculation has a loss-of-significance error. The loss of accuracy was a by-product of

- The form of $f(x)$.
- The finite precision six-digit decimal arithmetic being used.

For this particular $f(x)$, there is a simple way to reformulate it and avoid the loss-of-significance error.

Let us take

$$f(x) = \frac{x}{\sqrt{x+1} - \sqrt{x}}$$

which on a six-significant digit decimal computation will imply

$$f(100) = 4.98756$$

which is correct to six significant digits.

TABLE 1.2
Results Obtained Using Six Significant Digits Decimal Computation

x	Computed $f(x)$	True $f(x)$
1	0.414210	0.414214
10	1.54340	1.54347
100	4.99000	4.98756
1000	15.8000	15.8074
10,000	50.0000	49.9988
100,000	100.000	158.113

Example 1.10

Consider the equation $x^2 - 22x + 1 = 0$ whose roots are

$$r_T^{(1)} = 11 + \sqrt{120} \quad \text{and} \quad r_T^{(2)} = 11 - \sqrt{120} \tag{1.16}$$

Taking $\sqrt{120} = 10.954$, we obtain

$$r_A^{(1)} = 11 + 10.954 = 21.954 \text{ and } r_A^{(2)} = 11 - 10.954 = 0.045$$

Using the exact answers,

$$E_r\left(r_A^{(1)}\right) = 2.054 \times 10^{-5}, \quad E_r\left(r_A^{(2)}\right) = 1.20 \times 10^{-2}$$

Let us take

$$x_T = x_A = 11, \quad y_T = \sqrt{120}, \quad y_A = 10.954$$

$$E_r(x_A) = 0, \quad E_r(y_A) = 4.11 \times 10^{-5}$$

The accuracy in $r_A^{(2)}$ is much less than that of x_A and y_A entering into the calculation. So the significant digits have been lost in the subtraction $r_A^{(2)} = x_A - y_A$. Thus, we have had a *loss-of-significance error* in calculating $r_A^{(2)}$. In $r_A^{(1)}$, the accuracy is of five significant digits, whereas we have only two significant digits in case of $r_A^{(2)}$.

To overcome this particular problem of accuracy, we calculate $r_A^{(2)}$ converting Equation 1.16 to

$$r_A^{(2)} = \frac{11 - \sqrt{120}}{1} \cdot \frac{11 + \sqrt{120}}{11 + \sqrt{120}} = \frac{1}{11 + \sqrt{120}}$$

Then we use

$$\frac{1}{11 + \sqrt{120}} = \frac{1}{21.954} = 0.0455498 = r_A^{(2)}$$

There are two errors here: that of $\sqrt{120} = 10.954$ and in the final division. But each of these will have small relative errors, and the new value of $r_A^{(2)}$ will be more accurate than the preceding one.

By exact calculations, we have $E_r\left(r_A^{(2)}\right) = 2.085 \times 10^{-5}$ which is clearly much better than the preceding one.

This new computation of $r_A^{(2)}$ exhibits the *loss-of-significance error* is due to the form of the calculation, not to errors in the data of the computation.

1.10 NUMERICAL STABILITY, CONDITION NUMBER, AND CONVERGENCE

In numerical analysis, mathematical problems may have solutions that are quite sensitive to small computational errors. In numerical computation, one usually starts with initial data, then one computes in turn all intermediate values, finally arriving at the results. If the initial data involves errors, they will generally affect the final result. Even each arithmetic operation performed by the computer introduces round-off error. There are also errors arising out because of conversion of numbers during machine representation. Sometimes these errors grow and the cumulative effect of the round-off errors becomes unbounded. In such cases, computation is said to be unstable. To reckon with this phenomenon, we now present the concept of condition number and stability.

1.10.1 Condition of a Problem

Every mathematical problem that we try to solve numerically is based on an expression of some form or another. In order to have reliable solution, we first need to ensure that the expression is continuous in its inputs, so that we would not get completely different results from slight changes in the input. Further, we also need to ensure that the expression is well-conditioned. If it is well-conditioned, then small changes in the input to the expression will lead to small changes in the results. If small changes in the input lead to large changes in the output, then we call the problem *ill-conditioned.* The exact cutoff between well- and ill-conditioned depends on the context of the problem and the uses of the results.

Example 1.11

Let us consider the following system of equations:

$$x+y=1$$

$$1.1x+y=2$$

The above equations represent the intersection of the nearly two parallel lines, and it has the solution $x=10$, $y=-9$. Now, if we change the coefficient 1.1 of x to 1.05 and solve again, then we get $x=20$ and $y=-19$. That means 5% change in input causes 100% change in output. Hence, we would say that it is an ill-conditioned problem.

Example 1.12

Suppose we want to evaluate the expression $y = x/(1-x)$. With $x=0.93$, we get $y=13.29$ correct to four significant figures, but with $x=0.94$ we get $y=15.67$ correct to four significant figures. So, we would say that this expression is ill-conditioned when evaluated for x near 0.93. On the other hand, if we use $x=-0.93$ and $x=-0.94$, we get values of −0.4819 and −0.4845 correct to four significant figures, respectively, and we would say that it is well-conditioned for x near −0.93.

To deal with this difficulty, we introduce a measure of stability called a *condition number.* The condition number is used to describe the sensitivity of a function $f(x)$ to small changes in the argument x. It is measured by the maximum relative change in the function value $f(x)$ caused by a unit change in x.

For many types of problems, we can compute a condition number that indicates the magnification of the changes. Thus, the condition number is defined by

Relative error in the output ≈ condition number × relative error in the input

For example, consider evaluating a function $f(x)$ at a point $x=x_0$. The input is x_0 and the output is $f(x_0)$. If we perturb the input to $x=x_0+\varepsilon$, then the output is $f(x_0+\varepsilon)$ and by applying the mean value theorem, we get

$$\frac{f(x_0+\varepsilon)-f(x_0)}{f(x_0)} = \frac{\varepsilon f'(\xi)}{f(x_0)} \approx \frac{x_0 f'(x_0)}{f(x_0)}\frac{\varepsilon}{x_0} \tag{1.17}$$

where ξ lies between x_0 and $x_0+\varepsilon$. The condition number of $f(x)$ at x_0 is given by

$$C_f(x) = \max\left|\frac{\text{realtive error in } f(x)}{\text{realtive error in } x}\right|$$

$$= \max\left|\frac{\left[f(x_0+\varepsilon)-f(x_0)\right]/f(x_0)}{(\varepsilon/x_0)}\right|$$

$$\approx \left|\frac{x_0 f'(x_0)}{f(x_0)}\right|, \quad \text{using Equation 1.17}$$

A large value of the condition number indicates that $f(x)$ is highly sensitive near x and the corresponding problem is called ill-conditioned. If the condition number is less than 1, then it is well-conditioned, and if the condition number is arbitrarily large, then it is ill-conditioned. To illustrate this, in case of previous example 1.12, for the function $f(x) = x/(1-x)$, we get the condition number $C_f(x) = 1/|1-x|$. Then clearly, $f(x)$ is ill-conditioned for x near 1 and well-conditioned for $x<0$ and $x>2$.

1.10.2 Stability of an Algorithm

In numerical analysis, numerical stability is a generally desirable property of numerical algorithms. In a stable algorithm, the cumulative effect of the round-off errors remain bounded. When we study an algorithm, our interest is the same as for an expression. We want small changes in the input to only produce small changes in the output. An algorithm or numerical process is called *stable* if this is true, and it is called *unstable* if large changes in the output are produced. Analyzing an algorithm for stability is more complicated than determining the condition of an expression, even if the algorithm simply evaluates the expression. This is because an algorithm consists of many basic calculations and each one must be analyzed and, due to round-off error, we must consider the possibility of small errors being introduced in every computed value. For example, evaluating $y = x/(1-x)$ may be accomplished into two steps: $t = 1-x$ and $y = x/t$. Also, we may consider both x and t to have small errors. An algorithm is stable if every step is well-conditioned. It is unstable if any step is ill-conditioned.

Example 1.13

Let us consider evaluation of the function $f(x) = \sqrt{1+x} - 1$ for x near 0. The condition number $C_f(x) = \left(\sqrt{1+x} + 1\right)/2\sqrt{1+x}$ yields $C_f(0) = 1$. This implies that it is well-conditioned. To compute this we adopt three steps: (1) $t_1 = 1+x$, (2) $t_2 = \sqrt{t_1}$, and (3) $f(x) = t_2 - 1$. Steps (1) and (2) are well-conditioned, since their condition numbers are 0 and ½, respectively. But the step (3) is ill-conditioned. Therefore, we may say that this algorithm is unstable.

The problem in the previous example is well-conditioned, but the algorithm applied to evaluate the function $f(x)$ is unstable. However, there is a different algorithm for evaluating the function $f(x)$, which would be stable. In this case, we can re-formulate the given function as $f(x) = x/\left(\sqrt{1+x} + 1\right)$. This new expression for the function $f(x)$ is equivalent to the original function, but by evaluating it in the obvious way, we have a stable algorithm.

1.11 BRIEF IDEA OF CONVERGENCE

While the concepts of stability and condition are associated with the propagation of round-off errors, the concepts of convergence is closely related to truncated errors. In numerical analysis, we frequently use iterative methods for which the investigation of convergence is very important. It is associated with the concept of rate of convergence, assuming that the iteration method converges. By analyzing the rate of convergence, it is possible to compare the speed of convergence of different iterative methods.

EXERCISES

1. Round-off the following numbers to three decimal places:
   a. 2.47235, b. 0.003568, c. 42.3085, and d. 9.77345.
2. Round-off the following numbers to four significant figures:
   a. 38.46235, b. 0.70029, c. 0.0022218, d. 19.235101, and e. 2.36425.
3. Round-off the following numbers to four significant digits
   a. 46.2356, b. 0.72679, c. 12.00678, and d. 0.025845.
4. Round-off the following numbers correct to four significant figures:
   a. 53908, b. 4.4995001, c. 10.62501, and d. 638150.
5. Calculate the value of $\sqrt{102} - \sqrt{101}$ correct to four significant figure.
6. Give exact ways of avoiding loss-of-significance errors in the following computations:
   a. $\log(x+1) - \log(x)$ Large x, b. $\sin(x) - \sin(y)$ $x \approx y$, c. $\tan(x) - \tan(y)$ $x \approx y$.

7. Convert the following binary numbers to a. decimal, b. octal, and c. hexadecimal: a. $(1010111)_2$, b. $(1110101)_2$, and c. $(1011101)_2$.
8. Determine the upper bound on the error for the function $f(x)=(x+1)^{1/2}$ using a polynomial approximation with third-order Taylor series (computed about $x_0=0$) for all $x\in[0, 1]$.
9. Convert to octal and hexadecimal: a. $(10101\ .01110)_2$, b. $(1100111\ .1011010)_2$, and c. $(1110\ .0111)_2$.
10. Show that if 5*D* accuracy is required to calculate the value of $\log(1+x)$ at $x=0.1, 0.5$ by using the following equation:

$$\log(1+x)=x-\frac{1}{2}x^2+\frac{1}{3}x^3-\cdots+\frac{1}{n}(-1)^{n-1}x^n+R_{n+1}$$

then the value of n is 4 and 13, respectively.
11. If $u=8xy^2/z^3$ and errors in x, y, z be 0.001, compute the maximum absolute and relative error when $x=y=z=1$.
12. For the following numbers x_A and x_T, how many significant digits are there in x_A with respect to x_T?
    a. $x_A=451.023$ and $x_T=451.01$
    b. $x_A=-0.045113$ and $x_T=-0.04518$
    c. $x_A=23.4213$ and $x_T=23.4604$
13. (a) Find the value of $\sqrt{102}+\sqrt{101}$ correct to three significant figures.
    (b) Calculate $\sqrt{25.11}-\sqrt{25.1001}$ correct to three significant figures, giving necessary steps.
14. Assume that $x_A=0.937$ has three significant digits with respect to x_T. Bound the relative error in x_A. For $f(x)=\sqrt{1-x}$, bound the error and relative error in $f(x_A)$ with respect to $f(x_T)$.
15. If $z=(1/8)xy^3$, find the percentage error in z when $x=3.14\pm0.0016$ and $y=4.5\pm0.05$.
16. Show that if 5*D* accuracy is required to calculate the values of $\sin x$ at $x=0.1, 1.0$ by using the following equation:

$$\sin x=x-\frac{1}{3!}x^3+\frac{1}{5!}x^5+\cdots+\frac{1}{(2n-1)!}(-1)^{n-1}x^{2n-1}+R_{n+1}$$

then the value of n is 2 and 4, respectively.
17. Find the relative maximum error in F, where $F=5x^2y/z^3$. Given $\Delta x=\Delta y=\Delta z=0.001$, where Δx, Δy, and Δz denote the errors in x, y, and z, respectively, such that $x=y=z=1$.
18. Show that the series $\sin x=x-(x^3/3!)+(x^5/5!)+\cdots$, cannot be used for computing sin 100 to 10 significant figures. What is the difficulty and how would you avoid it?
19. The Maclaurin expansion of $\sin x$ is given by

$$\sin x=x-\frac{x^3}{3!}+\frac{x^5}{5!}-\frac{x^7}{7!}+\cdots$$

where x is in radians. Use the series to compute the value of $\sin 25°$ to an accuracy of 0.001.
20. Given the equation $x^2-40x+1=0$, find its roots to five significant digits. Use $\sqrt{399}=19.975$, correctly rounded to five digits.
21. Find the relative error in the quotient $4.536/1.32$, the numbers being correct to the digits given.

22. Sometimes the loss of significance error can be avoided by rearranging terms in the function using a known identity from trigonometry or algebra. Find an equivalent formula for the following functions that avoids a loss of significance:
    a. $\ln(x+1)-\ln(x)$ for large x
    b. $\sqrt{x^2+1}-x$ for large x
    c. $\cos^2(x)-\sin^2(x)$ for $x \approx \frac{\pi}{4}$
    d. $\sqrt{\frac{1+\cos(x)}{2}}$ for $x \approx \pi$

2 Numerical Solutions of Algebraic and Transcendental Equations

2.1 INTRODUCTION

In this chapter, we shall consider the numerical computation problem of real root of a given equation $f(x)=0$, which may be algebraic, trigonometric, or transcendental. It will be assumed that the function $f(x)$ is continuously differentiable sufficient number of times.

All methods for numerical solution of equations will consist of two stages. In the first stage, called *location of root*, rough values of the root are obtained, and the second stage consists in improvement of rough value of each root to any desired degree of accuracy.

In the second stage, a method of improvement of the rough value of a root will generate a sequence of successive approximation or iterates $\{x_n | n \geq 0\}$, starting with initial rough value x_0 of the root α obtained in the first stage, such that $x_n \to \alpha$ as $n \to \infty$.

2.2 BASIC CONCEPTS AND DEFINITIONS

2.2.1 Sequence of Successive Approximations

Let $\{x_n\}$ be a sequence of successive approximations for a desired root α of the equation $f(x)=0$.

The error ε_n at the nth iteration is defined by

$$\varepsilon_n = \alpha - x_n \tag{2.1}$$

In addition, we defined h_n by

$$h_n = x_{n+1} - x_n = \varepsilon_n - \varepsilon_{n+1} \tag{2.2}$$

which may be considered as an approximation of ε_n.

The iteration process converges if and only if $\varepsilon_n \to 0$ as $n \to \infty$.

2.2.2 Order of Convergence

Definition:

If an iterative method converges and two constants $p \geq 1$ and $C > 0$ exists such that

$$\lim_{n\to\infty} \left| \frac{\varepsilon_{n+1}}{\varepsilon_n^p} \right| = C \tag{2.3}$$

then p is called the *order of convergence* of the method, and C is called *asymptotic error constant.*

A sequence of iterates $\{x_n | n \geq 0\}$ is said to converge with order of convergence $p \geq 1$ to a root α if

$$|\varepsilon_{n+1}| \leq k |\varepsilon_n|^p, \quad n \geq 0 \tag{2.4}$$

for some $k > 0$. If $p = 1$, then the sequence of iterates $\{x_n | n \geq 0\}$ is said to be linearly convergent. If $p = 2$, then the iterative method is said to have quadratic convergence.

2.3 INITIAL APPROXIMATION

2.3.1 Graphical Method

In this method, we plot the graph of the curve $y = f(x)$ on the graph paper; the point at which the curve crosses the x-axis gives the root of the equation $f(x) = 0$. Any value in the neighborhood of this point may be taken as an initial approximation to the required root.

Sometimes, the equation $f(x) = 0$ can be written in the form $g(x) = h(x)$, where the graphs of $y = g(x)$ and $y = h(x)$ may be conveniently drawn. In that case, the abscissae of the point of intersection of the two graphs gives the required root of $f(x) = 0$, and therefore any value in the neighborhood of this point can be taken as initial approximation to the required root. Figure 2.1 shows the graph of $y = \cos x - xe^x$ and Figure 2.2 shows the graphs of $y = x$ and $y = \cos x/e^x$. The abscissae of the point of intersection of these two graphs give the required root of the $f(x) = 0$.

Another commonly used method is based upon the *intermediate mean value theorem*, which states that

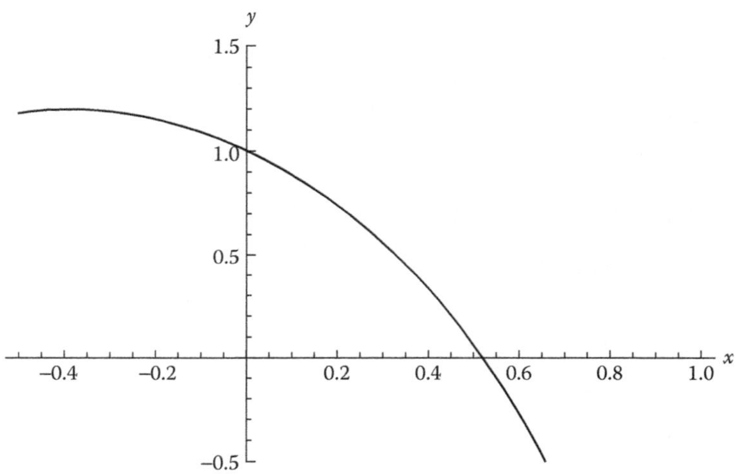

FIGURE 2.1 Graph of $y = \cos x - xe^x$.

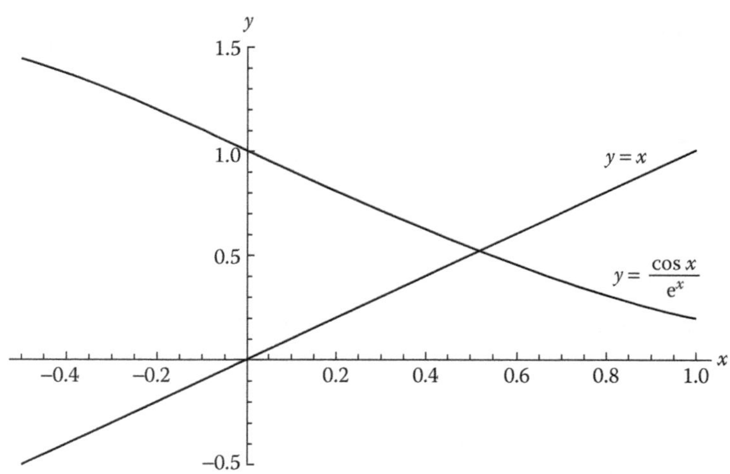

FIGURE 2.2 Graphs of $y = x$ and $y = \cos x/e^x$.

Theorem 2.1

If $f(x)$ be continuous function in the closed interval $[a,b]$ and c be any number such that $f(a) \le c \le f(b)$, then there is at least one number $\alpha \in [a,b]$ such that $f(\alpha) = c$.

2.3.2 Incremental Search Method

The incremental search method is a numerical method that is used when it is needed to find an interval of two values of x where the root is supposed to be existed. The incremental search method starts with an initial value x_0 and a sufficiently small interval Δx. It is supposed that we are going to search the location of the root in the x-axis from left to right.

We can find the value of x_1 easily with this following equation

$$x_1 = x_0 + \Delta x$$

If we convert that equation into an iterative one, we get

$$x_n = x_{n-1} + \Delta x$$

If $f(x_{n-1})f(x_n) < 0$, we can assure that there exists a root between the interval $[x_{n-1}, x_n]$.

We construct a table of values of $f(x)$ for various values of x and choose a suitable initial approximation to the root. This method is also known as *method of tabulation*.

Example 2.1

In order to obtain an initial approximation to the root of the equation $f(x) = \cos x - xe^x = 0$, we prepare the following table of values of $f(x)$ for known values of x:

x	0	0.5	1	1.5	2
$f(x)$	1	0.0532	−2.1780	−6.6518	−15.1942

From this table, we find that the equation $f(x)=0$ has at least one roct in the interval (0.5,1).

Example 2.2

Find real root of the equation $f(x)=10^x - x - 4 = 0$ correct to two significant digits by the method of tabulation.

Solution:

Let us tabulate the values of $f(x)$ in [0,1] with step size 0.1.

x	$f(x)$
0	−3
0.1	−2.841
0.2	−2.615
0.3	−2.305
0.4	−1.888
0.5	−1.338
0.6	−0.6189
0.7	0.3119
0.8	1.510
0.9	3.043
1	5

So the root of the given equation lies in (0.6,0.7). Since there is only one change of sign between 0.6 and 0.7, there is only one real root between 0.6 and 0.7.

We tabulate again between 0.6 and 0.7 with step length 0.01.

x	$f(x)$
0.6	−0.6189
0.61	−0.5362
0.62	−0.4513
0.63	−0.3642
0.64	−0.2748
0.65	−0.1832
0.66	−0.0891
0.67	0.007351
0.68	0.1063
0.69	0.2078
0.7	0.3119

From the above table, we can observe that the root lies between 0.66 and 0.67. Therefore, we may conclude that the value of the required root is 0.67 correct to two significant figures.

2.4 ITERATIVE METHODS

2.4.1 Method of Bisection

This is the simplest iterative method based on the repeated application of following *Bolzano's theorem* on continuity, which is a particular case of the intermediate value theorem.

Theorem 2.2: Bolzano's Theorem

If $f(x)$ be continuous in the closed interval $[a,b]$ and $f(a)$, $f(b)$ are of opposite signs, then there exists a number $\alpha \in [a,b]$ such that $f(\alpha)=0$, that is, there exists a real root α of the equation $f(x)=0$.

We first find a sufficiently small interval $[a_0,b_0]$ containing α by the method of tabulation or graphical method such that $f(a_0)$, $f(b_0)$ are of opposite signs.

Next, we proceed to generate the sequence $\{x_n\}$ of successive approximations as follows:

We set $x_0 = a_0$ or b_0 and $x_1 = (a_0+b_0)/2$ and compute $f(x_1)$.

Now, if $f(a_0)$, $f(x_1)$ are of opposite signs, we set $a_1 = a_0$ and $b_1 = x_1$ or $[a_1,b_1]=[a_0,x_1]$. Otherwise, if $f(x_1)$, $f(b_0)$ are of opposite signs, we set $a_1 = x_1$ and $b_1 = b_0$ or $[a_1,b_1]=[x_1,b_0]$ so that in either case $[a_1,b_1]$ contains the root α and $b_1 - a_1 = (b_0-a_0)/2$.

We again set $x_2 = (a_1+b_1)/2$ and repeat the above steps and so on.

In general, if the interval $[a_n,b_n]$ containing α has been obtained so that $f(a_n)f(b_n)<0$, then we set $x_{n+1} = (a_n+b_n)/2$.

If $f(a_n)f(x_{n+1})<0$, we set $a_{n+1}=a_n$, $b_{n+1}=x_{n+1}$.

Otherwise, if $f(x_{n+1})f(b_n)<0$, we set $a_{n+1}=x_{n+1}$, $b_{n+1}=b_n$.

So in either case $[a_{n+1},b_{n+1}]$ contains α that is, $f(a_{n+1})f(b_{n+1})<0$.

Therefore,

$$b_{n+1}-a_{n+1} = \frac{b_n-a_n}{2}$$

This implies

$$b_n - a_n = \frac{b_{n-1} - a_{n-1}}{2}$$
$$= \frac{b_{n-2} - a_{n-2}}{2^2} \tag{2.5}$$
$$\vdots$$
$$= \frac{b_0 - a_0}{2^n}$$

Since $x_n = a_n$ or b_n, approximation of ε_n that is, $|h_n| = |x_{n+1} - x_n| = (b_n - a_n)/2$.

Now error at nth iteration $|\varepsilon_n| = |\alpha - x_n| \leq |b_n - a_n|$

Therefore, we have

$$|\varepsilon_n| \leq |b_n - a_n| = \frac{|b_0 - a_0|}{2^n} \tag{2.6}$$

Consequently, $\varepsilon_n \to 0$ as $n \to \infty$.

Therefore, the iteration of bisection method invariably converges, that is, bisection method converges unconditionally.

2.4.1.1 Order of Convergence of the Bisection Method

We know

$$b_n - a_n = \frac{b_{n-1} - a_{n-1}}{2} = \frac{b_{n-2} - a_{n-2}}{2^2} = \cdots = \frac{b_0 - a_0}{2^n}$$

Therefore,

$$|\varepsilon_n| = |\alpha - x_n| \leq |b_n - a_n| \leq \frac{|b_0 - a_0|}{2^n} \tag{2.7}$$

This implies

$$|\varepsilon_{n-1}| \leq \frac{|b_0 - a_0|}{2^{n-1}}$$

Therefore,

$$\left|\frac{\varepsilon_{n+1}}{\varepsilon_n}\right| \cong \frac{1}{2} \tag{2.8}$$

Hence, the order of convergence for bisection method is 1.

Example 2.3

Find the real root of the equation $x\log_{10} x - 1.2 = 0$ correct to five significant figures using method of bisection.

Solution:

We first apply method of tabulation in order to find the location of rough value of the root (Table 2.1).

We note that $f(2) < 0$ and $f(3) > 0$. Thus, the given equation changes its sign within the interval [2,3]. Therefore, there exists at least one real root of the equation within [2,3].

Now, for the bisection method, we compute $x_{n+1} = (a_n + b_n)/2$.

If $f(a_n)f(x_{n+1}) < 0$, then the root lies between $[a_n, x_{n+1}]$; that is, we set $a_{n+1} = a_n$ and $b_{n+1} = x_{n+1}$.

Otherwise, if $f(x_{n+1})f(b_n) < 0$, then the root lies between $[x_{n+1}, b_n]$; that is, we set $a_{n+1} = x_{n+1}$ and $b_{n+1} = b_n$. The successive iterations have been presented in Table 2.2.

At the 14th step, the root lies between 2.740662 and 2.740631. Hence, the required real root is 2.7406 correct to five significant figures.

2.4.1.2 Advantage and Disadvantage of the Bisection Method

- *Advantage*: The bisection method is very simple because the iteration in each stage does not depend on the function values $f(x_n)$ but only on their signs. Also, the convergence of the method is unconditional. This method converges invariably, so it is surely convergent.
- *Disadvantage*: The method is very slow since it converges linearly. Consequently, it requires a large number of iteration to obtain moderate result up to certain degree of accuracy and thus this method is very laborious.

TABLE 2.1
Location of the Root

x	$f(x)$
1	−1.2
2	−0.6
3	0.23

TABLE 2.2
Table for Finding Real Root

n	a_n	b_n	$f(a_n)$	$f(b_n)$	$x_{n+1} = \frac{a_n + b_n}{2}$	$f(x_{n+1})$
0	2	3	−0.59794	0.231364	2.5	−0.20515
1	2.5	3	−0.20515	0.231364	2.75	0.00816491
2	2.5	2.75	−0.20515	0.00816491	2.625	−0.0997856
3	2.625	2.75	−0.0997856	0.00816491	2.6875	−0.046126
4	2.6875	2.75	−0.046126	0.00816491	2.718750	−0.0190585
5	2.71875	2.75	−0.0190585	0.00816491	2.734375	−0.0054662
6	2.734375	2.75	−0.0054662	0.00816491	2.742188	0.00134452
7	2.734375	2.742188	−0.0054662	0.00134452	2.738281	−0.00206205
8	2.738281	2.742188	−0.00206205	0.00134452	2.740234	−0.000359068
9	2.740234	2.742188	−0.000359068	0.00134452	2.741211	0.00049265
10	2.740234	2.741211	−0.000359068	0.00049265	2.740723	0.0000667723
11	2.740234	2.740723	−0.000359068	0.0000667723	2.740479	−0.000146153
12	2.740479	2.740723	−0.000146153	0.0000667723	2.740601	−0.0000396913
13	2.740601	2.740723	−0.0000396913	0.0000667723	2.740662	0.0000135402
14	2.740601	2.740662	−0.0000396913	0.0000135402	2.740631	−0.0000130756

2.4.1.3 Algorithm for the Bisection Method

Step 1: Start the program.
Step 2: Define the function $f(x)$.
Step 3: Enter the interval $[a,b]$ in which the root lies.
Step 4: If $f(a)f(b)<0$. Then go to Step 5 else Step 9.
Step 5: Calculate $x=(a+b)/2$.
Step 6: If $f(a)f(x)<0$, set $b=x$, otherwise if $f(x)f(b)<0$ set $a=x$.
Step 7: If $|a-b|<\varepsilon$, ε being the prescribed accuracy then go to Step 8 else Step 5.
Step 8: Print the value of x which is required root.
Step 9: Stop the program.

MATHEMATICA® Program for the Bisection Method (Chapter 2, Example 2.3)

```
f[x_]:=x²-10*Log[10,x]-3
Clear[a,b,x,n];
ε=0.000001;
a=2;
b=3;
x=a;
Print["n        an        bn        xn+1        f(an)        f(bn)        f(xn+1)"]
k=0;
While[Abs[f[x]] > ε, y=x; x=(a+b)/2;Print[k,"        ",N[a],"        ",
N[b],"        ",N[x,7],"        ",N[f[a]],"        ",N[f[b]],"        ",N[f[x]]];
If[f[a]*f[x]<0,b=x,If[f[x]*f[b]<0,a=x]];k=k+1];
```

Output:

n	an	bn	xn+1	f(an)	f(bn)	f(xn+1)
0	2.	3.	2.500000	-2.0103	1.22879	-0.7294
1	2.5	3.	2.750000	-0.7294	1.22879	0.169173
2	2.5	2.75	2.625000	-0.7294	0.169173	-0.300668
3	2.625	2.75	2.687500	-0.300668	0.169173	-0.0708285
4	2.6875	2.75	2.718750	-0.0708285	0.169173	0.0479088
5	2.6875	2.71875	2.703125	-0.0708285	0.0479088	-0.0117765
6	2.70313	2.71875	2.710938	-0.0117765	0.0479088	0.0179871
7	2.70313	2.71094	2.707031	-0.0117765	0.0179871	0.0030855
8	2.70313	2.70703	2.705078	-0.0117765	0.0030855	-0.00435046
9	2.70508	2.70703	2.706055	-0.00435046	0.0030855	-0.00063372
10	2.70605	2.70703	2.706543	-0.00063372	0.0030855	0.00122558
11	2.70605	2.70654	2.706299	-0.00063372	0.00122558	0.000295852
12	2.70605	2.7063	2.706177	-0.00063372	0.000295852	-0.000168953
13	2.70618	2.7063	2.706238	-0.000168953	0.000295852	0.0000634448
14	2.70618	2.70624	2.706207	-0.000168953	0.0000634448	-0.0000527553
15	2.70621	2.70624	2.706223	-0.0000527553	0.0000634448	5.34444*10^-6
16	2.70621	2.70622	2.706215	-0.0000527553	5.34444*10^-6	-0.0000237055
17	2.70621	2.70622	2.706219	-0.0000237055	5.34444*10^-6	-9.18055*10^-6
18	2.70622	2.70622	2.706221	-9.18055*10^-6	5.34444*10^-6	-1.91806*10^-6
19	2.70622	2.70622	2.706222	-1.91806*10^-6	5.34444*10^-6	1.71319*10^-6
20	2.70622	2.70622	2.706221	-1.91806*10^-6	1.71319*10^-6	-1.02437*10^-7

2.4.2 Regula-Falsi Method or Method of False Position

This method is the oldest method for finding the real root of an equation $f(x)=0$. It is also known as *method of chords* or *method of linear interpolation*. Like the bisection method, the false position method starts with two points a_0 and b_0 such that $f(a_0)$ and $f(b_0)$ are of opposite signs, which implies by the *intermediate value theorem* that the function f has a root in the interval $[a_0,b_0]$, assuming continuity of the function f.

We first determine by the method of tabulation a sufficiently small interval $[a_0,b_0]$ containing the only root α of the equation $f(x)=0$ such that $f(a_0)f(b_0)<0$.

Then, we approximate the portion of the curve $y=f(x)$ between the two points $(a_0,f(a_0))$ and $(b_0,f(b_0))$ by a straight line. It has been shown in Figure 2.3.

Now the equation of the chord joining the two points $(a_0,f(a_0))$ and $(b_0,f(b_0))$ is given by

$$y-f(b_0)=\frac{f(b_0)-f(a_0)}{b_0-a_0}(x-b_0) \tag{2.9}$$

The method consists in replacing the portion of the curve between the points $(a_0,f(a_0))$ and $(b_0,f(b_0))$ by means of the chord joining these two points, and taking the point of intersection of the chord with the x-axis as an approximation to the root. In this case, the point of intersection is obtained by putting $y=0$ in Equation 2.9. Thus, the first approximation x_1 to the root α is obtained as

$$x_1=\frac{a_0f(b_0)-b_0f(a_0)}{f(b_0)-f(a_0)} \tag{2.10}$$

Equivalently we may write

$$x_1=\frac{a_0\left|f(b_0)\right|+b_0\left|f(a_0)\right|}{\left|f(b_0)\right|+\left|f(a_0)\right|} \tag{2.11}$$

Next, we compute $f(x_1)$.

If $f(a_0)$ and $f(x_1)$ are of opposite signs, that is, $f(a_0)f(x_1)<0$, then the root lies in $[a_0,x_1]$, so we set $a_1=a_0$ and $b_1=x_1$.

Otherwise, if $f(x_1)$ and $f(b_0)$ are of opposite signs, that is, $f(x_1)f(b_0)<0$, then the root lies in $[x_1,b_0]$, we set $a_1=x_1$ and $b_1=b_0$, so that in either case $[a_1,b_1]$ contains α, that is, $f(a_1)f(b_1)<0$.

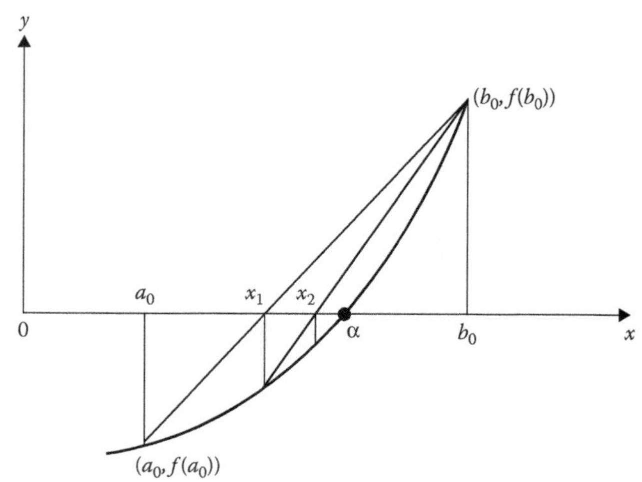

FIGURE 2.3 Graphical representation of the regula-falsi method.

We repeat the above step to obtain second approximation x_2, third approximation x_3, and so on.

In general, having obtained the interval $[a_n, b_n]$ containing α such that $f(a_n)f(b_n) < 0$. We obtain the $(n + 1)$th approximation x_{n+1} as

$$x_{n+1} = a_n - \frac{b_n - a_n}{f(b_n) - f(a_n)} f(a_n)$$

$$= \frac{b_n |f(a_n)| + a_n |f(b_n)|}{|f(a_n)| + |f(b_n)|}$$

Now, we set $a_{n+1} = a_n$ and $b_{n+1} = x_{n+1}$ if $f(a_n)f(x_{n+1}) < 0$.

Otherwise, we set $a_{n+1} = x_{n+1}$ and $b_{n+1} = b_n$ if $f(x_{n+1})f(b_n) < 0$, so that $[a_{n+1}, b_{n+1}]$ contains α, that is, $f(a_{n+1})f(b_{n+1}) < 0$.

Example 2.4

Find the real root of the equation $x\log_{10}x - 1.2 = 0$ correct to four decimal places using the regula-falsi method.

Solution:

We first apply method of tabulation in order to find the location of rough value of the root (Table 2.3).

We note that $f(2) < 0$ and $f(3) > 0$. Thus, the given equation changes its sign within the interval [2,3]. Therefore, there exists at least one real root of the equation within [2,3].

Now, for the regula-falsi method, we compute $x_{n+1} = \left(b_n|f(a_n)| + a_n|f(b_n)|\right) / \left(|f(a_n)| + |f(b_n)|\right)$. If $f(a_n)f(x_{n+1}) < 0$, then the root lies between $[a_n, x_{n+1}]$; that is, we set $a_{n+1} = a_n$ and $b_{n+1} = x_{n+1}$.

Otherwise, if $f(x_{n+1})f(b_n) < 0$, then the root lies between $[x_{n+1}, b_n]$; that is, we set $a_{n+1} = x_{n+1}$ and $b_{n+1} = b_n$. The successive iterations have been presented in Table 2.4.

Hence, the required real root is 2.7406 correct to four decimal places.

TABLE 2.3
Location of the Root

x	$f(x)$
1	−1.2
2	−0.6
3	0.23

TABLE 2.4
Table for Finding Real Root

n	a_n	b_n	$f(a_n)$	$f(b_n)$	$x_{n+1} = \frac{b_n\lvert f(a_n)\rvert + a_n\lvert f(b_n)\rvert}{\lvert f(a_n)\rvert + \lvert f(b_n)\rvert}$	$f(x_{n+1})$
0	2	3	−0.59794	0.231364	2.72101	−0.0170911
1	2.72101	3	−0.0170911	0.231364	2.74021	−0.000384056
2	2.74021	3	−0.000384056	0.231364	2.74064	−8.58134E-6
3	2.74064	3	−8.58134E-6	0.231364	2.74065	−1.91717E-7

Example 2.5

Find the real root of the equation $x\,e^{x}-\cos x=0$ correct to four significant figures using the regula-falsi method.

Solution:

We first apply method of tabulation in order to find the location of rough value of the root (Table 2.5).

We note that $f(0.5)<0$ and $f(0.6)>0$. Thus, the given equation changes its sign within the interval [0.5,0.6]. Therefore, there exists at least one real root of the equation within [0.5,0.6].

Now, for the regula-falsi method, we compute $x_{n+1}=\left(b_n|f(a_n)|+a_n|f(b_n)|\right)/\left(|f(a_n)|+|f(b_n)|\right)$. If $f(a_n)f(x_{n+1})<0$, then the root lies between $[a_n,x_{n+1}]$; that is, we set $a_{n+1}=a_n$ and $b_{n+1}=x_{n+1}$.

Otherwise, if $f(x_{n+1})f(b_n)<0$, then the root lies between $[x_{n+1},b_n]$; that is, we set $a_{n+1}=x_{n+1}$ and $b_{n+1}=b_n$. The successive iterations have been presented in Table 2.6. Hence, the required real root is 0.5178 correct to four significant figures.

2.4.2.1 Order of Convergence of the Regula-Falsi Method

It may be shown that the error at the $(n+1)$th step is related to the error in the nth step by the expression

$$\frac{\varepsilon_{n+1}}{\varepsilon_n}\cong A \tag{2.12}$$

where A is a constant depending on the function f. This shows that the sequence of successive iteration $\{x_n\}$ converges linearly to the root.

2.4.2.2 Advantage and Disadvantage of the Regula-Falsi Method

- *Advantage*: This method is very simple and does not require to calculate the derivative of $f(x)$. Moreover, this method is certainly convergent. Since the solution remains bracketed at each step, convergence is guaranteed, as was the case for the bisection method.

TABLE 2.5
Location of the Root

x	$f(x)$
0	−1
0.5	−0.0532219
0.6	0.267936
1	2.17798

TABLE 2.6
Table for Finding Real Root

n	a_n	b_n	$f(a_n)$	$f(b_n)$	$x_{n+1}=\dfrac{b_n\lvert f(a_n)\rvert+a_n\lvert f(b_n)\rvert}{\lvert f(a_n)\rvert+\lvert f(b_n)\rvert}$	$f(x_{n+1})$
0	0.5	0.6	−0.0532219	0.267936	0.516572	−0.00360274
1	0.516572	0.6	−0.00360274	0.267936	0.517679	−0.000238932
2	0.517679	0.6	−0.000238932	0.267936	0.517752	−0.0000158242
3	0.517752	0.6	−0.0000158242	0.267936	0.517757	−1.04793E-6

- *Disadvantage*: The method is first order and is exact for linear f. The method is very slow since it converges linearly. Also, the initial interval in which the root lies is to be chosen very small.

2.4.2.3 Algorithm for the Regula-Falsi Method

Step 1: Start the program.
Step 2: Define the function $f(x)$.
Step 3: Enter the interval $[a,b]$ in which the root lies.
Step 4: If $f(a)f(b)<0$. Then go to Step 5 else Step 9.
Step 5: Calculate $x=\left(a\left|f(b)\right|+b\left|f(a)\right|\right)/\left(\left|f(a)\right|+\left|f(b)\right|\right)$.
Step 6: If $f(a)f(x)<0$, set $b=x$, otherwise if $f(x)f(b)<0$ set $a=x$.
Step 7: If $\left|a-b\right|<\varepsilon$, ε being the prescribed accuracy. Then go to Step 8 else Step 5.
Step 8: Print the value of x which is required root.
Step 9: Stop the program.

MATHEMATICA® Program Implementing the Regula-Falsi Method (Chapter 2, Example 2.4)

```
f[x_]:=x*Log[10,x]-1.2
Clear[a,b,x,n];
ε=0.00000001;
a=2;
b=3;
x=a;
Print["n       an      bn      xn+1    f(an)   f(bn)   f(xn+1)"];
n=0;
While[Abs[f[x]]> ε, y=x; x=(b*Abs[f[a]]+a*Abs[f[b]])/(Abs[f[a]]+Abs[f[b]]);
Print[n,"      ",N[a,7],"            ",N[b,7],"                       ",N[x,7],"
",N[f[a]],"                        ",N[f[b]],"                        ",N[f[x]]];
If[f[a]*f[x]<0,b=x,If[f[x]*f[b]<0,a=x]];n++];
```

Output:

n	an	bn	xn+1	f(an)	f(bn)	f(xn+1)
0	2.000000	3.000000	2.72101	-0.59794	0.231364	-0.0170911
1	2.72101	3.000000	2.74021	-0.0170911	0.231364	-0.000384056
2	2.74021	3.000000	2.74064	-0.000384056	0.231364	-8.58134*10^-6
3	2.74064	3.000000	2.74065	-8.58134*10^-6	0.231364	-1.91717*10^-7
4	2.74065	3.000000	2.74065	-1.91717*10^-7	0.231364	-4.28317*10^-9

2.4.3 Fixed-Point Iteration

Let $[a_0,b_0]$ be an initial small interval containing the only root α of the given equation $f(x)=0$.

Let us rewrite given equation as

$$x=\phi(x) \tag{2.13}$$

so that the root α satisfies the equation

$$\alpha=\phi(\alpha) \tag{2.14}$$

We assume that, $\phi(x)$ is continuously differentiable sufficient number of times in $[a_0, b_0]$ such that for $x \in [a_0, b_0]$, $\phi(x) \in [a_0, b_0]$.

We take the successive approximation using the formula

$$x_{n+1} = \phi(x_n), \quad n \geq 0 \tag{2.15}$$

starting with $x_0 = a_0$ or b_0.

Substracting Equation 2.15 from Equation 2.14, we get

$$\alpha - x_{n+1} = \phi(\alpha) - \phi(x_n) = (\alpha - x_n)\phi'(\xi_n) \text{ (by using Lagrange's mean value theorem)}$$

where $\min\{\alpha, x_n\} < \xi_n < \max\{\alpha, x_n\}$.

Therefore, $\varepsilon_{n+1} = \varepsilon_n \phi'(\xi_n)$, where $\min\{\alpha, x_n\} < \xi_n < \max\{\alpha, x_n\}$. This is called *error equation.*

2.4.3.1 Condition of Convergence for the Fixed-Point Iteration Method

Theorem 2.3

Let α be the root of the equation $f(x) = 0$, that is, $x = \alpha$ be a solution of $x = \phi(x)$ and suppose $\phi(x)$ has a continuous derivative in some interval $[a_0, b_0]$ containing the root α. If $|\phi'(x)| \leq K < 1$ for all $x \in [a_0, b_0]$, then the fixed-point iteration process $x_{n+1} = \phi(x_n)$ converges with any initial approximation $x_0 \in [a_0, b_0]$.

Proof:

According to the hypothesis of the theorem

$$\alpha = \phi(\alpha) \tag{2.16}$$

We know the iteration scheme of the fixed-point method is

$$x_{n+1} = \phi(x_n) \tag{2.17}$$

By subtracting (2.17) from (2.16), we have

$$\alpha - x_{n+1} = \phi(\alpha) - \phi(x_n)$$

Since, $\varepsilon_{n+1} = \alpha - x_{n+1}$ is error at the $(n + 1)$th approximation.

So, $\varepsilon_{n+1} = (\alpha - x_n)\phi'(\xi)$, applying Lagrange's mean value theorem where $\min\{\alpha, x_n\} < \xi < \max\{\alpha, x_n\}$.

Therefore,

$$|\varepsilon_{n+1}| = |\alpha - x_n||\phi'(\xi)| \leq K|\alpha - x_n| \leq K|\varepsilon_n| \tag{2.18}$$

Again, it can be written as

$$|\varepsilon_n| \leq K|\varepsilon_{n-1}|$$

Therefore,

$$|\varepsilon_n| \leq K^2|\varepsilon_{n-2}| \leq K^3|\varepsilon_{n-3}| \cdots \leq K^n|\varepsilon_0|$$

Now, since $K < 1$, $K^n \to 0$ as $n \to \infty$

So, $|\varepsilon_n| \to 0$ as $n \to \infty$

This implies

$$|\alpha - x_n| \to 0 \quad \text{as} \quad n \to \infty$$

Therefore, $x_n \to \alpha$ as $n \to \infty$.

Hence, the condition of convergence of the fixed-point iteration is that $K < 1$ or $|\phi'(x)| < 1$ in $[a_0, b_0]$.

Corollary: It can be observed that the condition of convergence of the fixed-point iteration method is $|\phi'(x)| < 1$ in the neighborhood of α. This fact can be easily examined from Figure 2.4.

Example 2.6

Find the real root of the equation $x^3 + x - 1 = 0$ correct to three significant figures using the fixed-point iteration method.

Solution:

Let $f(x) = x^3 + x - 1 = 0$.

We first apply method of tabulation in order to find the location of rough value of the root (Table 2.7).

We note that $f(0) < 0$ and $f(1) > 0$. Thus the given equation changes its sign within the interval [0,1]. Therefore, there exists at least one real root of the equation within [0,1].

Now, we rewrite the equation $x^3 + x - 1 = 0$ as

$$x = \frac{1}{x^2 + 1} = \phi(x), \quad \text{say}$$

Hence, $\phi'(x) = -2x/(x^2 + 1)^2$, so that $|\phi'(x)| < 1$ in $0 < x < 1$. Therefore, according to Theorem 2.3, iteration in the fixed-point method certainly converges.

We choose initial approximation $x_0 = 0$. The successive iterations generated by the Equation 2.15 have been presented in Table 2.8, and it has been cited graphically in Figure 2.5.

Hence, the required real root of the given equation is 0.682 correct to three significant digits.

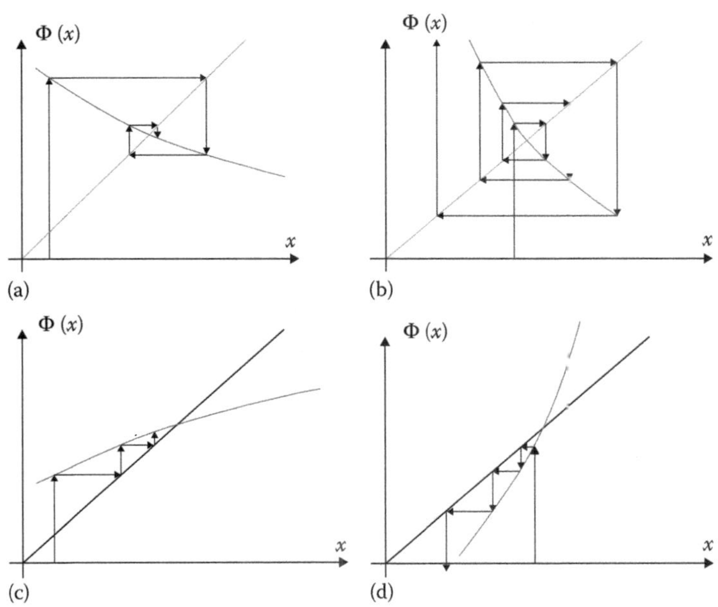

FIGURE 2.4 The conditional convergence of the fixed-point iteration method: (a) $-1 < \phi'(x) \le 0 \Rightarrow$ convergence, (b) $-1 < \phi'(x) \le 0 \Rightarrow$ divergence, (c) $0 < \phi'(x) \le 1 \Rightarrow$ convergence, and (d) $1 < \phi'(x) \Rightarrow$ divergence.

TABLE 2.7
Location of the Root

x	$f(x)$
0	−1
1	1

TABLE 2.8
Table for Finding Real Root

n	x_n	$x_{n+1} = \phi(x_n)$	$f(x_{n+1})$
0	0	1	1
1	1	0.5	−0.375
2	0.5	0.8	0.312
3	0.8	0.60976	−0.163535
4	0.60976	0.72897	0.116337
5	0.72897	0.653	−0.0685556
6	0.653	0.70106	0.045624
7	0.70106	0.67047	−0.0281294
8	0.67047	0.68988	0.0182119
9	0.68988	0.67754	−0.011432
10	0.67754	0.68537	0.0073189
11	0.68537	0.68039	−0.00462747
12	0.68039	0.68356	0.00294911
13	0.68356	0.68155	−0.00187003
14	0.68155	0.68282	0.00118959
15	0.68282	0.68201	−0.000755204

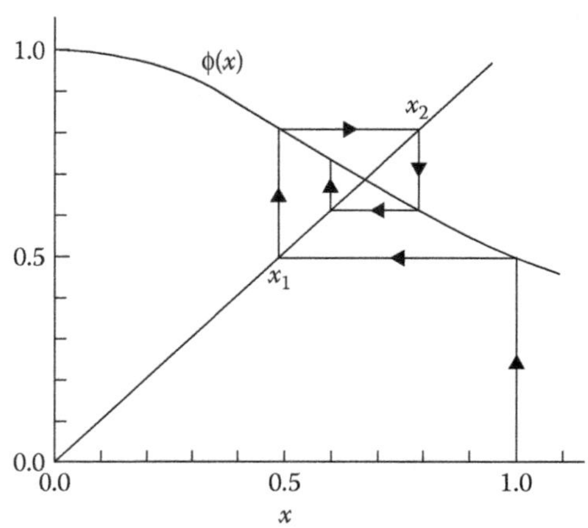

FIGURE 2.5 Successive iterates in the fixed-point iteration.

Example 2.7

Find the real root of the equation $10^x + x - 4 = 0$ correct to six decimal places using the fixed-point iteration method.

Solution:

Let $f(x) = 10^x + x - 4 = 0$.

We first apply method of tabulation in order to find the location of rough value of the root (Table 2.9).

We note that $f(0) < 0$ and $f(1) > 0$. Thus, the given equation changes its sign within the interval [0,1]. Therefore, there exists at least one real root of the equation within [0,1].

Now, we rewrite the equation $10^x + x - 4 = 0$ as $x = \log_{10}(4 - x) = \phi(x)$, say. Hence, $\phi'(x) = -1/(4 - x)\ln 10$ so that $|\phi'(x)| < 1$ in $0 < x < 1$. Therefore, according to Theorem 2.3, iteration in the fixed-point method certainly converges.

We choose initial approximation $x_0 = 0$. The successive iterations generated by the Equation 2.15 have been presented in Table 2.10.

Hence, the required real root of the given equation is 0.539179 correct to six decimal places.

2.4.3.2 Acceleration of Convergence: Aitken's Δ^2-Process

From the Equation 2.18, we have $|\alpha - x_{n+1}| = |\phi(\alpha) - \phi(x_n)| \le K|\alpha - x_n|$, where $K < 1$. Therefore, it is clear that the fixed-point iteration method is linearly convergent. This slow rate of convergence can be accelerated by using Aitken's method.

Now, from Equation 2.18, we have $\alpha - x_{n+1} = (\alpha - x_n)\phi'(\xi_n)$, where $\min\{\alpha, x_n\} < \xi_n < \max\{\alpha, x_n\}$. As $n \to \infty$, $\xi_n \to \alpha$ and therefore we may approximate $\phi'(\xi_n)$ by $\phi'(\alpha)$ for large n that is, $\phi'(\xi_n) \approx \phi'(\alpha)$

TABLE 2.9
Location of the Root

x	$f(x)$
0	−3
1	7

TABLE 2.10
Table for Finding Real Root

n	x_n	$x_{n+1} = \phi(x_n)$	$f(x_{n+1})$
0	0	0.60205999	0.60206
1	0.60205999	0.53121571	−0.0708443
2	0.53121571	0.54017729	0.00896159
3	0.54017729	0.53905384	−0.00112345
4	0.53905384	0.53919484	0.000140998
5	0.53919484	0.53917715	−0.0000176934
6	0.53917715	0.53917937	2.22033E-6
7	0.53917937	0.53917909	−2.78627E-7

Therefore, $\alpha - x_{n+1} \cong (\alpha - x_n)\phi'(\alpha)$ and consequently, $\alpha - x_{n+2} = (\alpha - x_{n+1})\phi'(\alpha)$.
Dividing, we get

$$\frac{\alpha - x_{n+1}}{\alpha - x_{n+2}} \cong \frac{\alpha - x_n}{\alpha - x_{n+1}}$$

which gives on simplification

$$\alpha \cong x_{n+2} - \frac{(x_{n+2} - x_{n+1})^2}{x_{n+2} - 2x_{n+1} + x_n} \tag{2.19}$$

Therefore, Equation 2.19 can be written in the simpler form

$$\alpha \cong x_{n+2} - \frac{(\Delta x_{n+1})^2}{\Delta^2 x_n} \tag{2.20}$$

where $\Delta x_{n+1} = x_{n+2} - x_{n+1}$ and $\Delta^2 x_n = \Delta(\Delta x_n) = x_{n+2} - 2x_{n+1} + x_n$.

This gives the approximate value of the root in terms of the previously known successive approximations. This modified iterative method is known as Aitken's Δ^2-process.

Example 2.8

Find the real root of the equation $\cos x - 2x + 3 = 0$ correct to three decimal places using (1) the fixed-point iteration method and (2) Aitken's Δ^2-method.

Solution:

Let $f(x) = \cos x - 2x + 3 = 0$.

We first apply method of tabulation in order to find the location of rough value of the root (Table 2.11).

We note that $f(1) > 0$ and $f(2) < 0$. Thus, the given equation changes its sign within the interval [1,2]. Therefore, there exists at least one real root of the equation within [1,2].

1. *Iterative method*

   Now, we rewrite the equation $\cos x - 2x + 3 = 0$ as

$$x = \frac{\cos x + 3}{2} = \phi(x), \text{ say}$$

   Hence, $\phi'(x) = (-1/2)\sin x$ so that $|\phi'(x)| < 1$ in $0 < x < 1$. Therefore, according to Theorem 2.3, iteration in the fixed-point method certainly converges.

We choose initial approximation $x_0 = 1$. The successive iterations generated by the Equation 2.15 have been presented in Table 2.12.

Hence, the required real root of the given equation is 1.524 correct to three decimal places.

TABLE 2.11
Location of the Root

x	$f(x)$
0	4
1	1.5403
2	−1.41615

TABLE 2.12
Table for Finding Real Root

n	x_n	$x_{n+1} = \phi(x_n)$	$f(x_{n+1})$
0	1	1.7702	−0.738339
1	1.7702	1.4010	0.367037
2	1.4010	1.5845	−0.182703
3	1.5845	1.4931	0.0912731
4	1.4931	1.5388	−0.045564
5	1.5388	1.5160	0.0227601
6	1.5160	1.5274	−0.0113662
7	1.5274	1.5217	0.00567704
8	1.5217	1.5245	−0.00283529
9	1.5245	1.5231	0.00141608
10	1.5231	1.5238	−0.000707249

2. *Aitken's Δ^2-method*
   As above we have
   $x_2 = 1.4010$, $x_3 = 1.5845$, and $x_4 = 1.4931$. Now, we construct the following difference table:

x_n	Δx_n	$\Delta^2 x_n$
$x_2 = 1.4010$		
	0.1835	
$x_3 = 1.5845$		−0.2749
	−0.0914	
$x_4 = 1.4931$		

Using Equation 2.20, we obtain

$$\begin{aligned}\alpha &\cong x_4 - \frac{(\Delta x_3)^2}{\Delta^2 x_2} \\ &= 1.4931 - \frac{(-0.0914)^2}{-0.2749} \\ &= 1.5235\end{aligned}$$

which corresponds to nine normal iterations in the iteration method.

Hence, the required real root is 1.524 correct to three decimal places.

2.4.3.3 Advantage and Disadvantage of the Fixed-Point Iteration Method

- *Advantage*: This method converges rapidly if the initial approximation x_0 is chosen very close to the desired root. Like other iterative methods, this method is also self-correcting; that is, if a computational error occurs at any iteration, the error is corrected in the next iteration.
- *Disadvantage*: The method is conditionally convergent. Sometimes, it is very difficult to express a given equation $f(x) = 0$ in the form $x = \phi(x)$.

2.4.3.4 Algorithm of the Fixed-Point Iteration Method

Step 1: Start the program.
Step 2: Define the function $x = \phi(x)$.
Step 3: Enter the initial guess of the root (say a).
Step 4: Calculate $b = \phi(a)$.
Step 5: If $|a - b| < \varepsilon$, where ε is a prescribed accuracy, then go to step 7.
Step 6: Set $a = b$ and go to step 4.
Step 7: Print the value of a.
Step 8: Stop the program.

MATHEMATICA® Program for the Fixed-Point Iteration Method (Chapter 2, Example 2.7)

```
f[x_]:=10^x+x - 4;
Clear[a,b,x,n];
ε=0.000001;
a=0;
b=1;
x=a;
Print["n       xn+1    f(xn+1)"]
n=0;
While[Abs[f[x]]>ε, y=x; x=Log[10,4-y];Print[n,"      ",N[x,8],"        ",
N[f[x]]];n++];
```

Output:

```
n              xn+1                    f(xn+1)
0           0.60205999               0.60206
1           0.53121571              -0.0708443
2           0.54017729               0.00896159
3           0.53905384              -0.00112345
4           0.53919484               0.000140998
5           0.53917715              -0.0000176934
6           0.53917937               2.22033*10^-6
7           0.53917909              -2.78627*10^-7
```

2.4.4 Newton–Raphson Method

Let, $[a_0, b_0]$ be an initial interval containing the only root α of the given equation $f(x) = 0$ and let $f(x)$ be continuously differentiable sufficient number of times in $[a_0, b_0]$.

Let $x_0 \in [a_0, b_0]$ be an initial approximation to α. We set initial approximation of α as $x_0 = a_0$ or b_0. Let $x_1 = x_0 + h$ be the exact root closer to x_0 so that $f(x_1) = f(x_0 + h) = 0$, where h is sufficiently small.

Then, expanding $f(x_0 + h)$ by Taylor's series about x_0, we obtain

$$f(x_0 + h) = f(x_0) + hf'(x_0) + \frac{h^2}{2!} f''(x_0) + \cdots = 0.$$

Now, neglecting the terms containing h^2 and higher powers of h, we have

$$f(x_0) + hf'(x_0) \cong 0$$

This implies

$$h \cong -\frac{f(x_0)}{f'(x_0)}$$

Therefore, we take the first approximation to α as

$$\begin{aligned} x_1 &= x_0 + h \\ &= x_0 - \frac{f(x_0)}{f'(x_0)} \end{aligned}$$

Continuing similarly, we get successive approximations (second approximation, third approximation, and so on) to α as

$$x_2 = x_1 - \frac{f(x_1)}{f'(x_1)}$$

$$x_3 = x_2 - \frac{f(x_2)}{f'(x_2)}$$

and so on.

In general, the $(n+1)$th approximation to α is given by

$$x_{n+1} = x_n - \frac{f(x_n)}{f'(x_n)}, \quad n = 0,1,2\ldots \tag{2.21}$$

This is the iteration scheme of Newton–Raphson method.

2.4.4.1 Condition of Convergence

The iteration scheme shows that the Newton–Raphson method is only a particular case of the fixed-point iteration method. In this case, the equation $f(x) = 0$ can to be rewritten as

$$x = \phi(x) = x - \frac{f(x)}{f'(x)}$$

Therefore,

$$\phi'(x) = 1 - \left[\frac{\{f'(x)\}^2 - f(x)f''(x)}{\{f'(x)\}^2}\right] = \frac{f(x)f''(x)}{\{f'(x)\}^2}$$

Thus, the condition of convergence of the Newton–Raphson method is

$$|\phi'(x)| < 1, \quad \text{for all} \quad x \in [a_0, b_0]$$

This implies

$$\left|\frac{f(x)f''(x)}{\{f'(x)\}^2}\right| < 1, \quad \text{for} \quad x \in [a_0, b_0]$$

Hence, the condition of convergence of the Newton–Raphson method is $|f(x)f''(x)| < |f'(x)|^2$, for all $x \in [a_0, b_0]$.

2.4.4.2 Order of Convergence for the Newton–Raphson Method

We know that, error at nth iteration $\varepsilon_n = \alpha - x_n$, that is, $\alpha = x_n + \varepsilon_n$.

Therefore,

$$0 = f(\alpha) = f(x_n + \varepsilon_n) = f(x_n) + \varepsilon_n f'(x_n) + \left(\varepsilon_n^2/2\right) f''(\xi_n), \text{ applying Taylor series expansion}$$

where $\min\{\alpha, x_n\} < \xi_n < \max\{\alpha, x_n\}$.

Now, iteration formula is $x_{n+1} = x_n - \left[f(x_n)/f'(x_n)\right]$

Therefore,

$$\alpha - x_{n+1} = \alpha - x_n + \frac{f(x_n)}{f'(x_n)}$$

This implies

$$\varepsilon_{n+1} = \varepsilon_n - \left[\varepsilon_n + \frac{\varepsilon_n^2}{2}\frac{f''(\xi_n)}{f'(x_n)}\right]$$

Hence,

$$\varepsilon_{n+1} = -\frac{\varepsilon_n^2}{2}\frac{f''(\xi_n)}{f'(x_n)} \tag{2.22}$$

where $\min\{\alpha, x_n\} < \xi_n < \max\{\alpha, x_n\}$.

This is the error equation. If the iteration converges, then $x_n, \xi_n \to \alpha$.

Therefore,

$$\lim_{n\to\infty}\left|\frac{\varepsilon_{n+1}}{\varepsilon_n^2}\right| = \frac{1}{2}\left|\frac{f''(\alpha)}{f'(\alpha)}\right|$$

which shows that the order of convergence of the Newton–Raphson method is 2 and the asymptotic error constant is

$$\frac{1}{2}\left|\frac{f''(\alpha)}{f'(\alpha)}\right|$$

2.4.4.3 Geometrical Significance of the Newton–Raphson Method

The geometrical meaning of the Newton–Raphson method is that the point at which the tangent to the curve $y = f(x)$ at the points $(x_n, f(x_n))$ cuts the x-axis is x_{n+1} (Figure 2.6).

The equation of the tangent is $y - f(x_n) = f'(x_n)(x - x_n)$. It intersects x-axis, that is,

$$y = 0 \quad \text{at} \quad x = x_n - \frac{f(x_n)}{f'(x_n)} = x_{n+1}$$

Accordingly, the Newton–Raphson method is also known as the *method of tangents*.

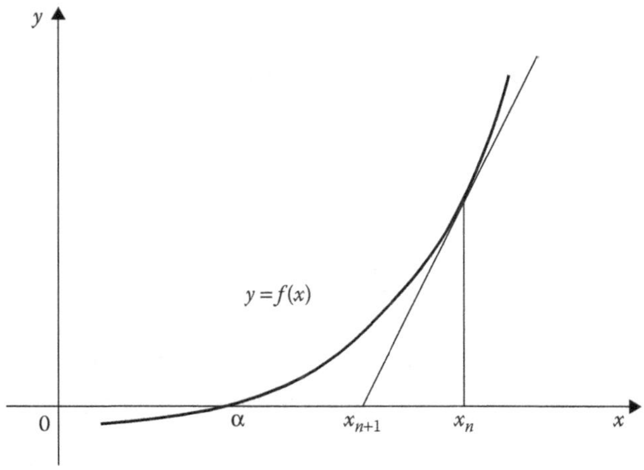

FIGURE 2.6 Graphical representation of the Newton–Raphson method.

2.4.4.4 Advantage and Disadvantage of the Newton–Raphson Method

- *Advantage*: The basic advantage of the Newton–Raphson method is that it converges very rapidly, since the order of convergence of the method is quadratic.
- *Disadvantage*: The Newton–Raphson method fails if $f'(x)=0$ or very small in the neighborhood of the desired root. In such cases, the regula-falsi method should be used. The initial approximation should be taken very close to the desired root, otherwise the method may diverge. Sometimes, this method may not be suitable for a function $f(x)$ whose derivative is difficult to calculate.

Example 2.9

Find the real root of the equation $10^x + x - 4 = 0$ correct to six significant digits using the Newton–Raphson method.

Solution:

Let $f(x) = 10^x + x - 4 = 0$.

We first apply method of tabulation in order to find the location of rough value of the root (Table 2.13).

We note that $f(0) < 0$ and $f(1) > 0$. Thus, the given equation changes its sign within the interval [0,1]. Therefore, there exists at least one real root of the equation within [0,1].

Now, $f'(x) = 10^x \ln 10 + 1$. We choose initial approximation $x_0 = 0$.

The successive iterations generated by the Equation 2.21 have been presented in Table 2.14.

TABLE 2.13
Location of the Root

x	$f(x)$
0	−3
1	7

TABLE 2.14
Table for Finding Real Root

n	$x_{n+1} = x_n - \frac{f(x_n)}{f'(x_n)}$	$f(x_{n+1})$
0	0.90837932	5.00641
1	0.65355360	1.15709
2	0.55178472	0.11453
3	0.53934062	0.00144869
4	0.53917915	2.39265E-7
5	0.53917912	6.18949E-15

Hence, the required real root of the given equation is 0.5391791 correct to six significant digits.

Example 2.10

Find the real root of the equation $x^2 - 10\log_{10} x - 3 = 0$ correct to five decimal places using the Newton–Raphson method.

Solution:

Let $f(x) = x^2 - 10\log_{10} x - 3 = 0$

We first apply method of tabulation in order to find the location of rough value of the root (Table 2.15).

We note that $f(2) < 0$ and $f(3) > 0$. Thus, the given equation changes its sign within the interval [2,3]. Therefore, there exists at least one real root of the equation within [2,3].

Now, $f'(x) = 2x - (10/x\ln 10)$. We choose initial approximation $x_0 = 2$.

The successive iterations generated by the Equation 2.21 have been presented in Table 2.16.

Hence, the required real root of the given equation is 2.70622 correct to five decimal places.

TABLE 2.15
Location of the Root

x	$f(x)$
2	−2.0103
3	1.22879

TABLE 2.16
Table for Finding Real Root

n	$x_{n+1} = x_n - \frac{f(x_n)}{f'(x_n)}$	$f(x_{n+1})$
0	3.0994091	1.69355
1	2.7464103	0.155115
2	2.7067541	0.00202975
3	2.7062212	3.68133E-7
4	2.7062211	1.24345E-14

2.4.4.5 Algorithm for the Newton–Raphson Method

Step 1: Start the program.
Step 2: Define the functions $f(x)$, $f'(x)$.
Step 3: Enter the initial guess of the root (say x_0) and set $n = 0$.
Step 4: Calculate $x_{n+1} = x_n - \left[f(x_n)/f'(x_n) \right]$.
Step 5: If $|x_{n+1} - x_n| < \varepsilon$, where ε is a prescribed accuracy, then go to Step 7.
Step 6: Set $n = n+1$ and go to Step 4.
Step 7: Print the value of x_n which is the required root.
Step 8: Stop the program.

MATHEMATICA® Program for Newton–Raphson Method (Chapter 2, Example 2.9)

```
f[x_]:=10^x+x - 4
N[f[0]]
N[f[1]]
Df[x_]:=1 + 10^x * Log[10]
Clear[a,b,x,n];
ε=0.000000001;
a=0;
b=1;
x=a;
Print["n                xn+1                    f(xn+1)"]
n=0;
While[Abs[f[x]]>ε, y=x; x=y-(f[y]/Df[y]);Print[n,"      ",N[x,8],"      ",
N[f[x]]];n++];
```

Input:

```
-3.
7.
```

Output:

```
n           xn+1                    f(xn+1)
0        0.90837932              5.00641
1        0.65355360              1.15709
2        0.55178472              0.11453
3        0.53934062              0.00144869
4        0.53917915              2.39265*10^-7
5        0.53917912              6.18949*10^-15
```

2.4.5 Secant Method

It has been already mentioned in the disadvantage of the Newton–Raphson method that, although this method is very powerful, sometimes the computation of derivative may be difficult. Particularly in the case of functions arising in practical problems, the evolution of derivatives of the function is not always possible. This suggests the idea of replacing the derivative $f'(x_n)$ by the difference quotient.

In secant method, the derivative at x_n is approximated by the following difference quotient:

$$f'(x_n) \approx \frac{f(x_n) - f(x_{n-1})}{x_n - x_{n-1}}$$

Hence, the iteration scheme of the Newton–Raphson method given in Equation 2.21 reduces to

$$x_{n+1} = x_n - \frac{(x_n - x_{n-1})f(x_n)}{f(x_n) - f(x_{n-1})} \tag{2.23}$$

This is the iteration scheme for secant method. It is to be noted that the iteration scheme in Equation 2.23 requires two initial approximations to find the root.

2.4.5.1 Geometrical Significance of the Secant Method

We approximate the curve $y = f(x)$ between the two points $(x_n, f(x_n))$ and $(x_{n-1}, f(x_{n-1}))$ by a straight line. The equation of this straight line is $[y - f(x_n)]/(x - x_n) = [f(x_{n-1}) - f(x_n)]/(x_{n-1} - x_n)$. It cuts the x-axis, that is, $y = 0$ at $x = x_{n+1}$ (Figure 2.7).

Therefore, the $(n+1)$th approximation x_{n+1} is $x = x_n - [(x_n - x_{n-1})/(y_n - y_{n-1})]\, y_n = x_{n+1}$.

In case, y_n, y_{n-1} are of the same sign, the required point of intersection falls outside the range of interpolation; that is, we have a case of extrapolation and in that case there is no guarantee that the iteration will converge or not. In fact, the secant method may or may not converge. But in case the iteration converges, this renders an efficient method for computing the required real root.

Example 2.11

Find the real root of the equation $10^x + x - 4 = 0$ correct to five decimal places using the secant method.

Solution:

Let $f(x) = 10^x + x - 4 = 0$.

We first apply method of tabulation in order to find the location of rough value of the root (Table 2.17).

We note that $f(0) < 0$ and $f(1) > 0$. Thus, the given equation changes its sign within the interval [0,1]. Therefore, there exists at least one real root of the equation within [0,1].

The successive iterations generated by Equation 2.23 have been presented in Table 2.18. Hence, the required real root of the given equation is 0.53918 correct to five decimal places.

Example 2.12

Find the real root of the equation $xe^x - 2 = 0$ correct to five significant digits using secant method.

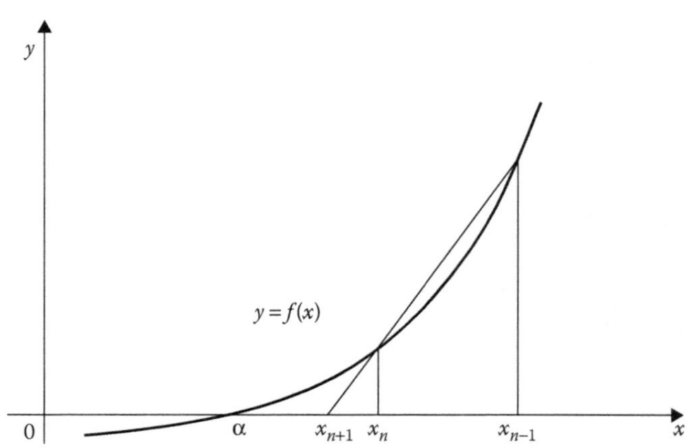

FIGURE 2.7 Graphical representation of the secant method.

TABLE 2.17
Location of the Root

x	$f(x)$
0	−3
1	7

TABLE 2.18
Table for Finding Real Root

n	x_{n-1}	x_n	y_{n-1}	y_n	$x_{n+1} = x_n - \frac{(x_n - x_{n-1})y_n}{(y_n - y_{n-1})}$	$f(x_{n+1})$
0	0	1	−3	7	0.3000000	−1.70474
1	1	0.3	7	−1.70474	0.4370881	−0.827088
2	0.3	0.4370881	−1.70474	−0.827088	0.5662786	0.24993
3	0.4370881	0.5662786	−0.827088	0.24993	0.5362989	−0.0257559
4	0.5662786	0.5362989	0.24993	−0.0257559	0.5390998	−0.000711539
5	0.5362989	0.5390998	−0.0257559	−0.000711539	0.5391794	2.0981E-6
6	0.5390998	0.5391794	−0.000711539	2.0981E-6	0.5391791	−1.70286E-10

Solution:

Let $f(x) = xe^x - 2 = 0$.

We first apply method of tabulation in order to find the location of rough value of the root (Table 2.19).

We note that $f(0) < 0$ and $f(1) > 0$. Thus, the given equation changes its sign within the interval [0,1]. Therefore, there exists at least one real root of the equation within [0,1].

The successive iterations generated by the Equation 2.23 have been presented in Table 2.20. Hence, the required real root of the given equation is 0.85261 correct to five significant digits.

2.4.5.2 Order of Convergence for the Secant Method

The error at nth iteration is given by

$$\varepsilon_n = \alpha - x_n$$

Now, $f(x_n) = f(\alpha - \varepsilon_n) = f(\alpha) - \varepsilon_n f'(\alpha) + \left(\varepsilon_n^2/2!\right) f''(\alpha) + O(\varepsilon_n^3)$, applying Taylor series expansion about α.

TABLE 2.19
Location of the Root

x	$f(x)$
0	−2
1	0.718282

TABLE 2.20
Table for Finding Real Root

n	x_{n-1}	x_n	y_{n-1}	y_n	$x_{n+1} = x_n - \frac{(x_n - x_{n-1})y_n}{(y_n - y_{n-1})}$	$f(x_{n+1})$
0	0	1	−2	0.718282	0.7357589	−0.464423
1	1	0.7357589	0.718282	−0.464423	0.8395208	−0.0562935
2	0.7357589	0.8395208	−0.464423	−0.0562935	0.8538327	0.00533812
3	0.8395208	0.8538327	−0.0562935	0.00533812	0.8525931	−0.0000539276
4	0.8538327	0.8525931	0.00533812	−0.0000539276	0.8526055	−5.09317E-8
5	0.8525931	0.8526055	−0.0000539276	−5.09317E-8	0.8526055	4.86722E-13

Then,

$$\begin{aligned}
\varepsilon_{n+1} &= \alpha - x_{n+1} \\
&= \alpha - x_n + \frac{(x_n - x_{n-1})}{(y_n - y_{n-1})} y_n \\
&= \frac{\alpha y_n - \alpha y_{n-1} - x_{n-1} y_n + x_n y_{n-1}}{(y_n - y_{n-1})} \\
&= \frac{\varepsilon_{n-1} y_n - \varepsilon_n y_{n-1}}{y_n - y_{n-1}} \\
&= \frac{\varepsilon_{n-1}\left[-\varepsilon_n f'(\alpha) + \left(\varepsilon_n^2/2!\right) f''(\alpha) + O(\varepsilon_n^3)\right] - \varepsilon_n\left[-\varepsilon_{n-1} f'(\alpha) + \left(\varepsilon_{n-1}^2/2!\right) f''(\alpha) + O(\varepsilon_{n-1}^3)\right]}{-\varepsilon_n f'(\alpha) + \left(\varepsilon_n^2/2!\right) f''(\alpha) + O(\varepsilon_n^3) + \varepsilon_{n-1} f'(\alpha) - \left(\varepsilon_{n-1}^2/2!\right) f''(\alpha) + O(\varepsilon_{n-1}^3)} \\
&= \frac{\left[\varepsilon_n \varepsilon_{n-1}(\varepsilon_n - \varepsilon_{n-1}) f''(\alpha)\right]/2 + \cdots}{-(\varepsilon_n - \varepsilon_{n-1}) f'(\alpha) + \cdots} \\
&= -\frac{\varepsilon_n \varepsilon_{n-1} f''(\alpha)}{2 f'(\alpha)} + \cdots
\end{aligned}$$

This implies,

$$\varepsilon_{n+1} \cong -\frac{\varepsilon_n \varepsilon_{n-1} f''(\alpha)}{2 f'(\alpha)} \quad \text{(by neglecting the higher order term)} \tag{2.24}$$

Now, we want to find p such that

$$|\varepsilon_{n+1}| = C |\varepsilon_n|^p$$

Using Equation 2.24, we obtain

$$\left|\frac{\varepsilon_n \varepsilon_{n-1} f''(\alpha)}{2 f'(\alpha)}\right| = C |\varepsilon_n|^p$$

Therefore, $|\varepsilon_n|^{p-1} = \tilde{C} |\varepsilon_{n-1}|$,

where

$$\tilde{C} = \left|\frac{f''(\alpha)}{2Cf'(\alpha)}\right|$$

So, $|\varepsilon_{n+1}|^{p-1} = \tilde{C}|\varepsilon_n|$

This implies, $|\varepsilon_{n+1}|^{p(p-1)} = \tilde{C}^p|\varepsilon_n|^p$

Therefore, if the iteration method; that is, secant method converges then $p(p-1)$ must be equal to 1.

Therefore,

$$p(p-1) = 1$$

or,

$$p^2 - p - 1 = 0$$

So,

$$p = \frac{1 \pm \sqrt{5}}{2}$$

Hence, the only positive solution is $\left(1+\sqrt{5}\right)/2$. Therefore, $p = \left(1+\sqrt{5}\right)/2 = 1.618$ (Golden ratio or Golden mean).

Furthermore, the asymptotic error constant

$$C = \tilde{C}^p = \left|\frac{f''(\alpha)}{2Cf'(\alpha)}\right|^p$$

This implies

$$C = \left|\frac{f''(\alpha)}{2f'(\alpha)}\right|^{p/(p+1)}$$

Consequently,

$$\lim_{n\to\infty}\left|\frac{\varepsilon_{n+1}}{\varepsilon_n{}^p}\right| = \left|\frac{f''(\alpha)}{2f'(\alpha)}\right|^{p/(p+1)}$$

where

$$p = \frac{1+\sqrt{5}}{2}$$

is order of convergence for the secant method.

2.4.5.3 Advantage and Disadvantage of the Secant Method

- *Advantage*: The advantage of the secant method is that if it converges, then it converges more rapidly than the regula-falsi method, since the order of convergence of secant method is 1.61803.
- *Disadvantage*: This method does not always converge in contrast to the regula-falsi method. If it converges, then it is faster than the regula-falsi method, but lesser than the Newton–Raphson method. If at any instance $f(x_{n-1}) \cong f(x_n)$, then the method fails.

2.4.5.4 Algorithm for the Secant Method

Step 1: Start the program.
Step 2: Define the function $f(x)$.
Step 3: Enter the initial interval $[a,b]$.
Step 4: Calculate $x = a - [(b-a)/(f(b)-f(a))]f(a)$.
Step 5: If $|a-b| < \varepsilon$, ε being the prescribed accuracy, then go to Step 7 else Step 6.
Step 6: Set $a = b, \quad b = x$ and go to Step 4.
Step 7: Print the value of x which is the required root.
Step 8: Stop the program.

MATHEMATICA® Program for Root Finding by Using the Secant Method (Chapter 2, Example 2.11)

```
f[x_] := 10^x + x - 4;
Clear[a,b,x,n];
ε=0.0000001;
a=0;
b=1;
x=a;
fa=f[a];
fb=f[b];
Print["n        xn-1        xn        yn-1        yn        xn+1        f(xn+1)"];
n=0;
While[Abs[f[x]]>ε, y=x; x=(b*fa-a*fb)/(fa-fb);Print[n,"      ",N[a],"      ",
N[b],"      ",N[f[a]],"      ",N[f[b]],"      ",SetPrecision[N[x],7],"      ",
N[f[x]]];
fa=fb;fb=f[x];a=b;b=x;n++];
```

Output:

n	xn-1	xn	yn-1	yn	xn+1	f(xn+1)
1	0.	1.	-3.	7.	0.3000000	-1.70474
2	1.	0.3	7.	-1.70474	0.4370881	-0.827088
3	0.3	0.437088	-1.70474	-0.827088	0.5662786	0.24993
4	0.437088	0.566279	-0.827088	0.24993	0.5362989	-0.0257559
5	0.566279	0.536299	0.24993	-0.0257559	0.5390998	-0.000711539
6	0.536299	0.5391	-0.0257559	-0.000711539	0.5391794	2.0981*10^-6
7	0.5391	0.539179	-0.000711539	2.0981*10^-6	0.5391791	-1.70286*10^-10

2.5 GENERALIZED NEWTON'S METHOD

A function $f(x)$ is said to have a root α of multiplicity $r(>1)$ if

$$f(x) = (x-\alpha)^r g(x), \quad \text{where } g(\alpha) \neq 0 \tag{2.25}$$

We suppose that $r \in Z^+$, that is, r is a positive integer and $g(x)$ is sufficiently differentiable at $x = \alpha$. Then we have

$$f(\alpha) = f'(\alpha) = f''(\alpha) = \cdots = f^{(r-1)}(\alpha) = 0 \quad \text{and} \quad f^{(r)}(\alpha) \neq 0 \tag{2.26}$$

Now according to the fixed-point iteration method, let us consider

$$\phi(x) = x - \frac{f(x)}{f'(x)}, \quad x \neq \alpha \tag{2.27}$$

Differentiating Equation 2.25 with respect to x, we get

$$f'(x) = (x-\alpha)^r g'(x) + r(x-\alpha)^{r-1} g(x)$$

Therefore, from Equation 2.27, we obtain

$$\phi(x) = x - \frac{(x-\alpha)g(x)}{rg(x)+(x-\alpha)g'(x)} \tag{2.28}$$

Again differentiating Equation 2.28 with respect to x, we have

$$\phi'(x) = 1 - \frac{g(x)}{rg(x)+(x-\alpha)g'(x)} - (x-\alpha)\frac{d}{dx}\left[\frac{g(x)}{rg(x)+(x-\alpha)g'(x)}\right] \tag{2.29}$$

so that $\phi'(\alpha) = 1 - (1/r) \neq 0$ for $r > 1$.

We therefore proceed to find a function $\phi(x)$ for which $\phi'(\alpha) = 0$. Based on Equation 2.27, we get

$$\phi(x) = x - r\frac{f(x)}{f'(x)} \quad \text{so that} \quad \phi'(\alpha) = 0$$

Now,

$$\begin{aligned}\alpha - x_{n+1} &= \phi(\alpha) - \phi(x_n)\\ &= \phi(\alpha) - \phi(\alpha) - (x_n - \alpha)\phi'(\alpha) - \frac{1}{2}(x_n-\alpha)^2\phi''(\xi_n), \text{ applying Taylor's series expansion}\end{aligned}$$

about α, where $\min\{\alpha, x_n\} < \xi_n < \max\{\alpha, x_n\}$.

$$= -\frac{1}{2}(x_n-\alpha)^2\phi''(\xi_n) \tag{2.30}$$

Hence the new iteration scheme is

$$x_{n+1} = x_n - r\frac{f(x_n)}{f'(x_n)}, \quad n = 0,1,2\ldots \tag{2.31}$$

From Equation 2.30, it shows that this iteration method has order of convergence two, the same as the original Newton method for simple roots.

Remarks: If α be a root of $f(x)=0$ with multiplicity r, then it is also a root of $f'(x)=0$ with multiplicity $r-1$ and so on. Hence, if x_0 is an initial approximation to the root α, then the expressions

$$x_0 - r\frac{f(x_0)}{f'(x_0)}, \; x_0 - (r-1)\frac{f'(x_0)}{f''(x_0)}, \ldots$$

will have same values.

Example 2.13

Find the double root of the equation $x^3+x^2-16x+20=0$.

Solution:

Let $f(x)=x^3+x^2-16x+20$.

Therefore, $f'(x)=3x^2+2x-16$ and $f''(x)=6x+2$. Let us choose the initial approximation of the root be 1.5, that is, $x_0=1.5$. Then using the iteration scheme in Equation 2.31 of generalized Newton's method, we obtain

$$x_1 = x_0 - 2\frac{f(x_0)}{f'(x_0)} = 2.02 \quad (\text{here, } r=2)$$

Again

$$x_1 = x_0 - (2-1)\frac{f'(x_0)}{f''(x_0)} = 2.06818$$

The closeness of these values implies that there is a double root near $x=1.5$.

Now,

$$x_2 = x_1 - 2\frac{f(x_1)}{f'(x_1)} = 2.00003$$

and

$$x_2 = x_1 - (2-1)\frac{f'(x_1)}{f''(x_1)} = 2.00008$$

Similarly,

$$x_3 = x_2 - 2\frac{f(x_2)}{f'(x_2)} = 2.00003$$

and

$$x_3 = x_2 - (2-1)\frac{f'(x_2)}{f''(x_2)} = 2.00005$$

Therefore, we conclude that there is a double root at $x = 2.00003$, which is sufficiently close to the actual root $x = 2$.

2.5.1 Numerical Solution of Simultaneous Nonlinear Equations

2.5.1.1 Newton's Method

The Newton–Raphson method can be extended to find the solutions of simultaneous equations in several unknowns. Let us consider a system of two nonlinear equations in two unknowns

$$f(x,y)=0, \quad g(x,y)=0 \tag{2.32}$$

Let (x_0, y_0) be an initial approximation to the roots of the system given in Equation 2.32. If $(x_0+\Delta x,\quad y_0+\Delta y)$ is the root of the system, then we must have

$$f(x_0+\Delta x, y_0+\Delta y)=0, \quad g(x_0+\Delta x, y_0+\Delta y)=0 \tag{2.33}$$

Assuming the functions f and g to be sufficiently differentiable, expanding f and g by Taylor's series in the neighborhood (x_0, y_0), we have

$$f(x_0,y_0)+\left(\Delta x\frac{\partial}{\partial x}+\Delta y\frac{\partial}{\partial y}\right)f(x_0,y_0)+\frac{1}{2!}\left(\Delta x\frac{\partial}{\partial x}+\Delta y\frac{\partial}{\partial y}\right)^2 f(x_0,y_0)+\cdots=0$$

$$g(x_0, y_0) + \left(\Delta x \frac{\partial}{\partial x} + \Delta y \frac{\partial}{\partial y} \right) g(x_0, y_0) + \frac{1}{2!} \left(\Delta x \frac{\partial}{\partial x} + \Delta y \frac{\partial}{\partial y} \right)^2 g(x_0, y_0) + \cdots = 0$$

Neglecting second and higher powers of Δx and Δy, we have

$$\begin{aligned} f(x_0, y_0) + \Delta x \frac{\partial f(x_0, y_0)}{\partial x} + \Delta y \frac{\partial f(x_0, y_0)}{\partial y} &= 0 \\ g(x_0, y_0) + \Delta x \frac{\partial g(x_0, y_0)}{\partial x} + \Delta y \frac{\partial g(x_0, y_0)}{\partial y} &= 0 \end{aligned} \tag{2.34}$$

Solving Equation 2.34 for Δx and Δy, we obtain

$$\Delta x = -\frac{1}{J_0} \begin{vmatrix} f(x_0, y_0) & \dfrac{\partial f(x_0, y_0)}{\partial y} \\ g(x_0, y_0) & \dfrac{\partial g(x_0, y_0)}{\partial y} \end{vmatrix} \quad \text{and} \quad \Delta y = -\frac{1}{J_0} \begin{vmatrix} \dfrac{\partial f(x_0, y_0)}{\partial x} & f(x_0, y_0) \\ \dfrac{\partial g(x_0, y_0)}{\partial x} & g(x_0, y_0) \end{vmatrix}$$

where the Jacobian

$$J_0 = \begin{vmatrix} \dfrac{\partial f(x_0, y_0)}{\partial x} & \dfrac{\partial f(x_0, y_0)}{\partial y} \\ \dfrac{\partial g(x_0, y_0)}{\partial x} & \dfrac{\partial g(x_0, y_0)}{\partial y} \end{vmatrix} \neq 0$$

Thus, the first approximation to the required root is given by

$$x_1 = x_0 + \Delta x = x_0 - \frac{1}{J_0} \begin{vmatrix} f(x_0, y_0) & \dfrac{\partial f(x_0, y_0)}{\partial y} \\ g(x_0, y_0) & \dfrac{\partial g(x_0, y_0)}{\partial y} \end{vmatrix}$$

$$y_1 = y_0 + \Delta y = y_0 - \frac{1}{J_0} \begin{vmatrix} \dfrac{\partial f(x_0, y_0)}{\partial x} & f(x_0, y_0) \\ \dfrac{\partial g(x_0, y_0)}{\partial x} & g(x_0, y_0) \end{vmatrix}$$

Repeating the above procedure, the $(n+1)$th approximation to the roots is given by

$$x_{n+1} = x_n + \Delta x = x_n - \frac{1}{J_n} \begin{vmatrix} f(x_n, y_n) & \dfrac{\partial f(x_n, y_n)}{\partial y} \\ g(x_n, y_n) & \dfrac{\partial g(x_n, y_n)}{\partial y} \end{vmatrix}$$

$$y_{n+1} = y_n + \Delta y = y_n - \frac{1}{J_n} \begin{vmatrix} \dfrac{\partial f(x_n, y_n)}{\partial x} & f(x_n, y_n) \\ \dfrac{\partial g(x_n, y_n)}{\partial x} & g(x_n, y_n) \end{vmatrix}$$

where, in each iteration, it is assumed that the Jacobian

$$J_n = \begin{vmatrix} \dfrac{\partial f(x_n, y_n)}{\partial x} & \dfrac{\partial f(x_n, y_n)}{\partial y} \\ \dfrac{\partial g(x_n, y_n)}{\partial x} & \dfrac{\partial g(x_n, y_n)}{\partial y} \end{vmatrix} \neq 0, \quad (n = 0,1,2,\ldots)$$

This iteration process will continue until $|x_{n+1} - x_n| < \varepsilon$, $|y_{n+1} - y_n| < \varepsilon$, where ε is the given accuracy.

2.5.1.1.1 Generalization of Newton's Method

In general, we can usually find solutions to a system of equations when the number of unknowns matches the number of equations. Now, Newton's method can be generalized for solving a system of n equations in n unknowns. Thus, we wish to find solutions to systems that have the following form:

$$\begin{aligned} &f_1(x_1, x_2, \ldots, x_n) = 0 \\ &f_2(x_1, x_2, \ldots, x_n) = 0 \\ &\vdots \\ &f_n(x_1, x_2, \ldots, x_n) = 0 \end{aligned}$$

In n-dimensional vector form,

$$\boldsymbol{f}(\boldsymbol{x}) = \boldsymbol{0}$$

where:

$$\boldsymbol{x} = [x_1, x_2, \ldots, x_n]^T$$

$$\boldsymbol{f} = [f_1, f_2, \ldots, f_n]^T$$

If $\boldsymbol{x}^{(0)} = [x_1^{(0)}, x_2^{(0)}, \ldots, x_n^{(0)}]^T$ be an initial approximation to the solution vector x.

In the single variable case, Newton's method was derived by considering the linear approximation of the function f at the initial guess x_0. From calculus, the following is the linear approximation of $\boldsymbol{f}$ at $\boldsymbol{x}^{(0)}$, for vectors and vector-valued functions:

$$\boldsymbol{f}(\boldsymbol{x}) \approx \boldsymbol{f}(\boldsymbol{x}^{(0)}) + J(\boldsymbol{x}^{(0)})(\boldsymbol{x} - \boldsymbol{x}^{(0)})$$

where $J(\boldsymbol{x})$ is Jacobian matrix of $\boldsymbol{f}$

$$\{J(\boldsymbol{x})\}_{ij} = \frac{\partial}{\partial x_j} f_i(\boldsymbol{x})$$

Therefore, the first approximation is given by

$$\boldsymbol{x}^{(1)} = \boldsymbol{x}^{(0)} - J(\boldsymbol{x}^{(0)})^{-1} \boldsymbol{f}(\boldsymbol{x}^{(0)})$$

provided that the inverse of Jacobian matrix exists.

In general, the $(k+1)$th approximation is given by

$$\boldsymbol{x}^{(k+1)} = \boldsymbol{x}^{(k)} - J(\boldsymbol{x}^{(k)})^{-1} \boldsymbol{f}(\boldsymbol{x}^{(k)})$$

The formula is the vector equivalent of the Newton's method formula we learned earlier. However, in practice, we never use the inverse of a matrix for computations, so we cannot use this formula directly. Rather, we can do the following. First, we shall solve the linear system of equations

$$J(\boldsymbol{x}^{(k)})\Delta\boldsymbol{x}^{(k)} = -\boldsymbol{f}(\boldsymbol{x}^{(k)})$$

Since, $J(\boldsymbol{x}^{(k)})$ is a known matrix and $\boldsymbol{f}(\boldsymbol{x}^{(k)})$ is a known vector, this system of equations can be solved efficiently and accurately for $\Delta\boldsymbol{x}$. Once we have the solution vector $\Delta\boldsymbol{x}$, we can obtain next improved approximate $\boldsymbol{x}^{(k+1)}$ by the following formula:

$$\boldsymbol{x}^{(k+1)} = \boldsymbol{x}^{(k)} + \Delta\boldsymbol{x}^{(k)}$$

The convergence of the method depends on the initial approximate vector $x^{(0)}$. A sufficient condition for convergence is that

$$\left\| J(\boldsymbol{x}^{(k)})^{-1} \right\| < 1, \quad \text{for each } k.$$

Moreover, a necessary and sufficient condition for convergence is

$$\rho\left[J(\boldsymbol{x}^{(k)})^{-1}\right] < 1$$

where $\|.\|$ is matrix row sum norm and $\rho\left[J(\boldsymbol{x}^{(k)})^{-1}\right]$ is the spectral radius for the inverse of Jacobian matrix, that is, $J(\boldsymbol{x}^{(k)})^{-1}$.

The iteration process will continue until $\left\|\boldsymbol{x}^{(k+1)} - \boldsymbol{x}^{(k)}\right\| < \varepsilon$,
where ε is the given tolerance level for error and in this case $\|.\|$ is the L_∞ norm.

Example 2.14

Solve the system of equations

$$x^2 + y^2 - 1 = 0$$

$$x^3 - y = 0$$

correct to four decimal places, using Newton's method, given that $x_0 = 1$ and $y_0 = 0.5$.

Solution:

Let $f(x,y)=x^2+y^2-1=0$, $g(x,y)=x^3-y=0$. Here, the Jacobian is given by

$$J_n = \begin{vmatrix} \dfrac{\partial f(x_n,y_n)}{\partial x} & \dfrac{\partial f(x_n,y_n)}{\partial y} \\ \dfrac{\partial g(x_n,y_n)}{\partial x} & \dfrac{\partial g(x_n,y_n)}{\partial y} \end{vmatrix}$$

$$= \begin{vmatrix} 2x_n & 2y_n \\ 3x_n^2 & -1 \end{vmatrix}$$

$$= -2x_n - 6x_n^2 y_n$$

Now, according to Newton's method, we have

$$x_{n+1} = x_n - \frac{1}{J_n}\begin{vmatrix} f(x_n,y_n) & \dfrac{\partial f(x_n,y_n)}{\partial y} \\ g(x_n,y_n) & \dfrac{\partial g(x_n,y_n)}{\partial y} \end{vmatrix} = x_n - \frac{1}{J_n}\begin{vmatrix} x_n^2 + y_n^2 - 1 & 2y_n \\ x_n^3 - y_n & -1 \end{vmatrix}$$

$$y_{n+1} = y_n - \frac{1}{J_n}\begin{vmatrix} \dfrac{\partial f(x_n, y_n)}{\partial x} & f(x_n, y_n) \\ \dfrac{\partial g(x_n, y_n)}{\partial x} & g(x_n, y_n) \end{vmatrix} = y_n - \frac{1}{J_n}\begin{vmatrix} 2x_n & x_n^2 + y_n^2 - 1 \\ 3x_n^2 & x_n^3 - y_n \end{vmatrix}$$

In matrix form, we have

$$\begin{bmatrix} x_{n+1} \\ y_{n+1} \end{bmatrix} = \begin{bmatrix} x_n \\ y_n \end{bmatrix} - \frac{1}{J_n}\begin{bmatrix} 1 - x_n^2 - 2x_n^3 y_n + y_n^2 \\ 3x_n^2 - x_n^4 - 2x_n y_n - 3x_n^2 y_n^2 \end{bmatrix}$$

Starting with $x_0 = 1$ and $y_0 = 0.5$, we get

$$\begin{bmatrix} x_1 \\ y_1 \end{bmatrix} = \begin{bmatrix} x_0 \\ y_0 \end{bmatrix} - \frac{1}{J_0}\begin{bmatrix} 1 - x_0^2 - 2x_0^3 y_0 + y_0^2 \\ 3x_0^2 - x_0^4 - 2x_0 y_0 - 3x_0^2 y_0^2 \end{bmatrix} = \begin{bmatrix} 1 \\ 0.5 \end{bmatrix} - \frac{1}{-5}\begin{bmatrix} -0.75 \\ 0.25 \end{bmatrix} = \begin{bmatrix} 0.85 \\ 0.55 \end{bmatrix}$$

$$\begin{bmatrix} x_2 \\ y_2 \end{bmatrix} = \begin{bmatrix} x_1 \\ y_1 \end{bmatrix} - \frac{1}{J_1}\begin{bmatrix} 1 - x_1^2 - 2x_1^3 y_1 + y_1^2 \\ 3x_1^2 - x_1^4 - 2x_1 y_1 - 3x_1^2 y_1^2 \end{bmatrix}$$

$$= \begin{bmatrix} 0.85 \\ 0.55 \end{bmatrix} - \frac{1}{-4.08425}\begin{bmatrix} -0.0955375 \\ 0.054825 \end{bmatrix} = \begin{bmatrix} 0.826608 \\ 0.563424 \end{bmatrix}$$

Similarly,

$$\begin{bmatrix} x_3 \\ y_3 \end{bmatrix} = \begin{bmatrix} 0.826032 \\ 0.563624 \end{bmatrix}$$

$$\begin{bmatrix} x_4 \\ y_4 \end{bmatrix} = \begin{bmatrix} 0.826031 \\ 0.563624 \end{bmatrix}$$

Therefore, the required roots of the given equations are $x = 0.826$ and $y = 0.5636$ correct to four decimal places.

Example 2.15

Using Newton's method, find the roots of the following system of equations

$$x^2 y + y^3 = 10$$

$$xy^2 - x^2 = 3$$

given that $x_0 = 0.8$ and $y_0 = 2.2$.

Solution:

Let $f(x,y) = x^2 y + y^3 - 10$, $g(x,y) = xy^2 - x^2 - 3$.
We can write the system of equations in the following matrix form:

$$\mathbf{f}(\mathbf{x}) = \begin{bmatrix} x^2 y + y^3 - 10 \\ xy^2 - x^2 - 3 \end{bmatrix} = 0$$

Here, the Jacobian matrix is given by

$$J_f(\mathbf{x}) = \begin{bmatrix} 2xy & x^2 + 3y^2 \\ y^2 - 2x & 2xy \end{bmatrix}$$

First iteration:

According to initial approximation $\mathbf{x}^{(0)} = \begin{bmatrix} 0.8 \\ 2.2 \end{bmatrix}$, we have

$$\mathbf{f}(\mathbf{x}^{(0)}) = \begin{bmatrix} 2.056 \\ 0.232 \end{bmatrix}, \quad J_f(\mathbf{x}^{(0)}) = \begin{bmatrix} 3.52 & 15.16 \\ 3.24 & 3.52 \end{bmatrix}$$

Now, from equation $J_f(\mathbf{x}^{(0)})\,\Delta\mathbf{x}^{(0)} = -\mathbf{f}(\mathbf{x}^{(0)})$, we have

$$\begin{bmatrix} 3.52 & 15.16 \\ 3.24 & 3.52 \end{bmatrix} \Delta\mathbf{x}^{(0)} = \begin{bmatrix} -2.056 \\ -0.232 \end{bmatrix}$$

Solving the above system, we have

$$\Delta\mathbf{x}^{(0)} = \begin{bmatrix} 0.101285 \\ -0.159137 \end{bmatrix}$$

Therefore, we obtain the first approximation

$$\mathbf{x}^{(1)} = \mathbf{x}^{(0)} + \Delta\mathbf{x}^{(0)} = \begin{bmatrix} 0.901285 \\ 2.040863 \end{bmatrix}$$

Second iteration:

$$\mathbf{f}(\mathbf{x}^{(1)}) = \begin{bmatrix} 0.158266 \\ -0.0583529 \end{bmatrix}, \quad J_f(\mathbf{x}^{(1)}) = \begin{bmatrix} 3.6788 & 13.3077 \\ 2.36255 & 3.6788 \end{bmatrix}$$

Now, from equation $J_f(\mathbf{x}^{(1)})\,\Delta\mathbf{x}^{(1)} = -\mathbf{f}(\mathbf{x}^{(1)})$, we have

$$\begin{bmatrix} 3.6788 & 13.3077 \\ 2.36255 & 3.6788 \end{bmatrix} \Delta\mathbf{x}^{(1)} = \begin{bmatrix} -0.158266 \\ 0.0583529 \end{bmatrix}$$

Solving the above system, we have

$$\Delta\mathbf{x}^{(1)} = \begin{bmatrix} 0.0758813 \\ -0.0328696 \end{bmatrix}$$

Therefore, we obtain the second approximation

$$\mathbf{x}^{(2)} = \mathbf{x}^{(1)} + \Delta\mathbf{x}^{(1)} = \begin{bmatrix} 0.977166 \\ 2.00799 \end{bmatrix}$$

Third iteration:

$$f(x^{(2)})=\begin{bmatrix} 0.0135996 \\ -0.0148968 \end{bmatrix}, \quad J_f(x^{(2)})=\begin{bmatrix} 3.92428 & 13.0509 \\ 2.07769 & 3.92428 \end{bmatrix}$$

Now, from equation $J_f(\boldsymbol{x}^{(2)})\,\Delta\boldsymbol{x}^{(2)} = -\boldsymbol{f}(\boldsymbol{x}^{(2)})$, we have

$$\begin{bmatrix} 3.92428 & 13.0509 \\ 2.07769 & 3.92428 \end{bmatrix}\Delta x^{(2)} = \begin{bmatrix} -0.0135996 \\ 0.0148968 \end{bmatrix}$$

Solving the above system, we have

$$\Delta x^{(2)}=\begin{bmatrix} 0.0211496 \\ -0.00740152 \end{bmatrix}$$

Therefore, we obtain the third approximation

$$x^{(3)}=x^{(2)}+\Delta x^{(2)}=\begin{bmatrix} 0.998316 \\ 2.00059 \end{bmatrix}$$

Therefore, the required roots of the given equations are $x = 1$ and $y = 2$, which can be verified by substituting these values into the given equations.

2.5.1.1.2 *Algorithm of Newton's Method for System of Nonlinear Equations*

Step 1: Start the program.

Step 2: Define the vector function $\boldsymbol{f}(\boldsymbol{x})$ where $\boldsymbol{x} = [x_1, x_2, \ldots, x_n]^T$ and $\boldsymbol{f} = [f_1(x_1, x_2, \ldots, x_n), f_2(x_1, x_2, \ldots, x_n), \ldots, f_n(x_1, x_2, \ldots, x_n)]^T$.

Step 3: Enter the initial approximation $\boldsymbol{x}^{(0)} = [x_1^{(0)}, x_2^{(0)}, \ldots, x_n^{(0)}]^T$ of the solution vector. Set $k = 0$.

Step 4: Solve the linear system of equations $J(\boldsymbol{x}^{(k)})\Delta\boldsymbol{x}^{(k)} = -\boldsymbol{f}(\boldsymbol{x}^{(k)})$, where the kth approximation $x^{(k)} = [x_1^{(k)}, x_2^{(k)}, \ldots, x_n^{(k)}]^T$ for the unknown vector $\Delta\boldsymbol{x}^{(k)}$.

Step 5: Calculate $\boldsymbol{x}^{(k+1)} = \boldsymbol{x}^{(k)} + \Delta\boldsymbol{x}^{(k)}$.

Step 6: If $\left\|\boldsymbol{x}^{(k+1)} - \boldsymbol{x}^{(k)}\right\| < \varepsilon$, where ε is prescribed accuracy, then go to Step 8.

Step 7: Set $k = k + 1$. Then go to Step 4.

Step 8: Print the solution vector $\boldsymbol{x}^{(k+1)}$ which gives the roots of the given simultaneous nonlinear equations.

Step 9: Stop the program.

MATHEMATICA® Program Implementing Newton's Method for Nonlinear System (Chapter 2, Example 2.14)

```
f[x_, y_]:= x² + y² - 1;
g[x_, y_]:= x³ - y;
ε=0.00001;
xnew=1;
ynew=0.5;
J[x,y]=Det[{{D[f[x,y],x],D[f[x,y],y]},{D[g[x,y],x],D[g[x,y],y]}}];
temp1[x,y]=N[Det[{{f[x,y],D[f[x,y],y]},{g[x,y],D[g[x,y],y]}}]];
temp2[x,y]=N[Det[{{D[f[x,y],x],f[x,y]},{D[g[x,y],x],g[x,y]}}]];
Do[Print["Iteration ",n,":"];
```

```
xold=xnew;
yold=ynew;
xnew=xold-1/(J[x,y]/.x->xold/.y->yold)*(temp1[x,y]/.x->xold/.y->yold);
ynew=yold-1/(J[x,y]/.x->xold/.y->yold)*(temp2[x,y]/.x->xold/.y->yold);
Print[xnew];
Print[ynew];

Print[Abs[xnew-xold]];
Print[Abs[ynew-yold]];

Print["-------------------------------------------"];
If[Max[Abs[xnew-xold],Abs[ynew-yold]]<ε,Break[]],{n,1,100}];
Print["Max Error=",Max[Abs[xnew-xold],Abs[ynew-yold]]];
Print["x=",xnew];     Print["y=",ynew];
```

Output:

```
Iteration 1:
0.85
0.55
0.15
0.05
-------------------------------------------
Iteration 2:
0.826608
0.563424
0.0233917
0.0134235
-------------------------------------------
Iteration 3:
0.826032
0.563624
0.000576626
0.000200494
-------------------------------------------
Iteration 4:
0.826031
0.563624
3.28811*10^-7
1.51272*10^-7
-------------------------------------------
Max Error=3.28811*10^-7
x=0.826031
y=0.563624
```

2.5.1.2 Fixed-Point Iteration Method

Let us consider a system of two nonlinear equations in two unknowns

$$f(x,y)=0, \quad g(x,y)=0 \tag{2.35}$$

Suppose that Equation 2.35 may be written as

$$x=\phi(x,y), \quad y=\psi(x,y) \tag{2.36}$$

where the functions ϕ and ψ satisfy the following conditions in a closed neighborhood D of the desired root (ξ,η):

(1) ϕ and Ψ and their first-order partial derivatives are continuous in D, and

(2) $$\left|\frac{\partial\phi}{\partial x}\right|+\left|\frac{\partial\phi}{\partial y}\right|<1 \quad \text{and} \quad \left|\frac{\partial\psi}{\partial x}\right|+\left|\frac{\partial\psi}{\partial y}\right|<1 \quad \text{for all}\,(x,y)\,\text{in}\,D \tag{2.37}$$

Then, if (x_0, y_0) be an initial approximation to the desired root (ξ, η) of Equation 2.35, then the first approximation is given by

$$x_1 = \phi(x_0, y_0), \quad y_1 = \psi(x_0, y_0)$$

Repeating this process, we construct the successive approximations as follows:

$$x_2 = \phi(x_1, y_1), \quad y_2 = \psi(x_1, y_1)$$

$$x_3 = \phi(x_2, y_2), \quad y_3 = \psi(x_2, y_2)$$

$$\vdots$$

$$x_{n+1} = \phi(x_n, y_n), \quad y_{n+1} = \psi(x_n, y_n)$$

It may be noted that the convergence of the sequences $\{x_n\}$ and $\{y_n\}$ depends on the suitable choice of the functions $\phi(x, y)$ and $\psi(x, y)$ and also on the initial approximation (x_0, y_0) of the desired root.

If the sequences $\{x_n\}$ and $\{y_n\}$ converges to ξ and η, respectively, then we must have

$$\xi = \phi(\alpha,\beta), \quad \eta = \psi(\alpha,\beta)$$

which shows that α, β are the roots of Equation 2.35. The conditions given in Equation 2.37 are the sufficient conditions for the convergence of iteration process.

Therefore, a sufficient condition for convergence of this method is that for each n, $\|J_n\|_\infty < 1$, where $\|\,.\,\|_\infty$ is matrix row sum norm and

$$J_n = \begin{vmatrix} \phi_x(x_n, y_n) & \phi_y(x_n, y_n) \\ \psi_x(x_n, y_n) & \psi_y(x_n, y_n) \end{vmatrix}$$

is the Jacobian matrix evaluated at (x_n, y_n).

This method can be easily generalized to a system of n equations in n unknowns.

Example 2.16:

Using the fixed-point iteration method, find the roots of the following system of equations

$$x + 3\log_{10} x - y^2 = 0$$

$$2x^2 - xy - 5x + 1 = 0$$

with $x_0 = 3.4$ and $y_0 = 2.2$ as the initial approximation.

Solution:

The given equations can be written as

$$x = \sqrt{\frac{x(y+5)-1}{2}} = \phi(x, y)$$

$$y = \sqrt{x + 3\log_{10} x} = \psi(x, y)$$

Using initial approximation $x_0 = 3.4$ and $y_0 = 2.2$, from the fixed-point iteration scheme, we have the first approximation

$$x_1 = 3.42637, \quad y_1 = 2.23482$$

Similarly, we obtain the successive approximations as follows:

$$x_2 = 3.44885, \quad y_2 = 2.24296$$

$$x_3 = 3.46265, \quad y_3 = 2.24986$$

$$x_4 = 3.47158, \quad y_4 = 2.25408$$

$$x_5 = 3.47729, \quad y_5 = 2.2568$$

Therefore, the required roots of the given equations are $x = 3.48$, $y = 2.26$ correct to three significant figures.

2.5.1.2.1 Algorithm of the Fixed-Point Iteration Method for a System of Nonlinear Equations

Step 1: Start the program.
Step 2: Define the suitable vector function $\boldsymbol{F}(\boldsymbol{x})$
where $\boldsymbol{x} = [x_1, x_2, \ldots, x_n]^T$ and $\boldsymbol{F} = [f_1(x_1, x_2, \ldots, x_n), f_2(x_1, x_2, \ldots, x_n), \ldots, f_n(x_1, x_2, \ldots, x_n)]^T$.
Step 3: Enter the initial approximation $\boldsymbol{x}^{(0)} = [x_1^{(0)}, x_2^{(0)}, \ldots, x_n^{(0)}]^T$ of the solution vector. Set $k = 0$.
Step 4: Calculate $\boldsymbol{x}^{(k+1)} = \boldsymbol{F}(\boldsymbol{x}^{(k)})$, where the kth approximation $\boldsymbol{x}^{(k)} = [x_1^{(k)}, x_2^{(k)}, \ldots, x_n^{(k)}]^T$.
Step 5: If $\left\| \boldsymbol{x}^{(k+1)} - \boldsymbol{x}^{(k)} \right\| < \varepsilon$, where ε is prescribed accuracy, then go to Step 7.
Step 6: Set $k = k + 1$. Then go to Step 4.
Step 7: Print the solution vector $\boldsymbol{x}^{(k+1)}$ which gives the roots of the given simultaneous nonlinear equations.
Step 8: Stop the program.

MATHEMATICA® Program Implementing the Fixed-Point Iteration Method for Nonlinear System (Chapter 2, Example 2.16)

```
F[x, y] = Sqrt[(x * y + 5 * x - 1)/2];
G[x, y] = Sqrt[x + 3 * Log[10, x]];
ε=0.001;
xnew=3.4;
ynew=2.2;
Do[
Print["Iteration ",n,":"];
xold=xnew;
yold=ynew;
Print[N[Abs[D[F[x,y],x]]+Abs[D[F[x,y],y]]/.x->xold/.y->yold]];
Print[N[Abs[D[G[x,y],x]]+Abs[D[G[x,y],y]]/.x->xold/.y->yold]];
xnew=F[x,y]/.x->xold/.y->yold;
Print[xnew];
ynew=G[x,y]/.x->xold/.y->yold;
Print[ynew];
Print[Abs[xnew-xold]];
Print[Abs[ynew-yold]];
Print["-----------------------------------------"];
If[Max[Abs[xnew-xold],Abs[ynew-yold]]<ε,Break[]],{n,1,100}];
Print["Max Error=",Max[Abs[xnew-xold],Abs[ynew-yold]]];
Print["x=",xnew];
Print["y=",ynew];
```

Output:

```
Iteration 1:
0.773414
0.309465
3.42637
2.23482
0.0263683
0.0348237
---------------------------------------------
Iteration 2:
0.772807
0.307685
3.44885
2.24296
0.0224844
0.00813655
---------------------------------------------
Iteration 3:
0.771938
0.306191
3.46265
2.24986
0.0137982
0.00690128
---------------------------------------------
Iteration 4:
0.771444
0.305284
3.47158
2.25408
0.00892936
0.00421861
---------------------------------------------
Iteration 5:
0.771122
0.304701
3.47729
2.2568
0.00571185
0.00272338
---------------------------------------------
Iteration 6:
0.770917
0.30433
3.48095
2.25854
0.00365784
0.00173935
---------------------------------------------
Iteration 7:
0.770786
0.304093
3.48329
2.25966
0.00234044
0.00111276
```

```
-------------------------------------------
Iteration 8:
0.770703
0.303941
3.48479
2.26037
0.00149713
0.000711534
-------------------------------------------
Iteration 9:
0.77065
0.303845
3.48575
2.26082
0.000957473
0.000454967
-------------------------------------------
Max Error=0.000957473
x=3.48575
y=2.26082
```

2.6 GRAEFFE'S ROOT SQUARING METHOD FOR ALGEBRAIC EQUATIONS

This direct method is used to find the roots of any polynomial equation with real coefficients. The basic idea behind this method is to separate the roots of the equations by squaring the roots. This can be done by separating even and odd powers of x in

$$P_n(x) = a_0x^n + a_1x^{n-1} + a_2x^{n-2} + \cdots + a_{n-1}x + a_n = 0, \quad a_0 \neq 0 \tag{2.38}$$

where $a_0, a_1, \ldots, a_n$ are real coefficients.

Separate even and odd terms of Equation (2.38) and then squaring on both sides, we obtain

$$(a_0x^n + a_2x^{n-2} + a_4x^{n-4} + \cdots)^2 = (a_1x^{n-1} + a_3x^{n-3} + \cdots)^2 \tag{2.39}$$

On simplification, we obtain

$$a_0^2x^{2n} - \left(a_1^2 - 2a_0a_2\right)x^{2n-2} + \left(a_2^2 - 2a_1a_3 + 2a_0a_4\right)x^{2n-4} - \cdots + (-1)^n a_n^2 = 0 \tag{2.40}$$

Now substituting y for $-x^2$, we have

$$b_0y^n + b_1y^{n-1} + \cdots + b_{n-1}y + b_n = 0 \tag{2.41}$$

where:

$$\begin{aligned}
b_0 &= a_0^2 \\
b_1 &= \left(a_1^2 - 2a_0a_2\right) \\
b_2 &= \left(a_2^2 - 2a_1a_3 + 2a_0a_4\right) \\
&\vdots \\
b_n &= a_n^2
\end{aligned}$$

Thus, all the b_i's are known in terms of a_i's. The roots of Equation 2.41 are $-\alpha_1^2, -\alpha_2^2, \ldots, -\alpha_n^2$, where $\alpha_1, \alpha_2, \ldots, \alpha_n$ are the roots of Equation 2.38.

The coefficients b_is can be determined from the following table:

a_0	a_1	a_2	a_3	…	a_n
a_0^2	a_1^2	a_2^2	a_3^2	…	a_n^2
	$-2a_0a_2$	$-2a_1a_3$	$-2a_0a_4$		
		$+2a_0a_4$	$+2a_1a_5$		
			$-2a_0a_6$		
b_0	b_1	b_2	b_3	…	b_n

A typical coefficient b_i $(i = 0,1,2,\ldots,n)$ in the $(i+1)$th column can be obtained by the following:

The terms alternate in sign starting with a positive sign. The first term is the square of the coefficient a_i. The second term is twice the product of the nearest neighboring coefficients a_{i-1} and a_{i+1}. The third term is twice the product of the next nearest neighboring coefficients a_{i-2} and a_{i+2}. This procedure is continued until there are no available coefficients to form the cross products.

This procedure can be repeated m times, so that we obtain the equation

$$A_0x^n + A_1x^{n-1} + A_2x^{n-2} + \cdots + A_{n-1}x + A_n = 0 \tag{2.42}$$

which has the roots $\xi_1, \xi_2, \ldots, \xi_n$ such that

$$\xi_i = -\alpha_i^{2^m}, \quad i = 1,2,\ldots,n \tag{2.43}$$

If we assume $|\alpha_1| > |\alpha_2| > \cdots > |\alpha_n|$, then $|\xi_1| >> |\xi_2| >> \cdots >> |\xi_n|$.

Thus, the roots ξ_i are very widely separated for large m.

Case I: Roots are real and unequal

Now from Equation 2.42, we have

$$-\frac{A_1}{A_0} = \sum \xi_i \cong \xi_1$$

$$\frac{A_2}{A_0} = \sum \xi_i\xi_j \cong \xi_1\xi_2$$

$$-\frac{A_3}{A_0} = \sum \xi_i\xi_j\xi_k \cong \xi_1\xi_2\xi_3$$

$$\vdots$$

$$(-1)^n\frac{A_n}{A_0} = \xi_1\xi_2\ldots\xi_n$$

which yields

$$\xi_i = -\frac{A_i}{A_{i-1}}, \quad i = 1,2,\ldots,n \tag{2.44}$$

Since

$$|\alpha_i|^{2^m} = |\xi_i| = \left|\frac{A_i}{A_{i-1}}\right|, \quad i = 1,2,\ldots,n$$

This implies,

$$\log|\alpha_i| = 2^{-m}\left(\log|A_i| - \log|A_{i-1}|\right), \quad i = 1,2,\ldots,n \tag{2.45}$$

This determines the absolute values of the roots and substitution in the original Equation 2.38 will give the sign of the roots.

Case II: Roots are real and two are numerically equal

We have

$$\xi_i \cong -\frac{A_i}{A_{i-1}} \quad \text{and} \quad \xi_{i+1} \cong -\frac{A_{i+1}}{A_i}$$

Therefore,

$$\xi_i\xi_{i+1} \cong \xi_i^2 \cong \left|\frac{A_{i+1}}{A_{i-1}}\right|$$

Hence,

$$\left|\xi_i^2\right| = \left|\alpha_i^{2^m}\right|^2 \cong \left|\frac{A_{i+1}}{A_{i-1}}\right| \tag{2.46}$$

Equation 2.46 gives the magnitude of the double root. By substituting in the given Equation 2.38, we can find its sign.

Case III: One pair of complex roots and others are real and distinct

Let us assume α_r, $\alpha_{r+1} = \rho e^{\pm i\theta} = \alpha \pm i\beta$ form a complex pair and all other roots to be real and distinct. It will cause the coefficient of x^{n-r} in the successive squaring fluctuate both in magnitude and sign.

Now, we suppose that

$$|\alpha_1| > |\alpha_2| > \cdots > |\alpha_r| = |\alpha_{r+1}| = \rho > \cdots > |\alpha_n|$$

Since the roots of the final equation are widely separately in magnitude, we have

$$|\xi_1| >> |\xi_2| >> \cdots >> |\xi_r| = |\xi_{r+1}| >> \cdots >> |\xi_n|$$

Now, proceeding in the same way as in Case I and Case II, we have up to the prescribed level of accuracy

$$-\frac{A_1}{A_0} \cong \xi_1$$

$$\frac{A_2}{A_0} \cong \xi_1\xi_2$$

$$\vdots$$

$$(-1)^{r-1}\frac{A_{r-1}}{A_0} \cong \xi_1\xi_2\ldots\xi_{r-1}$$

$$(-1)^r\frac{A_r}{A_0} \cong 2\xi_1\xi_2\ldots\xi_{r-1}\rho^{2^m}\cos(2^m\theta)$$

$$(-1)^{r+1}\frac{A_{r+1}}{A_0} \cong \xi_1\xi_2\ldots\xi_{r-1}\rho^{2(2^m)}$$

$$\vdots$$

$$(-1)^n\frac{A_n}{A_0} = \xi_1\xi_2\ldots\xi_n$$

For sufficiently large m, ρ can be determined from the relation

$$\rho^{2(2^m)} \cong \left|\frac{A_{r+1}}{A_{r-1}}\right|$$

and θ can also be determined from the relation

$$2\rho^{2^m}\cos(2^m\theta) = \left|\frac{A_r}{A_{r-1}}\right|$$

Now, Equation 2.38 gives

$$\alpha_1 + \alpha_2 + \alpha_3 + \cdots + \alpha_{r-1} + 2\alpha + \alpha_{r+2} + \cdots + \alpha_n = -\frac{a_1}{a_0}$$

which yields the value of α and the relation $\beta = \sqrt{\rho^2 - \alpha^2}$ determines β. Hence, all the roots are determined.

Example 2.17

Find the roots of the equation

$$x^3 - 8x^2 + 17x - 10 = 0$$

correct to four significant figures using the Graeffe's root squaring method.

Solution:

The coefficients of the successive root squaring are given in Table 2.21.

Let $\alpha_1, \alpha_2, \alpha_3$ are the roots of the given equation.

Then, from Table 2.21, we have

$$\alpha_1^2 = 30, \quad \alpha_2^2 = \frac{129}{30}, \quad \alpha_3^2 = \frac{100}{129}$$

TABLE 2.21
Coefficients in the Root Squaring by Graeffe's Method

m	2^m	x^3	x^2	x	1
0	1	1	−8	17	−10
		1	64	289	100
			−34	−160	
1	2	1	30	129	100
		1	900	16641	10,000
			−258	−6000	
2	4	1	642	10641	10,000
		1	412,164	113,230,881	10^8
			−21282	−12,840,000	
3	8	1	3.90882E5	1.00391E8	10^8
		1	1.52789E11	1.00783E16	10^{16}
			−2.00782E8	−7.81764E13	
4	16	1	1.52588E11	1.00001E16	10^{16}

Therefore,

$$|\alpha_1| = 5.47723, \quad |\alpha_2| = 2.07364, \quad |\alpha_3| = 0.880451$$

Again,

$$\alpha_1^4 = 642, \quad \alpha_2^4 = \frac{10641}{642}, \quad \alpha_3^4 = \frac{10,000}{10,641}$$

Therefore,

$$|\alpha_1| = 5.03366, \quad |\alpha_2| = 2.01772, \quad |\alpha_3| = 0.984588$$

and so on.

Therefore, the successive approximations to the roots are given in Table 2.22.

By Descarte's rule of sign, the given equation has three positive real roots, so the roots of the equation are 5, 2, and 1.

Example 2.18

Solve the equation

$$x^3 - 2x + 4 = 0$$

using the Graeffe's root squaring method.

Solution:

The coefficients of the successive root squaring are given in Table 2.23.

TABLE 2.22
Approximations to the Roots

m	α_1	α_2	α_3
2	5.47723	2.07364	0.880451
4	5.03366	2.01772	0.984588
8	5.00041	2.00081	0.999512
16	5.00000	2.00000	0.999999

TABLE 2.23
Coefficients in the Root Squaring by Graeffe's Method

m	2^m	x^3	x^2	x	1
0	1	1	0	−2	4
		1	0	4	16
			4	0	
1	2	1	4	4	16
		1	16	16	256
			−8	−128	
2	4	1	8	−112	256
		1	64	12,544	65,536
			224	−4,096	
3	8	1	288	8,448	65,536
		1	82,944	71,368,704	4,294,967,296
			−16,896	−37,748,736	
4	16	1	66,048	3,3619,968	4,294,967,296

Since the coefficients of x in the successive equations fluctuate both in magnitude and sign, α_2 and α_3 are complex conjugate roots.

Let $|\alpha_2| = |\alpha_3| = \rho$, then

$$\rho^{32} \cong \left|\frac{A_{r+1}}{A_{r-1}}\right| = \left|\frac{A_3}{A_1}\right| = \frac{4294967296}{66048}$$

Therefore, $\rho = 1.41387$

Now,

$$\alpha_1 = \pm 2.00097$$

Since -2 satisfies the given equation, thus $\alpha_1 = -2$.

Let $\alpha_2 = \alpha + i\beta$ and $\alpha_3 = \alpha - i\beta$, then the sum of the roots

$$2\alpha + \alpha_1 = 0$$

Therefore, $\alpha = 1$. Since $\rho^2 = \alpha^2 + \beta^2$, we have $\beta = \pm 1$

Hence, the roots of the given equation are -2 and $1 \pm i$.

EXERCISES

1. Find the approximate value of the smallest real root of the following equations using (i) graphical method and (ii) tabulation method:
   a. $e^{-x} - \sin x = 0$
   b. $3x - \cos x - 1 = 0$
   c. $\log x = \cos x$
2. Locate real roots of each of the following equations correct to two significant figures by the (i) tabulation method and (ii) graphical method:
   a. $x^2 - 10\log x - 3 = 0$
   b. $x^2 + \ln x = 2$
   c. $\sqrt{x} = 2\cos(\pi x/2)$
3. Obtain the root, correct to three decimal places, of each of the following equations using the bisection method:
   a. $5x\log_{10} x - 6 = 0$
   b. $x^2 + x - \cos x = 0$
4. Find the interval in which the smallest positive root of the following equations lies:
   a. $\tan x + \tanh x = 0$
   b. $x^3 - x - 4 = 0$

   Determine the roots correct to two decimals using the bisection method.
5. By using the following methods (i) bisection, (ii) fixed-point iteration, using Δ^2-process whenever necessary, (iii) Newton–Raphson, (iv) modified Newton–Raphson, (v) inverse interpolation, (vi) secant, and (vii) regula-falsi, compute correct to six significant figures the root of
   a. $x = \tan 2(x-1)$, which lies between 1 and 2
   b. $x^2 = \sin x$, which lies between $1/2$ and 1
   c. $x + \ln x = 2$, which lies between 1 and 2
   d. $x^x + 2x - 6 = 0$, which lies between 1 and 2
6. Explain the Newton–Raphson method to compute a real root of the equation $f(x) = 0$ and find the condition of convergence. Hence, find a nonzero root of the equation $x^2 + 4\sin x = 0$

7. Find the real root, which lies between 2 and 3, of the following equation:

$$x\log_{10} x - 1.2 = 0$$

using the methods of bisection and false-position to a tolerance of 0.5%.
8. Compute a real root of the following equations by the bisection method, correct to five significant figures
   a. $x = (1/2) + \sin x$
   b. $x^6 - x^4 - x^3 - 1 = 0$
   c. $x + \log x - 2 = 0$
   d. $\cos x = xe^x$
   e. $\tan x + x = 0$
9. Use the method of false position to find a real root, correct to three decimal places, of each of the following equations:
   a. $x^3 + x^2 + x + 7 = 0$
   b. $x^3 - x - 4 = 0$
   c. $x = 3e^{-x}$
   d. $x\tan x + 1 = 0$
10. Find the iterative functions of the following equations for which the fixed-point iteration method is convergent in the given interval
    a. $2x^3 + x^2 - 1 = 0$ in [0,1]
    b. $3x - \sqrt{1+\sin x} = 0$ in $\left[0, \pi/2\right]$
    c. $e^x - 4x = 0$ in [2,3]
    d. $x^2 + 2x - 2 = 0$ in [0,1]
11. Find a real root of each of the following equations correct to four decimal places using (i) the fixed-point iteration method and (ii) Aitken's Δ^2 method
    a. $2x - \log_{10} x = 7$ in [3,4]
    b. $x^3 + 3x - 5 = 0$ in [1,2]
    c. $x^x + 2x - 6 = 0$ in [1,2]
    d. $\sin x = 10(x-1)$ in [1,2]
    e. $\log x = \cos x$ in [1,2]
12. Using the Newton–Raphson method, find a real root of the following equations correct to five decimal places:
    a. $2x - 3\sin x - 5 = 0$
    b. $x^3 + x^2 + 3x + 4 = 0$
    c. $x\log_{10} x - 1.2 = 0$
    d. $10^x + x - 4 = 0$
    e. $x\sin x + \cos x = 0$
13. Use the method of iteration to find a real root of each of the following equations correct to four significant figures
    a. $e^x = 3x$
    b. $x = 1/(x+1)^2$
    c. $1 + x^2 = x^3$
    d. $x - \sin x = 1/2$
14. Use the Newton–Raphson method to obtain a root, correct to three decimal places, of each of the following equations:
    a. $x^{\sin 2} - 4 = 0$
    b. $e^x = 4x$

c. $xe^x = \cos x$

d. $\cot x = -x$

15. Show that an iterative method for computing $\sqrt[k]{a}$ is given by

$$x_{n+1} = \frac{1}{k}\left[(k-1)x_n + \frac{a}{x_n^{k-1}}\right] \quad \text{and that} \quad \varepsilon_{n+1} \cong -\frac{k-1}{2\sqrt[k]{a}}\varepsilon_n^2$$

16. Find the iterative methods based on the Newton–Raphson method for finding $\sqrt{N}$, $1/N$, $N^{1/3}$, where N is a positive real number. Apply the methods to $N = 18$ to obtain the results correct to two decimals.
17. Construct an iterative formula to evaluate the following using the Newton–Raphson method and hence evaluate

a. $\sqrt[9]{15}$

b. $\sqrt[7]{125}$

c. $\sqrt[3]{21}$

d. $\sqrt[4]{13}$

18. Find a double root of the following equations:

a. $x^3 - x^2 - x + 1 = 0$

b. $x^4 - 6x^3 + 9x^2 + 4x - 12 = 0$

c. $x^3 - 5x^2 + 8x - 4 = 0$

19. The root of the equation $x = (1/2) + \sin x$, by using the iteration method

$$x_{k+1} = (1/2) + \sin x_k, \quad x_0 = 1$$

correct to six decimal places is $x = 1.497300$. Determine the number of iteration steps required to reach the root by the linear iteration. If the Aitken-Δ^2 process is used after three approximations are available, how many iterations are required?

20. Use the regula-falsi method to evaluate the smallest real root of each of the following equations:

a. $x^3 + x^2 - 1 = 0$

b. $x^2 + 4\sin x = 0$

c. $xe^x = \cos x$

d. $2x^3 - 3x - 6 = 0$

21. Using the secant method, find a real root of the following equations correct to four significant figures:

a. $\sinh x - x = 0$

b. $3x^2 + 5x - 40 = 0$

c. $2(x+1) = e^x$

d. $x = \sin^{-1} 10(x-1)$

e. $x^2 - \ln x - 2 = 0$

22. Write a program implementing the algorithm for the bisection method. Use the program to calculate the real roots of the following equations. Use an error tolerance of $\varepsilon = 10^{-5}$:

a. $e^x - 3x^2 = 0$

b. $e^x = 1/1 + x^2$

c. $x = 1 + 0.3\cos x$

d. $xe^x = \cos x$

23. Use the program from Problem 22 to calculate the following:

(a) the smallest positive root of $x - \tan x = 0$ and (b) the root of this equation that is closest to $x = 100$

24. Use the Newton's method to find a solution of the following simultaneous equations:
    a. $x^2 + y - 11 = 0$, $x + y^2 - 7 = 0$ given $x_0 = 3$, $y_0 = -2$ as initial approximation.
    b. $y - \sin x = 1.32$, $x - \cos y = -0.85$ given $x_0 = 0.5$, $y_0 = 2.0$ as initial approximation.
25. Consider Newton's method for finding the positive square root of $a > 0$. Derive the following results, assuming $x_0 > 0$, $x_0 \neq \sqrt{a}$
    a. $x_{n+1} = (x_n + (a/x_n))/2$
    b. $x_{n+1}^2 - a = \left[(x_n^2 - a)/2x_n\right]^2/2$, $n \geq 0$, and thus $x_n > \sqrt{a}$ for all $n > 0$
26. Show that the following two sequences, both have convergence of the second order with limit $x_{n+1} = x_n\left(1 + a/x_n^2\right)/2$, $x_{n+1} = x_n\left(3 - x_n^2/a\right)/3$.
27. Show that $x = 1 + \tan^{-1}(x)$ has a solution α. Find an interval $[a,b]$ containing α such that for every $x_0 \in [a,b]$, the iteration

$$x_{n+1} = 1 + \tan^{-1}(x_n), \quad n \geq 0$$

    will converge to α. Calculate the first few iterates and estimate the rate of convergence.
28. Solve the following equations by Graeffe's root squaring method
    a. $x^4 + x^3 - 29x^2 - 63x - 90 = 0$
    b. $x^4 - 2x^3 - 6x^2 + 6x + 9 = 0$
    c. $x^3 - 9x^2 + 18x - 6 = 0$
    d. $x^3 - 7x^2 + 16x - 12 = 0$
29. Solve by Graeffe's root squaring method correct to four significant figures:
    a. $x^3 - 2x^2 - 5x + 6 = 0$
    b. $x^4 - 5x^3 + 20x^2 - 40x + 60 = 0$
    c. $x^3 + 6.09510x^2 - 35.3942x - 25.7283 = 0$
    d. $x^3 - 6x^2 + 11x - 6 = 0$
    e. $32x^3 - 6x - 1 = 0$
    f. $x^3 - 3x^2 + 6x - 8 = 0$
    g. $x^4 - 12x^3 + 49x^2 - 78x + 40 = 0$
    h. $x^3 - x - 1 = 0$
30. Find the real roots of the following equations using Graeffe's root-squaring method
    a. $x^3 - 4x^2 + 5x - 2 = 0$
    b. $x^3 - 2x^2 - 5x - 6 = 0$
31. Using the fixed-point iteration method, solve the following system of equations

$$x^2y + y^3 = 10$$
$$xy^2 - x^2 = 3$$

    with initial approximation $x = 0.8$, $y = 2.2$. Also, perform two iterations of Newton's method to obtain this root.
32. The following system of equations

$$y\cos(xy) + 1 = 0$$
$$\sin(xy) + x - y = 0$$

    has one solution close to $x = 1$, $y = 2$. Calculate this solution correct to two decimal places.

33. The following system of equations

$$\log_e(x^2+y)-1+y=0$$
$$\sqrt{x}+xy=0$$

has one approximate solution $(x_0, y_0)=(2.4, -0.6)$. Improve this solution and estimate the accuracy of the result.

34. Using Newton's method, solve the solution of the following system of equations:

$$x^3+y^3=53$$
$$2y^3+z^4=69$$
$$3x^5+10z^2=770$$

which is close to $x=3,\ y=3,\ z=2$.

35. Show that the equation $f(x)=\cos\left[\pi(x+1)/8\right]+0.148x-0.9062=0$ has one root in the interval $(-1,0)$ and one in $(0,1)$. Calculate the negative root correct to four decimals.

36. Solve the following systems by the Newton's method:
    a. $x^2+y^2=1,\quad y=x^3$
    b. $x^2-y^2=4,\quad x^2+y^2=16$
    c. $x^2+y^2=11,\quad y^2+x=7$

37. Obtain the Newton–Raphson extended formula

$$x_1=x_0-\frac{f(x_0)}{f'(x_0)}-\frac{1}{2}\frac{\{f(x_0)\}^2 f''(x_0)}{\{f'(x_0)\}^3}$$

for the root of the equation $f(x)=0$.

38. Let $x=\xi$ be a simple root of the equation $f(x)=0$. We try to find the root by means of the iteration formula

$$x_{i+1}=x_i-\left[\left(f(x_i)\right)^2/\left(f(x_i)-f\left(x_i-f(x_i)\right)\right)\right]$$

Find the order of convergence and compare the convergence properties with those of the Newton–Raphson method.

39. Show that

$$x_{n+1}=\frac{x_n\left(x_n^2+3a\right)}{3x_n^2+a},\quad n\geq 0$$

is a third-order method for computing $\sqrt{a}$. Calculate

$$\lim_{n\to\infty}\frac{\sqrt{a}-x_{n+1}}{(\sqrt{a}-x_n)^3}$$

assuming x_0 has been chosen sufficiently close to α.

40. The equation $x^2+ax+b=0$ has two real roots α and β. Show that the iteration method $x_{k+1}=-(ax_k+b)/x_k$ is convergent near $x=\alpha$ if $|\alpha|>|\beta|$ and that $x_{k+1}=-b/(x_k+a)$ is convergent near $x=\alpha$ if $|\alpha|<|\beta|$. Show also that the iteration method $x_{k+1}=-(x_k^2+b)/a$ is convergent near $x=\alpha$ if $2|\alpha|<|\alpha+\beta|$.

41. Using Newton's method for nonlinear systems, solve the following nonlinear system:

$$x^2 + y^2 = 4, \quad x^2 - y^2 = 1$$

The true solutions are easily determined to be $(\pm\sqrt{2.5}, \pm\sqrt{1.5})$. As an initial guess, use $(x_0, y_0) = (1.6, 1.2)$.

42. Determine the order of the convergence of the iterative method

$$x_{k+1} = \frac{x_0 f(x_k) - x_k f(x_0)}{f(x_k) - f(x_0)}$$

for finding a simple root of the equation $f(x) = 0$.

43. Solve the following system:

$$x^2 + xy^3 = 9, \quad 3x^2 y - y^3 = 4$$

using Newton's method for nonlinear systems. Use each of the initial guesses $(x_0, y_0) = (1.2, 2.5)$, $(-2, 2.5)$, $(-1.2, -2.5)$, $(2, -2.5)$. Observe from which root to which the method converges, the number of iterates required, and the speed of convergence

44. Using Newton's method for nonlinear systems, solve for all roots of the following nonlinear system:
    a. $x^2 + y^2 - 2x - 2y + 1 = 0, \quad x + y - 2xy = 0$
    b. $x^2 + 2xy + y^2 - x + y - 4 = 0, \quad 5x^2 - 6xy + 5y^2 + 16x - 16y + 12 = 0$

3 Interpolation

3.1 INTRODUCTION

Let us assume that $f(x)$ be a function defined in $(-\infty,\infty)$, in which it is continuously differentiable a sufficient number of times. Here, we are concerned with the function $f(x)$ such that the analytical formula representing the function is unknown, but the values of $f(x)$ are known for a given set of $n+1$ distinct values of x, say, x_0, x_1, x_2, …, x_n.

x	$f(x)$
x_0	$f(x_0)$
x_1	$f(x_1)$
x_2	$f(x_2)$
$\vdots$	$\vdots$
x_n	$f(x_n)$

Our problem is to find the value of $f(x)$ for a given value of x in the vicinity of the above-tabulated values of the arguments, but different from the above tabulated values of x. Since the formula for $f(x)$ is unknown, the precise value of $f(x)$ cannot be obtained. We try to find an approximate value of the same by the principle of interpolation. In interpolation, the unknown function $f(x)$ is approximated by a simple function $\varphi_n(x)$, so that it takes the same values as $f(x)$ for the given argument values x_0, x_1, x_2, …, x_n. In case, if the given value of x lies slightly outside the interval $\left[\min\{x_0,x_1,\ldots,x_n\},\max\{x_0,x_1,\ldots,x_n\}\right]$, the corresponding process is often called *extrapolation*.

Now the function $\varphi_n(x)$ may be in a variety of forms. When $\varphi_n(x)$ is a polynomial, then the process of approximating $f(x)$ by $\varphi_n(x)$ is called *polynomial interpolation*. If $\varphi_n(x)$ is a piecewise polynomial, the interpolation is called *piecewise polynomial interpolation*; if $\varphi_n(x)$ is a finite trigonometric series, then interpolation is called *trigonometric interpolation*. Likewise, $\varphi_n(x)$ may be a series of Legendre polynomials, Bessel functions, and Chebyshev polynomials.

First, we shall be familiar with polynomial interpolation. It is concerned with following famous theorem put forth by Weierstrass.

Theorem 3.1: Weierstrass Approximation Theorem

Suppose $f(x)$ be a continuous real-valued function defined on the real interval $[a, b]$. For every $\varepsilon > 0$, there exists a polynomial $\varphi_n(x)$ such that for all x in $[a, b]$, we have $\left\|f(x)-\varphi_n(x)\right\|_\infty < \varepsilon$.

Proof:

Please see Chapter 9.

3.2 POLYNOMIAL INTERPOLATION

In polynomial interpolation, $f(x)$ is approximated by a polynomial $\varphi_n(x)$ of degree $\le n$ such that

$$f(x_i) \cong \varphi_n(x_i) \text{ for all } i = 0,1,2,3,\ldots,n \tag{3.1}$$

Let $\varphi_n(x) = a_0 + a_1x + \cdots + a_nx^n$.

Then from Equation 3.1, we get

$$a_0 + a_1x_i + \cdots + a_nx_i^n = f\left(x_i\right) \text{ for all } i = 0,1,2,3, \ldots, n \tag{3.2}$$

This is a set of $(n+1)$ linear equations in $(n+1)$ unknowns of $a_0, a_1, a_2, \ldots, a_n$. The coefficient determinant of the system Equation 3.2 is

$$\begin{vmatrix} 1 & x_0 & \cdots & x_0^n \\ 1 & x_1 & \cdots & x_1^n \\ . & . & . & . \\ . & . & . & . \\ 1 & x_n & \cdots & x_n^n \end{vmatrix} = \prod_{\substack{0 \le i,j \le n \\ i \ne j}} \left(x_i - x_j\right) \ne 0$$

This determinant is known as *Vandermonde's determinant.* The value of this determinant is different from zero because $x_0, x_1, x_2, \ldots, x_n$ are distinct.

Therefore, by Cramer's rule, the values of $a_0, a_1, a_2, \ldots, a_n$ can be uniquely determined, so that the polynomial $\varphi_n(x)$ exists and is unique. This polynomial $\varphi_n(x)$ is called the *interpolation polynomial.* The given points $x_0, x_1, x_2, \ldots, x_n$ are called *interpolating points* or *nodes.*

3.2.1 Geometric Interpretation of Interpolation

Geometrically, the curve representing the unknown function $y = f(x)$ passes through the points (x_i, y_i), $(i = 0,1, \ldots, n)$. This unknown function is approximated by a unique nth degree parabola $y = \varphi_n(x)$, which passes through the above points. It has been depicted in Figure 3.1. The parabola $y = \varphi_n(x)$ is called *interpolation parabola* or *interpolation polynomial.* In this context, polynomial interpolation is also referred to as *parabolic interpolation.*

3.2.2 Error in Polynomial Interpolation

Let us assume that the unknown function $f(x)$ be $(n+1)$ times continuously differentiable on $[x_0, x_n]$. Let $f(x)$ be approximated by a polynomial $\varphi_n(x)$ of degree less than equal to n such that

$$f(x_i) = \varphi_n(x_i) \text{ for all } i = 0,1,2,3, \ldots, n \tag{3.3}$$

Since $f(x) - \varphi_n(x)$ vanishes at the interpolating points $x_0, x_1, x_2, \ldots, x_n$, we may write

$$f(x) = \varphi_n(x) + \Omega(x)R(x) \tag{3.4}$$

where $\Omega(x) = (x - x_0)(x - x_1)\ldots(x - x_n)$, which is a polynomial of degree $n+1$.

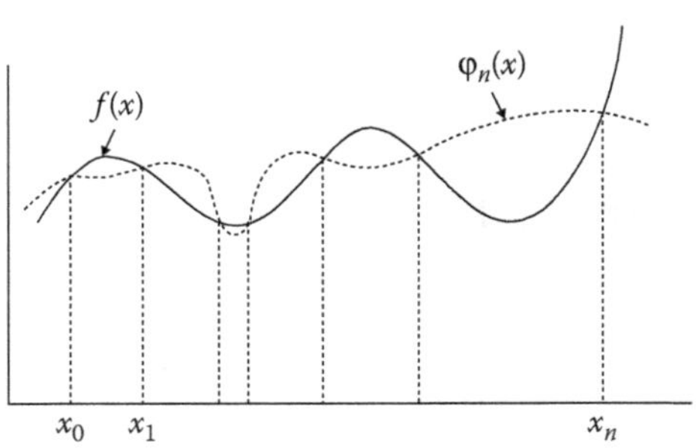

FIGURE 3.1 Geometrical representation of interpolation.

Let x' be any point in $[x_0, x_n]$ that is distinct from interpolating points $x_0, x_1, x_2, \ldots, x_n$.
Now let us construct a function

$$F(x) = f(x) - \varphi_n(x) - \Omega(x)R(x') \tag{3.5}$$

where $R(x')$ is a constant and from Equation 3.4, it is given by $R(x') = (f(x') - \varphi_n(x'))/\Omega(x')$.
It is obviously clear that

$$F(x_0) = F(x_1) = \ldots = F(x_n) = F(x') = C$$

Therefore, $F(x)$ vanishes at the $(n+2)$ points, that is, $x = x', x_0, x_1, \ldots, x_n$. Now the function $F(x)$ is continuously differentiable everywhere $(n+1)$ times. Consequently, by the repeated application of Rolle's theorem, $F'(x)$ must vanish $(n+1)$ times in $[x_0, x_n]$, $F''(x)$ must vanish n times in $[x_0, x_n]$ and so on. Finally, $F^{(n+1)}(x)$ must vanish once in $[x_0, x_n]$, say, at the point ξ, so that

$$F^{(n+1)}(\xi) = 0 \tag{3.6}$$

where $\min\{x', x_0, x_1, \ldots, x_n\} < \xi < \max\{x', x_0, x_1, \ldots, x_n\}$.
On differentiating Equation 3.5, $(n+1)$ times with respect to x and substituting $x = \xi$, we obtain

$$R(x') = \frac{f^{(n+1)}(\xi)}{(n+1)!}$$

since $\varphi_n(x)$ is a polynomial of degree less than equal to n, its $(n+1)$th derivative vanishes identically. Moreover, since $\Omega(x) = (x - x_0)(x - x_1)\ldots(x - x_n)$ is a polynomial of degree $n+1$, $\Omega^{(n+1)}(x) = (n+1)!$.
Thus, from Equation 3.4, we have

$$f(x') - \varphi_n(x') = \Omega(x')\frac{f^{(n+1)}(\xi)}{(n+1)!}$$

the quantity on the right-hand side gives the error at any point x' other than the interpolating points.
Since x' is an arbitrary point in $[x_0, x_n]$, on dropping the prime, we may write

$$f(x) - \varphi_n(x) = \Omega(x)\frac{f^{(n+1)}(\xi)}{(n+1)!} \tag{3.7}$$

where $\min\{x, x_0, x_1, \ldots, x_n\} < \xi < \max\{x, x_0, x_1, \ldots, x_n\}$.

3.2.3 Finite Differences

Let $y = f(x)$ be a real-valued function of x and $y_0, y_1, \ldots, y_n$ be the values of $y = f(x)$ corresponding to the values $x_0, x_1, \ldots, x_n$ of x. The values of x, that is, $x_i (i = 0, 1, \ldots, n)$ are called the *nodes* or *arguments*. The argument values $x_0, x_1, \ldots, x_n$ may or may not be equidistant or equally spaced.

3.2.3.1 Forward Differences

Let $y_0, y_1, \ldots, y_n$ be a given set of values of y corresponding to the equidistant values $x_0, x_1, \ldots, x_n$ of x, that is, $x_i = x_0 + ih, i = 0, 1, \ldots, n$. The differences $y_1 - y_0, y_2 - y_1, \ldots, y_n - y_{n-1}$ are called *first forward differences* if these are denoted by $\Delta y_0, \Delta y_1, \ldots, \Delta y_{n-1}$, respectively. Thus we have

$$\Delta y_i = y_{i+1} - y_i, \quad i = 0, 1, \ldots, n-1 \tag{3.8}$$

The operator Δ is called *first forward difference operator.*

In general, the first forward difference operator is defined by

$$\Delta f(x) = f(x+h) - f(x) \tag{3.9}$$

Similarly, we can define the second order, third order, fourth order, and many more forward differences formulae, respectively, as follows:

$$\begin{aligned} \Delta^2 y_i &= \Delta(\Delta y_i) = \Delta y_{i+1} - \Delta y_i \\ \Delta^3 y_i &= \Delta(\Delta^2 y_i) = \Delta^2 y_{i+1} - \Delta^2 y_i \\ &\dots \\ \Delta^k y_i &= \Delta(\Delta^{k-1} y_i) = \Delta^{k-1} y_{i+1} - \Delta^{k-1} y_i \end{aligned} \tag{3.10}$$

where $i = 0, 1, \dots, n-1$ and $k\,(k = 1, \dots, n)$ is a positive integer.

Thus, we have

$$\begin{aligned} \Delta y_0 &= y_1 - y_0 \\ \Delta^2 y_0 &= \Delta(\Delta y_0) = \Delta y_1 - \Delta y_0 = y_2 - 2y_1 + y_0 \\ \Delta^3 y_0 &= \Delta(\Delta^2 y_0) = \Delta^2 y_1 - \Delta^2 y_0 = y_3 - 3y_2 + 3y_1 - y_0 \\ \Delta^4 y_0 &= \Delta(\Delta^3 y_0) = \Delta^3 y_1 - \Delta^3 y_0 = y_4 - 4y_3 + 6y_2 - 4y_1 + y_0 \end{aligned} \tag{3.11}$$

and so on.

In general,

$$\Delta^k y_i = \sum_{r=0}^{k} (-1)^r \binom{k}{r} y_{i+k-r} \tag{3.12}$$

where $i = 0, 1, \dots, n-1$ and $k\,(k = 1, \dots, n)$ is a positive integer.

3.2.3.1.1 Forward Difference Table

We can calculate the above forward differences very easily with the help of Table 3.1, which is called *forward difference table.*

3.2.3.1.2 Some Properties of Forward Differences

1. The first order forward difference of a constant is zero.
2. The first order forward difference of a polynomial of degree n is a polynomial of degree $n-1$.

Proof:

Let $P(x) = a_0 + a_1 x + \cdots + a_n x^n$, where $a_n \neq 0$ be a polynomial of degree n.

Now, we have

$$\Delta x^n = (x+h)^n - x^n = \sum_{i=1}^{n} \binom{n}{i} x^{n-i} h^i$$

which is a polynomial of degree $n-1$.

TABLE 3.1
Forward Difference Table

x	y	Δy	$\Delta^2 y$	$\Delta^3 y$	$\Delta^4 y$	$\Delta^5 y$
x_0	y_0					
		Δy_0				
x_1	y_1		$\Delta^2 y_0$			
		Δy_1		$\Delta^3 y_1$		
x_2	y_2		$\Delta^2 y_1$		$\Delta^4 y_0$	
		Δy_2		$\Delta^3 y_2$		$\Delta^5 y_0$
x_3	y_3		$\Delta^2 y_2$		$\Delta^4 y_1$	
		Δy_3		$\Delta^3 y_3$		
x_4	y_4		$\Delta^2 y_3$			
		Δy_4				
x_5	y_5					

Therefore,

$$\Delta P(x) = \Delta(a_0 + a_1 x + \cdots + a_n x^n) = a_1 \Delta x + a_2 \Delta x^2 + \cdots + a_n \Delta x^n$$

which is a polynomial of degree $n-1$. ■

3. The kth order difference of a polynomial of degree $n(\geq k)$ is a polynomial of degree $n-k$. In particular, The nth order difference of a polynomial of degree n is constant and so the $(n+1)$th order difference is zero.

Proof:

Let $P(x) = a_0 + a_1 x + \cdots + a_n x^n$, where $a_n \neq 0$ be a polynomial of degree n.

According to property (2), $\Delta P(x) = \Delta(a_0 + a_1 x + \cdots + a_n x^n) = a_1 \Delta x + a_2 \Delta x^2 + \cdots + a_n \Delta x^n$ is a polynomial of degree $n-1$.

Thus, $\Delta P(x) = b_0 + b_1 x + \cdots + b_{n-1} x^{n-1}$, where $b_{n-1} \neq 0$.

Similarly, $\Delta^2 P(x) = b_1 \Delta x + b_2 \Delta x^2 + \cdots + b_{n-1} \Delta x^{n-1} = c_0 + c_1 x + \cdots + c_{n-2} x^{n-2}$, (where $c_{n-2} \neq 0$) say, is a polynomial of degree $n-2$.

Using method of induction, it may be easily proved that $\Delta^k P(x)$ is a polynomial of degree $n-k\,(k \leq n)$. Therefore, $\Delta^n P(x)$ is a polynomial of degree 0, that is, $\Delta^n P(x)$ is a constant, say c.

Then according to property (1), $\Delta^{n+1} P(x) = c - c = 0$.

This establishes the results. ■

Corollary: The converse of the property (3) is also true. Thus if the nth order forward difference of a polynomial is constant, then it is of degree n.

3.2.3.1.3 Propagation Error in Forward Difference Table

Let $y_0, y_1, \ldots, y_n$ be the actual values of a function, and suppose that the value of y_4 has been affected with an error ε, so that the erroneous value of y_4 is $y_4 + \varepsilon$. Then effect of error propagation in successive forward differences has been shown in Table 3.2.

TABLE 3.2
The Effect of an Error in the Forward Difference Table

x	y	Δy	$\Delta^2 y$	$\Delta^3 y$	$\Delta^4 y$
x_0	y_0				
		Δy_0			
x_1	y_1		$\Delta^2 y_0$		
		Δy_1		$\Delta^3 y_0$	
x_2	y_2		$\Delta^2 y_1$		$\Delta^4 y_0 + \varepsilon$
		Δy_2		$\Delta^3 y_1 + \varepsilon$	
x_3	y_3		$\Delta^2 y_2 + \varepsilon$		$\Delta^4 y_1 - 4\varepsilon$
		$\Delta y_3 + \varepsilon$		$\Delta^3 y_2 - 3\varepsilon$	
x_4	$y_4 + \varepsilon$		$\Delta^2 y_3 - 2\varepsilon$		$\Delta^4 y_2 + 6\varepsilon$
		$\Delta y_4 - \varepsilon$		$\Delta^3 y_3 + 3\varepsilon$	
x_5	y_5		$\Delta^2 y_4 + \varepsilon$		$\Delta^4 y_3 - 4\varepsilon$
		Δy_5		$\Delta^3 y_4 - \varepsilon$	
x_6	y_6		$\Delta^2 y_5$		$\Delta^4 y_4 + \varepsilon$
		Δy_6		$\Delta^3 y_5$	
x_7	y_7		$\Delta^2 y_6$		
		Δy_7			
x_8	y_8				

Table 3.2 shows that

1. The effect of error increases with the successive forward differences.
2. The coefficients of the ε's are the binomial coefficients with alternating signs.
3. The algebraic sum of the errors in any order of difference column is zero.
4. The maximum error in the forward differences occurs in the same horizontal line as the erroneous tabular value.

3.2.3.1.4 Relation between nth Order Forward Difference and Derivative of a Function

Theorem 3.2

The nth order forward difference of a function $f(x)$, which is continuously differentiable sufficient number of times, is related with its nth order derivative by the following relation:

$$\Delta^n f(x) = h^n f^{(n)}(x) + O(h^n) \tag{3.13}$$

where:

h is the step length

$O(h^n)$ is a function of the form $h^n R_n(x,h)$ such that $R_n(x,h) \to 0$ as $h \to 0$

Proof:

We have

$$\Delta f(x) = f(x+h) - f(x)$$

$$= hf'(x) + \frac{1}{2!}h^2 f''(x+\theta h), \quad 0<\theta<1, \text{ applying Taylor's series expansion}$$

$$= hf'(x) + hR_1(x,h)$$

where $R_1(x,h) = (1/2)hf''(x+\theta h) \to 0$ as $h \to 0$.

We shall prove the result by the method of induction on n.

Let us assume that the result is true for $n = k$. Therefore,

$$\Delta^k f(x) = h^k f^{(k)}(x) + h^k R_k(x,h)$$

where $R_k(x,h) \to 0$ as $h \to 0$.

Now,

$$\begin{aligned}\Delta^k f(x+h) &= h^k f^{(k)}(x+h) + h^k R_k(x+h,h) \\ &= h^k\left(f^{(k)}(x) + hf^{(k+1)}(x) + \frac{1}{2}h^2 f^{(k+2)}(x+\theta_1 h)\right) \\ &\quad + h^k\left(R_k(x,h) + h\frac{\partial R_k(x+\theta_2 h,h)}{\partial x}\right), \quad 0<\theta_1,\theta_2<1\end{aligned}$$

applying Taylor's series expansion of $f^{(k)}(x+h)$ and $R_k(x+h,h)$ about x.

Therefore,

$$\Delta^{k+1} f(x) = \Delta^k f(x+h) - \Delta^k f(x) = h^{k+1} f^{(k+1)}(x) + h^{k+1} R_{k+1}(x,h)$$

where

$$R_{k+1}(x,h) = \frac{1}{2}hf^{(k+2)}(x+\theta_1 h) + \frac{\partial R_k(x+\theta_2 h,h)}{\partial x} \tag{3.14}$$

Now, we have to prove that $R_{k+1}(x,h) \to 0$ as $h \to 0$.

The first term on the right-hand side of Equation 3.14 clearly tends to be zero as $h \to 0$. Regarding the second term on the right-hand side of Equation 3.14, due to induction hypothesis $R_k(x,h) \to 0$ as $h \to 0$ and consequently $R_k(x,h) \to R_k(x,0)$, so that $R_k(x,0) = 0$ and this is true for all x.

Therefore, we have

$$\frac{\partial R_k(x,0)}{\partial x} = 0$$

and so

$$\frac{\partial R_k(x+\theta_2 h,h)}{\partial x} \to \frac{\partial R_k(x,0)}{\partial x} = 0 \quad \text{as} \quad h \to 0$$

Hence, by the method of induction, the result in Equation 3.13 holds for all values of n. ■

3.2.4 Shift, Differentiation, and Averaging Operators

3.2.4.1 Shift Operator

The shift operator or shifting operator E is defined by

$$Ef(x) = f(x+h) \tag{3.15}$$

Now, we know that

$$\Delta f(x) = f(x+h) - f(x)$$
$$= Ef(x) - f(x)$$
$$= (E - I)f(x)$$

where I is the identity operator.

Therefore, we have

$$\Delta = E - I \tag{3.16}$$

Equation 3.16 represents a relation between the shift operator and the forward difference operator Δ.

Now, we have

$$Ef(x) = f(x+h)$$
$$E^2 f(x) = E.Ef(x) = Ef(x+h) = f(x+2h)$$
$$E^3 f(x) = E.E^2 f(x) = Ef(x+2h) = f(x+3h)$$
$$\ldots$$

By the method of induction, it may be easily proved that

$$E^n f(x) = f(x+nh) \tag{3.17}$$

where n is a positive integer.

The inverse operator is defined by

$$E^{-1} f(x) = f(x-h) \tag{3.18}$$

Proceeding in the similar manner as above, we may obtain

$$E^{-n} f(x) = f(x-nh) \tag{3.19}$$

where n is a positive integer.

3.2.4.2 Differentiation Operator

The differentiation operator D is defined by

$$Df(x) = \frac{d}{dx} f(x)$$

Let $f(x)$ be a function that is continuously differentiable sufficient number of times in a finite interval $[a,b]$. If h is the step length, then by Taylor's expansion, we have

$$f(x+h) = f(x) + hf'(x) + \frac{h^2}{2!} f''(x) + \frac{h^3}{3!} f'''(x) + \cdots$$

which may be written in terms of operators as

$$Ef(x) = f(x) + hDf(x) + \frac{h^2}{2!} D^2 f(x) + \frac{h^3}{3!} D^3 f(x) + \cdots$$
$$= \left(1 + hD + \frac{h^2}{2!} D^2 + \frac{h^3}{3!} D^3 + \cdots\right) f(x)$$
$$= e^{hD} f(x)$$

This implies that

$$E \equiv e^{hD} \tag{3.20}$$

Therefore, from Equation 3.16, we get

$$\Delta + I \equiv e^{hD} \tag{3.21}$$

Thus,

$$D \equiv \frac{1}{h}\log(I+\Delta) \equiv \frac{1}{h}\left(\Delta - \frac{\Delta^2}{2} + \frac{\Delta^3}{3} - \cdots\right) \tag{3.22}$$

3.2.4.3 Averaging Operator

The averaging operator μ is defined by

$$\begin{aligned}\mu f(x) &= \frac{1}{2}\left[f\left(x+\frac{h}{2}\right)+f\left(x-\frac{h}{2}\right)\right]\\ &= \frac{1}{2}\left[E^{1/2}f(x)+E^{-1/2}f(x)\right]\\ &= \frac{1}{2}\left[E^{1/2}+E^{-1/2}\right]f(x)\end{aligned} \tag{3.23}$$

This implies that

$$\mu = \frac{1}{2}\left[E^{1/2}+E^{-1/2}\right] \tag{3.24}$$

Equation 3.24 represents a relation between the averaging operator and the shift operator E.

Example 3.1

Prove that

$$\left(\frac{\Delta^2}{E}\right)x^4 = 12x^2+2$$

where the interval of differencing being unity.

Solution:

$$\left(\frac{\Delta^2}{E}\right)x^4 = \frac{(E-1)^2}{E}x^4 = \frac{(E^2-2E+1)}{E}x^4 = (E-2+E^{-1})x^4 = (x+1)^4 - 2x^4 + (x-1)^4 = 12x^2+2$$

Example 3.2

Prove that

$$\Delta \log f(x) = \log\left\{1+\frac{\Delta f(x)}{f(x)}\right\}$$

Solution:

Let,

$$g(x) = \log f(x)$$

Then,

$$g(x+h) = \log f(x+h)$$

Now,

$$\Delta \log f(x) = \Delta g(x) = g(x+h) - g(x)$$

$$= \log f(x+h) - \log f(x)$$

$$= \log \frac{f(x+h) - f(x) + f(x)}{f(x)}$$

$$= \log \frac{f(x) + \Delta f(x)}{f(x)}$$

$$= \log \left\{ 1 + \frac{\Delta f(x)}{f(x)} \right\}$$

Example 3.3

Prove that

$$\Delta^n \left(\frac{1}{x} \right) = \frac{(-1)^n n! h^n}{x(x+h)(x+2h)\ldots(x+nh)}$$

Solution:

$$\Delta \left(\frac{1}{x} \right) = \frac{1}{x+h} - \frac{1}{x} = \frac{-h}{x(x+h)} = \frac{(-1)1!h}{x(x+h)}$$

$$\Delta^2 \left(\frac{1}{x} \right) = -h \left[\frac{1}{(x+h)(x+2h)} - \frac{1}{x(x+h)} \right]$$

$$= \frac{2h^2}{x(x+h)(x+2h)}$$

$$= \frac{(-1)^2 2! h^2}{x(x+h)(x+2h)}$$

Thus, the result is true for $n = 1, 2$. Let us suppose that the result is valid for $n = m$, that is,

$$\Delta^m \left(\frac{1}{x} \right) = \frac{(-1)^m m! h^m}{x(x+h)(x+2h)\ldots(x+mh)}$$

Now,

$$\Delta^{m+1} \left(\frac{1}{x} \right) = (-1)^m m! h^m \left[\frac{1}{(x+h)(x+2h)\ldots(x+\overline{m+1}h)} - \frac{1}{x(x+h)(x+2h)\ldots(x+mh)} \right]$$

$$= \frac{(-1)^m m! h^m}{(x+h)(x+2h)\ldots(x+mh)} \cdot \frac{-(m+1)h}{x(x+\overline{m+1}h)} = \frac{(-1)^{m+1}(m+1)! h^{m+1}}{x(x+h)(x+2h)\ldots(x+\overline{m+1}h)}$$

Therefore, the result also holds for $n = m + 1$.

Hence, by induction, the result holds for any positive integer n, that is, $n = 1,2,\ldots$.

Example 3.4

Prove that

$$\Delta^n \cos(a+bx) = \left(2\sin\frac{b}{2}\right)^n \cos\left[a+bx+\frac{n}{2}(b+\pi)\right]$$

where $h = 1$ and a and b are constant.

Solution:

$$\Delta\cos(a + bx) = \cos(a + b(x + 1)) - \cos(a + bx)$$

$$= -2\sin\left(a + bx + \frac{b}{2}\right)\sin\frac{b}{2}$$

$$= 2\sin\frac{b}{2}\cos\left[a + bx + \frac{1}{2}(b+\pi)\right]$$

$$\Delta^2\cos(a + bx) = 2\sin\frac{b}{2}\Delta\cos\left[a + bx + \frac{1}{2}(b+\pi)\right]$$

$$= \left(2\sin\frac{b}{2}\right)^2\cos\left[a + bx + 2\frac{(b+\pi)}{2}\right]$$

Thus, the result is true for $n = 1, 2$. Let us suppose that the result is valid for $n = m$, that is,

$$\Delta^m\cos(a+bx) = \left(2\sin\frac{b}{2}\right)^m\cos\left[a + bx + \frac{m}{2}(b+\pi)\right]$$

Now,

$$\Delta^{m+1}\cos(a + bx) = \Delta\left(\Delta^m\cos(a + bx)\right)$$

$$= \left(2\sin\frac{b}{2}\right)^m \Delta\cos\left[a + bx + \frac{m}{2}(b + \pi)\right]$$

$$= \left(2\sin\frac{b}{2}\right)^m\left(2\sin\frac{b}{2}\right)\cos\left[a + bx + \frac{m+1}{2}(b + \pi)\right]$$

$$= \left(2\sin\frac{b}{2}\right)^{m+1}\cos\left[a + bx + \frac{m+1}{2}(b + \pi)\right]$$

Therefore, the result also holds for $n = m+1$.

Hence by induction, the result holds for any positive integer n, that is, $n = 1,2,\ldots$.

Example 3.5

Find the missing terms in the following table

x	0	1	2	3	4	5
$f(x)$	0	–	8	15	–	35

Solution:

Here, $f(x)$ has four given values. Let us assume that $f(x)$ be a polynomial of degree 3. Hence the fourth forward difference of $f(x)$ is

$$\Delta^4 f(x) = 0, \quad \text{for all } x$$

This implies

$$(E - I)^4 f(x) = 0$$

Thus $E^4 f(x) - 4E^3 f(x) + 6E^2 f(x) - 4Ef(x) + f(x) = 0$
which yields

$$f(x+4) - 4f(x+3) + 6f(x+2) - 4f(x+1) + f(x) = 0, \quad \text{since } h = 1 \tag{3.25}$$

Putting $x = 0$ in Equation 3.25, we get

$$f(4) - 4f(3) + 6f(2) - 4f(1) + f(0) = 0 \tag{3.26}$$

Substituting the values of $f(0)$, $f(2)$ and $f(3)$ in Equation 3.26, we obtain

$$f(4) - 4f(1) = 12 \tag{3.27}$$

Again, putting $x = 1$ in Equation 3.25, we get

$$f(5) - 4f(4) + 6f(3) - 4f(2) + f(1) = 0 \tag{3.28}$$

Substituting the values of $f(2)$, $f(3)$ and $f(5)$ in Equation 3.28, we obtain

$$4f(4) - f(1) = 93 \tag{3.29}$$

Solving Equations 3.27 and 3.29, we get

$$f(1) = 3 \text{ and } f(4) = 24$$

3.2.5 Factorial Polynomial

If n is any positive integer, then the factorial nth power of x is denoted by $[x]^n$ or $x^{(n)}$ and is defined by

$$x^{(n)} = x(x-h)(x-2h)\ldots(x-\overline{n-1}h) \tag{3.30}$$

which is also called a *factorial polynomial* of degree n.

In particular, $x^{(0)} = 1$ and $x^{(1)} = x$.

On the other hand, the function $1/[(x+h)(x+2h)\ldots(x+nh)]$ is called a *reciprocal factorial polynomial* of order n.

3.2.5.1 Forward Differences of Factorial Polynomial

$$\begin{aligned}\Delta x^{(n)} &= (x+h)^{(n)} - x^{(n)} \\ &= (x+h)x(x-h)\ldots(x-\overline{n-2}h) - x(x-h)(x-2h)\ldots(x-\overline{n-1}h) \\ &= x(x-h)\ldots(x-\overline{n-2}h)\left[(x+h)-(x-\overline{n-1}h)\right] \\ &= nhx(x-h)\ldots(x-\overline{n-2}h) \\ &= nhx^{(n-1)}\end{aligned}$$

Again,

$$\Delta^2 x^{(n)} = \Delta\left(\Delta x^{(n)}\right)$$
$$= nh\Delta x^{(n-1)}$$
$$= n(n-1)h^2 x^{(n-2)}$$

Proceeding in this manner, by the method of induction, we can obtain

$$\Delta^k x^{(n)} = n(n-1)(n-2)\ldots(n-k+1)h^k x^{(n-k)}, \quad \text{for } k = 1,2,\ldots,n \tag{3.31}$$

It follows that

$$\Delta^n x^{(n)} = n!h^n, \text{which is a constant} \tag{3.32}$$

Consequently,

$$\Delta^{n+1} x^{(n)} = 0 \tag{3.33}$$

Thus, in particular, if $h = 1$

$$\Delta^n x^{(n)} = n! \tag{3.34}$$

Corollary: From Equation 3.31, we can have

$$\Delta x^{(n)} = nhx^{(n-1)}$$

Thus,

$$\Delta x^{(n+1)} = (n+1)hx^{(n)} \tag{3.35}$$

Therefore,

$$x^{(n)} = \frac{\Delta x^{(n+1)}}{(n+1)h} \tag{3.36}$$

Hence,

$$\Delta^{-1} x^{(n)} = \frac{x^{(n+1)}}{(n+1)h} \tag{3.37}$$

Example 3.6

Express the function $f(x) = 2x^3 + x^2 + 3x + 4$ in terms of factorial polynomials, taking $h = 3$ and hence, find its forward differences.

Solution:

Let

$$f(x) = a_0 x^{(3)} + a_1 x^{(2)} + a_2 x^{(1)} + a_3 \tag{3.38}$$

Thus,

$$a_0 x(x-3)(x-6) + a_1 x(x-3) + a_2 x + a_3 = 2x^3 + x^2 + 3x + 4 \tag{3.39}$$

Now, equating the coefficients of like powers in x on both sides of Equation 3.39, we get

$$\text{Coefficients of } x^0: a_3 = 4 \tag{3.40}$$

$$\text{Coefficients of } x^1: 18a_0 - 3a_1 + a_2 = 3 \tag{3.41}$$

$$\text{Coefficients of } x^2: -9a_0 + a_1 = 1 \tag{3.42}$$

$$\text{Coefficients of } x^3: a_0 = 2 \tag{3.43}$$

Solving Equations 3.40 through 3.43, we obtain

$$a_0 = 2,\ a_1 = 19,\ a_2 = 24 \text{ and } a_3 = 4$$

Hence, the required factorial form the given function is

$$f(x) = 2x^{(3)} + 19x^{(2)} + 24x^{(1)} + 4$$

Therefore,

$$\Delta f(x) = 6 \times 3x^{(2)} + 38 \times 3x^{(1)} + 24 \times 3$$
$$= 18x(x-3) + 114x + 72$$
$$= 18x^2 + 60x + 72$$
$$\Delta^2 f(x) = 36 \times 3x^{(1)} + 114 \times 3$$
$$= 108x + 342$$
$$\Delta^3 f(x) = 108 \times 3 = 324$$

Example 3.7

Obtain a function whose first difference is $f(x) = x^3 + 3x^2 + 5x + 12$.

Solution:

Let

$$f(x) = a_0 x^{(3)} + a_1 x^{(2)} + a_2 x^{(1)} + a_3 \tag{3.44}$$

Thus,

$$a_0 x(x-1)(x-2) + a_1 x(x-1) + a_2 x + a_3 = x^3 + 3x^2 + 5x + 12 \tag{3.45}$$

Now, equating the coefficients of like powers in x on both sides of Equation 3.45, we get

$$\text{Coefficients of } x^0: a_3 = 12 \tag{3.46}$$

$$\text{Coefficients of } x^1: 2a_0 - a_1 + a_2 = 5 \tag{3.47}$$

$$\text{Coefficients of } x^2: -3a_0 + a_1 = 3 \tag{3.48}$$

$$\text{Coefficients of } x^3: a_0 = 1 \tag{3.49}$$

Solving Equations 3.46 through 3.49, we obtain

$$a_0 = 1,\ a_1 = 6,\ a_2 = 9, \text{ and } a_3 = 12$$

Hence, the factorial form of the given function is

$$f(x) = x^{(3)} + 6x^{(2)} + 9x^{(1)} + 12$$

Let, $g(x)$ be the required function whose first difference is $f(x)$.
Using Equation 3.37, we get

$$\begin{aligned} g(x) &= \Delta^{-1}f(x) = \Delta^{-1}x^{(3)} + 6\Delta^{-1}x^{(2)} + 9\Delta^{-1}x^{(1)} + \Delta^{-1}(12) \\ &= \frac{x^{(4)}}{4} + 6\frac{x^{(3)}}{3} + 9\frac{x^{(2)}}{2} + 12x^{(1)} \\ &= \frac{x(x-1)(x-2)(x-3)}{4} + 6\frac{x(x-1)(x-2)}{3} + 9\frac{x(x-1)}{2} + 12x \\ &= \frac{1}{4}\left(x^4 + 2x^3 + 5x^2 + 40x\right) \end{aligned}$$

Hence, the required function is

$$g(x) = \frac{1}{4}\left(x^4 + 2x^3 + 5x^2 + 40x\right)$$

3.2.6 Backward Differences

Let $y_0, y_1, \ldots, y_n$ be a given set of values of y corresponding to the equidistant values $x_0, x_1, \ldots, x_n$ of x, that is, $x_i = x_0 + ih$, $i = 0, 1, \ldots, n$. The differences $y_1 - y_0$, $y_2 - y_1$, $\ldots$, $y_n - y_{n-1}$ are called *first backward differences*, if these are denoted by $\nabla y_1, \nabla y_2, \ldots, \nabla y_n$, respectively. Thus, we have

$$\nabla y_i = y_i - y_{i-1}, \quad i = 1, \ldots, n \tag{3.50}$$

The operator ∇ is called the *first backward difference operator.*
In general, the first backward difference operator is defined by

$$\begin{aligned} \nabla f(x) &= f(x) - f(x-h) \\ &= f(x) - E^{-1}f(x) \\ &= (I - E^{-1})f(x), \quad \text{where } I \text{ is the identity operator} \end{aligned} \tag{3.51}$$

Therefore, we have

$$\nabla = I - E^{-1} \tag{3.52}$$

Equation 3.51 represents a relation between the shift operator and the backward difference operator ∇.
Similarly, we can define the second order, third order, fourth order, and many more backward differences formulae, respectively, as follows:

$$\begin{aligned} \nabla^2 y_i &= \nabla(\nabla y_i) = \nabla y_i - \nabla y_{i-1} = y_i - 2y_{i-1} + y_{i-2} \\ \nabla^3 y_i &= \nabla(\nabla^2 y_i) = \nabla^2 y_i - \nabla^2 y_{i-1} = y_i - 3y_{i-1} + 3y_{i-2} - y_{i-3} \\ &\cdots \\ \nabla^k y_i &= \nabla(\nabla^{k-1} y_i) = \nabla^{k-1} y_i - \nabla^{k-1} y_{i-1} = \sum_{r=0}^{k} (-1)^r \binom{k}{r} y_{i-r} \end{aligned} \tag{3.53}$$

where $i = 1, \ldots, n$ and $k(k = 1, \ldots, n)$ is a positive integer.

3.2.6.1 Relation between the Forward and Backward Difference Operators

We have

$$\nabla y_i = y_i - y_{i-1} = \Delta y_{i-1}$$

$$\nabla^2 y_i = y_i - 2y_{i-1} + y_{i-2} = \Delta^2 y_{i-2}$$

$$\nabla^3 y_i = y_i - 3y_{i-1} + 3y_{i-2} - y_{i-3} = \Delta^3 y_{i-3}$$

In general,

$$\nabla^k y_i = \Delta^k y_{i-k} \tag{3.54}$$

where $i = 1, \ldots, n$ and $k(k = 1, \ldots, n)$ is a positive integer.

Equation 3.54 represents a relation between the forward and the backward difference operators.

3.2.6.2 Backward Difference Table

We can calculate the above backward differences very easily with the help of Table 3.3, which is called *backward difference table.* However, according to the result in Equation 3.54, the backward differences may also be derived from the forward difference table. In that case, Table 3.3 is not required.

3.2.7 Newton's Forward Difference Interpolation

Let $y = f(x)$ be a function that takes the value $y_0, y_1, \ldots, y_n$ for the equidistant values $x_0, x_1, x_2, \ldots, x_n$, that is, $x_i = x_0 + ih$ for all $i = 0,1,2,\ldots, n$.

Let, $\varphi_n(x)$ be a polynomial of degree n. This polynomial may be written as

$$\varphi_n(x) = a_0 + a_1(x - x_0) + a_2(x - x_1)(x - x_0) + \ldots + a_n(x - x_0)(x - x_1)(x - x_2)\ldots(x - x_{n-1}) \tag{3.55}$$

We shall now determine the coefficient $a_0, a_1, \ldots, a_n$ so as to make

$$\varphi_n(x_0) = y_0, \varphi_n(x_1) = y_1, \ldots, \varphi_n(x_n) = y_n$$

Substituting in Equation 3.55 the successive values $x_0, x_1, x_2, \ldots, x_n$ for x at the same time putting $\varphi_n(x_0) = y_0$, $\varphi_n(x_1) = y_1$, $\ldots$, $\varphi_n(x_n) = y_n$, we have

$$a_0 = y_0;\ a_1 = \frac{y_1 - y_0}{x_1 - x_0} = \frac{\Delta y_0}{h};\ a_2 = \frac{\Delta^2 y_0}{2!h^2};\ a_3 = \frac{\Delta^3 y_0}{3!h^3};\ \cdots;\ a_n = \frac{\Delta^n y_0}{n!h^n}$$

TABLE 3.3
Backward Difference Table

x	y	∇y	$\nabla^2 y$	$\nabla^3 y$	$\nabla^4 y$
x_0	y_0				
x_1	y_1	∇y_1			
x_2	y_2	∇y_2	$\nabla^2 y_2$		
x_3	y_3	∇y_3	$\nabla^2 y_3$	$\nabla^3 y_3$	
x_4	y_4	∇y_4	$\nabla^2 y_4$	$\nabla^3 y_4$	$\nabla^4 y_4$

Substituting these values of $a_0, a_1, \ldots, a_n$ Equation 3.55, we have

$$\varphi_n(x) = y_0 + \frac{(x-x_0)}{1!h}\Delta y_0 + \frac{(x-x_1)(x-x_0)}{2!h^2}\Delta^2 y_0 + \cdots + \frac{(x-x_0)(x-x_1)(x-x_2)\ldots(x-x_{n-1})}{n!h^n}\Delta^n y_0 \quad (3.56)$$

which is Gregory–Newton's forward difference interpolation formula, and it is useful to interpolate near the beginning of a set of tabular values.

Now, settting $u = (x-x_0)/h$, from Equation 3.56, we obtain

$$\begin{aligned}\varphi_n(x) &= y_0 + u\Delta y_0 + \frac{u(u-1)}{2!}\Delta^2 y_0 + \cdots + \frac{u(u-1)\ldots(u-n+1)}{n!}\Delta^n y_0 \\ &= y_0 + \binom{u}{1}\Delta y_0 + \binom{u}{2}\Delta^2 y_0 + \cdots + \binom{u}{n}\Delta^n y_0\end{aligned} \quad (3.57)$$

In practical numerical computation, instead of Equation 3.56, we should use Equation 3.57 in order to ease the calculation involved with it.

3.2.7.1 Error in Newton's Forward Difference Interpolation

To find the error committed in approximating $f(x)$ by the polynomial $\varphi_n(x)$, we have from Equation 3.7, the remainder or truncation error or simply error is

$$R_{n+1}(x) = \frac{(x-x_0)(x-x_1)(x-x_2)\ldots(x-x_n)}{(n+1)!} f^{(n+1)}(\xi) \quad (3.58)$$

where $\min\{x, x_0, x_1, \ldots, x_n\} < \xi < \max\{x, x_0, x_1, \ldots, x_n\}$.

According to Equation 3.13, Equation 3.58 can be written as

$$\begin{aligned}R_{n+1}(x) &= \frac{(x-x_0)(x-x_1)(x-x_2)\ldots(x-x_n)}{(n+1)!h^{n+1}}\Delta^{n+1} f(\xi) \\ &= \frac{u(u-1)\ldots(u-n)}{(n+1)!}\Delta^{n+1} f(\xi)\end{aligned}$$

Hence,

$$|R_{n+1}(x)| = \left|\frac{u(u-1)\ldots(u-n)}{(n+1)!}\right| |\Delta^{n+1} f(\xi)| \leq |\Delta^{n+1} f(\xi)|, \quad \text{if } |u| \leq 1$$

Therefore, the error is about of the order of magnitude of the next difference not appeared in the expression of $\varphi_n(x)$ given in Equation 3.57.

3.2.7.2 Algorithm for Newton's Forward Difference Interpolation

Input: Enter the number of given data N (where $N = n + 1$) and interpolating point x. Enter the data $x_i, y_i, i = 0(1)n$.

Output: Write the value of the function $y = f(x)$ for given value of x

Initial step: Initialize $sum = 0$, $p = 1$.

Step 1: compute $h = x_1 - x_0$, $u = (x - x_0)/h$ and set $sum = y_0$

Step 2: for $j = 0(1)n$ do

set $f_{0,j} = y_j$.

Step 3: for $i = 1(1)n$ do
for $j = 0(1)n - i$ do
compute $f_{i,j} = f_{i-1,j+1} - f_{i-1,j}$.
Step 4: for $i = 1(1)n$ do
$p = p * \left(\frac{u - i + 1}{i}\right)$;
$sum = sum + p * f_{i,0}$.
Step 5: Print the value of *sum*.
Step 6: Stop. ■

MATHEMATICA® Program for Newton's Forward Difference Interpolation (Chapter 3, Example 3.8)

```
x[0]=200;x[1]=250;x[2]=300;x[3]=350;x[4]=400;
y[0]=15.04;y[1]=16.81;y[2]=18.42;y[3]=19.90;y[4]=21.27;
n=4;
h=x[1]-x[0];
u=(x-x[0])/h;
sum=y[0];
For[j=0,j<=n,j++,
f[0,j]=y[j]];
For[i=1,i<=n,i++,
For[j=0,j<=n-i,j++,
f[i,j]=f[i-1,j+1]-f[i-1,j]]];
For[i=1;p=1,i<=n,i++,
p=p*(u-i+1)/i;
sum=sum+p*f[i,0]];
Print["p[x]=",Simplify[sum]];
sum/.x->218
sum/.x->410
```

Output:

```
p[x]=5.41 +0.0625167 x-0.0000918333 x^2+1.13333*10^-7 x^3-6.66667*10^-11
x^4
15.6979
21.5319
```

3.2.8 Newton's Backward Difference Interpolation

Let $y = f(x)$ be a function that takes the value $y_0, y_1, \ldots, y_n$ for the equidistant values $x_0, x_1, x_2, \ldots, x_n$, that is, $x_i = x_0 + ih$ for all $i = 0,1,2,3, \ldots, n$.

Let, $\varphi_n(x)$ be a polynomial of degree n. This polynomial may be written as

$$\varphi_n(x) = a_0 + a_1(x - x_n) + a_2(x - x_n)(x - x_{n-1}) + \cdots + a_n(x - x_n)(x - x_{n-1})\ldots(x - x_1) \quad (3.59)$$

We shall now determine the coefficients $a_0, a_1, \ldots, a_n$ so as to make

$$\varphi_n(x_n) = y_n, \varphi_n(x_{n-1}) = y_{n-1}, \ldots, \varphi_n(x_0) = y_0$$

Substituting in Equation 3.59 the successive values x_n, x_{n-1}, … for x at the same time putting $\varphi_n(x_n) = y_n$, $\varphi_n(x_{n-1}) = y_{n-1}$, …, $\varphi_n(x_0) = y_0$, we have

$$a_0 = y_n;\ a_1 = \frac{y_n - y_{n-1}}{x_n - x_{n-1}} = \frac{\nabla y_n}{h};\ a_2 = \frac{\nabla^2 y_n}{2!h^2};\ a_3 = \frac{\nabla^3 y_n}{3!h^3};\ldots;\ a_n = \frac{\nabla^n y_n}{n!h^n}$$

Substituting these values of $a_0, a_1, \ldots, a_n$ in Equation 3.59, we have

$$\varphi_n(x) = y_n + \frac{(x-x_n)}{h}\nabla y_n + \frac{(x-x_n)(x-x_{n-1})}{2!h^2}\nabla^2 y_n + \cdots + \frac{(x-x_n)(x-x_{n-1})\ldots(x-x_1)}{n!h^n}\nabla^n y_n \quad (3.60)$$

which is Gregory–Newton's backward difference interpolation formula, and it is useful to interpolate near the end of a set of tabular values.

Now, setting $u = (x - x_n)/h$, we can obtain

$$\varphi_n(x) = y_n + u\nabla y_n + \frac{u(u+1)}{2!}\nabla^2 y_n + \cdots + \frac{u(u+1)\ldots(u+n-1)}{n!}\nabla^n y_n \quad (3.61)$$

In the case of practical numerical computation, instead of Equation 3.60, we should use Equation 3.61 in order to ease the calculation involved with it.

3.2.8.1 Error in Newton's Backward Difference Interpolation

To find the error committed in approximating $f(x)$ by the polynomial $\varphi_n(x)$, we have from Equation 3.7, the remainder or truncation error or simply error is

$$R_{n+1}(x) = \frac{(x-x_0)(x-x_1)(x-x_2)\ldots(x-x_n)}{(n+1)!} f^{(n+1)}(\xi) \quad (3.62)$$

where $\min\{x, x_0, x_1, \ldots, x_n\} < \xi < \max\{x, x_0, x_1, \ldots, x_n\}$.

According to Equation 3.13, Equation 3.62 can be written as

$$R_{n+1}(x) = \frac{h^{n+1}u(u+1)\ldots(u+n)}{(n+1)!} f^{(n+1)}(\xi) \quad (3.63)$$

3.2.8.2 Algorithm for Newton's Backward Difference Interpolation

Input: Enter the number of given data N (where $N = n + 1$) and interpolating point x. Enter the data x_i, y_i, $i = 0(1)n$.

Output: Write the value of the function $y = f(x)$ for given value of x.

Initial step: Initialize $sum = 0$, $p = 1$.

Step 1: compute $h = x_1 - x_0$, $u = (x - x_n)/h$ and set $sum = y_n$

Step 2: for $j = 0(1)n$ do

  set $f_{0,j} = y_j$.

Step 3: for $i = 1(1)n$ do

  for $j = i(1)n$ do

    compute $f_{i,j} = f_{i-1,j} - f_{i-1,j-1}$.

Step 4: for $i = 1(1)n$ do

  $p = p * \left(\frac{u+i-1}{i}\right)$;

  $sum = sum + p * f_{i,n}$.

Step 5: Print the value of sum.

Step 6: Stop. ■

MATHEMATICA® Program for Newton's Backward Difference Interpolation (Chapter 3, Example 3.8)

```
x[0]=200;x[1]=250;x[2]=300;x[3]=350;x[4]=400;
y[0]=15.04;y[1]=16.81;y[2]=18.42;y[3]=19.90;y[4]=21.27;
n=4;
h=x[1]-x[0];
u=(x-x[n])/h;
sum=y[n];
For[j=0,j<=n,j++,
f[0,j]=y[j]];
For[i=1,i<=n,i++,
For[j=i,j<=n,j++,
f[i,j]=f[i-1,j]-f[i-1,j-1]]];
For[i=1;p=1,i<=n,i++,
p=p*(u+i-1)/i;
sum=sum+p*f[i,n]];
Print["p[x]=",Simplify[sum]];
sum/.x->218
sum/.x->410
```

Output:

```
p[x]=5.41 +0.0625167 x-0.0000918333 x^2+1.13333*10^-7 x^3-6.66667*10^-11
x^4
15.6979
21.5319
```

Example 3.8

Calculate the values of $f(218)$ and $f(410)$ from the following data

x	200	250	300	350	400
y	15.04	16.81	18.42	19.90	21.27

Solution:

The forward difference table is

x	y	Δy	Δ²y	Δ³y	Δ⁴y
200	<u>15.04</u>				
		<u>1.77</u>			
250	16.81		<u>−0.16</u>		
		1.61		<u>0.03</u>	
300	18.42		−0.13		<u>−0.01</u>
		1.48		<u>0.02</u>	
350	19.90		<u>−0.11</u>		
		<u>1.37</u>			
400	<u>21.27</u>				

Since, $x = 218$ is near the beginning of the table, we use Newton's forward interpolation formula. From the above forward difference table, only the underlined upper portion values will be used in the Newton's forward interpolation formula.

Here,

$$u = \frac{x - x_0}{h} = \frac{218 - 200}{50} = \frac{18}{50} = 0.36$$

Using Newton's forward interpolation formula, we get

$$f(218) \cong y_0 + u\Delta y_0 + \frac{u(u-1)}{2!}\Delta^2 y_0 + \frac{u(u-1)(u-2)}{3!}\Delta^3 y_0 + \frac{u(u-1)(u-2)(u-3)}{4!}\Delta^4 y_0$$

$$= 15.04 + 0.36 \times 1.77 + \frac{0.36(0.36-1)}{2!} \times (-0.16) + \frac{0.36(0.36-1)(0.36-2)}{3!} \times (0.03)$$

$$+ \frac{0.36(0.36-1)(0.36-2)(0.36-3)}{4!} \times (-0.01)$$

$$= 15.04 + 0.6372 + 0.018432 + 0.00188928 + 0.0004156416$$

$$= 15.6979$$

Since, $x = 410$ is near the end of the table, we use Newton's backward interpolation formula. From the above forward difference table, only the underlined lower portion values will be used in the Newton's backward interpolation formula.

Here,

$$u = \frac{x - x_n}{h} = \frac{410 - 400}{50} = \frac{10}{50} = 0.2$$

Using Newton's backward interpolation formula, we get

$$f(410) \cong y_n + u\nabla y_n + \frac{u(u+1)}{2!}\nabla^2 y_n + \frac{u(u+1)(u+2)}{3!}\nabla^3 y_n + \frac{u(u+1)(u+2)(u+3)}{4!}\nabla^4 y_n$$

$$= 21.27 + 0.2 \times 1.37 + \frac{0.2 \times 1.2}{2!} \times (-0.11)$$

$$+ \frac{0.2 \times 1.2 \times 2.2}{3!} \times 0.02 + \frac{0.2 \times 1.2 \times 2.2 \times 3.2}{4!} \times (-0.01)$$

$$= 21.27 + 0.274 - 0.0132 + 0.00176 - 0.000704$$

$$= 21.532$$

3.2.9 Lagrange's Interpolation Formula

Let us consider $y_0, y_1, \ldots, y_n$ be the values of $y = f(x)$ corresponding to the values $x_0, x_1, x_2, \ldots, x_n$ of x. In this case, the values of x need not necessarily be equispaced.

Let, $\varphi_n(x)$ be a polynomial of degree n. Let us take the polynomial in the following form:

$$\begin{aligned}\varphi_n(x) &= c_0(x-x_1)(x-x_2)\ldots(x-x_n) + c_1(x-x_0)(x-x_2)\ldots(x-x_n) + \cdots \\ &+ c_r(x-x_0)(x-x_1)\ldots(x-x_{r-1})(x-x_{r+1})\ldots(x-x_n) + \cdots \\ &+ c_n(x-x_0)(x-x_1)(x-x_2)\ldots(x-x_{n-1})\end{aligned} \tag{3.64}$$

We shall determine the coefficients $c_0, c_1, \ldots, c_n$ so as to make

$$\varphi_n(x_0) = y_0, \varphi_n(x_1) = y_1, \ldots, \varphi_n(x_n) = y_n$$

Now, substituting in Equation 3.64 the successive values $x_0, x_1, \ldots, x_n$ for x at the same time putting $\varphi_n(x_0) = y_0$, $\varphi_n(x_1) = y_1$, $\ldots$, $\varphi_n(x_n) = y_n$, we have

$$y_0 = \varphi_n(x_0) = c_0(x_0 - x_1)(x_0 - x_2)\ldots(x_0 - x_n)$$

This implies

$$c_0 = \frac{y_0}{(x_0 - x_1)(x_0 - x_2)\ldots(x_0 - x_n)}$$

Similarly,

$$c_1 = \frac{y_1}{(x_1 - x_0)(x_1 - x_2)\ldots(x_1 - x_n)}$$

$$\ldots$$

$$c_n = \frac{y_n}{(x_n - x_0)(x_n - x_1)\ldots(x_n - x_{n-1})}$$

Substituting these values for $c_0, c_1, \ldots, c_n$ in Equation 3.64, we get

$$\varphi_n(x) = \frac{(x - x_1)(x - x_2)\ldots(x - x_n)y_0}{(x_0 - x_1)(x_0 - x_2)\ldots(x_0 - x_n)} + \frac{(x - x_0)(x - x_2)\ldots(x - x_n)y_1}{(x_1 - x_0)(x_1 - x_2)\ldots(x_1 - x_n)} + \cdots + \frac{(x - x_0)(x - x_1)\ldots(x - x_{n-1})y_n}{(x_n - x_0)(x_n - x_1)\ldots(x_n - x_{n-1})} \quad (3.65)$$

This is known as *Lagrange's interpolation formula*.

Now, the Equation 3.65 may be written in the form

$$\varphi_n(x) = \sum_{i=0}^{n} \omega_i(x) y_i \quad (3.66)$$

where $\omega_i(x)$ is a polynomial of degree n for each i.

Since, $\varphi_n(x_r) = f(x_r)$ for $r = 0, 1, 2, \ldots, n$. From Equation 3.66, we have

$$\sum_{i=0}^{n} \omega_i(x_r) y_i = f(x_r) \quad \text{for } r = 0, 1, 2, \ldots, n$$

It holds if

$$\begin{aligned} \omega_i(x_r) &= 0 \quad \text{for } r \neq i \\ &= 1 \quad \text{for } r = i \end{aligned} \quad (3.67)$$

Now, let us set

$$\Pi(x) = (x - x_0)(x - x_1)\ldots(x - x_{i-1})(x - x_i)(x - x_{i+1})\ldots(x - x_n) \quad (3.68)$$

then

$$\Pi'(x_i) = (x_i - x_0)(x_i - x_1)\ldots(x_i - x_{i-1})(x_i - x_{i+1})\ldots(x_i - x_n) \quad (3.69)$$

Therefore, Equation 3.66 may be written in the following form:

$$\varphi_n(x) = \sum_{i=0}^{n} \frac{\Pi(x)}{(x - x_i)\Pi'(x_i)} y_i \quad (3.70)$$

where

$$\omega_i(x) = \frac{\Pi(x)}{(x - x_i)\Pi'(x_i)} \tag{3.71}$$

which is called *Lagrange's fundamental polynomial.*

Hence, from Equation 3.70, we may obtain

$$f(x) \approx \varphi_n(x) = \sum_{i=0}^{n} \frac{\Pi(x)}{(x - x_i)\Pi'(x_i)} y_i = \sum_{i=0}^{n} \prod_{\substack{j=0 \\ j \neq i}}^{n} \frac{(x - x_j)}{(x_i - x_j)} y_i \tag{3.72}$$

This is the alternate form of Lagrange's interpolation formula.

Now if we treat x (dependent variable) as a function of y (independent variable), then interchanging x and y in Equation 3.72, we may obtain

$$x = \sum_{i=0}^{n} \frac{\Pi(y)}{(y - y_i)\Pi'(y_i)} x_i \tag{3.73}$$

This relation is useful for inverse interpolation, and it is sometimes referred to as *Lagrange's inverse interpolation formula.*

3.2.9.1 Error in Lagrange's Interpolation

To find the error committed in approximating $f(x)$ by the polynomial $\varphi_n(x)$, we have from Equation 3.7, the remainder or error is

$$f(x) - \varphi_n(x) = R_{n+1}(x) = \frac{(x - x_0)(x - x_1)(x - x_2)\ldots(x - x_n)}{(n+1)!} f^{(n+1)}(\xi) \tag{3.74}$$

where $\min\{x, x_0, x_1, \ldots, x_n\} < \xi < \max\{x, x_0, x_1, \ldots, x_n\}$.

If there exists a positive integer K such that

$$\left| f^{(n+1)}(\xi) \right| \leq K$$

where $\min\{x, x_0, x_1, \ldots, x_n\} < \xi < \max\{x, x_0, x_1, \ldots, x_n\}$,

then the error in Lagrange's interpolation formula is

$$\varepsilon_L = \left\| R_{n+1}(x) \right\| \leq \frac{K}{(n+1)!} \underset{\alpha \leq x \leq \beta}{Sup} \left| (x - x_0)(x - x_1)(x - x_2)\ldots(x - x_n) \right|$$

where $\alpha \equiv \min\{x_0, x_1, \ldots, x_n\}$ and $\beta \equiv \max\{x_0, x_1, \ldots, x_n\}$.

3.2.9.2 Advantages and Disadvantages of Lagrange's Interpolation

- *Advantages*: The main advantage of this method is that it is applicable for both equispaced and unequispaced arguments. Another advantage is that the value of x for which corresponding value of $y = f(x)$ is to be determined may lie anywhere in the tabulated values.
- *Disadvantages*: The numerical computation in Lagrange's interpolation is laborious. For any computation, the whole data is taken into calculation. Also, if a new node or interpolation point is added causing the increase in the degree of the interpolation polynomial, then the whole computation has to be done afresh. For these reasons, Lagrange's interpolation is less suitable from the practical point of view. However, Lagrange's interpolation is an important theoretical tool.

3.2.9.3 Algorithm for Lagrange's Interpolation

Input: Enter the number of given data N (where $N = n + 1$) and interpolating point x. Enter the data x_i, y_i, $i = 0(1)n$.

Output: Print the value of the function $y = f(x)$ for given value of x.

Initial step: Initialize $sum = 0$.

Step 1: for $i = 0(1)n$ do

set $p = 1$.

for $j = 0(1)n$ do

if $j \neq i$ then compute

$$p = p * \frac{x - x_j}{x_i - x_j};$$

$sum = sum + p * y_i$

Step 2: Print the value of sum.

Step 3: Stop. ■

MATHEMATICA® Program for Lagrange's Interpolation (Chapter 3, Example 3.10)

```
x[0]=110;x[1]=130;x[2]=160;x[3]=190;
y[0]=10.8;y[1]=8.1;y[2]=5.5;y[3]=4.8;
n=3;
sum=0;
For[i=0,i<=n,i++,
p=1;
For[j=0,j<=n,j++,
If[j!=i,
p=p*(x-x[j])/(x[i]-x[j]),True]];
sum=sum+p*y[i]];
Print["p[x]=",Simplify[sum]];
sum/.x->140
```

Output:

```
p[x]=36.9311 -0.308444 x+0.000522222 x^2+1.11111*10^-6 x^3
7.03333
```

3.2.10 Divided Difference

Let us consider $y_0 = f(x_0)$, $y_1 = f(x_1)$, …, $y_n = f(x_n)$ be the values of the function $y = f(x)$ corresponding to the values x_0, x_1, x_2, …, x_n, which are not necessarily equally spaced.

The first-order divided differences of $f(x)$ for two arguments x_0, x_1 denoted by $f[x_0, x_1]$ and is defined by

$$f[x_0, x_1] = \frac{f(x_0) - f(x_1)}{x_0 - x_1} = \frac{f(x_1) - f(x_0)}{x_1 - x_0}$$

Similarly,

$$f[x_1, x_2] = \frac{f(x_1) - f(x_2)}{x_1 - x_2} = \frac{f(x_2) - f(x_1)}{x_2 - x_1}$$

$$f[x_2, x_3] = \frac{f(x_2) - f(x_3)}{x_2 - x_3} = \frac{f(x_3) - f(x_2)}{x_3 - x_2}$$

and so on.

The second-order divided differences of $f(x)$ for three arguments x_0, x_1, x_2 denoted by $f[x_0,x_1,x_2]$ and is defined by

$$f[x_0,x_1,x_2]=\frac{f[x_0,x_1]-f[x_1,x_2]}{x_0-x_2}$$

In general, the nth order divided differences of $f(x)$ with $n+1$ arguments x_0, x_1, x_2, …, x_n is defined by

$$f[x_0,x_1,x_2,\ldots,x_n]=\frac{f[x_0,x_1,x_2,\ldots,x_{n-1}]-f[x_1,x_2,\ldots,x_n]}{x_0-x_n}$$

3.2.10.1 Some Properties of Divided Differences

1. The divided differences are symmetric with respect to their arguments.
   Proof:

$$f[x_0,x_1]=\frac{f(x_0)-f(x_1)}{x_0-x_1}=\frac{f(x_1)-f(x_0)}{x_1-x_0}=f[x_1,x_0]$$

Also, we have

$$f[x_0,x_1]=\frac{f(x_0)}{x_0-x_1}+\frac{f(x_1)}{x_1-x_0}=\sum_{i=0}^{1}\frac{f(x_i)}{\prod_{\substack{j=0\\ j\neq i}}^{1}(x_i-x_j)}$$

Again,

$$\begin{aligned}f[x_0,x_1,x_2]&=\frac{f[x_0,x_1]-f[x_1,x_2]}{x_0-x_2}\\&=\frac{1}{(x_0-x_2)}\left[\frac{f(x_0)}{x_0-x_1}+\frac{f(x_1)}{x_1-x_0}-\frac{f(x_1)}{x_1-x_2}-\frac{f(x_2)}{x_2-x_1}\right]\\&=\frac{f(x_0)}{(x_0-x_1)(x_0-x_2)}+\frac{f(x_1)}{(x_1-x_0)(x_1-x_2)}+\frac{f(x_2)}{(x_2-x_0)(x_2-x_1)}\\&=\sum_{i=0}^{2}\frac{f(x_i)}{\prod_{\substack{j=0\\ j\neq i}}^{2}(x_i-x_j)}\end{aligned}$$

Now, we shall prove that

$$f[x_0,x_1,\ldots,x_n]=\sum_{i=0}^{n}\frac{f(x_i)}{\prod_{\substack{j=0\\ j\neq i}}^{n}(x_i-x_j)} \tag{3.75}$$

It has been already shown that the above result is true for $n=1,2$. Let us assume that the result in Equation 3.75 is true for $n=k$. Therefore,

$$f[x_0,x_1,\ldots,x_k]=\sum_{i=0}^{k}\frac{f(x_i)}{\prod_{\substack{j=0\\ j\neq i}}^{k}(x_i-x_j)} \tag{3.76}$$

Now,

$$
\begin{aligned}
f[x_0,x_1,\ldots,x_{k+1}] &= \frac{f[x_0,x_1,\ldots,x_k]-f[x_1,x_2,\ldots,x_{k+1}]}{x_0-x_{k+1}} \\
&= \frac{1}{(x_0-x_{k+1})}\left[\frac{f(x_0)}{(x_0-x_1)\ldots(x_0-x_k)}+\frac{f(x_1)}{(x_1-x_0)\ldots(x_1-x_k)}\right. \\
&\quad \left.-\frac{f(x_1)}{(x_1-x_2)\ldots(x_1-x_{k+1})}-\cdots-\frac{f(x_{k+1})}{(x_{k+1}-x_1)\ldots(x_{k+1}-x_k)}\right], \qquad (3.77)
\end{aligned}
$$

using induction hypothesis

$$
\begin{aligned}
&= \frac{f(x_0)}{(x_0-x_1)\ldots(x_0-x_k)(x_0-x_{k+1})}+\frac{f(x_1)}{(x_1-x_0)\ldots(x_1-x_{k+1})} \\
&\quad +\cdots+\frac{f(x_{k+1})}{(x_{k+1}-x_0)(x_{k+1}-x_1)\ldots(x_{k+1}-x_k)} \\
&= \sum_{i=0}^{k+1}\frac{f(x_i)}{\prod_{\substack{j=0\\ j\neq i}}^{k+1}(x_i-x_j)}
\end{aligned}
$$

Therefore, the result is also true for $n=k+1$. Hence, by the method of induction, the result in Equation 3.75 holds for all positive integer n, that is, $n=1, 2, \ldots$.

From Equation 3.75, it is clear that $f[x_0,x_1,\ldots,x_n]$ is symmetrical with respect to their arguments. Hence, divided differences of all orders are symmetrical in their arguments. ■

2. The divided differences of $f(x)\pm g(x)$ is the sum (or difference) of the corresponding divided differences of $f(x)$ and $g(x)$, that is, the divided difference is linear.
3. The divided difference of a constant is zero.
4. The divided difference of $kf(x)$, where k is a constant, is k times the divided difference of $f(x)$.
5. The nth order divided difference of a polynomial is constant.

Proof:

Let us consider the case when $f(x)=x^n$, where n is a positive integer.

Now,

$$
\begin{aligned}
f[x_0,x_1] &= \frac{f(x_1)-f(x_0)}{x_1-x_0}=\frac{x_1^n-x_0^n}{x_1-x_0} \\
&= x_1^{n-1}+x_1^{n-2}x_0+x_1^{n-3}x_0^2+\cdots+x_0^{n-1} \\
&= \text{a symmetric polynomial of degree } (n-1) \text{ in } x_0, x_1
\end{aligned}
$$

Similarly, it may be shown that $f[x_0,x_1,x_2]=$ a symmetric polynomial of degree $(n-2)$ in x_0, x_1, x_2.

Proceeding in the similar manner, we can prove that $f[x_0,x_1,x_2,\ldots,x_n]$ is a symmetric polynomial of degree $(n-n)=0$ in x_0, x_1, x_2, $\ldots$, x_n.

Hence, the nth order divided difference of a polynomial is constant. ■

Theorem 3.3: Relation between Divided Differences and Forward Differences

Let the arguments $x_0, x_1, \dots, x_n$ be equally spaced with step length h, that is, $x_i = x_0 + ih$ $(i = 0, 1, 2, \dots, n)$, then a divided difference reduces to a finite difference given by

$$f[x_0, x_1, \dots, x_n] = \frac{\Delta^n f(x_0)}{n!h^n} \tag{3.78}$$

Proof:

We shall prove the result in Equation 3.78 by the method of induction on n.

For $n = 1$

$$f[x_0, x_1] = \frac{f(x_0) - f(x_1)}{x_0 - x_1} = \frac{f(x_1) - f(x_0)}{x_1 - x_0} = \frac{\Delta f(x_0)}{1!h^1}$$

For $n = 2$

$$f[x_0, x_1, x_2] = \frac{f[x_0, x_1] - f[x_1, x_2]}{x_0 - x_2} = \frac{[\Delta f(x_1) - \Delta f(x_0)]}{2h^2} = \frac{\Delta^2 f(x_0)}{2!h^2}$$

Thus, the result is true for $n = 1, 2$. Let us assume that the result is true for $n = k$, so

$$f[x_0, x_1, x_2, \dots, x_k] = \frac{\Delta^k f(x_0)}{k!h^k}$$

Now, for $n = k + 1$, we have

$$f[x_0, x_1, \dots, x_{k+1}] = \frac{f[x_0, x_1, \dots, x_k] - f[x_1, x_2, \dots, x_{k+1}]}{x_0 - x_{k+1}}$$

$$= \frac{[\Delta^k f(x_1) - \Delta^k f(x_0)]}{(k+1)!h^{k+1}}, \text{ using induction hypothesis}$$

$$= \frac{\Delta^{k+1} f(x_0)}{(k+1)!h^{k+1}}$$

Therefore, the result is also true for $n = k + 1$. Hence, by the method of induction, the result holds for all positive integer k, that is, $k = 1, 2, 3, \dots$ ■

3.2.10.2 Newton's Divided Difference Interpolation Formula

Let us consider $y_0 = f(x_0)$, $y_1 = f(x_1)$, …, $y_n = f(x_n)$ be the values of the function $y = f(x)$ corresponding to the arguments x_0, x_1, x_2, …, x_n, which are not necessarily equally spaced.

Then from the definition of divided differences, we have

$$f[x, x_0] = \frac{f(x) - f(x_0)}{x - x_0}$$

This implies

$$f(x) = f(x_0) + (x - x_0) f[x, x_0] \tag{3.79}$$

Again,

$$f[x,x_0,x_1]=\frac{f[x,x_0]-f[x_0,x_1]}{x-x_1}$$

Therefore,

$$f[x,x_0]=f[x_0,x_1]+\left(x-x_1\right)f[x,x_0,x_1] \tag{3.80}$$

Substituting this value of $f[x,x_0]$ in Equation 3.79, we get

$$f(x)=f(x_0)+\left(x-x_0\right)f[x_0,x_1]+\left(x-x_0\right)\left(x-x_1\right)f[x,x_0,x_1] \tag{3.81}$$

Also,

$$f[x,x_0,x_1,x_2]=\frac{f[x,x_0,x_1]-f[x_0,x_1,x_2]}{x-x_2}$$

Therefore,

$$f[x,x_0,x_1]=f[x_0,x_1,x_2]+\left(x-x_2\right)f[x,x_0,x_1,x_2] \tag{3.82}$$

Again, substituting this value of $f[x,x_0,x_1]$ in Equation 3.81, we get

$$f(x)=f(x_0)+\left(x-x_0\right)f[x_0,x_1]+\left(x-x_0\right)\left(x-x_1\right)f[x_0,x_1,x_2]+\left(x-x_0\right)\left(x-x_1\right)\left(x-x_2\right)f[x,x_0,x_1,x_2]$$

Proceeding in this manner, we can obtain

$$\begin{aligned} y=f\left(x\right)&=y_0+\left(x-x_0\right)f[x_0,x_1]+\left(x-x_0\right)\left(x-x_1\right)f[x_0,x_1,x_2]+\cdots \\ &+\left(x-x_0\right)\left(x-x_1\right)\ldots\left(x-x_{n-1}\right)f[x_0,x_1,\ldots,x_n] \\ &+\left(x-x_0\right)\left(x-x_1\right)\ldots\left(x-x_n\right)f[x,x_0,x_1,\ldots,x_n] \end{aligned} \tag{3.83}$$

Since $(n+1)$ values of $f(x)$ are given, $f(x)$ can be approximated by a polynomial of degree n. Hence, $f(x)$ can be approximated by

$$\begin{aligned} f\left(x\right)\approx y_0&+\left(x-x_0\right)f[x_0,x_1]+\left(x-x_0\right)\left(x-x_1\right)f[x_0,x_1,x_2]+\cdots \\ &+\left(x-x_0\right)\left(x-x_1\right)\ldots\left(x-x_{n-1}\right)f[x_0,x_1,\ldots,x_n] \end{aligned} \tag{3.84}$$

which is called *Newton's divided difference interpolation formula* or *Newton's general interpolation formula with divided differences*, the last term in Equation 3.83 being the error or remainder term after $(n+1)$ terms.

3.2.10.2.1 Alternate Approach

Let $\varphi_n(x)$ be interpolating polynomial interpolating at the $n+1$ distinct points $x_0, x_1, \ldots, x_n$. Let us consider this polynomial as

$$\varphi_n(x)=c_0+c_1(x-x_0)+c_2(x-x_0)(x-x_1)+\cdots+c_n(x-x_0)(x-x_1)\ldots(x-x_{n-1}) \tag{3.85}$$

Substituting successively $x=x_0,x_1,\ldots,x_n$ in Equation 3.85, we obtain

$$\varphi_n(x_0)=f(x_0)=c_0,\varphi_n(x_1)=f(x_1)=c_0+c_1(x_1-x_0)=f(x_0)+c_1(x_1-x_0)$$

yields to

$$c_1 = \frac{f(x_1)-f(x_0)}{x_1-x_0} = f[x_0,x_1]$$

$$\varphi_n(x_2) = f(x_2) = c_0 + c_1(x_2-x_0) + c_2(x_2-x_0)(x_2-x_1)$$

$$= f(x_0) + \frac{f(x_1)-f(x_0)}{x_1-x_0}(x_2-x_0) + c_2(x_2-x_0)(x_2-x_1)$$

yields to

$$c_2 = \frac{1}{(x_2-x_0)(x_2-x_1)}\left[f(x_2)-f(x_0)-\frac{f(x_1)-f(x_0)}{x_1-x_0}(x_2-x_0)\right]$$

$$= \frac{f(x_0)}{(x_0-x_1)(x_0-x_2)} + \frac{f(x_1)}{(x_1-x_0)(x_1-x_2)} + \frac{f(x_2)}{(x_2-x_0)(x_2-x_1)}$$

$$= f[x_0,x_1,x_2]$$

Using induction, the following can be obtained

$$c_n = f[x_0,x_1,\ldots,x_n] \tag{3.86}$$

Therefore, the divided difference interpolating polynomial becomes

$$f(x) \cong \phi_n(x) = f(x_0) + (x-x_0)f[x_0,x_1] + (x-x_0)(x-x_1)f[x_0,x_1,x_2]$$
$$+\cdots+(x-x_0)(x-x_1)\ldots(x-x_{n-1})f[x_0,x_1,\ldots,x_n]$$

3.2.10.3 Divided Difference Table

Table 3.4 shows the divided differences where the differences used in Equation 3.84 have been underlined.

3.2.10.4 Algorithm for Newton's Divided Difference Interpolation

Input: Enter the number of given data N (where $N = n + 1$) and interpolating point x. Enter the data $x_i,\ y_i,\ i = 0(1)n$.

Output: Print the value of the function $y = f(x)$ for given value of x.

Initial step: Initialize $sum = 0,\ p = 1$.

Step 1: set $sum = y_0$

Step 2: for $j = 0(1)n$ do

set $d_{0,j} = y_j$.

Step 3: for $i = 1(1)n$ do

for $j = 0(1)n - i$ do

compute $d_{i,j} = \dfrac{d_{i-1,j+1} - d_{i-1,j}}{x_{i+j} - x_j}$.

Step 4: for $i = 0(1)n - 1$ do

$p = p * (x - x_i)$;

$sum = sum + p * d_{i+1,0}$.

Step 5: Print the value of sum.

Step 6: Stop. ■

TABLE 3.4
Divided Difference Table

x	y	First-Order Divided Difference	Second-Order Divided Difference	nth-Order Divided Difference
x_0	y_0			
		$f[x_0, x_1]$		
x_1	y_1		$f[x_0, x_1, x_2]$	
		$f[x_1, x_2]$		
x_2	y_2		.	
		.		
.	.		.	$f[x_0, x_1, x_2, \ldots, x_n]$
		.		
.	.		.	
		.		
.	.		.	
		.		
x_{n-1}	y_{n-1}		$f[x_{n-2}, x_{n-1}, x_n]$	
		$f[x_{n-1}, x_n]$		
x_n	y_n			

MATHEMATICA® Program for Newton's Divided Difference Interpolation (Chapter 3, Example 3.9)

```
x[0]=654;x[1]=658;x[2]=659;x[3]=661;
y[0]=2.8156;y[1]=2.8182;y[2]=2.8189;y[3]=2.8202;
n=3;
sum=y[0];
For[j=0,j<=n,j++,
d[0,j]=y[j]];
For[i=1,i<=n,i++,
For[j=0,j<=n-i,j++,
d[i,j]=(d[i-1,j+1]-d[i-1,j])/(x[i+j]-x[j])]];
For[i=0;p=1,i<=n-1,i++,
p=p*(x-x[i]);
sum=sum+p*d[i+1,0]];
Print["p[x]=",Simplify[sum]];
sum/.x->652
sum/.x->656
sum/.x->663
```

Output:

```
p[x]=1087.03 -4.94557 x+0.00751857 x^2-3.80952*10^-6 x^3
2.81474
2.81681
2.82121
```

3.2.10.5 Some Important Relations

1. Comparing Equations 3.7 and 3.83, we get

$$f[x, x_0, x_1, \ldots, x_n] = \frac{f^{(n+1)}(\xi)}{(n+1)!} \tag{3.87}$$

where $\min\{x, x_0, x_1, \ldots, x_n\} < \xi < \max\{x, x_0, x_1, \ldots, x_n\}$. From Equation 3.87, we have

$$f[x_0, x_1, \ldots, x_n] = \frac{f^{(n)}(\xi)}{n!} \tag{3.88}$$

where $\min\{x_0, x_1, \ldots, x_n\} < \xi < \max\{x_0, x_1, \ldots, x_n\}$.

2. In case of equidistant arguments given by $x_i = x_0 + ih$, $i = 0, 1, \ldots, n$, we have from Equation 3.78

$$f[x_0, x_1, \ldots, x_n] = \frac{\Delta^n f(x_0)}{n! h^n} = \frac{f^{(n)}(\xi)}{n!} \tag{3.89}$$

which yields

$$\Delta^n f(x_0) = h^n f^{(n)}(\xi), x_0 < \xi < x_n \tag{3.90}$$

3. If two or more arguments concide, then the previous definition of divided difference becomes indeterminate. In such a case, we extend the definition by a limiting process, called *confluent divided difference*.
We define

$$f[x_0, x_0] = \lim_{x \to x_0} f[x, x_0] \tag{3.91}$$

More generally,

$$f[x_0, x_0, x_1, \ldots, x_n] = \lim_{x \to x_0} f[x, x_0, x_1, \ldots, x_n] \tag{3.92}$$

Now, from Equation 3.91, we have

$$f[x_0, x_0] = \lim_{x \to x_0} f[x, x_0] = \lim_{x \to x_0} \frac{f(x) - f(x_0)}{x - x_0} = f'(x_0) \tag{3.93}$$

provided $f'(x_0)$ exists. Again,

$$\begin{aligned} f[x_0, x_0, x_1, \ldots, x_n] &= \lim_{x \to x_0} f[x, x_0, x_1, \ldots, x_n] = \lim_{x \to x_0} f[x, x_1, \ldots, x_n, x_0] \\ &= \lim_{x \to x_0} \frac{f[x, x_1, \ldots, x_n] - f[x_1, \ldots, x_n, x_0]}{x - x_0} \\ &= \lim_{x \to x_0} \frac{f[x, x_1, \ldots, x_n] - f[x_0, x_1, \ldots, x_n]}{x - x_0} \\ &= \frac{d}{dx} f[x, x_1, \ldots, x_n]\Big|_{x = x_0} \end{aligned} \tag{3.94}$$

Now, from Equation 3.76, we have

$$f[x, x_1, \ldots, x_n] = \frac{f(x)}{(x - x_1)(x - x_2)\ldots(x - x_n)} + \sum_{i=1}^{n} \frac{f(x_i)}{(x_i - x)(x_i - x_1)\ldots(x_i - x_{i-1})(x_i - x_{i+1})\ldots(x_i - x_n)}$$

Therefore, the confluent divided difference $f[x_0, x_0, x_1, \ldots, x_n]$ exists and Equation 3.94 holds if $f'(x_0)$ exists.

4. For the confluent divided difference, we can obtain from Equations 3.89 and 3.90

$$f[x_0, x_0, \ldots, x_0] = \frac{f^{(n)}(x_0)}{n!} \tag{3.95}$$

Also, from Equation 3.89 and 3.90, we have

$$f[x_0, x_0, \ldots, x_0, x_1, \ldots, x_n] = \frac{f^{(n+k)}(\xi)}{(n+k)!} \tag{3.96}$$

where x_0 occurs $(k+1)$ times in the argument of $f[x_0, x_0, \ldots, x_0, x_1, \ldots, x_n]$ and $\min\{x_0, x_1, \ldots, x_n\} < \xi < \max\{x_0, x_1, \ldots, x_n\}$.

5. Moreover, we have

$$f[x, x, \ldots, x, x_0, x_1, \ldots, x_n] = \frac{1}{k!}\frac{d^k}{dx^k} f[x, x_0, x_1, \ldots, x_n] \tag{3.97}$$

where x occurs $(k+1)$ times in the argument of $f[x, x, \ldots, x, x_0, x_1, \ldots, x_n]$.

Proof:
Let

$$g(x) = f[x, x_0, x_1, \ldots, x_n]$$

Then,

$$\begin{aligned} f[x, x', x_0, x_1, \ldots, x_n] = f[x, x_0, x_1, \ldots, x_n, x'] &= \frac{f[x, x_0, x_1, \ldots, x_n] - f[x_0, x_1, \ldots, x_n, x']}{x - x'} \\ &= \frac{f[x, x_0, x_1, \ldots, x_n] - f[x', x_0, x_1, \ldots, x_n]}{x - x'} \\ &= \frac{g(x) - g(x')}{x - x'} \\ &= g[x, x'] \end{aligned}$$

In general, it can be proved by induction that

$$f[x, x_1', x_2', \ldots, x_k', x_0, x_1, \ldots, x_n] = g[x, x_1', x_2', \ldots, x_k']$$

Now making $(x_1', x_2', \ldots, x_k') \to (x, x, \ldots, x)$, we obtain

$$f[x, x, \ldots, x, x_0, x_1, \ldots, x_n] = g[x, x, \ldots, x]$$

Hence, from Equation 3.95, we get

$$f[x, x, \ldots, x, x_0, x_1, \ldots, x_n] = g[x, x, \ldots, x] = \frac{g^{(k)}(x)}{k!} = \frac{1}{k!}\frac{d^k}{dx^k} f[x, x_0, x_1, \ldots, x_n]$$ ■

Example 3.9

Given that $\log_{10}654 = 2.8156$, $\log_{10}658 = 2.8182$, $\log_{10}659 = 2.8189$, and $\log_{10}661 = 2.8202$, find $\log_{10}652$, $\log_{10}656$, and $\log_{10}663$, using Newton's divided difference interpolation formula.

Solution:

To use the Newton's divided difference interpolation formula, we first construct the divided difference table

x	y	First-Order Divided Difference	Second-Order Divided Difference	Third-Order Divided Difference
654	2.8156			
		0.00065		
658	2.8182		0.00001	
		0.0007		−0.000004
659	2.8189		−0.000017	
		0.00065		
661	2.8202			

From this divided difference table, only the underlined values will be used in the Newton's divided difference interpolation formula.

Now, we obtain the Newton's divided difference interpolating polynomial as

$$f(x) \cong 2.8156 + 0.00065 \times (x-654) + (x-654)(x-658) \times (0.00001)$$
$$+ (x-654)(x-658)(x-659) \times (-0.000004)$$

Putting x = 652, 656, and 663, we obtain, respectively,

$$\log_{10} 652 = f(652) \cong 2.8156 + 0.00065 \times (-2) + (-2) \times (-6) \times (0.00001) + (-2) \times (-6) \times (-7)$$
$$\times (-0.000004)$$
$$= 2.8148$$

Since $x = 652$ lies outside the tabular values, this process is called *extrapolation*.

$$\log_{10} 656 = f(656) \cong 2.8156 + 0.00065 \times 2 + 2 \times (-2) \times (0.00001) + 2 \times (-2) \times (-3) \times (-0.000004)$$
$$= 2.81681$$

Here, the process of finding value of y is called *interpolation* because $x = 656$ lies inside the tabular values.

$$\log_{10} 664 = f(664) \cong 2.8156 + 0.00065 \times 10 + 10 \times 6 \times (0.00001) + 10 \times 6 \times 5 \times (-0.000004)$$
$$= 2.8215$$

In this case also, this process is called extrapolation because $x = 664$ lies outside the tabular values.

Example 3.10

The following table gives the viscosity of an oil as a function of temperature. Use Lagrange's formula to find viscosity of oil at a temperature of 140°.

Temperature (°C)	110	130	160	190
Viscosity	10.8	8.1	5.5	4.8

Solution:

Let x denotes temperature and y denotes viscosity, respectively.Using Lagrange's interpolation formula, we have

$$f(x) \cong \frac{(x-130)(x-160)(x-190)}{(110-130)(110-160)(110-190)} \times 10.8 + \frac{(x-110)(x-160)(x-190)}{(130-110)(130-160)(130-190)} \times 8.1$$

$$+ \frac{(x-110)(x-130)(x-190)}{(160-110)(160-130)(160-190)} \times 5.5 + \frac{(x-110)(x-130)(x-160)}{(190-110)(190-130)(190-160)} \times 4.8$$

which gives

$$f(140) \cong 7.03$$

Example 3.11

Use Lagrange's and Newton's divided difference formulae to calculate $f(3)$ from the following table:

x	0	1	2	4	5	6
f(*x*)	1	14	15	5	6	19

Solution:

Using Lagrange's interpolation formula, we have

$$f(x) \cong \frac{(x-1)(x-2)(x-4)(x-5)(x-6)}{(0-1)(0-2)(0-4)(0-5)(0-6)} \times 1 + \frac{x(x-2)(x-4)(x-5)(x-6)}{1.(1-2)(1-4)(1-5)(1-6)} \times 14$$

$$+ \frac{x(x-1)(x-4)(x-5)(x-6)}{2.(2-1)(2-4)(2-5)(2-6)} \times 15 + \frac{x(x-1)(x-2)(x-5)(x-6)}{4.(4-1)(4-2)(4-5)(4-6)} \times 5$$

$$+ \frac{x(x-1)(x-2)(x-4)(x-6)}{5.(5-1)(5-2)(5-4)(5-6)} \times 6 + \frac{x(x-1)(x-2)(x-4)(x-5)}{6.(6-1)(6-2)(6-4)(6-5)} \times 19$$

which gives

$$f(3) \cong 10$$

To use the Newton's divided difference interpolation formula, we first construct the divided difference table

x	y	First-Order Divided Difference	Second-Order Divided Difference	Third-Order Divided Difference	Fourth-Order Divided Difference	Fifth-Order Divided Difference
0	1					
		13				
1	14		−6			
		1		1		
2	15		−2		0	
		−5		1		0
4	5		2		0	
		1		1		
5	6		6			
		13				
6	19					

From this divided difference table, only the underlined values will be used in the Newton's divided difference interpolation formula. Now, we obtain the Newton's divided difference interpolating polynomial as

$$f(x) \cong 1+13x + x(x-1)\times(-6) + x(x-1)(x-2)\times 1$$

$$= 1+13x - 6x(x-1) + x(x-1)(x-2)$$

$$= x^3 - 9x^2 + 21x + 1$$

which gives

$$f(3) \cong 27 - 81 + 63 + 1 = 10$$

3.2.11 Gauss's Forward Interpolation Formula

Let …, y_{-2}, y_{-1}, y_0, y_1, y_2, … be the values of $y = f(x)$ corresponding to the equidistant values of x, that is, …, x_{-2}, x_{-1}, x_0, x_1, x_2, …, with step size h. We know Newton's forward interpolation formula is

$$y = f(x) = y_0 + \binom{u}{1}\Delta y_0 + \binom{u}{2}\Delta^2 y_0 + \binom{u}{3}\Delta^3 y_0 + \cdots \tag{3.98}$$

where

$$\binom{u}{r} = \frac{u(u-1)\ldots(u-r+1)}{r!} \quad \text{and} \quad u = \frac{(x-x_0)}{h}$$

Now,

$$\Delta^2 y_0 = \Delta^2 E y_{-1} = \Delta^2(1+\Delta)y_{-1} = \Delta^2 y_{-1} + \Delta^3 y_{-1}$$

Similarly,

$$\Delta^3 y_0 = \Delta^3 E y_{-1} = \Delta^3(1+\Delta)y_{-1} = \Delta^3 y_{-1} + \Delta^4 y_{-1}$$

$$\Delta^4 y_0 = \Delta^4 E y_{-1} = \Delta^4(1+\Delta)y_{-1} = \Delta^4 y_{-1} + \Delta^5 y_{-1}$$

and so on.

Substituting these values of $\Delta^2 y_0$, $\Delta^3 y_0$, and $\Delta^4 y_0$ in Equation 3.98, we obtain

$$y = f(x) = y_0 + \binom{u}{1}\Delta y_0 + \binom{u}{2}\left(\Delta^2 y_{-1} + \Delta^3 y_{-1}\right) + \binom{u}{3}\left(\Delta^3 y_{-1} + \Delta^4 y_{-1}\right) + \binom{u}{4}\left(\Delta^4 y_{-1} + \Delta^5 y_{-1}\right) + \cdots$$

$$= y_0 + \binom{u}{1}\Delta y_0 + \binom{u}{2}\Delta^2 y_{-1} + \left[\binom{u}{2} + \binom{u}{3}\right]\Delta^3 y_{-1} + \left[\binom{u}{3} + \binom{u}{4}\right]\Delta^4 y_{-1} + \cdots$$

$$= y_0 + \binom{u}{1}\Delta y_0 + \binom{u}{2}\Delta^2 y_{-1} + \binom{u+1}{3}\Delta^3 y_{-1} + \binom{u+1}{4}\Delta^4 y_{-1}$$

$$+ \binom{u+1}{5}\Delta^5 y_{-1} + \cdots, \text{ using Pascal's identity}$$

$$= y_0 + \binom{u}{1}\Delta y_0 + \binom{u}{2}\Delta^2 y_{-1} + \binom{u+1}{3}\Delta^3 y_{-1} + \binom{u+1}{4}\left(\Delta^4 y_{-2} + \Delta^5 y_{-2}\right) + \binom{u+1}{5}\left(\Delta^5 y_{-2} + \Delta^6 y_{-2}\right) + \cdots$$

$$= y_0 + \binom{u}{1}\Delta y_0 + \binom{u}{2}\Delta^2 y_{-1} + \binom{u+1}{3}\Delta^3 y_{-1} + \binom{u+1}{4}\Delta^4 y_{-2} + \left[\binom{u+1}{4} + \binom{u+1}{5}\right]\Delta^5 y_{-2} + \cdots$$

$$= y_0 + \binom{u}{1}\Delta y_0 + \binom{u}{2}\Delta^2 y_{-1} + \binom{u+1}{3}\Delta^3 y_{-1} + \binom{u+1}{4}\Delta^4 y_{-2} + \binom{u+2}{5}\Delta^5 y_{-2} + \cdots$$

Proceeding in the similar manner and grouping the terms, we have

$$\begin{aligned} y = y_0 &+ \left[\binom{u}{1}\Delta y_0 + \binom{u}{2}\Delta^2 y_{-1}\right] + \left[\binom{u+1}{3}\Delta^3 y_{-1} + \binom{u+1}{4}\Delta^4 y_{-2}\right] \\ &+ \left[\binom{u+2}{5}\Delta^5 y_{-2} + \binom{u+2}{5}\Delta^6 y_{-2}\right] + \cdots \end{aligned} \tag{3.99}$$

which is called *Gauss's forward interpolation formula.*

Note: Gauss's forward interpolation formula is used when the interpolation point is near the center of the tabular values. Moreover, this formula gives best result if the starting argument x_0 is such that $0 < u < 1$.

3.2.12 Gauss's Backward Interpolation Formula

Let $\ldots, y_{-2}, y_{-1}, y_0, y_1, y_2, \ldots$ be the values of $y = f(x)$ corresponding to the equidistant values of x, that is, $\ldots, x_{-2}, x_{-1}, x_0, x_1, x_2, \ldots$, with step size h. We know Newton's forward interpolation formula is

$$y = f(x) = y_0 + \binom{u}{1}\Delta y_0 + \binom{u}{2}\Delta^2 y_0 + \binom{u}{3}\Delta^3 y_0 + \cdots \tag{3.100}$$

where

$$\binom{u}{r} = \frac{u(u-1)\ldots(u-r+1)}{r!} \quad \text{and} \quad u = \frac{(x-x_0)}{h}$$

Now,

$$\Delta y_0 = \Delta E y_{-1} = \Delta(1+\Delta)y_{-1} = \Delta y_{-1} + \Delta^2 y_{-1}$$

Similarly,

$$\Delta^2 y_0 = \Delta^2 E y_{-1} = \Delta^2(1+\Delta)y_{-1} = \Delta^2 y_{-1} + \Delta^3 y_{-1}$$

$$\Delta^3 y_0 = \Delta^3 E y_{-1} = \Delta^3(1+\Delta)y_{-1} = \Delta^3 y_{-1} + \Delta^4 y_{-1}$$

$$\Delta^4 y_0 = \Delta^4 E y_{-1} = \Delta^4(1+\Delta)y_{-1} = \Delta^4 y_{-1} + \Delta^5 y_{-1}$$

and so on.

Substituting these values of Δy_0, $\Delta^2 y_0$, $\Delta^3 y_0$, and $\Delta^4 y_0$ in Equation 3.100, we obtain

$$y = f(x) = y_0 + \binom{u}{1}\left(\Delta y_{-1} + \Delta^2 y_{-1}\right) + \binom{u}{2}\left(\Delta^2 y_{-1} + \Delta^3 y_{-1}\right) + \binom{u}{3}\left(\Delta^3 y_{-1} + \Delta^4 y_{-1}\right)$$

$$+ \binom{u}{4}\left(\Delta^4 y_{-1} + \Delta^5 y_{-1}\right) + \cdots$$

$$= y_0 + \binom{u}{1}\Delta y_{-1} + \left[\binom{u}{1} + \binom{u}{2}\right]\Delta^2 y_{-1} + \left[\binom{u}{2} + \binom{u}{3}\right]\Delta^3 y_{-1} + \left[\binom{u}{3} + \binom{u}{4}\right]\Delta^4 y_{-1} + \cdots$$

$$= y_0 + \binom{u}{1}\Delta y_{-1} + \binom{u+1}{2}\Delta^2 y_{-1} + \binom{u+1}{3}\Delta^3 y_{-1} + \binom{u+1}{4}\Delta^4 y_{-1} + \binom{u+1}{5}\Delta^5 y_{-1}$$

$+\cdots$, using Pascal's identity

$$= y_0 + \binom{u}{1}\Delta y_{-1} + \binom{u+1}{2}\Delta^2 y_{-1} + \binom{u+1}{3}\left(\Delta^3 y_{-2} + \Delta^4 y_{-2}\right) + \binom{u+1}{4}\left(\Delta^4 y_{-2} + \Delta^5 y_{-2}\right)$$

$$+ \binom{u+1}{5}\left(\Delta^5 y_{-2} + \Delta^6 y_{-2}\right) + \cdots$$

$$= y_0 + \binom{u}{1}\Delta y_{-1} + \binom{u+1}{2}\Delta^2 y_{-1} + \binom{u+1}{3}\Delta^3 y_{-2} + \left[\binom{u+1}{3} + \binom{u+1}{4}\right]\Delta^4 y_{-2} + \cdots$$

$$= y_0 + \binom{u}{1}\Delta y_{-1} + \binom{u+1}{2}\Delta^2 y_{-1} + \binom{u+1}{3}\Delta^3 y_{-2} + \binom{u+2}{4}\Delta^4 y_{-2} + \cdots$$

Proceeding in the similar manner and grouping the terms, we have

$$y = y_0 + \binom{u}{1}\Delta y_{-1} + \left[\binom{u+1}{2}\Delta^2 y_{-1} + \binom{u+1}{3}\Delta^3 y_{-2}\right] + \left[\binom{u+2}{4}\Delta^4 y_{-2} + \binom{u+2}{5}\Delta^5 y_{-3}\right] + \cdots \quad (3.101)$$

which is called *Gauss's backward interpolation formula.*

Note: Gauss's backward interpolation formula is used when the interpolation point is near the center of the tabular values. Moreover, this formula gives best result if the starting argument x_0 is such that $-1 < u < 0$.

Example 3.12

Find the values of y when the values of $x = 3.3$ and 2.8, respectively, from the following data using

1. Gauss's forward interpolation formula
2. Gauss's backward interpolation formula

x	2.0	2.5	3.0	3.5	4.0
y	246.2	409.3	537.2	636.3	715.9

Solution:

To use the Gauss's interpolation formulae, we first construct the difference table

x	y	Δy	$\Delta^2 y$	$\Delta^3 y$	$\Delta^4 y$
2.0	246.2				
		163.1			
2.5	409.3		−35.2		
		<u>127.9</u>		<u>6.4</u>	
3.0	**<u>537.2</u>**		**<u>−28.8</u>**		**<u>2.9</u>**
		99.1		**9.3**	
3.5	636.3		−19.5		
		79.6			
4.0	715.9				

Now in order to determine the value of y for the given value of $x = 3.3$, if we choose $x_0 = 3$, then $u = [(3.3 - 3)/0.5] = 0.6$ lies between 0 and 1. Therefore, we can use Gauss's forward interpolation formula.

From the above difference table, we identify $y_0 = 537.2$, $\Delta y_0 = 99.1$, $\Delta^2 y_{-1} = -28.8$, $\Delta^3 y_{-1} = 9.3$, and $\Delta^4 y_{-2} = 2.9$, which are to be used in the Gauss's forward interpolation formula. These values have been shown in bold digits.

Using Gauss's forward interpolation formula, we have

$$\varphi(x) \cong y_0 + \left[\binom{u}{1}\Delta y_0 + \binom{u}{2}\Delta^2 y_{-1}\right] + \left[\binom{u+1}{3}\Delta^3 y_{-1} + \binom{u+1}{4}\Delta^4 y_{-2}\right] + \cdots$$

where $u = [(x - x_0)/h]$. Therefore,

$$y(3.3) \cong \varphi(3.3) = 537.2 + \left[0.6 \times 99.1 + \frac{0.6(0.6-1)}{2} \times (-28.8)\right]$$

$$+ \left[\frac{1.6(1.6-1)(1.6-2)}{6} \times 9.3 + \frac{1.6(1.6-1)(1.6-2)(1.6-3)}{24} \times 2.9\right]$$

$$= 599.586$$

Again, in order to determine y for given $x = 2.8$, if we choose $x_0 = 3$, then $u = [(2.8 - 3)/0.5] = -0.4$ lies between −1 and 0. Therefore, we can use Gauss's backward interpolation formula. From the above difference table, we identify $y_0 = 537.2$, $\Delta y_{-1} = 127.9$, $\Delta^2 y_{-1} = -28.8$, $\Delta^3 y_{-2} = 6.4$, and $\Delta^4 y_{-2} = 2.9$, which are to be used in the Gauss's backward interpolation formula. These values have been underlined in the difference table.

Using Gauss's backward interpolation formula, we have

$$\varphi(x) \cong y_0 + \binom{u}{1}\Delta y_{-1} + \left[\binom{u+1}{2}\Delta^2 y_{-1} + \binom{u+1}{3}\Delta^3 y_{-2}\right] + \left[\binom{u+2}{4}\Delta^4 y_{-2} + \binom{u+2}{5}\Delta^5 y_{-3}\right] + \cdots$$

where $u = (x - x_0)/h$.

Therefore,

$$y(2.8) \cong \varphi(2.8) = 537.2 + (-0.4)\times 127.9 + \frac{(1-0.4)(-0.4)}{2}\times(-28.8)$$

$$+\frac{(1-0.4)(-0.4)(-0.4-1)}{6}\times 6.4$$

$$+\frac{(1-0.4)(-0.4)(-0.4-1)(-0.4-2)}{24}\times 2.9$$

$$= 489.757$$

3.2.13 Central Difference

The central difference operator δ is defined by

$$\delta f(x) = f\left(x+\frac{h}{2}\right) - f\left(x-\frac{h}{2}\right)$$

Now, we note that

$$\delta f(x) = E^{1/2} f(x) - E^{-(1/2)} f(x)$$
$$= \left(E^{1/2} - E^{-(1/2)}\right) f(x)$$

Therefore,

$$\delta = E^{1/2} - E^{-(1/2)} \tag{3.102}$$

Also,

$$\delta f\left(x+\frac{h}{2}\right) = f(x+h) - f(x) = \Delta f(x)$$

and

$$\delta f(x) = f\left(x+\frac{h}{2}\right) - f\left(x-\frac{h}{2}\right) = \Delta f\left(x-\frac{h}{2}\right)$$

Thus, we may obtain the following results:

$$\delta E^{1/2} = E^{1/2}\delta = \Delta = E - I,\ \delta = \Delta E^{-(1/2)} = E^{-(1/2)}\Delta,\ \delta E^{-(1/2)} = E^{-(1/2)}\delta = I - E^{-1} = \nabla \tag{3.103}$$

In general, we can obtain

$$\delta^k = \Delta^k E^{-(k/2)} = E^{-(k/2)}\Delta^k,\ \Delta^k = \delta^k E^{k/2} = E^{k/2}\delta^k,\ \nabla^k = \delta^k E^{-(k/2)} = E^{-(k/2)}\delta^k \tag{3.104}$$

Therefore, it can be shown that

$$\delta^k y_i = \Delta^k y_{i-(k/2)},\ \Delta^k y_i = \delta^k y_{i+(k/2)},\ \nabla^k y_i = \delta^k y_{i-(k/2)} \tag{3.105}$$

Therefore,

$$\Delta y_0 = y_1 - y_0 = \delta y_{1/2},\ \Delta y_1 = y_2 - y_1 = \delta y_{3/2}, \ldots, \Delta y_{n-1} = y_n - y_{n-1} = \delta y_{n-(1/2)}$$

Similarly,

$$\Delta^2 y_0 = \Delta y_1 - \Delta y_0 = \delta y_{3/2} - \delta y_{1/2} = \delta^2 y_1$$

$$\Delta^2 y_1 = \Delta y_2 - \Delta y_1 = \delta y_{5/2} - \delta y_{3/2} = \delta^2 y_2$$

$$\cdots$$

$$\Delta^3 y_0 = \Delta^2 y_1 - \Delta^2 y_0 = \delta^2 y_2 - \delta^2 y_1 = \delta^3 y_{3/2}$$

and so on.

3.2.13.1 Central Difference Table

The central differences are displayed in Table 3.5, which is called *central difference table*, in which the differences $\delta^k y_i$, for a given i and for $k = 1, 2, \ldots$ lie along the same horizontal line as y_i.

3.2.13.2 Stirling's Interpolation Formula

Let the values of the function $y = f(x)$ be known for the following odd number of equispaced arguments

$$x_{-n} = x_0 - nh, \ldots, x_{-2} = x_0 - 2h, x_{-1} = x_0 - h, x_0 = x_0, x_1 = x_0 + h, x_2 = x_0 + 2h, \ldots, x_n = x_0 + nh$$

Using Gauss's forward interpolation formula,

$$y = f(x) = y_0 + \left[\binom{u}{1}\Delta y_0 + \binom{u}{2}\Delta^2 y_{-1}\right] + \left[\binom{u+1}{3}\Delta^3 y_{-1} + \binom{u+1}{4}\Delta^4 y_{-2}\right] + \left[\binom{u+2}{5}\Delta^5 y_{-2} + \binom{u+2}{6}\Delta^6 y_{-3}\right] + \cdots \tag{3.106}$$

TABLE 3.5
Central Difference Table

x	y	δ	δ^2	δ^3	δ^4
x_{-2}	y_{-2}				
		$\delta y_{-\frac{3}{2}} (= \Delta y_{-2})$			
x_{-1}	y_{-1}		$\delta^2 y_{-1} (= \Delta^2 y_{-2})$		
		$\delta y_{-\frac{1}{2}} (= \Delta y_{-1})$		$\delta^3 y_{-\frac{1}{2}} (= \Delta^3 y_{-2})$	
x_0	y_0		$\delta^2 y_0 (= \Delta^2 y_{-1})$		$\delta^4 y_0 (= \Delta^4 y_{-2})$
		$\delta y_{\frac{1}{2}} (= \Delta y_0)$		$\delta^3 y_{\frac{1}{2}} (= \Delta^3 y_{-1})$	
x_1	y_1		$\delta^2 y_1 (= \Delta^2 y_0)$		
		$\delta y_{\frac{3}{2}} (= \Delta y_1)$			
x_2	y_2				

Again, using Gauss's backward interpolation formula,

$$y=f(x)=y_0+\binom{u}{1}\Delta y_{-1}+\left[\binom{u+1}{2}\Delta^2 y_{-1}+\binom{u+1}{3}\Delta^3 y_{-2}\right]+\left[\binom{u+2}{4}\Delta^4 y_{-2}+\binom{u+2}{5}\Delta^5 y_{-3}\right]+\cdots \quad (3.107)$$

Adding Equations 3.106 and 3.107 such that differences of the same order are grouped and dividing by 2, we have

$$y=f(x)=y_0+\binom{u}{1}\left(\frac{\Delta y_{-1}+\Delta y_0}{2}\right)+\left[\binom{u}{2}+\binom{u+1}{2}\right]\frac{\Delta^2 y_{-1}}{2}+\binom{u+1}{3}\left(\frac{\Delta^3 y_{-1}+\Delta^3 y_{-2}}{2}\right)$$

$$+\left[\binom{u+1}{4}+\binom{u+2}{4}\right]\frac{\Delta^4 y_{-2}}{2}+\cdots+\frac{\binom{u+n-1}{2n}+\binom{u+n}{2n}}{2}\Delta^{2n} y_{-n}$$

where $u=[(x-x_0)/h]$

This implies

$$y=f(x)=y_0+\frac{u(\Delta y_0+\Delta y_{-1})}{2}+\frac{u^2}{2}\Delta^2 y_{-1}+\frac{u(u^2-1^2)}{3!}\frac{\Delta^3 y_{-2}+\Delta^3 y_{-1}}{2}+\frac{u^2(u^2-1^2)}{4!}\Delta^4 y_{-2}+\cdots$$

$$+\frac{u^2(u^2-1^2)(u^2-2^2)\ldots[u^2-(n-1)^2]}{2n!}\Delta^{2n} y_{-n} \quad (3.108)$$

which is called *Stirling's interpolation formula*, and it is used when the interpolating point x is near the center of the table and the number of arguments or nodes is odd.

Using Equation 3.7, the remainder or error is given by

$$R_{2n+1}(x)=\frac{u\left(u^2-1^2\right)\left(u^2-2^2\right)\ldots\left(u^2-n^2\right)}{(2n+1)!}h^{2n+1}f^{(2n+1)}(\xi) \quad (3.109)$$

where $\min\{x_{-n},x,x_n\}<\xi<\max\{x_{-n},x,x_n\}$. In terms of the central differences, Stirling's formula in Equation 3.108 becomes symmetrical form with respect to the starting argument x_0, that is,

$$y=f(x)=y_0+u\left(\frac{\delta y_{-(1/2)}+\delta y_{1/2}}{2}\right)+\frac{u^2}{2}\delta^2 y_0+\frac{u(u^2-1^2)}{3!}\left(\frac{\delta^3 y_{-(1/2)}+\Delta^3 y_{1/2}}{2}\right)+\cdots$$

$$+\frac{u^2(u^2-1^2)(u^2-2^2)\ldots[u^2-(n-1)^2]}{2n!}\delta^{2n} y_0 \quad (3.110)$$

Remarks: Stirling's formula yields best approximation for $f(x)$ if the initial argument x_0 is such that $-0.25<u<0.25$.

3.2.13.3 Bessel's Interpolation Formula

Let the values of the function $y=f(x)$ be known for the following even number of equispaced arguments

$$x_{-(n-1)}=x_0-(n-1)h,\ldots,x_{-2}=x_0-2h,\; x_{-1}=x_0-h,\; x_0=x_0,\; x_1=x_0+h,\; x_2=x_0+2h,\ldots,x_n=x_0+nh$$

Using Gauss's forward interpolation formula

$$y = f(x) = y_0 + \left[\binom{u}{1}\Delta y_0 + \binom{u}{2}\Delta^2 y_{-1}\right] + \left[\binom{u+1}{3}\Delta^3 y_{-1} + \binom{u+1}{4}\Delta^4 y_{-2}\right]$$
$$+ \left[\binom{u+2}{5}\Delta^5 y_{-2} + \binom{u+2}{6}\Delta^6 y_{-3}\right] + \cdots \tag{3.111}$$

Now,

$$y_0 = y_1 - \Delta y_0 \tag{3.112}$$

Similarly,

$$y_{-1} = y_0 - \Delta y_{-1}$$

Therefore,

$$\Delta^2 y_{-1} = \Delta^2 y_0 - \Delta^3 y_{-1} \tag{3.113}$$

Again,

$$y_{-2} = y_{-1} - \Delta y_{-2}$$

This implies

$$\Delta^4 y_{-2} = \Delta^4 y_{-1} - \Delta^5 y_{-2} \tag{3.114}$$

and so on.

Now, Equation 3.111 can be written as

$$y = \left(\frac{y_0}{2} + \frac{y_0}{2}\right) + \binom{u}{1}\Delta y_0 + \frac{1}{2}\frac{u(u-1)}{2!}\Delta^2 y_{-1} + \frac{1}{2}\frac{u(u-1)}{2!}\Delta^2 y_{-1} + \binom{u+1}{3}\Delta^3 y_{-1}$$
$$+ \frac{1}{2}\frac{(u+1)u(u-1)(u-2)}{4!}\Delta^4 y_{-2} + \frac{1}{2}\frac{(u+1)u(u-1)(u-2)}{4!}\Delta^4 y_{-2} + \cdots \tag{3.115}$$

Using Equations 3.112 through 3.114, for the second split terms in Equation 3.115, we have

$$y = \frac{y_0}{2} + \frac{y_1 - \Delta y_0}{2} + \binom{u}{1}\Delta y_0 + \frac{1}{2}\frac{u(u-1)}{2!}\Delta^2 y_{-1} + \frac{1}{2}\frac{u(u-1)}{2!}\left(\Delta^2 y_0 - \Delta^3 y_{-1}\right) + \binom{u+1}{3}\Delta^3 y_{-1}$$
$$+ \frac{1}{2}\frac{(u+1)u(u-1)(u-2)}{4!}\Delta^4 y_{-2} + \frac{1}{2}\frac{(u+1)u(u-1)(u-2)}{4!}\left(\Delta^4 y_{-1} - \Delta^5 y_{-2}\right) + \cdots$$
$$= \left(\frac{y_0}{2} + \frac{y_1}{2}\right) + \left(u - \frac{1}{2}\right)\Delta y_0 + \frac{u(u-1)}{2!}\left(\frac{\Delta^2 y_{-1} + \Delta^2 y_0}{2}\right) + \frac{u(u-1)}{2!}\left(-\frac{1}{2} + \frac{u+1}{3}\right)\Delta^3 y_{-1} + \cdots$$

Thus,

$$y = \left(\frac{y_0 + y_1}{2}\right) + \left(u - \frac{1}{2}\right)\Delta y_0 + \frac{u(u-1)}{2!}\left(\frac{\Delta^2 y_{-1} + \Delta^2 y_0}{2}\right) + \frac{\left(u - (1/2)\right)u(u-1)}{3!}\Delta^3 y_{-1}$$
$$+ \frac{(u+1)u(u-1)(u-2)}{4!}\left(\frac{\Delta^4 y_{-2} + \Delta^4 y_{-1}}{2}\right) + \cdots$$

Therefore,

$$f(x)=\frac{y_0+y_1}{2}+\left(u-\frac{1}{2}\right)\Delta y_0+\binom{u}{2}\left(\frac{\Delta^2 y_{-1}+\Delta^2 y_0}{2}\right)+\frac{u(u-1)(u-(1/2))}{3!}\Delta^3 y_{-1}$$
$$+\binom{u+1}{4}\left(\frac{\Delta^4 y_{-2}+\Delta^4 y_{-1}}{2}\right)+\cdots$$
$$+\binom{u+n-2}{2n-2}\left(\frac{\Delta^{2n-2} y_{-n+2}+\Delta^{2n-2} y_{-n+1}}{2}\right)+\frac{\binom{u+n-1}{2n-1}+\binom{u+n-2}{2n-1}}{2}\Delta^{2n-1} y_{-n+1} \quad (3.116)$$

which is called *Bessel's interpolation formula*, and it is used when the interpolating point x is near the center of the table and the number of arguments or nodes is even. Setting $v = u-(1/2)$, from Equation 3.116 yields

$$f(x)=\frac{y_0+y_1}{2}+v\Delta y_0+\frac{\left[v^2-(1/4)\right]}{2!}\frac{\left(\Delta^2 y_{-1}+\Delta^2 y_0\right)}{2}+\frac{v\left[v^2-(1/4)\right]}{3!}\Delta^3 y_{-1}$$
$$+\frac{\left[v^2-(1/4)\right]\left[v^2-(9/4)\right]}{4!}\frac{\left(\Delta^4 y_{-2}+\Delta^4 y_{-1}\right)}{2}+\frac{v\left[v^2-(1/4)\right]\left[v^2-(9/4)\right]}{5!}\Delta^5 y_{-2}+\cdots$$
$$+\frac{v\left[v^2-(1/4)\right]\left[v^2-(9/4)\right]\ldots\left\{v^2-\left[(2n-3)^2/4\right]\right\}}{(2n-1)!}\Delta^{2n-1} y_{-n+1} \quad (3.117)$$

Using Equation 3.7, the remainder or error is given by

$$R_{2n}(x)=\frac{\left[v^2-(1/4)\right]\left[v^2-(9/4)\right]\ldots\left\{v^2-\left[(2n-1)^2/4\right]\right\}}{2n!}h^{2n}f^{(2n)}(\xi) \quad (3.118)$$

where $\min\{x_{-(n-1)},x,x_n\}<\xi<\max\{x_{-(n-1)},x,x_n\}$. In terms of the central differences, Bessel's formula in Equation 3.118 becomes

$$f(x)=\frac{y_0+y_1}{2}+v\delta y_{1/2}+\frac{\left[v^2-(1/4)\right]}{2!}\frac{\left(\delta^2 y_0+\delta^2 y_1\right)}{2!}+\frac{v\left[v^2-(1/4)\right]}{3!}\delta^3 y_{1/2}+\cdots$$
$$+\frac{v\left[v^2-(1/4)\right]\left[v^2-(9/4)\right]\ldots\left\{v^2-\left[(2n-3)^2/4\right]\right\}}{(2n-1)!}\delta^{2n-1} y_{1/2} \quad (3.119)$$

Remarks: Bessel's formula results in best approximation for $f(x)$ if the initial argument x_0 is such that $0.25 < u < 0.75$, that is, $-0.25 < v = u-(1/2) < 0.25$.

Example 3.13

Apply Stirling's formula to find the value of $f(1.58)$ from the following table:

x	1.3	1.4	1.5	1.6	1.7	1.8	1.9
$f(x)$	0.26236	0.33647	0.40547	0.47	0.53063	0.58779	0.64185

Solution:

Here, we choose $x_0 = 1.6$ so that $u = [(x - x_0)/h] = [(1.58 - 1.6)/0.1] = -0.2$. Therefore, the value of u satisfies the condition $-0.25 < u < 0.25$. Moreover, the number of given arguments is odd. Hence, we apply Stirling's interpolation formula. Using Stirling's interpolation formula, we have

$$\varphi(x) = y_0 + \frac{u(\Delta y_0 + \Delta y_{-1})}{2} + \frac{u^2}{2}\Delta^2 y_{-1} + \frac{u(u^2 - 1^2)}{3!}\frac{\Delta^3 y_{-2} + \Delta^3 y_{-1}}{2} + \frac{u^2(u^2 - 1^2)}{4!}\Delta^4 y_{-2} + \cdots$$
$$+ \frac{u^2(u^2 - 1^2)(u^2 - 2^2)\ldots[u^2 - (n-1)^2]}{2n!}\Delta^{2n} y_{-n}$$

Now, we construct the difference table

x	y	Δy	$\Delta^2 y$	$\Delta^3 y$	$\Delta^4 y$	$\Delta^5 y$	$\Delta^6 y$
1.3	0.26236						
		0.07411					
1.4	0.33647		−0.00518				
		0.06893		0.00085			
1.5	0.4054		−0.00433		−0.00049		
		0.0646		**0.00036**		**0.00063**	
1.6	**0.47**		**−0.00397**		**0.00014**		**−0.0009**
		0.06063		**0.0005**		**−0.00027**	
1.7	0.53063		−0.00347		−0.00013		
		0.05716		0.00037			
1.8	0.58779		−0.0031				
		0.05406					
1.9	0.64185						

From the above difference table, we identify $y_0 = 0.47$, $\Delta y_{-1} = 0.0646$, $\Delta y_0 = 0.06063$, $\Delta^2 y_{-1} = -0.00397$, $\Delta^3 y_{-2} = 0.00036$, $\Delta^3 y_{-1} = 0.0005$, $\Delta^4 y_{-2} = 0.00014$, $\Delta^5 y_{-3} = 0.00063$, $\Delta^5 y_{-2} = -0.00027$, and $\Delta^6 y_{-3} = -0.0009$, which are to be used in the Stirling's interpolation formula. These values are shown in bold digits. Using Stirling's interpolation formula, we have

$$f(1.58) \cong \varphi(1.58) = 0.47 + \frac{(-0.2)(0.0646 + 0.06063)}{2} + \frac{(-0.2)^2}{2} \times (-0.00397)$$
$$+ \frac{(-0.2)\{(-0.2)^2 - 1^2\}}{3!} \times \left(\frac{0.00036 + 0.0005}{2}\right)$$
$$+ \frac{(-0.2)^2\{(-0.2)^2 - 1^2\}}{4!} \times 0.00014 + \frac{(-0.2) \times \{(-0.2)^2 - 1^2\}\{(-0.2)^2 - 2^2\}}{5!}$$
$$\times \left(\frac{0.00063 - 0.00027}{2}\right)$$
$$+ \frac{(-0.2)^2\{(-0.2)^2 - 1^2\}\{(-0.2)^2 - 2^2\}}{6!} \times (-0.0009)$$
$$= 0.457418$$

Example 3.14

Use Bessel's formula to find the value of $f(x)$ at $x = 1.95$, given that

x	1.7	1.8	1.9	2.0	2.1	2.2	2.3
$f(x)$	2.979	3.144	3.283	3.391	3.463	3.997	4.491

Which other interpolation formula can be used? Which is more appropriate? Give reasons.

Solution:

x	y	Δ	Δ^2	Δ^3	Δ^4	Δ^5	Δ^6
1.7	2.979						
		0.165					
1.8	3.144		−0.026				
		0.139		−0.005			
1.9	3.283		−0.031		0		
		0.108		−0.005		0.503	
2.0	3.391		−0.036		0.503		−2.006
		0.072		0.498		−1.503	
2.1	3.463		0.462		−1		
		0.534		−0.502			
2.2	3.997		−0.04				
		0.494					
2.3	4.491						

Here, $u = [(x - x_0)/h] = [(1.95 - 1.9)/0.1] = 0.5$. Therefore, $u > 0.25$. Since $u > 0.25$, Stirling's formula is not appropriate. Now, $v = u - 0.5 = 0.5 - 0.5 = 0$. Therefore, $v > -0.25$. Since $v > -0.25$, that is, $u > 0.25$, Bessel's formula should be used. Using Bessel's interpolation formula, we get

$$y_{1.95} \cong \frac{3.283 + 3.391}{2} + 0 \times 0.108 + \frac{[0 - (1/4)]}{2!} \times \frac{(-0.031)(-0.036)}{2} + \frac{0 \times [0 - (1/4)]}{3!} \times (-0.005)$$

$$+ \frac{[0 - (1/4)] \times [0 - (9/4)]}{4!} \times \frac{(0 + 0.503)}{2} + \frac{0 \times [0 - (1/4)] \times [0 - (9/4)]}{5!} \times 0.503$$

$$= 3.337 + 0.0041875 + 0.0057945$$

$$= 3.347082$$

3.2.13.4 Everette's Interpolation Formula

Let the values of the function $y = f(x)$ be known for the following even number of equispaced arguments

$$x_{-(n-1)} = x_0 - (n-1)h, \ \ldots, \ x_{-2} = x_0 - 2h, \ x_{-1} = x_0 - h, x_0 = x_0, x_1 = x_0 + h, x_2 = x_0 + 2h, \ldots, x_n = x_0 + nh$$

Using Gauss's forward interpolation formula

$$y=f(x)=y_0+\left[\binom{u}{1}\Delta y_0+\binom{u}{2}\Delta^2 y_{-1}\right]+\left[\binom{u+1}{3}\Delta^3 y_{-1}+\binom{u+1}{4}\Delta^4 y_{-2}\right]$$
$$+\left[\binom{u+2}{5}\Delta^5 y_{-2}+\binom{u+2}{6}\Delta^6 y_{-3}\right]+\cdots \tag{3.120}$$

Now, we have

$$\Delta y_0 = y_1 - y_0$$
$$\Delta^3 y_{-1} = \Delta^2 y_0 - \Delta^2 y_{-1}$$
$$\Delta^5 y_{-2} = \Delta^4 y_{-1} - \Delta^4 y_{-2}$$

and so on.

Substituting these values of Δy_0, $\Delta^3 y_{-1}$, and $\Delta^5 y_{-2}$ in Equation 3.120, we get

$$y = f(x) = y_0 + \binom{u}{1}(y_1 - y_0) + \binom{u}{2}\Delta^2 y_{-1} + \binom{u+1}{3}\left(\Delta^2 y_0 - \Delta^2 y_{-1}\right) + \binom{u+1}{4}\Delta^4 y_{-2}$$
$$+\binom{u+2}{5}\left(\Delta^4 y_{-1} - \Delta^4 y_{-2}\right) + \binom{u+2}{6}\Delta^6 y_{-3} + \cdots$$
$$= (1-u)y_0 + uy_1 + \left[\binom{u}{2} - \binom{u+1}{3}\right]\Delta^2 y_{-1} + \binom{u+1}{3}\Delta^2 y_0$$
$$+\left[\binom{u+1}{4} - \binom{u+2}{5}\right]\Delta^4 y_{-2} + \binom{u+2}{5}\Delta^4 y_{-1} + \cdots \tag{3.121}$$

According to Pascal's identity,

$$\binom{u}{r} + \binom{u}{r+1} = \binom{u+1}{r+1}$$

This implies

$$\binom{u}{r} - \binom{u+1}{r+1} = -\binom{u}{r+1}$$

Using this in Equation 3.121, we have

$$y = (1-u)y_0 + uy_1 - \binom{u}{3}\Delta^2 y_{-1} + \binom{u+1}{3}\Delta^2 y_0 - \binom{u+1}{5}\Delta^4 y_{-2} + \binom{u+2}{5}\Delta^4 y_{-1} - \cdots \tag{3.122}$$

Now, setting

$$v = 1-u$$

Equation 3.122 yields

$$y = \left[vy_0 + \binom{v+1}{3}\Delta^2 y_{-1} + \binom{v+2}{5}\Delta^4 y_{-2} + \cdots\right] + \left[uy_1 + \binom{u+1}{3}\Delta^2 y_0 + \binom{u+2}{5}\Delta^4 y_{-1} + \cdots\right]$$

Thus,

$$y=\left[vy_0+\frac{v\left(v^2-1^2\right)}{3!}\Delta^2 y_{-1}+\frac{v\left(v^2-1^2\right)\left(v^2-2^2\right)}{5!}\Delta^4 y_{-2}+\cdots\right]$$
$$+\left[uy_1+\frac{u\left(u^2-1^2\right)}{3!}\Delta^2 y_0+\frac{u\left(u^2-1^2\right)\left(u^2-2^2\right)}{5!}\Delta^4 y_{-1}+\cdots\right] \tag{3.123}$$

Equation 3.123 is called *Everett's interpolation formula.* It is also known as *Laplace–Everett's interpolation formula.*

This formula is used when the interpolating point x is near the middle of the table and the number of equispaced arguments or nodes is even. Using Equation 3.7, the remainder or error is given by

$$R_{2n}(x)=\frac{u\left(u^2-1^2\right)\left(u^2-2^2\right)\ldots\left\{u^2-(n-1)^2\right\}(u-n)}{(2n-1)!}h^{2n}f^{(2n)}(\xi) \tag{3.124}$$

where $\min\{x_{-(n-1)},x,x_n\}<\xi<\max\{x_{-(n-1)},x,x_n\}$.

Remarks: The Everett's formula yields best approximation for $f(x)$ if the initial argument x_0 is such that $0.25<u<0.75$.

Example 3.15

Using Laplace–Everett's formula, find the value of log 335.5 from the following data:

x	310	320	330	340	350	360
$\log_{10} x$	2.491	2.505	2.518	2.531	2.544	2.556

Solution:

We first construct the difference table

x	y	Δy	$\Delta^2 y$	$\Delta^3 y$	$\Delta^4 y$	$\Delta^5 y$
310	2.491					
		0.014				
320	2.505		−0.001			
		0.013		0.001		
330	**2.518**		**0**		**−0.001**	
		0.013		**0**		**0**
340	**2.531**		**0**		**−0.001**	
		0.013		−0.001		
350	2.544		−0.001			
		0.012				
360	2.556					

Here, we choose $x_0=330$, so that $u=[(x-x_0)/h]=[(335.5-330)/10]=0.55$. Therefore, the value of u satisfies the condition $0.25<u<0.75$. Hence, we apply Everett's interpolation formula. From the above difference table, we identify $y_0=2.518$, $y_1=2.531$, $\Delta^2 y_{-1}=0$, $\Delta^2 y_0=0$, $\Delta^3 y_{-1}=0$, $\Delta^4 y_{-2}=-0.001$, $\Delta^4 y_{-1}=-0.001$, and $\Delta^5 y_{-2}=0$, which are to be used in the Everett's interpolation formula. These values have been shown in bold digits.

Using Everett's interpolation formula, we have

$$y = \left[vy_0 + \frac{v(v^2-1^2)}{3!}\Delta^2 y_{-1} + \frac{v(v^2-1^2)(v^2-2^2)}{5!}\Delta^4 y_{-2} + \cdots \right] + \left[uy_1 + \frac{u(u^2-1^2)}{3!}\Delta^2 y_0 + \frac{u(u^2-1^2)(u^2-2^2)}{5!}\Delta^4 y_{-1} + \cdots \right]$$

where $v = 1-u$. Therefore,

$$\begin{aligned}\log 335.5 &\cong 0.45\times 2.518 + \frac{0.45(0.45^2-1^2)}{6}\times 0 + \frac{0.45(0.45^2-1^2)(0.45^2-2^2)}{120}\times(-0.001)\\ &\quad + 0.55\times 2.531 + \frac{0.55(0.55^2-1^2)}{6}\times 0 + \frac{0.55(0.55^2-1^2)(0.55^2-2^2)}{120}\times(-0.001)\\ &= 2.52513\end{aligned}$$

3.2.14 Hermite's Interpolation Formula

The interpolation formulae so far discussed involve only a certain number of function values. Now we proceed to derive an interpolation formula in which not only the values of the unknown function $f(x)$ are given at $(n+1)$ distinct points $x = x_0, x_1, \ldots, x_n$, but the values of the derivative $f'(x)$ are also known at these points. The data is represented by the following table:

x	$f(x)$	$f'(x)$
x_0	$f(x_0)$	$f'(x_0)$
x_1	$f(x_1)$	$f'(x_1)$
x_2	$f(x_2)$	$f'(x_2)$
$\vdots$	$\vdots$	$\vdots$
x_n	$f(x_n)$	$f'(x_n)$

Let us determine a polynomial $H_{2n+1}(x)$ of degree less than equal to $2n+1$, such that

$$H_{2n+1}(x_i) = f(x_i),\ H'_{2n+1}(x_i) = f'(x_i), \quad i = 0,1,2,\ldots,n \tag{3.125}$$

which forms a system of $2n+2$ linear equations for determination of the $2n+2$ coefficients of the polynomial $H_{2n+1}(x)$ and the degree of the polynomial has been assumed as $2n+1$. Let

$$H_{2n+1}(x) = \sum_{i=0}^{n} A_i(x)y_i + \sum_{i=0}^{n} B_i(x)y_i' \tag{3.126}$$

where $A_i(x)$ and $B_i(x)$ are polynomials in x of degree at most $2n+1$ such that

$$A_i(x_j) = \delta_{ij},\ A_i'(x_j) = 0 \text{ for all } i \tag{3.127}$$

$$B_i(x_j) = 0 \text{ for all } i,\ B_i'(x_j) = \delta_{ij} \tag{3.128}$$

where $i, j = 0,1,2,\ldots,n$.

Then the conditions $H_{2n+1}(x_j)=y_j$ and $H'_{2n+1}(x_j)=y'_j$ are satisfied since from Equation 3.126, we have

$$H_{2n+1}(x_j)=\sum_{i=0}^{n}A_i(x_j)y_i+\sum_{i=0}^{n}B_i(x_j)y'_i=A_j(x_j)y_j=y_j$$

and

$$H'_{2n+1}(x_j)=\sum_{i=0}^{n}A'_i(x_j)y_i+\sum_{i=0}^{n}B'_i(x_j)y'_i=B'_j(x_j)y'_j=y'_j$$

Let

$$\omega_i(x)=\frac{(x-x_0)(x-x_1)\ldots(x-x_{i-1})(x-x_{i+1})\ldots(x-x_n)}{(x_i-x_0)(x_i-x_1)\ldots(x_i-x_{i-1})(x_i-x_{i+1})\ldots(x_i-x_n)} \tag{3.129}$$

be a polynomial of degree n used in the Lagrange's interpolation formula (called *Lagrange's fundamental polynomial*).

For fixed $i\neq j$, $A_i(x_j)=A'_i(x_j)=0$, so that $A_i(x)$ has a factor $(x-x_j)^2, j\neq i$ and so $A_i(x)$ has a factor $\{\omega_i(x)\}^2$, which is a polynomial of degree $2n$. Thus, $A_i(x)$ is of the form

$$A_i(x)=(a_ix+b_i)\{\omega_i(x)\}^2 \tag{3.130}$$

where a_i and b_i are constants. Similarly,

$$B_i(x)=(c_ix+d_i)\{\omega_i(x)\}^2 \tag{3.131}$$

where c_i and d_i are constants. Using conditions in Equations 3.127 and 3.128, we obtain

$$a_i=-2\omega'_i(x_i),\ b_i=1+2x_i\omega'_i(x_i),\ c_i=1,\text{ and } d_i=-x_i \tag{3.132}$$

Substituting Equation 3.132 in 3.130 and 3.131, the Equation 3.126 becomes

$$H_{2n+1}(x)=\sum_{i=0}^{n}\left[1-2(x-x_i)\omega'_i(x_i)\right]\{\omega_i(x)\}^2y_i+\sum_{i=0}^{n}(x-x_i)\{\omega_i(x)\}^2y'_i \tag{3.133}$$

which is called *Hermite's interpolation polynomial.*

3.2.14.1 Uniqueness of Hermite Polynomial

To prove that this polynomial is the unique polynomial of degree $2n+1$, let us assume that there is another polynomial $\tilde{H}_{2n+1}(x)$ of degree $2n+1$ that satisfies the same constraints given in Equation 3.125. Since $H_{2n+1}(x_i)=\tilde{H}_{2n+1}(x_i)=f(x_i)$ for $i=0,1,2,\ldots,n$, therefore, $H_{2n+1}(x)-\tilde{H}_{2n+1}(x)$ has at least $n+1$ zeros. It follows from Rolle's theorem that $H'_{2n+1}(x)-\tilde{H}'_{2n+1}(x)$ has n zeros that lie within the intervals (x_{i-1},x_i) for $i=0,1,2,\ldots,n$.

Again, since $H'_{2n+1}(x_i)=\tilde{H}'_{2n+1}(x_i)=f'(x_i)$ for $i=0,1,2,\ldots,n$, we have $H_{2n+1}(x)-\tilde{H}_{2n+1}(x)$ has $n+1$ additional zeros. However, $H_{2n+1}(x)-\tilde{H}_{2n+1}(x)$ is polynomial of degree $2n+1$ and the only way that a polynomial of degree $2n+1$ can have $2n+1$ zeros if it is identically zero. Therefore, $H_{2n+1}(x)=\tilde{H}_{2n+1}(x)$, and hence, the Hermite polynomial is unique.

3.2.14.2 The Error in Hermite Interpolation

Let $H_{2n+1}(x)$ denote the Hermite polynomial that approximate $f(x)$ with interpolation points $x_0, x_1, \ldots, x_n$ in $[a,b]$ and let $f(x)$ be $2n+2$ times continuously differentiable on $[a,b]$, that is, $f(x) \in C^{2n+2}[a,b]$. Let us construct a function

$$\varphi_n(x) = f(x) - H_{2n+1}(x) - \lambda w(x) \tag{3.134}$$

where

$$w(x) = \prod_{i=0}^{n} (x - x_i)^2 \tag{3.135}$$

and λ is selected such that $\varphi_n(x') = 0$. Therefore,

$$\lambda = \frac{f(x') - H_{2n+1}(x')}{w(x')} \tag{3.136}$$

Therefore, $\varphi_n(x)$ has (at least) $n+2$ zeros, that is, $x', x_0, x_1, \ldots, x_n$ in $[a,b]$. By Rolle's theorem, we know that $\varphi_n'(x)$ has (at least) $n+1$ zeros that are different from $x', x_0, x_1, \ldots, x_n$. In addition, $\varphi_n'(x)$ vanishes at $x_0, x_1, \ldots, x_n$, which implies that $\varphi_n'(x)$ has at least $2n+2$ zeros in $[a,b]$. Again, Rolle's theorem implies that $\varphi_n''(x)$ has at least $2n+1$ zeros in (a,b) and by induction $\varphi_n^{(2n+2)}(x)$ at least one zero in (a,b), say ξ. Therefore,

$$0 = \varphi^{(2n+2)}(\xi) = f^{(2n+2)}(\xi) - H_{2n+1}^{(2n+2)}(\xi) - \lambda w^{(2n+2)}(\xi) \tag{3.137}$$

Since the leading term in $w^{(2n+2)}(x)$ is $x^{(2n+2)}$, $w^{(2n+2)}(\xi) = (2n+2)!$ In addition, since $H_{2n+1}^{(2n+2)}(x)$ is a polynomial of degree $2n+1$, $H_{2n+1}^{(2n+2)}(\xi) = 0$. Therefore, from Equation 3.137, we have

$$\lambda = \frac{f^{(2n+2)}(\xi)}{(2n+2)!} \tag{3.138}$$

Substituting this value of λ in Equation 3.136, we obtain

$$f(x') - H_{2n+1}(x') = \frac{f^{(2n+2)}(\xi)}{(2n+2)!} \prod_{i=0}^{n} (x' - x_i)^2 \tag{3.139}$$

On dropping the prime, we may write

$$f(x) - H_{2n+1}(x) = \frac{f^{(2n+2)}(\xi)}{(2n+2)!} \prod_{i=0}^{n} (x - x_i)^2 \tag{3.140}$$

where $a < \xi < b$. Equation 3.140 represents the error in the Hermite interpolation.

3.2.15 Piecewise Interpolation

If we use interpolating polynomials of high degree, then we may obtain reasonably accurate results using interpolation. However, if the degree of the interpolating polynomial becomes higher, the computation will become more unwieldy and the computational overhead will be very high. In addition, it will result in higher round-off errors. In order to maintain the degree of the interpolating polynomial small and also to achieve desirable accurate results, we employ piecewise interpolation. We divide the given interval $[a,b]$ into a finite number of subintervals $[x_{i-1},x_i]$, $i=1,2,\ldots,n$, and the function $y=f(x)$ is approximated by a lower degree polynomial in each subinterval (x_{i-1},x_i), $i=1,2,\ldots,n$.

3.2.15.1 Piecewise Linear Interpolation

In this method, the interpolation polynomial is assumed to be linear in each subinterval (x_{i-1},x_i), $i=1,2,\ldots,n$, and it agrees with function $f(x)$ at the $n+1$ nodal points $x_0,x_1,\ldots,x_n$. The subintervals are called *finite elements* in one space dimension and the nodal points are called *knots*.

Using the linear Lagrange's interpolation formula, the piecewise linear interpolation polynomial $P_i(x)$ is given by

$$P_{i,1}(x)=\frac{x-x_i}{x_{i-1}-x_i}y_{i-1}+\frac{x-x_{i-1}}{x_i-x_{i-1}}y_i \tag{3.141}$$

where

$$x_{i-1}\le x\le x_i,\quad i=1,2,\ldots,n$$

For $x\in[x_i,x_{i+1}]$, we have

$$P_{i+1,1}(x)=\frac{x-x_{i+1}}{x_i-x_{i+1}}y_i+\frac{x-x_i}{x_{i+1}-x_i}y_{i+1} \tag{3.142}$$

Therefore, the overall interpolation polynomial is given by

$$\varphi(x)=\sum_{i=1}^{n}P_{i,1}(x) \tag{3.143}$$

3.2.15.2 Piecewise Quadratic Interpolation

Let the given interval $[a,b]$ be divided into $2n$ number of subintervals by $2n+1$ nodal points such that

$$a=x_0<x_1<\cdots<x_{2n}=b$$

We consider the groups of three contiguous nodal points as $[x_0,x_1],[x_1,x_2],\ldots,[x_{i-1},x_{i+1}],\ldots,$ $[x_{2n-2},x_{2n}]$.

In this method, we shall approximate the function $f(x)$ in two contiguous subintervals by a quadratic polynomial.

For $x\in[x_{i-1},x_{i+1}]$, the quadratic polynomial is given by

$$P_{i,2}(x)=\frac{(x-x_i)(x-x_{i+1})}{(x_{i-1}-x_i)(x_{i-1}-x_{i+1})}y_{i-1}+\frac{(x-x_{i-1})(x-x_{i+1})}{(x_i-x_{i-1})(x_i-x_{i+1})}y_i+\frac{(x-x_{i-1})(x-x_i)}{(x_{i+1}-x_{i-1})(x_{i+1}-x_i)}y_{i+1} \tag{3.144}$$

where $i=1,2,\ldots,2n-1$. Therefore, the overall interpolation polynomial is given by

$$\varphi_n(x)=\sum_{i=1}^{2n-1}P_{i,2}(x) \tag{3.145}$$

3.2.15.3 Piecewise Cubic Interpolation

In this method, the function $f(x)$ is approximated by a cubic polynomial $P_{i,3}(x)$ in each subinterval (x_{i-1}, x_i), which satisfies the following Hermite type of conditions:

$$\begin{aligned} P_{i,3}(x_{i-1}) &= y_{i-1}, \quad P_{i,3}(x_i) = y_i \\ P'_{i,3}(x_{i-1}) &= y'_{i-1}, \quad P'_{i,3}(x_i) = y'_i \end{aligned} \tag{3.146}$$

The polynomial thus obtained is called *piecewise cubic Hermite interpolation polynomial.* Using Equations 3.126 and 3.133, we can write this polynomial in the following form:

$$P_{i,3}(x) = A_{i-1}(x)y_{i-1} + A_i(x)y_i + B_{i-1}(x)y'_{i-1} + B_i(x)y'_i \tag{3.147}$$

where $x_{i-1} \le x \le x_i$, $i = 1, 2, \ldots, n$.

Here,

$$A_{i-1}(x) = \left[1 - 2(x - x_{i-1})\omega'_{i-1}(x_{i-1})\right]\left\{\omega_{i-1}(x)\right\}^2$$

where

$$\omega_{i-1}(x) = \frac{x - x_i}{x_{i-1} - x_i} \quad \text{and} \quad \omega'_{i-1}(x) = \frac{1}{x_{i-1} - x_i}$$

Therefore,

$$A_{i-1}(x) = \left[1 + \frac{2(x_{i-1} - x)}{x_{i-1} - x_i}\right]\left(\frac{x - x_i}{x_{i-1} - x_i}\right)^2 \tag{3.148}$$

In addition,

$$B_{i-1}(x) = (x - x_{i-1})\left\{\omega_{i-1}(x)\right\}^2$$

This implies

$$B_{i-1}(x) = \frac{(x - x_{i-1})(x - x_i)^2}{(x_{i-1} - x_i)^2} \tag{3.149}$$

Similarly,

$$A_i(x) = \left[1 + \frac{2(x_i - x)}{x_i - x_{i-1}}\right]\left(\frac{x - x_{i-1}}{x_i - x_{i-1}}\right)^2 \tag{3.150}$$

and

$$B_i(x) = \frac{(x - x_i)(x - x_{i-1})^2}{(x_i - x_{i-1})^2} \tag{3.151}$$

Thus, the Hermite interpolation polynomial in Equation 3.147 becomes

$$\begin{aligned} P_{i,3}(x) = &\left[1 + \frac{2(x_{i-1} - x)}{x_{i-1} - x_i}\right]\left(\frac{x - x_i}{x_{i-1} - x_i}\right)^2 y_{i-1} + \left[1 + \frac{2(x_i - x)}{(x_i - x_{i-1})}\right]\left(\frac{x - x_{i-1}}{x_i - x_{i-1}}\right)^2 y_i \\ &+ \frac{(x - x_{i-1})(x - x_i)^2}{(x_{i-1} - x_i)^2} y'_{i-1} + \frac{(x - x_i)(x - x_{i-1})^2}{(x_i - x_{i-1})^2} y'_i \end{aligned} \tag{3.152}$$

where $x_{i-1} \le x \le x_i$, $i = 1, 2, \ldots, n$. Therefore, the overall interpolation polynomial is given by

$$\varphi_n(x) = \sum_{i=1}^{n} P_{i,3}(x) \tag{3.153}$$

Example 3.16

Using Hermite interpolation formula, express y as a polynomial in x of degree 5 from the following data. Hence, compute the value of y corresponding to $x = -0.5$ and $x = 1$.

x	−1	0	2
y	4	5	37
y'	5	0	80

Solution:

Here, $n = 2$. Using Hermite interpolation formula, we have

$$H_5(x) = \sum_{i=0}^{2} [1 - 2(x - x_i)\omega_i'(x_i)]\{\omega_i(x)\}^2 y_i + \sum_{i=0}^{2} (x - x_i)\{\omega_i(x)\}^2 y_i'$$

where

$$\omega_i(x) = \frac{(x - x_0)(x - x_1)\ldots(x - x_{i-1})(x - x_{i+1})\ldots(x - x_n)}{(x_i - x_0)(x_i - x_1)\ldots(x_i - x_{i-1})(x_i - x_{i+1})\ldots(x_i - x_n)}$$

Therefore,

$$\begin{aligned} H_5(x) &= [1 - 2(x - x_0)\omega_0'(x_0)]\{\omega_0(x)\}^2 y_0 + [1 - 2(x - x_1)\omega_1'(x_1)]\{\omega_1(x)\}^2 y_1 \\ &\quad + [1 - 2(x - x_2)\omega_2'(x_2)]\{\omega_2(x)\}^2 y_2 \\ &\quad + (x - x_0)\{\omega_0(x)\}^2 y_0' + (x - x_1)\{\omega_1(x)\}^2 y_1' + (x - x_2)\{\omega_2(x)\}^2 y_2' \end{aligned} \tag{3.154}$$

Now,

$$\omega_0(x) = \frac{(x - x_1)(x - x_2)}{(x_0 - x_1)(x_0 - x_2)} = \frac{(x - 0)(x - 2)}{(-1 - 0)(-1 - 2)} = \frac{x(x - 2)}{3}$$

$$\omega_0'(x) = \frac{2x - 2}{3}$$

$$\omega_0'(x_0) = -\frac{4}{3}$$

$$\omega_1(x) = \frac{(x - x_0)(x - x_2)}{(x_1 - x_0)(x_1 - x_2)} = \frac{(x + 1)(x - 2)}{(0 + 1)(0 - 2)} = \frac{(x + 1)(x - 2)}{-2}$$

$$\omega_1'(x) = \frac{2x - 1}{-2}$$

$$\omega_1'(x_1) = \frac{1}{2}$$

$$\omega_2(x) = \frac{(x - x_0)(x - x_1)}{(x_2 - x_0)(x_2 - x_1)} = \frac{(x + 1)(x - 0)}{(2 + 1)(2 - 0)} = \frac{x(x + 1)}{6}$$

$$\omega_2'(x) = \frac{2x + 1}{6}$$

$$\omega_2'(x_2) = \frac{5}{6}$$

Thus, from Equation 3.154, we obtain the required Hermite interpolating polynomial as follows:

$$H_5(x) = \left[1 - 2(x + 1) \times \left(-\frac{4}{3}\right)\right]\left\{\frac{x(x - 2)}{3}\right\}^2 \times 4 + \left[1 - 2x \times \left(\frac{1}{2}\right)\right]\left\{\frac{(x + 1)(x - 2)}{-2}\right\}^2 \times 5$$

$$+ \left[1 - 2(x - 2) \times \left(\frac{5}{6}\right)\right]\left\{\frac{x(x + 1)}{6}\right\}^2 \times 37$$

$$+ (x + 1)\left\{\frac{x(x - 2)}{3}\right\}^2 \times 5 + x \times \left\{\frac{(x + 1)(x - 2)}{-2}\right\}^2 \times 0 + (x - 2)\left\{\frac{x(x + 1)}{6}\right\}^2 \times 80$$

$$= x^5 + 5$$

Therefore,

$$f(-0.5) = 4.96875 \quad \text{and} \quad f(1) = 6$$

Example 3.17

Find the interpolating polynomial for the following data using (1) piecewise linear interpolation and (2) piecewise cubic Hermite interpolation.

x	−2	0	1
y	3	1	−2
y'	−1	0	1

Solution:

1. The piecewise linear interpolating polynomial is given by

$$P_{i,1}(x) = \frac{x - x_i}{x_{i-1} - x_i} y_{i-1} + \frac{x - x_{i-1}}{x_i - x_{i-1}} y_i,$$

for $x_{i-1} \le x \le x_i$ and $i = 1, 2, \ldots, n$.
Thus, if $-2 \le x \le 0$,

$$P_{1,1}(x) = \frac{x - x_1}{x_0 - x_1} y_0 + \frac{x - x_0}{x_1 - x_0} y_1$$

$$= \frac{x - 0}{-2 - 0} \times 3 + \frac{x + 2}{0 + 2} \times 1$$

$$= 1 - x$$

If $0 \le x \le 1$,

$$P_{2,1}(x) = \frac{x - x_2}{x_1 - x_2} y_1 + \frac{x - x_1}{x_2 - x_1} y_2$$

$$= \frac{x - 1}{0 - 1} \times 1 + \frac{x - 0}{1 - 0} \times (-2)$$

$$= 1 - 3x$$

Hence, the required piecewise linear interpolating polynomial is

$$y \cong \varphi(x) = \begin{cases} 1 - x, & -2 \le x \le 0 \\ 1 - 3x, & 0 \le x \le 1 \end{cases}$$

2. From Equation 3.152, the piecewise Hermite interpolating polynomial is given by

$$P_{i,3}(x) = \left[1 + \frac{2(x_{i-1} - x)}{x_{i-1} - x_i}\right]\left(\frac{x - x_i}{x_{i-1} - x_i}\right)^2 y_{i-1} + \left[1 + \frac{2(x_i - x)}{(x_i - x_{i-1})}\right]\left(\frac{x - x_{i-1}}{x_i - x_{i-1}}\right)^2 y_i$$

$$+ \frac{(x - x_{i-1})(x - x_i)^2}{(x_{i-1} - x_i)^2} y'_{i-1} + \frac{(x - x_i)(x - x_{i-1})^2}{(x_i - x_{i-1})^2} y'_i$$

where $x_{i-1} \le x \le x_i$, $i = 1, 2, \ldots, n$.
Thus, if $-2 \le x \le 0$,

$$P_{1,3}(x) = \left[1 + \frac{2 \times (-2 - x)}{-2 - 0}\right]\left(\frac{x - 0}{-2 - 0}\right)^2 \times 3 + \left[1 + \frac{2 \times (0 - x)}{(0 + 2)}\right]\left(\frac{x + 2}{0 + 2}\right)^2 \times 1$$

$$+ \frac{(x + 2)(x - 0)^2}{(-2 - 0)^2} \times (-1) + \frac{(x - 0)(x + 2)^2}{(0 + 2)^2} \times 0$$

$$= 1 + x^2 + \frac{x^3}{4}$$

If $0 \le x \le 1$,

$$P_{2,3}(x) = \left[1 + \frac{2 \times (0 - x)}{(0 - 1)}\right]\left(\frac{x - 1}{0 - 1}\right)^2 \times 1 + \left[1 + \frac{2(1 - x)}{(1 - 0)}\right]\left(\frac{x - 0}{1 - 0}\right)^2 \times (-2)$$

$$+ \frac{(x - 0)(x - 1)^2}{(0 - 1)^2} \times 0 + \frac{(x - 1)(x - 0)^2}{(1 - 0)^2} \times 1$$

$$= 1 - 10x^2 + 7x^3$$

Hence, the required piecewise Hermite cubic interpolating polynomial is

$$y \cong \varphi(x) = \begin{cases} 1 + x^2 + \dfrac{x^3}{4}, & -2 \le x \le 0 \\ 1 - 10x^2 + 7x^3, & 0 \le x \le 1 \end{cases}$$

3.2.16 Cubic Spline Interpolation

Spline interpolation is a piecewise polynomial interpolation. Let $f(x)$ be a function defined in $a \le x \le b$, and we partition $[a, b]$ into finite number of subintervals by the node points

$$a = x_0 < x_1 < \cdots < x_n = b \tag{3.155}$$

Now, we shall find a cubic spline function $\varphi(x)$ that approximates $f(x)$ such that

$$\varphi(x_0) = f(x_0) = f_0,\ \varphi(x_1) = f(x_1) = f_1, \ldots, \varphi(x_n) = f(x_n) = f_n \tag{3.156}$$

3.2.16.1 Cubic Spline

A cubic spline $\varphi(x)$ defined on $a \le x \le b$ corresponding to the partition Equation 3.155 is a continuous function. $\varphi(x)$ has continuous first and second derivatives everywhere in that interval. In addition, in each subinterval $[x_{i-1}, x_i]$, $i = 1, 2, \ldots, n$, of that partition Equation 3.155, $\varphi(x)$ is represented by a polynomial $s_i(x)$ of degree not exceeding three. Hence $\varphi(x)$ consists of n such polynomials, one in each subinterval. Therefore, in each subinterval $[x_{i-1}, x_i]$, the spline $\varphi(x)$ must agree with the polynomial $s_i(x)$ of degree not exceeding three, such that

1. $s_i(x_{i-1}) = y_{i-1}$ and $s_i(x_i) = y_i$, $i = 1, 2, \ldots, n$ (3.157)
2. $s_i(x)$, $s_i'(x)$ and $s_i''(x)$ are continuous functions in $[a, b]$. This implies that

$$\begin{aligned} s_i(x_i) &= s_{i+1}(x_i), \\ s_i'(x_i) &= s_{i+1}'(x_i), \\ s_i''(x_i) &= s_{i+1}''(x_i), \quad i = 1, 2, \ldots, n-1 \end{aligned} \tag{3.158}$$

From Figure 3.2, the above conditions can be easily perceived.

3.2.16.1.1 *Method I*

As $s_i(x)$ is a cubic polynomial in $[x_{i-1}, x_i]$, $s_i''(x)$ is a linear polynomial in that interval. Then by Lagrange's interpolation formula,

$$s_i''(x) = \left(\frac{x - x_i}{x_{i-1} - x_i}\right) s_i''(x_{i-1}) + \left(\frac{x - x_{i-1}}{x_i - x_{i-1}}\right) s_i''(x_i), \quad i = 1, 2, \ldots, n \tag{3.159}$$

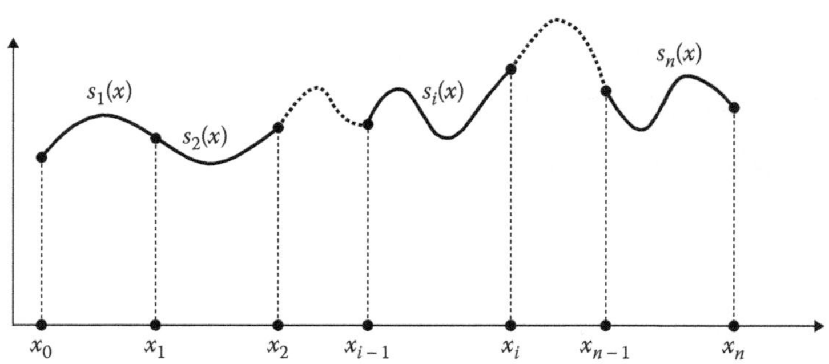

FIGURE 3.2 Geometrical representation of spline interpolation.

Integrating Equation 3.159 twice with respect to x, we have

$$s_i(x) = \frac{(x_i - x)^3}{6h_i} M_{i-1} + \frac{(x - x_{i-1})^3}{6h_i} M_i + c_1 x + c_2, \quad i = 1, 2, \ldots, n \tag{3.160}$$

where $x_i - x_{i-1} = h_i$, $M_{i-1} = s_i''(x_{i-1})$, $M_i = s_i''(x_i)$, and c_1 and c_2 are arbitrary constants to be determined.

Since $s_i(x)$ agrees with $y = f(x)$ at the points x_{i-1} and x_i, we have $s_i(x_{i-1}) = y_{i-1}$ and $s_i(x_i) = y_i$ using these conditions in Equation 3.160, we have

$$\frac{h_i^2}{6} M_{i-1} + c_1 x_{i-1} + c_2 = y_{i-1} \tag{3.161}$$

and

$$\frac{h_i^2}{6} M_i + c_1 x_i + c_2 = y_i \tag{3.162}$$

Solving Equations 3.161 and 3.162 for c_1 and c_2, we obtain

$$c_1 = \frac{(y_i - y_{i-1})}{h_i} + \frac{(M_{i-1} - M_i)}{6} h_i \tag{3.163}$$

and

$$c_2 = \frac{(x_i y_{i-1} - x_{i-1} y_i)}{h_i} + \frac{(x_{i-1} M_i - x_i M_{i-1})}{6} h_i \tag{3.164}$$

Substituting these values of c_1 and c_2 in Equation 3.160, we have

$$\begin{aligned} s_i(x) = {} & \frac{(x_i - x)^3}{6h_i} M_{i-1} + \frac{(x - x_{i-1})^3}{6h_i} M_i + \frac{(y_i - y_{i-1})x}{h_i} - \frac{(M_i - M_{i-1})}{6} h_i x \\ & + \frac{(x_i y_{i-1} - x_{i-1} y_i)}{h_i} - \frac{(x_i M_{i-1} - x_{i-1} M_i)}{6} h_i \end{aligned} \tag{3.165}$$

$$\begin{aligned} = {} & \left[\frac{(x_i - x)^3}{6h_i} - \frac{h_i (x_i - x)}{6} \right] M_{i-1} + \left[\frac{(x - x_{i-1})^3}{6h_i} - \frac{h_i (x - x_{i-1})}{6} \right] M_i \\ & + \frac{(x_i - x) y_{i-1}}{h_i} + \frac{(x - x_{i-1})}{h_i} y_i, \quad x_{i-1} \le x \le x_i, \quad i = 1, 2, \ldots, n \end{aligned} \tag{3.166}$$

where M_{i-1} and M_i are unknown to be determined. These values can be found by using the condition of continuity of $s_i'(x)$ at the point x_i, that is,

$$s_i'(x_i - 0) = s_{i+1}'(x_i + 0) \tag{3.167}$$

Differentiating Equation 3.165 with respect to x, we get

$$s_i'(x) = \frac{-(x_i - x)^2}{2h_i} M_{i-1} + \frac{(x - x_{i-1})^2}{2h_i} M_i - \frac{(M_i - M_{i-1})}{6} h_i + \frac{(y_i - y_{i-1})}{h_i}, \quad x_{i-1} < x < x_i \tag{3.168}$$

and setting $i = i+1$, we have

$$s'_{i+1}(x) = \frac{-(x_{i+1}-x)^2}{2h_{i+1}}M_i + \frac{(x-x_i)^2}{2h_{i+1}}M_{i+1} - \frac{(M_{i+1}-M_i)}{6}h_{i+1} + \frac{(y_{i+1}-y_i)}{h_{i+1}}, \quad x_i < x < x_{i+1} \tag{3.169}$$

From Equations 3.167, 3.168, and 3.169, we have

$$\frac{h_i}{2}M_i - \frac{(M_i - M_{i-1})}{6}h_i + \frac{(y_i - y_{i-1})}{h_i} = \frac{-h_{i+1}}{2}M_i - \frac{(M_{i+1}-M_i)}{6}h_{i+1} + \frac{(y_{i+1}-y_i)}{h_{i+1}}$$

This implies that

$$\frac{h_i}{6}M_{i-1} + \left(\frac{h_i + h_{i+1}}{3}\right)M_i + \frac{h_{i+1}}{6}M_{i+1} = \frac{(y_{i+1}-y_i)}{h_{i+1}} - \frac{(y_i - y_{i-1})}{h_i} \quad \text{for } i = 1,2,\ldots,n-1 \tag{3.170}$$

Equation 3.170 forms a system of $(n-1)$ linear equations in $(n+1)$ unknowns $M_0, M_1, \ldots, M_n$.

The two additional conditions may be taken in one of the following forms:

Case I:

$$\varphi''(x_0) = s''_1(x_0) = M_0 = 0, \quad \varphi''(x_n) = s''_n(x_n) = M_n = 0 \tag{3.171}$$

These conditions are called *free* or *natural boundary conditions.*

Case II:

$$\varphi'(x_0) = s'_1(x_0) = y'_0, \quad \varphi'(x_n) = s'_n(x_n) = y'_n \tag{3.172}$$

These conditions are called *clamped boundary conditions.*

In case I, substituting the values of M_0 and M_n, that is, $M_0 = M_n = 0$ in Equation 3.170, we can obtain an $(n-1)\times(n-1)$ tridiagonal system of equations for finding $M_i, i = 1,2,\ldots,n-1$.

In case II, using Equation 3.172, from Equation 3.168, we can obtain

$$2M_0 + M_1 = \frac{6}{h_1}\left(\frac{y_1 - y_0}{h_1} - y'_0\right) \tag{3.173}$$

$$M_{n-1} + 2M_n = \frac{6}{h_n}\left(y'_n - \frac{y_n - y_{n-1}}{h_n}\right) \tag{3.174}$$

These two equations (3.173) and (3.175) along with Equation 3.170 form a system of $(n+1)$ linear equations in $(n+1)$ unknowns $M_0, M_1, \ldots, M_n$. This tridiagonal system of equations can be solved for $M_i, i = 0,1,2,\ldots,n$.

- *Determination of splines for equispaced nodes or knots*: For equidistant knots $h_i = h$ for all i, Equations 3.166 and 3.170 yield, respectively,

$$s_i(x) = \frac{1}{6h}\left[(x_i - x)^3 M_{i-1} + (x - x_{i-1})^3 M_i\right] + \frac{1}{h}(x_i - x)\left(y_{i-1} - \frac{h^2}{6}M_{i-1}\right)$$
$$+ \frac{1}{h}(x - x_{i-1})\left(y_i - \frac{h^2}{6}M_i\right), \quad x_{i-1} \le x \le x_i, \quad i = 1,2,\ldots,n \tag{3.175}$$

and

$$M_{i-1}+4M_i+M_{i+1}=\frac{6}{h^2}\left(y_{i-1}-2y_i+y_{i+1}\right),\quad i=1,2,\ldots,n-1 \tag{3.176}$$

In case of free or natural boundary conditions, Equations 3.176 along with Equations 3.171 may be solved for unknowns $M_0, M_1, \ldots, M_n$. Otherwise, in case of clamped boundary conditions, Equations 3.176 together with Equation 3.173 and 3.174 may be solved for unknowns $M_0, M_1, \ldots, M_n$. Solutions obtained for M_i, $i=0,1,2,\ldots,n$ are substituted in Equation 3.175 to find the cubic spline interpolation.

3.2.16.1.2 Method II

In this method, we consider piecewise cubic Hermite interpolation. According to Equation 3.152, in subinterval $[x_{i-1}, x_i]$, let us consider the piecewise cubic Hermite interpolation polynomial

$$\begin{aligned} s_i(x)=&\left[1+\frac{2(x_{i-1}-x)}{x_{i-1}-x_i}\right]\left(\frac{x-x_i}{x_{i-1}-x_i}\right)^2 y_{i-1}+\left[1+\frac{2(x_i-x)}{(x_i-x_{i-1})}\right]\left(\frac{x-x_{i-1}}{x_i-x_{i-1}}\right)^2 y_i \\ &+\frac{\left(x-x_{i-1}\right)\left(x-x_i\right)^2}{\left(x_{i-1}-x_i\right)^2}y'_{i-1}+\frac{\left(x-x_i\right)\left(x-x_{i-1}\right)^2}{\left(x_i-x_{i-1}\right)^2}y'_i \end{aligned} \tag{3.177}$$

where $x_{i-1}\le x\le x_i$, $i=1,2,\ldots,n$. Differentiating Equation 3.177 twice with respect to x, we get

$$\begin{aligned} s''_i(x)=&\,2y_{i-1}\left[\frac{1}{\left(x_{i-1}-x_i\right)^2}-\frac{2}{(x_{i-1}-x_i)}\left[\frac{\left(x-x_{i-1}\right)}{\left(x_{i-1}-x_i\right)^2}+\frac{2\left(x-x_i\right)}{\left(x_{i-1}-x_i\right)^2}\right]\right] \\ &+2y_i\left[\frac{1}{\left(x_i-x_{i-1}\right)^2}-\frac{2}{(x_i-x_{i-1})}\left[\frac{\left(x-x_i\right)}{\left(x_i-x_{i-1}\right)^2}+\frac{2\left(x-x_{i-1}\right)}{\left(x_i-x_{i-1}\right)^2}\right]\right] \\ &+2y'_{i-1}\left[\frac{\left(x-x_{i-1}\right)}{\left(x_{i-1}-x_i\right)^2}+\frac{2\left(x-x_i\right)}{\left(x_{i-1}-x_i\right)^2}\right]+2y'_i\left[\frac{\left(x-x_i\right)}{\left(x_i-x_{i-1}\right)^2}+\frac{2\left(x-x_{i-1}\right)}{\left(x_i-x_{i-1}\right)^2}\right] \end{aligned}$$

where $x_{i-1}<x<x_i$, $i=1,2,\ldots,n$. Therefore,

$$\begin{aligned} s''_i(x)=&\,2y_{i-1}\left[\frac{1}{h_i^2}+\frac{2}{h_i^3}\left[\left(x-x_{i-1}\right)+2\left(x-x_i\right)\right]\right] \\ &+2y_i\left[\frac{1}{h_i^2}-\frac{2}{h_i^3}\left[\left(x-x_i\right)+2\left(x-x_{i-1}\right)\right]\right] \\ &+2y'_{i-1}\left[\frac{\left(x-x_{i-1}\right)}{h_i^2}+\frac{2\left(x-x_i\right)}{h_i^2}\right]+2y'_i\left[\frac{\left(x-x_i\right)}{h_i^2}+\frac{2\left(x-x_{i-1}\right)}{h_i^2}\right] \end{aligned} \tag{3.178}$$

where $h_i=x_i-x_{i-1}$, $i=1,2,\ldots,n$. Replacing $i=i+1$ in Equation 3.178, we have

$$s''_{i+1}(x) = 2y_i\left[\frac{1}{h_{i+1}^{\,2}} + \frac{2}{h_{i+1}^{\,3}}\left[(x-x_i)+2(x-x_{i+1})\right]\right]$$
$$+2y_{i+1}\left[\frac{1}{h_{i+1}^{\,2}} - \frac{2}{h_{i+1}^{\,3}}\left[(x-x_{i+1})+2(x-x_i)\right]\right] \tag{3.179}$$
$$+2y'_i\left[\frac{(x-x_i)}{h_{i+1}^{\,2}} + \frac{2(x-x_{i+1})}{h_{i+1}^{\,2}}\right] + 2y'_{i+1}\left[\frac{(x-x_{i+1})}{h_{i+1}^{\,2}} + \frac{2(x-x_i)}{h_{i+1}^{\,2}}\right]$$

Using the condition of continuity of $s''_i(x)$ at the point x_i, that is, $s''_i(x_i-0) = s''_{i+1}(x_i+0)$ given in Equation 3.158, we obtain

$$3\frac{y_{i-1}}{h_i^2} - 3\frac{y_i}{h_i^2} + \frac{y'_{i-1}}{h_i} + \frac{2y'_i}{h_i} = -3\frac{y_i}{h_{i+1}^{\,2}} + 3\frac{y_{i+1}}{h_{i+1}^{\,2}} - \frac{2y'_i}{h_{i+1}} - \frac{y'_{i+1}}{h_{i+1}}$$

This implies that

$$\frac{y'_{i-1}}{h_i} + 2y'_i\left(\frac{1}{h_i} + \frac{1}{h_{i+1}}\right) + \frac{y'_{i+1}}{h_{i+1}} = -3\frac{y_{i-1}}{h_i^2} + 3y_i\left(\frac{1}{h_i^2} - \frac{1}{h_{i+1}^{\,2}}\right) + 3\frac{y_{i+1}}{h_{i+1}^{\,2}} \tag{3.180}$$

for $i = 1, 2, \ldots, n-1$. Now, setting $s'_i(x_i) = y'_i = k_i$, $s'_i(x_{i-1}) = y'_{i-1} = k_{i-1}$, and also $s'_{i+1}(x_{i+1}) = y'_{i+1} = k_{i+1}$, Equation 3.180 can be written as

$$\frac{k_{i-1}}{h_i} + 2k_i\left(\frac{1}{h_i} + \frac{1}{h_{i+1}}\right) + \frac{k_{i+1}}{h_{i+1}} = -3\frac{y_{i-1}}{h_i^2} + 3y_i\left(\frac{1}{h_i^2} - \frac{1}{h_{i+1}^{\,2}}\right) + 3\frac{y_{i+1}}{h_{i+1}^{\,2}} \tag{3.181}$$

Now, Equation 3.181 constitutes a system of $n-1$ linear equations in $n+1$ unknowns $k_0, k_1, \ldots, k_n$. The remaining two equations may be derived using either the free or clamped boundary conditions as discussed earlier.

- In case of free or natural boundary conditions,

$$s''_1(x_0) = 0, \quad s''_n(x_n) = 0 \tag{3.182}$$

from Equation 3.178, we have

$$2y'_0 + y'_1 = \frac{3(y_1 - y_0)}{h_1} \tag{3.183}$$

$$y'_{n-1} + 2y'_n = \frac{3(y_n - y_{n-1})}{h_n} \tag{3.184}$$

Equations 3.183 and 3.184 can be written as

$$2k_0 + k_1 = \frac{3(y_1 - y_0)}{h_1} \tag{3.185}$$

$$k_{n-1} + 2k_n = \frac{3(y_n - y_{n-1})}{h_n} \tag{3.186}$$

Solving Equations 3.181, 3.185, and 3.186, we can determine the unknowns $k_0, k_1, \ldots, k_n$.

For equispaced points $h_i = h$ for all i, then Equations 3.181, 3.185, and 3.186 become, respectively,

$$k_{i-1} + 4k_i + k_{i+1} = \frac{3}{h}(f_{i+1} - f_{i-1}), \quad i = 1, 2, \ldots, n-1 \tag{3.187}$$

$$2k_0 + k_1 = \frac{3(y_1 - y_0)}{h} \tag{3.188}$$

$$k_{n-1} + 2k_n = \frac{3(y_n - y_{n-1})}{h} \tag{3.189}$$

Solving Equations 3.187 through 3.189, we can determine the unknowns $k_i = y_i'$, $i = 0, 1, 2, \ldots, n$. It may be noted that, we need to solve an $(n+1) \times (n+1)$ tridiagonal system of equations in order to find the values of k_i, $i = 0, 1, 2, \ldots, n$.

Substituting the values of y_i and y_i', $i = 0, 1, 2, \ldots, n$ in the piecewise cubic Hermite interpolation polynomial Equation 3.177, we can obtain the required cubic spline interpolation.

- In case of clamped boundary conditions,

$$s_1'(x_0) = y_0' = k_0, \quad s_n'(x_n) = y_n' = k_n \tag{3.190}$$

are known explicitly. Substituting these values of k_0 and k_n in Equation 3.181, a system of $n-1$ linear equations in $n-1$ unknowns k_i, $i = 1, 2, \ldots, n-1$ are obtained. Solving this $(n-1) \times (n-1)$ tridiagonal system of equations, we can determine the values of k_i, $i = 1, 2, \ldots, n-1$.

In many practical applications, it may be convenient to use the spline first derivatives. In those cases, method II is inevitably suitable in compared to method I.

- Another approach for method II in the case of equidistant nodes

In the subinterval $[x_{i-1}, x_i]$, the spline $\varphi(x)$ is given by a cubic polynomial that can be written in the following form:

$$s_i(x) = a_{i0} + a_{i1}(x - x_i) + a_{i2}(x - x_i)^2 + a_{i3}(x - x_i)^3, \quad i = 1, 2, \ldots, n \tag{3.191}$$

We shall now determine the coefficient of the spline $s_i(x)$.

In Equation 3.178, substituting $x = x_i$, we have

$$\frac{s_i''(x)}{2} = 3\frac{y_{i-1}}{h^2} - 3\frac{y_i}{h^2} + \frac{y_{i-1}'}{h} + \frac{2y_i'}{h}, \quad \text{since, } h_i = h \text{ for all } i \tag{3.192}$$

Again, differentiating Equation 3.178 with respect to x and then substituting $x = x_i$, we can obtain

$$\frac{s_i'''(x)}{6} = 2\frac{y_{i-1}}{h^3} - 2\frac{y_i}{h^3} + \frac{y_{i-1}'}{h^2} + \frac{y_i'}{h^2} \tag{3.193}$$

Now, using Taylor's formula, from Equation 3.191 with the help of Equations 3.192 and 3.193, we get

$$\begin{aligned}
a_{i0} &= s_i(x_i) = y_i \\
a_{i1} &= s_i'(x_i) = k_i \\
a_{i2} &= \frac{1}{2}s_i''(x_i) = \frac{3}{h^2}(y_{i-1} - y_i) + \frac{1}{h}(k_{i-1} + 2k_i) \\
a_{i3} &= \frac{1}{6}s_i'''(x_i) = \frac{2}{h^3}(y_{i-1} - y_i) + \frac{1}{h^2}(k_{i-1} + k_i)
\end{aligned} \tag{3.194}$$

Hence, we can obtain the required cubic spline $\varphi(x)$ by determining $s_i(x)$ in each subinterval $[x_{i-1}, x_i]$, $i = 1, 2, \ldots, n$.

Cubic spline interpolation provides continuity and smoothness at the knot points of the subintervals. Splines are very useful for interpolation with a large number of data points. They perform better than higher order interpolating polynomials, since higher order polynomials may exhibit strong oscillatory behavior.

3.2.16.2 Error in Cubic Spline

In subinterval $[x_{i-1}, x_i]$, from the Equation 3.177, the spline equation is

$$\begin{aligned} s_i(x) &= \left[1 + \frac{2(x_{i-1} - x)}{x_{i-1} - x_i}\right]\left(\frac{x - x_i}{x_{i-1} - x_i}\right)^2 y_{i-1} + \left[1 + \frac{2(x_i - x)}{(x_i - x_{i-1})}\right]\left(\frac{x - x_{i-1}}{x_i - x_{i-1}}\right)^2 y_i \\ &+ \frac{(x - x_{i-1})(x - x_i)^2}{(x_{i-1} - x_i)^2} y'_{i-1} + \frac{(x - x_i)(x - x_{i-1})^2}{(x_i - x_{i-1})^2} y'_i \end{aligned} \tag{3.195}$$

where $x_{i-1} \le x \le x_i$, $i = 1, 2, \ldots, n$. Moreover, from Equations 3.178, we have

$$\begin{aligned} s''_i(x) &= 2y_{i-1}\left[\frac{1}{h_i^2} + \frac{2}{h_i^3}\left[(x - x_{i-1}) + 2(x - x_i)\right]\right] \\ &+ 2y_i\left[\frac{1}{h_i^2} - \frac{2}{h_i^3}\left[(x - x_i) + 2(x - x_{i-1})\right]\right] \\ &+ 2y'_{i-1}\left[\frac{(x - x_{i-1})}{h_i^2} + \frac{2(x - x_i)}{h_i^2}\right] + 2y'_i\left[\frac{(x - x_i)}{h_i^2} + \frac{2(x - x_{i-1})}{h_i^2}\right] \end{aligned} \tag{3.196}$$

where $h_i = x_i - x_{i-1}$, $i = 1, 2, \ldots, n$. Consequently, according to Equation 3.179,

$$\begin{aligned} s''_{i+1}(x) &= 2y_i\left[\frac{1}{h_{i+1}^2} + \frac{2}{h_{i+1}^3}\left[(x - x_i) + 2(x - x_{i+1})\right]\right] \\ &+ 2y_{i+1}\left[\frac{1}{h_{i+1}^2} - \frac{2}{h_{i+1}^3}\left[(x - x_{i+1}) + 2(x - x_i)\right]\right] \\ &+ 2y'_i\left[\frac{(x - x_i)}{h_{i+1}^2} + \frac{2(x - x_{i+1})}{h_{i+1}^2}\right] + 2y'_{i+1}\left[\frac{(x - x_{i+1})}{h_{i+1}^2} + \frac{2(x - x_i)}{h_{i+1}^2}\right] \end{aligned} \tag{3.197}$$

Using the condition of continuity of $s''_i(x)$ at the point x_i, that is, $s''_i(x_i - 0) = s''_{i+1}(x_i + 0)$, we can obtain

$$s''_i(x_i - 0) = 2\left(\frac{y'_{i-1}}{h_i} + \frac{2y'_i}{h_i}\right) - 6\left(\frac{y_i - y_{i-1}}{h_i^2}\right) \tag{3.198}$$

and

$$s''_{i+1}(x_i + 0) = 2\left(-\frac{2y'_i}{h_{i+1}} - \frac{y'_{i+1}}{h_{i+1}}\right) + 6\left(\frac{y_{i+1} - y_i}{h_{i+1}^2}\right) \tag{3.199}$$

To determine the error estimate involved in spline approximation at the equidistant interior points, we take the average of Equations 3.198 and 3.199 yielding

$$\varphi''(x_i) = \frac{1}{2}\left[s_i''(x_i - 0) + s_{i+1}''(x_i + 0)\right]$$

$$= \frac{3}{h^2}(y_{i+1} - 2y_i + y_{i-1}) - \frac{1}{h}(y_{i+1}' - y_{i-1}')$$

Thus,

$$\varphi''(x_i) = \frac{3}{h^2}(y_{i+1} - 2y_i + y_{i-1}) - \frac{1}{h}(s_{i+1}'(x_{i+1}) - s_i'(x_{i-1})) \tag{3.200}$$

when $1 \le i \le n-1$. From Equations 3.168 and 3.169, we have

$$s_i'(x) = \frac{-(x_i - x)^2}{2h_i} s_i''(x_{i-1}) + \frac{(x - x_{i-1})^2}{2h_i} s_i''(x_i) - \frac{(s_i''(x_i) - s_i''(x_{i-1}))}{6} h_i + \frac{(y_i - y_{i-1})}{h_i} \tag{3.201}$$

where:

$$x_{i-1} < x < x_i$$
$$h_i = x_i - x_{i-1}$$
$$i = 1, 2, \ldots, n$$

and

$$s_{i+1}'(x) = \frac{-(x_{i+1} - x)^2}{2h_{i+1}} s_i''(x_i) + \frac{(x - x_i)^2}{2h_{i+1}} s_{i+1}''(x_{i+1}) - \frac{(s_{i+1}''(x_{i+1}) - s_i''(x_i))}{6} h_{i+1} + \frac{(y_{i+1} - y_i)}{h_{i+1}} \tag{3.202}$$

where $x_i < x < x_{i+1}$, and $h_{i+1} = x_{i+1} - x_i$, $i = 0, 1, 2, \ldots, n-1$. Using the condition of continuity of $s_i'(x)$ at the point x_i, that is, $s_i'(x_i - 0) = s_{i+1}'(x_i + 0)$, we can obtain

$$s_i'(x_i - 0) = \frac{h_i}{2} s_i''(x_i) - \frac{s_i''(x_i) - s_i''(x_{i-1})}{6} h_i + \frac{(y_i - y_{i-1})}{h_i} \tag{3.203}$$

and

$$s'_{i+1}(x_i + 0) = \frac{-h_{i+1}}{2} s_i''(x_i) - \frac{(s_{i+1}''(x_{i+1}) - s_i''(x_i))}{6} h_{i+1} + \frac{(y_{i+1} - y_i)}{h_{i+1}} \tag{3.204}$$

Again, taking the average of Equations 3.203 and 3.204, we get at the equidistant interior points

$$\begin{aligned} \varphi'(x_i) &= \frac{1}{2}\left[s_i'(x_i - 0) + s'_{i+1}(x_i + 0)\right] \\ &= \frac{1}{2h}(y_{i+1} - y_{i-1}) - \frac{h}{12}(s_{i+1}''(x_{i+1}) - s_i''(x_{i-1})) \end{aligned} \tag{3.205}$$

when $1 \le i \le n-1$. Let us assume that $y = f(x)$ be continuously differentiable sufficient number of times and h is sufficiently small. Then, $s_i'(x_i)$ and $s_i''(x_i)$ can be approximated by Taylor series expansion as follows:

$$\begin{aligned} \varphi'(x_i) &= s_i'(x_i) \approx s_i'(x_{i+1}) \sim y_i' + A_1 h y_i'' + A_2 h^2 y_i''' + A_3 h^3 y_i^{iv} + A_4 h^4 y_i^{v} + \cdots \\ \varphi''(x_i) &= s_i''(x_i) \approx s_i''(x_{i+1}) \sim y_i'' + B_1 h y_i''' + B_2 h^2 y_i^{iv} + B_3 h^3 y_i^{v} + B_4 h^4 y_i^{vi} + \cdots \end{aligned} \tag{3.206}$$

Now, from Equations 3.200 and 3.205 together with the relations in Equation 3.206, we can obtain

$$y_i''+B_1hy_i'''+B_2h^2y_i^{iv}+B_3h^3y_i^{v}+B_4h^4y_i^{vi}+O(h^6)\sim y_i''+h^2\left(-\frac{1}{12}-2A_2\right)y_i^{iv}$$
$$+h^4\left(-\frac{1}{120}-\frac{1}{3}A_2-2A_4\right)y_i^{vi}+O(h^6) \tag{3.207}$$

and

$$y_i'+A_1hy_i''+A_2h^2y_i'''+A_3h^3y_i^{iv}+A_4h^4y_i^{v}+O(h^6)\sim y_i'+h^4\left(-\frac{7}{360}-\frac{1}{6}B_2\right)y_i^{v}+O(h^6) \tag{3.208}$$

By equating the coefficients of corresponding powers of h in Equations 3.207 and 3.208, we then obtain a set of equations relating the A's and B's, which yields

$$A_1=A_2=A_3=0, A_4=-\frac{1}{180},\ldots$$

$$B_1=0, B_2=-\frac{1}{12},\; B_3=0, B_4=\frac{1}{360},\ldots$$

Accordingly, from Equation 3.206, we can obtain

$$y_i'-s_i'(x_i)\sim\frac{h^4}{180}y_i^{v}-\frac{h^6}{1512}y_i^{vii}+O(h^8) \tag{3.209}$$

$$y_i''-s_i''(x_i)\sim\frac{h^2}{12}y_i^{iv}-\frac{h^4}{360}y_i^{vi}+O(h^6) \tag{3.210}$$

Similarly, we can deduce other such relations at the interior nodes. Hence, the errors in other derivatives are given by

$$y_i'''-\frac{1}{2}[s'''(x_i+)+s'''(x_i-)]\sim-\frac{h^2}{12}y_i^{v}+O(h^4) \tag{3.211}$$

$$y_i^{iv}-\frac{1}{h}[s'''(x_i+)-s'''(x_i-)]\sim\frac{67h^4}{30240}y_i^{viii}+O(h^6) \tag{3.212}$$

Now, in any subinterval $[x_{i-1},x_i]$, we may write

$$f(x)-s_i(x)=[f(x)-P_{i,3}(x)]+[P_{i,3}(x)-s_i(x)] \tag{3.213}$$

where $P_{i,3}(x)$ is the cubic Hermite polynomial approximating $f(x)$ on that subinterval. $P_{i,3}(x)$ agrees with $f(x)$ at x_{i-1} and x_i and also $f'(x)$ and $P_{i,3}'(x)$ agree with each other at x_{i-1} and x_i. If on that subinterval we write $x=x_{i-1}+\theta h$, where $0\le\theta\le1$ and $i=1,2,\ldots,n$, Equation 3.140 takes the form

$$f(x)-P_{i,3}(x)=\frac{(x-x_{i-1})^2(x-x_i)^2}{4!}f^{iv}(\xi_i)$$

This implies that

$$f(x_{i-1}+\theta h)-P_{i,3}(x_{i-1}+\theta h)=\frac{h^4}{24}\theta^2(1-\theta)^2f^{iv}(\xi_i) \tag{3.214}$$

where $x_{i-1} < \xi_i < x_i$ and $i = 1, 2, \ldots, n$. Furthermore, since $P_{i,3}(x)$ and $s_i(x)$ agree at x_{i-1} and x_i, we must have

$$P_{i,3}(x) - s_i(x) = \left[p'(x_{i-1}) - s_i'(x_{i-1})\right]\frac{(x - x_{i-1})(x_i - x)^2}{h^2} - \left[p'(x_i) - s_i'(x_i)\right]\frac{(x - x_{i-1})^2(x_i - x)}{h^2} \quad (3.215)$$

In addition, since $p'(x_{i-1}) = f'(x_{i-1})$ and $p'(x_i) = f'(x_i)$, we have

$$P_{i,3}(x_{i-1} + \theta h) - s_i(x_{i-1} + \theta h) = h\theta(1-\theta)^2\left[f'(x_{i-1}) - s_i'(x_{i-1})\right] - h\theta^2(1-\theta)\left[f'(x_i) - s_i'(x_i)\right] \quad (3.216)$$

Finally, from Equation 3.209, we have

$$y_i' - s_i'(x_i) \sim O(h^4) \quad (3.217)$$

It follows from Equation 3.216 that the spline approximation tends to differ from the cubic Hermite approximation only by terms that are small of order h^5 in interior subintervals. Consequently, from Equation 3.213, we have

$$f(x) - s_i(x) \sim \frac{h^4}{24}\theta^2(1-\theta)^2 f^{iv}(x_i) + O(h^5) \quad (3.218)$$

in any interior point of the subinterval $[x_{i-1}, x_i]$, where $\theta = (x - x_{i-1})/h$. It can be verified that the factor $\theta^2(1-\theta)^2 \le 1/16$ when $0 \le \theta \le 1$.

Example 3.18

Obtain the cubic spline approximation for the function $y = f(x)$ from the following data, given that $y_0'' = y_3'' = 0$

x	−1	0	1	2
y	−1	1	3	35

Solution:

Since the values of x are equispaced with $h = 1$, from Equation 3.176, we have

$$M_{i-1} + 4M_i + M_{i+1} = 6(y_{i-1} - 2y_i + y_{i+1}) \quad (3.219)$$

for $i = 1, 2$ and $M_0 = y_0'' = 0, M_3 = y_3'' = 0$. Putting $i = 1$ in Equation 3.219, we get

$$M_0 + 4M_1 + M_2 = 6(y_0 - 2y_1 + y_2) = 6(-1 - 2 + 3) = 0 \quad (3.220)$$

Putting $i = 2$, we get

$$M_1 + 4M_2 + M_3 = 6(y_1 - 2y_2 + y_3) = 6(1 - 6 + 35) = 180 \quad (3.221)$$

Therefore, from Equations 3.220 and 3.221, we have

$$4M_1 + M_2 = 0 \quad (3.222)$$

$$M_1 + 4M_2 = 180 \quad (3.223)$$

since $M_0 = M_3 = 0$. Solving Equations 3.222 and 3.223, we get $M_1 = -12$ and $M_2 = 48$.

The cubic spline in $x_{i-1} \le x \le x_i$ is given by

$$y \cong s_i(x) = \frac{1}{6}\left[(x_i - x)^3 M_{i-1} + (x - x_{i-1})^3 M_i\right] + (x_i - x)$$
$$\left(y_{i-1} - \frac{1}{6}M_{i-1}\right) + (x - x_{i-1})\left(y_i - \frac{1}{6}M_i\right) \tag{3.224}$$

for $i = 1,2,3$. Putting $i = 1$ in Equation 3.224, the cubic spline in $-1 \le x \le 0$ is given by

$$s_1(x) = \frac{1}{6}\left[(x+1)^3(-12)\right] + (0 - x)(-1) + (x+1)\left(1 - \frac{1}{6}(-12)\right)$$
$$= -2x^3 - 6x^2 - 2x + 1$$

Putting $i = 2$ in Equation 3.224, the cubic spline for $0 \le x \le 1$ is given by

$$s_2(x) = \frac{1}{6}\left[(1-x)^3(-12) + (x-0)^3 \times 48\right] + (1-x)\left\{1 - \frac{1}{6}(-12)\right\} + (x - 0)\left(3 - \frac{1}{6} \times 48\right)$$
$$= 10x^3 - 6x^2 - 2x + 1$$

Putting $i = 3$ in Equation 3.224, the cubic spline for $1 \le x \le 2$ is given by

$$s_3(x) = \frac{1}{6}\left[(2-x)^3 \times 48\right] + (2 - x)\left(3 - \frac{1}{6} \times 48\right) + (x - 1) \times 35$$
$$= -8x^3 + 48x^2 - 56x + 19$$

Hence, the required cubic spline approximation for the given function is

$$y = \begin{cases} -2x^3 - 6x^2 - 2x + 1, & \text{for } -1 \le x \le 0 \\ 10x^3 - 6x^2 - 2x + 1, & \text{for } 0 \le x \le 1 \\ -8x^3 + 486x^2 - 56x + 19, & \text{for } 1 \le x \le 2 \end{cases}$$

Example 3.19

Interpolate $f(x) = x^4$ on the interval $-1 \le x \le 1$ by the cubic spline $g(x)$ corresponding to the partition $x_0 = -1$, $x_1 = 0$, and $x_2 = 1$ and satisfying the clamped conditions $g'(-1) = f'(-1)$ and $g'(1) = f'(1)$.

Solution:

We write $f(-1) = f_0 = 1$, $f(0) = f_1 = 0$, and $f(1) = f_2 = 1$. Here, $x_0 = -1$, $x_1 = 0$, and $x_2 = 1$.

The given interval $-1 \le x \le 1$ is partitioned into $n = 2$ parts, each of length $h = 1$. Hence the spline $g(x)$ consists of $n = 2$ cubic polynomials. From Equation 3.191, we have

$$s_i(x) = a_{i0} + a_{i1}(x - x_i) + a_{i2}(x - x_i)^2 + a_{i3}(x - x_i)^3, \quad i = 1,2 \tag{3.225}$$

Putting $i = 1$ in Equation 3.225, we get

$$s_1(x) = a_{10} + a_{11}(x - x_1) + a_{12}(x - x_1)^2 + a_{13}(x - x_1)^3$$
$$= a_{10} + a_{11}x + a_{12}x^2 + a_{13}x^3, \quad -1 \le x \le 0 \tag{3.226}$$

Again, putting $i = 2$ in Equation 3.225, we get

$$\begin{aligned} s_2(x) &= a_{20} + a_{21}(x - x_2) + a_{22}(x - x_2)^2 + a_{23}(x - x_2)^3 \\ &= a_{20} + a_{21}(x - 1) + a_{22}(x - 1)^2 + a_{23}(x - 1)^3, \quad 0 \le x \le 1 \end{aligned} \tag{3.227}$$

From Equation 3.187, we have

$$k_0 + 4k_1 + k_2 = \frac{3}{h}(f_2 - f_0) = \frac{3(1-1)}{1} = 0 \tag{3.228}$$

where $s_1'(x_0) = k_0$, $s_1'(x_1) = k_1$ and $s_2'(x_2) = k_2$. Now, given that, $g' = f'$ at $x = \pm 1$. Therefore,

$$f'(-1) = -4 = g'(-1) = s_1'(-1) = k_0$$

$$f'(1) = 4 = g'(1) = s_1'(1) = k_2$$

Substituting k_0 and k_2 into Equation 3.228, we get $k_1 = 0$. Now, from Equation 3.194, we obtain

$$a_{10} = f_1 = 0$$

$$a_{11} = k_1 = 0$$

$$a_{12} = \frac{3}{h^2}(f_0 - f_1) + \frac{(k_0 + 2k_1)}{h} = \frac{3(1-0)}{1^2} + \frac{(-4+0)}{1} = -1$$

$$a_{03} = \frac{2}{1^3}(f_0 - f_1) + \frac{1}{1^2}(k_0 + k_1) = \frac{2}{1^3}(1-0) + \frac{1}{1^2}(-4+0) = -2$$

Similarly,

$$a_{20} = f_2 = 1$$

$$a_{21} = k_2 = 4$$

$$a_{22} = \frac{3}{1^2}(f_1 - f_2) + \frac{1}{1}(k_1 + 2k_2) = \frac{3}{1^2}(0-1) + \frac{1}{1}(0+8) = 5$$

$$a_{23} = \frac{2}{1^3}(f_1 - f_2) + \frac{1}{1^2}(k_1 + k_2) = \frac{2}{1^3}(0-1) + \frac{1}{1^2}(0+4) = 2$$

Thus, the cubic polynomials are

$$\begin{aligned} s_1(x) &= a_{10} + a_{11}x + a_{12}x^2 + a_{13}x^3, \quad \text{for } -1 \le x \le 0 \\ &= -x^2 - 2x^3, \quad \text{for } -1 \le x \le 0 \end{aligned}$$

and

$$\begin{aligned} s_2(x) &= a_{20} + a_{21}(x-1) + a_{22}(x-1)^2 + a_{23}(x-1)^3, \quad \text{for } 0 \le x \le 1 \\ &= 1 + 4(x-1) + 5(x-1)^2 + 2(x-1)^3, \quad \text{for } 0 \le x \le 1 \\ &= -x^2 + 2x^3, \quad \text{for } 0 \le x \le 1 \end{aligned}$$

Therefore, the required cubic spline $g(x)$ is given by

$$g(x) = -x^2 - 2x^3 \quad \text{if } -1 \le x \le 0$$
$$= -x^2 + 2x^3 \quad \text{if } 0 \le x \le 1$$

Example 3.20

Interpolate $f(x) = \sin x$ on the interval $[-(\pi/2),(\pi/2)]$ by the cubic spline $g(x)$ corresponding to the partition $x_0 = -(\pi/2)$, $x_1 = 0$, $x_2 = (\pi/2)$ and satisfying the clamped conditions $g' = f'$ at $x = \pm(\pi/2)$.

Solution:

We write $f(-(\pi/2)) = f_0 = -1$, $f(0) = f_1 = 0$, and $f(\pi/2) = f_2 = 1$. Here, $x_0 = -(\pi/2)$, $x_1 = 0$, and $x_2 = (\pi/2)$.

The given interval $-(\pi/2) \le x \le (\pi/2)$ is partitioned into $n = 2$ parts, each of length $h = (\pi/2)$. Hence, the spline $g(x)$ consists of $n = 2$ cubic polynomials. From Equation 3.191, we have

$$s_i(x) = a_{i0} + a_{i1}(x-x_i) + a_{i2}(x-x_i)^2 + a_{i3}(x-x_i)^3, \quad i=1,2 \tag{3.229}$$

Putting $i = 1$ in Equation 3.229, we get

$$\begin{aligned} s_1(x) &= a_{10} + a_{11}(x-x_1) + a_{12}(x-x_1)^2 + a_{13}(x-x_1)^3 \\ &= a_{10} + a_{11}x + a_{12}x^2 + a_{13}x^3, \quad -1 \le x \le 0 \end{aligned} \tag{3.230}$$

Again, putting $i = 2$ in Equation 3.229, we get

$$\begin{aligned} s_2(x) &= a_{20} + a_{21}(x-x_2) + a_{22}(x-x_2)^2 + a_{23}(x-x_2)^3 \\ &= a_{20} + a_{21}(x-1) + a_{22}(x-1)^2 + a_{23}(x-1)^3, \quad 0 \le x \le 1 \end{aligned} \tag{3.231}$$

From Equation 3.187, we have

$$k_0 + 4k_1 + k_2 = \frac{6}{\pi}(1+1) = \frac{12}{\pi} \tag{3.232}$$

where $s_1'(x_0) = k_0$, $s_1'(x_1) = k_1$, and $s_2'(x_2) = k_2$. Now, given that, $g' = f'$ at $x = \pm(\pi/2)$. Therefore,

$$f'\left(-\frac{\pi}{2}\right) = 0 = g'\left(-\frac{\pi}{2}\right) = s_1'\left(-\frac{\pi}{2}\right) = k_0$$

$$f'\left(\frac{\pi}{2}\right) = 0 = g'\left(\frac{\pi}{2}\right) = s_1'\left(\frac{\pi}{2}\right) = k_2$$

Substituting k_0 and k_2 into Equation 3.232, we get $k_1 = (3/\pi)$. Now, from Equation 3.194, we obtain

$$a_{10} = f_1 = 0$$

$$a_{11} = k_1 = \frac{3}{\pi}$$

$$a_{12} = \frac{3}{h^2}(f_0 - f_1) + \frac{(k_0 + 2k_1)}{h} = \frac{12}{\pi^2}(-1-0) + \frac{2(0+6)}{\pi^2} = 0$$

$$a_{03} = \frac{2}{h^3}(f_0 - f_1) + \frac{1}{h^2}(k_0 + k_1) = \frac{16}{\pi^3}(-1-0) + \frac{4}{\pi^2}\left(0 + \frac{3}{\pi}\right) = -\frac{4}{\pi^3}$$

Similarly,

$$a_{20} = f_2 = 1$$

$$a_{21} = k_2 = 0$$

$$a_{22} = \frac{3}{h^2}(f_1 - f_2) + \frac{1}{h}(k_1 + 2k_2) = \frac{12}{\pi^2}(0-1) + \frac{2}{\pi}\left(\frac{3}{\pi} + 0\right) = -\frac{6}{\pi^2}$$

$$a_{23} = \frac{2}{h^3}(f_1 - f_2) + \frac{1}{h^2}(k_1 + k_2) = \frac{16}{\pi^3}(0-1) + \frac{4}{\pi^2}\left(\frac{3}{\pi} + 0\right) = -\frac{4}{\pi^3}$$

Thus, the cubic polynomials are

$$s_1(x) = a_{10} + a_{11}x + a_{12}x^2 + a_{13}x^3, \quad \text{for } -\frac{\pi}{2} \le x \le 0$$

$$= \frac{3x}{\pi} - \frac{4x^3}{\pi^3}, \quad \text{for } -\frac{\pi}{2} \le x \le 0$$

and

$$s_2(x) = a_{20} + a_{21}\left(x - \frac{\pi}{2}\right) + a_{22}\left(x - \frac{\pi}{2}\right)^2 + a_{23}\left(x - \frac{\pi}{2}\right)^3, \quad \text{for } 0 \le x \le \frac{\pi}{2}$$

$$= 1 - \frac{6}{\pi^2}\left(x - \frac{\pi}{2}\right)^2 - \frac{4}{\pi^3}\left(x - \frac{\pi}{2}\right)^3, \quad \text{for } 0 \le x \le \frac{\pi}{2}$$

$$= \frac{3x}{\pi} - \frac{4x^3}{\pi^3}, \quad \text{for } 0 \le x \le \frac{\pi}{2}$$

Therefore, the required cubic spline $g(x)$ is given by

$$g(x) = \frac{3x}{\pi} - \frac{4x^3}{\pi^3}, \quad -\frac{\pi}{2} \le x \le \frac{\pi}{2}$$

Example 3.21

Fit a cubic natural spline from the following given data:

x	0	1	2
y	4	1	2

Solution:

Since the values of x are equispaced with $h = 1$ from Equation 3.176, we have

$$M_0 + 4M_1 + M_2 = 6(y_0 - 2y_1 + y_2) = 6(4 - 2 + 2) = 24 \tag{3.233}$$

and $M_0 = y_0'' = 0$, $M_2 = y_2'' = 0$. Therefore, from Equation 3.233, we get

$$M_1 = 6 \tag{3.234}$$

The cubic spline in $x_i - 1 \le x \le x_i$ is given by

$$y = s_i(x) = \frac{1}{6}\left[(x_i - x)^3 M_{i-1} + (x - x_{i-1})^3 M_i\right] + (x_i - x)\left(y_{i-1} - \frac{1}{6}M_{i-1}\right) + (x - x_{i-1})\left(y_i - \frac{1}{6}M_i\right) \tag{3.235}$$

for $i = 1, 2$. Putting $i = 1$ in Equation 3.221, the cubic spline in $0 \le x \le 1$ is given by

$$s_1(x) = \frac{1}{6}\left[(1-x)^3 \times 0 + x^3 \times 6\right] + (1-x)(4-0) + (x-0)\left(1 - \frac{1}{6} \times 6\right)$$
$$= x^3 - 4x + 4$$

Putting $i = 2$ in Equation 3.224, the cubic spline for $1 \le x \le 2$ is given by

$$s_2(x) = \frac{1}{6}\left[(2-x)^3 \times 6 + (x-1)^3 \times 0\right] + (2-x)\left\{1 - \frac{1}{6} \times 6\right\} + (x-1)\left(2 - \frac{1}{6} \times 0\right)$$
$$= 6 - 10x + 6x^2 - x^3$$

Hence, the required cubic spline approximation for the given function is

$$y = \begin{cases} 4 - 4x + x^3, & \text{for } 0 \le x \le 1 \\ 6 - 10x + 6x^2 - x^3, & \text{for } 1 \le x \le 2 \end{cases}$$

Example 3.22

Find the cubic spline $g(x)$ corresponding to the given data $f_0 = f(0) = 1$, $f_1 = f(1) = 0$, $f_2 = f(2) = -1$, $f_3 = f(3) = 0$, $k_0 = 0$, and $k_3 = -6$.

Solution:

We write $f_0 = f(0) = 1$, $f_1 = f(1) = 0$, $f_2 = f(2) = -1$, and $f_3 = f(3) = 0$. Here, $x_0 = 0$, $x_1 = 1$, $x_2 = 2$, and $x_3 = 3$.

The given interval $0 \le x \le 3$ is partitioned into $n = 3$ parts, each of length $h = 1$. Hence, the spline $g(x)$ consists of $n = 3$ cubic polynomials. From Equation 3.191, we have

$$s_i(x) = a_{i0} + a_{i1}(x - x_i) + a_{i2}(x - x_i)^2 + a_{i3}(x - x_i)^3, \quad i = 1,2,3 \tag{3.236}$$

Putting $i = 1$ in Equation 3.236, we get

$$\begin{aligned} s_1(x) &= a_{10} + a_{11}(x - x_1) + a_{12}(x - x_1)^2 + a_{13}(x - x_1)^3 \\ &= a_{10} + a_{11}(x-1) + a_{12}(x-1)^2 + a_{13}(x-1)^3, \quad 0 \le x \le 1 \end{aligned} \tag{3.237}$$

Putting $i = 2$ in Equation 3.236, we get

$$\begin{aligned} s_2(x) &= a_{20} + a_{21}(x - x_2) + a_{22}(x - x_2)^2 + a_{23}(x - x_2)^3 \\ &= a_{20} + a_{21}(x - 2) + a_{22}(x - 2)^2 + a_{23}(x - 2)^3, \quad 1 \le x \le 2 \end{aligned} \tag{3.238}$$

Again, putting $i = 3$ in Equation 3.236, we get

$$\begin{aligned} s_3(x) &= a_{30} + a_{31}(x - x_3) + a_{32}(x - x_3)^2 + a_{33}(x - x_3)^3 \\ &= a_{30} + a_{31}(x - 3) + a_{32}(x - 3)^2 + a_{33}(x - 3)^3, \quad 2 \le x \le 3 \end{aligned} \tag{3.239}$$

From Equation 3.187, we have

$$k_0 + 4k_1 + k_2 = \frac{3}{h}\left(f_2 - f_0\right) = \frac{3(-1-1)}{1} = -6 \tag{3.240}$$

and

$$k_1 + 4k_2 + k_3 = \frac{3}{h}\left(f_3 - f_1\right) = \frac{3(0-0)}{1} = 0 \tag{3.241}$$

where $s_1'(x_0) = k_0$, $s_1'(x_1) = k_1$, $s_2'(x_2) = k_2$, and $s_3'(x_3) = k_3$. Now, given that, $k_0 = 0$ and $k_3 = -6$. Therefore, from Equations 3.240 and 3.241, we get

$$4k_1 + k_2 = -6 \tag{3.242}$$

and

$$k_1 + 4k_2 = 6 \tag{3.243}$$

Solving Equations 3.242 and 3.243, we get

$$k_1 = -2 \quad \text{and} \quad k_2 = 2 \tag{3.244}$$

Now, from Equation 3.194, we obtain

$$a_{10} = f_1 = 0$$

$$a_{11} = k_1 = -2$$

$$a_{12} = \frac{3}{h^2}\left(f_0 - f_1\right) + \frac{\left(k_0 + 2k_1\right)}{h} = \frac{3(1-0)}{1^2} + \frac{(0-4)}{1} = -1$$

$$a_{03} = \frac{2}{1^3}\left(f_0 - f_1\right) + \frac{1}{1^2}\left(k_0 + k_1\right) = \frac{2}{1^3}(1-0) + \frac{1}{1^2}(0-2) = 0$$

Similarly,

$$a_{20} = f_2 = -1$$

$$a_{21} = k_2 = 2$$

$$a_{22} = \frac{3}{1^3}\left(f_1 - f_2\right) + \frac{1}{1}\left(k_1 + 2k_2\right) = \frac{3}{1^2}(0+1) + \frac{1}{1}(-2+4) = 5$$

$$a_{23} = \frac{2}{1^3}\left(f_1 - f_2\right) + \frac{1}{1^2}\left(k_1 + k_2\right) = \frac{2}{1^3}(0+1) + \frac{1}{1^2}(-2+2) = 2$$

Again, we have

$$a_{30} = f_3 = 0$$

$$a_{31} = k_3 = -6$$

$$a_{32} = \frac{3}{1^2}\left(f_2 - f_3\right) + \frac{1}{1}\left(k_2 + 2k_3\right) = 3(-1-0) + (2-12) = -13$$

$$a_{33} = \frac{2}{1^3}\left(f_2 - f_3\right) + \frac{1}{1^2}\left(k_2 + k_3\right) = 2(-1-0) + (2-6) = -6$$

Thus, the cubic polynomials are

$$
\begin{aligned}
s_1(x) &= a_{10} + a_{11}(x-1) + a_{12}(x-1)^2 + a_{13}(x-1)^3, \quad 0 \le x \le 1 \\
&= 1 - x^2, \quad 0 \le x \le 1 \\
s_2(x) &= a_{20} + a_{21}(x-2) + a_{22}(x-2)^2 + a_{23}(x-2)^3, \quad 1 \le x \le 2 \\
&= -1 + 2(x-2) + 5(x-2)^2 + 2(x-2)^3, \quad 1 \le x \le 2 \\
s_3(x) &= a_{30} + a_{31}(x-x_3) + a_{32}(x-x_3)^2 + a_{33}(x-x_3)^3, \quad 2 \le x \le 3 \\
&= -6(x-3) - 13(x-3)^2 - 6(x-3)^3, \quad 2 \le x \le 3
\end{aligned}
$$

Therefore, the required cubic spline $g(x)$ is given by

$$
\begin{aligned}
g(x) &= 1 - x^2 \quad \text{if } 0 \le x \le 1 \\
&= -1 + 6x - 7x^2 + 2x^3 \quad \text{if } 1 \le x \le 2 \\
&= 63 - 90x + 41x^2 - 6x^3 \quad \text{if } 2 \le x \le 3
\end{aligned}
$$

3.2.17 Interpolation by Iteration

3.2.17.1 Aitken's Interpolation Formula

Newton's general interpolation formula generates successively higher order interpolation formulae as discussed in Section 3.2.10.2. We shall now derive another formula of this kind, called *Aitken's interpolation formula*, which can be easily used to set computer programming code.

Let $y_0 = f(x_0)$ and $y_1 = f(x_1), \ldots, y_n = f(x_n)$ be the values of the function $y = f(x)$ corresponding to the values $x_0, x_1, x_2, \ldots, x_n$, which are not necessarily equally spaced.

We now proceed iteratively to find the value of $y = f(x)$ corresponding to a given value of x by the method of iteration as follows:

A first approximation to $f(x)$ is obtained by considering the first two points x_0 and x_1 only. Then we obtain the second approximation by considering the first three points x_0, x_1, x_2 and so on.

We denote the interpolation polynomial in the first iteration by $p_{0,1}(x)$. Thus, for the first stage of approximation, we have

$$
\begin{aligned}
p_{0,1}(x) &= f(x_0) + (x - x_0) f[x_0, x_1] \\
&= \frac{(x_1 - x) f(x_0) - (x_0 - x) f(x_1)}{x_1 - x_0} \\
&= \frac{1}{x_1 - x_0} \begin{vmatrix} f(x_0) & x_0 - x \\ f(x_1) & x_1 - x \end{vmatrix}
\end{aligned}
\tag{3.245}
$$

Similarly, we can obtain $p_{0,2}(x)$, $p_{0,3}(x)$, and so on.

Next, we construct the polynomial $p_{0,1,2}(x)$ of second degree through the first three points: (x_0, y_0), (x_1, y_1), and (x_2, y_2), and then the second approximation is given by

$$
p_{0,1,2}(x) = \frac{1}{x_2 - x_1} \begin{vmatrix} p_{0,1}(x) & x_1 - x \\ p_{0,2}(x) & x_2 - x \end{vmatrix} \tag{3.246}
$$

TABLE 3.6
Aitken's Scheme for Iterated Polynomials

x	y					
x_0	y_0					
		$p_{0,1}(x)$				
x_1	y_1		$p_{0,1,2}(x)$			
		$p_{0,2}(x)$		$p_{0,1,2,3}(x)$		
x_2	y_2		$p_{0,1,3}(x)$		$p_{0,1,2,3,4}(x)$	
		$p_{0,3}(x)$		$p_{0,1,2,4}(x)$		$p_{0,1,2,3,4,5}(x)$
x_3	y_3		$p_{0,1,4}(x)$		$p_{0,1,2,3,5}(x)$	
		$p_{0,4}(x)$		$p_{0,1,2,5}(x)$		
x_4	y_4		$p_{0,1,5}(x)$			
		$p_{0,5}(x)$				
x_5	y_5					

Similarly, we can obtain $p_{0,1,3}(x)$, $p_{0,1,4}(x)$, and so on.

Proceeding in this manner, at the nth stage of approximation, we can obtain

$$p_{0,1,\ldots,n}(x) = \frac{1}{x_n - x_{n-1}} \begin{vmatrix} p_{0,1,\ldots,n-2,n-1}(x) & x_{n-1} - x \\ p_{0,1,\ldots,n-2,n}(x) & x_n - x \end{vmatrix} \tag{3.247}$$

Table 3.6 may be constructed to facilitate the computation of the iterated polynomials.

3.2.17.1.1 Algorithm for Aitken's Interpolation

Input: Enter the number of given data N (where $N = n + 1$) and interpolating point x. Enter the data $x_i, y_i, i = 0(1)n$.

Output: Print the value of the function $y = f(x)$ for a given value of x.

Initial step: Initialize $\tilde{p}_{i,0} = y_i, i = 0(1)n$.

Step 1: Set $k = 1$;

for $i = 0(1)\overline{n-1}$ do

Set, $j = i + k$;

Compute $\tilde{p}_{i,j} = \frac{1}{x_j - x_0}\left((x_j - x) * \tilde{p}_{0,0} - (x_0 - x) * \tilde{p}_{j,0}\right)$

Step 2: for $k = 2(1)n$ do

for $i = 0(1)n - k$ do

Set, $j = i + k$

Compute $\tilde{p}_{i,j} = \frac{1}{x_j - x_{j-i-1}}\left((x_j - x) * \tilde{p}_{0,j-i-1} - (x_{j-i-1} - x) * \tilde{p}_{i+1,j}\right)$

Step 3: Print the value of $s = \tilde{p}_{0,n}$.

Step 4: Stop. ■

Table 3.7 shows the execution of Aitken's algorithm.

TABLE 3.7
Execution of Aitken's Algorithm

x	$f(x)$				
x_0	$y_0 = \tilde{p}_{0,0}$				
		$p_{0,1}(x) \equiv \tilde{p}_{0,1}$			
x_1	$y_1 = \tilde{p}_{1,0}$		$p_{0,1,2}(x) \equiv \tilde{p}_{0,2}$		
		$p_{0,2}(x) \equiv \tilde{p}_{1,2}$		$p_{0,1,2,3}(x) \equiv \tilde{p}_{0,3}$	
x_2	$y_2 = \tilde{p}_{2,0}$		$p_{0,1,3}(x) \equiv \tilde{p}_{1,3}$		$p_{0,1,2,3,4}(x) \equiv \tilde{p}_{0,4}$
		$p_{0,3}(x) \equiv \tilde{p}_{2,3}$		$p_{0,1,2,4}(x) \equiv \tilde{p}_{1,4}$	
x_3	$y_3 = \tilde{p}_{3,0}$		$p_{0,1,4}(x) \equiv \tilde{p}_{2,4}$		
		$p_{0,4}(x) \equiv \tilde{p}_{3,4}$			
x_4	$y_4 = \tilde{p}_{4,0}$				

*MATHEMATICA® Program Implementing Aitken's Interpolation for the Following Data

```
        x[0]=300        x[1]=304        x[2]=305        x[3]=307;
        y[0]=2.4771     y[1]=2.4829     y[2]=2.4843     y[3]=2.4871*
n=3;
x[0]=300;
x[1]=304;
x[2]=305;
x[3]=307;
y[0]=2.4771;
y[1]=2.4829;
y[2]=2.4843;
y[3]=2.4871;
For[i=0,i<=n,i++,
p[i,0]=y[i];
Print[p[i,0]]];
k=1;
For[i=0,i<=n-1,i++,
j=i+k;
p[i,j]=1/(x[j]-x[j-i-1])*Det[{{p[0,0],x[0]-x},{p[j,0],x[j]-x}}];
Print["p[",i,",",j,"]=",p[i,j]/.x->301]];
Print["..................."];
For[k=2,k<=n,k++,
For[i=0,i<=n-k,i++,
j=i+k;
p[i,j]=1/
(x[j]-x[j-i-1])*Det[{{p[0,j-i-1],x[j-i-1]-x},{p[i+1,j],x[j]-x}}];
Print["p[",i,",",j,"]=",p[i,j]/.x->301]]];
Print["..................."];
Print["p[","0",",",n,"]=",p[0,n]/.x->301];
```

Output:

```
2.4771
2.4829
2.4843
2.4871
```

```
p[0,1]=2.47855
p[1,2]=2.47854
p[2,3]=2.47853
...................
p[0,2]=2.47858
p[1,3]=2.47857
p[0,3]=2.4786
...................
p[0,3]=2.4786
```

3.2.17.2 Neville's Interpolation Formula

The above Aitken's interpolation scheme has been modified by Neville, which is described as follows. In this scheme, the first approximations are as follows:

$$p_{0,1}(x) = \frac{1}{x_1 - x_0}\begin{vmatrix} f(x_0) & x_0 - x \\ f(x_1) & x_1 - x \end{vmatrix} \tag{3.248}$$

$$p_{1,2}(x) = \frac{1}{x_2 - x_1}\begin{vmatrix} f(x_1) & x_1 - x \\ f(x_2) & x_2 - x \end{vmatrix} \tag{3.249}$$

Similarly, we can obtain $p_{2,3}(x)$, $p_{3,4}(x)$, and so on. The second approximations are given by

$$p_{0,1,2}(x) = \frac{1}{x_2 - x_0}\begin{vmatrix} p_{0,1}(x) & x_0 - x \\ p_{1,2}(x) & x_2 - x \end{vmatrix} \tag{3.250}$$

$$p_{1,2,3}(x) = \frac{1}{x_3 - x_1}\begin{vmatrix} p_{1,2}(x) & x_1 - x \\ p_{2,3}(x) & x_3 - x \end{vmatrix} \tag{3.251}$$

Similarly, we can obtain $p_{2,3,4}(x)$, $p_{3,4,5}(x)$, and so on. Proceeding in this manner, at the nth stage of approximation, we can obtain

$$\begin{aligned} p_{0,1,\ldots,n}(x) &= p_{0,1,\ldots,n-1}(x) + \frac{(x - x_0)}{(x_n - x_0)}\left(p_{1,2,\ldots,n}(x) - p_{0,1,\ldots,n-1}(x)\right) \\ &= \frac{(x - x_0)\,p_{1,2,\ldots,n}(x) - (x - x_n)\,p_{0,1,\ldots,n-1}(x)}{(x_n - x_0)} \\ &= \frac{1}{x_n - x_0}\begin{vmatrix} p_{0,1,\ldots,n-1}(x) & x_0 - x \\ p_{1,2,\ldots,n}(x) & x_n - x \end{vmatrix} \end{aligned} \tag{3.252}$$

The computations for the iterated polynomials may be carried out as shown in Table 3.8.

3.2.17.2.1 Algorithm for Neville's Interpolation

Input: Enter the number of given data N (where $N = n + 1$) and interpolating point x. Enter the data $x_i, y_i, i = 0(1)n$.

Output: Print the value of the function $y = f(x)$ for a given value of x.

Initial step: Initialize $\tilde{p}_{i,0} = y_i, i = 0(1)n$.

Step 1: set, $k = 1$;

for $i = 0(1)n-1$ do

Set, $j = i + k$;

Compute $\tilde{p}_{i,j} = \dfrac{1}{x_j - x_i}\left((x_j - x) * \tilde{p}_{i,0} - (x_i - x) * \tilde{p}_{j,0}\right)$

Step 2: for $k = 2(1)n$ do

for $i = 0(1)n-k$ do

set, $j = i + k$

compute $\tilde{p}_{i,j} = \dfrac{1}{x_j - x_i}\left((x_j - x) * \tilde{p}_{i,j-1} - (x_i - x) * \tilde{p}_{i+1,j}\right)$

Step 3: Print the value of $s = \tilde{p}_{0,n}$.

Step 4: Stop. ■

Table 3.9 shows the execution of Neville's algorithm.

TABLE 3.8
Neville's Scheme for Iterated Polynomials

x	y					
x_0	y_0					
		$p_{0,1}(x)$				
x_1	y_1		$p_{0,1,2}(x)$			
		$p_{1,2}(x)$		$p_{0,1,2,3}(x)$		
x_2	y_2		$p_{1,2,3}(x)$		$p_{0,1,2,3,4}(x)$	
		$p_{2,3}(x)$		$p_{1,2,3,4}(x)$		$p_{0,1,2,3,4,5}(x)$
x_3	y_3		$p_{2,3,4}(x)$		$p_{1,2,3,4,5}(x)$	
		$p_{3,4}(x)$		$p_{2,3,4,5}(x)$		
x_4	y_4		$p_{3,4,5}(x)$			
		$p_{4,5}(x)$				
x_5	y_5					

TABLE 3.9
Execution of Neville's Algorithm

x	y				
x_0	$y_0 \equiv \tilde{p}_{0,0}$				
		$p_{0,1}(x) \equiv \tilde{p}_{0,1}$			
x_1	$y_1 \equiv \tilde{p}_{1,0}$		$p_{0,1,2}(x) \equiv \tilde{p}_{0,2}$		
		$p_{1,2}(x) \equiv \tilde{p}_{1,2}$		$p_{0,1,2,3}(x) \equiv \tilde{p}_{0,3}$	
x_2	$y_2 \equiv \tilde{p}_{2,0}$		$p_{1,2,3}(x) \equiv \tilde{p}_{1,3}$		$p_{0,1,2,3,4}(x) \equiv \tilde{p}_{0,4}$
		$p_{2,3}(x) \equiv \tilde{p}_{2,3}$		$p_{1,2,3,4}(x) \equiv \tilde{p}_{1,4}$	
x_3	$y_3 \equiv \tilde{p}_{3,0}$		$p_{2,3,4}(x) \equiv \tilde{p}_{2,4}$		
		$p_{3,4}(x) \equiv \tilde{p}_{3,4}$			
x_4	$y_4 \equiv \tilde{p}_{4,0}$				

*MATHEMATICA® Program Implementing Neville's Interpolation for the Following Data

```
        x[0]=0.38      x[1]=0.42      x[2]=0.35      x[3]=0.44      x[4]=0.33;
        y[0]=0.39941  y[1]=0.44657  y[2]=0.36503  y[3]=0.47078  y[4]=0.34252*
n=4;
x[0]=0.38;
x[1]=0.42;
x[2]=0.35;
x[3]=0.44;
x[4]=0.33;
y[0]=0.39941;
y[1]=0.44657;
y[2]=0.36503;
y[3]=0.47078;
y[4]=0.34252;
For[i=0,i<=n,i++,
p[i,0]=y[i];
Print[p[i,0]]];
k=1;
For[i=0,i<=n-1,i++,
j=i+k;
p[i,j]=1/(x[j]-x[i])*Det[{{p[i,0],x[i]-x},{p[j,0],x[j]-x}}];
Print["p[",i,",",j,"]=",p[i,j]/.x->0.36]];
Print["..................."];
For[k=2,k<=n,k++,
For[i=0,i<=n-k,i++,
j=i+k;
p[i,j]=1/(x[j]-x[i])*Det[{{p[i,j-1],x[i]-x},{p[i+1,j],x[j]-x}}];
Print["p[",i,",",j,"]=",p[i,j]/.x->0.36]]];
Print["..................."];
Print["p[","0",",",n,"]=",p[0,n]/.x->0.36];
```

Output:

```
0.39941
0.44657
0.36503
0.47078
0.34252
p[0,1]=0.37583
p[1,2]=0.376679
p[2,3]=0.37678
p[3,4]=0.3775
.........................
p[0,2]=0.376396
p[1,3]=0.376374
p[2,4]=0.37642
p[0,3]=0.376403
p[1,4]=0.376405
p[0,4]=0.376404
.........................
p[0,4]=0.376404
```

Example 3.23

Use Aitken's iterated interpolation to find the value of $\log_{10} 4.3$ from the following table:

x	4	4.2	4.4	4.6	4.8
$\log_{10} x$	0.60206	0.62325	0.64345	0.66276	0.68124

Solution:

We construct Table 3.10 according to Table 3.6.

Therefore, $\log_{10} 4.3 = 0.633467$, which is correct up to five decimal place with the exact value.

Example 3.24

Use Neville's iterated interpolation to find the value of tan 0.36 from the following table:

x	0.33	0.35	0.38	0.42	0.44
$\tan x$	0.34252	0.36503	0.39941	0.44657	0.47078

Solution:

We construct Table 3.11 according to Table 3.8. Therefore, $\tan(0.36) = 0.376404$, which is correct up to five decimal place with the exact value.

3.2.18 Inverse Interpolation

In the case of inverse interpolation, we have to compute the value of x for a given value of y. We assume that the function $y = f(x)$ has an inverse $x = F(y)$. Here, the interpolating points are $y_0, y_1, \ldots, y_n$ and the corresponding function values are $x_0, x_1, x_2, \ldots, x_n$. If the values of x are unequispaced, the most easy way of finding interpolating polynomial is interchange of x and y in Lagrange's interpolation formula or Aitken's iterated interpolating polynomials. Use of Lagrange's interpolation formula has been already mentioned in Equation 3.73.

TABLE 3.10
Aitken's Iteration

x	y				
4	0.60206				
		0.633845			
4.2	0.62325		0.633474		
		0.633103		0.633467	
4.4	0.64345		0.633486		0.633467
		0.63241		0.633468	
4.6	0.66276		0.633496		
		0.631752			
4.8	0.68124				

TABLE 3.11
Neville's Iteration

x	*y*				
0.33	0.34252				
		0.376285			
0.35	0.36503		0.376408		
		0.37649		0.376404	
0.38	0.39941		0.376396		0.376404
		0.37583		0.376403	
0.42	0.44657		0.37646		
		0.37394			
0.44	0.47078				

Now, Newton's divided difference formula for inverse interpolation is

$$x = x_0 + (y - y_0)F[y_0, y_1] + (y - y_0)(y - y_1)F[y_0, y_1, y_2] + \cdots + (y - y_0)(y - y_1)\ldots(y - y_{n-1})F[y_0, y_1, \ldots, y_n] \quad (3.253)$$

Example 3.25

Find the value of x when $y = f(x) = 15$ from the following table:

x	5	6	9	11
y	12	13	14	16

Solution:

We first construct Newton's divided difference table as follows:

y	*x*	First-Order Divided Difference	Second-Order Divided Difference	Third-Order Divided Difference
12	5			
		1		
13	6		1	
		3		−0.4167
14	9		−0.6667	
		1		
16	11			

Using Newton's divided difference formula in Equation 3.253, we obtain

$$x = 5 + (15 - 12) + (15 - 12)(15 - 13) + (15 - 12)(15 - 13)(15 - 14)(-0.4167)$$
$$= 11.4998$$

EXERCISES

1. Prove that
   a. $\Delta \cdot \nabla \equiv \Delta - \nabla \equiv \delta^2$
   b. $\Delta + \nabla \equiv \frac{\Delta}{\nabla} - \frac{\nabla}{\Delta}$
   c. $\delta \equiv E^{1/2}\nabla$
   d. $E \equiv e^{hD}$
   e. $D \equiv \frac{1}{h}\left[\Delta - \frac{\Delta^2}{2} + \frac{\Delta^3}{3} \cdots\right]$
   f. $E\nabla \equiv \nabla E \equiv \Delta$
   g. $\mu\delta = \frac{\Delta}{2} + \frac{\Delta E^{-1}}{2}$
   h. $1 - e^{-hD} = \nabla$
   i. $\delta\left(E^{1/2} + E^{-(1/2)}\right) \equiv \Delta E^{-1} + \Delta$
   j. $\left(\Delta - \frac{1}{2}\delta^2\right) \equiv \delta\left(1 + \frac{\delta^2}{4}\right)^{1/2}$

   Where the symbols have their usual meaning.
2. Evaluate
   a. $\Delta^2 x^3$
   b. $\Delta^2(\cos x)$
   c. $\Delta\left[(x+1)(x+2)\right]$
   d. $\Delta\left(\tan^{-1} x\right)$
   e. $\Delta\left[\frac{f(x)}{g(x)}\right]$
3. Locate and correct the error in the following table.

x	1.00	1.05	1.10	1.15	1.20	1.25	1.30
e^x	2.7183	2.8577	3.0042	3.1528	3.3201	3.4903	3.6693

4. Prove the following
   a. $\left(\frac{\Delta^2}{E}\right)x^3 = 6x$
   b. $\Delta^3(1-x)(1-2x)(1-3x) = -36$ taking $h = 1$
   c. $\Delta(x + \cos x) = \pi - 2\cos x$, taking $h = \pi$
   d. $\Delta\left(\frac{1}{1+x^2}\right) = -\frac{1+2x}{\left(1+x^2\right)\left(x^2+2x+2\right)}$
   e. $\Delta\left(\frac{x^2}{\cos 2x}\right) = \frac{h(2x+h)\cos 2x + 2x^2 \sin h \sin(2x+h)}{\cos(2x+2h)\cos 2x}$
   f. $\frac{\Delta^2 x^2}{E(x+\log x)} = \frac{2}{x+1+\log(x+1)}$

g. $\Delta^n x^{(n)} = n!h^n$
h. $\Delta^{n+1} x^{(n)} = 0$

5. Evaluate
   a. $\Delta^2\left(\dfrac{5x+12}{x^2+5x+6}\right)$
   b. $\Delta^{10}\left[(1-x)(1-x^2)(1-3x^3)(1-4x^4)\right]$
   c. $\Delta\left(\dfrac{2^x}{x!}\right)$
6. Prove the following relations:
   a. $\delta^3 E = \Delta^2$
   b. $E^{-1/2} = \mu - \dfrac{\delta}{2}$
   c. $\nabla = \delta E^{-1/2}$
   d. $\mu = \cos h\dfrac{hD}{2}$
7. Prove that $\delta \equiv \Delta E^{-1/2}$ and hence prove that $E \equiv (\Delta/\delta)^2$.
8. Prove that $hD \equiv \log(1+\Delta) \equiv -\log(1-\nabla) = \sin h^{-1}(\mu\delta)$.
9. Prove that $(\Delta^2/E)e^x \cdot [E(e^x)/\Delta^2(e^x)] = e^x$, taking the interval of differencing as h.
10. Prove that the following relations:
   a. $\sum_{k=0}^{n-1} \Delta^2 f_k = \Delta f_n - \Delta f_0$
   b. $\Delta(f_i g_i) = f_i \Delta g_i + g_{i+1}\Delta f_i$
   c. $\Delta f_i^2 = (f_i + f_{i+1})\Delta f_i$
   d. $\Delta(f_i/g_i) = (g_i \Delta f_i - f_i \Delta g_i)/(g_i g_{i+1})$
   e. $\Delta(1/f_i) = -\Delta f_i/(f_i f_{i+1})$
11. Prove that for equally spaced arguments with space length h
   a. $\delta^{2m+1} y_{1/2} + (2m+1)!h^{2m+1} f(x_{-m}, \ldots, x_0, \ldots, x_m)$
   b. $\delta^{2m} y_0 + (2m)!h^{2m} f(x_{-m}, \ldots, x_{-1}, x_1, \ldots, x_m)$

   where δ is a central difference operator.
12. If $f(x) = e^{ax}$, show that $\Delta^n f(x) = (e^{ah} - 1)^n e^{ax}$.
13. Find the missing terms in the following:

x	0	5	10	15	20	25	30
y	1	3	?	73	225	?	1153

14. The following table gives the values of y, which is a polynomial in x of degree 3. If one value of y is incorrect, identify and correct it:

x	0	1	2	3	4	5	6
y	14	20	40	85	170	304	500

15. Find the missing term/terms in the following tables:
   a.

x	0	1	2	3	4
$f(x)$	1	3	9	—	81

b.

x	0	1	2	3	4	5
$f(x)$	0	—	8	15	—	35

c.

x	10	15	20	25	30	35
$f(x)$	19.97	21.51	—	23.52	24.65	—

d.

x	10	15	20	25	30	35	40
$f(x)$	270	—	222	200	—	164	148

16. Given $y_0 = 2$, $y_1 = 11$, $y_2 = 80$, $y_3 = 200$, $y_4 = 100$, and $y_5 = 8$ find $\nabla^5 y_5$.
    a. Without constructing the difference table
    b. By constructing the difference table
17. Find the functions whose first differences are as follows:
    a. $6x^2 + 2$
    b. $x^3 + 3x^2 + 5x + 12$
    c. $2x^3 - 3x^2 + 4x + 10$
18. If y is a polynomial of degree 3 and the values are as follows. Locate and correct the wrong value of y.

x	0	1	2	3	4	5	6	7
y	25	21	18	18	27	45	76	123

19. Locate and correct the error in the following table:

x	2.5	3.0	3.5	4.0	4.5	5.0	5.5
y	4.32	4.83	5.27	5.47	6.26	6.79	7.23

20. Find and correct an error in the following table:

x	0	1	2	3	4	5	6	7
y	0	0	1	6	24	60	120	210

21. The values of $f(x)$ given below are those of a polynomial of degree 4. Find $f(x)$ and hence evaluate $f(6)$.

x	0	1	2	3	4	5
$f(x)$	1	5	31	121	341	781

22. If $u_r(x) = (x - x_0)(x - x_1)(x - x_2)\ldots(x - x_{r-1})$ and $u_0(x) = 1$ prove that $\Delta^i u_r(x) = r(r-1)(r-2)\ldots(r-i+1)h^i u_{r-i}(x)$ $(1 \le i \le r)$ where $x_r = x_0 + rh$ and $h > 0$ $(r = 0, 1, 2, \ldots, n)$. Hence deduce $\Delta^n u_n(x) = n!h^n$.
23. The population of a town in the decennial census is given below. Estimate the population for the year 1905 and 1935.

Years (x)	1901	1911	1921	1931	1941
Population in thousands (y)	56	76	91	103	111

24. From the following table, find the number of students who obtained marks between 60 and 70.

Marks obtained	0–40	40–60	60–80	80–100	100–120
No. of students	250	120	100	70	50

25. Given the following table of values of $y = x^{1/2}$, find $(30)^{1/2}$ using (1) Lagrange's interpolation and (2) Newton's backward interpolation formula. Compare with the exact value. Also, calculate the truncation error.

x	25	27	29	31
y	5	5.196	5.385	5.568

26. Find $f(0.23)$ and $f(0.29)$ from the following table:

x	0.20	0.22	0.24	0.26	0.28	0.30
$f(x)$	1.6596	1.6698	1.6804	1.6912	1.7024	1.7139

27. From the following data

x	0	1	2	3	4
y	1	2	4	8	16

a. Find y at $x = 2.2$, using (1) Stirling, (2) Bessel, and (3) Everett interpolation formulae
b. Also, find x for $y = 3$. If the above table of values be those of $y = 2^x$, find an estimate of the truncation error

28. The following table gives the distance in nautical miles of the visible horizon for the given heights in feet above the earth's surface:

Height (x)	100	150	200	250	300	350	400
Distance (d)	10.66	13.06	15.07	16.84	18.45	19.93	21.30

Find the value of d when $x = 410$ feet.

29. Using Gauss forward formula find $f(30)$ from the following data

x	21	25	29	33	37
y	18.4708	17.8144	17.1070	16.3432	15.5154

30. Using Gauss's forward formula, find the value of $f(32)$ given that $f(25) = 0.2707$, $f(30) = 0.3027$, $f(35) = 0.3386$, and $f(40) = 0.3794$.

31. Find the value of cos $51^\circ 42'$ by using the Gauss backward interpolation formula from the following table:

x	50°	51°	52°	53°	54°
cos x	0.6428	0.6293	0.6157	0.6018	0.5878

32. State Gauss's backward formula and use it to find the value of $\sqrt{12525}$, given that $\sqrt{12500} = 111.8034$, $\sqrt{12510} = 111.8481$, $\sqrt{12520} = 111.8982$, $\sqrt{12530} = 111.937$, and $\sqrt{12540} = 111.9822$.

33. Construct a table of divided difference for the following data

x_i	0	1	3	4
y_i	1	−1	1	−1

34. Apply Newton's general interpolation formula to obtain the polynomial of degree ≤3 through the above points. Compare it with that obtained by using Lagrange's interpolation formula.

35. From the following table determine (1) $y(0.2)$ and (2) $y(0.75)$ using Everett's formula and Newton's formula. Also calculate the corresponding values using Stirling's and Bessel's interpolation formulae.

x	0	0.5	1.0	1.5
y	1.0	1.41421	2.0	2.82843

36. State Stirling's formula for interpolation at the middle of a table of values and find $e^{1.91}$ from the following table:

x	1.7	1.8	1.9	2.0	2.1	2.2
e^x	5.4739	6.0496	6.6859	7.3891	8.1662	9.0250

37. Show that $\sum_{i=0}^{n} \frac{\Pi_{n+1}(x)}{(x-x_i)\Pi'_{n+1}(x_i)} = 1$ where $\Pi_{n+1}(x) = (x-x_0)(x-x_1)(x-x_2)\ldots(x-x_n)$.
38. Use Lagrange's interpolation formula to express $(x^2+x+2)/[x(x-1)(x+1)]$ as a sum of partial fractions.
39. Find the interpolation polynomial for the following data:

x	0	2	4	6	8
y	100	113	152	256	524

by using
a. Newton's forward interpolation formula
b. Newton's backward interpolation formula and
c. Stirling's interpolation formula.
40. Using Bessel's formula find $y(62.5)$ given

x	60	61	62	63	64	65
y	7782	7853	7924	7993	8062	8129

41. Employ Bessel's formula to find the value $f(1.95)$ given that

x	1.7	1.8	1.9	2.0	2.1	2.2	2.3
$f(x)$	2.979	3.144	3.283	3.391	3.463	3.997	4.491

which other interpolation formula can be used here? Which is more appropriate? Give reasons.
42. Find y_{34} using Everett's formula, given that $y_{20} = 11.4699$, $y_{25} = 12.7834$, $y_{30} = 13.7648$, $y_{35} = 14.4982$, and $y_{40} = 15.0463$.
43. Using Everett's formula find $f(23.75)$ from the following data:

x	21	22	23	24	25	26
$f(x)$	1.3222	1.3424	1.3617	1.3802	1.3979	1.4150

44. Using Everett's formula, evaluate $f(25)$ from the set of values $f(20) = 2854$, $f(24) = 3162$, $f(28) - 3544$, and $f(32) = 3992$.
45. Deduce Everett's formula from Bessel's formula and show that Everett's formula truncated after second difference is equivalent to Bessel's formula truncated after third difference. Use Everett's formula to find $\cos(12.5°)$ given that $\cos(0°) = 1$, $\cos(5°) = 0.9962$, $\cos(10°) = 0.9848$, $\cos(15°) = 0.9659$, and $\cos(20°) = 09397$.
46. Establish Newton's divided difference interpolation formula and give an estimate of the remainder term. Deduce Newton's forward and backward difference interpolation formula as particular cases.
47. Given the set of tabulated point (0, 2), (1, 3), (2, 12), and (15, 3587) satisfying the function $y = f(x)$, compute $f(4)$ using Newton's divided difference formula.
48. Find $f(8)$ using Newton's divided difference formula given that

x	4	5	7	10	11	13
$f(x)$	48	100	294	900	1210	2028

49. Using inverse divided difference formula find x for which $y = 0$, given that

x	1	2	2.5	3
y	−6	−1	5.625	16

50. Using appropriate inverse interpolation formula find ln (1.2461) from the following data

x	0.0	0.1	0.2	0.3	0.4
e^x	1.0000	1.1052	1.2214	1.3499	1.4919

51. Using (1) Stirling's formula and (2) Bessel's formula, find $\log_{10} 337.5$ from the following table:

x	310	320	330	340	350	360
$\log_{10} x$	2.491362	2.505150	2.518514	2.531479	2.544068	2.556303

52. Estimate $\sqrt{1.12}$ using Stirling's formula from the following data:

x	1.00	1.05	1.10	1.15	1.20	1.25	1.30
$\sqrt{x}$	1.0000	1.0247	1.0488	1.0724	1.0954	1.1180	1.1402

53. Using Stirling's formula, find $\cos(17)$, given that $\cos(0) = 1$, $\cos(0.05) = 0.9988$, $\cos(0.10) = 0.9950$, $\cos(0.15) = 0.9888$, $\cos(0.20) = 0.9801$, $\cos(0.25) = 0.9689$, and $\cos(0.30) = 0.9553$.
54. State Bessel's formula for interpolation and mention its limitations. Use this formula to solve the problems in 17.
55. Following are the marks obtained by 590 students in a certain examination

Marks	Below 20	20–40	40–60	60–80	80–100
No. of students	250	120	100	70	50

Find out the number of candidate who has failed in the examination where pass mark is 30.
56. From the data given below, find the number of students whose weight is between 60 and 70.

Weight	0–40	40–60	60–80	80–100	100–120
No. of students	250	120	100	70	50

57. Find $\sin^{-1}(0.43837)$ using inverse interpolation formula, given that

x	0	10	20	30	40
$\sin x$	0	0.17365	0.34202	0.50000	0.64279

58. Verify the function defined by

$$s(x) = \begin{cases} -x^3 + \dfrac{17}{2}x^2 - 9x + \dfrac{3}{2}, & -1 \le x \le 1 \\ 2x^3 - \dfrac{1}{2}x^2 - \dfrac{3}{2}, & 1 \le x \le 2 \\ x^3 + \dfrac{11}{2}x^2 - 12x + \dfrac{13}{2}, & 2 \le x \le 4 \end{cases}$$

is a cubic spline.
59. Construct the Hermite interpolation polynomial that fits the data

x	$f(x)$	$f'(x)$
2	29	50
3	105	105

Interpolate $f(x)$ at $x = 2.5$.

60. Given the following table, compute $f(0.30)$ by Hermite interpolation formula.

x	$f(x)$	$f'(x)$
0.20	0.08371	2.5000
0.25	0.30685	2.0000
0.35	0.64333	1.4286
0.45	0.77686	1.2500

61. Construct the Hermite interpolation polynomial that fits the data

x	$f(x)$	$f'(x)$
0	0	1.0000
0.5	0.4794	0.8776
1.0	0.8415	0.5403

Estimate the value of $f(0.75)$. Find a bound on the error. If the data represents the function $f(x)=\sin x$, then find the actual error at $x=0.75$.

62. Obtain the cubic spline fit for the data

x	0	1	2	3
$f(x)$	1	4	10	8

under the end conditions $f''(0)=0=f''(3)$ and valid in the interval [1, 2]. Hence, obtain the estimate of $f(1.5)$.

63. Find whether the following functions are splines or not.

a. $f(x)=\begin{cases} x^2-x+1, & 1\le x\le 2 \\ 3x+3, & 2\le x\le 3 \end{cases}$

b. $f(x)=\begin{cases} -x^2-2x^3, & -1\le x\le 0 \\ -x^2+2x^3, & 0\le x\le 1 \end{cases}$

c. $f(x)=\begin{cases} -x^2-2x^3, & -1\le x\le 0 \\ x^2+2x^3, & 0\le x\le 1 \end{cases}$

64. Find the Hermite polynomial of the third degree approximating the function $y(x)$ such that

$$y(0)=1,\quad y'(0)=0$$

$$y(1)=3,\quad y'(1)=5$$

65. Fit the following four points by the cubic splines:

x	1	2	3	4
y	1	5	11	8

Use the end conditions $y''(1)=0=y''(4)$. Hence, compute (1) $y(1.5)$ and (2) $y'(2.5)$.

66. The following table gives the normal weight of babies during the first 12 months of life.

Age (m) (in months)	0	2	5	8	10	12
Weights (w) (in lbs)	7.5	10.25	15	16	18	21

Estimate the weight of baby at the age of 7 months.

67. Obtain the natural cubic spline that satisfies the following data:

x	2	3	4
y	11	49	123

68. Obtain the cubic spline approximations that fit the given data.

x	−1	0	1	2
$f(x)$	5	−2	−7	2

Assume $f'(-1)=1$ and $f'(2)=1$.

69. From the table given below interpolate by Lagrange's formula the number of persons under the age of 30.

Age (in years)	10–15	15–20	20–25	25–35	35–45	45–55
Person (in thousands)	193.5	880	923	1636	1221	830

70. Construct the interpolating cubic spline with (1) free and (2) clamped boundary conditions to the function $f(x)=x^4$ using the following data:

x_i	−1	−0.9	−0.7	−0.1	0.2	0.6	0.8	1
$f(x_i)$	1	0.6561	0.2400	0.0001	0.0016	0.1296	0.4096	1

71. Find the Hermite polynomial for the following data and hence find the approximate value at $x=1.5$.

x	−1	0	1
$f(x)$	1	1	3
$f'(x)$	−5	1	7

72. Using Hermite interpolation formula find the value of $f(3.2)$ from the following data:

x	3.0	3.5	4.0
y	1.09861	1.25276	1.38629
y'	0.33333	0.28571	0.25000

73. Obtain the cubic spline approximation valid in [3,4], for the function given in the tabular form

X	1	2	3	4
$f(x)$	3	10	29	65

under the natural spline conditions $f''(1)=0$ and $f''(4)=0$.

74. Fit a cubic spline, $s(x)$ to the function $f(x)=x^4$ on the interval $-1\le x\le 1$ corresponding to the partition $x_0=-1$, $x_1=0$, $x_2=1$ and satisfying the conditions $s'(-1)=f'(-1)$ and $s'(1)=f'(1)$.

4 Numerical Differentiation

4.1 INTRODUCTION

By the problem of numerical differentiation, we mean computing the values of the different-order derivatives of the function $f(x)$ for a given value of x in terms of a given set of tabulated values. Since using the method of polynomial interpolation for a function $f(x)$, which is continuous and differentiable in any interval $[a, b]$, we can find an interpolating polynomial $\phi_n(x)$, which can approximate values of $f(x)$ at any point within $[a, b]$ in terms of some given values of $f(x)$ in that interval. The approximated values of $f'(x)$, $f''(x)$, and so on would be the values of the derivatives of $\phi_n(x)$ for required number of times. The choice of the interpolating polynomial $\phi_n(x)$, which is to be differentiated, depends on the position of the point in the given set of values, at which the value of the derivatives is to be obtained.

Let (x_i, y_i), $i = 0(1)n$, be a given set of values of a function defined over an interval $[a, b]$ in which it is continuous and differentiable.

If the interpolating points are equally spaced, then we may use either of Newton's, Stirling's, or Bessel's interpolation formula for numerical differentiation. On the other hand, if the interpolating points are unequally spaced, then we may use either Newton's divided difference or Lagrange's interpolation formula.

The choice of the formula is the same as in the case of interpolation. For example, Newton's forward formula may be used when derivatives are to be obtained at a point near the beginning of the set of tabulated values. On the other hand, Newton's backward interpolation formula may be used when derivatives are to be obtained at a point near the end of the set of tabulated values. If the derivatives are required near the middle of the set of tabulated values, then we shall employ Stirling's or Bessel's interpolation formula.

4.2 ERRORS IN COMPUTATION OF DERIVATIVES

Let

$$f(x) = \phi_n(x) + R_{n+1}(x) \tag{4.1}$$

Then from Equation 3.7 of Chapter 3, we have

$$R_{n+1}(x) = \Omega(x)\frac{f^{(n+1)}(\xi)}{(n+1)!}$$

where:

$\Omega(x) = (x - x_0)(x - x_1)\ldots(x - x_n)$, $\min\{x, x_0, x_1, \ldots, x_n\} < \xi < \max\{x, x_0, x_1, \ldots, x_n\}$

$\xi = \xi(x)$

Differentiating both sides of Equation 4.1 with respect to x, we get

$$f'(x) = \phi'_n(x) + R'_{n+1}(x) \tag{4.2}$$

where:

$$R'_{n+1}(x) = \Omega'(x)\frac{f^{(n+1)}(\xi)}{(n+1)!} + \Omega(x)\frac{f^{(n+2)}(\xi)}{(n+1)!}\xi'(x) \tag{4.3}$$

where $\min\{x, x_0, x_1, \ldots, x_n\} < \xi < \max\{x, x_0, x_1, \ldots, x_n\}$ and $\xi = \xi(x)$ is unknown.

In practical, the function $\xi(x)$ being unknown, its bounds cannot be determined. But if we want to find out the value of the error term at a node $x = x_i$, $i = 0,1,\ldots,n$, then the second term in Equation 4.3 vanishes (since $\Omega(x_i) = 0$, $i = 0,1,\ldots,n$), and we get

$$R'_{n+1}(x_i) = \Omega'(x_i)\frac{f^{(n+1)}(\xi_i)}{(n+1)!}, \quad i = 0,1,\ldots,n \tag{4.4}$$

where $\xi_i = \xi(x_i)$ lies between the smallest and the largest values of $x_0, x_1, \ldots, x_n$.

A more convenient formula can be determined by taking $R_{n+1}(x)$ in the following divided difference form as

$$R_{n+1}(x) = \Omega(x) f[x, x_0, \ldots, x_n] \tag{4.5}$$

Differentiating both sides of Equation 4.5 with respect to x, we get

$$R'_{n+1}(x) = \Omega'(x) f[x, x_0, \ldots, x_n] + \Omega(x) f[x, x, x_0, \ldots, x_n] \tag{4.6}$$

since $f[x, x, x_0, \ldots, x_n] = \dfrac{d}{dx} f[x, x_0, \ldots, x_n]$

Now, differentiating both sides of Equation 4.5 k times with respect to x, we get by Leibnitz's theorem

$$R_{n+1}^{(k)}(x) = \sum_{i=0}^{k} \binom{k}{i} \Omega^{(i)}(x) \frac{d^{k-i}}{dx^{k-i}} f[x, x_0, \ldots, x_n] \tag{4.7}$$

According to Equation 3.97 of Chapter 3, we have

$$\frac{d^{k-i}}{dx^{k-i}} f[x, x_0, \ldots, x_n] = (k-i)!\, f[x, x, \ldots, x, x_0, \ldots, x_n]$$

where x occurs $(k-i+1)$ times in the argument of $f[x, x, \ldots, x, x_0, x_1, \ldots, x_n]$.

Therefore, from Equation 4.6, we get

$$R_{n+1}^{(k)}(x) = \sum_{i=0}^{k} \frac{k!}{i!} \Omega^{(i)}(x) f[x, x, \ldots, x, x_0, \ldots, x_n] \tag{4.8}$$

Again according to Equation 3.96 of Chapter 3, we have

$$f[x, x, \ldots, x, x_0, \ldots, x_n] = \frac{f^{(n+k-i+1)}(\xi)}{(n+k-i+1)!}$$

where $\min\{x, x_0, x_1, \ldots, x_n\} < \xi < \max\{x, x_0, x_1, \ldots, x_n\}$.

Hence, from Equation 4.7, we obtain

$$R_{n+1}^{(k)}(x) = \sum_{i=0}^{k} \frac{k!}{i!(n+k-i+1)!} \Omega^{(i)}(x) f^{(n+k-i+1)}(\xi) \tag{4.9}$$

where $\min\{x, x_0, x_1, \ldots, x_n\} < \xi < \max\{x, x_0, x_1, \ldots, x_n\}$.

4.3 NUMERICAL DIFFERENTIATION FOR EQUISPACED NODES

Let $y = f(x)$ be a continuous function defined over a closed interval $[a,b]$ having continuous derivatives, and let values $y_i = f(x_i)$ of this function are known at the equispaced nodes $x_i \in [a,b]$, $i = 0,1,\ldots,n$. We now propose to find a numerical formula on the basis of the given values to find the value of the derivatives of $f(x)$ at points close to x_0 and x_n, respectively.

4.3.1 Formulae Based on Newton's Forward Interpolation

We consider Newton's forward interpolation formula, which gives values of $f(x)$ at points near x_0 in terms of the given tabulated values. This interpolation formula without error term is

$$f(x) \approx \varphi_n(x) = y_0 + u\Delta y_0 + \frac{u(u-1)}{2!}\Delta^2 y_0 + \cdots + \frac{u(u-1)(u-2)(u-n+1)}{n!}\Delta^n y_0 \tag{4.10}$$

where $u = (x - x_0)/h$ and $x_i = x_0 + ih$, $i = 0,1,\ldots,n$.

Since Equation 4.10 is a polynomial in u, that is, in x, it is differentiable any number of times. Then differentiating both sides of Equation 4.10 with respect to x, we get

$$\frac{dy}{dx} = f'(x) \approx \varphi_n'(x) = \frac{d}{du}\left[y_0 + u\Delta y_0 + \frac{u(u-1)}{2!}\Delta^2 y_0 + \cdots + \frac{u(u-1)(u-2)}{3!}\Delta^3 y_0 + \cdots\right]\frac{du}{dx}$$

$$= \frac{1}{h}\left[\Delta y_0 + \frac{2u-1}{2!}\Delta^2 y_0 + \frac{3u^2-6u+2}{3!}\Delta^3 y_0 + \frac{(4u^3-18u^2+22u-6)}{4!}\Delta^4 y_0 + \cdots\right], \tag{4.11}$$

since $\dfrac{du}{dx} = \dfrac{1}{h}$

This formula may be used for computing approximate values of the derivatives of $f(x)$ at a point near x_0.

If we want to find the approximate value of $f'(x_0)$, then as $u = \left[(x - x_0)/h\right] = 0$, we obtain from Equation 4.11

$$\left.\frac{dy}{dx}\right|_{x=x_0} = f'(x_0) \approx \varphi_n'(x_0) = \frac{1}{h}\left[\Delta y_0 - \frac{1}{2}\Delta^2 y_0 + \frac{1}{3}\Delta^3 y_0 - \frac{1}{4}\Delta^4 y_0 + \frac{1}{5}\Delta^5 y_0 - \cdots\right] \tag{4.12}$$

Similarly, differentiating both sides of Equation 4.11, we get

$$\frac{d^2y}{dx^2} = f''(x) \approx \varphi_n''(x) = \frac{1}{h^2}\left[\Delta^2 y_0 + (u-1)\Delta^3 y_0 + \frac{6u^2-18u+11}{12}\Delta^4 y_0 + \cdots\right] \tag{4.13}$$

If $x = x_0$, that is, $u = 0$, then from Equation 4.13, we get

$$\left.\frac{d^2y}{dx^2}\right|_{x=x_0} = f''(x_0) \approx \varphi_n''(x_0)$$

$$= \frac{1}{h^2}\left[\Delta^2 y_0 - \Delta^3 y_0 + \frac{11}{12}\Delta^4 y_0 - \frac{5}{6}\Delta^5 y_0 + \frac{137}{180}\Delta^6 y_0 - \frac{7}{10}\Delta^7 y_0 + \frac{363}{560}\Delta^8 y_0 - \cdots\right] \tag{4.14}$$

Again, differentiating both sides of Equation 4.13, we get

$$\frac{d^3y}{dx^3} = f'''(x) \approx \varphi_n'''(x) = \frac{1}{h^3}\left[\Delta^3 y_0 + \frac{12u-18}{12}\Delta^4 y_0 + \cdots\right] \tag{4.15}$$

In particular, if $x = x_0$, that is, $u = 0$, then from Equation 4.15, we get

$$\left.\frac{d^3y}{dx^3}\right|_{x=x_0} = f'''(x_0) \approx \varphi'''_n(x_0) = \frac{1}{h^3}\left[\Delta^3 y_0 - \frac{3}{2}\Delta^4 y_0 + \frac{7}{4}\Delta^5 y_0 - \cdots\right] \tag{4.16}$$

It is possible to find a numerical formula for finding approximate values of the derivatives of any order for $f(x)$ in terms of the given tabulated values. The above formulae for numerical differentiation are used only at the points near the beginning of the set of tabulated values.

4.3.1.1 Error Estimate

From Equation 4.3, we get the general error term for the first-order differentiation formula as follows:

$$R'_{n+1}(x) = \Omega'(x)\frac{f^{(n+1)}(\xi)}{(n+1)!} + \Omega(x)\frac{f^{(n+2)}(\xi)}{(n+1)!}\xi'(x) \tag{4.17}$$

where $\min\{x, x_0, x_1, \ldots, x_n\} < \xi < \max\{x, x_0, x_1, \ldots, x_n\}$.

Now,

$$\begin{aligned}\Omega'(x) &= \frac{d}{dx}\left[(x-x_0)(x-x_1)\ldots(x-x_n)\right]\\ &= \frac{d}{du}\left[uh(u-1)h\ldots(u-n)h\right]\frac{du}{dx}, \quad \text{where} \quad u = \frac{x-x_0}{h}\\ &= h^{n+1}\frac{d}{du}\left[u(u-1)\ldots(u-n)\right]\frac{1}{h}\\ &= h^n\frac{d}{du}\left[u(u-1)\ldots(u-n)\right]\end{aligned}$$

and

$$\frac{f^{(n+2)}(\xi)}{(n+1)!}\xi'(x) = \frac{d}{dx}\left[\frac{f^{(n+1)}(\xi)}{(n+1)!}\right] = \frac{1}{(n+1)!h}\frac{d}{du}\left[f^{(n+1)}(\xi)\right]$$

Therefore, the general error term for the first-order numerical differentiation formula is given by

$$\begin{aligned}R'_{n+1}(x) = {} & \frac{h^n}{(n+1)!}f^{(n+1)}(\xi)\frac{d}{du}\{u(u-1)\ldots(u-n)\}\\ & + \frac{u(u-1)(u-2)\ldots(u-n)}{(n+1)!}h^{n+1}f^{(n+2)}(\xi)\xi'(x)\end{aligned} \tag{4.18}$$

$$= \frac{h^n}{(n+1)!}\left[f^{(n+1)}(\xi)\frac{d}{du}\{u(u-1)\ldots(u-n)\} + u(u-1)(u-2)\ldots(u-n)\frac{d}{du}\left[f^{(n+1)}(\xi)\right]\right] \tag{4.19}$$

This is the error term for the formula in Equation 4.11.

We can obtain another form of general error term for the first-order numerical differentiation formula. Using Equation 3.89 of Chapter 3 and Equation 4.5, we get

$$\begin{aligned}R'_{n+1}(x) &= \Omega'(x)f[x, x_0, \ldots, x_n] + \Omega(x)f[x, x, x_0, \ldots, x_n]\\ &= h^n\frac{d}{du}\left[u(u-1)\ldots(u-n)\right]\frac{f^{(n+1)}(\xi)}{(n+1)!} + u(u-1)(u-2)\ldots(u-n)h^{n+1}\frac{f^{(n+2)}(\xi')}{(n+2)!}\end{aligned} \tag{4.20}$$

where:

$$f[x, x_0, \ldots, x_n] = \frac{f^{(n+1)}(\xi)}{(n+1)!}$$

$$f[x, x, x_0, \ldots, x_n] = \frac{f^{(n+2)}(\xi')}{(n+2)!}$$

$$\min\{x, x_0, x_1, \ldots, x_n\} < \xi, \xi' < \max\{x, x_0, x_1, \ldots, x_n\}$$

Now, assuming $f^{(n+1)}(x)$ and $f^{(n+2)}(x)$ do not vary strongly and h is very small, by the use of Equation 3.90 of Chapter 3, from Equation 4.20 we obtain

$$\begin{aligned} R'_{n+1}(x) &\approx \frac{h^n}{(n+1)!}\frac{\Delta^{n+1}y_0}{h^{n+1}}\frac{d}{du}\{u(u-1)\ldots(u-n)\} + \frac{u(u-1)(u-2)\ldots(u-n)}{(n+2)!}h^{n+1}\frac{\Delta^{n+2}y_0}{h^{n+2}} \\ &= \frac{1}{h}\left[\frac{d}{du}\binom{u}{n+1}\Delta^{n+1}y_0 + \binom{u}{n+1}\frac{\Delta^{n+2}y_0}{(n+2)}\right] \end{aligned} \tag{4.21}$$

This is another form of error term for the formula in Equation 4.11.

In particular, at $x = x_0$, that is, $u = 0$, the second term in Equation 4.21 vanishes, and thus, we obtain

$$R'_{n+1}(x_0) \approx \frac{(-1)^n \Delta^{n+1} y_0}{(n+1)h} \tag{4.22}$$

In a similar manner, we can find error terms for higher-order numerical differentiation formulae.

4.3.2 Formulae Based on Newton's Backward Interpolation

Let us now find out a numerical formula for finding approximate values of the derivatives of $f(x)$ at a point near the end of the given values, that is, near x_n. For this purpose, we consider Newton's backward interpolation formula without error term as

$$\begin{aligned} f(x) \approx \varphi_n(x) &= y_n + u\nabla y_n + \frac{u(u+1)}{2!}\nabla^2 y_n + \frac{u(u+1)(u+2)}{3!}\nabla^3 y_n \\ &+ \cdots + \frac{u(u+1)\ldots(u+n-1)}{n!}\nabla^n y_n \end{aligned} \tag{4.23}$$

where $u = (x - x_n)/h$ so that $(du/dx) = (1/h)$.

Next differentiating both sides of Equation 4.23 with respect to x, we get the differentiation formula as

$$\begin{aligned} \frac{dy}{dx} &= f'(x) \approx \varphi'_n(x) \\ &= \frac{1}{h}\left[\nabla y_n + \frac{2u+1}{2!}\nabla^2 y_n + \frac{3u^2+6u+2}{3!}\nabla^3 y_n + \frac{4u^3+18u^2+22u+6}{4!}\nabla^4 y_n + \cdots\right] \end{aligned} \tag{4.24}$$

At $x = x_n$, that is, $u = 0$, the above formula reduces to

$$\left.\frac{dy}{dx}\right|_{x=x_n} = f'(x_n) \approx \varphi_n'(x_n) = \frac{1}{h}\left[\nabla y_n + \frac{1}{2}\nabla^2 y_n + \frac{1}{3}\nabla^3 y_n + \frac{1}{4}\nabla^4 y_n + \frac{1}{5}\nabla^5 y_n + \cdots\right] \tag{4.25}$$

Again differentiating both sides of Equation 4.24 with respect to x, we get

$$\frac{d^2 y}{dx^2} = f''(x) \approx \varphi''(x) = \frac{1}{h^2}\left[\nabla^2 y_n + (u+1)\nabla^3 y_n + \frac{6u^2+18u+11}{12}\nabla^4 y_n + \cdots\right] \tag{4.26}$$

At $x = x_n$, that is, $u = 0$, the above formula reduces to

$$\begin{aligned}\left.\frac{d^2y}{dx^2}\right|_{x=x_n} &= f''(x_n) \approx \varphi''(x_n) \\ &= \frac{1}{h^2}\left[\nabla^2 y_n + \nabla^3 y_n + \frac{11}{12}\nabla^4 y_n + \frac{5}{6}\nabla^5 y_n + \frac{137}{180}\nabla^6 y_n + \frac{7}{10}\nabla^7 y_n + \frac{363}{560}\nabla^8 y_n + \cdots\right]\end{aligned} \tag{4.27}$$

Similarly, we may obtain formulae for higher-order differentiations. The above formulae for numerical differentiation are used only at the points near the end of the set of tabulated values.

4.3.2.1 Error Estimate

From Equation 4.3, we get the general error term for the first-order differentiation formula as follows:

$$R_{n+1}'(x) = \Omega'(x)\frac{f^{(n+1)}(\xi)}{(n+1)!} + \Omega(x)\frac{f^{(n+2)}(\xi)}{(n+1)!}\xi'(x) \tag{4.28}$$

where $\min\{x, x_0, x_1, \ldots, x_n\} < \xi < \max\{x, x_0, x_1, \ldots, x_n\}$

Now,

$$\begin{aligned}\Omega'(x) &= \frac{d}{dx}\left[(x-x_0)(x-x_1)\ldots(x-x_n)\right] \\ &= \frac{d}{du}\left[uh(u+1)h\ldots(u+n)h\right]\frac{du}{dx}, \quad \text{where} \quad u = \frac{x-x_n}{h} \\ &= h^{n+1}\frac{d}{du}\left[u(u+1)\ldots(u+n)\right]\frac{1}{h} \\ &= h^n\frac{d}{du}\left[u(u+1)\ldots(u+n)\right]\end{aligned}$$

and

$$\frac{f^{(n+2)}(\xi)}{(n+1)!}\xi'(x) = \frac{d}{dx}\left[\frac{f^{(n+1)}(\xi)}{(n+1)!}\right] = \frac{1}{(n+1)!h}\frac{d}{du}\left[f^{(n+1)}(\xi)\right]$$

Therefore, the general error term for the first-order numerical differentiation formula is given by

$$\begin{aligned}R_{n+1}'(x) &= \frac{h^n}{(n+1)!}f^{(n+1)}(\xi)\frac{d}{du}\{u(u+1)\ldots(u+n)\} \\ &\quad + \frac{u(u+1)(u+2)\ldots(u+n)}{(n+1)!}h^{n+1}f^{(n+2)}(\xi)\xi'(x)\end{aligned} \tag{4.29}$$

$$= \frac{h^n}{(n+1)!}\left[f^{(n+1)}(\xi)\frac{d}{du}\{u(u+1)\ldots(u+n)\} + u(u+1)(u+2)\ldots(u+n)\frac{d}{du}\left[f^{(n+1)}(\xi)\right]\right] \quad (4.30)$$

This is the error term for the formula in Equation 4.24.

We can obtain another form of general error term for the first-order numerical differentiation formula. Using Equation 3.89 of Chapter 3 and Equation 4.5, we get

$$R'_{n+1}(x) = \Omega'(x) f[x, x_0, \ldots, x_n] + \Omega(x) f[x, x, x_0, \ldots, x_n]$$

$$= h^n \frac{d}{du}\left[u(u-1)\ldots(u-n)\right]\frac{f^{(n+1)}(\xi)}{(n+1)!} + u(u-1)(u-2)\ldots(u-n)h^{n+1}\frac{f^{(n+2)}(\xi')}{(n+2)!} \quad (4.31)$$

where:

$$f[x, x_0, \ldots, x_n] = \frac{f^{(n+1)}(\xi)}{(n+1)!}$$

$$f[x, x, x_0, \ldots, x_n] = \frac{f^{(n+2)}(\xi')}{(n+2)!}$$

$$\min\{x, x_0, x_1, \ldots, x_n\} < \xi, \xi' < \max\{x, x_0, x_1, \ldots, x_n\}$$

Now, assuming $f^{(n+1)}(x)$ and $f^{(n+2)}(x)$ do not vary strongly and h is very small, by the use of Equation 3.90 of Chapter 3, from Equation 4.31 we obtain

$$R'_{n+1}(x) \approx \frac{h^n}{(n+1)!}\frac{\Delta^{n+1}y_0}{h^{n+1}}\frac{d}{du}\{u(u+1)\ldots(u+n)\} + \frac{u(u+1)(u+2)\ldots(u+n)}{(n+2)!}h^{n+1}\frac{\Delta^{n+2}y_0}{h^{n+2}}$$

$$= \frac{1}{h}\left[\frac{d}{du}\binom{u+n}{n+1}\Delta^{n+1}y_0 + \binom{u+n}{n+1}\frac{\Delta^{n+2}y_0}{(n+2)}\right] \quad (4.32)$$

This is another form of error term for the formula in Equation 4.24.

In particular, at $x = x_n$, that is, $u = 0$, the second term in Equation 4.32 vanishes, and thus, we obtain

$$R'_{n+1}(x_n) \approx \frac{\Delta^{n+1}y_0}{(n+1)h} \quad (4.33)$$

In a similar manner, we can find error terms for higher-order numerical differentiation formulae.

Example 4.1

Determine the derivatives of the function $y = f(x)$ at the points $x = 0$, 1.2, 3.4, and 4, respectively, from the following table:

x	0	1	2	3	4
y	0.3010	0.4771	0.6020	0.6990	0.7782

Solution:

The forward difference table is as follows:

x	y	Δy	$\Delta^2 y$	$\Delta^3 y$	$\Delta^4 y$
0	<u>0.3010</u>				
		<u>0.1761</u>			
1	0.4771		<u>−0.0512</u>		
		0.1249		<u>0.0233</u>	
2	0.6020		−0.0279		<u>−0.0132</u>
		0.0970		<u>0.0101</u>	
3	0.6990		<u>−0.0178</u>		
		<u>0.0792</u>			
4	<u>0.7782</u>				

Since, $x = 0$ and 1.2 are near the beginning of the table, we use Newton's forward interpolation formula. From the above forward difference table, only underlined upper portion values will be used in the Newton's forward interpolation formula.

1. When $x = 0$, then

$$u = \frac{x - x_0}{h} = 0$$

Using Equation 4.12, we get

$$\left.\frac{dy}{dx}\right|_{x=0} = f'(0) \approx \frac{1}{h}\left[\Delta y_0 - \frac{1}{2}\Delta^2 y_0 + \frac{1}{3}\Delta^3 y_0 - \frac{1}{4}\Delta^4 y_0 - \cdots\right]$$

$$= \frac{1}{1}\left[0.1761 - \frac{1}{2}\times(-0.0512) + \frac{1}{3}\times(0.0233) - \frac{1}{4}\times(-0.0132)\right]$$

$$= 0.212767$$

Again, using Equation 4.14, we get

$$\left.\frac{d^2y}{dx^2}\right|_{x=0} = f''(0) \approx \frac{1}{h^2}\left[\Delta^2 y_0 - \Delta^3 y_0 + \frac{11}{12}\Delta^4 y_0 - \cdots\right]$$

$$= \frac{1}{1^2}\left[-0.0512 - 0.0233 + \frac{11}{12}\times(-0.0132)\right]$$

$$= -0.0866$$

2. When $x = 1.2$, then

$$u = \frac{x - x_0}{h} = 1.2$$

Using Equation 4.11, we get

$$\left.\frac{dy}{dx}\right|_{x=1.2} = f'(1.2) \approx \frac{1}{h}\left[\Delta y_0 + \frac{2u-1}{2!}\Delta^2 y_0 + \frac{3u^2-6u+2}{3!}\Delta^3 y_0 + \frac{(4u^3-18u^2+22u-6)}{4!}\Delta^4 y_0 + \cdots\right]_{u=1.2}$$

$$= \frac{1}{1}\left[0.1761 + \frac{2\times1.2-1}{2}\times(-0.0512) + \frac{3\times1.44-6\times1.2+2}{6}\times(0.0233)\right.$$

$$\left. + \frac{4\times1.728-18\times1.44+22\times1.2-6}{24}\times(-0.0132)\right]$$

$$= 0.136077$$

Again, using Equation 4.13, we get

$$\left.\frac{d^2y}{dx^2}\right|_{x=1.2} = f''(1.2) \approx \frac{1}{h^2}\left[\Delta^2 y_0 + (u-1)\Delta^3 y_0 + \frac{6u^2-18u+11}{12}\Delta^4 y_0 + \cdots\right]_{u=1.2}$$

$$= \frac{1}{1^2}\left[-0.0512 + (1.2-1)\times0.0233 + \frac{6\times1.44-18\times1.2+11}{12}\times(-0.0132)\right]$$

$$= -0.044384$$

Since, $x = 3.4$ and 4 are near the end of the table, we use Newton's backward interpolation formula. From the above forward difference table, only underlined lower portion values will be used in the Newton's backward interpolation formula.

3. When $x = 4$, then

$$u = \frac{x-x_n}{h} = 0$$

Using Equation 4.25, we get

$$\left.\frac{dy}{dx}\right|_{x=4} = f'(4) \approx \frac{1}{h}\left[\nabla y_n + \frac{1}{2}\nabla^2 y_n + \frac{1}{3}\nabla^3 y_n + \frac{1}{4}\nabla^4 y_n + \cdots\right]$$

$$= \frac{1}{h}\left[\Delta y_{n-1} + \frac{1}{2}\Delta^2 y_{n-2} + \frac{1}{3}\Delta^3 y_{n-3} + \frac{1}{4}\Delta^4 y_{n-4} + \cdots\right], \quad \text{since } \nabla^r y_n = \Delta^r y_{n-r}$$

$$= \frac{1}{1}\left[0.0792 + \frac{1}{2}\times(-0.0178) + \frac{1}{3}\times(0.0101) + \frac{1}{4}\times(-0.0132)\right]$$

$$= 0.0703667$$

Again, using Equation 4.27, we get

$$\left.\frac{d^2y}{dx^2}\right|_{x=4} = f''(4) \approx \frac{1}{h^2}\left[\nabla^2 y_n + \nabla^3 y_n + \frac{11}{12}\nabla^4 y_n + \cdots\right]$$

$$= \frac{1}{1^2}\left[-0.0178 + 0.0101 + \frac{11}{12}\times(-0.0132)\right]$$

$$= -0.0198$$

4. When $x = 3.4$, then

$$u = \frac{x-x_n}{h} = \frac{3.4-4}{1} = -0.6$$

Using Equation 4.24, we get

$$\left.\frac{dy}{dx}\right|_{x=3.4} = f'(3.4) \approx \frac{1}{h}\left[\nabla y_n + \frac{2u+1}{2!}\nabla^2 y_n + \frac{3u^2+6u+2}{3!}\nabla^3 y_n + \frac{4u^3+18u^2+22u+6}{4!}\nabla^4 y_n + \cdots\right]_{u=-0.6}$$

$$= \frac{1}{h}\left[\Delta y_{n-1} + \frac{2u+1}{2!}\Delta^2 y_{n-2} + \frac{3u^2+6u+2}{3!}\Delta^3 y_{n-3} + \frac{4u^3+18u^2+22u+6}{4!}\Delta^4 y_{n-4} + \cdots\right]_{u=-0.6}$$

$$= \frac{1}{1}\left[0.0792 + \frac{2\times(-0.6)+1}{2}\times(-0.0178) + \frac{3\times 0.36 - 6\times 0.6 + 2}{6}\times(0.0101)\right.$$

$$\left. + \frac{4\times(-0.216)+18\times 0.36 - 22\times 0.6 + 6}{24}\times(-0.0132)\right]$$

$$= 0.0809759$$

Again, using Equation 4.26, we get

$$\left.\frac{d^2y}{dx^2}\right|_{x=3.4} = f''(3.4) \approx \frac{1}{h^2}\left[\nabla^2 y_n + (u+1)\nabla^3 y_n + \frac{6u^2+18u+11}{12}\nabla^4 y_n + \cdots\right]_{u=-0.6}$$

$$= \frac{1}{h^2}\left[\Delta^2 y_{n-2} + (u+1)\Delta^3 y_{n-3} + \frac{6u^2+18u+11}{12}\Delta^4 y_{n-4} + \cdots\right]_{u=-0.6}$$

$$= \frac{1}{1^2}\left[(-0.0178) + (-0.6+1)\times 0.0101 + \frac{6\times 0.36 - 18\times 0.6 + 11}{12}\times(-0.0132)\right]$$

$$= -0.016356$$

4.3.3 Formulae Based on Stirling's Interpolation

Let the values of the function $y = f(x)$ be known for the following odd number of equispaced arguments:

$$x_{-n} = x_0 - nh, \ldots,\ x_{-2} = x_0 - 2h,\ x_{-1} = x_0 - h, x_0 = x_0, x_1 = x_0 + h, x_2 = x_0 + 2h, \ldots, x_n = x_0 + nh$$

Then, from Equation 3.108 of Chapter 3, Stirling's interpolation formula is given by

$$\begin{aligned} y = f(x) \cong y_0 &+ \frac{u(\Delta y_0 + \Delta y_{-1})}{2} + \frac{u^2}{2}\Delta^2 y_{-1} + \frac{u(u^2-1^2)}{3!}\frac{\Delta^3 y_{-2} + \Delta^3 y_{-1}}{2} + \frac{u^2(u^2-1^2)}{4!}\Delta^4 y_{-2} \\ &+ \cdots + \frac{u^2(u^2-1^2)(u^2-2^2)\ldots(u^2-(n-1)^2)}{2n\,!}\Delta^{2n} y_{-n} \end{aligned} \tag{4.34}$$

where $u = (x - x_0)/h$.

Differentiating both sides of Equation 4.34 with respect to x, we have

$$\begin{aligned} \frac{dy}{dx} = f'(x) \cong \frac{1}{h}\Bigg[& \frac{(\Delta y_{-1} + \Delta y_0)}{2} + u\Delta^2 y_{-1} + \frac{3u^2-1}{3!}\frac{\Delta^3 y_{-2} + \Delta^3 y_{-1}}{2} \\ & + \frac{4u^3-2u}{4!}\Delta^4 y_{-2} + \frac{5u^4-15u^2+4}{5!}\frac{\Delta^5 y_{-3} + \Delta^5 y_{-2}}{2} + \cdots\Bigg] \end{aligned} \tag{4.35}$$

Again, differentiating both sides of Equation 4.35 with respect to x, we get

$$\frac{d^2y}{dx^2} = f''(x) \cong \frac{1}{h^2}\left[\Delta^2 y_{-1} + u\frac{\Delta^3 y_{-2} + \Delta^3 y_{-1}}{2} + \frac{12u^2 - 2}{4!}\Delta^4 y_{-2} + \frac{20u^3 - 30u}{5!}\frac{\Delta^5 y_{-3} + \Delta^5 y_{-2}}{2} + \cdots\right] \quad (4.36)$$

and so on.

The above formulae for numerical differentiation are used when the point x is near the middle of the set of tabulated values.

In particular, when $x = x_0$, that is, $u = 0$, we get

$$\left.\frac{dy}{dx}\right|_{x=x_0} = f'(x_0) \cong \frac{1}{h}\left[\frac{(\Delta y_{-1} + \Delta y_0)}{2} - \frac{1}{3!}\frac{\Delta^3 y_{-2} + \Delta^3 y_{-1}}{2} + \frac{4}{5!}\frac{\Delta^5 y_{-3} + \Delta^5 y_{-2}}{2} + \cdots\right] \quad (4.37)$$

$$\left.\frac{d^2y}{dx^2}\right|_{x=x_0} = f''(x_0) \cong \frac{1}{h^2}\left[\Delta^2 y_{-1} - \frac{1}{12}\Delta^4 y_{-2} + \cdots\right] \quad (4.38)$$

and so on.

4.3.3.1 Error Estimate

From Equation 3.109 of Chapter 3, we have the error term as follows:

$$R_{2n+1}(x) = \frac{u(u^2 - 1^2)(u^2 - 2^2)\ldots(u^2 - n^2)}{(2n+1)!}h^{2n+1}f^{(2n+1)}(\xi) \quad (4.39)$$

where $\min\{x_{-n}, x, x_n\} < \xi < \max\{x_{-n}, x, x_n\}$ and $\xi = \xi(x)$.

Differentiating both sides of Equation 4.39 with respect to x, we get

$$R'_{2n+1}(x) = \frac{h^{2n}f^{(2n+1)}(\xi)}{(2n+1)!}\frac{d}{du}\left\{u(u^2 - 1^2)(u^2 - 2^2)\ldots(u^2 - n^2)\right\} + \frac{u(u^2 - 1^2)(u^2 - 2^2)\ldots(u^2 - n^2)}{(2n+1)!}h^{2n+1}f^{(2n+2)}(\xi)\xi'(x) \quad (4.40)$$

This is the error term for the formula in Equation 4.35.

In particular, when $x = x_0$, that is, $u = 0$, the second term in Equation 4.40 vanishes, consequently using Equation 3.90 of Chapter 3, we get

$$R'_{2n+1}(x_0) = \frac{(-1)^n (n!)^2}{(2n+1)!h}\Delta^{2n+1}y_0 \quad (4.41)$$

In a similar manner, we can find error terms for higher-order numerical differentiation formulae.

Example 4.2

Compute the values of dy/dx and d^2y/dx^2 at $x = 2.72$ from the following table:

x	2.5	2.6	2.7	2.8	2.9
y	0.4938	0.4953	0.4965	0.4974	0.4981

Solution:

First, we construct the following difference table:

x	y	Δy	$\Delta^2 y$	$\Delta^3 y$	$\Delta^4 y$
2.5	0.4938				
		0.0015			
2.6	0.4953		−0.0003		
		0.0012		**0.0000**	
2.7	**0.4965**		**−0.0003**		**0.0001**
		0.0009		**0.0001**	
2.8	0.4974		−0.0002		
		0.0007			
2.9	0.4981				

Since the point $x = 2.72$ lies near the middle of the set of tabular values, we shall use Stirling's interpolation differentiation formula.

We choose $x_0 = 2.7$ so that $u = (x - x_0)/h = (2.72 - 2.7)/0.1 = 0.2$.

From the above difference table, we identify $y_0 = 0.4965$, $\Delta y_{-1} = 0.0012$, $\Delta y_0 = 0.0009$, $\Delta^2 y_{-1} = -0.0003$, $\Delta^3 y_{-2} = 0.0000$, $\Delta^3 y_{-1} = 0.0001$, and $\Delta^4 y_{-2} = 0.0001$, which are to be used in the Stirling's interpolation differentiation formula. These values have been shown in bold digits.

Using Equation 4.35, we have

$$\frac{dy}{dx} = f'(x) \cong \frac{1}{h}\left[\frac{(\Delta y_{-1} + \Delta y_0)}{2} + u\Delta^2 y_{-1} + \frac{3u^2 - 1}{3!}\frac{\Delta^3 y_{-2} + \Delta^3 y_{-1}}{2} + \frac{4u^3 - 2u}{4!}\Delta^4 y_{-2} + \cdots\right]$$

Thus,

$$\left.\frac{dy}{dx}\right|_{x=2.72} = f'(2.72) \cong \frac{1}{0.1}\left[\frac{(0.0012 + 0.0009)}{2} + 0.2 \times (-0.0003) + \frac{3 \times (0.2)^2 - 1}{3!} \times \left(\frac{0.0000 + 0.0001}{2}\right) + \frac{4 \times (0.2)^3 - 2 \times 0.2}{4!} \times 0.0001\right]$$

$$= 0.00981133$$

Again, using Equation 4.36, we have

$$\frac{d^2 y}{dx^2} = f''(x) \cong \frac{1}{h^2}\left[\Delta^2 y_{-1} + u\frac{\Delta^3 y_{-2} + \Delta^3 y_{-1}}{2} + \frac{12u^2 - 2}{4!}\Delta^4 y_{-2} + \cdots\right]$$

Therefore,

$$\left.\frac{d^2 y}{dx^2}\right|_{x=2.72} = f''(2.72) \cong \frac{1}{(0.1)^2}\left[(-0.0003) + 0.2 \times \frac{0.0000 + 0.0001}{2} + \frac{12 \times (0.2)^2 - 2}{4!} \times 0.0001\right]$$

$$= -0.0296333$$

4.3.4 Formulae Based on Bessel's Interpolation

Let the values of the function $y = f(x)$ be known for the following even number of equispaced arguments:

$$x_{-(n-1)} = x_0 - (n-1)h, \ldots, x_{-2} = x_0 - 2h, x_{-1} = x_0 - h, x_0 = x_0, x_1 = x_0 + h, x_2 = x_0 + 2h, \ldots, x_n = x_0 + nh$$

Then, from Equation 3.117 of Chapter 3, Bessel's interpolation formula is given by

$$\begin{aligned} f(x) = {} & \frac{y_0 + y_1}{2} + v\Delta y_0 + \frac{\left(v^2 - \frac{1}{4}\right)}{2!}\frac{\left(\Delta^2 y_{-1} + \Delta^2 y_0\right)}{2!} + \frac{v\left(v^2 - \frac{1}{4}\right)}{3!}\Delta^3 y_{-1} \\ & + \frac{\left(v^2 - \frac{1}{4}\right)\left(v^2 - \frac{9}{4}\right)}{4!}\frac{\left(\Delta^4 y_{-2} + \Delta^4 y_{-1}\right)}{2} + \frac{v\left(v^2 - \frac{1}{4}\right)\left(v^2 - \frac{9}{4}\right)}{5!}\Delta^5 y_{-2} \\ & + \cdots + \frac{v\left(v^2 - \frac{1}{4}\right)\left(v^2 - \frac{9}{4}\right)\ldots\left(v^2 - \frac{(2n-3)^2}{4}\right)}{(2n-1)!}\Delta^{2n-1} y_{-n+1} \end{aligned} \tag{4.42}$$

where $v = (x - x_0)/h - (1/2)$.

Differentiating both sides of Equation 4.42 with respect to x, we have

$$\frac{dy}{dx} = f'(x) \cong \frac{1}{h}\left[\Delta y_0 + v\frac{\left(\Delta^2 y_{-1} + \Delta^2 y_0\right)}{2!} + \frac{3v^2 - \frac{1}{4}}{3!}\Delta^3 y_{-1} + \frac{4v^3 - 5v}{4!}\frac{\left(\Delta^4 y_{-2} + \Delta^4 y_{-1}\right)}{2} + \cdots\right] \tag{4.43}$$

Again, differentiating both sides of Equation 4.43 with respect to x, we get

$$\frac{d^2 y}{dx^2} = f''(x) \cong \frac{1}{h^2}\left[\frac{\left(\Delta^2 y_{-1} + \Delta^2 y_0\right)}{2!} + v\Delta^3 y_{-1} + \frac{12v^2 - 5}{4!}\frac{\left(\Delta^4 y_{-2} + \Delta^4 y_{-1}\right)}{2} + \cdots\right] \tag{4.44}$$

and so on.

The above formulae for numerical differentiation are used when the point x is near the middle of the set of tabulated values.

In particular, when $x = x_0$, that is, $v = -(1/2)$, we get

$$\left.\frac{dy}{dx}\right|_{x=x_0} = f'(x_0) \cong \frac{1}{h}\left[\Delta y_0 - \frac{\left(\Delta^2 y_{-1} + \Delta^2 y_0\right)}{4} + \frac{1}{12}\Delta^3 y_{-1} + \frac{\left(\Delta^4 y_{-2} + \Delta^4 y_{-1}\right)}{24} - \cdots\right] \tag{4.45}$$

$$\left.\frac{d^2 y}{dx^2}\right|_{x=x_0} = f''(x_0) \cong \frac{1}{h^2}\left[\frac{\left(\Delta^2 y_{-1} + \Delta^2 y_0\right)}{2} - \frac{1}{2}\Delta^3 y_{-1} - \frac{\left(\Delta^4 y_{-2} + \Delta^4 y_{-1}\right)}{24} + \cdots\right] \tag{4.46}$$

and so on.

4.3.4.1 Error Estimate

From Equation 3.118 of Chapter 3, we have the error term as follows:

$$R_{2n}(x) = \frac{\left(v^2 - \frac{1}{4}\right)\left(v^2 - \frac{9}{4}\right)\ldots\left(v^2 - \frac{(2n-1)^2}{4}\right)}{2n!}h^{2n} f^{(2n)}(\xi) \tag{4.47}$$

where $\min\{x_{-(n-1)}, x, x_n\} < \xi < \max\{x_{-(n-1)}, x, x_n\}$ and $\xi = \xi(x)$.

Differentiating both sides of Equation 4.47 with respect to x, we get

$$R'_{2n}(x) = \frac{h^{2n-1} f^{(2n)}(\xi)}{(2n)!} \frac{d}{dv}\left\{\left(v^2 - \tfrac{1}{4}\right)\left(v^2 - \tfrac{9}{4}\right)\ldots\left(v^2 - \tfrac{(2n-1)^2}{4}\right)\right\}$$
$$+ \frac{\left(v^2 - \tfrac{1}{4}\right)\left(v^2 - \tfrac{9}{4}\right)\ldots\left(v^2 - \tfrac{(2n-1)^2}{4}\right)}{(2n)!} h^{2n} f^{(2n+1)}(\xi)\xi'(x) \tag{4.48}$$

This is the error term for the formula in Equation 4.43.

In particular, when $x = x_0$, that is, $v = -(1/2)$, the second term in Equation 4.48 vanishes, consequently using Equation 3.90 of Chapter 3, we get

$$R'_{2n}(x_0) = \frac{(-1)^n (n-1)!n!}{(2n)!h} \Delta^{2n} y_0 \tag{4.49}$$

In a similar manner, we can find error terms for higher-order numerical differentiation formulae.

Example 4.3

A rod is rotating in a plane. The angle θ (in radians) through which the rod has turned for various values of time *t* (in seconds) is given in the following table:

t	0.0	0.2	0.4	0.6	0.8	1.0
θ	0.00	0.12	0.49	1.12	2.02	3.20

Calculate the angular velocity and the angular acceleration of the rod when $t = 0.54$ s.

Solution:

First, we construct the following difference table:

t	θ	$\Delta\theta$	$\Delta^2\theta$	$\Delta^3\theta$	$\Delta^4\theta$	$\Delta^5\theta$
0.0	0.00					
		0.12				
0.2	0.12		0.25			
		0.37		0.01		
0.4	**0.49**		**0.26**		**0.00**	
		0.63		**0.01**		**0.00**
0.6	**1.12**		**0.27**		**0.00**	
		0.90		0.01		
0.8	2.02		0.28			
		1.18				
1.0	3.20					

Since the point $t = 0.54$ lies near the middle of the set of tabular values, we apply Bessel's interpolation differentiation formula (Stirling's interpolation differentiation formula may also be used).

We choose $t_0 = 0.4$ so that $v = [(t-t_0)/h] - (1/2) = [(0.54-0.4)/0.2] - 0.5 = 0.2$.

From the above difference table, we identify $\Delta\theta_0 = 0.63$, $\Delta^2\theta_{-1} = 0.26$, $\Delta^2\theta_0 = 0.27$, $\Delta^3\theta_{-1} = 0.01$, $\Delta^4\theta_{-2} = 0.00$, $\Delta^4\theta_{-1} = 0.00$, and $\Delta^5\theta_{-2} = 0.00$, which are to be used in the Bessel's interpolation differentiation formula. These values have been shown in bold digits.

Using Equation 4.35, we have

$$\frac{d\theta}{dt} \cong \frac{1}{h}\left[\Delta\theta_0 + v\frac{\left(\Delta^2\theta_{-1}+\Delta^2\theta_0\right)}{2!} + \frac{3v^2-\frac{1}{4}}{3!}\Delta^3\theta_{-1} + \frac{4v^3-5v}{4!}\frac{\left(\Delta^4\theta_{-2}+\Delta^4\theta_{-1}\right)}{2} + \cdots\right]$$

Thus,

$$\left.\frac{d\theta}{dt}\right|_{t=0.54} \cong \frac{1}{0.2}\left[0.63 + 0.2\times\frac{(0.26+0.27)}{2!} + \frac{3\times(0.2)^2-\frac{1}{4}}{3!}\times 0.01\right]$$

$$= 3.41392 \text{radians/s}$$

Again, using Equation 4.44, we have

$$\frac{d^2\theta}{dt^2} \cong \frac{1}{h^2}\left[\frac{\left(\Delta^2\theta_{-1}+\Delta^2\theta_0\right)}{2!} + v\Delta^3\theta_{-1} + \frac{12v^2-5}{4!}\frac{\left(\Delta^4\theta_{-2}+\Delta^4\theta_{-1}\right)}{2} + \cdots\right]$$

Therefore,

$$\left.\frac{d^2\theta}{dt^2}\right|_{t=0.54} \cong \frac{1}{(0.2)^2}\left[\frac{(0.26+0.27)}{2!} + 0.2\times 0.01\right]$$

$$= 6.675 \text{ radians/s}^2$$

4.4 NUMERICAL DIFFERENTIATION FOR UNEQUALLY SPACED NODES

Let $y = f(x)$ be a continuous function defined over a closed interval $[a,b]$ having continuous derivatives, and let values $y_i = f(x_i)$ of this function are known at the unequally spaced nodes $x_i \in [a,b]$, $i = 0,1,\ldots,n$. We now propose to find a numerical formula on the basis of the given values to find the value of the derivatives of $f(x)$ at points close to x_0 and x_n, respectively.

4.4.1 Formulae Based on Lagrange's Interpolation

Differentiating both sides of Equation 3.72 of Chapter 3 with respect to x, we get

$$\frac{dy}{dx} = f'(x) \approx \varphi_n'(x) = \sum_{i=0}^{n}\frac{\Pi'(x)}{(x-x_i)\Pi'(x_i)}y_i - \sum_{i=0}^{n}\frac{\Pi(x)}{(x-x_i)^2\Pi'(x_i)}y_i \tag{4.50}$$

For a node point $x = x_j$, $j = 0,1,\ldots,n$, the Equation 4.50 becomes indeterminate and so we proceed as follows:

$$f(x) \approx \varphi_n(x) = \sum_{i=0}^{n}\frac{\Pi(x)}{(x-x_i)\Pi'(x_i)}y_i = \sum_{i\neq j}^{n}\frac{\Pi(x)}{(x-x_i)\Pi'(x_i)}y_i + \omega_j(x)y_j \tag{4.51}$$

where:

$$\omega_j(x) = \frac{\Pi(x)}{(x-x_j)\Pi'(x_j)} = \frac{(x-x_0)(x-x_1)\ldots(x-x_{j-1})(x-x_{j+1})\ldots(x-x_n)}{(x_j-x_0)(x_j-x_1)\ldots(x_j-x_{j-1})(x_j-x_{j+1})\ldots(x_j-x_n)} \tag{4.52}$$

Therefore,

$$\left.\frac{dy}{dx}\right|_{x=x_j} = f'(x_j) \approx \varphi_n'(x_j) = \sum_{i \neq j}^{n} \frac{\Pi'(x_j)}{(x_j - x_i)\Pi'(x_i)} y_i + \omega_j'(x_j) y_j \tag{4.53}$$

where:

$$\omega_j'(x_j) = \sum_{i \neq j} \frac{1}{x_j - x_i} \tag{4.54}$$

Hence,

$$\left.\frac{dy}{dx}\right|_{x=x_j} = f'(x_j) \approx \varphi_n'(x_j) = \sum_{i \neq j}^{n} \frac{\Pi'(x_j)}{(x_j - x_i)\Pi'(x_i)} y_i + \sum_{i \neq j} \frac{1}{x_j - x_i} y_j, \quad j = 0,1,\ldots,n \tag{4.55}$$

Similarly, we can obtain the derivatives of higher order.

4.4.1.1 Error Estimate

Using Equation 3.89 of Chapter 3 and Equation 4.5, we get

$$\begin{aligned} R_{n+1}'(x) &= \Omega'(x) f[x, x_0, \ldots, x_n] + \Omega(x) f[x, x, x_0, \ldots, x_n] \\ &= \Omega'(x) \frac{f^{(n+1)}(\xi)}{(n+1)!} + \Omega(x) \frac{f^{(n+2)}(\xi')}{(n+2)!} \end{aligned} \tag{4.56}$$

where:

$$f[x, x_0, \ldots, x_n] = \frac{f^{(n+1)}(\xi)}{(n+1)!}$$

$$f[x, x, x_0, \ldots, x_n] = \frac{f^{(n+2)}(\xi')}{(n+2)!}$$

$$\min\{x, x_0, x_1, \ldots, x_n\} < \xi, \xi' < \max\{x, x_0, x_1, \ldots, x_n\}$$

At a node $x = x_i$, $i = 0,1,\ldots,n$, then second term in Equation 4.56 vanishes (since $\Omega(x_i) = 0$, $i = 0,1,\ldots,n$) and we get

$$R_{n+1}'(x_i) = \Omega'(x_i) \frac{f^{(n+1)}(\xi_i)}{(n+1)!}, \quad i = 0,1,\ldots,n \tag{4.57}$$

where $\xi_i = \xi(x_i)$ lies between the smallest and the largest values of $x_0, x_1, \ldots x_n$.

4.4.2 Formulae Based on Newton's Divided Difference Interpolation

Differentiating both sides of Equation 3.84 of Chapter 3 with respect to x, we get

$$\begin{aligned} \frac{dy}{dx} &= f'(x) \approx f[x_0, x_1] + (2x - x_0 - x_1) f[x_0, x_1, x_2] \\ &+ \left\{3x^2 - 2(x_0 + x_1 + x_2)x + (x_0x_1 + x_1x_2 + x_2x_0)\right\} f[x_0, x_1, x_2, x_3] + \cdots \end{aligned} \tag{4.58}$$

$$\frac{d^2y}{dx^2} = f''(x) \approx 2f[x_0,x_1,x_2] + \{6x - 2(x_0 + x_1 + x_2)\} f[x_0,x_1,x_2,x_3] + \cdots \quad (4.59)$$

and so on.

4.4.2.1 Error Estimate

The errors $R'_{n+1}(x)$ and $R'_{n+1}(x_i)$ are same as given by Equations 4.56 and 4.57, respectively.

Example 4.4

Find the values of $f'(6)$ and $f''(8)$ from the following table, using an appropriate interpolation formula:

x	3	4	7	9	10
y	48	100	295	900	1210

Solution:

Since the values of x are unequally spaced, we may use either the Newton's divided interpolation formula or Lagrange's interpolation formula.

Case I:

To use the Newton's divided difference interpolation formula, we construct the following divided difference table:

x	y	First-Order Divided Difference	Second-Order Divided Difference	Third-Order Divided Difference	Fourth-Order Divided Difference
3	<u>48</u>				
		<u>52</u>			
4	100		<u>3.25</u>		
		65		<u>7.375</u>	
7	295		47.5		−<u>2.125</u>
		302.5		−7.5	
9	900		2.5		
		310			
10	1210				

From this divided difference table, only underlined values will be used in the Newton divided difference interpolation formula.

Using Newton's divided interpolation formula, from Equation 3.84 of Chapter 3, we have

$$y = f(x) \approx y_0 + (x - x_0)f[x_0,x_1] + (x - x_0)(x - x_1)f[x_0,x_1,x_2] + \cdots$$

$$+ (x - x_0)(x - x_1)\ldots(x - x_{n-1})f[x_0,x_1,\ldots,x_n]$$

$$= 48 + (x-3)\times 52 + (x-3)(x-4)\times 3.25 + (x-3)(x-4)(x-7)\times 7.375$$

$$+ (x-3)(x-4)(x-7)(x-9)\times(-2.125)$$

$$= -2.125x^4 + 56.25x^3 - 497.375x^2 + 1824.25x - 2295$$

Therefore,

$$\frac{dy}{dx} = -8.5x^3 + 168.75x^2 - 994.75x + 1824.25$$

and

$$\frac{d^2y}{dx^2} = -25.5x^2 + 337.5x - 994.75$$

Hence,

$$\left.\frac{dy}{dx}\right|_{x=6} = -8.5\times 6^3 + 168.75\times 6^2 - 994.75\times 6 + 1824.25 = 94.75$$

$$\left.\frac{d^2y}{dx^2}\right|_{x=8} = -25.5\times 8^2 + 337.5\times 8 - 994.75 = 73.25$$

Case II:
Using Lagrange's interpolation formula, we have

$$y = f(x) \cong \frac{(x-4)(x-7)(x-9)(x-10)}{(3-4)(3-7)(3-9)(3-10)}\times 48 + \frac{(x-3)(x-7)(x-9)(x-10)}{(4-3)(4-7)(4-9)(4-10)}\times 100$$

$$+\frac{(x-3)(x-4)(x-9)(x-10)}{(7-3)(7-4)(7-9)(7-10)}\times 295 + \frac{(x-3)(x-4)(x-7)(x-10)}{(9-3)(9-4)(9-7)(9-10)}\times 900$$

$$+\frac{(x-3)(x-4)(x-7)(x-9)}{(10-3)(10-4)(10-7)(10-9)}\times 1210$$

$$= -2.125x^4 + 56.25x^3 - 497.375x^2 + 1824.25x - 2295$$

It may be noted significantly that the same polynomial has been obtained as in the case of Newton's divided interpolation formula.
Hence,

$$\left.\frac{dy}{dx}\right|_{x=6} = 94.75$$

$$\left.\frac{d^2y}{dx^2}\right|_{x=8} = 73.25$$

Example 4.5

Find the maximum and minimum values of $y = f(x)$ from the following table:

x	0	1	2	5
y	2	3	12	147

Solution:

Since the values of x are unequally spaced, we may use either the Newton's divided interpolation formula or Lagrange's interpolation formula.

Let us apply Lagrange's interpolation formula.
Using Lagrange's interpolation formula, we have

$$y = f(x) \cong \frac{(x-1)(x-2)(x-5)}{(0-1)(0-2)(0-5)} \times 2 + \frac{(x-0)(x-2)(x-5)}{(1-0)(1-2)(1-5)} \times 3$$

$$+ \frac{(x-0)(x-1)(x-5)}{(2-0)(2-1)(2-5)} \times 12 + \frac{(x-0)(x-1)(x-2)}{(5-0)(5-1)(5-2)} \times 147$$

$$= x^3 + x^2 - x + 2$$

Now, for maxima or minima, we have

$$\frac{dy}{dx} = 0$$

This implies

$$3x^2 + 2x - 1 = 0 \quad \text{yielding} \quad x = -1, \frac{1}{3}$$

At

$$x = -1, \quad \left.\frac{d^2y}{dx^2}\right|_{x=-1} = -4 < 0$$

Again, at

$$x = \frac{1}{3}, \quad \left.\frac{d^2y}{dx^2}\right|_{x=\frac{1}{3}} = 4 > 0$$

Therefore, $y = f(x)$ attains its maximum value 3 at $x = -1$ and also attains its minimum value $(49/27) \cong 1.8148$ at $x = (1/3)$.

4.5 RICHARDSON EXTRAPOLATION

The technique of combining two computed approximate values obtained by using the same formula or method with two different step sizes, to obtain a higher-order method which provides closer approximation of a certain quantity, is called Richardson's extrapolation method.

To illustrate the method, let us consider

$$f'(x_0) \cong \frac{f_1 - f_{-1}}{2h} \tag{4.60}$$

where $f_1 = f(x_0 + h)$ and $f_{-1} = f(x_0 - h)$.

Now, from Equation 4.60, we get

$$f'(x_0) \cong \frac{1}{2h}\left[f(x_0 + h) - f(x_0 - h)\right]$$

$$= \frac{1}{2h}\left[\left\{f(x_0) + \frac{h}{1!}f'(x_0) + \frac{h^2}{2!}f''(x_0) + \frac{h^3}{3!}f'''(x_0) + \cdots\right\}\right.$$

$$\left. - \left\{f(x_0) - \frac{h}{1!}f'(x_0) + \frac{h^2}{2!}f''(x_0) - \frac{h^3}{3!}f'''(x_0) + \cdots\right\}\right]$$

Taylor's series expansion about x_0.

$$= f'(x_0) + \frac{h^2}{3!} f'''(x_0) + \frac{h^4}{5!} f^{v}(x_0) + \cdots$$

$$= f'(x_0) + c_1 h^2 + c_2 h^4 + \cdots = g(h), \quad \text{say}$$

Thus,

$$g(h) = f'(x_0) + c_1 h^2 + c_2 h^4 + \cdots \tag{4.61}$$

is an approximation of $f'(x_0)$, when the step size is taken as h. Here, $c_1 h^2 + c_2 h^4 + \cdots$ is the local truncation error associated with Equation 4.61.

A closer approximation of $f'(x_0)$ can be obtained by the following procedure.

Taking the step size as $h/2^r$, $r = 1, 2, \ldots$, we have from Equation 4.61

$$g\left(\frac{h}{2}\right) = f'(x_0) + c_1 \frac{h^2}{4} + c_2 \frac{h^4}{16} + \cdots \tag{4.62}$$

$$g\left(\frac{h}{2^2}\right) = f'(x_0) + c_1 \frac{h^2}{16} + c_2 \frac{h^4}{256} + \cdots \tag{4.63}$$

and so on.

Eliminating c_1 from Equations 4.61 and 4.62, we obtain

$$g^{(1)}(h) = \frac{4g(h/2) - g(h)}{4-1} = f'(x_0) - c_2 \frac{h^4}{4} - \frac{5}{16} c_3 h^6 - \cdots \tag{4.64}$$

Again, eliminating c_1 from Equations 4.62 and 4.63, we obtain

$$g^{(1)}\left(\frac{h}{2}\right) = \frac{4g(h/2^2) - g(h/2)}{4-1} = f'(x_0) - c_2 \frac{h^4}{64} - \frac{5}{1024} c_3 h^6 - \cdots \tag{4.65}$$

Since the truncation error is of $O(h^4)$, $g^{(1)}(h)$, $g^{(1)}(h/2)$, … are closer approximations to $f'(x_0)$.

Similarly, eliminating c_2 from Equations 4.64 and 4.65, we obtain

$$g^{(2)}(h) = \frac{4^2 g^{(1)}(h/2) - g^{(1)}(h)}{4^2 - 1} = f'(x_0) + c_3 \frac{h^6}{64} + \cdots$$

Again, since the truncation error is of $O(h^6)$, $g^{(2)}(h)$ is a more closer approximation to $f'(x_0)$.

Proceeding in the similar way as before, we can obtain the successive closer approximations to $f'(x_0)$ from the given formula

$$g^{(r)}(h) = \frac{4^r g^{(r-1)}(h/2) - g^{(r-1)}(h)}{4^r - 1}, \quad r = 1, 2, \ldots \tag{4.66}$$

where $g^{(0)}(h) = g(h)$.

We thus obtain a table of successive values of $g^{(r)}(h)$ for various values of r from Table 4.1.

From Table 4.1, it may be easily observed that the successive entries in a given table for a particular column results in better approximation than those of the preceding ones. Also,

TABLE 4.1
Extrapolation Method

$O(h^2)$	$O(h^4)$	$O(h^6)$	$O(h^8)$	$O(h^{10})$
$g(h)$				
	$g^{(1)}(h)$			
$g\left(\frac{h}{2}\right)$		$g^{(2)}(h)$		
	$g^{(1)}\left(\frac{h}{2}\right)$		$g^{(3)}(h)$	
$g\left(\frac{h}{2^2}\right)$		$g^{(2)}\left(\frac{h}{2}\right)$		$g^{(4)}(h)$
	$g^{(1)}\left(\frac{h}{2^2}\right)$		$g^{(3)}\left(\frac{h}{2}\right)$	
$g\left(\frac{h}{2^3}\right)$		$g^{(2)}\left(\frac{h}{2^2}\right)$		
	$g^{(1)}\left(\frac{h}{2^3}\right)$			
$g\left(\frac{h}{2^4}\right)$				

the successive columns give better approximations than preceding ones. Moreover, the lower diagonal entries yield best results.

This method of extrapolation will be continued until

$$\left|g^{(r)}(h)-g^{(r)}\left(\frac{h}{2}\right)\right|<\varepsilon \tag{4.67}$$

for a given tolerance error ε.

Note:

1. The recurrence formula in Equation 4.66 also holds if we wish to approximate

$$f''(x_0)\cong\frac{f_1-2f_0+f_{-1}}{h^2}$$

2. When $f'(x_0)\cong(f_1-f_0)/h$ or $(f_0-f_{-1})/h$ and $f''(x_0)\cong(f_2-2f_1+f_0)/h^2$ or $(f_0-2f_{-1}+f_{-2})/h^2$, it can be shown that the recurrence formula for Richardson's extrapolation method will be of the following form:

$$g^{(r)}(h)=\frac{2^r g^{(r-1)}(h/2)-g^{(r-1)}(h)}{2^r-1},\quad r=1,2,\ldots \tag{4.68}$$

where $g^{(0)}(h)=g(h)$.

Example 4.6

Find the values of $f'(19)$ and $f''(19)$ from the following table, using appropriate initial values based on finite differences and Richardson's extrapolation method:

x	15	17	19	21	23	25
$y = f(x)$	3.873	4.123	4.359	4.583	4.796	5

Solution:

Using the formula $f'(x_0) \cong (f_1 - f_{-1})/2h$, we can obtain

$$g(h) = f'(x_0) \cong \frac{1}{2h}\left[f(x_0+h) - f(x_0-h)\right] = f'(x_0) + \frac{h^2}{3!}f'''(x_0) + \frac{h^4}{5!}f^{(v)}(x_0) + \cdots$$

using Taylor's series expansion about x_0.

Therefore,

$$g(h) = f'(19) \cong \frac{f_1 - f_{-1}}{2h}, \quad \text{taking} \quad x_0 = 19$$

The computed values of successive approximations have been tabulated in the following table:

h \ g	$g(h)$	$g^{(1)}(h)$
4	0.115375	
		0.114875
2	0.115	

Hence, $f'(19) \cong g^{(1)}(4) = 0.114875$.

Again, using the formula $f''(x_0) \cong (f_1 - 2f_0 + f_{-1})/h^2$, we can obtain

$$g(h) = f''(x_0) \cong \frac{f(x_0+h) - 2f(x_0) + f(x_0-h)}{h^2} = f''(x_0) + \frac{h^2}{12}f^{(iv)}(x_0) + \cdots$$

using Taylor's series expansion about x_0.

Therefore,

$$g(h) = f''(19) \cong \frac{f_1 - 2f_0 + f_{-1}}{h^2}, \quad \text{taking} \quad x_0 = 19$$

The computed values of successive approximations have been tabulated in the following table:

h \ g	$g(h)$	$g^{(1)}(h)$
4	−0.0030625	
		−0.002979167
2	−0.003	

Hence, $f''(19) \cong g^{(1)}(4) = -0.002979167$.

EXERCISES

1. Compute the first and second derivatives of the function $y = f(x)$ at $x = 1.5$ and 5.8, using appropriate differentiation formula.

x	1	2	3	4	5	6
$y = f(x)$	1	8	27	65	120	210

2. The following data give corresponding values of pressure (p) and the specific volume (v) of a superheated steam:

v	2	4	6	8	10
p	105	42.7	25.3	16.7	13

Find the rate of change of

a. Pressure with respect to volume when $v = 2$.
b. Volume with respect to pressure when $p = 105$.

3. A curve is expressed by the following values of x and y. Find the slope at the point $x = 1.5$.

x	0.0	0.5	1.0	1.5	2.0
y	0.4	0.35	0.24	0.13	0.05

4. In a machine, a slider moves along a fixed straight rod. Its distance x cm along the rod is given below for various values of time t seconds. Find the velocity and acceleration of the slider when $t = 0.3$.

t	0	0.1	0.2	0.3	0.4	0.5	0.6
x	30.13	31.62	32.87	33.64	33.95	33.81	33.24

5. The table below gives the results of an observation: θ is the observed temperature, in celsius, of a vessel of cooling water and t is the time, in minutes, from the beginning of observation.

t	1	3	5	7	9
θ	85.3	74.5	67.0	60.5	54.3

Find the approximate rate of cooling at $t = 3$ and $t = 3.5$.

6. Compute the value of $y'(1.2)$ and $y''(1.2)$ from the following table:

x	1	1.1	1.2	1.3	1.4
y	0.0254	0.0437	0.0587	0.0670	0.0780

7. Using Bessel's formula, calculate $f'(1.35)$ and $f''(1.35)$ from the following data:

x	1.1	1.2	1.3	1.4	1.5	1.6
y	0.9916	0.9857	0.9781	0.9691	0.9584	13.83072

8. From the following table of values, evaluate $\left.\frac{dy}{dx}\right|_{x=0.9}$, respectively, with $h = 0.1$ and 0.2. Apply Richardson's formula to obtain a more accurate value of the first derivative.

x	0.6	0.7	0.8	0.9	1.0	1.1
y	−0.51083	−0.35667	−0.22314	−0.10536	0.0	0.09531

9. The distances (x cm) traversed by a particle at different times (t seconds) are given below. Find the velocity and acceleration of the particle at $t = 0.3$ s.

t	0.0	0.1	0.2	0.3	0.4	0.5	0.6
x	3.01	3.16	3.29	3.36	3.40	3.38	3.32

10. A particle is moving along a straight line. The displacement x at some time instance t is given below. Find the velocity and acceleration of the particle at $t = 4$.

t	1	3	5	7	9	11
x	0.1405	0.7676	3.5135	9.9351	21.5892	40.0324

11. A rod is rotating in a plane about one of its ends. The angle θ (in radians) at different times t (seconds) is given below. Find its angular velocity and angular acceleration when $t = 0.6$ s.

t	0	0.2	0.4	0.6	0.8	1.0
θ	0.0	0.15	0.50	1.15	2.0	3.20

12. Find x for which y is maximum and also find the corresponding value of y, from the table given below:

x	1	1.5	2	2.5	3	3.5
y	2.7	−5.5188	−27.8	−75.4688	−163.3	−309.5188

13. The following table gives angular displacements θ (in radians) at different times t (seconds). Calculate the angular velocity at $t = 0.06$.

t	0	0.02	0.04	0.06	0.08	0.10	0.12
θ	0.052	0.105	0.168	0.242	0.327	0.408	0.489

14. The population of certain town is given below in the following table. Estimate the rate of growth of population in 1981 and 1990.

Year	1951	1961	1971	1981	1991
Population (in thousands)	19.96	39.65	58.81	77.21	94.61

15. Find the first three derivatives of the function tabulated below at the point $x = 2.5$ by using Lagrange's interpolation formula or Newton's divided interpolation formula.

x	1.5	1.9	2.5	3.2	4.3	5.9
$f(x)$	3.375	6.059	13.625	29.368	73.907	196.579

16. Find the values of $f'(0)$ and $f''(0)$ from the following table, using appropriate initial values based on finite differences and Richardson's extrapolation method:

x	0	1	2	3	4	5	6	7	8
$y = f(x)$	−1	0	5	20	51	104	185	300	455

17. Find the values of $f'(3)$ and $f''(3)$ from the following table, using appropriate initial values based on finite differences and Richardson's extrapolation method:

x	**−1**	**1**	**2**	**3**	**4**	**5**	**7**
$y = f(x)$	**1**	**1**	**16**	**81**	**256**	**625**	**2401**

18. Find the value of x for which $f(x)$ is maximum in the range of x given in the following data. Find also the maximum value of $f(x)$.

x	60	75	90	105	120
$y = f(x)$	28.2	38.2	43.2	40.9	37.7

19. From the below table find the minimum value of $f(x)$.

x	0	1	2	3	4	5
y	58	43	40	45	52	60

20. Using Newton's divided difference formula, construct the interpolating polynomial and hence compute (dy/dx) and (d^2y/dx^2) at $x = 5$ by using the following table:

x	0	2	3	4	7	9
$f(x)$	4	26	58	112	466	922

5 Numerical Integration

5.1 INTRODUCTION

To find the value of the definite integral $\int_a^b f(x)dx$ standard methods exist, when $f(x)$ is known and integrable. But if $f(x)$ is not known excepting at some desirable points or if $f(x)$ is not integrable analytically, we may try to find numerical value of the integral up to a desired degree of accuracy. This problem is called *numerical integration* or *mechanical quadrature*. In case where $f(x)$ is known but the form of $f(x)$ is so complicated that it is very difficult to integrate it and find a value, then we may take help of numerical quadrature.

In this method, we first choose a suitable set of nodes $x_i, i = 0(1)n$ corresponding to the interval of integration, find the values $f(x_i)$ of the function and then obtain a suitable interpolation polynomial for $f(x)$. We then integrate it to get an approximate value of the given definite integral.

Geometrically, we determine $(n + 1)$ suitable points $(x_i, y_i), i = 0(1)n$ on the curve $y = f(x)$, and then replace the curve by the nth degree parabola represented by the interpolating polynomial passing through those $(n + 1)$ points, then the area bounded by this parabola, x-axis and the ordinates at $x = a$ and $x = b$, is taken to be an approximation to the area under the curve $y = f(x)$.

We may get various numerical integration formula changing the nodes x_i, varying the number of nodes, and also changing the interpolation polynomial for $f(x)$.

Definition 5.1:

A numerical integration formula is said to be closed or open type accordingly as the limits of integration *a, b* are taken as nodes, that is, interpolating points or not.

5.2 NUMERICAL INTEGRATION FROM LAGRANGE'S INTERPOLATION

Let $x_i, i = 0(1)n$ be a suitable set of interpolating points of the closed interval $[a, b]$. Let $y_i, = f(x_i), i = 0(1)n$ are known. Let $f(x)$ be approximated by a polynomial $\varphi_n(x)$ of degree less than equal to n. Then Lagrange's interpolation formula for $f(x)$ is given by

$$f(x) = \varphi_n(x) + \frac{\Pi(x) f^{(n+1)}(\xi)}{(n+1)!}, \quad \text{where } a < \xi < b$$

This implies that

$$f(x) = \sum_{i=0}^{n} \frac{\Pi(x)}{(x - x_i)\Pi'(x_i)} y_i + \frac{\Pi(x) f^{(n+1)}(\xi)}{(n+1)!} \tag{5.1}$$

where ξ lies between the smallest and largest value of $x, x_0, \ldots x_n$. Then,

$$\int_a^b f(x)dx = \int_a^b \sum_{i=0}^{n} \frac{\Pi(x)}{(x - x_i)\Pi'(x_i)} y_i dx + \int_a^b \frac{\Pi(x) f^{(n+1)}(\xi)}{(n+1)!} dx$$

$$= \sum_{i=0}^{n} H_i^{(n)} y_i + R_{n+1}(f) \tag{5.2}$$

where:

$$H_i^{(n)} = \int_a^b \frac{\Pi(x)dx}{(x - x_i)\Pi'(x_i)} \tag{5.3}$$

and

$$R_{n+1}(f) = \int_a^b \frac{\Pi(x) f^{(n+1)}(\xi)}{(n+1)!} dx \tag{5.4}$$

which is the error term in the numerical integration formula (5.2).

Corollary:

1. For any given set of nodes and the values of the function there, the coefficients $H_i^{(n)}$ are independent of $f(x)$.
2. For any polynomial of degree less than equal to n, formula (5.2) is exact, so that, $f(x) = \varphi_n(x)$ and $R_n(f) = 0$.
3. Taking $f(x) = 1$, we get from (5.3)

$$\sum_{i=0}^{n} H_i^{(n)} y_i = \int_a^b 1dx, \quad \text{as } R_n(f) = 0$$

$$\text{that is,} \quad \sum_{i=0}^{n} H_i^{(n)} = b - a, \quad \text{as } y_i = 1 \quad \text{for all } i.$$

Definition 5.2: Degree of Precision

A mechanical quadrature formula is said to have degree of precision k, a positive integer, if the formula is exact for any arbitrary polynomial of degree less than or equal to k and there exists at least one polynomial of degree $(k + 1)$ for which the formula is not exact.

Theorem 5.1

If the degree of precision of an $(n + 1)$-point interpolation formula is k, then $n \le k \le 2n + 1$.

Proof:

We have seen that any mechanical quadrature formula is exact for an arbitrary polynomial of degree $\le n$ if there are $(n + 1)$ nodes. Therefore, for $(n + 1)$ nodes any quadrature formula has degree of precision $\ge n$.

Therefore,

$$k \ge n$$

We now want to show that, whatever be the nodes and the quadrature formula, degree of precision cannot exceed 2n + 1 for (n + 1) nodes.

For this let us take $f(x)=\{\Pi(x)\}^2$, so that $f(x_i)=y_i=0,\quad i=0(1)n$.

$$\int_a^b f(x)dx=\sum_{i=0}^{n}H_i^{(n)}y_i+R_{n+1}(f)$$

$$=0+R_{n+1}(f)$$

Therefore,

$$\int_a^b \{\Pi(x)\}^2dx=R_{n+1}(f)>0$$

As error term is nonzero, the formula is not exact for a polynomial $\{\Pi(x)\}^2$ of degree $2n+2$, where $(n+1)$ nodes are taken. Hence, degree of precision $k\le 2n+1$.

Hence, $n\le k\le 2n+1$. ■

5.3 NEWTON–COTES FORMULA FOR NUMERICAL INTEGRATION (CLOSED TYPE)

We are to find out the value of the integral $I=\int_a^b f(x)dx$ for a given function $y=f(x)$.

We first divide the closed interval $[a,b]$ into n equal parts by the points x_i such that, $x_0=a$, $x_n=b$, and $x_i=x_0+ih=a+ih$, where $h=(b-a)/n\equiv$ width of each subinterval.

We choose the points $x_i, i=0(1)n$ as interpolating points, then $y_i=f(x_i)$.

We then replace $f(x)$ by the Lagrangian interpolating polynomial, so that

$$f(x)=\sum_{i=0}^{n}\frac{\Pi(x)}{(x-x_i)\Pi'(x_i)}y_i+\frac{\Pi(x)f^{(n+1)}(\xi)}{(n+1)!}$$

and

$$\int_a^b f(x)dx=\sum_{i=0}^{n}H_i^{(n)}y_i+R_{n+1}(f)$$

where $H_i^{(n)}$ and $R_{n+1}(f)$ are given by (5.3) and (5.4).

To evaluate $H_i^{(n)}$, we take $x=x_0+th$ where $0\le t\le n$.

Now,

$$\Pi(x)=(x-x_0)(x-x_1)\dots(x-x_n)=\prod_{i=0}^{n}(x-x_i)=\prod_{i=0}^{n}(x_0+th-x_0-ih)$$

$$=\prod_{i=0}^{n}h\,(t-i)=h^{n+1}t(t-1)(t-2)\dots(t-n)$$

$$\begin{aligned}\Pi'(x_i) &= (x_i - x_0)(x_i - x_1)\ldots(x_i - x_{i-1})(x_i - x_{i+1})\ldots(x_i - x_n)\\ &= \{ih\cdot(i-1)h\cdot\ldots\cdot h\}\{(-h)(-2h)\ldots(-(n-i)h)\}\\ &= i!h^i(-1)^{n-i}(n-i)!h^{n-i}\\ &= (-1)^{n-i}h^n i!(n-i)!,\quad i = 0(1)n\end{aligned}$$

From Equation 5.3,

$$H_i^{(n)} = \int_a^b \frac{\Pi(x)dx}{(x-x_i)\Pi'(x_i)}$$

$$= \int_0^n \frac{h^{n+1}t(t-1)(t-2)\ldots(t-n)}{(t-i)h(-1)^{n-i}h^n i!(n-i)!}hdt \tag{5.5}$$

$$= \frac{(-1)^{n-i}h}{i!(n-i)!}\int_0^n \frac{t(t-1)(t-2)\ldots(t-n)}{(t-i)}dt$$

$$= \frac{(-1)^{n-i}(b-a)}{ni!(n-i)!}\int_0^n \frac{t(t-1)(t-2)\ldots(t-n)}{(t-i)}dt \tag{5.6}$$

$$= (b-a)K_i^{(n)},\quad i = 0(1)n$$

where:

$$K_i^{(n)} = \frac{(-1)^{n-i}}{ni!\,(n-i)!}\int_0^n \frac{t(t-1)(t-2)\ldots(t-n)}{(t-i)}dt,\quad i = 0(1)n \tag{5.7}$$

Therefore, the numerical integration formula for equispaced points reduces to

$$\int_a^b f(x)dx = \sum_{i=0}^{n} H_i^{(n)}y_i + R_{n+1}(f) = (b-a)\sum_{i=0}^{n} K_i^{(n)}y_i + R_{n+1}(f) \tag{5.8}$$

where:

$$\begin{aligned}R_{n+1}(f) &= \int_a^b \frac{\Pi(x)f^{(n+1)}(\xi)}{(n+1)!}dx = \int_0^n \frac{h^{n+1}t(t-1)\ldots(t-n)f^{(n+1)}(\xi)}{(n+1)!}hdt\\ &= \frac{h^{n+2}}{(n+1)!}\int_0^n t(t-1)\ldots(t-n)f^{(n+1)}(\xi)dt\end{aligned} \tag{5.9}$$

where ξ is a function of t and $0 < \xi < n$.

Equation 5.8 gives Newton–Cotes numerical integration formula with error term.

Note: The coefficients $K_i^{(n)}$ are called Newton–Cotes coefficient for $(n+1)$ points and are independent of the function f.

Theorem 5.2

a. If the function $f(x)$ be $n+2$ times continuously differentiable on $[a,b]$, then for $a<\xi<b$ the error committed in the closed-type Newton–Cotes formula is

$$E_{NC} \cong \frac{h^{n+3} f^{(n+2)}(\xi)}{(n+2)!} \int_0^n u^2(u-1)\ldots(u-n)du, \quad \text{if } n \text{ is even}$$

b. If the function $f(x)$ be $n+1$ times continuously differentiable on $[a,b]$, then for $a<\xi<b$ the error committed in the closed-type Newton–Cotes formula is

$$E_{NC} \cong \frac{h^{n+2} f^{(n+1)}(\xi)}{(n+1)!} \int_0^n u(u-1)\ldots(u-n)du, \quad \text{if } n \text{ is odd}$$

Proof:

From Equation 5.9, we have the error in the closed-type of Newton–Cotes quadrature formula as

$$E_{n+1}(f) = \int_a^b \frac{(x-x_0)(x-x_1)\ldots(x-x_n)f^{(n+1)}(\xi)}{(n+1)!}dx \tag{5.10}$$

Now, comparing the error terms in Equations 3.7 and 3.83, we find that

$$f[x,x_0,x_1,\ldots,x_n] = \frac{f^{(n+1)}(\xi)}{(n+1)!} \tag{5.11}$$

where $\min\{x,x_0,x_1,\ldots,x_n\} < \xi < \max\{x,x_0,x_1,\ldots,x_n\}$. $a<\xi<b$.

Let us define,

$$w_n(x) = \int_a^x (\zeta - x_0)(\zeta - x_1)\ldots(\zeta - x_n)d\zeta$$

a. For, n even, it is clear that $w_n(a) = w_n(b) = 0$, and also it can be shown that $w_n(x) > 0$ for $a<x<b$

Now from Equations 5.1 and 5.11, we have

$$E_{n+1}(f) = \int_a^b w_n'(x) f[x,x_0,x_1,\ldots,x_n]dx \tag{5.12}$$

$$= \left[w_n(x) f[x,x_0,x_1,\ldots,x_n]\right]_a^b - \int_a^b w_n(x)\frac{d}{dx} f[x,x_0,x_1,\ldots,x_n]dx$$

$$= -\int_a^b w_n(x)\frac{d}{dx} f[x,x_0,x_1,\ldots,x_n]dx \tag{5.13}$$

Again,

$$\frac{d}{dx}f[x,x_0,x_1,\ldots,x_n]=\lim_{h\to 0}\frac{f[x+h,x_0,x_1,\ldots,x_n]-f[x,x_0,x_1,\ldots,x_n]}{h}$$

$$=\lim_{h\to 0}\frac{f[x+h,x_0,x_1,\ldots,x_n]-f[x_0,x_1,\ldots,x_n,x]}{h}$$

$$=\lim_{h\to 0}f[x+h,x_0,x_1,\ldots,x_n,x]$$

$$=f[x,x_0,x_1,\ldots,x_n,x]$$

Therefore,

$$\frac{d}{dx}f[x,x_0,x_1,\ldots,x_n]=f[x,x,x_0,x_1,\ldots,x_n] \tag{5.14}$$

Thus, from Equation 5.13, we obtain

$$E_{n+1}(f)=-\int_a^b w_n(x)f[x,x,x_0,x_1,\ldots,x_n]dx$$

Using integral mean value theorem and Equation 5.13, we have

$$\begin{aligned}E_{n+1}(f)&=-f[\eta,\eta,x_0,x_1,\ldots,x_n]\int_a^b w_n(x)dx, \quad \text{where} \quad a<\eta<b\\&=-\frac{f^{(n+2)}(\xi)}{(n+2)!}\int_a^b\int_a^x(\zeta-x_0)(\zeta-x_1)\ldots(\zeta-x_n)d\zeta\,dx\end{aligned} \tag{5.15}$$

where $a<\xi<b$.

Now, we change the order of integration and then substituting $\zeta=x_0+\mu h$, $0\le\mu\le n$, we obtain

$$E_{n+1}(f)=-\frac{f^{(n+2)}(\xi)}{(n+2)!}\int_a^b\int_\zeta^b(\zeta-x_0)(\zeta-x_1)\ldots(\zeta-x_n)dx\,d\zeta$$

$$=-\frac{f^{(n+2)}(\xi)}{(n+2)!}\int_{x_0}^{x_n}(\zeta-x_0)(\zeta-x_1)\ldots(\zeta-x_n)(x_n-\zeta)d\zeta$$

$$=\frac{h^{n+3}f^{(n+2)}(\xi)}{(n+2)!}\int_0^n\mu(\mu-1)\ldots(\mu-n+1)(\mu-n)^2d\mu$$

$$=\frac{h^{n+3}f^{(n+2)}(\xi)}{(n+2)!}\int_0^n u^2(1-u)\ldots(n-u)du, \quad \text{where} \quad u=n-\mu$$

$$=\frac{h^{n+3}f^{(n+2)}(\xi)}{(n+2)!}\int_0^n u^2(u-1)\ldots(u-n)du, \quad \text{if } n \text{ is even}$$

Hence, the error in the closed-type Newton–Cotes quadrature formula is

$$E_{NC} \cong \frac{h^{n+3} f^{(n+2)}(\xi)}{(n+2)!} \int_0^n u^2(u-1)\ldots(u-n)du, \quad \text{if } n \text{ is even}$$

b. For n odd, it is clear that $w_n(a) = 0$, $w_n(b) = 2w_n(x_{n/2})$ and also it can be shown that $w_n(x) < 0$ for $a < x \leq b$.
Now, from Equation 5.12, we have

$$E_{n+1}(f) = \int_a^b w_n'(x) f[x, x_0, x_1, \ldots, x_n]dx$$

$$= \int_a^{b-h} w_n'(x) f[x, x_0, x_1, \ldots, x_n]dx + \int_{b-h}^{b} w_n'(x) f[x, x_0, x_1, \ldots, x_n]dx$$

where $w_n'(x) = (x-x_0)(x-x_1)\ldots(x-x_n)$ does not change sign in $[b-h, b]$.
Then by the properties of divided differences and using Equation 5.11, we get

$$E_{n+1}(f) = \int_a^{b-h} w_{n-1}'(x)\big(f[x, x_0, x_1, \ldots, x_{n-1}] - f[x_0, x_1, \ldots, x_n]\big)dx$$
$$+ \frac{f^{(n+1)}(\xi_1)}{(n+1)!} \int_{b-h}^{b} w_n'(x)dx, \quad \text{where} \quad a < \xi_1 < b \tag{5.16}$$

Now, $n-1$ is even, and so $w_{n-1}(a) = w_{n-1}(b-h) = 0$. In addition, $w_{n-1}(x) > 0$ for $a < x < b$. Thus,

$$\int_a^{b-h} w_{n-1}'(x)dx = 0$$

Therefore, Equation 5.16 yields

$$E_{n+1}(f) = \int_a^{b-h} w_{n-1}'(x) f[x, x_0, x_1, \ldots, x_{n-1}]dx + \frac{f^{(n+1)}(\xi_1)}{(n+1)!} \int_{b-h}^{b} w_n'(x)dx \tag{5.17}$$

For the first integral, an integration by parts and application of the integral mean value theorem as before yield

$$E_{n+1}(f)=-\frac{f^{(n+1)}(\xi_2)}{(n+1)!}\int_a^{b-h} w_{n-1}(x)\,dx+\frac{f^{(n+1)}(\xi_1)}{(n+1)!}\int_{b-h}^{b} w_n'(x)\,dx,\quad \text{where } a<\xi_2<b \tag{5.18}$$

Now,

$$\int_a^{b-h} w_n'(x)\,dx=\int_a^{b-h} w_{n-1}'(x)(x-b)\,dx$$

$$=w_{n-1}(x)(x-b)\Big|_a^{b-h}-\int_a^{b-h} w_{n-1}(x)\,dx$$

$$=-\int_a^{b-h} w_{n-1}(x)\,dx$$

Then, Equation 5.18 reduces to

$$E_{n+1}(f)=\frac{f^{(n+1)}(\xi_2)}{(n+1)!}\int_a^{b-h} w_n'(x)\,dx+\frac{f^{(n+1)}(\xi_1)}{(n+1)!}\int_{b-h}^{b} w_n'(x)\,dx$$

Since $f^{(n+1)}(x)$ is continuous on $[a,b]$, according to intermediate value theorem, there exists a point ξ, where $\min\{\xi_1,\xi_2\}<\xi<\max\{\xi_1,\xi_2\}$ such that

$$E_{n+1}(f)=\frac{f^{(n+1)}(\xi)}{(n+1)!}\int_a^{b}(x-x_0)(x-x_1)\ldots(x-x_n)\,dx \tag{5.19}$$

Now, we change the order of integration and then substituting $x=x_0+uh$, $0\le u\le n$, we obtain

$$E_{n+1}(f)=\frac{f^{(n+1)}(\xi)h^{n+2}}{(n+1)!}\int_0^{n} u(u-1)\ldots(u-n)\,du,\quad \text{if } n \text{ is odd}$$

Hence, the error in the closed-type Newton–Cotes quadrature formula is

$$E_{\text{NC}}\cong\frac{h^{n+2}f^{(n+1)}(\xi)}{(n+1)!}\int_0^{n} u(u-1)\ldots(u-n)\,du,\quad \text{if } n \text{ is odd}$$ ■

Properties:

1. $K_i^{(n)}=K_{n-i}^{(n)},\quad i=0(1)n$ (5.20)

Proof:

$$K_{n-i}^{(n)} = \frac{(-1)^{n-(n-i)}}{n(n-i)!(n-n+i)!}\int_0^n \frac{t(t-1)(t-2)\ldots(t-n)}{t-(n-i)}dt$$

$$= \frac{(-1)^i}{n\,i!\,(n-i)!}\int_0^n \frac{t(t-1)\ldots(t-n)}{(t-n+i)}dt$$

$$= \frac{(-1)^i}{n\,i!\,(n-i)!}\int_n^0 \frac{(n-u)(n-u-1)\ldots(-u)}{(-u+i)}(-1)\,du$$

$$= \frac{(-1)^i(-1)^n}{n\,i!(n-i)!}\int_0^n \frac{u(u-1)\ldots(u-n)}{u-i}du$$

$$= \frac{(-1)^{2i}(-1)^{n-i}}{ni!\,(n-i)!}\int_0^n \frac{u(u-1)\ldots(u-n)}{u-i}du$$

$$= \frac{(-1)^{n-i}}{ni!\,(n-i)!}\int_0^n \frac{t(t-1)\ldots(t-n)}{t-i}dt = K_i^{(n)} \quad \text{for} \quad i = 0(1)n$$

■

2. $\sum_{i=0}^{n} K_i^{(n)} = 1$ (5.21)

Proof: Taking $f(x) = 1$ in Equation 5.2, we get

$$\int_a^b f(x)dx = \sum_{i=0}^{n} H_i^{(n)} y_i \quad \text{as} \quad R_{n+1}(f) = 0 \quad \text{when} \quad f(x) = 1 \quad \text{and} \quad y_i = f(x_i) = 1$$

If follows that

$$\int_a^b 1dx = \sum_{i=0}^{n} H_i^{(n)}$$

Thus,

$$\sum_{i=0}^{n} H_i^{(n)} = (b-a)$$

Again, from Equation 5.6,

$$H_i^{(n)} = (b-a)K_i^{(n)} \quad \text{for} \quad i = 0(1)n$$

This implies that

$$(b-a)=\sum_{i=0}^{n} H_i^{(n)}=(b-a)\sum_{i=0}^{n} K_i^{(n)}$$

Hence,

$$\sum_{i=0}^{n} K_i^{(n)}=1$$

■

5.3.1 Deduction of Trapezoidal, Simpson's One-Third, Weddle's, and Simpson's Three-Eighth Rules from the Newton–Cotes Numerical Integration Formula

5.3.1.1 Trapezoidal Rule and Its Error Estimate

Taking $n=1$, we get two-point Newton–Cotes formula as

$$\int_a^b f(x)dx=(b-a)\sum_{i=0}^{1} K_i^{(1)}y_i+R_2(f)$$

which is called *trapezoidal rule* for numerical integration.

Now,

$$K_0^{(1)}=-\int_0^1 \frac{t(t-1)}{t}dt=-\int_0^1 (t-1)dt=-\frac{1}{2}\left[(t-1)^2\right]_0^1=\frac{1}{2}$$

and from Equation 5.21, we have

$$K_1^{(1)}=1-K_0^{(1)}=1-\frac{1}{2}=\frac{1}{2}$$

$$\int_a^b f(x)dx=(b-a)\left(\frac{1}{2}y_0+\frac{1}{2}y_1\right)+R_2(f) \tag{5.22}$$

$$=\frac{h}{2}\left(y_0+y_1\right)+R_2(f),\quad \text{where}\quad h=b-a$$

- *Truncation error in trapezoidal rule*: The truncation error in trapezoidal rule is given by

$$R_2(f)=\int_a^b f(x)\,dx-\frac{h}{2}\left(y_0+y_1\right) \tag{5.23}$$

which is obviously a function of h where $x_0=a, x_1=b$, $h=b-a=x_1-x_0$, $y_0=f(x_0)$, $y_1=f(x_1)=f(x_0+h)$.

Now, let us assume that $f(x)$ and its first, second and higher order derivatives are continuous in $[x_0,x_1]$ that is, $[x_0,x_0+h]$ so that primitive of $f(x)$ exits and let $f(x)=(d/dx)F(x)$ where $F(x)$ is the primitive of $f(x)$.

Therefore,

$$R_2(f) = \int_{x_0}^{x_0+h} f(x)dx - \frac{h}{2}\left[f(x_0) + f(x_0+h)\right]$$

$$= F(x_0+h) - F(x_0) - \frac{h}{2}\left[f(x_0) + f(x_0+h)\right]$$

$$= h\,F'(x_0) + \frac{h^2}{2!}F''(x_0) + \frac{h^3}{3!}F'''(x_0) + \cdots - \frac{h}{2}\left[f(x_0) + f(x_0+h)\right]$$

by Taylor series expansion of $F(x)$ about x_0

$$R_2(f) = hf(x_0) + \frac{h^2}{2!}f'(x_0) + \frac{h^3}{3!}f''(x_0) + \frac{h^4}{4!}f'''(x_0)\cdots$$

$$-\frac{h}{2}\left[f(x_0) + f(x_0+h)\right]$$

$$= hf(x_0) + \frac{h^2}{2!}f'(x_0) + \frac{h^3}{3!}f''(x_0) + \frac{h^4}{4!}f'''(x_0)\cdots$$

$$-\frac{h}{2}\left[f(x_0) + f(x_0) + hf'(x_0) + \frac{h^2}{2!}f''(x_0) + \cdots\right],$$ by Taylor series expansion of $f(x)$ about x_0

$$= \left(\frac{h^3}{6} - \frac{h^3}{4}\right)f''(x_0) + h^4\left(\frac{1}{24} - \frac{1}{12}\right)f'''(x_0) + \cdots$$

$$\approx -\frac{h^3}{12}f''(x_0), \quad \text{if } h \text{ is small} \tag{5.24}$$

Since $f''(x)$ is assumed to be continuous in $[a,b]$ and $h = x_1 - x_0 = b - a$ is small, we get approximately

$$R_2(f) \approx -\frac{h^3}{12}f''(\xi) \quad \text{where} \quad a < \xi < b \tag{5.25}$$

It can be verified from the Theorem 5.2 for the error of the closed-type Newton–Cotes formula when $n = 1$.

Hence, the trapezoidal rule with error term is given by

$$\int_a^b f(x)dx = \frac{h}{2}(y_0 + y_1) - \frac{h^3}{12}f''(\xi) \tag{5.26}$$

5.3.1.1.1 *Composite Trapezoidal Rule*

If h is not small so that $f(x)$ varies much in the interval $[a,b]$, we divide $[a,b]$ into n parts by the node points $x_i = x_0 + ih$ where $x_0 = a$, $x_n = b$ and $h = \left[(b-a)/n\right] \equiv$ width of each subinterval.

We then apply trapezoidal rule for each of the subintervals $[x_{i-1}, x_i]$, $i = 1,2,\ldots,n$, yielding

$$\int_a^b f(x)dx = \int_{x_0}^{x_1} f(x)dx + \int_{x_1}^{x_2} f(x)dx + \cdots + \int_{x_{n-1}}^{x_n} f(x)dx = \sum_{i=1}^{n} \int_{x_{i-1}}^{x_i} f(x)dx$$

Now, according to Equation 5.24, the leading term of truncation error or local error in subinterval $[x_0, x_1]$ is

$$-\frac{h^3}{12} f''(x_0) \sim O(h^3) \tag{5.27}$$

Similarly, the local errors in the remaining subintervals $[x_i, x_{i+1}]$, $(i = 1,2,\cdots,n-1)$ are given, respectively, by

$$-\frac{h^3}{12} f''(x_1), -\frac{h^3}{12} f''(x_2), \cdots, -\frac{h^3}{12} f''(x_{n-1})$$

Thus,

$$\begin{aligned} \int_a^b f(x)dx &= \sum_{i=1}^{n} \frac{h}{2}\left[y_{i-1} + y_i\right] - \frac{h^3}{12} \sum_{i=0}^{n-1} f''(x_i) \\ &= \frac{h}{2}\left[y_0 + y_n + 2\left(y_1 + y_2 + \cdots + y_{n-1}\right)\right] - \frac{h^3}{12} \sum_{i=0}^{n-1} f''(x_i) \end{aligned} \tag{5.28}$$

Therefore, the total error is

$$E_T^C \cong -\frac{h^3}{12}\left[f''(x_0) + f''(x_1) + \cdots + f''(x_{n-1})\right] \tag{5.29}$$

Let $f''(\xi) = \underset{a \le x \le b}{Sup} |f''(x)|$, where $a < \xi < b$

Therefore, from Equation 5.29, we get the error estimate

$$E_T^C \cong -\frac{nh^3}{12} f''(\xi) = -\frac{(b-a)h^2}{12} f''(\xi) \sim O(h^2) \tag{5.30}$$

which is called *global truncation error* in composite trapezoidal rule.

Now, we may observe that global error in trapezoidal rule is $O(h^2)$, whereas the corresponding local error is $O(h^3)$.

Thus,

$$\int_a^b f(x)dx = \frac{h}{2}\left[y_0 + y_n + 2\left(y_1 + y_2 + \cdots + y_{n-1}\right)\right] - \frac{(b-a)h^2}{12} f''(\xi), \quad a < \xi < b \tag{5.31}$$

This is called *composite trapezoidal rule* with error term.

If $f''(x)$ does not vary strongly in $(a, b+h)$, a practical estimate of E_T^C is given by

$$E_T^C \cong -\frac{h}{12}\left[\Delta^2 f\left(x_0\right) + \Delta^2 f(x_2) + \cdots + \Delta^2 f(x_{n-1})\right]$$

It follows that

$$E_T^C \cong -\frac{h}{12}\left[(y_0 + y_{n+1}) - (y_1 + y_n)\right].$$

5.3.1.1.2 Geometrical Interpretation

Let the curve $y = f(x)$ cuts the ordinates $x = a$ and $x = b$ at the points A and B with coordinates $(a, f(a))$ and $(b, f(b))$, respectively. It has been shown in Figure 5.1.

Then to find the area under the curve $y = f(x)$ bounded by ordinates at $x = a$, $x = b$ and x-axis, we replace the curve between the ordinates by the line AB and find the area of the trapezium formula with x-axis, as

$$\frac{1}{2}(x_1 - x_0)\left[f(a) + f(b)\right] = \frac{1}{2}h(y_0 + y_1)$$

From Figure 5.2, it can be observed that the geometrical significance of trapezoidal rule is that the curve $y = f(x)$ is replaced by n straight line segments joining the points $(x_0, y_0), (x_1, y_1), \ldots, (x_{n-1}, y_{n-1})$ and (x_n, y_n). The area bounded by the curve $y = f(x)$, the ordinates $x = x_0 = a$, $x = x_n = b$ and the x-axis is then approximately equal to the sum of the areas of the n trapeziums obtained as shown in Figure 5.2.

For this reason, trapezoidal rule is also known as *trapezium rule*. This formula has degree of precision 1.

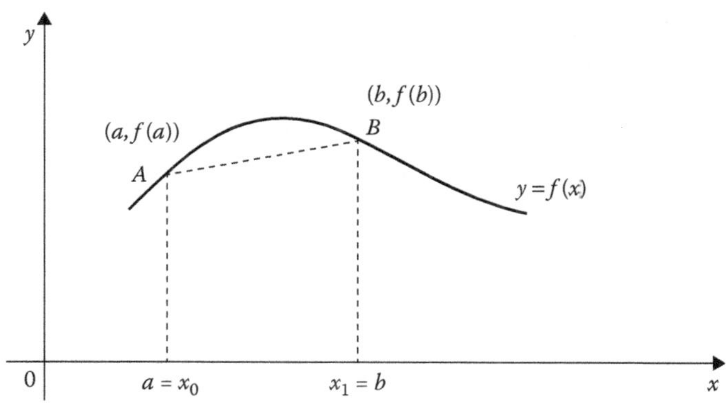

FIGURE 5.1 Geometrical interpretation of the trapezoidal rule.

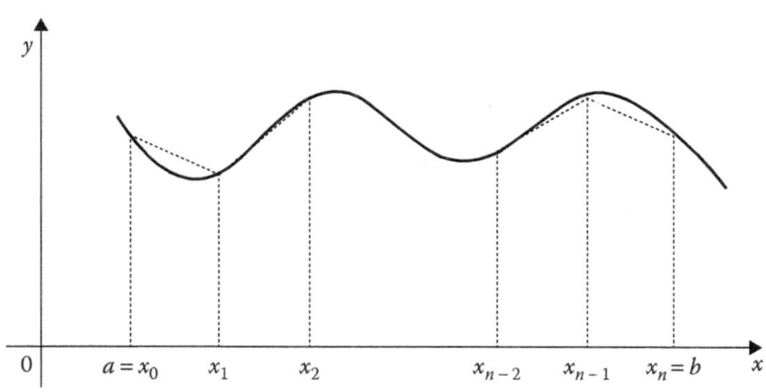

FIGURE 5.2 Geometrical interpretation of the composite trapezoidal rule.

5.3.1.1.3 Algorithm for Trapezoidal Rule

Input: Enter the given step size h, lower and upper limits of the given integral a, b, and the integrand $f(x)$.

Output: Print the value of the integral I obtained by using trapezoidal rule.

Initial step: compute

$$n = \frac{(b-a)}{h};$$

Step 1: for $i = 0 \ldots n$ do
calculate $x_i = a + i\,h$;
next compute
$$f_i = f(x_i);$$
end

Step 2: sum $= 0$;
for $j = 1 \ldots n-1$ do
sum $=$ sum $+ f_j$;
end

Step 3: compute
$$I = \frac{h}{2}(f_0 + f_n + 2*\text{sum});$$

Step 4: Print the value of the integral I.

Step 5: Stop. ■

MATHEMATICA® Program for Numerical Integration by Trapezoidal Rule (Chapter 5, Example 1)

```
a=0;b=N[Pi/2];n=6;h=N[(b-a)/n];
For[x=0,x<=b, x+=h, y[x]=Sqrt[Sin[x]];Print[y[x]]];
For[i=1;s=0,i<=n-1,i++,s+=2*y[a+i*h]];
```

$$I_T = \frac{h}{2} * (y[a] + y[b] + s)$$

Output:

```
0
0.508743
0.707107
0.840896
0.930605
0.982815
1.
1.17029
```

5.3.1.2 Simpson's One-Third Rule or Parabolic Rule with Error Term

The three-point or two-interval Newton–Cotes numerical integration formula is known as *Simpson's one-third rule* or *parabolic rule.*

Taking $n = 2$, from Equation 5.8, we get

$$\int_a^b f(x)dx = (b-a)\sum_{i=0}^{2} K_i^{(2)} y_i + R_3(f) \tag{5.32}$$

which is called Simpson's one-third rule with error term.

Now,

$$K_0^{(2)} = \frac{1}{4}\int_0^2 (t-1)(t-2)dt = \frac{1}{4}\int_0^2 (t^2-3t+2)dt = \frac{1}{6}$$

Similarly,

$$K_1^{(2)} = -\frac{1}{2}\int_0^2 t(t-2)dt = \frac{2}{3}$$

and

$$K_2^{(2)} = 1 - K_0^{(2)} - K_1^{(2)} = \frac{1}{6}$$

From (5.32) we get,

$$\int_a^b f(x)dx = (b-a)\left[\frac{1}{6}y_0 + \frac{2}{3}y_1 + \frac{1}{6}y_2\right] + R_3(f)$$

$$= \frac{(b-a)}{6}[y_0 + 4y_1 + y_2] + R_3(f) \tag{5.33}$$

$$= \frac{h}{3}[y_0 + 4y_1 + y_2] + R_3(f)$$

This formula has degree of precision 3.

- *Truncation error in Simpson's one-third rule*: The truncation error term in this formula is given by

$$R_3(f) = \int_a^b f(x)dx - \frac{h}{3}[y_0 + 4y_1 + y_2] \tag{5.34}$$

  where $y_0 = f(x_0), y_1 = f(x_0 + h), y_2 = f(x_0 + 2h)$

  Now, let $f(x)$ and its derivatives up to fourth order are all continuous in $[a,b]$, that is, in $[x_0, x_2]$ and $F'(x) = f(x)$.

  Then, by fundamental theorem of integral calculus

$$\int_a^b f(x)dx = F(b) - F(a) = F(x_0 + 2h) - F(x_0)$$

$$= F(x_0) + 2hF'(x_0) + \frac{(2h)^2}{2!}F''(x_0) + \frac{(2h)^3}{3!}F'''(x_0) + \frac{(2h)^4}{4!}F^{iv}(x_0) + \cdots - F(x_0)$$

using Taylor series expansion about x_0

$$= 2hF'(x_0) + 2h^2F''(x_0) + \frac{4}{3}h^3F'''(x_0) + \frac{2}{3}h^4F^{iv}(x_0) + \frac{4}{15}h^5F^{v}(x_0) + \cdots$$

$$= 2hf(x_0) + 2h^2f'(x_0) + \frac{4}{3}h^3f''(x_0) + \frac{2}{3}h^4f'''(x_0) + \frac{4}{15}h^5f^{iv}(x_0) + \cdots$$

Again,

$$y_0 + 4y_1 + y_2 = f(x_0) + 4f(x_0 + h) + f(x_0 + 2h)$$

$$= f(x_0) + 4\left[f(x_0) + hf'(x_0) + \frac{h^2}{2!}f''(x_0) + \frac{h^3}{3!}f'''(x_0) + \frac{h^4}{4!}f^{iv}(x_0) + \cdots\right]$$

$$+ \left[f(x_0) + 2hf'(x_0) + \frac{(2h)^2}{2!}f''(x_0) + \frac{(2h)^3}{3!}f'''(x_0) + \frac{(2h)^4}{4!}f^{iv}(x_0) + \cdots\right]$$

$$= 6f(x_0) + 6hf'(x_0) + 4h^2f''(x_0) + 2h^3f'''(x_0) + \frac{5}{6}h^4f^{iv}(x_0) + \cdots$$

Therefore, from Equation 5.34, we obtain

$$R_3(f) = \int_a^b f(x)dx - \frac{h}{3}[y_0 + 4y_1 + y_2] = -\frac{h^5}{90}f^{iv}(x_0) + \cdots$$

Therefore,

$$R_3(f) \approx -\frac{h^5}{90}f^{iv}(x_0) \tag{5.35}$$

Since $f^{iv}(x)$ is assumed to be continuous in $[a,b]$ and h is small, we get approximately

$$R_3(f) \approx -\frac{h^5}{90}f^{iv}(\xi), \quad a < \xi < b \tag{5.36}$$

if $f^{iv}(x)$ does not vary strongly in $[a,b]$.

The error term in Equation 5.36 can be verified from Theorem 5.2 for the error of closed-type Newton–Cotes formula when $n = 2$.

Therefore, Simpson's one-third rule for numerical integration is given by

$$\int_a^b f(x)dx = \frac{h}{3}[y_0 + 4y_1 + y_2] - \frac{h^5}{90}f^{iv}(\xi), \quad a < \xi < b \tag{5.37}$$

5.3.1.2.1 Composite Simpson's One-Third Rule

If h is not small, then it is not wise to take only two subintervals of $[a,b]$, since $f^{iv}(x)$ may vary much in this interval and error may be quite significant.

In this case, we divide the interval $[a,b]$ into an even number of equal subintervals by the points $a = x_0, x_1, x_2, \ldots, x_n = b$, where $n = 2k$ is even and length of each subinterval is $h = (b-a)/n = (b-a)/2k$.

Then, we apply Simpson's one-third rule to each of $n/2$ subintervals $[x_{i-2}, x_i]$, $i = 2,4,\ldots, n = 2k$ and get the value of the definite interval as

$$\int_a^b f(x)dx = \int_{x_0}^{x_2} f(x)dx + \int_{x_2}^{x_4} f(x)dx + \cdots + \int_{x_{n-2}}^{x_n} f(x)dx$$

Now, according to Equation 5.36, the leading term of truncation error or local error in subinterval $[x_0, x_2]$ is

$$-\frac{h^5}{90} f^{iv}(x_0) \sim O(h^5) \tag{5.38}$$

Similarly, the local errors in the remaining subintervals $[x_i, x_{i+2}]$, $(i = 2,4,\ldots,n-2)$ are given, respectively, by

$$-\frac{h^5}{90} f^{iv}(x_2), -\frac{h^5}{90} f^{iv}(x_4), \ldots, -\frac{h^5}{90} f^{iv}(x_{n-2})$$

Therefore,

$$\int_{x_0}^{x_2} f(x)dx = \frac{h}{3}[y_0 + 4y_1 + y_2] - \frac{h^5}{90} f^{iv}(x_0)$$

$$\int_{x_2}^{x_4} f(x)dx = \frac{h}{3}[y_2 + 4y_3 + y_4] - \frac{h^5}{90} f^{iv}(x_2)$$

$$\vdots$$

Finally,

$$\int_{x_{n-2}}^{x_n} f(x)dx = \frac{h}{3}[y_{n-2} + 4y_{n-1} + y_n] - \frac{h^5}{90} f^{iv}(x_{n-2})$$

Adding both sides of the above equations, we have

$$\begin{aligned}\int_a^b f(x)dx &= \frac{h}{3}[y_0 + 4y_1 + y_2] + \frac{h}{3}[y_2 + 4y_3 + y_4] + \cdots + \frac{h}{3}[y_{n-2} + 4y_{n-1} + y_n] \\ &\quad - \frac{h^5}{90}\left[f^{iv}(x_0) + f^{iv}(x_2) + \cdots + f^{iv}(x_{n-2})\right] \\ &= \frac{h}{3}[y_0 + 4(y_1 + y_3 + y_5 + \cdots y_{n-1}) + 2(y_2 + y_4 + \cdots + y_{n-2}) + y_n] \\ &\quad - \frac{h^5}{90}\left[f^{iv}(x_0) + f^{iv}(x_2) + \cdots + f^{iv}(x_{n-2})\right]\end{aligned} \tag{5.39}$$

Therefore, the total error is

$$E_S^C \cong -\frac{h^5}{90}\left[f^{iv}(x_0)+f^{iv}(x_2)+\cdots+f^{iv}(x_{n-2})\right] \tag{5.40}$$

Let $f^{iv}(\xi)=\underset{a\le x\le b}{Sup}\left|f^{iv}(x)\right|$, where $a<\xi<b$

Therefore, from Equation 5.40, we get the error estimate

$$E_S^C \cong -\frac{nh^5}{180}f^{iv}(\xi)=-\frac{(b-a)h^4}{180}f^{iv}(\xi)\sim O(h^4) \tag{5.41}$$

which is called the *global truncation error* in composite Simpson's one-third rule.

Now we may observe that global error in Simpson's one-third rule is $O(h^4)$, whereas the corresponding local error is $O(h^5)$.

Thus,

$$\int_a^b f(x)\,dx=\frac{h}{3}\left[y_0+4(y_1+y_3+y_5+\cdots y_{n-1})+2(y_2+y_4+\cdots+y_{n-2})+y_n\right]$$
$$-\frac{(b-a)h^4}{180}f^{iv}(\xi),\quad a<\xi<b \tag{5.42}$$

This is called composite Simpson's one-third rule for numerical integration with error term.

If $f^{iv}(x)$ does not vary strongly in (x_0,x_{n+2}), a practical estimate of E_S^C is given by

$$E_S^C \cong -\frac{h}{90}\left[\Delta^4 f(x_0)+\Delta^4 f(x_2)+\cdots+\Delta^4 f(x_{n-2})\right]$$

It follows that

$$E_S^C \cong -\frac{h}{90}\left[y_0+y_{n+2}-4(y_1+y_{n+1})+7(y_2+y_n)-8(y_3+y_5+y_{n-1})+8(y_4+y_6+\cdots+y_{n-2})\right],\quad n\ge 6$$
$$=-\frac{h}{90}\left[y_0+y_6-4(y_1+y_5)+7(y_2+y_4)-8(y_3+y_5)\right],\quad n=4$$

5.3.1.2.2 *Geometrical Significance*

In Figure 5.3, the curve $y=f(x)$ is intersected by the ordinates at $x=x_0,x_1,x_2$ at A, C, and B, respectively.

Let $y=a_0+a_1x+a_2x^2$ be a parabola passing through these points.

Therefore,

$$y_0=a_0+a_1x_0+a_2x_0^2$$
$$y_1=a_0+a_1(x_0+h)+a_2(x_0+h)^2$$
$$y_2=a_0+a_1(x_0+2h)+a_2(x_0+2h)^2$$

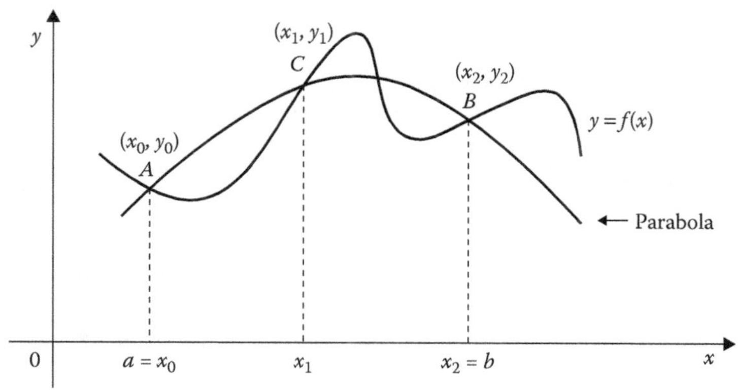

FIGURE 5.3 Geometrical interpretation of Simpson's one-third rule.

Now,

$$y_0 + 4y_1 + y_2 = 6a_0 + 6a_1(x_0 + h) + a_2(6x_0^2 + 12hx_0 + 8h^2)$$

Thus,

$$\frac{h}{3}[y_0 + 4y_1 + y_2] = h\left[2a_0 + 2a_1(x_0 + h) + a_2\left(2x_0^2 + 4x_0h + \frac{8}{3}h^2\right)\right]$$

Now, area under the parabola bounded by the ordinates at x_0 and x_2 and x-axis is given by

$$\int_{x_0}^{x_0+h} (a_0 + a_1x + a_2x^2)dx = \left[a_0x + a_1\frac{x^2}{2} + a_2\frac{x^3}{3}\right]_{x_0}^{x_0+2h}$$

$$= h\left[2a_0 + 2a_1(x_0 + h) + a_2\left(2x_0^2 + 4x_0h + \frac{8}{3}h^2\right)\right]$$

Thus, we find that the area under parabola bounded by the ordinates at $x_0 = a,\ x_0 + 2h = b$ and x-axis is same as the value of the integral given by Simpson's one-third rule excepting the error term.

From Figure 5.4, it can be observed that the geometrical significance of Simpson's one-third is that the curve $y = f(x)$ is replaced by $n/2$ parabolic chains joining three consecutive points $(x_0, y_0),\ (x_1, y_1),\ (x_2, y_2);\ (x_2, y_2),\ (x_3, y_3),\ (x_4, y_4);\ \ldots;\ (x_{n-2}, y_{n-2}),\ (x_{n-1}, y_{n-1}),\ (x_n, y_n)$, where $n = 2m,\ m \in Z^+$.

The sum of these $n/2$ areas thus obtained as shown in Figure 5.4 is then approximately equal to the area bounded by the curve $y = f(x)$, the ordinate $x = x_0 = a,\ x = x_n = b$ and x-axis.

5.3.1.2.3 Algorithm for Simpson's One-Third Rule

Input: Enter the given step size h, lower upper limits of the given integral a, b, and the integrand $f(x)$.

Output: Print the value of the integral I obtained by using Simpson's one-third rule.

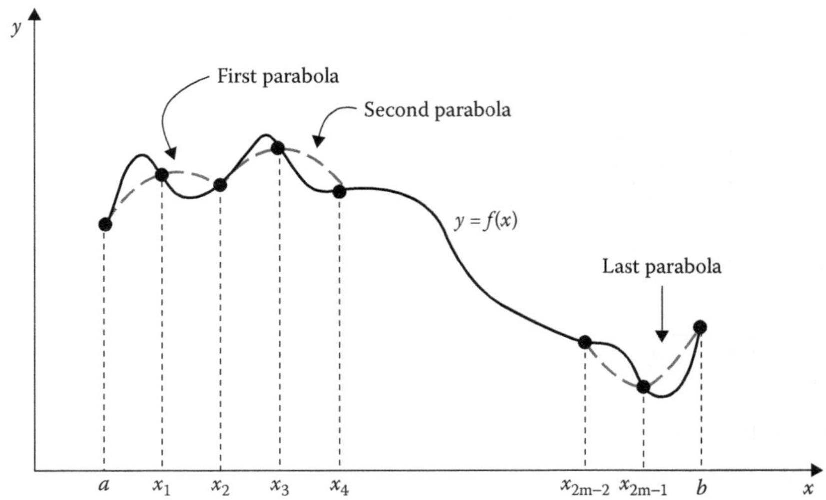

FIGURE 5.4 Geometrical interpretation of composite Simpson's one-third rule.

Initial step: compute
$n=(b-a)/h$;
Step 1: for $i=0,\ldots,n$ do
calculate $x_i = a+ih$;
next compute
$f_i = f(x_i)$;
end
Step 2: sum1 = 0;
sum2 = 0;
for $j=1,\ldots,n-1$ do
If (j mod 2)==0
$\text{sum1} = \text{sum1} + f_j$;
else
$\text{sum2} = \text{sum2} + f_j$;
end
Step 3: compute

$$I = \frac{h}{3}(f_0 + f_n + 2*\text{sum1} + 4*\text{sum2});$$

Step 4: Print the value of the integral I.
Step 5: Stop. ■

MATHEMATICA® Program for Numerical Integration by Simpson One-Third Rule (Chapter 5, Example 1)

```
a=0;b=N[Pi/2];n=6;h=N[(b-a)/n];
For[x=0,x<=b, x+=h, y[x]=Sqrt[Sin[x]];Print[y[x]]];
For[i=1;s1=0;s2=0,i<=n-1,i++,If[Mod[i,2]==0,s1+=2*y[a+i*h],
 s2+=4*y[a+i*h]]];
```

$$I_S = \frac{h}{3}*(y[a]+y[b]+s1+s2)$$

Output:

```
0
0.508743
0.707107
0.840896
0.930605
0.982815
1.
1.18728
```

5.3.1.3 Weddle's Rule

Taking $n = 6$, we get the seven-point Newton–Cotes formula, which is called Weddle's rule. Then, from Equation 5.8, we get

$$\int_a^b f(x)dx = (b-a)\sum_{i=0}^{6} K_i^{(6)} y_i + R_7(f) \tag{5.43}$$

This is called Weddle's rule with error term.

Using Equation 5.7, the coefficients $K_i^{(6)}$ are obtained as

$$K_0^{(6)} = \frac{41}{840}, K_1^{(6)} = \frac{216}{840}, K_2^{(6)} = \frac{27}{840}, K_3^{(6)} = \frac{272}{840}, K_4^{(6)} = \frac{27}{840}, K_5^{(6)} = \frac{216}{840}, K_6^{(6)} = \frac{41}{840}$$

Then, Equation 5.43 reduce to

$$\begin{aligned}\int_a^b f(x)dx &= \frac{b-a}{840}\left[41y_0 + 216y_1 + 27y_2 + 272y_3 + 27y_4 + 216y_5 + 41y_6\right] + R_7(f) \\ &= \frac{h}{140}\left[41y_0 + 216y_1 + 27y_2 + 272y_3 + 27y_4 + 216y_5 + 41y_6\right] + R_7(f)\end{aligned} \tag{5.44}$$

where $h = (b-a)/6$.

Again, we note that

$$\begin{aligned}\Delta^6 y_0 &= (E-1)^6 y_0 \\ &= E^6 y_0 - 6E^5 y_0 + {}^6C_2 E^4 y_0 - {}^6C_3 E^3 y_0 + {}^6C_4 E^2 y_0 - {}^6C_5 E y_0 + y_0 \\ &= y_0 - 6y_1 + 15y_2 - 20y_3 + 15y_4 - 6y_5 + y_6\end{aligned}$$

Adjusting $(h/140)\Delta^6 y_0$ with Equation 5.44, we get

$$\begin{aligned}\int_a^b f(x)dx &= \frac{h}{140}\left[42y_0 + 210y_1 + 42y_2 + 252y_3 + 42y_4 + 210y_5 + 42y_6\right] + R_7(f) - \frac{h}{140}\Delta^6 y_0 \\ &= \frac{3h}{10}\left[y_0 + 5y_1 + y_2 + 6y_3 + y_4 + 5y_5 + y_6\right] + E_W\end{aligned} \tag{5.45}$$

where $E_W = R_7(f) - h/140\ \Delta^6 y_0$ is the error term.

Now, from Theorem 5.2, according to the error of closed-type Newton–Cotes formula in case of $n = 2$, the following can be obtained:

$$E_W = -\frac{h^7}{140} f^{vi}(\xi_1) - \frac{9h^9}{1400} f^{viii}(\xi_2), \quad \text{where } a < \xi_1, \xi_2 < b$$

In fact, the first term being the dominant part, we may take the error term approximately as

$$E_W \approx -\frac{h^7}{140} f^{vi}(\xi) \sim O(h^7), \quad a < \xi < b \tag{5.46}$$

Therefore, Weddle's rule together with the error term is given by

$$\int_a^b f(x)dx = \frac{3h}{10}\left[y_0 + 5y_1 + y_2 + 6y_3 + y_4 + 5y_5 + y_6\right] - \frac{h^7}{140} f^{vi}(\xi), \quad a < \xi < b \tag{5.47}$$

Therefore, the degree of precision of Weddle's rule is 5.

A practical estimate of E_W is given by

$$\begin{aligned} E_W &\approx -\frac{h}{140}\Delta^6 y_0 \\ &= -\frac{h}{140}\left[y_0 - 6y_1 + 15y_2 - 20y_3 + 15y_4 - 6y_5 + y_6\right] \end{aligned}$$

5.3.1.3.1 Composite Weddle's Rule

If h is not small so that $f^{vi}(x)$ vary strongly in the interval $[a,b]$, then error in the previous formula becomes significant. We then subdivide the closed interval $[a,b]$ by n points, $a = x_0, x_1, x_2, \ldots, x_n = b$, where $n \geq 12$ and is a multiple of 6. We then consider the subintervals $[x_0, x_6], [x_6, x_{12}], \ldots, [x_{n-6}, x_n]$, where $x_i = x_0 + ih$, $i = 0, 1, 2, \ldots, n$ with $h = (b-a)/n \equiv$ length of each subinterval and then we apply Weddle's rule in each of these subintervals.

We then get the composite Weddle's rule with error term as

$$\begin{aligned} \int_a^b f(x)dx = \frac{3h}{10}&[y_0 + y_n + 5(y_1 + y_5 + y_7 + y_{11} + \cdots + y_{n-5} + y_{n-1}) \\ &+ (y_2 + y_4 + y_8 + y_{10} + \cdots + y_{n-4} + y_{n-2}) \\ &+ 6(y_3 + y_9 + \cdots y_{n-3}) + 2(y_6 + y_{12} + \cdots y_{n-6})] - \frac{nh^7}{840} f^{vi}(\xi_1) - \frac{9nh^9}{8400} f^{viii}(\xi_2) \end{aligned} \tag{5.48}$$

where $a < \xi_1, \xi_2 < b$

Thus, the error estimate in composite Weddle's rule is given by

$$E_W^C \cong -\frac{nh^7}{840} f^{vi}(\xi_1) - \frac{9nh^9}{8400} f^{viii}(\xi_2)$$

$$\cong -\frac{nh^7}{840} f^{vi}(\xi_1), \quad \text{if } h \text{ is sufficiently small} \tag{5.49}$$

$$= -\frac{(b-a)h^6}{840} f^{vi}(\xi_1) \sim O(h^6)$$

which is called the global error in composite Weddle's rule.

For practical purpose, an estimate of error term is given by

$$E_W^C \cong -\frac{h}{140}\left[\Delta^6 y_0 + \Delta^6 y_6 + \ldots + \Delta^6 y_{n-6}\right]$$

5.3.1.3.2 Algorithm for Weddle's Rule

Input: Enter the given step size h, lower and upper limits of the given integral a, b, and the integrand $f(x)$.

Output: Print the value of the integral I obtained by using Weddle's rule.

Initial step: compute

$$n = \frac{b-a}{h};$$

Step 1: for $i = 0(1)n$ do
 calculate $x_i = a + ih$;
 next compute
 $f_i = f(x_i)$;
end

Step 2: for $i = 0(6)\,\overline{n-6}$ do

$$T_i = \frac{3*h}{10} * \left(f_i + 5f_{i+1} + f_{i+2} + 6f_{i+3} + f_{i+4} + 5f_{i+5} + f_{i+6}\right);$$

end

Step 3: sum = 0;
 for $j = 0\,(6)\,\overline{n-6}$ do
 sum = sum + T_j;
end

Step 4: assign I = sum;

Step 5: Print the value of the integral I.

Step 6: Stop. ■

MATHEMATICA® Program for Numerical Integration by Weddle's Rule (Chapter 5, Example 3)

```
a=0;
b=Pi/2;
h=Pi/24;
n=(b-a)/h;
f[x_]:=Exp[Sin[x]];
For[i=0,i<=n,i++,
x[i]=a+i*h];
For[i=0,i<=n-6,i=i+6,
T[i]=((3*h)/10)*(f[x[i]]+5*f[x[i+1]]+f[x[i+2]]+6*f[x[i+3]]+f[x[i+4]]+5*f
[x[i+5]]+f[x[i+6]])];
sum=0;
For[j=0,j<=n-6,j=j+6,
sum=sum+T[j]];
Print["sum=",N[sum]];
```

Output:

```
sum=3.10438
```

5.3.1.4 Simpson's Three-Eighth Rule with Error Term

The four-point or three-interval Newton–Cotes numerical integration formula is known as *Simpson's three-eighth rule.*

Taking $n = 3$, from Equation 5.8, we get

$$\int_a^b f(x)dx = (b-a)\sum_{i=0}^{3} K_i^{(3)} y_i + R_4(f) \tag{5.50}$$

which is called Simpson's three-eighth rule with error term.

Now,

$$K_0^{(3)} = -\frac{1}{18}\int_0^3 (t-1)(t-2)(t-3)dt = \frac{1}{8}$$

Similarly,

$$K_1^{(3)} = \frac{1}{6}\int_0^3 t(t-2)(t-3)dt = \frac{3}{8}$$

$$K_2^{(3)} = \frac{1}{6}\int_0^3 t(t-1)(t-3)dt = \frac{3}{8}$$

and

$$K_3^{(3)} = 1 - K_0^{(3)} - K_1^{(3)} - K_2^{(3)} = \frac{1}{8}$$

From (5.48) we get,

$$\begin{aligned}\int_a^b f(x)dx &= \frac{(b-a)}{8}[y_0 + 3y_1 + 3y_2 + y_3] + R_4(f) \\ &= \frac{3h}{8}[y_0 + 3y_1 + 3y_2 + y_3] + R_4(f)\end{aligned} \tag{5.51}$$

This formula has degree of precision 3.

Now, from Theorem 5.2, according to the error of closed-type Newton–Cotes formula in case of $n = 3$, the following truncation error can be obtained:

$$E_{S_1} = -\frac{3h^5}{80} f^{iv}(\xi) \sim O(h^5) \quad \text{where } a < \xi < b$$

5.3.1.4.1 *Composite Simpson's Three-Eighth Rule*

If h is not small so that $f^{iv}(x)$ vary strongly in the interval $[a,b]$, then error in the previous formula becomes significant. We then subdivide the closed interval $[a,b]$ by the n points, $a = x_0, x_1, x_2, \ldots, x_n = b$, where $n \geq 3$ and is a multiple of 3. We then consider the subintervals $[x_0, x_3], [x_3, x_6], \ldots, [x_{n-3}, x_n]$, where $x_i = x_0 + ih$, $i = 0,1,2,\ldots,n$ with $h = (b-a)/n \equiv$ length of each subinterval and then we apply Simpson's three-eighth rule in each of these subintervals.

We then get the composite Simpson's three-eighth rule with error term as

$$\int_a^b f(x)dx = \frac{3h}{8}\left[y_0 + y_n + 3\left(y_1 + y_2 + y_4 + y_5 + \cdots + y_{n-2} + y_{n-1}\right) + 2\left(y_3 + y_6 + \cdots y_{n-3}\right)\right] - \frac{nh^5}{80} f^{iv}(\xi) \quad \text{where } a < \xi < b \tag{5.52}$$

Thus, the error estimate in composite Simpson's three-eighth rule is given by

$$\begin{aligned} E_{S_1}^C &\cong -\frac{nh^5}{80} f^{iv}(\xi), \quad \text{if } h \text{ is sufficiently small} \\ &= -\frac{(b-a)h^4}{80} f^{iv}(\xi) \sim O(h^4) \end{aligned} \tag{5.53}$$

which is called the global error in composite Simpson's three-eighth rule.

5.3.1.4.2 *Algorithm for Simpson's Three-Eighth Rule*

Input: Enter the given step size h, lower and upper limits of the given integral a, b, and the integrand $f(x)$.

Output: Print the value of the integral I obtained by using Simpson's three-eighth rule.

Initial step: compute

$$n = \frac{b-a}{h};$$

Step 1: for $i = 0,\ldots,n$ do
  calculate $x_i = a + i\,h$;
  next compute
    $f_i = f(x_i)$;
 end

Step 2: sum1 = 0;
  sum2 = 0;
 for $j = 1,\ldots,n-1$ do
   If (j mod 3)==0
    sum1 = sum1 + f_j;
   else
    sum2 = sum2 + f_j;
 end

Step 3: compute

$$I = \frac{3h}{8}(f_0 + f_n + 2*\text{sum1} + 3*\text{sum2});$$

Step 4: Print the value of the integral I.

Step 5: Stop. ■

MATHEMATICA® Program for Integration by Trapezoidal and Simpson's Three-Eighth Rule (Chapter 5, Example-2)

```
a=0;b=N[1];n=6;h=N[(b-a)/n];
For[x=0,x<=b,x+=h,y[x]=Exp[-x*x];Print[y[x]]];
For[i=1;s=0,i<=n-1,i++,s+=2*y[a+i*h]];
```

$$I_T = \frac{h}{2} * \left(y[a] + y[b] + s\right)$$

```
For[i=1;s1 = 0;s2 = 0,i<=n-1,i++,If[Mod[i,3]==0,s1+=2*y[a+i*h],
s2+=3*y[a+i*h]]];
```

$$I_S = \frac{3*h}{8} * \left(y[a] + y[b] + s1 + s2\right)$$

Output:

```
1
0.972604
0.894839
0.778801
0.64118
0.499352
0.367879
0.745119
0.746838
```

5.4 NEWTON–COTES QUADRATURE FORMULA (OPEN TYPE)

In Newton–Cotes quadrature formula of open type, the interval of integration $[a,b]$ is divided into $n+2$ equal subintervals by the node points $x_i = x_0 + ih$, $i = -1,0,\ldots,n+1$, where $x_{-1} = a$, $x_{n+1} = b$ and step size $h = (b-a)/(n+2)$.

If we set $x = x_0 + th$, where $0 \le t \le n$, using Equation 5.8 of Section 5.3, we get

$$\int_a^b f(x)dx = \sum_{i=0}^{n} H_i^{(n)} y_i + R_{n+1}(f) = (b-a)\sum_{i=0}^{n} K_i^{(n)} y_i + R_{n+1}(f) \tag{5.54}$$

where:

$$\begin{aligned} H_i^{(n)} &= \int_a^b \frac{\Pi(x)dx}{(x-x_i)\Pi'(x_i)} \\ &= \int_{-1}^{n+1} \frac{h^{n+1}t(t-1)(t-2)\ldots(t-n)}{(t-i)h(-1)^{n-i}h^n i!(n-i)!} hdt \\ &= \frac{(-1)^{n-i}(b-a)}{(n+2)i!(n-i)!} \int_{-1}^{n+1} \frac{t(t-1)(t-2)\ldots(t-n)}{(t-i)} dt \\ &= (b-a)K_i^{(n)}, \quad K_i^{(n)} = \frac{(-1)^{n-i}}{(n+2)i!\,(n-i)!} \int_{-1}^{n+1} \frac{t(t-1)(t-2)\ldots(t-n)}{(t-i)} dt, \quad i = 0(1)n \end{aligned} \tag{5.55}$$

and

$$R_{n+1}(f) = \int_a^b \frac{\Pi(x) f^{(n+1)}(\zeta)}{(n+1)!} dx \tag{5.56}$$

where ζ is a function of x and $a < \zeta < b$.

Therefore, using Equations 5.55 and 5.56, Equation 5.54 becomes

$$\int_a^b f(x)dx = (b-a)\sum_{i=0}^{n} K_i^{(n)} y_i + E_{NC} \tag{5.57}$$

where:

$$E_{NC} \cong \begin{cases} \dfrac{h^{n+3} f^{(n+2)}(\xi)}{(n+2)!} \displaystyle\int_{-1}^{n+1} u(u-1)\ldots(u-n)(u-n-1)du, & \text{if } n \text{ is even} \\ \dfrac{h^{n+2} f^{(n+1)}(\xi)}{(n+1)!} \displaystyle\int_{-1}^{n+1} u(u-1)\ldots(u-n)du, & \text{if } n \text{ is odd} \end{cases}, \quad \text{for } a < \xi < b$$

Equation 5.57 gives Newton–Cotes numerical integration formula of open type with error term. The coefficients $K_i^{(n)}$ also satisfy the Equations 5.20 and 5.21.

5.5 NUMERICAL INTEGRATION FORMULA FROM NEWTON'S FORWARD INTERPOLATION FORMULA

Let the values of the function $f(x)$ be given at the $(n+1)$ equispaced points $a = x_0, x_1, x_2, \ldots, x_n = b$, where $x_i = x_0 + ih, i = 0(1)n$ and $h = (b-a)/n = (x_n - x_0)/n$.

Now to find the definite integral $\int_a^b f(x)dx$, where $f(x)$ is not known everywhere in $[a,b]$, we are to use an interpolating polynomial for $f(x)$ over $[a,b]$ and then integrate it.

We first divide the range of integration $[a,b]$ into n equal parts, each of length h by the $(n+1)$ equidistant points $a = x_0, x_1, x_2, \ldots, x_n = b$.

Instead of Lagrange's interpolating polynomial as in the previous case, we may use Newton's forward interpolating formula as

$$f(x) = \varphi_n(x) + \frac{\Pi(x) f^{(n+1)}(\xi)}{(n+1)!}, \quad \text{where } a < \xi < b$$

$$= y_0 + u\Delta y_0 + \frac{u(u-1)}{2!}\Delta^2 y_0 + \cdots \frac{u(u-1)\ldots(u-n+1)}{n!}\Delta^n y_0 + \frac{\Pi(x) f^{(n+1)}(\xi)}{(n+1)!}$$

where, $u = (x - x_0)/h$, so that $dx = hdu$.

Thus,

$$\int_a^b f(x)dx = h\int_0^n \left[y_0 + u.\Delta y_0 + \frac{u(u-1)}{2!}\Delta^2 y_0 + \cdots + \frac{u(u-1)\ldots(u-n+1)}{n!}\Delta^n y_0 \right] du + R_{n+1}(f) \tag{5.58}$$

where:

$$R_{n+1}(f)=\int_a^b \frac{\Pi(x)f^{(n+1)}(\xi)}{(n+1)!}dx$$

Now, we may now deduce different quadrature formula from this Equation 5.58.

1. *Trapezoidal Rule*
   Putting $n=1$ in Equation 5.58 we get

$$\int_a^b f(x)dx = h\int_0^1 [y_0+u\Delta y_0]du + R_2(f)$$

$$= h\left[y_0+\frac{1}{2}(y_1-y_0)\right]+R_2(f)$$

$$= \frac{h}{2}[y_0+y_1]+R_2(f)$$

where:

$$R_2(f)\cong -\frac{h^3}{12}f''(\xi), \quad a<\xi<b$$

- *Composite trapezoidal rule*: We divide the range of integration $[a,b]$ into n equal subintervals, each of width h by the $n+1$ equidistant points of x, say $a=x_0$, $x_1=x_0+h$, $x_2=x_0+2h$, …, $x_n=x_0+nh$ so that $b-a=nh$. Let the values of the integrand at these equidistant points be $y_i=f(x_i), i=0,1,2,\ldots,n$.

  Let

$$I=\int_a^b ydx=\int_{x_0}^{x_n} ydx=\int_{x_0}^{x_1} ydx+\int_{x_1}^{x_2} ydx+\cdots+\int_{x_{n-1}}^{x_n} ydx$$

Taking the first subinterval $[x_0,x_1]$ and using Newton's forward interpolation formula, from Equation 5.58, we get

$$I_1=\int_{x_0}^{x_1} ydx=\int_{x_0}^{x_1}[y_0+u\Delta y_0]dx, \quad \text{where } x=x_0+uh$$

$$= h\int_0^1 (y_0+u\Delta y_0)du$$

$$= \frac{h}{2}(y_0+y_1)$$

For the next subinterval $[x_1, x_2]$, we can obtain

$$I_2 = \int_{x_1}^{x_2} y dx = \frac{h}{2}(y_1 + y_2)$$

and so on.

Finally, for the last subinterval $[x_{n-1}, x_n]$, we have

$$I_n = \int_{x_{n-1}}^{x_n} y dx = \frac{h}{2}(y_{n-1} + y_n)$$

Adding $I_1, I_2, \ldots, I_n$, we have

$$I_T^C = I_1 + I_2 + \cdots + I_n$$

$$I_T^C = \frac{h}{2}[y_0 + y_n + 2(y_1 + y_2 + y_{n-1})]$$

which is known as composite trapezoidal rule for numerical integration.

Proceeding in the similar way as before, from Equation 5.30, the error estimate is given by

$$E_T^C \cong -\frac{nh^3}{12} f''(\xi) = -\frac{(b-a)h^2}{12} f''(\xi) \sim O(h^2) \tag{5.59}$$

Hence, the composite trapezoidal rule with error term is

$$\int_a^b f(x)dx = \frac{h}{2}\left[y_0 + y_n + 2(y_1 + y_2 + \cdots + y_{n-1})\right] - \frac{(b-a)h^2}{12} f''(\xi), \quad a < \xi < b \tag{5.60}$$

2. *Simpson's one-third rule*
   Putting $n = 2$ in Equation 5.58, we get

$$\int_a^b f(x)dx = h\int_0^2 \left[y_0 + u\Delta y_0 + \frac{u(u-1)}{2!}\Delta^2 y_0 \right] du + R_3(f)$$

$$= h\left[2y_0 + 2(y_1 - y_0) + \frac{1}{3}(y_0 - 2y_1 + y_2) \right] + R_3(f) \tag{5.61}$$

$$= \frac{h}{3}[y_0 + 4y_1 + y_2] + R_3(f)$$

where:

$$R_3(f) \cong -\frac{h^5}{90} f^{iv}(\xi), \quad a = x_0 < \xi < x_2 = b$$

- *Composite Simpson's one-third rule*: We divide the range of integration $[a,b]$ into even number of equal subintervals, that is, $n = 2m, m \in Z^+$ say, by the $n+1$ equidistant points of x, say $a = x_0$, $x_1 = x_0 + h$, $x_2 = x_0 + 2h$, …, $x_n = x_0 + nh$, so that $b - a = nh$. Let the values of the integrand at these equidistant points be $y_i = f(x_i)$, $i = 0,1,2,\ldots n$. Now, we apply Simpson's one-third rule Equation 5.61 in each of the subintervals $[x_0, x_2]$, $[x_2, x_4]$, $[x_4, x_6]$,…, $[x_{n-2}, x_n]$ so that

$$I = \int_a^b y dx = \int_{x_0}^{x_n} y dx = \int_{x_0}^{x_2} y dx + \int_{x_2}^{x_4} y dx + \int_{x_4}^{x_6} y dx + \cdots + \int_{x_{n-2}}^{x_n} y dx$$

Taking the subinterval $[x_0, x_2]$ and using Newton's forward interpolation formula, from Equation 5.58, we get

$$I_1 = \int_{x_0}^{x_2} y dx$$

$$= h \int_0^2 \left[y_0 + u\Delta y_0 + \frac{u(u-1)}{2!} \Delta^2 y_0 \right] du$$

$$= h \left[2y_0 + 2\Delta y_0 + \frac{1}{3} \Delta^2 y_0 \right]$$

$$= \frac{h}{3} \left[y_0 + 4y_1 + y_2 \right]$$

Similarly, for the interval $[x_2, x_4]$, we can obtain

$$I_2 = \int_{x_2}^{x_4} y dx = \frac{h}{3} \left[y_2 + 4y_3 + y_4 \right]$$

and finally, for the last interval $[x_{n-2}, x_n]$, we have

$$I_{\frac{n}{2}} = \int_{x_{n-2}}^{x_n} y dx = \frac{h}{3} \left[y_{n-2} + 4y_{n-1} + y_n \right]$$

Adding $I_1, I_2, \ldots, I_{\frac{n}{2}}$, we get

$$I_S^C = I_1 + I_2 + \cdots + I_{\frac{n}{2}}$$

$$I_S^C = \frac{h}{3} \left[y_0 + y_n + 4\left(y_1 + y_3 + \cdots + y_{n-1} \right) + 2\left(y_2 + y_4 + \cdots + y_{n-2} \right) \right]$$

which is known as composite Simpson's one-third rule for numerical integration.

Proceeding in the similar way as before, from Equation 5.41, the error estimate is given by

$$E_S^C \cong -\frac{nh^5}{180} f^{iv}(\xi) = -\frac{(b-a)h^4}{180} f^{iv}(\xi) \sim O(h^4) \tag{5.62}$$

Hence, the composite Simpson's one-third rule with error term is

$$\int_a^b f(x)dx = \frac{h}{3}\left[y_0 + y_n + 4(y_1 + y_3 + y_5 + \cdots y_{n-1}) + 2(y_2 + y_4 + \cdots + y_{n-2})\right] \tag{5.63}$$

$$-\frac{(b-a)h^4}{180} f^{iv}(\xi), \quad a < \xi < b$$

3. *Simpson's three-eighth rule*
   Putting $n = 3$ in Equation 5.58, we get

$$\int_a^b f(x)dx = h\int_0^3 \left[y_0 + u\Delta y_0 + \frac{u(u-1)}{2!}\Delta^2 y_0 + \frac{u(u-1)(u-2)}{3!}\Delta^3 y_0 \right] du + R_4(f)$$

$$= h\left[3y_0 + \frac{9}{2}\Delta y_0 + \frac{9}{4}\Delta^2 y_0 + \frac{3}{8}\Delta^3 y_0 \right] + R_4(f) \tag{5.64}$$

$$= \frac{3h}{8}\left[y_0 + 3y_1 + 3y_2 + y_3\right] + R_4(f)$$

where:

$$R_4(f) \cong -\frac{3h^5}{80} f^{iv}(\xi), \quad a = x_0 < \xi < x_3 = b.$$

- *Composite Simpson's three-eighth rule*: We divide the range of integration $[a,b]$ into $n = 3m$, $m \in Z^+$ number of equal subintervals by the $n+1$ equidistant points of x, say $a = x_0$, $x_1 = x_0 + h$, $x_2 = x_0 + 2h, \ldots$, $x_n = x_0 + nh$ so that $b - a = nh$. Let the values of the integrand at these equidistant points be $y_i = f(x_i)$, $i = 0,1,2,\ldots n$. Now, we apply Simpson's three-eighth rule to Equation 5.64 in each of the subintervals $[x_0, x_3]$, $[x_3, x_6], \ldots, [x_{n-3}, x_n]$ so that

$$I = \int_a^b y dx = \int_{x_0}^{x_n} y dx = \int_{x_0}^{x_3} y dx + \int_{x_3}^{x_6} y dx + \cdots + \int_{x_{n-3}}^{x_n} y dx$$

  Taking the subinterval $[x_0, x_3]$ and using Newton's forward interpolation formula, from Equation 5.58, we get

$$I_1 = \int_{x_0}^{x_3} y dx$$

$$= h\int_0^3 \left[y_0 + u\Delta y_0 + \frac{u(u-1)}{2!}\Delta^2 y_0 + \frac{u(u-1)(u-2)}{3!}\Delta^3 y_0 \right] du$$

$$= \frac{3h}{8}\left[y_0 + 3y_1 + 3y_2 + y_3\right]$$

  Similarly, for the interval $[x_3, x_6]$, we can obtain

$$I_2 = \int_{x_3}^{x_6} y dx = \frac{3h}{8}\left[y_3 + 3y_4 + 3y_5 + y_6\right]$$

and finally, for the last interval $[x_{n-3}, x_n]$, we have

$$I_{\frac{n}{3}} = \int_{x_{n-3}}^{x_n} y dx = \frac{3h}{8}\left[y_{n-3} + 3y_{n-2} + 3y_{n-1} + y_n\right]$$

Adding $I_1, I_2, \ldots, I_{\frac{n}{3}}$, we get

$$I_{S1}^{C} = I_1 + I_2 + \cdots + I_{\frac{n}{3}}$$

$$I_{S1}^{C} = \frac{3h}{8}\left[y_0 + y_n + 3\left(y_1 + y_2 + y_4 + y_5 + \cdots + y_{n-2} + y_{n-1}\right) + 2\left(y_3 + y_6 + \cdots y_{n-3}\right)\right]$$

which is known as composite Simpson's three-eighth rule for numerical integration.

Proceeding in the similar way as before, from Equation 5.53, the error estimate is given by

$$E_{S1}^{C} \cong -\frac{nh^5}{80} f^{iv}(\xi) = -\frac{(b-a)h^4}{80} f^{iv}(\xi) \sim O(h^4) \tag{5.65}$$

Hence, the composite Simpson's three-eighth rule with error term is

$$\int_a^b f(x)dx = \frac{3h}{8}\left[y_0 + y_n + 3\left(y_1 + y_2 + y_4 + y_5 + \cdots + y_{n-2} + y_{n-1}\right) + 2\left(y_3 + y_6 + \cdots y_{n-3}\right)\right] - \frac{(b-a)h^4}{80} f^{iv}(\xi), \quad a < \xi < b. \tag{5.66}$$

4. *Weddle's rule*

Putting $n = 6$ in Equation 5.58, we get

$$\begin{aligned} \int_a^b f(x)dx &= h\int_0^6 \left[y_0 + u\Delta y_0 + \frac{u(u-1)}{2!}\Delta^2 y_0 + \cdots + \frac{u(u-1)\ldots(u-5)}{6!}\Delta^6 y_0 \right] du + R_7(f) \\ &= h\left[6y_0 + 18\Delta y_0 + 27\Delta^2 y_0 + 24\Delta^3 y_0 + \frac{123}{10}\Delta^4 y_0 + \frac{33}{10}\Delta^5 y_0 + \frac{31}{140}\Delta^6 y_0 \right] + R_7(f) \\ &= \frac{h}{140}\left[41y_0 + 216y_1 + 27y_2 + 272y_3 + 27y_4 + 216y_5 + 41y_6\right] + R_7(f) \\ &= \frac{h}{140}\left[42y_0 + 210y_1 + 42y_2 + 252y_3 + 42y_4 + 210y_5 + 42y_6\right] + R_7(f) - \frac{h}{140}\Delta^6 y_0 \\ &= \frac{3h}{10}\left[y_0 + 5y_1 + y_2 + 6y_3 + y_4 + 5y_5 + y_6\right] + E_W \end{aligned} \tag{5.67}$$

Where using Equation 5.46, E_W is given by

$$E_W = R_7(f) - \frac{h}{140}\Delta^6 y_0 \approx -\frac{h^7}{140} f^{vi}(\xi) \sim O(h^7), \quad a < \xi < b$$

- *Composite Weddle's rule*: We divide the range of integration $[a,b]$ into $n = 6m,\ m \in Z^+$ number of equal subintervals by the $n+1$ equidistant points of x, say $a = x_0$, $x_1 = x_0 + h$, $x_2 = x_0 + 2h$, …, $x_n = x_0 + nh$ so that $b - a = nh$. Let the values of the integrand at these equidistant points be $y_i = f(x_i)$, where $i = 0,1,2,\ldots,n$. Now, we apply Weddle's rule to Equation 5.67 in each of the subintervals $[x_0, x_6]$, $[x_6, x_{12}]$, … , $[x_{n-6}, x_n]$ so that

$$I = \int_a^b y dx = \int_{x_0}^{x_n} y dx = \int_{x_0}^{x_6} y dx + \int_{x_6}^{x_{12}} y dx + \int_{x_{12}}^{x_{18}} y dx + \cdots + \int_{x_{n-6}}^{x_n} y dx$$

Taking the subinterval $[x_0, x_6]$ and using Newton's forward interpolation formula, from Equation 5.58, we get

$$I_1 = \int_{x_0}^{x_6} y dx$$

$$= \frac{3h}{10}\left[y_0 + 5y_1 + y_2 + 6y_3 + y_4 + 5y_5 + y_6\right]$$

Similarly, for the interval $[x_6, x_{12}]$, we can obtain

$$I_2 = \int_{x_6}^{x_{12}} y dx = \frac{3h}{10}\left[y_6 + 5y_7 + y_8 + 6y_9 + y_{10} + 5y_{11} + y_{12}\right]$$

and finally, for the last interval $[x_{n-6}, x_n]$, we have

$$I_{\frac{n}{6}} = \int_{x_{n-6}}^{x_n} y dx = \frac{3h}{10}\left[y_{n-6} + 5y_{n-5} + y_{n-4} + 6y_{n-3} + y_{n-2} + 5y_{n-1} + y_n\right]$$

Adding $I_1, I_2, \ldots, I_{\frac{n}{6}}$, we get

$$I_W^C = I_1 + I_2 + \cdots + I_{\frac{n}{6}}$$

$$I_W^C = \frac{3h}{10}\Big[y_0 + y_n + 5\left(y_1 + y_5 + y_7 + y_{11} + \cdots + y_{n-5} + y_{n-1}\right)$$

$$+ \left(y_2 + y_4 + y_8 + y_{10} + \cdots + y_{n-4} + y_{n-2}\right)$$

$$+ 6\left(y_3 + y_9 + \cdots y_{n-3}\right) + 2\left(y_6 + y_{12} + \cdots y_{n-6}\right)\Big]$$

which is known as composite Weddle's rule for numerical integration.

Proceeding in the similar way as before, from Equation 5.49, the error estimate is given by

$$E_W^C = -\frac{(b-a)h^6}{840} f^{vi}(\xi) \sim O(h^6), \quad a<\xi<b \tag{5.68}$$

Hence, the composite Weddle's rule with error term is

$$\begin{aligned}\int_a^b f(x)dx = \frac{3h}{10}\Big[& y_0 + y_n + 5\left(y_1 + y_5 + y_7 + y_{11} + \cdots + y_{n-5} + y_{n-1}\right) \\ & + \left(y_2 + y_4 + y_8 + y_{10} + \cdots + y_{n-4} + y_{n-2}\right) \\ & + 6\left(y_3 + y_9 + \cdots y_{n-3}\right) + 2\left(y_6 + y_{12} + \cdots y_{n-6}\right)\Big] \\ & - \frac{(b-a)h^6}{840} f^{vi}(\xi), \quad a<\xi<b\end{aligned} \tag{5.69}$$

Example 5.1

Evaluate $\int_0^{\pi/2}\sqrt{\sin x}\,dx$ correct to five significant digits taking $n = 6$ subintervals, using following:

1. Trapezoidal rule
2. Simpson's one-third rule

Solution:

Here, step size $h = \pi/12$. We first tabulate the functional values of $f(x)$ in the following table:

x	y
0	0
$\frac{\pi}{12}$	0.5087426
$\frac{\pi}{6}$	0.7071068
$\frac{\pi}{4}$	0.8408964
$\frac{\pi}{3}$	0.9306048
$\frac{5\pi}{12}$	0.9828152
$\frac{\pi}{2}$	1

1. Using trapezoidal rule, we get

$$I_T = \frac{\pi}{24}\left[0+1+2\times(0.5087426+0.7071068+0.8408964+0.9306048+0.9828152)\right]$$

$$= 1.1703 \quad \text{(rounding off up to five significant digits)}$$

2. Using Simpson's one-third rule, we get

$$I_s = \frac{\pi}{36}\begin{bmatrix}0+1+4\times(0.5087426+0.8408964+0.9828152) \\ +2\times(0.7071068+0.9306048)\end{bmatrix}$$

$$= 1.1873 \quad \text{(rounding off up to five significant digits)}$$

Example 5.2

Evaluate $\int_0^1 e^{-x^2}\,dx$ by using Simpson's 3/8 rule correct to five significant digits using $n=6$ subintervals.

Solution:

Here, step size $h = 1/6 = 0.1666667$. We first tabulate the functional values of $f(x)$ in the following table:

x	y
0	1
0.1666667	0.972604
0.3333334	0.894839
0.5	0.778801
0.6666667	0.641180
0.8333334	0.499352
1	0.367879

Using Simpson's three-eighth rule, we get

$$I_{s_1} = \frac{3}{48}\left[1+0.367879+3\times(0.972604+0.894839+0.641180+0.499352)+2\times 0.778801\right]$$

$$=0.74684 \quad \text{(rounding off up to five significant digits)}$$

Example 5.3

Evaluate $\int_0^{\pi/2} e^{\sin x}\,dx$ by using Weddle's rule correct to six significant digits using $n = 12$ subintervals.

Solution:

Here, step size $h=\pi/24$. We first tabulate the functional values of $f(x)$ in the following table:

x	y
0	1
$\frac{\pi}{24}$	1.13943
$\frac{\pi}{12}$	1.2954
$\frac{\pi}{8}$	1.46621
$\frac{\pi}{6}$	1.64872
$\frac{5\pi}{24}$	1.83815
$\frac{\pi}{4}$	2.02811
$\frac{7\pi}{24}$	2.2108

x	y
0	1
$\frac{\pi}{3}$	2.37744
$\frac{3\pi}{8}$	2.51904
$\frac{5\pi}{12}$	2.62722
$\frac{11\pi}{24}$	2.69513
$\frac{\pi}{2}$	2.71828

Using Weddle's rule, we get

$$I_W = \frac{3\pi}{240}\big[1+2.71828+5\times(1.13943+1.83815+2.2108+2.69513)$$
$$+(1.2954+1.64872+2.37744+2.62722)+6\times(1.46621+2.51904)+2\times 2.02811\big]$$
$$= 3.10438 \quad \text{(rounding off up to six significant digits)}$$

5.6 RICHARDSON EXTRAPOLATION

Let us consider a given composite quadrature formula with equally spaced nodes in which the error is of the form $ch^m f^{(m)}(\xi)$, where the constant c is independent of the step length h, m is a fixed positive integer and $\xi = \xi(h)$ such that $a < \xi < b$. Thus, the composite quadrature formula can be written as

$$I = \int_a^b f(x)dx = I(h) + ch^m f^{(m)}(\xi)$$

where $I(h)$ is the computed value of the integral.

Now if we carry out the computation for two different step-lengths h_1 and h_2, then letting $I_1 = I(h_1)$ and $I_2 = I(h_2)$, we have

$$\int_a^b f(x)dx = I_1 + ch_1^m f^{(m)}(\xi_1) = I_2 + ch_2^m f^{(m)}(\xi_2), \quad \text{where } a < \xi_1, \xi_2 < b \tag{5.70}$$

If $f^{(m)}(\xi)$ does not vary strongly in (a,b), we may write

$$f^{(m)}(\xi_1) \cong f^{(m)}(\xi_2)$$

Then from Equation 5.70, we obtain

$$c\left(h_1^m - h_2^m\right) f^{(m)}(\xi_2) \cong I_2 - I_1$$

Therefore,

$$ch_2^m f^{(m)}(\xi_2) \cong \frac{I_2 - I_1}{\left[\left(h_1^m / h_2^m\right) - 1\right]}$$

This gives an estimate of the error for step length h_2.

Therefore,

$$\int_a^b f(x)dx \cong I_{12}$$

where:

$$I_{12} = I_2 + \frac{I_2 - I_1}{\left(\alpha^m - 1\right)}, \quad \text{where } \alpha = \frac{h_1}{h_2} \tag{5.71}$$

It may be noted that I_2 is more accurate than I_1 if $h_2 < h_1$ and therefore, I_{12} is an improved approximation over I_2. Since $\alpha > 1$, we see that $I_{12} > I_2$ if $I_1 < I_2$ and $I_{12} < I_2$ if $I_2 < I_1$. Thus, in either case, I_{12} lies outside interval $[I_1, I_2]$ or $[I_2, I_1]$ as the case may be, that is, I_{12} is computed from I_1 and I_2 with the help of extrapolation. This process is known as *Richardson extrapolation.*

In particular, if we take $h_2 = h_1/2$ so that $\alpha = 2$, then Equation 5.71 becomes

$$I_{12} = I_2 + \frac{I_2 - I_1}{\left(2^m - 1\right)} \tag{5.72}$$

For the composite trapezoidal rule $m = 2$, Equation 5.72 reduces to

$$I_{12} = I_2 + \frac{I_2 - I_1}{3}$$

For the composite Simpson's rule $m = 4$ Equation 5.72 yields

$$I_{12} = I_2 + \frac{I_2 - I_1}{15}$$

This extrapolation process can be continued recursively.

Example 5.4

Compute $\int_0^1 e^{-x^2} dx$ by Richardson extrapolation method.

Solution:

x	y
0	1
0.1	0.99005
0.2	0.960789
0.3	0.913931
0.4	0.852144
0.5	0.7788001
0.6	0.697676
0.7	0.612626
0.8	0.527292
0.9	0.444858
1	0.367879

Taking $h_1 = 0.2$, using trapezoidal rule, we get

$$I_1 = \frac{0.2}{2}\left[1+0.367879+2\times(0.960789+0.852144+0.697676+0.527292)\right]$$
$$= 0.744368$$

Taking $h_2 = 0.1$, using trapezoidal rule, we get

$$I_2 = \frac{0.1}{2}[1+0.367879+2\times(0.99005+0.960789+0.913931+0.852144+0.7788001$$
$$+0.697676+0.612626+0.527292+0.444858)]$$
$$= 0.746211$$

Therefore,

$$I_{12} = I_2+\frac{I_2-I_1}{3} = 0.746825$$

Taking $h_1 = 0.2$, using Simpson's one-third rule, we get

$$I_1 = \frac{0.2}{3}[1+0.367879+2\times(0.852144+0.527292) + 4 \times (0.960789 + 0.697676)]$$
$$= 0.717374$$

Taking $h_2 = 0.1$, using Simpson's one-third rule, we get

$$I_2 = \frac{0.2}{3}[1+0.367879+2\times(0.960789+0.852144+0.697676+0.527292)$$
$$+ 4\times(0.99005 + 0.913931 + 0.7788001 + 0.612626 + 0.444858)]$$
$$= 0.746825$$

Therefore,

$$I_{12} = I_2 + \frac{I_2 - I_1}{15} = 0.748788$$

Note: The efficiency of the method of extrapolation is that it gives a fairly good result by computing a relatively small number of ordinates and hence the effect of round-off error is less.

MATHEMATICA® Program for Integration by Richardson Extrapolation (Chapter 5, Example 4)

```
a=0.0;b=1.0;n=10;h=0.1;
For[x=a, x<=b, x+=h, y[x]=Exp[-x^2];Print["y[",x,"]=",y[x]]];
For[i=1;s=0,i<=n-1,i++,s+=2*Exp[-(a+i*h)^2]];
```

$I_{T_2} = \frac{h}{2} * \left(Exp[-a^2] + Exp[-b^2] + s\right)$

```
For[i=1;s1=0;s2=0,i<=n-1,i++,If[Mod[i,2]==0,s1+=2*Exp[-(a+i*h)^2],
s2+=4*Exp[-(a+i*h)^2]]];
```

$I_{S_2} = \frac{h}{3} * \left(Exp[-a^2] + Exp[-b^2] + s1 + s2\right)$

```
a=0.0;b=1.0;n=5;h=0.2;
For[x=a, x<=b, x+=h, y[x]=Exp[-x^2];Print["y[",x,"]=",y[x]]];
For[i=1;s=0,i<=n-1,i++,s+=2*Exp[-(a+i*h)^2]];
```

$I_{T_1} = \frac{h}{2} * \left(Exp[-a^2] + Exp[-b^2] + s\right)$

```
For[i=1;s1=0;s2=0,i<=n-1,i++,If[Mod[i,2]==0,s1+=2*Exp[-(a+i*h)^2],
s2+=4*Exp[-(a+i*h)^2]]];
```

$I_{S_1} = \frac{h}{3} * \left(Exp[-a^2] + Exp[-b^2] + s1 + s2\right)$

$I_{TR} = I_{T2} + \frac{I_{T2} - I_{T1}}{3}$

$I_{SR} = I_{S2} + \frac{I_{S2} - I_{S1}}{15}$

Output:

```
y[0.]=1.
y[0.1]=0.99005
y[0.2]=0.960789
y[0.3]=0.913931
y[0.4]=0.852144
y[0.5]=0.778801
y[0.6]=0.697676
y[0.7]=0.612626
y[0.8]=0.527292
y[0.9]=0.444858
y[1.]=0.367879
0.746211
0.746825
y[0.]=1.
y[0.2]=0.960789
y[0.4]=0.852144
y[0.6]=0.697676
```

```
y[0.8]=0.527292
y[1.]=0.367879
0.744368
0.717374
0.746825
0.748788
```

5.7 ROMBERG INTEGRATION

When Richardson's extrapolation procedure is applied to the numerical integration, then it is called *Romberg's integration method.*

Let us consider $I = \int_a^b y dx$

Since the error in composite trapezoidal rule is of order h^2, so the composite trapezoidal rule for I with error term can be written as

$$I = I_T(h) + c_1h^2 + c_2h^4 + c_3h^6 + \cdots \tag{5.73}$$

where $I_T(h)$ is the composite trapezoidal formula for I with the subinterval of width h.

From Equation 5.73, we get

$$I = I_T\left(\frac{h}{2}\right) + c_1\frac{h^2}{4} + c_2\frac{h^4}{16} + c_3\frac{h^6}{64} + \cdots \tag{5.74}$$

$$I = I_T\left(\frac{h}{2^2}\right) + c_1\frac{h^2}{16} + c_2\frac{h^4}{256} + c_3\frac{h^6}{4096} + \cdots \tag{5.75}$$

Eliminating c_1 from Equations 5.73 and 5.74, we have

$$\begin{aligned} I &= \frac{4I_T(h/2) - I_T(h)}{4-1} - \frac{1}{4}c_2h^4 - \frac{5}{16}c_3h^6 - \cdots \\ &= I_T^{(1)}(h) - \frac{1}{4}c_2h^4 - \frac{5}{16}c_3h^6 - \cdots \end{aligned} \tag{5.76}$$

It can be easily observed that $I_T^{(1)}(h)$ is a closer approximation to I than $I_T(h)$.

Replacing h by $(h/2)$ in Equation 5.76, we get

$$I = I_T^{(1)}\left(\frac{h}{2}\right) - \frac{1}{64}c_2h^4 - \frac{5}{1024}c_3h^6 - \cdots \tag{5.77}$$

From Equations 5.76 and 5.77, we see that error terms of both approximations are of $O(h^4)$.

Again, eliminating c_2 from Equations 5.76 and 5.77, we have

$$I = I_T^{(2)}(h) + \frac{1}{64}c_3h^6 + \cdots \tag{5.78}$$

where:

$$I_T^{(2)}(h) = \frac{4^2 I_T^{(1)}(h/2) - I_T^{(1)}(h)}{4^2 - 1}$$

Here, $I_T^{(2)}(h)$ is more closer approximation to I than $I_T^{(1)}(h)$.

Thus, the successive closer approximations to I are given by the formula

$$I_T^{(r)}(h) = \frac{4^r I_T^{(r-1)}(h/2) - I_T^{(r-1)}(h)}{4^r - 1}, \quad r = 1, 2, \ldots \tag{5.79}$$

where $I_T^{(0)}(h) = I_T(h)$.

Next, we consider the application of Simpson's one-third rule. Since the error in composite Simpson's one-third is of the order h^4, the composite Simpson's one-third rule for I with error term can be written as follows:

$$I = I_S(h) + k_1 h^4 + k_2 h^6 + k_3 h^8 + \cdots \tag{5.80}$$

where $I_S(h)$ is the composite Simpson's one-third formula for I with the subinterval of width h.

From Equation 5.80, we have

$$I = I_S\left(\frac{h}{2}\right) + k_1 \frac{h^4}{16} + k_2 \frac{h^6}{64} + k_3 \frac{h^8}{256} + \cdots \tag{5.81}$$

Eliminating k_1 from Equations 5.80 and 5.81, we have

$$\begin{aligned} I &= \frac{4^2 I_S(h/2) - I_S(h)}{4^2 - 1} - \frac{1}{20} k_2 h^6 - \cdots \\ &= I_S^{(1)}(h) - \frac{1}{20} k_2 h^6 - \cdots \end{aligned} \tag{5.82}$$

It can be easily observed that $I_S^{(1)}(h)$ is a closer approximation to I than $I_S(h)$.

Proceeding in the similar way as before, we can obtain the successive closer approximations to I in case of Simpson's one-third rule as follows:

$$I_S^{(r)}(h) = \frac{4^{r+1} I_S^{(r-1)}(h/2) - I_S^{(r-1)}(h)}{4^{r+1} - 1}, \quad r = 1, 2, \ldots \tag{5.83}$$

where $I_S^{(0)}(h) = I_S(h)$.

In the case of application of the trapezoidal rule, we thus obtain a table of successive approximations to the integral as shown in Table 5.1.

From Table 5.1, it may be easily observed that the given successive entries for a particular column result in better approximation than those of the preceding ones. Also, the successive columns give better approximations than preceding ones. Moreover, the lower diagonal entries yield best results.

This method of extrapolation will be continued until

$$\left| I_T^{(r)}(h) - I_T^{(r)}\left(\frac{h}{2}\right) \right| < \varepsilon \tag{5.84}$$

for a given tolerance error ε.

Similar results hold for the case of Simpson's one-third rule also.

TABLE 5.1
Computation in Romberg's Integration Method

$O(h^2)$	$O(h^4)$	$O(h^6)$	$O(h^8)$	$O(h^{10})$
$I_T(h)$				
	$I_T^{(1)}(h)$			
$I_T\left(\frac{h}{2}\right)$		$I_T^{(2)}(h)$		
	$I_T^{(1)}\left(\frac{h}{2}\right)$		$I_T^{(3)}(h)$	
$I_T\left(\frac{h}{2^2}\right)$		$I_T^{(2)}\left(\frac{h}{2}\right)$		$I_T^{(4)}(h)$
	$I_T^{(1)}\left(\frac{h}{2^2}\right)$		$I_T^{(3)}\left(\frac{h}{2}\right)$	
$I_T\left(\frac{h}{2^3}\right)$		$I_T^{(2)}\left(\frac{h}{2^2}\right)$		
	$I_T^{(1)}\left(\frac{h}{2^3}\right)$			
$I_T\left(\frac{h}{2^4}\right)$				

5.7.1 Algorithm for Romberg's Integration

Input: Enter a positive integer J, an interval $[a,b]$ and a function $f(x)$

Output: The following algorithm computes an approximation to $I(f)=\int_a^b f(x)dx$ that is accurate to

$O\left(h^{2J+2}\right)$

Initial step:

Set $h=b-a$

Compute

$$I_{0,0}=\frac{h}{2}\left[f(a)+f(b)\right]$$

Next, set $h=h/2$

Step 1: for $j=1,2,\ldots,J$ do

$$I_{j,0}=\frac{h}{2}\left[f(a)+2\sum_{r=1}^{2^j-1}f(a+rh)+f(b)\right] \quad \text{(Composite Trapezoidal Rule)}$$

set $h=h/2$

end

Step 2: for $k=1,2,\ldots,J$ do

for $l=k,k+1,\ldots,J$ do

$$I_{l,k}=I_{l,k-1}+\frac{I_{l,k-1}-I_{l-1,k-1}}{4^k-1} \quad \text{(Richardson Extrapolation)}$$

end

end

Step 3: Stop. ■

MATHEMATICA® Program for Romberg Integration (Chapter 5, Example 5)

```
J=4;
a=1;
b=2;
f[x_]:=1/x;
h=b-a;
T[0,0]=(h/2)*(f[a]+f[b]);
h=h/2;
For[j=1,j<=J, j++,
```

$$T[j,0]=\frac{h}{2}*\left(f[a]+f[b]+2*\sum_{r=1}^{2^j-1} f[a+r*h]\right);$$

$$h=\frac{h}{2}];$$

```
For[k=1,k<=J,k++,
For[i=k,i<=J,i++,
```

$$T[i,k]=T[i,k-1]+\frac{T[i,k-1]-T[i-1,k-1]}{4^k-1};$$

```
    Print["T[",i,",",k,"]=",N[T[i, k]]]]];
```

Output:

```
T[1,1]=0.694444
T[2,1]=0.693254
T[3,1]=0.693155
T[4,1]=0.693148
T[2,2]=0.693175
T[3,2]=0.693148
T[4,2]=0.693147
T[3,3]=0.693147
T[4,3]=0.693147
T[4,4]=0.693147
```

Example 5.5

Using Romberg's integration, calculate $\int_0^1 dx/(1+x)$ with step size $h = 0.5$.

Solution:

x	y
0	1
0.125	0.889
0.25	0.8
0.375	0.7273
0.5	0.6667
0.625	0.61538
0.75	0.57143
0.875	0.53333
1	0.5

$$I_T(0.5) = \frac{0.5}{2}[1+0.5+2\times 0.6667]$$
$$= 0.70835$$

$$I_T(0.25) = \frac{0.5}{2}[1+0.5+2\times(0.8 + 0.6667+0.57143)]$$
$$= 0.6970325$$

$$I_T(0.125) = \frac{0.125}{2}\left[1+0.5+2\times(0.889+0.8+0.7273+0.6667+0.61538+0.57143+0.53333)\right]$$
$$= 0.6941425$$

The above computed values have been tabulated in the following table:

h \ I_T	$I_T(h)$	$I_T^1(h)$	$I_T^2(h)$
0.5	0.70835		
0.25	0.6970325	0.69326	0.69317378
0.125	0.6941425	0.69317917	

Hence, the required result is 0.69317, correct to five significant figures.

Example 5.6

Evaluate $\int_0^1 dx/(1+x^2)$ by using Romberg's integral method correct to 4 decimal places. Hence deduce an approximate value of π

Solution:

Taking $h = 0.5$ using trapezoidal rule, we get $I(h) = 0.755$.

Next, taking $h = 0.25$ again using trapezoidal rule, we get $I(h/2) = 0.7828$.

Finally, taking $h = 0.125$ using trapezoidal rule, we have $I(h/4) = 0.78475$.

The above successive approximations have been tabulated in the following table:

h \ I_T	$I_T(h)$	$I_T^{(1)}(h)$
0.5	0.775	
0.25	0.7828	0.7854
0.125	0.78475	0.7854

Therefore, the required value of the given integral is 0.7854, correct to four significant figures. Hence, the approximate value of π is 3.1416 correct to four significant figures.

5.8 GAUSS QUADRATURE FORMULA

In the integration formulae derived so far to evaluate $I = \int_a^b f(x)dx$, we used the values of $f(x)$ at equally spaced points of the interval [a, b]. Gauss's quadrature formula uses the same number of functional values $f(x)$ with different spacing in order to obtain better accuracy.

Let

$$I = \int_a^b w(x)y\,dx = \int_a^b w(x)f(x)dx \tag{5.85}$$

be the integral to be evaluated, where $w(x) > 0$ $(a \le x \le b)$ is the weight function.

Let us introduce transformation $x = [(b-a)/2]u + [(b+a)/2]$, so that the interval $[a,b]$ is transformed to $[-1,1]$.

Now, we have

$$y = f(x) = f\left(\frac{b-a}{2}u + \frac{b+a}{2}\right) = \phi(u)$$

$$w(x) = w\left(\frac{b-a}{2}u + \frac{b+a}{2}\right) = \psi(u)$$

and

$$dx = \left(\frac{b-a}{2}\right)du$$

Therefore, from Equation 5.85, we obtain

$$I = \left(\frac{b-a}{2}\right)\int_{-1}^{1} \psi(u)\phi(u)du \tag{5.86}$$

The n-point Gauss's quadrature formula is expressed in the form

$$\int_{-1}^{1} \psi(u)\phi(u)du = w_1\phi(u_1) + w_2\phi(u_2) + \cdots + w_n\phi(u_n) = \sum_{k=1}^{n} w_k\phi(u_k) \tag{5.87}$$

where $u_1, u_2, \ldots, u_n$ are the points of subinterval of the interval $(-1,1)$. These points are known as nodes or abscissae and $w_1, w_2, \ldots, w_n$ are the weights to be determined.

Now, suppose that, $\psi(u) = 1$. To evaluate Equation 5.87, we have to determine $2n$ unknowns u_i and w_i, $i = 1,2,\ldots,n$. Thus, $2n$ equations are required and these are obtained such that the formula in Equation 5.87 is exact for all polynomials of degree less than or equal to $2n-1$.

Therefore, we can take

$$\phi(u) = c_0 + c_1u + \cdots + c_{2n-1}u^{2n-1} \tag{5.88}$$

and so

$$\int_{-1}^{1} \phi(u)du = 2c_0 + \frac{2}{3}c_2 + \cdots + \frac{2}{2n-1}c_{2n-2} = 2\sum_{k=1}^{n} \frac{1}{2k-1}c_{2k-2} \tag{5.89}$$

Thus, from Equations 5.87 and 5.89, we get

$$w_1\phi(u_1) + w_2\phi(u_2) + \cdots + w_n\phi(u_n) = 2\sum_{k=1}^{n} \frac{1}{2k-1}c_{2k-2}$$

This implies that

$$w_1(c_0+c_1u_1+\cdots+c_{2n-1}{u_1}^{2n-1})+w_2(c_0+c_1u_2+\cdots+c_{2n-1}{u_2}^{2n-1})+\cdots$$
$$+w_n(c_0+c_1u_n+\cdots+c_{2n-1}{u_n}^{2n-1})=2\sum_{k=1}^{n}\frac{1}{2k-1}c_{2k-2} \tag{5.90}$$

Since the formula in Equation 5.87 is to be exact for all polynomials of degree less than or equal to $2n-1$, Equation 5.90 must be an identity and consequently, it is true for all values of c_j, where $j=0,1,\ldots,2n-1$. Now, comparing the different coefficients of c_j, $j=0,1,\ldots,2n-1$ from both sides of Equation 5.78, we get

$$\begin{aligned}
w_1+w_2+\cdots+w_n&=2\\
w_1u_1+w_2u_2+\cdots+w_nu_n&=0\\
w_1u_1^2+w_2u_2^2+\cdots+w_nu_n^2&=\frac{2}{3}\\
&\vdots\\
w_1u_1^{2n-2}+w_2u_2^{2n-2}+\cdots+w_nu_n^{2n-2}&=\frac{2}{2n-1}\\
w_1u_1^{2n-1}+w_2u_2^{2n-1}+\cdots+w_nu_n^{2n-1}&=0
\end{aligned} \tag{5.91}$$

Equation 5.91 form a system of $2n$ nonlinear equations in $2n$ unknowns u_i and w_i, $i=1,2,\ldots,n$. By solving these $2n$ equations, we can obtain the values of the $2n$ unknowns u_i and w_i, $i=1,2,\ldots,n$. Though theoretically it would be possible to solve Equation 5.91, but practically it is extremely complicated to solve these equations, in general.

Now, we look for alternative methods for solving the nonlinear system of equations in (5.91).

5.8.1 Guass–Legendre Integration Method

In this method, the weight function be $\psi(u)=1$. Then, from Equation 5.87, we get

$$\int_{-1}^{1}\phi(u)du=w_1\phi(u_1)+w_2\phi(u_2)+\cdots+w_n\phi(u_n)=\sum_{k=1}^{n}w_k\phi(u_k) \tag{5.92}$$

This method uses Gauss's quadrature formula, in which $u_1,u_2,\ldots,u_n$ are assumed as the roots of the equation $P_n(x)=0$, where $P_n(x)$ is the Legendre polynomial of degree n.

It is found that when $u_1,u_2,\ldots,u_n$ occurring in Equation 5.91 are the roots of the equation $P_n(x)=0$, then the values of w_i s are easily obtained.

Now, by Rodrigues' formula, we have

$$P_n(u)=\frac{1}{2^n n!}\frac{d^n}{du^n}\left(u^2-1\right)^n$$

Thus, $u_1,u_2,\ldots,u_n$ are the roots of the equation

$$\frac{d^n}{du^n}\left(u^2-1\right)^n=0 \tag{5.93}$$

whose roots are symmetrical about $u = 0$, since the power of u is even.

Case I:

When $n = 1$, Equation 5.93 gives $u = 0$, that is, $u_1 = 0$. From Equation 5.79, we have $w_1 = 2$. Therefore, from Equation 5.87, we get

$$\int_{-1}^{1} \phi(u)du = 2\phi(u_1) = 2\phi(0)$$

so that

$$\int_{a}^{b} f(x)dx \cong (b-a)f\left(\frac{a+b}{2}\right) \tag{5.94}$$

This is called *Gauss–Legendre one-point formula.*

Case II:

When $n = 2$, Equation 5.93 gives $u = \pm\left(1/\sqrt{3}\right)$. We take $u_1 = -\left(1/\sqrt{3}\right)$ and $u_2 = \left(1/\sqrt{3}\right)$.

Now, from Equation 5.91, we have

$$w_1 + w_2 = 2$$

$$-\frac{1}{\sqrt{3}}w_1 + \frac{1}{\sqrt{3}}w_2 = 0$$

Solving the above equations, we obtain

$$w_1 = w_2 = 1$$

Therefore, from Equation 5.92, we get

$$\int_{-1}^{1} \phi(u)du = \phi\left(-\frac{1}{\sqrt{3}}\right) + \phi\left(\frac{1}{\sqrt{3}}\right)$$

so that

$$\int_{a}^{b} f(x)dx \cong \frac{(b-a)}{2}\left[f\left(-\frac{b-a}{2\sqrt{3}} + \frac{b+a}{2}\right) + f\left(\frac{b-a}{2\sqrt{3}} + \frac{b+a}{2}\right)\right] \tag{5.95}$$

This is called *Gauss–Legendre two-point formula.*

Case III:

When $n = 3$, Equation 5.93 gives $u = 0, \pm\sqrt{3/5}$. We take $u_1 = -\sqrt{3/5}$, $u_2 = 0$ and $u_3 = \sqrt{3/5}$.

Now, from Equation 5.91, we have

$$w_1 + w_2 + w_3 = 2$$

$$-\sqrt{\frac{3}{5}}w_1 + \sqrt{\frac{3}{5}}w_3 = 0$$

$$\frac{3}{5}w_1 + \frac{3}{5}w_3 = \frac{2}{3}$$

Solving these equations, we get $w_1 = 5/9, w_2 = 8/9$, and $w_3 = 5/9$.

Therefore, from Equation 5.92, we get

$$\int_{-1}^{1} \phi(u)du = \frac{5}{9}\phi\left(-\sqrt{\frac{3}{5}}\right) + \frac{8}{9}\phi(0) + \frac{5}{9}\phi\left(\sqrt{\frac{3}{5}}\right)$$

so that

$$\int_a^b f(x)dx \cong \frac{(b-a)}{18}\left[5f\left(-\frac{b-a}{2}\sqrt{\frac{3}{5}} + \frac{b+a}{2}\right) + 8f\left(\frac{b+a}{2}\right) + 5f\left(\frac{b-a}{2}\sqrt{\frac{3}{5}} + \frac{b+a}{2}\right)\right] \quad (5.96)$$

This is called *Gauss–Legendre three-point formula.*

Note: From the above cases, we may note that the values of u are symmetrically placed with respect to the midpoint of the interval of integration and that the values of w are the same for each symmetric pair of u values.

In general, in order to drive the n-point Gauss–Legendre quadrature formula, we can obtain the nodes by finding the zeros of Legendre polynomial $P_n(x)$ using Rodrigues' formula and the corresponding weights can be obtained by the following equation:

$$w_k = \int_{-1}^{1} \frac{P_n(x)dx}{(x-x_k)P_n'(x_k)} = \frac{2}{(1-x_k^2)\left[P_n'(x_k)\right]^2}, \quad k = 1,2,\ldots,n$$

5.9 GAUSSIAN QUADRATURE: DETERMINATION OF NODES AND WEIGHTS THROUGH ORTHOGONAL POLYNOMIALS

Theorem 5.3

For each $n \geq 1$, Gauss quadrature integration formula has a degree of precision $2n-1$. Let $f(x) \in C^{2n}[a,b]$, then the Gauss quadrature formula and its error is given by

$$\int_a^b w(x)f(x)dx = \sum_{i=1}^{n} w_i f(x_i) + \frac{f^{(2n)}(\xi)}{(2n)!}\int_a^b \left[\pi_n(x)\right]^2 w(x)dx, \quad a < \xi < b$$

where:

$$\pi_n(x) = (x-x_1)(x-x_2)\ldots(x-x_n)$$

The nodes $\{x_i\}$ are the zeros of an orthogonal polynomial $\pi_n(x)$, which is orthogonal with respect to the weight function $w(x)$ over $[a,b]$ and the positive weights $\{w_i\}$ are given by

$$w_i = \int_a^b \frac{w(x)\pi_n(x)}{(x-x_i)\pi_n'(x_i)}dx, \quad i = 1,2,\ldots,n$$

Proof:

Let $H_{2n-1}(x)$ be the Hermite interpolating polynomial of degree $2n-1$. Then,

$$H_{2n-1}(x) = \sum_{i=1}^{n}\left[1-2(x-x_i)\omega_i'(x_i)\right]\left\{\omega_i(x)\right\}^2 f(x_i) + \sum_{i=1}^{n}(x-x_i)\left\{\omega_i(x)\right\}^2 f'(x_i) \quad (5.97)$$

where:

$$\omega_i(x) = \frac{\pi_n(x)}{(x - x_i)\pi_n'(x_i)}, \quad i = 1, 2, \ldots, n$$

Since $f(x) \in C^{2n}[a, b]$, then from Equation 3.140 of Section 3.2.14, we have the error term as

$$f(x) - H_{2n-1}(x) = \frac{f^{(2n)}(\xi)}{(2n)!}\prod_{i=1}^{n}(x - x_i)^2 = \frac{f^{(2n)}(\xi)}{(2n)!}[\pi_n(x)]^2, \quad a < \xi < b \tag{5.98}$$

This implies that

$$\int_a^b w(x)f(x)dx = \int_a^b w(x)H_{2n-1}(x)dx + \frac{f^{(2n)}(\xi)}{(2n)!}\int_a^b [\pi_n(x)]^2 w(x)dx \tag{5.99}$$

Thus, the degree of precision of Gauss quadrature is $2n - 1$, that is, the Gaussian quadrature is exact for polynomials of degree less than or equal to $2n - 1$.

Now, from Equation 5.97, we get

$$\int_a^b w(x)H_{2n-1}(x)dx = \sum_{i=1}^{n} f(x_i)\int_a^b w(x)h_i(x)dx + \sum_{i=1}^{n} f'(x_i)\int_a^b w(x)\tilde{h}_i(x)dx \tag{5.100}$$

where:

$$h_i(x) = [1 - 2(x - x_i)\omega_i'(x_i)]\{\omega_i(x)\}^2$$

and

$$\tilde{h}_i(x) = (x - x_i)\{\omega_i(x)\}^2$$

Therefore, Equation 5.100 becomes

$$\int_a^b w(x)H_{2n-1}(x)dx = \sum_{i=1}^{n} w_i f(x_i) + \sum_{i=1}^{n} f'(x_i)\int_a^b w(x)\tilde{h}_i(x)dx \tag{5.101}$$

where:

$$w_i = \int_a^b w(x)h_i(x)dx$$

Now, since degree of $\omega_i(x) = n - 1$, $\pi_n(x)$ is orthogonal to $\omega_i(x)$ with respect to the weight function $w(x) \geq 0$, and then from Equation 5.101, we have

$$\int_a^b w(x)\tilde{h}_i(x)dx = \frac{1}{\pi_n'(x_i)}\int_a^b w(x)\pi_n(x)\omega_i(x)dx = 0 \tag{5.102}$$

Therefore, using Equations 5.101 and 5.102, Equation 5.99 becomes

$$\int_a^b w(x)f(x)dx = \sum_{i=1}^{n} w_i f(x_i) + \frac{f^{(2n)}(\xi)}{(2n)!}\int_a^b [\pi_n(x)]^2 w(x)dx \tag{5.103}$$

where:

$$w_i = \int_a^b w(x)h_i(x)dx, \quad i = 1,2,\ldots,n$$

Again,

$$w_i = \int_a^b w(x)h_i(x)dx = \int_a^b w(x)\left[1-2(x-x_i)\omega_i'(x_i)\right]\{\omega_i(x)\}^2 dx$$

$$= \int_a^b w(x)\{\omega_i(x)\}^2 dx - 2\omega_i'(x_i)\int_a^b w(x)(x-x_i)\{\omega_i(x)\}^2 dx$$

$$= \int_a^b w(x)\{\omega_i(x)\}^2 dx - 2\omega_i'(x_i)\int_a^b w(x)\tilde{h}_i(x)dx$$

Since from Equation 5.9, we have

$$\int_a^b w(x)\tilde{h}_i(x)dx = \frac{1}{\pi_n'(x_i)}\int_a^b w(x)\pi_n(x)\omega_i(x)dx = 0$$

Therefore,

$$w_i = \int_a^b w(x)\{\omega_i(x)\}^2 dx > 0, \quad i = 1,2,\ldots,n \tag{5.104}$$

Thus, all the weights are positive, for all n.

Now, from Equation 5.103, we have

$$\int_a^b w(x)f(x)dx = \sum_{k=1}^{n} w_k f(x_k) + E_n(f) \tag{5.105}$$

where:

$$E_n(f) = \frac{f^{(2n)}(\xi)}{(2n)!}\int_a^b \left[\pi_n(x)\right]^2 w(x)dx$$

To construct w_i, we substitute $f(x) = \omega_i(x)$ into Equation 5.93, yielding

$$\int_a^b w(x)\omega_i(x)dx = \sum_{k=1}^{n} w_k\omega_i(x_k) + E_n(f)$$

Since $\omega_i(x)$ is a polynomial of degree $n-1$, $E_n(f) = 0$.

Thus,

$$\int_a^b w(x)\omega_i(x)dx = \sum_{k=1}^{n} w_k\omega_i(x_k) = w_i, \quad \text{since } \omega_i(x_k) = \delta_{ik} \tag{5.106}$$

Therefore,

$$w_i = \int_a^b w(x)\omega_i(x)dx = \int_a^b \frac{w(x)\pi_n(x)}{(x-x_i)\pi'_n(x_i)}dx, \quad i = 1,2,\ldots,n \tag{5.107}$$

The Gauss–Legendre, Gauss–Chebyshev, Gauss–Laguerre, and Gauss–Hermite formulae can be derived using this Theorem 5.3.

5.9.1 Gauss–Legendre Quadrature Method

In the Gauss–Legendre quadrature method, the weight function is taken as $w(x) = 1$ over the interval $[-1,1]$.

Then, from Equation 5.105, we get

$$\int_{-1}^{1} f(x)dx \doteq \sum_{k=1}^{n} w_k f(x_k) \tag{5.108}$$

This method gives exact result for polynomial of degree less than or equal to $2n-1$. The nodes $x_1, x_2, \ldots, x_n$ are the zeros of Legendre polynomials

$$P_n(x) = \frac{1}{2^n n\,!}\frac{d^n}{dx^n}\left(x^2-1\right)^n$$

which satisfies the recurrence equation

$$\left(1-x^2\right)P'_n(x) = -nxP_n(x) + nP_{n-1}(x) = (n+1)xP_n(x) - (n+1)P_{n+1}(x)$$

The weights are given by

$$w_k = \int_{-1}^{1} \frac{P_n(x)dx}{(x-x_k)P'_n(x_k)}, \quad k = 1,2,\ldots,n$$

Therefore,

$$w_k = -\frac{2}{(n+1)P_{n+1}(x_k)P'_n(x_k)} = \frac{2}{nP_{n-1}(x_k)P'_n(x_k)}, \quad k = 1,2,\ldots,n$$

Using recurrence relation, we get

$$w_k = \frac{2}{(1-x_k)^2\left[P'_n(x_k)\right]^2}, \quad k = 1,2,\ldots,n \tag{5.109}$$

From Equation 5.105, the error of this formula is given by

$$E_n(f) = \frac{2^{2n+1}(n!)^4 f^{(2n)}(\xi)}{(2n+1)[(2n)!]^3}, \quad -1 < \xi < 1 \tag{5.110}$$

The nodes and the corresponding weights for the Gauss–Legendre integration method are given in Table 5.2.

TABLE 5.2
Nodes and Weights for the Gauss–Legendre Quadrature Method

n	Nodes x_k	Weights w_k
2	±0.5773502692	1.0000000000
3	0.0000000000	0.888888889
	±0.7745966692	0.555555556
4	±0.3399810436	0.6521451549
	±0.8611363116	0.3478548451
5	0.0000000000	0.5688888889
	±0.5384693101	0.4786286705
	±0.9061798459	0.2369268851
6	±0.2386191861	0.4679139346
	±0.6612093865	0.3607615730
	±0.9324695142	0.1713244924

Example 5.7

Evaluate the integral $I = \int_0^1 dx/(1+x^2)$ using

1. Two-point Gauss–Legendre quadrature formula
2. Three-point Gauss–Legendre quadrature formula

Solution:

Putting $x = (u/2) + (1/2)$ we get,

$$I = \int_{-1}^{1} \frac{2du}{u^2 + 2u + 5} = \int_{-1}^{1} \phi(u)\,du$$

where $\phi(u) = 2/(u^2 + 2u + 5)$

1. The two-point Gauss's quadrature formula is

$$I = w_1\phi(u_1) + w_2\phi(u_2)$$

where u_1 and u_2 are two roots of the equation

$$\frac{d^2(u^2-1)^2}{du^2} = 0$$

This implies

$$u = \pm\frac{1}{\sqrt{3}}$$

Therefore, $u_1 = -1/\sqrt{3}$ and $u_2 = 1/\sqrt{3}$
w_1 and w_2 are given by the equation

$$w_1 + w_2 = 2 \quad \text{and} \quad -\frac{1}{\sqrt{3}}w_1 + \frac{1}{\sqrt{3}}w_2 = 0$$

Solving the above equations, we get $w_1 = w_2 = 1$

Thus, the two-point Gauss–Legendre quadrature formula becomes

$$I = \phi\left(-\frac{1}{\sqrt{3}}\right) + \phi\left(\frac{1}{\sqrt{3}}\right) = 0.786885$$

2. The three-point Gauss–Legendre quadrature formula is

$$\begin{aligned} I &= w_1\varphi(u_1) + w_2\varphi(u_2) + w_3\varphi(u_3) \\ &= \frac{5}{9}\varphi\left(-\sqrt{\frac{3}{5}}\right) + \frac{8}{9}\varphi(0) + \frac{5}{9}\varphi\left(\sqrt{\frac{3}{5}}\right) \\ &= 0.785267 \end{aligned}$$

5.9.2 Gauss–Chebyshev Quadrature Method

In the Gauss–Chebyshev quadrature method, the weight function is taken as $w(x) = 1/\sqrt{1-x^2}$ over the interval [–1,1].

Then, from Equation 5.105, we get

$$\int_{-1}^{1} \frac{1}{\sqrt{1-x^2}} f(x)\,dx \doteq \sum_{k=1}^{n} w_k f(x_k) \tag{5.111}$$

This method gives exact result for polynomial of degree less than or equal to $2n-1$. The nodes $x_1, x_2, \ldots, x_n$ are the zeros of Chebyshev polynomials.

$$T_n(x) = \cos(n \cos^{-1} x) = 0$$

Thus, the abscissas are given explicitly by

$$x_k = \cos\left[\frac{(2k-1)\pi}{2n}\right], \quad k = 1,2,\ldots,n \tag{5.112}$$

Also, the weights are given by

$$w_k = \int_{-1}^{1} \frac{T_n(x)dx}{\sqrt{1-x^2}(x-x_k)T_n'(x_k)} = \frac{-\pi}{T_{n+1}(x_k)T_n'(x_k)} = \frac{\pi}{n}, \quad k = 1,2,\ldots,n \tag{5.113}$$

From Equation 5.105, the error of this formula is given by

$$E_n(f) = \frac{2\pi}{2^{2n}(2n)!} f^{(2n)}(\xi), \quad -1 < \xi < 1 \tag{5.114}$$

The nodes and the corresponding weights for the Gauss–Chebyshev integration method are given in Table 5.3:

TABLE 5.3
Nodes and Weights for the Gauss–Chebyshev Quadrature Method

n	Nodes x_k	Weights w_k
2	±0.7071068	1.5707963
3	0.0000000000	1.0471976
	±0.8660254	1.0471976
4	±0.3826834	0.7853982
	±0.9238795	0.7853982
5	0.0000000000	0.6283185
	±0.5877853	0.6283185
	±0.9510565	0.6283185

Example 5.8

Evaluate the integral $I=\int_{-1}^{1}(1-x^2)^{3/2}dx$ using the Gauss–Chebyshev quadrature method with $n = 5$.

Solution:

Let $f(x)=(1-x^2)^2$.

Now, from Table 5.3 of nodes and weights for the Gauss–Chebyshev quadrature method, we have for $n = 5$, $x_1=-0.9510565$, $x_2=-0.5877853$, $x_3=0$, $x_4=0.5877853$, and $x_5=0.9510565$.

We also have $w_1=0.6283185$, $w_2=0.6283185$, $w_3=0.6283185$, $w_4=0.6283185$, and $w_5=0.6283185$.

By the Gauss–Chebyshev quadrature method, we obtain

$$I=\sum_{i=1}^{5}w_if(x_i)=1.1781$$

5.9.3 Gauss–Laguerre Quadrature Method

The Gauss–Lagurre quadrature method is Gaussian quadrature over the interval $[0,\infty)$ with the weight function $w(x)=e^{-x}$. This quadrature formula is given by

$$\int_{0}^{\infty}e^{-x}f(x)dx \doteq \sum_{k=1}^{n}w_kf(x_k) \tag{5.115}$$

This method gives exact result for polynomial of degree less than or equal to $2n-1$. The nodes $x_1,x_2,\ldots,x_n$ are the zeros of Laguerre polynomials

$$L_n(x)=(-1)^n e^x\frac{d^n}{dx^n}(e^{-x}x^n)$$

which satisfy the recurrence equation

$$xL_n'(x)=nL_n(x)-nL_{n-1}(x)=(x-n-1)L_n(x)+(n+1)L_{n+1}(x)$$

The lower order Laguerre polynomials are

$$L_0(x)=1,\ L_1(x)=x-1,\ L_2(x)=x^2-4x+2 \text{ and } L_3(x)=x^3-9x^2+18x-6$$

The Laguerre polynomials $L_n(x)$ are orthogonal with respect to the weight function e^{-x} on $(0,\infty)$.

$$\int_0^\infty e^{-x} L_m(x) L_n(x) dx = 0, \quad \text{for } m \neq n \tag{5.116}$$

The weights are given by

$$w_k = \int_0^\infty \frac{e^{-x} L_n(x) dx}{(x - x_k) L_n'(x_k)}, \quad k = 1, 2, \ldots, n$$

$$= \frac{1}{(n+1) L_n'(x_k) L_{n+1}(x_k)} = -\frac{1}{n L_{n-1}(x_k) L_n'(x_k)}, \quad k = 1, 2, \ldots, n$$

Using recurrence relation, we get

$$w_k = \frac{1}{x_k \left[L_n'(x_k)\right]^2} = \frac{x_k}{(n+1)^2 \left[L_{n+1}(x_k)\right]^2}, \quad k = 1, 2, \ldots, n \tag{5.117}$$

From Equation 5.105, the error of this formula is given by

$$E_n(f) = \frac{(n!)^2}{(2n)!} f^{(2n)}(\xi), \quad 0 < \xi < \infty \tag{5.118}$$

The nodes and the corresponding weights for the Gauss–Laguerre integration method are given in Table 5.4.

TABLE 5.4
Nodes and Weights for the Gauss–Laguerre Quadrature Method

n	Nodes x_k	Weights w_k
2	0.5857864376	0.8535533906
	3.4142135624	0.1464466094
3	0.4157745568	0.7110930099
	2.2942803603	0.2785177336
	6.2899450829	0.0103892565
4	0.3225476896	0.6031541043
	1.7457611012	0.3574186924
	4.5366202969	0.0388879085
	9.3950709123	0.0005392947
5	0.2635603197	0.5217556106
	1.4134030591	0.3986668111
	3.5964257710	0.0759424497
	7.0858100059	0.0036117587
	12.6408008443	0.0000233700
6	0.2228466042	0.4589646740
	1.1889321017	0.4170008308
	2.9927363261	0.1133733821
	5.7751435691	0.0103991975
	9.8374674184	0.0002610172
	15.9828739806	0.0000008985

Example 5.9

Evaluate the integral $I = \int_0^{\infty} e^{-x} \sin x \, dx$ using the Gauss–Laguerre quadrature method with $n = 6$.

Solution:

Let $f(x) = \sin x$.

Now, from Table 5.4 of nodes and weights for the Gauss–Laguerre quadrature method, we have for $n = 6$, $x_1 = 0.2228466042$, $x_2 = 1.1889321017$, $x_3 = 2.9927363261$, $x_4 = 5.7751435691$, $x_5 = 9.8374674184$, and $x_6 = 15.9828739806$.

We also have $w_1 = 0.4589646740$, $w_2 = 0.4170008308$, $w_3 = 0.1133733821$, $w_4 = 0.0103991975$, $w_5 = 0.0002610172$, and $w_6 = 0.0000008985$.

By the Gauss–Laguerre quadrature method, we obtain

$$I = \sum_{i=1}^{6} w_i f(x_i) = 0.499979 \cong 0.5$$

5.9.4 Gauss–Hermite Quadrature Method

The Gauss–Hermite quadrature method is Gaussian quadrature over the interval $(-\infty, \infty)$ with the weight function $w(x) = e^{-x^2}$. This quadrature formula is given by

$$\int_{-\infty}^{\infty} e^{-x^2} f(x) dx \doteq \sum_{k=1}^{n} w_k f(x_k) \tag{5.119}$$

The nodes $x_1, x_2, \ldots, x_n$ are the zeros of the Hermite polynomial $H_n(x)$, where

$$H_n(x) = (-1)^n e^{x^2} \frac{d^n}{dx^n}\left(e^{-x^2}\right)$$

and these zeros are symmetrical about the origin.

These polynomials satisfy the following recurrence relation:

$$H_n'(x) = 2nH_{n-1}(x) = 2xH_n(x) - H_{n+1}(x)$$

The Hermite polynomials $H_n(x)$ are orthogonal with respect to the weight function e^{-x^2} on $(-\infty, \infty)$.

$$\int_{-\infty}^{\infty} e^{-x^2} H_m(x) H_n(x) dx = 0, \quad \text{for } m \neq n \tag{5.120}$$

The weights are given by

$$w_k = \int_{-\infty}^{\infty} \frac{e^{-x^2} H_n(x) dx}{(x - x_k) H_n'(x_k)} = -\frac{2^{n+1} n! \sqrt{\pi}}{H_{n+1}(x_k) H_n'(x_k)} = \frac{2^n (n-1)! \sqrt{\pi}}{H_{n-1}(x_k) H_n'(x_k)}, \quad k = 1, 2, \ldots, n$$

Using recurrence relation, we get

$$w_k = \frac{2^{n+1} n! \sqrt{\pi}}{\left[H_n'(x_k)\right]^2} = \frac{2^{n+1} n! \sqrt{\pi}}{\left[H_{n+1}(x_k)\right]^2} = \frac{2^{n-1} n! \sqrt{\pi}}{n^2 \left[H_{n-1}(x_k)\right]^2}, \quad k = 1, 2, \ldots, n \tag{5.121}$$

TABLE 5.5
Nodes and Weights for the Gauss–Hermite Quadrature Method

n	Nodes x_k	Weights w_k
2	±0.7071067812	0.8862269255
3	0.0000000000	1.1816359006
	±1.2247448714	0.2954089752
4	±0.5246476233	0.8049140900
	±1.6506801239	0.0813123354
5	0.0000000000	0.9453087205
	±0.9585724646	0.3936193232
	±2.0201828705	0.0199532421
6	±0.4360774119	0.7264295952
	±1.3358490740	0.1570673203
	±2.3506049737	0.0045300099

From Equation 5.105, the error of this formula is given by

$$E_n(f) = \frac{n!\sqrt{\pi}}{2^n(2n)!} f^{(2n)}(\xi), \quad -\infty < \xi < \infty \tag{5.122}$$

The nodes and the corresponding weights for the Gauss–Hermite integration method are given in Table 5.5.

Example 5.10

Evaluate the integral $I = \int_{-\infty}^{\infty} e^{-x^2}/(1+x^2)\,dx$ using the Gauss–Hermite quadrature method with $n = 5$.

Solution:

Let $f(x) = 1/(1+x^2)$.

Now, from Table 5.5 of nodes and weights for the Gauss–Hermite quadrature method, we have for $n = 5$, $x_1 = -2.0201828705$, $x_2 = -0.9585724646$, $x_3 = 0$, $x_4 = 0.9585724646$, and $x_5 = 2.0201828705$.

We also have $w_1 = 0.0199532421$, $w_2 = 0.3936193232$, $w_3 = 0.9453087205$, $w_4 = 0.3936193232$, and $w_5 = 0.0199532421$.

By the Gauss–Hermite quadrature method, we obtain

$$I = \sum_{i=1}^{5} w_i f(x_i) = 1.36343$$

5.10 LOBATTO QUADRATURE METHOD

The Gauss quadrature formula does not contain the value of the integrand at the end points. In certain type of problems, it is advantageous to utilize the end values of the integrand. In the case of the Lobatto quadrature method, both the ends of the interval $[-1, 1]$ are preassigned as abscissas and the weight function is considered as unity. Therefore, the endpoints of the interval $[-1, 1]$ are included in a total of n abscissas, giving $r = n-2$ free abscissas. Abscissas are symmetrical about the origin, and in this case, the general quadrature formula is

$$\int_{-1}^{1} f(x)dx = w_1 f(-1) + w_2 f(1) + \sum_{k=2}^{n-1} w_k f(x_k), \quad n \geq 3 \tag{5.123}$$

The free abscissas x_i for $i = 2,\ldots,n-1$ are the roots of the polynomial $P'_{n-1}(x)$, that is,

$$\frac{d^{n-2}}{dx^{n-2}}(x^2-1)^{n-1} = 0 \tag{5.124}$$

The weights of the free abscissas are given by

$$\begin{aligned} w_k &= -\frac{2n}{(1-x_i)^2 P''_{n-1}(x_i) P'_{n-1}(x_i)}, \quad x_i \neq \pm 1 \\ &= \frac{2}{n(n-1)\left[P_{n-1}(x_i)\right]^2}, \quad x_i \neq \pm 1 \end{aligned} \tag{5.125}$$

Again, the weights of the end nodes are given by

$$w_1 = w_n = \frac{2}{n(n-1)} \tag{5.126}$$

The error term can obtained as

$$E = -\frac{2^{2n-1} n(n-1)^3 \left[(n-2)!\right]^4}{(2n-1)\left[(2n-2)!\right]^3} f^{(2n-2)}(\xi), \quad -1 < \xi < 1 \tag{5.127}$$

The nodes and the corresponding weights for the Lobatto quadrature method are given in Table 5.6.

TABLE 5.6
Nodes and Weights for the Lobatto Quadrature Method

n	Nodes x_k	Weights w_k
3	0.00000	1.333333
	±1.00000	0.333333
4	±0.447214	0.833333
	±1.000000	0.166667
5	0.000000	0.711111
	±0.654654	0.544444
	±1.000000	0.100000
6	±0.285232	0.554858
	±0.765055	0.378475
	±1.000000	0.066667

Example 5.11

Evaluate the integral $I = \int_1^2 dx/\sqrt{1+x^4}$ using the Lobatto quadrature method with $n = 5$.

Solution:

Substituting $x = (1/2)u + (3/2)$, we get

$$I = \int_{-1}^{1} \frac{2du}{\sqrt{16+(u+3)^4}} \equiv \int_{-1}^{1} \phi(u)\,du$$

where:

$$\phi(u) = \frac{2}{\sqrt{16+(u+3)^4}}$$

Now, from Table 5.6, we have for $n = 5$,

$u_1 = -1$, $u_2 = -0.654654$, $u_3 = 0$, $u_4 = 0.654654$ and $u_5 = 1$.

Also, $w_1 = 0.1$, $w_2 = 0.544444$, $w_3 = 0.711111$, $w_4 = 0.544444$, and $w_5 = 0.1$.

Using the Lobatto quadrature method, we obtain

$$I = \sum_{i=1}^{5} w_i \phi(u_i) = 0.430086$$

5.11 DOUBLE INTEGRATION

Let us consider the double integral of the form

$$I = \int_a^b \int_c^d f(x,y)\,dx\,dy \tag{5.128}$$

over the rectangular region bounded by the lines $x = a$, $x = b$, $y = c$, and $y = d$. This integral can be evaluated numerically by two successive integrations in x and y directions, respectively, by considering one variable at one time.

5.11.1 Trapezoidal Method

Considering y as fixed, if we evaluate the inner integral of Equation 5.128 using the trapezoidal method, we get

$$I = \frac{b-a}{2}\int_c^d [f(a,y)+f(b,y)]\,dy = \frac{b-a}{2}\left[\int_c^d f(a,y)\,dy + \int_c^d f(b,y)\,dy\right] \tag{5.129}$$

Applying this method to the two integrals on the right-hand side of Equation 5.129, we obtain

$$I = \frac{(b-a)(d-c)}{4}\left[f(a,c)+f(a,d)+f(b,c)+f(b,d)\right] \tag{5.130}$$

From Equation 5.130, we may note that the evaluation of this integral is possible only if the values of the function $f(x,y)$ are known or available at the corner points of the rectangular region $a \le x \le b$, $c \le y \le d$.

We can use the composite rule to evaluate the integral in Equation 5.128. In order to apply the composite rule of integration, we divide the interval $[a,b]$ into n equal subintervals each of length h and the interval $[c,d]$ into m equal subintervals each of length k.

Thus, we have
$x_i = x_0 + ih$, $x_0 = a$, $x_n = b$, and $h = (b-a)/n$, where $i = 0,1,2,\ldots,n$
$y_j = x_0 + jk$, $y_0 = c$, $y_m = d$ and $k = (d-c)/m$, where $j = 0,1,2,\ldots,m$
Now,

$$\int_a^b f(x,y)dx = \frac{h}{2}\left[f(x_0,y) + 2\left(f(x_1,y) + f(x_2,y) + \cdots + f(x_{n-1},y)\right) + f(x_n,y)\right] \tag{5.131}$$

Again, integrating Equation 5.131 between c and d term by term using the trapezoidal rule yields

$$\begin{aligned} I = \int_c^d\int_a^b f(x,y)\,dxdy &= \frac{h}{2}\left[\frac{k}{2}\left[f(x_0,y_0) + 2\left(f(x_0,y_1) + f(x_0,y_2) + \cdots + f(x_0,y_{m-1})\right) + f(x_0,y_m)\right]\right. \\ &\quad + 2.\frac{k}{2}\left[f(x_1,y_0) + 2\left(f(x_1,y_1) + f(x_1,y_2) + \cdots + f(x_1,y_{m-1})\right) + f(x_1,y_m)\right] \\ &\quad + \cdots \\ &\quad + 2.\frac{k}{2}\left[f(x_{n-1},y_0) + 2\left(f(x_{n-1},y_1) + f(x_{n-1},y_2) + \cdots + f(x_{n-1},y_{m-1})\right) + f(x_{n-1},y_m)\right] \\ &\quad \left. + \frac{k}{2}\left[f(x_n,y_0) + 2\left(f(x_n,y_1) + f(x_n,y_2) + \cdots + f(x_n,y_{m-1})\right) + f(x_n,y_m)\right]\right] \\ &= \frac{hk}{4}\left[f_{0,0} + 2\left(f_{0,1} + f_{0,2} + \cdots + f_{0,m-1}\right) + f_{0,m}\right. \\ &\quad + 2\sum_{i=1}^{n-1} f_{i,0} + 2\left(f_{i,1} + f_{i,2} + \cdots + f_{i,m-1}\right) + f_{i,m} \\ &\quad \left. + f_{n,0} + 2\left(f_{n,1} + f_{n,2} + \cdots + f_{n,m-1}\right) + f_{n,m}\right] \end{aligned} \tag{5.132}$$

where $f_{i,j} \equiv f(x_i, y_j)$, $i = 0,1,2,\ldots,n$ and $j = 0,1,2,\ldots,m$.

This formula is of second order in both h and k. Table 5.7 tabulates the double integration form in the trapezoidal method.

TABLE 5.7
Tabular Form of Trapezoidal Rule for Double Integration

	y_0	y_1	...	y_m	Trapezoidal Formula
x_0	$f_{0,0}$	$f_{0,1}$	...	$f_{0,m}$	$I_0 = \frac{k}{2}\left[f_{0,0} + 2\left(f_{0,1} + f_{0,2} + \cdots + f_{0,m-1}\right) + f_{0,m}\right]$
x_1	$f_{1,0}$	$f_{1,1}$	...	$f_{1,m}$	$I_1 = \frac{k}{2}\left[f_{1,0} + 2\left(f_{1,1} + f_{1,2} + \cdots + f_{1,m-1}\right) + f_{1,m}\right]$
⋮	⋮	⋮	...	⋮	⋮
x_n	$f_{n,0}$	$f_{n,1}$	...	$f_{n,m}$	$I_n = \frac{k}{2}\left[f_{n,0} + 2\left(f_{n,1} + f_{n,2} + \cdots + f_{n,m-1}\right) + f_{n,m}\right]$

$I = \frac{h}{2}\left[I_0 + 2\left(I_1 + I_2 + \cdots I_{n-1}\right) + I_n\right]$

5.11.1.1 Algorithm for the Trapezoidal Method

Input: Enter the given step sizes h,k, upper and lower limits of integral a, b, c, d, and the integrand $f(x,y)$.

Output: Print the value of the double integral obtained by using trapezoidal method.

Initial step: compute

$$n=\frac{(b-a)}{h}$$

and

$$m=\frac{(d-c)}{k}$$

Step 1: for $i=0\ldots n$ do

calculate $x_i=a+i\,h$

for $j=0...m$ do

calculate $y_j=c+jk$

next compute

$$f_{i,j}=f(x_i,y_j)$$

end

$\text{sum1}=0$

for $j=1\ldots m-1$ do

$\text{sum1}=\text{sum1}+f_{i,j}$

end

$$I_i=\frac{k}{2}\left(f_{i,0}+f_{i,m}+2*\text{sum1}\right)$$

end

Step 2: $\text{sum2}=0$

for $i=1\ldots n-1$ do

$\text{sum2}=\text{sum2}+I_i$

end

Step 3: Compute

$$I=\frac{h}{2}\left(T_0+T_n+2*\text{sum2}\right)$$

Step 4: Print the value of the double integral I.

Step 5: Stop. ■

MATHEMATICA® Program for Double Integration by Trapezoidal Rule (Chapter 5, Example 12)

```
Clear[NI, NIT, sum1,h, k,i, j,l, r]
h=0.25;
k=0.25;
For[i=1.0,i<=2.0,i=i+0.25,
For[j=1.0,j<=2.0,j=j+0.25,
    f[i, j]= 1/(i^2+j^2);
    Print["f[",i,",",j,"]=",N[f[i, j]]]];
For[sum1 = 0.0;l=1,l<1.75,l=l+0.25;
sum1=sum1+f[i, l];
Print[l];
Print[f[i, l]];
];
Print[sum1];
NI[i]=(k/2)*(N[f[i,1.0]]+N[f[i,2.0]]+2*sum1);
Print["NI[",i,"]=",NI[i]]];
For[sum1 = 0.0;r=1,r<1.75,r=r+0.25;
```

```
sum1=sum1+NI[r]];
NIT=(h/2)*(N[NI[1.0]]+N[NI[2.0]]+2*sum1);
Print["NIT=",N[NIT]]
```

$$N\left[\int_1^2\int_1^2 \frac{1}{x^2+y^2}dxdy\right]$$

Output:

```
f[1.,1.]=0.5
f[1.,1.25]=0.390244
f[1.,1.5]=0.307692
f[1.,1.75]=0.246154
f[1.,2.]=0.2
1.25
0.390244
1.5
0.307692
1.75
0.246154
0.94409
NI[1.]=0.323523
f[1.25,1.]=0.390244
f[1.25,1.25]=0.32
f[1.25,1.5]=0.262295
f[1.25,1.75]=0.216216
f[1.25,2.]=0.179775
1.25
0.32
1.5
0.262295
1.75
0.216216
0.798511
NI[1.25]=0.27088
f[1.5,1.]=0.307692
f[1.5,1.25]=0.262295
f[1.5,1.5]=0.222222
f[1.5,1.75]=0.188235
f[1.5,2.]=0.16
1.25
0.262295
1.5
0.222222
1.75
0.188235
0.672753
NI[1.5]=0.22665
f[1.75,1.]=0.246154
f[1.75,1.25]=0.216216
f[1.75,1.5]=0.188235
f[1.75,1.75]=0.163265
f[1.75,2.]=0.141593
1.25
0.216216
1.5
```

```
0.188235
1.75
0.163265
0.567717
NI[1.75]=0.190398
f[2.,1.]=0.2
f[2.,1.25]=0.179775
f[2.,1.5]=0.16
f[2.,1.75]=0.141593
f[2.,2.]=0.125
1.25
0.179775
1.5
0.16
1.75
0.141593
0.481368
NI[2.]=0.160967
NIT=0.232543
0.231307 + 0. I
```

5.11.2 Simpson's One-Third Method

Let $x_0 = a$, $x_1 = x_0 + h$, and $x_2 = b$ be the three points on the interval $[a,b]$ and $y_0 = c$, $y_1 = y_0 + k$, and $y_2 = d$ be the three points on the interval $[c,d]$, where $h = (b-a)/2$ and $k = (d-c)/2$.

Applying Simpson's one-third rule on $\int_a^b f(x,y)\,dx$, we have

$$\int_a^b f(x,y)\,dx = \frac{h}{3}\left[f(x_0,y) + 4f(x_1,y) + f(x_2,y)\right] \tag{5.133}$$

Again, integrating Equation 5.133 between c and d term by term using Simpson's one-third rule yields

$$\begin{aligned} I = \int_c^d \int_a^b f(x,y)\,dx\,dy &= \frac{h}{3}\Bigg[\frac{k}{3}\left[f(x_0,y_0) + 4f(x_0,y_1) + f(x_0,y_2)\right] \\ &\quad + 4.\frac{k}{3}\left[f(x_1,y_0) + 4f(x_1,y_1) + f(x_1,y_2)\right] \\ &\quad + \frac{k}{3}\left[f(x_2,y_0) + 4f(x_2,y_1) + f(x_2,y_2)\right]\Bigg] \\ &= \frac{hk}{9}\left[f_{0,0} + f_{0,2} + f_{2,0} + 4\left(f_{0,1} + f_{1,0} + f_{1,2} + f_{2,1}\right) + 16f_{1,1}\right] \end{aligned} \tag{5.134}$$

where:

$$f_{i,j} \equiv f(x_i, y_j) \quad i = 0,1,2 \text{ and } j = 0,1,2$$

Table 5.8 shows the tabular form of double integration in Simpson's one-third method.

TABLE 5.8
Tabular Form of Simpson's One-Third Rule for Double Integration

	y_0	y_1	…	y_m	Simpson's One-Third Formula
x_0	$f_{0,0}$	$f_{0,1}$	…	$f_{0,m}$	$I_0=\frac{k}{3}\left[f_{0,0}+4\left(f_{0,1}+f_{0,3}+\cdots+f_{0,m-1}\right)+2\left(f_{0,2}+f_{0,4}+\cdots+f_{0,m-2}\right)+f_{0,m}\right]$
x_1	$f_{1,0}$	$f_{1,1}$	…	$f_{1,m}$	$I_1=\frac{k}{3}\left[f_{1,0}+4\left(f_{1,1}+f_{1,3}+\cdots+f_{1,m-1}\right)+2\left(f_{1,2}+f_{1,4}+\cdots+f_{1,m-2}\right)+f_{1,m}\right]$
⋮	⋮	⋮	…	⋮	⋮
x_n	$f_{n,0}$	$f_{n,1}$	…	$f_{n,m}$	$I_n=\frac{k}{3}\left[f_{n,0}+4\left(f_{n,1}+f_{n,3}+\cdots+f_{n,m-1}\right)+2\left(f_{n,2}+f_{n,4}+\cdots+f_{n,m-2}\right)+f_{n,m}\right]$

$$I=\frac{h}{3}\left[I_0+4\left(I_1+I_3+\cdots+I_{n-1}\right)+2\left(I_2+I_4+\cdots+I_{n-2}\right)+I_n\right]$$

5.11.2.1 Algorithm for Simpson's Method

Input: Enter the given step sizes h, k, upper and lower limits of integral a, b, c, d, and the integrand $f(x,y)$.

Output: Print the value of the double integral I obtained by using Simpson's method.

Initial step:

Compute

$$n=\frac{(b-a)}{h}$$

and

$$m=\frac{(d-c)}{k}$$

```
Step 1: for i = 0...n do
            calculate x_i = a + i h;
            for j = 0...m do
                    calculate y_j = c + j k;
                    next compute
                            f_{i,j} = f(x_i, y_j)
            end
            sum1 = 0
            sum2 = 0
            for j = 1...m-1 do
                    If (j mod 2)==0
                            sum1 = sum1 + f_{i,j}
                        else
                            sum2 = sum2 + f_{i,j}
            end
            I_i = k/3 (f_{i,0} + f_{i,m} + 2*sum1 + 4* sum2)
        end
Step 2:     sum3 = 0;
            sum4 = 0;
        for i = 1...n-1 do
            If (i mod 2)==0
                    sum3 = sum3 + I_i
```

else $\text{sum4} = \text{sum4} + I_i$;

end

Step 3: Compute

$$I = \frac{h}{3}\left(I_0 + I_n + 2*\text{sum3} + 4*\text{sum4}\right);$$

Step 4: Print the value of the double integral I.

Step 5: Stop. ■

MATHEMATICA® Program for Double Integration by Simpson's Rule (Chapter 5, Example 13)

```
Clear[NI, NIT, sum1,h, k,i, j,l, r]
h=0.25;
k=0.25;
n=4;
m=4;
For[x[i]=0.0;l=0,x[i]<=2.0;l<=n, x[i]=x[i]+0.25;l=l+1,
For[y[j]=0.0;r=0,y[j]<=2.0;r<=m, y[j]=y[j]+0.25;r=r+1,
f[l, r]=Exp[x[i]+y[j]];
    Print["f[",x[i],",",y[j],"]=",N[f[l, r]]]];
    For[sum1 = 0.0;sum2 = 0.0;r=1,r<m, r=r+1;
If[Mod[r,2]==0,sum1=sum1+f[l, r], sum2=sum2+f[l, r];]
];
Print[sum1];
Print[sum2];
NI[l]=(k/3)*(N[f[l,0.0]]+N[f[l, m.0]]+2*sum1 + 4*sum2);
Print["NI[",l,"]=",NI[l]]];
For[sum1 = 0.0;sum2 = 0.0;k=1,k<n, k=k+1;
If[Mod[k,2]==0,sum1=sum1+NI[k], sum2=sum2+NI[k]];
   ];
NIT=(h/3)*(N[NI[0]]+N[NI[n]]+2*sum1 + 4*sum2);
Print["NIT=",N[NIT]]
```

$$N\left[\int_0^1\int_0^1 \text{Exp}[x+y]\,dxdy\right]$$

Output:

```
f[0.,0.]=1.
f[0.,0.25]=1.28403
f[0.,0.5]=1.64872
f[0.,0.75]=2.117
f[0.,1.]=2.71828
4.367
2.117
NI[0]=1.60017
f[0.25,0.]=1.28403
f[0.25,0.25]=1.64872
f[0.25,0.5]=2.117
f[0.25,0.75]=2.71828
f[0.25,1.]=3.49034
5.60734
2.71828
NI[1]=2.05466
f[0.5,0.]=1.64872
```

```
f[0.5,0.25]=2.117
f[0.5,0.5]=2.71828
f[0.5,0.75]=3.49034
f[0.5,1.]=4.48169
7.19997
3.49034
NI[2]=2.63823
f[0.75,0.]=2.117
f[0.75,0.25]=2.71828
f[0.75,0.5]=3.49034
f[0.75,0.75]=4.48169
f[0.75,1.]=5.7546
9.24495
4.48169
NI[3]=3.38755
f[1.,0.]=2.71828
f[1.,0.25]=3.49034
f[1.,0.5]=4.48169
f[1.,0.75]=5.7546
f[1.,1.]=7.38906
11.8707
5.7546
NI[4]=4.34971
NIT=2.78966
2.95249
```

Example 5.12

Evaluate the integral $\int_1^2\int_1^2 dxdy/(x^2+y^2)$ using trapezoidal rule, taking $h = k = 0.25$.

Solution:

Given that $h = k = 0.25$. Then, we have the following table for values of the integrand $1/(x^2+y^2)$.

		y					
		1	**1.25**	**1.5**	**1.75**	**2**	**Trapezoidal Formula**
	1	0.5	0.390244	0.307692	0.246154	0.2	$I_0 = 0.323523$
	1.25	0.390244	0.32	0.262295	0.216216	0.179775	$I_1 = 0.27088$
x	1.5	0.307692	0.262295	0.222222	0.188235	0.16	$I_2 = 0.22665$
	1.75	0.246154	0.216216	0.188235	0.163265	0.141593	$I_3 = 0.190398$
	2	0.2	0.179775	0.16	0.141593	0.125	$I_4 = 0.160967$
							$I = 0.232543$

Using trapezoidal rule, we obtain

$$I = \frac{h}{2}\left[I_0 + 2(I_1 + I_2 + I_3) + I_4\right]$$

$$= \frac{0.25}{2}\left[0.323523 + 2(0.27088 + 0.22665 + 0.190398) + 0.160967\right]$$

$$= 0.232543$$

Example 5.13

Evaluate the integral $\int_0^1 \int_0^1 e^{x+y}\,dxdy$ using Simpson's one-third rule, taking $h = k = 0.25$.

Solution:

Given that $h = k = 0.25$. Then, we have the following table for values of the integrand e^{x+y}.

		y					
		0	0.25	0.5	0.75	1	Simpson's One-Third Formula
	0	1	1.28403	1.64872	2.117	2.71828	$I_0 = 1.71832$
	0.25	1.28403	1.64872	2.117	2.71828	3.49034	$I_1 = 2.20637$
x	0.5	1.64872	2.117	2.71828	3.49034	4.48169	$I_2 = 2.83303$
	0.75	2.117	2.71828	3.49034	4.48169	5.7546	$I_3 = 3.63768$
	1	2.71828	3.49034	4.48169	5.7546	7.38906	$I_4 = 4.67087$
							$I = 2.95262$

Using Simpson's one-third rule, we obtain

$$I = \frac{h}{3}\left[I_0 + 2I_2 + 4\left(I_1 + I_3\right) + I_4\right]$$

$$= \frac{0.25}{3}\left[1.71832 + 2 \times 2.83303 + 4 \times (2.20637 + 3.63768) + 4.67087\right]$$

$$= 2.95262$$

5.12 BERNOULLI POLYNOMIALS AND BERNOULLI NUMBERS

The Bernoulli polynomials $B_n(x)$ are defined by Equations 5.135 and 5.136:

$$B_n(x+1) - B_n(x) = nx^{n-1} \tag{5.135}$$

$$B_n'(x) = nB_{n-1}(x) \tag{5.136}$$

From these equations, the explicit forms of the Bernoulli polynomials may be successively deduced as follows:

From Equation 5.136, we can obtain

$$B_n^{(r)}(x) = r!\binom{n}{r}B_{n-r}(x)$$

and so by Taylor's theorem

$$B_n(x+h) = \sum_{r=0}^{n} \frac{h^r}{r!} B_n^{(r)}(x) = \sum_{r=0}^{n} \binom{n}{r} h^r B_{n-r}(x) = \sum_{r=0}^{n} \binom{n}{n-r} h^{n-r} B_r(x)$$

This implies

$$B_n(x+h) = \sum_{r=0}^{n} \binom{n}{r} h^{n-r} B_r(x) \tag{5.137}$$

Putting $h = 1$ and using Equation 5.135, we get the important relation

$$\sum_{r=0}^{n-1} \binom{n}{r} B_r(x) = nx^{n-1} \tag{5.138}$$

This equation uniquely determines the polynomials $B_0(x)$, $B_1(x)$,... successively.

Putting $n = 1,2,3,\ldots$ successively in Equation 5.138, we get

$$B_0(x) = 1$$

$$B_1(x) = x - \frac{1}{2}$$

$$B_2(x) = x^2 - x + \frac{1}{6}$$

$$B_3(x) = x\left(x - \frac{1}{2}\right)(x-1)$$

and so on.

The Bernoulli numbers $B_n\,(n = 0,1,\ldots)$ are defined by

$$B_n = B_n(0) \tag{5.139}$$

The first few Bernoulli numbers are then

$$B_0 = 1,\ B_1 = -\frac{1}{2},\ B_2 = \frac{1}{6},\ B_3 = 0,\ldots$$

Putting $x = 0$ in Equation 5.135, we have

$$B_n(1) = B_n \quad (n \neq 1), \quad B_1(1) = \frac{1}{2}$$

From Equation 5.136, we can obtain

$$\int_0^1 B_n(x)dx = \frac{1}{n+1}\int_0^1 B'_{n+1}(x)dx = \frac{B_{n+1}(1) - B_{n+1}}{n+1} = 0, \quad n \geq 1$$

Putting $x = 0$ in Equation 5.137 and replacing h by x, we have

$$B_n(x) = \sum_{r=0}^{n} \binom{n}{r} B_r x^{n-r} \tag{5.140}$$

This equation gives the explicit expression of the Bernoulli polynomials in terms of the Bernoulli numbers.

5.12.1 Some Properties of Bernoulli Polynomials

1. $B_n(1-x)=(-1)^n B_n(x)$
2. $\int_0^1 B_m(x)B_n(x)dx=(-1)^{m-1}\dfrac{m!\,n!}{(m+n)!}B_{m+n},\quad m,n\geq 1$
3. $\sup\limits_{x\in[0,1]}|B_{2n}(x)|=|B_{2n}|$
4. $\sup\limits_{x\in[0,1]}|B_{2n+1}(x)|\leq\dfrac{2n+1}{4}|B_{2n}|$
5. $B_{2n+1}=0,\quad n\geq 1$
6. $B_{2n}=B_{2n}(1),\quad n\geq 1$
7. $B_n\left(\dfrac{1}{2}\right)=(2^{1-n}-1)B_n$
8. $\sum\limits_{r=0}^{n-1}\binom{n}{r}B_r=0\ (n\geq 2)$

5.13 EULER–MACLAURIN FORMULA

Let the function $f(x)$ be continuous and continuously differentiable sufficient number of times on the interval $[a,b]$. We divide the interval $[a,b]$ into n equal subintervals by the points $x_0, x_1, x_2,\ldots,x_n$, where $x_i=x_0+ih$, $i=0,1,2,\ldots,n$, and $h=(b-a)/n$.

Thus,

$$\int_a^b f(x)\,dx=\sum_{i=0}^{n-1}\int_{x_i}^{x_{i+1}} f(x)\,dx \tag{5.141}$$

Then, setting $x=x_i+hu$, from Equation 5.141, we get

$$\int_a^b f(x)dx=\sum_{i=0}^{n-1}\int_{x_i}^{x_{i+1}} f(x)dx=\sum_{i=0}^{n-1}h\int_0^1 f(x_i+hu)du \tag{5.142}$$

Now, applying successive integration by parts, we obtain

$$\int f(x_i+hu)\,du=\int f(x_i+hu)B_0(u)\,du,\quad \text{since } B_0(u)=1$$

$$=f(x_i+hu)B_1(u)-h\int f'(x_i+hu)B_1(u)du,\ \text{using} \int B_n(x)dx=B_{n+1}(x)/(n+1) \text{ from}$$

Equation 5.136

$$
\begin{aligned}
&= f(x_i+hu)B_1(u) - \frac{h}{2!}f'(x_i+hu)B_2(u) + \frac{h^2}{2!}\int f''(x_i+hu)B_2(u)\,du \\
&= f(x_i+hu)B_1(u) - \frac{h}{2!}f'(x_i+hu)B_2(u) + \cdots + (-1)^{2m-1}\frac{h^{2m-1}}{2m!}f^{(2m-1)}(x_i+hu)B_{2m}(u) \\
&\quad + \frac{h^{2m}}{2m!}\int f^{(2m)}(x_i+hu)B_{2m}(u)\,du \\
&= \sum_{k=0}^{2m-1}\frac{(-1)^k h^k}{(k+1)!}f^{(k)}(x_i+hu)B_{k+1}(u) + \frac{h^{2m}}{2m!}\int f^{(2m)}(x_i+hu)B_{2m}(u)\,du
\end{aligned}
\tag{5.143}
$$

Now, substituting in Equation 5.143, $B_1(1)=1/2$, $B_1(0)=-1/2$, $B_n=B_n(0)=B_n(1)\,(n\geq 2)$ and $B_{2n+1}=0$ $(n\geq 1)$, we get

$$
\begin{aligned}
\int_0^1 f(x_i+hu)\,du &= \frac{1}{2}\left[f(x_{i+1})+f(x_i)\right] + \sum_{k=1}^{m}\frac{(-1)^{2k-1}h^{2k-1}}{2k!}B_{2k}(u)\left[f^{(2k-1)}(x_{i+1})-f^{(2k-1)}(x_i)\right] \\
&\quad + \frac{h^{2m}}{2m!}\int_0^1 f^{(2m)}(x_i+hu)B_{2m}(u)\,du
\end{aligned}
\tag{5.144}
$$

Since $B_{2m}(u)-B_{2m}$ does not change sign in (0,1), last term of Equation 5.144 can be written as

$$
\frac{h^{2m}}{2m!}\int_0^1 f^{(2m)}(x_i+hu)B_{2m}(u)\,du
$$

$$
= \frac{h^{2m}}{2m!}\int_0^1 f^{(2m)}(x_i+hu)(B_{2m}(u)-B_{2m})\,du + \frac{B_{2m}h^{2m-1}}{2m!}\left[f^{(2m-1)}(x_{i+1})-f^{(2m-1)}(x_i)\right]
$$

$= \dfrac{h^{2m}}{2m!}f^{(2m)}(x_i+\mu h)\displaystyle\int_0^1 (B_{2m}(u)-B_{2m})\,du + \dfrac{B_{2m}h^{2m-1}}{2m!}\left[f^{(2m-1)}(x_{i+1})-f^{(2m-1)}(x_i)\right]$, applying integral mean value theorem and $0<\mu<1$

$= -\dfrac{B_{2m}h^{2m}}{2m!}f^{(2m)}(\xi_i) + \dfrac{B_{2m}h^{2m-1}}{2m!}\left[f^{(2m-1)}(x_{i+1})-f^{(2m-1)}(x_i)\right]$, since $\int_0^1 B_{2m}(x)dx=0$ and $a<\xi_i<b$

Therefore, Equation 5.144 can be written as

$$
\int_0^1 f(x_i+hu)\,du = \frac{1}{2}\left[f(x_{i+1})+f(x_i)\right] - \sum_{k=1}^{m-1}\frac{B_{2k}h^{2k}}{2k!}\left[f^{(2k-1)}(x_{i+1})-f^{(2k-1)}(x_i)\right] - \frac{B_{2m}h^{2m}}{2m!}f^{(2m)}(\xi_i)
$$

Now, from Equation 5.142, we obtain

$$
\begin{aligned}
\int_a^b f(x)dx &= \frac{h}{2}\sum_{i=0}^{n-1}[f(x_{i+1})+f(x_i)] - \sum_{i=0}^{n-1}\sum_{k=1}^{m-1}\frac{B_{2k}h^{2k}}{2k!}\left[f^{(2k-1)}(x_{i+1})-f^{(2k-1)}(x_i)\right] \\
&\quad -\sum_{i=0}^{n-1}\frac{B_{2m}h^{2m+1}}{2m!}f^{(2m)}(\xi_i) \\
&= \frac{h}{2}[f(a)+f(b)] + h\sum_{i=1}^{n-1}f(x_i) - \sum_{k=1}^{m-1}\frac{B_{2k}h^{2k}}{2k!}\left[f^{(2k-1)}(b)-f^{(2k-1)}(a)\right] \\
&\quad -\frac{B_{2m}h^{2m+1}}{2m!}\sum_{i=0}^{n-1}f^{(2m)}(\xi_i) \\
&= h\sum_{i=0}^{n}f(x_i) - \frac{h}{2}[f(a)+f(b)] - \sum_{k=1}^{m-1}\frac{B_{2k}h^{2k}}{2k!}\left[f^{(2k-1)}(b)-f^{(2k-1)}(a)\right] \\
&\quad -\frac{B_{2m}h^{2m+1}}{2m!}nf^{(2m)}(\xi),
\end{aligned}
\tag{5.145}
$$

since by intermediate value theorem

$$\frac{1}{n}\sum_{i=0}^{n-1}f^{(2m)}(\xi_i) = f^{(2m)}(\xi), \quad a<\xi<b.$$

Therefore,

$$\sum_{i=0}^{n}f(x_i) = \frac{1}{h}\int_a^b f(x)dx + \frac{1}{2}\left[f(a)+f(b)\right] + \sum_{k=1}^{m-1}\frac{B_{2k}h^{2k-1}}{2k!}\left[f^{(2k-1)}(b)-f^{(2k-1)}(a)\right] + E \tag{5.146}$$

where:

$$E = \frac{B_{2m}h^{2m}}{2m!}nf^{(2m)}(\xi), \quad a<\xi<b \tag{5.147}$$

The Equation 5.146 is called the *Euler–Maclaurin sum formula* and Equation 5.147 is the corresponding error term.

Example 5.14

Compute the sum of the series

$$\frac{1}{11^3}+\frac{1}{12^3}+\cdots+\frac{1}{99^3}$$

correct to five significant figures by using the Euler–Maclaurin sum formula.

Solution:

It will be more convenient to compute $\sum_{n=10}^{100} 1/n^3$.

Let us take $f(x) = 1/x^3$, $a = 10$, $b = 100$, and $h = 1$.

Now,

$$f'(x) = -\frac{3}{x^4}, f''(x) = \frac{12}{x^5}, f'''(x) = -\frac{60}{x^6}, \text{ and so on.}$$

Therefore, from Equation 5.146, we get

$$\sum_{n=10}^{100} \frac{1}{n^3} = \int_{10}^{100} f(x)dx + \frac{1}{2}[f(10)+f(100)] + \frac{B_2 h}{2!}[f'(100) - f'(10)] + \frac{B_4 h^3}{4!}[f'''(100) - f'''(10)] + \cdots$$

$$= \frac{1}{2}\left(\frac{1}{10^2} - \frac{1}{100^2}\right) + \frac{1}{2}\left(\frac{1}{10^3} + \frac{1}{100^3}\right) - \frac{3}{12}\left(\frac{1}{100^4} - \frac{1}{10^4}\right) + \frac{60}{4!\times 30}\left[\frac{1}{100^6} - \frac{1}{10^6}\right]$$

$$= 0.00547541$$

Hence, the required sum is

$0.00547541 - 0.001 - 0.000001 = 0.00447441 \cong 0.00447$ correct to five significant figures.

Example 5.15

Evaluate the integral

$$\int_0^1 \frac{dx}{\sqrt{1+x^2}}$$

correct to five significant figures by using the Euler–Maclaurin sum formula. Hence determine the approximate value of $\sinh^{-1}1$.

Solution:

Let $f(x) = \dfrac{1}{\sqrt{1+x^2}}$

We divide the interval [0,1] into $n = 10$ equal subintervals so that $h = 1/10 = 0.1$.

Now,

$$f'(x) = -\frac{x}{(1+x^2)^{3/2}}, f'''(x) = \frac{9x - 6x^3}{\left(1+x^2\right)^{7/2}}, \text{ and so on.}$$

$$h\sum_{i=0}^{n} f(x_i) - \frac{h}{2}[f(a)+f(b)] - \sum_{k=1}^{m-1} \frac{B_{2k} h^{2k}}{2k!}\left[f^{(2k-1)}(b) - f^{(2k-1)}(a)\right] - \frac{B_{2m} h^{2m+1}}{2m!} n f^{(2m)}(\xi)$$

Therefore, from Equation 5.145, we get

$$\int_0^1 \frac{dx}{\sqrt{1+x^2}} \cong 0.1\times\left(\sum_{i=0}^{n} f(x_i)\right) - \frac{(0.1)}{2}[f(0)+f(1)] - \frac{B_2 h^2}{2!}\left[f'(1) - f'(0)\right] - \frac{B_4 h^4}{4!}\left[f'''(1) - f'''(0)\right] + \cdots$$

$$= 0.1\times(9.66434) - \frac{(0.1)}{2}\times 1.70711 - \frac{(0.1)^2}{2!\times 6}\left[-0.353553 - 0\right] + \frac{(0.1)^4}{4!\times 30}\left[0.265165 - 0\right]$$

$$= 0.881374$$

Hence, the required value of the integral is 0.881374.

Since

$$\int_0^1 \frac{dx}{\sqrt{1+x^2}} = \sinh^{-1}1.$$

Therefore, the required approximate value is $\sinh^{-1}1 \cong 0.881374$.

EXERCISES

1. Calculate $\int_0^1 (1+x^2)/(1+x^4)\,dx$ up to five significant figures by the trapezoidal, Simpson's, and Weddle's rule by taking six intervals and hence compare the obtained results.
2. Compute by Simpson's one-third rule with the integral

$$I = \int_{200}^{800} \frac{dx}{\log_{10} x}$$

taking 12 and 16 subintervals. Use the computed values to make an estimate of the accuracy.

3. The velocity v of a particle at distance s from a point on its path is given in the table below

s in meter	0	10	20	30	40	50	60
v meters per sec	47	58	64	65	61	52	38

Estimate the time taken to travel 60 meter by using Simpson's three-eighth rule.

4. The following table gives the velocity v of the particle at time t

t (seconds)	0	2	4	6	8	10	12
v meter per sec	4	6	16	34	60	94	136

Find (i) the distance moved by the particle in 12 s and also (ii) the acceleration at $t = 2$ s.

5. Find the value of $\log_e 2$ from $\int_0^1 x^2/(1+x^3)\,dx$, using Simpson's one-third rule by dividing the interval into six equal parts. Find the error also.
6. Compute

$$I_p = \int_0^1 \frac{x^p}{x^3+10}\,dx \quad \text{for } p = 0,1$$

using trapezoidal and Simpson's rules with the number of points 3, 5, and 9. Improve the results using the Romberg integration.

7. The velocity of a train which starts from rest is given by the following time being reckoned in minutes from the start and speed in miles per hour.

Minutes	2	4	6	8	10	12	14	16	18	20
Miles per hour	10	18	25	29	32	20	11	5	2	0

Estimate approximately the total distance traveled in 20 min.

8. Evaluate $\int_{4.0}^{5.2} \log_e x\,dx$ by (a) Simpson's one-third rule, (b) Simpson's three-eighth rule, and (c) Weddle's rule. Compare the results.
9. Compute $I = \int_0^1 dx/(1+x^2)^{3/2}$ by trapezoidal rule with $h = 0.25$ and $h = 0.125$. Improve it by Romberg rule. Also, compute I by Simpson's one-third rule with eight subintervals and compare the results.
10. Compute the value of $\int_{0.2}^{1.4} \left(\sin x - \log_e x + e^x\right) dx$, using Weddle's rule with 12 intervals.
11. Evaluate $\int_0^{1/2} dx/\sqrt{1-x^2}$ by Weddle's rule taking $h = 0.1$. Compare with the actual value and get the numerical difference between them.
12. The velocity of an electric train that starts from rest is given in the following data:

t (minutes)	0	1	2	3	4	5	6	7	8	9	10	11	12
v (km/hour)	0	10	25	40	55	60	62	57	42	30	20	13	0

Find the total distance covered in 12 min using Weddle's rule.
13. Evaluate $\int_0^1 e^{-x^2} \sin x\,dx$, by trapezoidal rule with subinterval lengths 0.2 and 0.1. Make an estimate of the error. Improve the result by Romberg rule.
14. A river is 80 ft wide. The depth d in feet at a distance x ft from one bank is given by the following table:

x	0	10	20	30	40	50	60	70	80
d	0	4	7	9	12	15	14	8	3

Estimate the cross-sectional area of the river using Simpson's three-eighth rule.
15. Evaluate $\int_2^6 (1/\log_e x)\,dx$ by using Simpson's three-eighth rule.
16. Compute $\int_{\pi/4}^{\pi/2} \cos x \ln(\sin x)/(\sin^2 x + 1)\,dx$ correct to three decimal places using the following:
    a. trapezoidal rule and Romberg integration
    b. Simpson's rule and Romberg integration
17. Evaluate $\int_0^{\pi/2} e^{\sin x}\,dx$ correct to four decimal places using Simpson's three-eighth rule.
18. Obtain the approximate value of

$$I = \int_{-1}^{1} e^{-x^2} \cos x\,dx$$

using
    a. The Gauss–Legendre integration method for $n = 2,3$
    b. The Lobatto integration method for $n = 3,5$
19. Use Simpson's one-third rule to evaluate the double integral $\int_{-2}^{2}\int_0^4 \left(x^2 - xy + y^2\right) dxdy$ and compare the results with the exact value.
20. Evaluate $\int_{-1}^{1} \sqrt{1-x^2} \cos x\,dx$ by Gaussian two- and three-point quadrature formulae.
21. Compute

$$I = \int_{-1}^{1} \left(1+x^2\right)^{\frac{3}{2}} \cos x\,dx,$$

by three-point Gauss–Legendre and Gauss–Chebyshev quadrature. Obtain a better result by four-point Gauss–Legendre formula.

22. Evaluate the function $I = \int_0^\infty \left((3+x)/\sqrt{x}\right) e^{-x} dx$ as accurately as possible using Gauss–Hermite quadrature.
23. Use Romberg's integration to evaluate $\int_0^{\sqrt{\pi}} 2x^2 \cos x^2 dx$.
24. Evaluate $I = \int_0^1 x^2/(1+x^3)\,dx$ using Lobatto's integration method with (a) $n = 3$ and (b) $n = 5$.
25. Evaluate $\int_0^{0.8} e^{-x^2}\, dx$ by Romberg's method with $h = 0.1$ and $h = 0.2$.
26. Evaluate $\int_0^1 dx/(1+x)\sqrt{1+2x-x^2}$ using Romberg's method correct up to five decimal places.
27. Use Romberg's method to compute $\int_{1.0}^{1.7} \log x/x\, dx$ correct up to four decimal places by taking $h = 0.1$.
28. Evaluate $\int_0^1 \sqrt{\sin x + \cos x}\, dx$ correct to two decimal places using Simpson's three-eighth rule.
29. Evaluate $\int_0^{1/2} dx/\sqrt{1-x^2}$ by Euler–Maclaurin's formula by taking $h = 0.1$.
30. Evaluate the double integral

$$\int_0^1 \left(\int_1^2 \frac{2xy}{(1+x^2)(1+y^2)} dy \right) dx$$

using

a. The trapezoidal rule with $h = k = 0.25$
b. The Simpson's rule with $h = k = 0.25$

Hence, compare the results obtained with the exact solution.

31. Using Euler–Maclaurin's quadrature formula, find the sum of the following series:

a. $\dfrac{1}{51^2} + \dfrac{1}{53^2} + \dfrac{1}{55^2} + \cdots + \dfrac{1}{99^2}$

b. $\dfrac{1}{11^2} + \dfrac{1}{12^2} + \dfrac{1}{13^2} + \cdots + \dfrac{1}{99^2}$

32. Using Weddle's rule, calculate $\int_{0.5}^{0.7} \sqrt{x}\, e^{-x} dx$, by taking suitable number of subintervals.
33. Using Simpson's one-third rule, compute the value of the following integral:

$$I = \int_0^{\pi/2} \sqrt{1 - 0.15\sin^2\phi}\; d\phi$$

with four and eight subintervals. From the computations, make a rough estimate of the truncation error.

34. Obtain an approximate value of

$$I = \int_{-1}^{1} (1-x^2)^{1/2} \cos x\; dx$$

using

a. The Gauss–Legendre integration method for $n = 2, 3$
b. The Gauss–Chebyshev integration method for $n = 2, 3$

35. Evaluate $\int_1^5 \int_1^5 1/\sqrt{x^2+y^2}\; dxdy$ using the trapezoidal rule with four subintervals.
36. Evaluate $\int_0^{\pi/2} \int_0^{\pi/2} \sin(x+y)\, dxdy$ using Simpson's rule with $h = k = \pi/4$.
37. Estimate $\int_0^{0.5} \int_0^{0.5} \sin xy/(1+xy)\, dxdy$ using Simpson's rule for double integrals with both step sizes = 0.25.

38. Compute the values of $I = \int_0^1 dx/1+x^2$ using the trapezoidal rule with $h = 0.5, 0.25$, and 0.125. Then obtain a better estimate using Romberg's method. Compare your results with the true value.
39. Use the Euler–Maclaurin formula to evaluate the integral $I = \int_1^2 (\cos x + \ln x - e^x)\,dx$ by taking $h = 0.1$.
40. Evaluate the integral

$$I = \int_{\pi/6}^{\pi/2} \log_{10} \sin x\,dx$$

by the trapezoidal rule with 4, 8, and 16 subintervals and make an estimate of the accuracy achieved.

Using the above-computed values, determine the integral by the Romberg rule.
41. Apply Simpson's three-eighth rule, calculate $\int_0^{\pi/2} e^{\sin x}\,dx$, correct to four decimal places.
42. Find $\int_{0.2}^{1.5} e^{x^2}\,dx$ by the Gauss three-point quadrature formula.
43. Compute the integral $\int_0^1 dx/\sqrt{1+x^4}$ by four-point and six-point Gauss–Legendre quadrature formula. How accurate is the result?
44. For the integral $I = \int_{-1}^{1} \sqrt{1-x^2}\, f(x)\,dx$ with weight $w(x) = \sqrt{1-x^2}$, find explicit formulas for the nodes and weights of the Gaussian quadrature formula. Also give the error formula.
45. Find the remainder of the Simpson's three-eighth rule

$$\int_{x_0}^{x_3} f(x)\,dx = \frac{3h}{8}\left[f(x_0) + 3f(x_1) + 3f(x_2) + f(x_3)\right]$$

for equally spaced points $x_i = x_0 + ih, \quad i = 1, 2, 3.$

Evaluate $\int_0^1 dx/(1+x^2)$ using the Simpson's three-eighth rule. Compare with the exact solution.
46. Derive the one- and two-point Gaussian quadrature formulae for

$$I = \int_0^1 x\, f(x)\,dx \cong \sum_{j=1}^{n} w_j\, f(x_j)$$

with weight function $w(x) = x$.
47. Determine the weights and abscissas in the quadrature formula

$$\int_{-1}^{1} f(x)dx = \sum_{k=1}^{4} w_k f(x_k)$$

with $x_1 = -1$ and $x_4 = 1$ so that the formula becomes exact for polynomials of highest possible degree.

6 Numerical Solution of System of Linear Algebraic Equations

6.1 INTRODUCTION

Linear systems of algebraic equations occur extensively in various fields of science and engineering. Most importantly, linear systems are involved in discretization of ordinary or partial differential equations or integral equations. The most common form of the system in n unknowns $x_1, x_2, \ldots, x_n$ is of the form

$$\begin{aligned} a_{11}x_1 + a_{12}x_2 + \ldots + a_{1n}x_n &= b_1 \\ a_{21}x_1 + a_{22}x_2 + \ldots + a_{2n}x_n &= b_2 \\ &\vdots \\ a_{n1}x_1 + a_{n2}x_2 + \ldots + a_{nn}x_n &= b_n \end{aligned} \tag{6.1}$$

Equation 6.1 can be written in the matrix form as

$$\boldsymbol{Ax} = \boldsymbol{b} \tag{6.2}$$

where

$$\boldsymbol{A} = [a_{ij}]_{n \times n} = \begin{bmatrix} a_{11} & a_{12} & \ldots & a_{1n} \\ a_{21} & a_{22} & \ldots & a_{2n} \\ \vdots & \vdots & \ldots & \vdots \\ a_{n1} & a_{n2} & \ldots & a_{nn} \end{bmatrix}, \quad \boldsymbol{x} = \begin{bmatrix} x_1 \\ x_2 \\ \vdots \\ x_n \end{bmatrix} \text{ and } \boldsymbol{b} = \begin{bmatrix} b_1 \\ b_2 \\ \vdots \\ b_n \end{bmatrix}$$

in which it is assumed that the matrix $\boldsymbol{A}$ is nonsingular, that is, $|\boldsymbol{A}| \neq 0$, so that the system in Equation 6.1 has a unique solution. If $\boldsymbol{b} = \boldsymbol{0}$, then the system of Equation 6.1 is called *homogeneous*. Otherwise, the system is called *nonhomogeneous*.

Many problems arising from engineering and sciences require the solution of systems of linear algebraic equations. Their most important application in engineering is in the analysis of linear systems in which the response or output is proportional to the input. The modeling of the linear systems invariably leads to equations of the form $\boldsymbol{Ax} = \boldsymbol{b}$, where $\boldsymbol{b}$ is the input and $\boldsymbol{x}$ represents the response of the system. The coefficient matrix $\boldsymbol{A}$, which reflects the characteristics of the system, is independent of the input. So, if the input is changed, the equations have to be solved again with a different $\boldsymbol{b}$, but the same $\boldsymbol{A}$.

The numerical methods for the solution of system of linear algebraic equations are, in general, of two types:

1. Direct methods
2. Iterative methods

The direct methods reduce the given system of equations into equivalent equations that can be solved more easily. This transformation is carried out by applying certain operators.

On the other hand, in iterative method, we start with an initial approximation to the solution $\boldsymbol{x}$ and repeatedly improve the approximation until a certain convergence criterion is reached. Iterative techniques are self-correcting causing round-off errors occurred in any iteration to be corrected in the subsequent iterations. A serious drawback of iterative methods is that they are not always convergent. The initial approximation affects only the number of iterations that are needed for convergence. The iterative method is most useful to solve a system of ill-conditioned equations.

In this chapter, we shall discuss five direct methods and three iterative methods:

1. *Direct methods*:
   a. Gauss elimination method
   b. Gauss–Jordan method
   c. Doolittle's method
   d. Crout's method
   e. Cholesky's method
2. *Iterative methods*:
   a. Gauss–Jacobi iteration method
   b. Gauss–Seidel iteration method
   c. Successive over-relaxation (SOR) method

6.2 VECTOR AND MATRIX NORM

We now introduce the general notion of the norm of a vector.

6.2.1 Vector Norm

Definition 6.1

Let V be a vector space. A vector norm for a column vector $\boldsymbol{x} = (x_1, x_2, \ldots, x_n)^T$ with n components is a generalized length or size or distance from the origin, is denoted by $\|\boldsymbol{x}\|$, and satisfies the following four properties:

i. $\|\boldsymbol{x}\|$ is a nonnegative real number, that is, $\|\boldsymbol{x}\| > 0$ for all $\boldsymbol{x} \in V$ (nonnegativity).
ii. $\|\boldsymbol{x}\| = 0$ if and only if $\boldsymbol{x} = \boldsymbol{0}$, where $\boldsymbol{0}$ denotes the zero vector (positive-definiteness).
iii. $\|k\boldsymbol{x}\| = |k|\,\|\boldsymbol{x}\|$, for all scalars k and for all $\boldsymbol{x} \in V$ (absolute homogeneity or absolute scalability).
iv. $\|\boldsymbol{x} + \boldsymbol{y}\| \le \|\boldsymbol{x}\| + \|\boldsymbol{y}\|$, for all $\boldsymbol{x}, \boldsymbol{y} \in V$ (triangle inequality or subadditivity).

Some examples of norms in the complex n-dimensional space $\boldsymbol{C}^n$ are given below:

a. $\|\boldsymbol{x}\|_1 = |x_1| + |x_2| + \cdots + |x_n| = \sum_{i=1}^{n} |x_i|, \quad \boldsymbol{x} \in \boldsymbol{C}^n \; (l_1 \text{ norm})$

b. $\|\boldsymbol{x}\|_2 = \sqrt{|x_1|^2 + |x_2|^2 + \cdots + |x_n|^2} = \left(\sum_{i=1}^{n} |x_i|^2 \right)^{1/2}, \quad \boldsymbol{x} \in \boldsymbol{C}^n \; (l_2 \text{ norm or Euclidean norm})$

c. $\|\boldsymbol{x}\|_p = \sqrt{|x_1|^p + |x_2|^p + \cdots + |x_n|^p} = \left(\sum_{i=1}^{n} |x_i|^p \right)^{1/p}, \quad \boldsymbol{x} \in \boldsymbol{C}^n \; (l_p \text{ norm or } p\text{-norm})$

d. $\|\boldsymbol{x}\|_\infty = \max_{1 \le i \le n} |x_i|, \quad \boldsymbol{x} \in \boldsymbol{C}^n \; (l_\infty \text{ norm or maximum norm})$

For $n = 3$, the l_2 norm or Euclidean norm is the usual length of a vector in three-dimensional space. The l_1 norm and l_∞ norm are generally more convenient in practical computation.

6.2.2 Matrix Norm

The set of all $n \times n$ matrices forms a vector space of dimension n^2. Thus, a matrix norm satisfies the conditions (i)–(iv) of a vector norm. In addition, we impose an additional condition on matrix norm.

Definition 6.2

A matrix norm $\|\boldsymbol{A}\|$ satisfies the following properties:

i. $\|\boldsymbol{A}\|$ is a nonnegative real number, that is, $\|\boldsymbol{A}\| > 0$ if $\boldsymbol{A} \neq \boldsymbol{0}$, where $\boldsymbol{0}$ denotes the zero vector.
ii. $\|\boldsymbol{A}\| = \boldsymbol{0}$ if and only if $\boldsymbol{A} = \boldsymbol{0}$.
iii. $\|k\boldsymbol{A}\| = |k|\|\boldsymbol{A}\|$, for all scalars k. (6.3)
iv. $\|\boldsymbol{A} + \boldsymbol{B}\| \le \|\boldsymbol{A}\| + \|\boldsymbol{B}\|$ (triangle inequality).
v. $\|\boldsymbol{AB}\| \le \|\boldsymbol{A}\|\|\boldsymbol{B}\|$.

The most commonly used matrix norms are as follows:

a. Frobenius or Euclidean norm

$$\|\boldsymbol{A}\|_F = \left(\sum_{i=1}^{n}\sum_{j=1}^{n}|a_{ij}|^2\right)^{1/2} = \sqrt{\text{trace}(A^*A)} \tag{6.4}$$

where $\boldsymbol{A}^*$ denotes the conjugate transpose of $\boldsymbol{A}$

b. Column *sum* matrix norm

$$\|A\|_1 = \max_j \sum_{i=1}^{n}|a_{ij}| \tag{6.5}$$

which is simply the maximum absolute column sum of the matrix

c. Row *sum* matrix norm

$$\|\boldsymbol{A}\|_\infty = \max_i \sum_{j=1}^{n}|a_{ij}| \tag{6.6}$$

which is simply the maximum absolute row sum of the matrix

d. Spectral norm

$$\|\boldsymbol{A}\|_2 = \sqrt{\rho(\boldsymbol{A}^*\boldsymbol{A})} \tag{6.7}$$

where $\boldsymbol{A}^*$ denotes the conjugate transpose of $\boldsymbol{A}$ and $\rho(\boldsymbol{A}^*\boldsymbol{A})$ is the spectral radius of $\boldsymbol{A}^*\boldsymbol{A}$

It can be proved that there is a number c such that

$$\|\boldsymbol{Ax}\| \le c\|\boldsymbol{x}\|, \quad \text{for all } \boldsymbol{x} \tag{6.8}$$

This implies

$$\frac{\|\boldsymbol{A}\,\boldsymbol{x}\|}{\|\boldsymbol{x}\|} \le c, \quad \text{for all } \boldsymbol{x} \neq \boldsymbol{0}.$$

The smallest possible c valid for all $\boldsymbol{x} \neq \boldsymbol{0}$ is called the matrix norm of $\boldsymbol{A}$ corresponding to the suitable vector norm and is denoted by $\|\boldsymbol{A}\|$. Thus,

$$\|\boldsymbol{A}\| = \sup_{\|\boldsymbol{x}\| \neq 0} \frac{\|\boldsymbol{A}\boldsymbol{x}\|}{\|\boldsymbol{x}\|} \tag{6.9}$$

which is called natural norm or the matrix norm induced by the vector norm $\|\cdot\|$. This is also known as the operator norm.

Since, for any $\boldsymbol{x} \neq \boldsymbol{0}$, we can define

$$\boldsymbol{u} = \frac{\boldsymbol{x}}{\|\boldsymbol{x}\|}$$

So that

$$\|\boldsymbol{u}\| = 1.$$

Thus, Equation 6.9 is equivalent to

$$\|\boldsymbol{A}\| = \sup_{\|\boldsymbol{u}\|=1} \|\boldsymbol{A}\boldsymbol{u}\| \tag{6.10}$$

By considering the smallest value of $c = \|\boldsymbol{A}\|$, we have from Equation 6.6

$$\|\boldsymbol{A}\boldsymbol{x}\| \leq \|\boldsymbol{A}\|\|\boldsymbol{x}\|, \quad \text{for all } \boldsymbol{x} \tag{6.11}$$

For two $n \times n$ matrices $\boldsymbol{A}$ and $\boldsymbol{B}$, the Equation 6.9 implies that

$$\|\boldsymbol{A}\boldsymbol{B}\| \leq \|\boldsymbol{A}\|\|\boldsymbol{B}\| \tag{6.12}$$

Hence, we have

$$\|\boldsymbol{A}^n\| \leq \|\boldsymbol{A}\|^n \tag{6.13}$$

6.2.3 Condition Number of a Matrix

The condition number of a nonsingular square matrix $\boldsymbol{A}$, denoted by $\kappa(\boldsymbol{A})$, is defined by

$$\kappa(\boldsymbol{A}) = \|\boldsymbol{A}\|\|\boldsymbol{A}^{-1}\| \tag{6.14}$$

6.2.4 Spectral Radius and Norm Convergence

Definition 6.3

Let $\boldsymbol{A}$ be any arbitrary $n \times n$ square matrix. The spectrum of $\boldsymbol{A}$ is defined as the set of all eigenvalues λ_i $(i = 1,2,\ldots,n)$ of $\boldsymbol{A}$. The spectral radius of $\boldsymbol{A}$, denoted by $\rho(\boldsymbol{A})$, is defined as

$$\rho(\boldsymbol{A}) = \max_{1 \leq i \leq n} |\lambda_i|$$

Theorem 6.1

Let $\boldsymbol{A}$ be any arbitrary $n \times n$ square matrix. Then, for any natural matrix norm

$$\rho(\boldsymbol{A}) \leq \|\boldsymbol{A}\|.$$

Proof:
Let λ be an eigenvalue in the spectrum of $\boldsymbol{A}$ such that

$$|\lambda| = \rho(\boldsymbol{A}).$$

Also, let $\boldsymbol{x}_s$ be the corresponding eigenvector of $\boldsymbol{A}$. The associated eigenvector $\boldsymbol{x}_s$ can be chosen to be normalized for any particular vector norm, that is, $\|\boldsymbol{x}_s\| = 1$. Then for the corresponding natural matrix norm

$$\rho(\boldsymbol{A}) = |\lambda| = |\lambda|\|\boldsymbol{x}_s\| = \|\lambda \boldsymbol{x}_s\| = \|\boldsymbol{A}\boldsymbol{x}_s\| \leq \|\boldsymbol{A}\|\|\boldsymbol{x}_s\| \leq \|\boldsymbol{A}\|$$

Hence, this completes the proof. ■

Corollary: For a square matrix $\boldsymbol{A}$, $\rho(\boldsymbol{A}) < 1$ if and only if $\|\boldsymbol{A}\| < 1$.

6.2.5 Jordan Block

Definition 6.4

A Jordan block $\boldsymbol{J}_n(\lambda) = [J_{ij}]_{n \times n}$ is a $n \times n$ square matrix of the following form:

$$J_{ij} = \begin{cases} \lambda, & \text{if } i = j, \\ 1, & \text{if } j = i+1, \\ 0, & \text{otherwise} \end{cases}$$

$$\text{that is,} \quad \boldsymbol{J}_n(\lambda) = \begin{pmatrix} \lambda & 1 & 0 & \dots & 0 \\ 0 & \lambda & 1 & \dots & 0 \\ \vdots & \vdots & \vdots & \dots & 1 \\ 0 & 0 & 0 & \dots & \lambda \end{pmatrix}, \quad n \geq 1 \tag{6.15}$$

So, Jordan block $\boldsymbol{J}_n(\lambda)$ has the single eigenvalue λ of algebraic multiplicity n and geometric multiplicity 1.

6.2.6 Jordan Canonical Form

Any square matrix $\boldsymbol{A}$ is similar to a block diagonal matrix with Jordan blocks on the *block* diagonal, if there is a nonsingular matrix $\boldsymbol{P}$ for which

$$\boldsymbol{A} = \boldsymbol{P}\boldsymbol{J}\boldsymbol{P}^{-1}$$

$$\text{that is,}\quad \boldsymbol{P}^{-1}\boldsymbol{A}\boldsymbol{P}=\boldsymbol{J}=\begin{pmatrix} \boldsymbol{J}_{n_1}(\lambda_1) & 0 & \cdots & 0 \\ 0 & \boldsymbol{J}_{n_2}(\lambda_2) & \cdots & 0 \\ \vdots & \vdots & \cdots & \vdots \\ 0 & \cdots & \cdots & \boldsymbol{J}_{n_k}(\lambda_k) \end{pmatrix} \tag{6.16}$$

Here, $\boldsymbol{J}_{n_i}(\lambda_i),\quad i=1,2,\ldots$ are the Jordan blocks of dimension $n_i \times n_i$ with λ_i on the diagonal so that $\sum_{i=1}^{k} n_i = n$. The eigenvalues $\lambda_1, \lambda_2, \ldots, \lambda_k$ are not necessarily to be distinct.

It is convenient to write Equation 6.16 as

$$\boldsymbol{P}^{-1}\boldsymbol{A}\boldsymbol{P}=\boldsymbol{D}+\boldsymbol{N} \tag{6.17}$$

where $\boldsymbol{D}=\text{diag}[\lambda_1,\lambda_2,\ldots,\lambda_k]$ with each λ_i appearing n_i times on the diagonal of $\boldsymbol{D}$. The matrix $\boldsymbol{N}$ has all zero entries, except for possible 1's on the superdiagonal. The matrix $\boldsymbol{N}$ is a nilpotent matrix that satisfies

$$\boldsymbol{N}^n=\boldsymbol{0} \tag{6.18}$$

Theorem 6.2

Let $\boldsymbol{A}$ be a square matrix of order n. A necessary and sufficient condition that the square matrix $\boldsymbol{A}$ be convergent is that the spectral radius of $\boldsymbol{A}$ is less than 1, that is,

$$\lim_{m\to\infty}\boldsymbol{A}^m=\boldsymbol{0}$$

if and only if $\rho(\boldsymbol{A})<1$.

Proof:

Let the matrix $\boldsymbol{A}$ is similar to a block diagonal matrix with Jordan blocks on the *block* diagonal such that

$$\boldsymbol{A}=\boldsymbol{P}\boldsymbol{J}\boldsymbol{P}^{-1} \tag{6.19}$$

where $\boldsymbol{P}$ is a nonsingular matrix and $\boldsymbol{J}$ is a block diagonal matrix with Jordan blocks on the diagonal.

Then,

$$\boldsymbol{A}^m=(\boldsymbol{P}\boldsymbol{J}\boldsymbol{P}^{-1})^m=\boldsymbol{P}\boldsymbol{J}^m\boldsymbol{P}^{-1} \tag{6.20}$$

Therefore,

$\boldsymbol{A}^m \to \boldsymbol{0}$ as $m\to\infty$ if and only if $\boldsymbol{J}^m\to\boldsymbol{0}$ as $m\to\infty$.

Now, from Equation 6.17, we have

$$\boldsymbol{J}=\boldsymbol{P}^{-1}\boldsymbol{A}\boldsymbol{P}=\boldsymbol{D}+\boldsymbol{N} \tag{6.21}$$

where $\boldsymbol{D}=\text{diag}[\lambda_1,\lambda_2,\ldots,\lambda_k]$, the eigenvalues $\lambda_1,\lambda_2,\ldots,\lambda_k$ of $\boldsymbol{A}$ need not be distinct, and $\boldsymbol{N}$ is a nilpotent matrix such that $\boldsymbol{N}^n=\boldsymbol{0}$.

Now,

$$\boldsymbol{J}^m=(\boldsymbol{D}+\boldsymbol{N})^m=\sum_{r=0}^{m}\binom{m}{r}\boldsymbol{D}^{m-r}\boldsymbol{N}^r \tag{6.22}$$

This implies that

$$J^m = \sum_{r=0}^{n-1} \binom{m}{r} D^{m-r} N^r, \quad \text{since } N^r = \mathbf{0} \text{ for } r \geq n$$

Therefore, for any matrix norm

$$\left\| J^m \right\| = \left\| \sum_{r=0}^{n-1} \binom{m}{r} D^{m-r} N^r \right\| \leq \sum_{r=0}^{n-1} \left\| \binom{m}{r} D^{m-r} N^r \right\| \leq \sum_{r=0}^{n-1} \frac{m^r}{r!} \left\| D^{m-r} \right\| \left\| N^r \right\| \leq \sum_{r=0}^{n-1} \frac{m^r}{r!} \left\| D \right\|^{m-r} \left\| N \right\|^r$$

Now, using the row sum norm, we have

$$\left\| J^m \right\| \leq \sum_{r=0}^{n-1} \frac{m^r}{r!} \left\| N \right\|_\infty^r \rho(A)^{m-r}, \quad \text{since } \left\| D \right\|_\infty = \rho(A)$$

If we assume that $\rho(A) < 1$, then J^m and consequently, according to Equation 6.20, A^m converges to the zero matrix as $m \to \infty$.

Conversely, let us suppose that $\rho(A) \geq 1$. Then, let λ be an eigenvalue of A such that $|\lambda| \geq 1$ and let $x\,(x \neq \mathbf{0})$ be the corresponding eigenvector of A. Then, we have

$$A^m x = \lambda^m x$$

Since, $|\lambda| \geq 1$, λ^m does not tend to zero as $m \to \infty$. Consequently, $A^m x$ does not converge to the zero matrix as $m \to \infty$. This implies, A^m also does not converge to the zero matrix as $m \to \infty$.

Hence, the theorem is proved. ■

Theorem 6.3

a. The geometric series

$$I + A + A^2 + A^3 + \ldots \tag{6.23}$$

converges if and only if A is convergent.

b. If A is convergent, then $I - A$ is nonsingular and

$$(I - A)^{-1} = I + A + A^2 + A^3 + \ldots \tag{6.24}$$

Proof:

A necessary condition for the series (6.23) in part (a) to converge is that A is convergent, that is, $\lim_{n\to\infty} A^n = \mathbf{0}$. The sufficiency follows from part (b).

Let A be convergent, then by Theorem 6.2 we know that $\rho(A) < 1$. Since, the eigenvalues of $I - A$ are $1 - \lambda_i$, $i = 1, 2, \ldots$, it follows that det $(I - A) \neq 0$, and hence, the matrix $I - A$ is nonsingular. So, $(I - A)^{-1}$ exists.

Now, we consider the identity

$$(\boldsymbol{I}-\boldsymbol{A})\left(\boldsymbol{I}+\boldsymbol{A}+\boldsymbol{A}^2+\boldsymbol{A}^3+\ldots+\boldsymbol{A}^n\right)=\boldsymbol{I}-\boldsymbol{A}^{n+1}$$

which is valid for all positive integer n. Multiplying both sides of the above equation by $(\boldsymbol{I}-\boldsymbol{A})^{-1}$, we get

$$(\boldsymbol{I}-\boldsymbol{A})^{-1}\left(\boldsymbol{I}-\boldsymbol{A}^{n+1}\right)=\left(\boldsymbol{I}+\boldsymbol{A}+\boldsymbol{A}^2+\boldsymbol{A}^3+\ldots+\boldsymbol{A}^n\right)$$

Letting the limit $n \to \infty$, both sides of the above equation yields

$$(\boldsymbol{I}-\boldsymbol{A})^{-1}=\boldsymbol{I}+\boldsymbol{A}+\boldsymbol{A}^2+\boldsymbol{A}^3+\ldots$$

■

Corollary: If for some natural norm, $\|\boldsymbol{A}\|<1$, then $\boldsymbol{I}-\boldsymbol{A}$ is non-singular and

$$\frac{1}{1+\|\boldsymbol{A}\|}\leq\left\|(\boldsymbol{I}-\boldsymbol{A})^{-1}\right\|\leq\frac{1}{1-\|\boldsymbol{A}\|} \tag{6.25}$$

Proof:

We know that if $\|\boldsymbol{A}\|<1$, then $\boldsymbol{A}$ is convergent. Therefore, by the result of Theorem 6.3, it follows that $\boldsymbol{I}-\boldsymbol{A}$ is nonsingular. For a natural norm, we have

$$\|\boldsymbol{I}\|=1$$

Now, from the identity, we get

$$\boldsymbol{I}=(\boldsymbol{I}-\boldsymbol{A})(\boldsymbol{I}-\boldsymbol{A})^{-1}$$

This implies that

$$\begin{aligned}1=\|\boldsymbol{I}\|&=\left\|(\boldsymbol{I}-\boldsymbol{A})(\boldsymbol{I}-\boldsymbol{A})^{-1}\right\|\\&\leq\|(\boldsymbol{I}-\boldsymbol{A})\|\left\|(\boldsymbol{I}-\boldsymbol{A})^{-1}\right\|\\&\leq\left(1+\|\boldsymbol{A}\|\right)\left\|(\boldsymbol{I}-\boldsymbol{A})^{-1}\right\|\end{aligned}$$

Thus,

$$\frac{1}{1+\|\boldsymbol{A}\|}\leq\left\|(\boldsymbol{I}-\boldsymbol{A})^{-1}\right\|$$

Again, we consider the following identity

$$(\boldsymbol{I}-\boldsymbol{A})^{-1}=\boldsymbol{I}+\boldsymbol{A}(\boldsymbol{I}-\boldsymbol{A})^{-1}$$

This implies that

$$\begin{aligned}\left\|(\boldsymbol{I}-\boldsymbol{A})^{-1}\right\| &= \left\|\boldsymbol{I}+\boldsymbol{A}(\boldsymbol{I}-\boldsymbol{A})^{-1}\right\| \\ &\leq 1+\|\boldsymbol{A}\|\left\|(\boldsymbol{I}-\boldsymbol{A})^{-1}\right\|\end{aligned}$$

which yields

$$\left\|(\boldsymbol{I}-\boldsymbol{A})^{-1}\right\| \leq \frac{1}{1-\|\boldsymbol{A}\|}, \quad \text{since} \|\boldsymbol{A}\|<1$$

It may be noted that if $\boldsymbol{A}$ is convergent, then $(-\boldsymbol{A})$ is also convergent, that is, $\|-\boldsymbol{A}\| = \|\boldsymbol{A}\| < 1$. Thus, Theorem 6.3 and its corollary are immediately applicable to $\boldsymbol{I}+\boldsymbol{A}$. Therefore, if in some natural norm, $\|-\boldsymbol{A}\| = \|\boldsymbol{A}\| < 1$, then

$$\frac{1}{1+\|\boldsymbol{A}\|} \leq \left\|(\boldsymbol{I}+\boldsymbol{A})^{-1}\right\| \leq \frac{1}{1-\|\boldsymbol{A}\|} \tag{6.26}$$

6.3 DIRECT METHODS

These are the methods that can find the solution of the system in a finite number of steps known a priori. Some of the important direct methods are

1. Elimination methods
2. Decomposition methods

6.3.1 Gauss Elimination Method

To solve a system of linear equations

$$\boldsymbol{A}\boldsymbol{x}=\boldsymbol{b}$$

$$\text{where } \boldsymbol{A}=[a_{ij}]_{n\times n}=\begin{bmatrix} a_{11} & a_{12} & \dots & a_{1n} \\ a_{21} & a_{22} & \dots & a_{2n} \\ \vdots & \vdots & \dots & \vdots \\ a_{n1} & a_{n2} & \dots & a_{nn} \end{bmatrix}, \quad \boldsymbol{x}=\begin{bmatrix} x_1 \\ x_2 \\ \vdots \\ x_n \end{bmatrix}, \quad \text{and } \boldsymbol{b}=\begin{bmatrix} b_1 \\ b_2 \\ \vdots \\ b_n \end{bmatrix}$$

we reduce it to an equivalent system $\boldsymbol{U}\boldsymbol{x}=\boldsymbol{g}$, where $\boldsymbol{U}$ is an upper triangular matrix. Then, this system can be easily solved by a process of back substitution.

Let us denote the original linear system of equations by

$$\boldsymbol{A}^{(1)}\boldsymbol{x} = \boldsymbol{b}^{(1)} \tag{6.27}$$

$$\text{where } \boldsymbol{A}^{(1)} = [a_{ij}^{(1)}]_{n\times n} \quad \text{and} \quad \boldsymbol{b}^{(1)} = \begin{bmatrix} b_1^{(1)} \\ b_2^{(1)} \\ \vdots \\ b_n^{(1)} \end{bmatrix}$$

Step 1:

Let us assume that $a_{11}^{(1)} \neq 0$. We define the row multipliers by

$$m_{i1} = \frac{a_{i1}^{(1)}}{a_{11}^{(1)}}, \quad i = 2,3,\ldots,n \tag{6.28}$$

These are used in eliminating the unknown x_1 term from the equations 2 through n.

We define

$$a_{ij}^{(2)} = a_{ij}^{(1)} - m_{i1}a_{1j}^{(1)}, \quad i,j = 2,3,\ldots,n$$

$$b_i^{(2)} = b_i^{(1)} - m_{i1}b_1^{(1)}, \quad i = 2,3,\ldots,n \tag{6.29}$$

Also, the first rows of $\boldsymbol{A}^{(1)}$ and $\boldsymbol{b}^{(1)}$ are left unchanged, and the first column of $\boldsymbol{A}^{(1)}$, below the diagonal, is set to zero. We obtain the system $\boldsymbol{A}^{(2)}\boldsymbol{x} = \boldsymbol{b}^{(2)}$ as follows:

$$\begin{bmatrix} a_{11}^{(1)} & a_{12}^{(1)} \cdots a_{1n}^{(1)} \\ 0 & a_{22}^{(2)} \cdots a_{2n}^{(2)} \\ \vdots & \vdots \cdots \vdots \\ 0 & a_{n2}^{(2)} \cdots a_{nn}^{(2)} \end{bmatrix} \begin{bmatrix} x_1 \\ x_2 \\ \vdots \\ x_n \end{bmatrix} = \begin{bmatrix} b_1^{(1)} \\ b_2^{(2)} \\ \vdots \\ b_n^{(2)} \end{bmatrix} \tag{6.30}$$

We continue to eliminate the other unknowns, going onto columns 2, 3, and so on, and this is expressed generally in the following.

Step k (General step):

Let $1 \leq k \leq n-1$. Let us assume $\boldsymbol{A}^{(k)}\boldsymbol{x} = \boldsymbol{b}^{(k)}$ has been constructed with $x_1, x_2, \ldots, x_{k-1}$ eliminated at successive previous stages, and $\boldsymbol{A}^{(k)}$ has the form

$$\boldsymbol{A}^{(k)} = \begin{bmatrix} a_{11}^{(1)} & a_{12}^{(1)} & & \cdots & & a_{1n}^{(1)} \\ 0 & a_{21}^{(2)} & & & & a_{21}^{(2)} \\ \vdots & & \ddots & & & \\ 0 & \cdots & 0 & a_{kk}^{(k)} & \cdots & a_{kn}^{(k)} \\ \vdots & & & \vdots & & \vdots \\ 0 & \cdots & 0 & a_{nk}^{(k)} & \cdots & a_{nn}^{(k)} \end{bmatrix} \tag{6.31}$$

Again, let us assume that $a_{kk}^{(k)} \neq 0$. We define the row multipliers by

$$m_{ik} = \frac{a_{ik}^{(k)}}{a_{kk}^{(k)}}, \quad i = k+1, k+2, \ldots, n \tag{6.32}$$

We use these to eliminate the unknown x_k from the equations $k+1$ through n.

We define

$$a_{ij}^{(k+1)} = a_{ij}^{(k)} - m_{ik} a_{kj}^{(k)}, \quad i, j = k+1, \ldots, n$$

$$b_i^{(k+1)} = b_i^{(k)} - m_{ik} b_k^{(k)}, \quad i = k+1, \ldots, n \tag{6.33}$$

The earlier rows of $\boldsymbol{A}^{(k)}$ and $\boldsymbol{b}^{(k)}$ starting from row 1 to row k are left unchanged and the zeros are introduced using the above elementary row operations into column k of $\boldsymbol{A}^{(k)}$ below the diagonal element $a_{kk}^{(k)}$.

Proceeding in this manner, after $n-1$ steps we obtain the system $\boldsymbol{A}^{(n)}\boldsymbol{x} = \boldsymbol{b}^{(n)}$ which is of the following form:

$$\begin{bmatrix} a_{11}^{(1)} & a_{12}^{(1)} \cdots a_{1n}^{(1)} \\ 0 & a_{22}^{(2)} \cdots a_{2n}^{(2)} \\ \vdots & \vdots \cdots \vdots \\ 0 & 0 \cdots a_{nn}^{(n)} \end{bmatrix} \begin{bmatrix} x_1 \\ x_2 \\ \vdots \\ x_n \end{bmatrix} = \begin{bmatrix} b_1^{(1)} \\ b_2^{(2)} \\ \vdots \\ b_n^{(n)} \end{bmatrix} \tag{6.34}$$

The leading coefficients of the above set of equations $a_{11}^{(1)}, a_{22}^{(2)}, \ldots, a_{nn}^{(n)}$, which are nonzero by assumption, are known as the pivots, and the corresponding equations (rows) are called the pivotal equations (rows).

Let $\boldsymbol{U} = \boldsymbol{A}^{(n)}$ and $\boldsymbol{g} = \boldsymbol{b}^{(n)}$

The system

$$\boldsymbol{U}\boldsymbol{x} = \boldsymbol{g} \tag{6.35}$$

is the upper triangular system, which can be easily solved by back-substitution method, where

$$\boldsymbol{U} = \begin{bmatrix} u_{11} & u_{12} \cdots u_{1n} \\ 0 & u_{22} \cdots u_{2n} \\ \vdots & \vdots \cdots \vdots \\ 0 & 0 \cdots u_{nn} \end{bmatrix} \quad \text{and} \quad \boldsymbol{g} = \begin{bmatrix} g_1 \\ g_2 \\ \vdots \\ g_n \end{bmatrix}$$

Thus, using the Gauss elimination method, we obtain

$$[\boldsymbol{A} \mid \boldsymbol{b}] \xrightarrow{\text{Gauss elimination method}} [\boldsymbol{U} \mid \boldsymbol{g}]$$

By the back-substitution method, we first find

$$x_n = \frac{b_n^{(n)}}{a_{nn}^{(n)}} = \frac{g_n}{u_{nn}} \tag{6.36}$$

and then we determine

$$\begin{aligned} x_r &= \frac{1}{a_{rr}^{(r)}}\left[b_r^{(r)} - \sum_{j=r+1}^{n} a_{rj}^{(r)} x_j\right], \quad r = n-1, n-2, \ldots, 1 \\ &= \frac{1}{u_{rr}}\left[g_r - \sum_{j=r+1}^{n} u_{rj} x_j\right], \quad r = n-1, n-2, \ldots, 1 \end{aligned} \tag{6.37}$$

This is the end of the Gauss elimination or the Gaussian elimination method.

6.3.1.1 Pivoting in the Gauss Elimination Method

We have assumed that in each step k of the Gaussian elimination method $a_{kk}^{(k)} \neq 0$, $k = 1, 2, \ldots, n$.

To remove this restriction, we begin with each step of elimination process by switching rows to put a nonzero element in the pivotal position. Since, $\boldsymbol{A}$ is nonsingular, this is always possible. However, sometimes it may happen that pivot element is very small.

To get rid of this situation and for other reason involving the propagation of rounding off errors, we use pivoting method. There are two types of pivoting. One is partial pivoting and another is complete pivoting. In practical computation, partial pivoting is mostly used, since in complete pivoting computational overhead is high. In view of operational time, complete pivoting is more expensive than partial pivoting.

- *Partial pivoting*: Let, at stage $k, 1 \leq k \leq n-1$

$$\left|a_{i_0 k}^{(k)}\right| = \max_{k \leq i \leq n}\left|a_{ik}^{(k)}\right| \tag{6.38}$$

  Let i_0 be the smallest row index for which the maximum in magnitude is attained. If $i_0 > k$, we first interchange the rows k and i_0 in $\boldsymbol{A}^{(k)}$ and $\boldsymbol{b}^{(k)}$ and then proceed with elementary row operations in step k of the Gaussian elimination process. All the row multipliers

$$|m_{ik}| \leq 1, \quad i = k+1, k+2, \ldots, n$$

  Thus, it helps in preventing the growth of the elements in $\boldsymbol{A}^{(k)}$ and consequently eliminates the possibility of the large loss of significant errors due to propagation of round-off errors.

  Thus, in the first stage of the Gauss elimination, the first column is searched for largest element in magnitude and brought as the first pivot by interchanging the first equation with the equation having the largest element in magnitude. Then proceed with elementary row operations in first step of Gauss elimination method. In the second elimination stage, the second column is searched for the largest element in magnitude among the $n-1$ elements leaving the first element, and this element is brought as the second pivot by interchanging the second equation with the equation having the largest element in magnitude. Next, we proceed with usual elementary row operations of the Gauss elimination method. This process is continued until we arrive at the upper triangular system $\boldsymbol{Ux} = \boldsymbol{g}$.

- *Complete pivoting*: In this process, the largest element in magnitude of the whole coefficient matrix $\boldsymbol{A}$ is brought as the first pivotal position of the coefficient matrix, and then leaving the first row and first column, the largest among the remaining elements is brought to the second pivotal position of the coefficient matrix and so on by interchanging both rows and columns is called complete pivoting.

  Let i_0 and j_0 be the smallest integers for which

$$\left|a_{i_0 j_0}^{(k)}\right| = \max_{k \le i, j \le n} \left|a_{ij}^{(k)}\right| \tag{6.39}$$

  and next interchange rows k and i_0 and also columns k and j_0.

  During the interchange of rows, the last column of the augmented matrix also has to be considered but this column is not considered to find the largest element in magnitude. Since the interchange of columns is also allowed in this process, there will be a change in the position of the individual elements of the unknown vector $\boldsymbol{x}$. Hence in the end, the elements of the unknown vector $\boldsymbol{x}$ has to be rearranged by applying column switching in reverse order to all the column transformations preformed.

Remarks: Without partial pivoting, the computational scheme of the Gauss elimination method may fail in case the pivotal coefficient vanishes at any step. Again if at any step, the pivot is zero or sufficiently small in magnitude, then the corresponding row multipliers will be numerically large so that on multiplying the pivotal equation by such a numerically large multiplier the round-off and other computational errors in the coefficients and the constants of that equation get considerably magnified thereby making the results of the Gauss elimination method affected with large errors.

6.3.1.2 Operation Count in the Gauss Elimination Method

Generally, the important factors involved in justifying the quality of a numerical method are

- Time complexity (amount of time which is estimated by number of operations performed)
- Space complexity (amount of storage required)
- Effect of round-off error

The operational count of the Gauss elimination method can be determined as follows:

In step k, we eliminate unknown x_k from $n-k$ equations. This requires $n-k$ divisions for computing the row multipliers m_{ik} in Equation 6.32 and $(n-k)(n-k+1)$ multiplications for computing row operations in Equation 6.33. Since, there are $n-1$ steps in the Gauss elimination method, k ranges from 1 to $n-1$.

Therefore, the total number of operations in forward elimination is

$$\sum_{k=1}^{n-1}(n-k) + \sum_{k=1}^{n-1}(n-k)(n-k+1) = \frac{n(n-1)(2n+5)}{6}$$

Now, for back substitution, the total number of multiplications is

$$\sum_{r=1}^{n-1}(n-r) = \frac{n(n-1)}{2}$$

and the total number of divisions is n.

Thus, the total number of multiplications and divisions in the back-substitution method is

$$\frac{n(n-1)}{2}+n=\frac{n(n+1)}{2}$$

Therefore, the total number of operations in the Gauss elimination is

$$\frac{n(n-1)(2n+5)}{6}+\frac{n(n+1)}{2}=\frac{n(n^2+3n-1)}{3}$$

For large n, the operational count is approximately $n^3/3$.

Therefore, the time complexity in the Gauss elimination method is $O(n^3)$.

6.3.1.3 Algorithm for the Gauss Elimination Method

Input: Enter the number of unknown variables n, the coefficient matrix $\boldsymbol{A}=[a_{ij}]_{n\times n}$, and the constant column vector $\boldsymbol{b}=[b_i]_{n\times 1}$.

Output: Solution $\boldsymbol{x}=[x_i]_{n\times 1}$ of the given system of equations.

Initial step: Create the augmented matrix $[\boldsymbol{A}\mid\boldsymbol{b}]=[\text{new}\_a_{ij}]_{n\times n+1}$, where $\text{new}\_a_{i,n+1}=b_i$.

Step 1: for $k=1,2,\ldots,n-1$ do

If i_0 be the smallest row index such that

$$\left|a_{i_0k}^{(k)}\right|=\max_{k\le i\le n}\left|a_{ik}^{(k)}\right|;$$

and $i_0>k$ then

interchange the rows k and i_0 in the augmented matrix $[\boldsymbol{A}\mid\boldsymbol{b}]=[\text{new}\_a_{ij}]_{n\times n+1}$;

Set, $\text{old}\_a_{ij}=\text{new}\_a_{ij};\quad i=1(1)n,\quad j=1(1)n+1.$

for $i=k+1,k+2,\ldots,n$ do

Compute

$$m_{ik}=(\text{old}\_a_{ik}/\text{old}\_a_{kk});\quad (m_{ik}\text{ is a row multiplier})$$

for $j=k+1,k+2,\ldots,n+1$ do

$\text{new}\_a_{ij}=\text{old}\_a_{ij}-m_{ik}*\text{old}\_a_{kj};$

end

end

end

Step 2: (Back-substitution method)

Compute $x_n=(\text{new}\_a_{n,n+1}/\text{new}\_a_{nn});$

for $r=n-1,n-2,\ldots,1$ do

$\text{sum}=0;$

for $j=r+1,r+2,\ldots,n$ do

$\text{sum}=\text{sum}+\text{new}\_a_{rj}*x_j;$

end

Compute

$x_r=(\text{new}\_a_{r,n+1}-\text{sum})/\text{new}\_a_{rr};$

end

Step 3: Print the value of x_i, $(i=1,2,\ldots,n)$.

Step 4: Stop the program. ■

MATHEMATICA® Program for Solving System of Equations by the Gauss Elimination Method without Pivoting (Chapter 6, Example 6.1)

```
n=3;
A={{1,2,1,0},{2,2,3,3},{-1,-3,0,2}};
For[i=1,i<=n, i++,
  For[j=1,j<=n+1,j++,
   a[i, j]=A[[i, j]]]];
For[k=1,k<=n-1,k++,
  For[i=k+1,i<=n, i++,
   m[i, k]=a[i, k]/a[k, k];
   Print["m[",i,",",k,"]=",m[i, k]];
   For[j=k+1,j<=n+1,j++,
    a[i, j]=a[i, j]-m[i, k]*a[k, j];
    Print["a[",i,",",j,"]=",a[i, j]]]]];
x[n]=a[n, n+1]/a[n, n];
For[r=n-1,r>=1,r--,
  sum=0;
  For[j=r+1,j<=n, j++,
   sum=sum+a[r, j]*x[j]];
  x[r]=(a[r, n+1]-sum)/a[r, r]];
For[i=1,i<=n, i++,
  Print["x[",i,"]=",x[i]]];
```

Output:

```
m[2,1]=2
a[2,2]=-2
a[2,3]=1
a[2,4]=3
m[3,1]=-1
a[3,2]=-1
a[3,3]=1
a[3,4]=2
m[3,2]=1/2
a[3,3]=1/2
a[3,4]=1/2
x[1]=1
x[2]=-1
x[3]=1
```

MATHEMATICA® Program for the Gauss Elimination Method with Partial Pivoting for Solving System of Equations (Chapter 6, Example 6.2)

```
ROWINDEX[M_, k_, n_]:=Module[{s, max, index},
   max=M[[k, k]];
   index=k;
   For[s=k+1,s<=n, s++,
    If[max<M[[k, s]], max=M[[k, s]];index=s]];
   Return [index]];

n=3;
A={{1,1,1,9},{2,-3,4,13},{3,4,5,40}};
```

```
For[i=1,i<=n, i++,
  For[j=1,j<=n+1,j++,
   a[i, j]=A[[i, j]]]];

For[k=1,k<=n-1,k++,
  i0=ROWINDEX[A, k,n];
  Print["i0=",i0];
  If[i0>k,
   For[j=1,j<=n+1,j++,
    temp1=a[k, j];
    a[k, j]=a[i0,j];
    a[i0,j]=temp1]];

  For[i=k+1,i<=n, i++,
   m[i, k]=a[i, k]/a[k, k];
   Print["m[",i,",",k,"]=",m[i, k]];
   For[j=k+1,j<=n+1,j++,
    a[i, j]=a[i, j]-m[i, k]*a[k, j];
    Print["a[",i,",",j,"]=",a[i, j]]]]];
x[n]=a[n, n+1]/a[n, n];
For[r=n-1,r>=1,r--,
  sum=0;
  For[j=r+1,j<=n, j++,
   sum=sum+a[r, j]*x[j]];
  x[r]=(a[r, n+1]-sum)/a[r, r]];
For[i=1,i<=n, i++,
  Print["x[",i,"]=",x[i]]];
```

Output:

```
i0 = 1
m[2,1]=2
a[2,2]=-5
a[2,3]=2
a[2,4]=-5
m[3,1]=3
a[3,2]=1
a[3,3]=2
a[3,4]=13
i0 = 3
m[3,2]=-5
a[3,3]=12
a[3,4]=60
x[1]=1
x[2]=3
x[3]=5
```

Example 6.1

Solve the following linear system of equations by the Gauss elimination method:

$$x_1+2x_2+x_3=0$$

$$2x_1+2x_2+3x_3=3$$

$$-x_1-3x_2=2$$

Solution:

First, we represent the given system of linear equations with the following augmented matrix:

$$[A|b] = \left[\begin{array}{ccc|c} 1 & 2 & 1 & 0 \\ 2 & 2 & 3 & 3 \\ -1 & -3 & 0 & 2 \end{array}\right]$$

Now, we shall apply the Gauss elimination method (without partial pivoting) in the following steps. Since, there are three unknowns x_1, x_2, x_3, the number of steps required in this method will be two.

Step 1:

$$\left[\begin{array}{ccc|c} 1 & 2 & 1 & 0 \\ 2 & 2 & 3 & 3 \\ -1 & -3 & 0 & 2 \end{array}\right] \xrightarrow[R'_3 \leftarrow R_3 + R_1]{R'_2 \leftarrow R_2 - 2R_1} \left[\begin{array}{ccc|c} 1 & 2 & 1 & 0 \\ 0 & -2 & 1 & 3 \\ 0 & -1 & 1 & 2 \end{array}\right]$$

Step 2:

$$\left[\begin{array}{ccc|c} 1 & 2 & 1 & 0 \\ 0 & -2 & 1 & 3 \\ 0 & -1 & 1 & 2 \end{array}\right] \xrightarrow{R'_3 \leftarrow R_3 - \frac{1}{2}R_2} \left[\begin{array}{ccc|c} 1 & 2 & 1 & 0 \\ 0 & -2 & 1 & 3 \\ 0 & 0 & 0.5 & 0.5 \end{array}\right]$$

Thus starting with augment matrix $[A|b]$, we arrive at the following upper triangular matrix $[U|g]$, viz.,

$$\left[\begin{array}{ccc|c} 1 & 2 & 1 & 0 \\ 0 & -2 & 1 & 3 \\ 0 & 0 & 0.5 & 0.5 \end{array}\right]$$

after completion of the Gauss elimination method in two steps.
Now, the equivalent system of equations is

$$x_1 + 2x_2 + x_3 = 0$$

$$-2x_2 + x_3 = 3$$

$$0.5x_3 = 0.5$$

We solve the above system of equations by the back-substitution method.
Hence, the required solution of the system of linear equations is

$$\begin{bmatrix} x_1 \\ x_2 \\ x_3 \end{bmatrix} = \begin{bmatrix} 1 \\ -1 \\ 1 \end{bmatrix}, \quad \text{that is,} \quad x_1 = 1, x_2 = -1, x_3 = 1$$

Example 6.2

Using the Gauss elimination method, solve the following linear system of equations:

$$x_1+x_2+x_3=9$$

$$2x_1-3x_2+4x_3=13$$

$$3x_1+4x_2+5x_3=40$$

Solution:

First, we represent the given system of linear equations with the following augmented matrix

$$[A|b]=\left[\begin{array}{ccc|c} 1 & 1 & 1 & 9 \\ 2 & -3 & 4 & 13 \\ 3 & 4 & 5 & 40 \end{array}\right]$$

Now, we shall apply the Gauss elimination method (with partial pivoting) in the following steps. Since, there are three unknowns x_1, x_2, x_3, the number of steps required in this method will be two.

Step 1:

$$\left[\begin{array}{ccc|c} 1 & 1 & 1 & 9 \\ 2 & -3 & 4 & 13 \\ 3 & 4 & 5 & 40 \end{array}\right] \xrightarrow{R_{13}} \left[\begin{array}{ccc|c} 3 & 4 & 5 & 40 \\ 2 & -3 & 4 & 13 \\ 1 & 1 & 1 & 9 \end{array}\right] \xrightarrow[R'_3 \leftarrow R_3-\frac{1}{3}R_1]{R'_2 \leftarrow R_2-\frac{2}{3}R_1} \left[\begin{array}{ccc|c} 3 & 4 & 5 & 40 \\ 0 & -\frac{17}{3} & \frac{2}{3} & -\frac{41}{3} \\ 0 & -\frac{1}{3} & -\frac{2}{3} & -\frac{13}{3} \end{array}\right]$$

Step 2:

$$\left[\begin{array}{ccc|c} 3 & 4 & 5 & 40 \\ 0 & -\frac{17}{3} & \frac{2}{3} & -\frac{41}{3} \\ 0 & -\frac{1}{3} & -\frac{2}{3} & -\frac{13}{3} \end{array}\right] \xrightarrow{R'_3 \leftarrow R_3-\frac{1}{17}R_2} \left[\begin{array}{ccc|c} 3 & 4 & 5 & 40 \\ 0 & -\frac{17}{3} & \frac{2}{3} & -\frac{41}{3} \\ 0 & 0 & -\frac{12}{17} & -\frac{180}{51} \end{array}\right]$$

Thus starting with augment matrix $[A|b]$, we arrive at the following upper triangular matrix $[U|g]$, viz.,

$$\left[\begin{array}{ccc|c} 3 & 4 & 5 & 40 \\ 0 & -\frac{17}{3} & \frac{2}{3} & -\frac{41}{3} \\ 0 & 0 & -\frac{12}{17} & -\frac{180}{51} \end{array}\right]$$

after completion of the Gauss elimination method in two steps.

Now, the equivalent system of equations is

$$3x_1+4x_2+5x_3=40$$

$$-\frac{17}{3}x_2+\frac{2}{3}x_3=-\frac{41}{3}$$

$$-\frac{12}{17}x_3=-\frac{180}{51}$$

We solve the above system of equations by the back-substitution method.
Hence, the required solution of the system of linear equations is

$$\begin{bmatrix} x_1 \\ x_2 \\ x_3 \end{bmatrix} = \begin{bmatrix} 1 \\ 3 \\ 5 \end{bmatrix}, \quad \text{that is,} \quad x_1=1, x_2=3, x_3=5$$

6.3.2 Gauss–Jordan Method

This method is much similar to the Gauss elimination method including the possible use of partial pivoting. In this method, elimination of unknowns is performed not in the equations below the diagonal but in the equations above the diagonal, as a result of which the coefficient matrix of the given system of equations reduces to a diagonal matrix or identity matrix form.

General step $k(1 \le k \le n)$: In step k of the elimination process, we choose the pivot element as discussed in Equation 6.38 of the Gauss elimination method. Then, we define

$$a_{kj}^{(k+1)} = \frac{a_{kj}^{(k)}}{a_{kk}^{(k)}}, \quad j = k, k+1, \ldots, n$$

$$b_k^{(k+1)} = \frac{b_k^{(k)}}{a_{kk}^{(k)}} \tag{6.40}$$

Eliminating the unknown x_k in equations, both the above and below equation k, we define

$$a_{ij}^{(k+1)} = a_{ij}^{(k)} - a_{ik}^{(k)} a_{kj}^{(k+1)}$$

$$b_i^{(k+1)} = b_i^{(k)} - a_{ik}^{(k)} b_k^{(k+1)}, \quad i = 1, 2, \ldots, n\,(i \neq k); \quad j = k, k+1, \ldots, n \tag{6.41}$$

This procedure will convert the augmented matrix $[A \mid b]$ to $[I \mid d]$.
Thus, using the Gauss–Jordan method, we obtain

$$[A \mid b] \xrightarrow{\text{Gauss–Jordan method}} [I \mid d]$$

Therefore, solution is given by

$$x_i = d_i, \quad i = 1, 2, \ldots, n$$

It can be shown that for large n, the operational count in the Gauss–Jordan method is approximately $n^3/2$.
Therefore, the time complexity in the Gauss–Jordan method is $O(n^3)$.

6.3.2.1 Algorithm for the Gauss–Jordan Method

Input: Enter the number of unknown variables n, the coefficient matrix $\boldsymbol{A} = [a_{ij}]_{n\times n}$, and the constant column vector $\boldsymbol{b} = [b_i]_{n\times 1}$.

Output: Solution $\boldsymbol{x} = [x_i]_{n\times 1}$ of the given system of equations.

Initial step: Create the augmented matrix $[\boldsymbol{A}\mid\boldsymbol{b}] = [\text{new}\_a_{ij}]_{n\times n+1}$, where $\text{new}\_a_{i,n+1} = b_i$.

Step 1: for $k = 1,2,\ldots,n$ do

If i_0 be the smallest row index such that

$$\left|a_{i_0k}^{(k)}\right| = \max_{k\le i\le n}\left|a_{ik}^{(k)}\right|;$$

and $i_0 > k$ then

interchange the rows k and i_0 in the augmented matrix $[\boldsymbol{A}\mid\boldsymbol{b}] = [\text{new}\_a_{i,j}]_{n\times n+1}$;

Set, $\text{old}\_a_{ij} = \text{new}\_a_{ij}$; $i = 1(1)n$, $j = 1(1)n+1$.

for $j = k, k+1,\ldots,n+1$ do

Compute

$$\text{new}\_a_{kj} = (\text{old}\_a_{kj} / \text{old}\_a_{kk});$$

for $i = 1,2,\ldots,n$ do

if $i \ne j$ then

$$\text{new}\_a_{ij} = \text{old}\_a_{ij} - \text{old}\_a_{ik} * \text{new}\_a_{kj};$$

end

end

end

Step 2: Set $x_i = \text{new}\_a_{i,n+1}$; $i = 1,2,\ldots,n$.

Step 3: Print the value of x_i, $(i = 1,2,\ldots,n)$.

Step 4: Stop the program. ■

MATHEMATICA® Program for Solving System of Equations by the Gauss–Jordan Method (Chapter 6, Example 6.3)

```
ROWINDEX[M_, k_, n_]:=Module[{s, max, index},
   max=M[[k, k]];
   index=k;
   For[s=k+1,s<=n, s++,
    If[max<M[[k, s]], max=M[[k, s]];index=s]];
   Return [index]];

n=3;
A={{2,4,1,3},{3,2,-2,-2},{1,-1,1,6}};
For[i=1,i<=n, i++,
  For[j=1,j<=n+1,j++,
   a[i, j]=A[[i, j]];
   ]];
Print[
  "......................................................................
   ............"];
For[k=1,k<=n, k++,
  Print["Step ",k,":"];
  Print[
    "....................................................................
     ..............."];
  i0=ROWINDEX[A, k,n];
  Print["i0=",i0];
  If[i0>k,
```

```
 For[j=1,j<=n+1,j++,
  temp1=a[k, j];
  a[k, j]=a[i0,j];
  a[i0,j]=temp1]];

For[i=1,i<=n, i++,
 For[j=1,j<=n+1,j++,
  temp[i, j]=a[i, j]]];

For[i=1,i<=n, i++,
  If[i!=k, For[j=k, j<=n+1,j++,
    a[k, j]=temp[k, j]/temp[k, k];
   a[i, j]=temp[i, j]-temp[i, k]*a[k, j];
   Print["a[",i,",",j,"]=",a[i, j]]]]]];
Print[
 "..............................................................
  ..........."]];

For[i=1,i<=n, i++,
  Print["x[",i,"]=",a[i, n+1]]];
```

Output:

```
................. .........................................
Step 1:
............................................................
i0=2
a[2,1]=0
a[2,2]=8/3
a[2,3]=7/3
a[2,4]=13/3
a[3,1]=0
a[3,2]=-(5/3)
a[3,3]=5/3
a[3,4]=20/3
............................................................
Step 2:
............................................................
i0=2
a[1,2]=0
a[1,3]=-(5/4)
a[1,4]=-(7/4)
a[3,2]=0
a[3,3]=25/8
a[3,4]=75/8
............................................................
Step 3:
............................................................
i0=3
a[1,3]=0
a[1,4]=2
a[2,3]=0
a[2,4]=-1
............................................................
x[1]=2
x[2]=-1
x[3]=3
```

Example 6.3

Using the Gauss–Jordan method, solve the following equations:

$$2x_1+4x_2+x_3=3$$
$$3x_1+2x_2-2x_3=-2$$
$$x_1-x_2+x_3=6$$

Solution:

First, we represent the given system of linear equations with the following augmented matrix:

$$\left[A|b\right]=\left[\begin{array}{ccc|c}2&4&1&3\\3&2&-2&-2\\1&-1&1&6\end{array}\right]$$

Now, we shall apply the Gauss–Jordan method in the following steps. Since, there are three unknowns x, y, z, the number of steps required in this method will be three.

Step 1:

$$\left[\begin{array}{ccc|c}2&4&1&3\\3&2&-2&-2\\1&-1&1&6\end{array}\right]\xrightarrow{\frac{1}{2}R_1}\left[\begin{array}{ccc|c}1&2&0.5&1.5\\3&2&-2&-2\\1&-1&1&6\end{array}\right]\xrightarrow[R_3'\leftarrow R_3-R_1]{R_2'\leftarrow R_2-3R_1}\left[\begin{array}{ccc|c}1&2&0.5&1.5\\0&-4&-3.5&-6.5\\0&-3&0.5&4.5\end{array}\right]$$

Step 2:

$$\left[\begin{array}{ccc|c}1&2&0.5&1.5\\0&-4&-3.5&-6.5\\0&-3&0.5&4.5\end{array}\right]\xrightarrow{-\frac{1}{4}R_2}\left[\begin{array}{ccc|c}1&2&0.5&1.5\\0&1&0.875&1.625\\0&-3&0.5&4.5\end{array}\right]\xrightarrow[R_3'\leftarrow R_3+3R_2]{R_1'\leftarrow R_1-2R_2}\left[\begin{array}{ccc|c}1&0&-1.25&-1.75\\0&1&0.875&1.625\\0&0&3.125&9.375\end{array}\right]$$

Step 3:

$$\left[\begin{array}{ccc|c}1&0&-1.25&-1.75\\0&1&0.875&1.625\\0&0&3.125&9.375\end{array}\right]\xrightarrow{\frac{8}{25}R_3}\left[\begin{array}{ccc|c}1&0&-1.25&-1.75\\0&1&0.875&1.625\\0&0&1&3\end{array}\right]\xrightarrow[R_2'\leftarrow R_2-0.875R_3]{R_1'\leftarrow R_1+1.25R_3}\left[\begin{array}{ccc|c}1&0&0&2\\0&1&0&-1\\0&0&1&3\end{array}\right]$$

Therefore, the required solution of the system of linear equations is

$$\begin{bmatrix}x_1\\x_2\\x_3\end{bmatrix}=\begin{bmatrix}2\\-1\\3\end{bmatrix},\quad \text{that is,}\quad x_1=2,\ x_2=-1,\ x_3=3$$

Example 6.4

Apply the Gauss–Jordan method to solve the following equations:

$$x + y + z = 9$$
$$2x - 3y + 4z = 13$$
$$3x + 4y + 5z = 40$$

Solution:

First, we represent the given system of linear equations with the following augmented matrix:

$$\left[A|b\right] = \left[\begin{array}{ccc|c} 1 & 1 & 1 & 9 \\ 2 & -3 & 4 & 13 \\ 3 & 4 & 5 & 40 \end{array}\right]$$

Now, we shall apply the Gauss–Jordan method in the following steps. Since, there are three unknowns x, y, z, the number of steps required in this method will be three.

Step 1:

$$\left[\begin{array}{ccc|c} 1 & 1 & 1 & 9 \\ 2 & -3 & 4 & 13 \\ 3 & 4 & 5 & 40 \end{array}\right] \xrightarrow[R_3' \leftarrow R_3 - 3R_1]{R_2' \leftarrow R_2 - 2R_1} \left[\begin{array}{ccc|c} 1 & 1 & 1 & 9 \\ 0 & -5 & 2 & -5 \\ 0 & 1 & 2 & 13 \end{array}\right]$$

Step 2:

$$\left[\begin{array}{ccc|c} 1 & 1 & 1 & 9 \\ 0 & -5 & 2 & -5 \\ 0 & 1 & 2 & 13 \end{array}\right] \xrightarrow{-\frac{1}{5}R_2} \left[\begin{array}{ccc|c} 1 & 1 & 1 & 9 \\ 0 & 1 & -0.4 & 1 \\ 0 & 1 & 2 & 13 \end{array}\right] \xrightarrow[R_3' \leftarrow R_3 - R_2]{R_1' \leftarrow R_1 - R_2} \left[\begin{array}{ccc|c} 1 & 0 & 1.4 & 8 \\ 0 & 1 & -0.4 & 1 \\ 0 & 0 & 2.4 & 12 \end{array}\right]$$

Step 3:

$$\left[\begin{array}{ccc|c} 1 & 0 & 1.4 & 8 \\ 0 & 1 & -0.4 & 1 \\ 0 & 0 & 2.4 & 12 \end{array}\right] \xrightarrow{\frac{5}{12}R_3} \left[\begin{array}{ccc|c} 1 & 0 & 1.4 & 8 \\ 0 & 1 & -0.4 & 1 \\ 0 & 0 & 1 & 5 \end{array}\right] \xrightarrow[R_2' \leftarrow R_2 + 0.4R_3]{R_1' \leftarrow R_1 - 1.4R_3} \left[\begin{array}{ccc|c} 1 & 0 & 0 & 1 \\ 0 & 1 & 0 & 3 \\ 0 & 0 & 1 & 5 \end{array}\right]$$

Therefore, the required solution of the system of linear equations is

$$\begin{bmatrix} x \\ y \\ z \end{bmatrix} = \begin{bmatrix} 1 \\ 3 \\ 5 \end{bmatrix}, \quad \text{that is,} \quad x = 1, y = 3, z = 5.$$

6.3.3 Triangularization Method

This method is also known as triangular decomposition or factorization method or ***LU***-decomposition method.

This method is based on the fact that every square matrix can be expressed as the product of a lower and an upper triangular matrix provided all the principal minors of the given square matrix $\boldsymbol{A} = [a_{ij}]_{n\times n}$ are nonsingular, that is,

$$a_{11} \neq 0, \begin{vmatrix} a_{11} & a_{12} \\ a_{21} & a_{22} \end{vmatrix} \neq 0, \ldots, \det \boldsymbol{A} = |\boldsymbol{A}| \neq 0$$

Further, if the matrix $\boldsymbol{A}$ can be factorized, then it is unique.

If it is possible, then in this method the coefficient matrix $\boldsymbol{A}$ of the system of equations $\boldsymbol{Ax} = \boldsymbol{b}$ is decomposed into the product of a lower triangular matrix $\boldsymbol{L}$ and an upper triangular matrix $\boldsymbol{U}$ so that

$$\boldsymbol{A} = \boldsymbol{LU} \tag{6.42}$$

where:

$$\boldsymbol{L} = \begin{bmatrix} l_{11} & 0 & 0 & \cdots & 0 \\ l_{21} & l_{22} & 0 & \cdots & 0 \\ l_{31} & l_{32} & l_{33} & \cdots & 0 \\ \vdots & \vdots & \vdots & \cdots & \vdots \\ l_{n1} & l_{n2} & l_{n3} & \cdots & l_{nn} \end{bmatrix} \text{ and } \boldsymbol{U} = \begin{bmatrix} u_{11} & u_{12} & u_{13} & \cdots & u_{1n} \\ 0 & u_{22} & u_{23} & \cdots & u_{2n} \\ 0 & 0 & u_{33} & \cdots & u_{3n} \\ \vdots & \vdots & \vdots & \cdots & \vdots \\ 0 & 0 & 0 & \cdots & u_{nn} \end{bmatrix}$$

Therefore, from Equation 6.48, we can obtain

$$l_{i1}u_{1j} + l_{i2}u_{2j} + \ldots + l_{in}u_{nj} = a_{ij}, \quad i, j = 1, 2, \ldots, n \tag{6.43}$$

where:

$$l_{ij} = 0, j > i \quad \text{and} \quad u_{ij} = 0, i > j$$

Thus, the system of equations $\boldsymbol{Ax} = \boldsymbol{b}$ becomes

$$\boldsymbol{LUx} = \boldsymbol{b} \tag{6.44}$$

Let us take

$$\boldsymbol{Ux} = \boldsymbol{z} \tag{6.45}$$

Substituting this in Equation 6.44, we get

$$\boldsymbol{Lz} = \boldsymbol{b} \tag{6.46}$$

The unknowns $z_1, z_2, \ldots, z_n$ in Equation 6.46 are determined by forward substitution and then consequently the unknowns $x_1, x_2, \ldots, x_n$ in Equation 6.45 are determined by back substitution.

To determine $\boldsymbol{L}$ and $\boldsymbol{U}$ in $\boldsymbol{LU}$-decomposition method, we can apply any of the following two methods.

6.3.3.1 Doolittle's Method

If we choose all the elements of principal diagonal of lower triangular matrix $\boldsymbol{L}$ as 1, that is, $l_{ii} = 1$, $i = 1, 2, \ldots, n$, then the corresponding method is called Doolittle's method.

From Equation 6.43, we get

$$
\begin{aligned}
l_{ii} &= 1 \\
u_{ii} &= a_{ii} - \sum_{k=1}^{i-1} l_{ik}u_{ki} \\
u_{ij} &= a_{ij} - \sum_{k=1}^{i-1} l_{ik}u_{kj}, \quad i < j \\
l_{ij} &= \frac{\left(a_{ij} - \sum_{k=1}^{j-1} l_{ik}u_{kj}\right)}{u_{jj}}, \quad i > j
\end{aligned}
\tag{6.47}
$$

6.3.3.1.1 *Algorithm for Doolittle's Decomposition Method*

Input: Enter the number of unknown variables n, the coefficient matrix $\boldsymbol{A} = [a_{ij}]_{n\times n}$, and the constant column vector $\boldsymbol{b} = [b_i]_{n\times 1}$.

Output: Solution $\boldsymbol{x} = [x_i]_{n\times 1}$ of the given system of equations.

Initial step: Read matrix $\boldsymbol{A}$ with elements a_{ij}, $i, j = 1, 2, \ldots n$.

Step 1: (Determination for elements l_{ij} and u_{ij} of lower and upper triangular matrices, respectively)

```
for i = 1(1)n do
    for j = 1(1)n do
        if (i == j)then
            l_ii = 1;
            sum = 0;
          for k = 1(1)(i-1) do
            sum = sum + l_ik * u_ki;
          end
            u_ii = a_ii - sum;
        else
          if (i < j) then
            sum1 = 0;
          for k = 1(1)(i-1) do
            sum1 = sum1 + l_ik * u_kj;
          end
            u_ij = a_ij - sum1;
          else
            if (i > j) then
                sum2 = 0;
              for k = 1(1)(j-1) do
                sum2 = sum2 + l_ik * u_kj;
              end
                l_ij = (a_ij - sum2) / u_jj;
    end
end
```

Step 2: Print lower triangular matrix $\boldsymbol{L} = [l_{ij}]_{n\times n}$;
and upper triangular matrix $\boldsymbol{U} = [u_{ij}]_{n\times n}$;

Step 3: (Forward substitution method for determining z)
Compute $z_1 = (b_1 / l_{11})$;

for $r = 2,3,\ldots,n$ do
    sum $= 0$;
  for $j = 1,2,\ldots,\overline{r-1}$ do
    sum $=$ sum $+ l_{rj} * z_j$;
  end
    Compute
      $z_r = (b_r - \text{sum}) / l_{rr}$;
end

Step 4: (Back-substitution method for determining the solution $\boldsymbol{x}$)
  Compute $x_n = (z_n / u_{nn})$;
  for $r = n-1, \quad n-2,\ldots,1$ do
    sum $= 0$;
  for $j = \overline{r+1}, \overline{r+2},\ldots,n$ do
    sum $=$ sum $+ u_{rj} * x_j$;
  end
    Compute
      $x_r = (z_r - \text{sum}) / u_{rr}$;
  end

Step 5: Print the value of x_i, $(i = 1,2,\ldots,n)$.
Step 6: Stop the program. ■

MATHEMATICA® Program for Solution of System of Algebraic Equations by Doolittle's Method (Chapter 6, Example 6.5)

```
A={{1,1,1},{4,3,-1},{3,5,3}};
b={1,6,4};
n=3;
For[i=1,i<=n, i++,
  For[j=1,j<=n, j++,
   If[i==j,
    l[i, i]=1;
    sum=0;
    For[k=1,k<=i-1,k++,
     sum=sum+l[i, k]*u[k, i]];
    u[i, i]=A[[i, i]]-sum;
    Print["l[",i,",",i,"]=",l[i, i]];
    Print["u[",i,",",i,"]=",u[i, i]]];

If[i<j,
 sum1 = 0;
 For[k=1,k<=i-1,k++,
  sum1=sum1+l[i, k]*u[k, j]];
 u[i, j]=A[[i, j]]-sum1;
 Print["u[",i,",",j,"]=",u[i, j]]];

If[i>j,
 sum2 = 0;
 For[k=1,k<=j-1,k++,
  sum2=sum2+l[i, k]*u[k, j]];
 l[i, j]=(A[[i, j]]-sum2)/u[j, j];
 Print["l[",i,",",j,"]=",l[i, j]]]]];

z[1]=b[[1]]/l[1,1];
Print["z[",1,"]=",z[1]];
```

```
For[r=2,r<=n, r++,
    sum3 = 0;
  For[j=1,j<=r-1,j++,
   sum3=sum3+l[r, j]*z[j]];
  z[r]=(b[[r]]-sum3)/l[r, r];Print["z[",r,"]=",z[r]]];

x[n]=z[n]/u[n, n];
Print["x[",n,"]=",x[n]];

For[r=n-1,r>=1,r--,
    sum4 = 0;
  For[j=r+1,j<=n, j++,
   sum4=sum4+u[r, j]*x[j]];
  x[r]=(z[r]-sum4)/u[r, r];Print["x[",r,"]=",x[r]]];
```

Output:

```
l[1,1]=1
u[1,1]=1
u[1,2]=1
u[1,3]=1
l[2,1]=4
l[2,2]=1
u[2,2]=-1
u[2,3]=-5
l[3,1]=3
l[3,2]=-2
l[3,3]=1
u[3,3]=-10
1
6
4
z[1]=1
z[2]=2
z[3]=5
x[3]=-(1/2)
x[2]=1/2
x[1]=1
```

6.3.3.2 Crout's Method

If we choose all the elements of principal diagonal of upper triangular matrix $\boldsymbol{U}$ as 1, that is, $u_{ii}=1$, $i=1,2,\ldots,n$, then the corresponding method is called Crout's method.

From Equation 6.43, we get

$$
\begin{aligned}
u_{ii} &= 1 \\
l_{ii} &= a_{ii} - \sum_{k=1}^{i-1} l_{ik}u_{ki} \\
l_{ij} &= a_{ij} - \sum_{k=1}^{j-1} l_{ik}u_{kj}, \quad i>j \\
u_{ij} &= \frac{\left(a_{ij} - \sum_{k=1}^{i-1} l_{ik}u_{kj}\right)}{l_{ii}}, \quad i<j
\end{aligned}
\tag{6.48}
$$

It can be shown that for large n, the operational count in ***LU***-decomposition method is approximately $n^3/3$.

Therefore, the time complexity in ***LU***-decomposition method is $O(n^3)$.

6.3.3.2.1 Algorithm for Crout's Decomposition Method

Input: Enter the number of unknown variables n, the coefficient matrix $\boldsymbol{A} = [a_{ij}]_{n\times n}$, and the constant column vector $\boldsymbol{b} = [b_i]_{n\times 1}$.

Output: Solution $\boldsymbol{x} = [x_i]_{n\times 1}$ of the given system of equations.

Initial step: Read matrix $\boldsymbol{A}$ with elements a_{ij}, $i, j = 1, 2, \ldots n$.

Step 1: (Determination for elements l_{ij} and u_{ij} of lower and upper triangular matrices, respectively)

```
for i = 1(1)n do
    for j = 1(1)n do
        if (i = j) then
            l_ii = 1;
            sum = 0;
          for k = 1(1)(i-1) do
            sum = sum + l_ik * u_ki;
          end
            l_ii = a_ii - sum;
        else
         if (i < j)then
           sum1 = 0;
          for k = 1(1)(i-1) do
           sum1 = sum1 + l_ik * u_kj;
          end
           u_ij = (a_ij - sum1) / l_ii;
         else
          if (i > j)then
              sum2 = 0;
            for k = 1(1)(j-1) do
              sum2 = sum2 + l_ik * u_kj;
            end
              l_ij = (a_ij - sum2);
    end
end
```

Step 2: Print lower triangular matrix $\boldsymbol{L} = [l_{ij}]_{n\times n}$;
and upper triangular matrix $\boldsymbol{U} = [u_{ij}]_{n\times n}$;

Step 3: (Forward substitution method for determining $\boldsymbol{z}$)

```
Compute z_1 = (b_1 / l_11);
  for r = 2,3,...,n do
      sum = 0;
    for j = 1,2,...,(r-1) do
      sum = sum + l_rj * z_j;
    end
      Compute
      z_r = (b_r - sum) / l_rr;
end
```

Step 4: (Back-substitution method for determining the solution $\boldsymbol{x}$)
Compute $x_n = (z_n / u_{nn})$;
for $r = n-1, n-2, \ldots, 1$ do
sum $= 0$;
for $j = \overline{r+1}, \overline{r+2}, \ldots, n$ do
sum $=$ sum $+ u_{rj} * x_j$;
end
Compute
$x_r = (z_r - \text{sum}) / u_{rr}$;
end

Step 5: Print the value of x_i, $(i = 1, 2, \ldots, n)$.

Step 6: Stop the program. ■

MATHEMATICA® Program for Solution of System of Algebraic Equations by Crout's Method (Chapter 6, Example 6.6)

```
A={{2,1,4},{8,-3,2},{4,11,-1}};
b={12,20,33};
n=3;
For[i=1,i<=n, i++,
  For[j=1,j<=n, j++,
   If[i==j,
    u[i, i]=1;
    sum=0;
    For[k=1,k<=i-1,k++,
     sum=sum+l[i, k]*u[k, i]];
    l[i, i]=A[[i, i]]-sum;
    Print["u[",i,",",i,"]=",u[i, i]];
    Print["l[",i,",",i,"]=",l[i, i]]];

   If[i<j,
    sum1 = 0;
    For[k=1,k<=i-1,k++,
     sum1=sum1+l[i, k]*u[k, j]];
    u[i, j]=(A[[i, j]]-sum1)/l[i, i];
    Print["u[",i,",",j,"]=",u[i, j]]];
   If[i>j,
    sum2 = 0;
    For[k=1,k<=j-1,k++,
     sum2=sum2+l[i, k]*u[k, j]];
    l[i, j]=A[[i, j]]-sum2;
    Print["l[",i,",",j,"]=",l[i, j]]]]];

z[1]=b[[1]]/l[1,1];
Print["z[",1,"]=",z[1]];

For[r=2,r<=n, r++,
    sum3 = 0;
  For[j=1,j<=r-1,j++,
   sum3=sum3+l[r, j]*z[j]];
  z[r]=(b[[r]]-sum3)/l[r, r];Print["z[",r,"]=",z[r]]];

x[n]=z[n]/u[n, n];
Print["x[",n,"]=",x[n]];
```

```
For[r=n-1,r>=1,r--,
    sum4 = 0;
  For[j=r+1,j<=n,j++,
   sum4=sum4+u[r, j]*x[j]];
  x[r]=(z[r]-sum4)/u[r, r];
  Print["x[",r,"]=",x[r]]];
```

Output:

```
u[1,1]=1
l[1,1]=2
u[1,2]=1/2
u[1,3]=2
l[2,1]=8
u[2,2]=1
l[2,2]=-7
u[2,3]=2
l[3,1]=4
l[3,2]=9
u[3,3]=1
l[3,3]=-27
z[1]=6
z[2]=4
z[3]=1
x[3]=1
x[2]=2
x[1]=3
```

Example 6.5

Find the solution of the system of equations

$$x_1+x_2+x_3=1$$

$$4x_1+3x_2-x_3=6$$

$$3x_1+5x_2+3x_3=4$$

by ***LU***-decomposition method.

Solution:

We apply Doolittle's method. However, Crout's method can also be applied. Using the Doolittle's method, the coefficient matrix $\boldsymbol{A}$ can be decomposed as follows:

$$\boldsymbol{A}=\begin{bmatrix}1&1&1\\4&3&-1\\3&5&3\end{bmatrix}=\begin{bmatrix}1&0&0\\l_{21}&1&0\\l_{31}&l_{32}&1\end{bmatrix}\begin{bmatrix}u_{11}&u_{12}&u_{13}\\0&u_{22}&u_{23}\\0&0&u_{33}\end{bmatrix}=\begin{bmatrix}u_{11}&u_{12}&u_{13}\\l_{21}u_{11}&l_{21}u_{12}+u_{22}&l_{21}u_{13}+u_{23}\\l_{31}u_{11}&l_{31}u_{12}+l_{32}u_{22}&l_{31}u_{13}+l_{32}u_{23}+u_{33}\end{bmatrix}$$

Therefore, we have

$$u_{11}=1,u_{12}=1,u_{13}=1,l_{21}u_{11}=4,l_{21}u_{12}+u_{22}=3,l_{21}u_{13}+u_{23}=-1,l_{31}u_{11}=3,l_{31}u_{12}+l_{32}u_{22}=5$$

$$\text{and } l_{31}u_{13}+l_{32}u_{23}+u_{33}=3$$

Solving the above equations, we get

$$u_{11} = u_{12} = u_{13} = 1,\ l_{21} = 4,\ u_{22} = -1,\ u_{23} = -5,\ u_{33} = -10$$

$$l_{31} = 3,\ l_{32} = -2$$

Now, using the forward substitution method, solution of the system of equations $\boldsymbol{Lz} = \boldsymbol{b}$, that is,

$$\begin{bmatrix} 1 & 0 & 0 \\ 4 & 1 & 0 \\ 3 & -2 & 1 \end{bmatrix} \begin{bmatrix} z_1 \\ z_2 \\ z_3 \end{bmatrix} = \begin{bmatrix} 1 \\ 6 \\ 4 \end{bmatrix}$$

yields $z_1 = 1$, $z_2 = 2$, and $z_3 = 5$.

Again, solving the system of equations $\boldsymbol{Ux} = \boldsymbol{z}$, that is,

$$\begin{bmatrix} 1 & 1 & 1 \\ 0 & -1 & -5 \\ 0 & 0 & -10 \end{bmatrix} \begin{bmatrix} x_1 \\ x_2 \\ x_3 \end{bmatrix} = \begin{bmatrix} 1 \\ 2 \\ 5 \end{bmatrix}$$

by the back-substitution method, we get

$$\boldsymbol{x} = \begin{bmatrix} 1 \\ 1/2 \\ -1/2 \end{bmatrix}$$

Therefore, the required solution is

$$x_1 = 1,\ x_2 = \frac{1}{2},\ \text{and}\ x_3 = -\frac{1}{2}$$

Example 6.6

Find the solution of the system of equations

$$2x_1 + x_2 + 4x_3 = 12$$

$$8x_1 - 3x_2 + 2x_3 = 20$$

$$4x_1 + 11x_2 - x_3 = 33$$

by Crout's decomposition method.

Solution:

Using Crout's method, the coefficient matrix $\boldsymbol{A}$ can be decomposed as follows:

$$\boldsymbol{A} = \begin{bmatrix} 2 & 1 & 4 \\ 8 & -3 & 2 \\ 4 & 11 & -1 \end{bmatrix} = \begin{bmatrix} l_{11} & 0 & 0 \\ l_{21} & l_{22} & 0 \\ l_{31} & l_{32} & l_{33} \end{bmatrix} \begin{bmatrix} 1 & u_{12} & u_{13} \\ 0 & 1 & u_{23} \\ 0 & 0 & 1 \end{bmatrix} = \begin{bmatrix} l_{11} & l_{11}u_{12} & l_{11}u_{13} \\ l_{21} & l_{21}u_{12} + l_{22} & l_{21}u_{13} + l_{22}u_{23} \\ l_{31} & l_{31}u_{12} + l_{32} & l_{31}u_{13} + l_{32}u_{23} + l_{33} \end{bmatrix}$$

Therefore, we have

$$l_{11}=2, l_{21}=8, l_{31}=4, u_{12}=\frac{1}{2}, u_{13}=2, l_{22}=-7, l_{21}u_{12}+l_{22}=-3, l_{21}u_{13}+l_{22}u_{23}=2, l_{31}u_{12}+l_{32}=11$$

and $l_{31}u_{13}+l_{32}u_{23}+l_{33}=-1$.

Solving the above equations, we get

$$l_{11}=2, l_{21}=8, l_{31}=4, u_{12}=\frac{1}{2}, u_{13}=2, l_{22}=-7, u_{23}=2, l_{32}=9, l_{33}=-27$$

Now, using the forward substitution method, solution of the system of equations $\boldsymbol{Lz}=\boldsymbol{b}$, that is,

$$\begin{bmatrix} 2 & 0 & 0 \\ 8 & -7 & 0 \\ 4 & 9 & -27 \end{bmatrix} \begin{bmatrix} z_1 \\ z_2 \\ z_3 \end{bmatrix} = \begin{bmatrix} 12 \\ 20 \\ 33 \end{bmatrix}$$

yields $z_1 = 6$, $z_2 = 4$, $z_3 = 1$.

Again, solving the system of equations $\boldsymbol{Ux}=\boldsymbol{z}$, that is,

$$\begin{bmatrix} 1 & 0.5 & 2 \\ 0 & 1 & 2 \\ 0 & 0 & 1 \end{bmatrix} \begin{bmatrix} x_1 \\ x_2 \\ x_3 \end{bmatrix} = \begin{bmatrix} 6 \\ 4 \\ 1 \end{bmatrix}$$

by the back-substitution method, we get

$$\boldsymbol{x} = \begin{bmatrix} 3 \\ 2 \\ 1 \end{bmatrix}$$

Therefore, the required solution is

$$x_1 = 3,\ x_2 = 2,\ x_3 = 1$$

6.3.3.3 Cholesky's Method

Let $\boldsymbol{A}=[a_{ij}]_{n\times n}$ be a symmetric positive definite matrix of order n. Thus, $\boldsymbol{A}=\boldsymbol{LL}^T$ and $\boldsymbol{x}^T\boldsymbol{Ax}>0$, that is, $\sum_{i=1}^{n}\sum_{j=1}^{n} a_{ij}x_i x_j > 0$.

The method states that there exists always a lower triangular matrix L such that

$$\boldsymbol{A}=\boldsymbol{LL}^T \tag{6.49}$$

where $\boldsymbol{L}=[l_{ij}]_{n\times n}$ and $l_{ij}=0, i<j$.

The equation $\boldsymbol{Ax}=\boldsymbol{b}$ gives

$$\boldsymbol{LL}^T\boldsymbol{x}=\boldsymbol{b} \tag{6.50}$$

Let us take

$$L^T x = z \tag{6.51}$$

then

$$Lz = b \tag{6.52}$$

The solutions $z_i\,(i=1,2,\ldots,n)$ are obtained from Equation 6.52 by the forward substitution method and then the solutions $x_i\,(i=1,2,\ldots,n)$ are determined from Equation 6.51 by the back-substitution process.

The elements $l_{ij}\,(i,j=1,2,\ldots,n)$ of lower triangular matrix $\boldsymbol{L}$ are given by

$$l_{11} = \sqrt{a_{11}},$$

$$l_{i1} = \frac{a_{i1}}{l_{11}}, \quad i=2,\ldots,n$$

$$l_{ii} = \sqrt{a_{ii} - \sum_{j=1}^{i-1} l_{ij}^2}, \quad i=2,\ldots,n \tag{6.53}$$

$$l_{ij} = \frac{1}{l_{jj}}\left(a_{ij} - \sum_{k=1}^{j-1} l_{ik} l_{jk}\right), \quad i=j+1,\ldots,n, j\geq 2 \text{ and } l_{ij}=0, i<j$$

This method is also known as square root method.

If the coefficient matrix $\boldsymbol{A}$ is symmetric but not positive definite, this method could be still applied. But in that case, it leads to a complex matrix $\boldsymbol{L}$, so that the method becomes impractical.

It can be shown that for large n, the operational count in Cholesky's method is approximately $n^3/6$.

Therefore, the time complexity in Cholesky's method is $O(n^3)$.

6.3.3.3.1 Algorithm for Cholesky's Decomposition Method

Input: Enter the number of unknown variables n, the coefficient matrix $\boldsymbol{A}=[a_{ij}]_{n\times n}$, and the constant column vector $\boldsymbol{b}=[b_i]_{n\times 1}$.

Output: Solution $\boldsymbol{x}=[x_i]_{n\times 1}$ of the given system of equations.

Initial step: Read matrix $\boldsymbol{A}$ with elements $a_{ij},\ i,j=1,2,\ldots n$.

Step 1: (Determination for elements l_{ij} of lower triangular matrix)

For $j = 1(1)n$ do
 $\text{sum1} = 0;$
 for $k = 1(1)\overline{j-1}$ do
  $\text{sum1} = \text{sum1} + l_{jk}^2;$
 end
 $l_{jj} = \text{sqrt}(a_{jj} - \text{sum1});$

 for $i = \overline{j+1}(1)n$ do
  $\text{sum2} = 0;$

for $k = 1(1)\overline{j-1}$ do

sum2 = sum2 + $l_{ik} * l_{jk}$;

end

$$l_{ij} = \frac{(a_{ij} - \text{sum2})}{l_{jj}};$$

end

end

Step 2: Print lower triangular matrix $\boldsymbol{L} = [l_{ij}]_{n\times n}$;

Step 3: (Forward substitution method for determining $\boldsymbol{z}$)

Compute $z_1 = (b_1 / l_{11})$;

for $r = 2,3,\ldots,n$ do

sum = 0;

for $j = 1,2,\ldots,\overline{r-1}$ do

sum = sum + $l_{rj} * z_j$;

end

Compute

$$z_r = \frac{(b_r - \text{sum})}{l_{rr}};$$

end

Step 4: (Back-substitution method for determining the solution $\boldsymbol{x}$)

Compute $x_n = (z_n / l_{nn})$;

for $r = n-1, n-2,\ldots,1$ do

sum = 0;

for $j = \overline{r+1}, \overline{r+2},\ldots,n$ do

sum = sum + $l_{jr} * x_j$;

end

Compute

$$x_r = \frac{(z_r - \text{sum})}{l_{rr}};$$

end

Step 5: Print the value of x_i, $(i = 1,2,\ldots,n)$.

Step 6: Stop the program. ■

MATHEMATICA® Program for Solution of System of Algebraic Equations by Cholesky's Method (Chapter 6, Example 6.7)

```
A={{4,6,8},{6,34,52},{8,52,129}};
b={0,-160,-452};
n=3;
For[j=1,j<=n, j++,
  sum=0;
  For[k=1,k<=j-1,k++,
   sum=sum+l[j, k]^2];
  l[j, j]=Sqrt[A[[j, j]]-sum];
  Print["l[",j,",",j,"]=",l[j, j]];
  For[i=j+1,i<=n, i++,
   sum1 = 0;
   For[s=1,s<=j-1,s++,
```

```
    sum1=sum1+l[i, s]*l[j, s]];
   l[i, j]=(A[[i, j]]-sum1)/l[j, j];
   Print["l[",i,",",j,"]=",l[i, j]]];
  ];

z[1]=b[[1]]/l[1,1];
Print["z[",1,"]=",z[1]];

For[r=2,r<=n, r++,
    sum3 = 0;
  For[j=1,j<=r-1,j++,
   sum3=sum3+l[r, j]*z[j]];
  z[r]=(b[[r]]-sum3)/l[r, r];Print["z[",r,"]=",z[r]]];

x[n]=z[n]/l[n, n];
Print["x[",n,"]=",x[n]];

For[r=n-1,r>=1,r--,
    sum4 = 0;
  For[j=r+1,j<=n, j++,
   sum4=sum4+l[j, r]*x[j]];
  x[r]=(z[r]-sum4)/l[r, r];Print["x[",r,"]=",x[r]]];
```

Output:

```
l[1,1]=2
l[2,1]=3
l[3,1]=4
l[2,2]=5
l[3,2]=8
l[3,3]=7
z[1]=0
z[2]=-32
z[3]=-28
x[3]=-4
x[2]=0
x[1]=8
```

Example 6.7

Solve the following system of equations by Cholesky's method:

$$4x_1+6x_2+8x_3=0$$

$$6x_1+34x_2+52x_3=-160$$

$$8x_1+52x_2+129x_3=-452$$

Solution:

The given system of equations can be written as $\boldsymbol{Ax}=\boldsymbol{b}$

$$\text{where } \boldsymbol{A}=\begin{bmatrix}4 & 6 & 8\\ 6 & 34 & 52\\ 8 & 52 & 129\end{bmatrix}, \boldsymbol{x}=\begin{bmatrix}x_1\\ x_2\\ x_3\end{bmatrix}, \boldsymbol{b}=\begin{bmatrix}0\\ -160\\ -452\end{bmatrix}$$

Since, the coefficient matrix $\boldsymbol{A}$ is symmetric, we can apply Cholesky's method.

We decompose the matrix $\boldsymbol{A}$ as $\boldsymbol{A}=\boldsymbol{L}\boldsymbol{L}^T$, where $\boldsymbol{L}$ is a 3×3 lower triangular matrix given by

$$L=\begin{bmatrix} l_{11} & 0 & 0 \\ l_{21} & l_{22} & 0 \\ l_{31} & l_{32} & l_{33} \end{bmatrix}$$

Using the formulae in Equation 6.53, we get

$$l_{11}=2, l_{21}=3, l_{22}=5, l_{31}=4, l_{32}=8, l_{33}=7.$$

Otherwise, we can determine l_{ij}'s value directly by solving the following equation $\boldsymbol{A}=\boldsymbol{L}\boldsymbol{L}^T$, that is,

$$\begin{bmatrix} l_{11} & 0 & 0 \\ l_{21} & l_{22} & 0 \\ l_{31} & l_{32} & l_{33} \end{bmatrix}\begin{bmatrix} l_{11} & l_{21} & l_{31} \\ 0 & l_{22} & l_{32} \\ 0 & 0 & l_{33} \end{bmatrix}=\begin{bmatrix} 4 & 6 & 8 \\ 6 & 34 & 52 \\ 8 & 52 & 129 \end{bmatrix}$$

Then, we solve the system $\boldsymbol{Lz}=\boldsymbol{b}$, that is,

$$\begin{bmatrix} 2 & 0 & 0 \\ 3 & 5 & 0 \\ 4 & 8 & 7 \end{bmatrix}\begin{bmatrix} z_1 \\ z_2 \\ z_3 \end{bmatrix}=\begin{bmatrix} 0 \\ -160 \\ -452 \end{bmatrix}$$

by the forward substitution method yielding $z_1=0, z_2=-32, \text{and}\, z_3=-28$.

The resulting solution is now obtained from $\boldsymbol{L}^T\boldsymbol{x}=\boldsymbol{z}$, that is,

$$\begin{bmatrix} 2 & 3 & 4 \\ 0 & 5 & 8 \\ 0 & 0 & 7 \end{bmatrix}\begin{bmatrix} x_1 \\ x_2 \\ x_3 \end{bmatrix}=\begin{bmatrix} 0 \\ -32 \\ -28 \end{bmatrix}$$

using the back-substitution method.

Therefore, the required solution is $x=8, x_2=0, x_3=-4$.

Example 6.8

Solve the following system of equations by Cholesky's method

$$x_1-x_2+3x_3+2x_4=15$$

$$-x_1+5x_2-5x_3-2x_4=-35$$

$$3x_1-5x_2+19x_3+3x_4=94$$

$$2x_1-2x_2+3x_3+21x_4=1$$

Solution:

The given system of equations can be written as $\boldsymbol{Ax}=\boldsymbol{b}$

$$\text{where } \boldsymbol{A}=\begin{bmatrix} 1 & -1 & 3 & 2 \\ -1 & 5 & -5 & -2 \\ 3 & -5 & 19 & 3 \\ 2 & -2 & 3 & 21 \end{bmatrix}, \boldsymbol{x}=\begin{bmatrix} x_1 \\ x_2 \\ x_3 \\ x_4 \end{bmatrix}, \boldsymbol{b}=\begin{bmatrix} 15 \\ -35 \\ 94 \\ 1 \end{bmatrix}$$

Since, the coefficient matrix $\boldsymbol{A}$ is symmetric, we can apply Cholesky's method.

We decompose the matrix $\boldsymbol{A}$ as $\boldsymbol{A}=\boldsymbol{L}\boldsymbol{L}^T$, where $\boldsymbol{L}$ is a 4×4 lower triangular matrix given by

$$\boldsymbol{L}=\begin{bmatrix} l_{11} & 0 & 0 & 0 \\ l_{21} & l_{22} & 0 & 0 \\ l_{31} & l_{32} & l_{33} & 0 \\ l_{41} & l_{42} & l_{43} & l_{44} \end{bmatrix}$$

Using the formulae in Equation 6.53, we get

$$l_{11}=1, l_{21}=-1, l_{22}=2, l_{31}=3, l_{32}=-1, l_{33}=3, l_{41}=2, l_{42}=0, l_{43}=-1, l_{44}=4$$

Then, we solve the system $\boldsymbol{Lz}=\boldsymbol{b}$, that is,

$$\begin{bmatrix} 1 & 0 & 0 & 0 \\ -1 & 2 & 0 & 0 \\ 3 & -1 & 3 & 0 \\ 2 & 0 & -1 & 4 \end{bmatrix}\begin{bmatrix} z_1 \\ z_2 \\ z_3 \\ z_4 \end{bmatrix}=\begin{bmatrix} 15 \\ -35 \\ 94 \\ 1 \end{bmatrix}$$

by the forward substitution method yielding $z_1=15, z_2=-10, z_3=13, z_4=-4$.

The resulting solution is now obtained from $\boldsymbol{L}^T\boldsymbol{x}=\boldsymbol{z}$, that is, from

$$\begin{bmatrix} 1 & -1 & 3 & 2 \\ 0 & 2 & -1 & 0 \\ 0 & 0 & 3 & -1 \\ 0 & 0 & 0 & 4 \end{bmatrix}\begin{bmatrix} x_1 \\ x_2 \\ x_3 \\ x_4 \end{bmatrix}=\begin{bmatrix} 15 \\ -10 \\ 13 \\ -4 \end{bmatrix}$$

using the back-substitution method.

Therefore, the required solution is $x_1=2, x_2=-3, x_3=4, x_4=-1$.

6.4 ITERATIVE METHOD

The direct methods so far discussed yield the solutions after an amount of computation that is known in advance. In contrast, in an indirect or iterative method we start with an approximation to the true solution, and if convergent, obtain a sequence of successive closer approximations from a computational cycle repeated till the required accuracy is obtained. Therefore, in a direct method the amount of computation is fixed, but in an iterative method the amount of computation depends on the accuracy required and varies from case to case.

We apply iterative methods if convergence is very rapid, so that we save operations compared to a direct method. In general, one would prefer a direct method for the solution of a linear system of equations. But in case of *sparse* matrix with large number of zero elements, it will be better to use iterative methods which preserve these elements.

In iterative methods, the process starts with an initial approximation to the unknown vector $\boldsymbol{x}$ of $\boldsymbol{A}\,\boldsymbol{x}=\boldsymbol{b}$ and then the successive approximations will be improved by an iterative process

$$\boldsymbol{x}^{(k+1)}=\boldsymbol{Q}\,\boldsymbol{x}^{(k)}+\boldsymbol{C} \tag{6.54}$$

where $\boldsymbol{x}^{(k+1)}$ and $\boldsymbol{x}^{(k)}$ are the $(k+1)$th and kth approximations of $\boldsymbol{x}$, respectively, $\boldsymbol{Q}$ and $\boldsymbol{C}$ are the iteration matrix and constant column vector of the corresponding scheme. There are various methods to generate $\boldsymbol{Q}$ and $\boldsymbol{C}$.

Next, we shall describe the following iterative methods:

- Gauss–Jacobi iterative method
- Gauss–Seidel iterative method
- SOR method

6.4.1 Gauss–Jacobi Iteration

Initially, the given equation of the system is so rearranging that $a_{ii} \neq 0$ for $i = 1,2,\ldots,n$. Suppose that this rearranging system of equation is

$$\begin{aligned} a_{11}x_1 + a_{12}x_2 + \ldots + a_{1n}x_n &= b_1 \\ a_{21}x_1 + a_{22}x_2 + \ldots + a_{2n}x_n &= b_2 \\ &\vdots \\ a_{n1}x_1 + a_{n2}x_2 + \ldots + a_{nn}x_n &= b_n \end{aligned} \tag{6.55}$$

Now the equations in (6.55) can be written as

$$\begin{aligned} x_1 &= -\frac{1}{a_{11}}\left(a_{12}x_2 + a_{13}x_3 + \ldots + a_{1n}x_n - b_1\right) \\ x_2 &= -\frac{1}{a_{22}}\left(a_{21}x_1 + a_{23}x_3 + \ldots + a_{2n}x_n - b_2\right) \\ &\vdots \\ x_n &= -\frac{1}{a_{nn}}\left(a_{n1}x_1 + a_{n2}x_2 + \ldots + a_{n,n-1}x_{n-1} - b_n\right) \end{aligned} \tag{6.56}$$

In the Gauss–Jacobi method, the iteration is generated by the formula

$$x_i^{(k+1)} = -\frac{1}{a_{ii}}\left[\sum_{\substack{j=1 \\ j\neq i}}^{n} a_{ij}x_j^{(k)} - b_i\right], \quad i = 1,2,\ldots,n \text{ and } k = 0,1,2,\ldots \tag{6.57}$$

where the initial guess $x_i^{(0)}\left(i = 1,2,\ldots,n\right)$ can be chosen arbitrarily.

Thus, we have

$$\begin{aligned} x_1^{(k+1)} &= -\frac{1}{a_{11}}\left(a_{12}x_2^{(k)} + a_{13}x_3^{(k)} + \ldots + a_{1n}x_n^{(k)} - b_1\right) \\ x_2^{(k+1)} &= -\frac{1}{a_{22}}\left(a_{21}x_1^{(k)} + a_{23}x_3^{(k)} + \ldots + a_{2n}x_n^{(k)} - b_2\right) \\ &\vdots \\ x_n^{(k+1)} &= -\frac{1}{a_{nn}}\left(a_{n1}x_1^{(k)} + a_{n2}x_3^{(k)} + \ldots + a_{n,n-1}x_{n-1}^{(k)} - b_n\right) \end{aligned} \tag{6.58}$$

where $k = 0,1,2,\ldots$

In matrix form, this method can be written as

$$\begin{aligned} \boldsymbol{x}^{(k+1)} &= -\boldsymbol{D}^{-1}[(\boldsymbol{L}+\boldsymbol{U})\boldsymbol{x}^{(k)} - \boldsymbol{b}] \\ &= -\boldsymbol{D}^{-1}(\boldsymbol{L}+\boldsymbol{U})\boldsymbol{x}^{(k)} + \boldsymbol{D}^{-1}\boldsymbol{b}, \quad k = 0,1,2,\ldots \end{aligned} \tag{6.59}$$

where $\boldsymbol{L}$ and $\boldsymbol{U}$ are, respectively, the lower and upper triangular matrices with zero diagonal entries, and $\boldsymbol{D}$ is the diagonal matrix such that

$$\boldsymbol{A} = \boldsymbol{L}+\boldsymbol{D}+\boldsymbol{U} \tag{6.60}$$

Therefore, from Equation 6.59, we have

$$\boldsymbol{x}^{(k+1)} = \boldsymbol{Q}_{GJ}\,\boldsymbol{x}^{(k)} + \boldsymbol{C}_{GJ} \tag{6.61}$$

where $\boldsymbol{Q}_{GJ} = -\boldsymbol{D}^{-1}\left(\boldsymbol{L}+\boldsymbol{U}\right)$ and $\boldsymbol{C}_{GJ} = \boldsymbol{D}^{-1}\boldsymbol{b}$.

Now, the Equation 6.59 can also be written as

$$\begin{aligned} \boldsymbol{x}^{(k+1)} &= \boldsymbol{x}^{(k)} - \left[\boldsymbol{I} + \boldsymbol{D}^{-1}\left(\boldsymbol{L}+\boldsymbol{U}\right)\right]\boldsymbol{x}^{(k)} + \boldsymbol{D}^{-1}\boldsymbol{b} \\ &= \boldsymbol{x}^{(k)} - \boldsymbol{D}^{-1}\left[\boldsymbol{D}+\boldsymbol{L}+\boldsymbol{U}\right]\boldsymbol{x}^{(k)} + \boldsymbol{D}^{-1}\boldsymbol{b} \\ &= \boldsymbol{x}^{(k)} + \boldsymbol{D}^{-1}[\boldsymbol{b} - \boldsymbol{A}\,\boldsymbol{x}^{(k)}] \end{aligned} \tag{6.62}$$

Therefore,

$$\boldsymbol{h}^{(k)} = \boldsymbol{D}^{-1}[\boldsymbol{b} - \boldsymbol{A}\,\boldsymbol{x}^{(k)}] = \boldsymbol{D}^{-1}\boldsymbol{r}^{(k)} \tag{6.63}$$

where $\boldsymbol{h}^{(k)} = \boldsymbol{x}^{(k+1)} - \boldsymbol{x}^{(k)}$ is the error in approximation and $\boldsymbol{r}^{(k)} = \boldsymbol{b} - \boldsymbol{A}\,\boldsymbol{x}^{(k)}$ is the residual vector.

We may solve the following equation

$$\boldsymbol{D}\,\boldsymbol{h}^{(k)} = \boldsymbol{r}^{(k)} \tag{6.64}$$

for the vector $\boldsymbol{h}^{(k)}$ and then we determine

$$\boldsymbol{x}^{(k+1)} = \boldsymbol{x}^{(k)} + \boldsymbol{h}^{(k)} \tag{6.65}$$

These equations constitute the Gauss–Jacobi method in an error format.

6.4.1.1 Convergence of the Gauss–Jacobi Iteration Method

To analyze the convergence of the Gauss–Jacobi iteration method, let $\varepsilon^{(k)} = \boldsymbol{x} - \boldsymbol{x}^{(k)}$, $k \geq 0$ be the error at the kth approximation.

Subtracting Equation 6.58 from Equation 6.56, we get

$$\varepsilon_i^{(k+1)} = -\sum_{\substack{j=1 \\ j\neq i}}^{n} \frac{a_{ij}}{a_{ii}}\varepsilon_i^{(k)}, \quad i = 1,2,\ldots,n \text{ and } k \geq 0 \tag{6.66}$$

So,

$$\left|\varepsilon_i^{(k+1)}\right| \le \sum_{\substack{j=1\\ j\ne i}}^{n} \left|\frac{a_{ij}}{a_{ii}}\right| \left|\varepsilon_i^{(k)}\right| \le \sum_{\substack{j=1\\ j\ne i}}^{n} \left|\frac{a_{ij}}{a_{ii}}\right| \left\|\varepsilon^{(k)}\right\|_\infty \le K \left\|\varepsilon^{(k)}\right\|_\infty \tag{6.67}$$

where

$$K = \operatorname*{Max}_{1\le i\le n} \sum_{\substack{j=1\\ j\ne i}}^{n} \left|\frac{a_{ij}}{a_{ii}}\right| \tag{6.68}$$

Now, from Equation 6.67, we get

$$\left\|\varepsilon^{(k+1)}\right\|_\infty \le K \left\|\varepsilon^{(k)}\right\|_\infty \tag{6.69}$$

This shows that the rate of convergence is linear.

This implies that

$$\left\|\varepsilon^{(k)}\right\|_\infty \le K^k \left\|\varepsilon^{(0)}\right\|_\infty \tag{6.70}$$

If $K < 1$, then $\varepsilon^{(k)} \to \mathbf{0}$ as $k \to \infty$, that is, the Gauss–Jacobi iteration method converges.

In order for $K < 1$ to be true, the coefficient matrix $\boldsymbol{A}$ must be diagonally dominant, that is,

$$|a_{ii}| > \sum_{\substack{j=1\\ j\ne i}}^{n} |a_{ij}|, \quad i = 1,2,\ldots,n \tag{6.71}$$

Thus, the Gauss–Jacobi iteration method converges if the given system of linear equations is strictly diagonally dominant. It may be noted that the above condition of convergence is sufficient but not necessary.

From Equation 6.65, we have

$$\boldsymbol{x}^{(k+1)} = \boldsymbol{x}^{(k)} + \boldsymbol{h}^{(k)}$$

This implies that

$$\boldsymbol{h}^{(k)} = \varepsilon^{(k)} - \varepsilon^{(k+1)}, \quad \text{since} \quad \varepsilon^{(k)} = \boldsymbol{x} - \boldsymbol{x}^{(k)}$$

Therefore, from Equation 6.69, we obtain

$$\left\|\varepsilon^{(k+1)}\right\|_\infty \le K \left\|\varepsilon^{(k)}\right\|_\infty = K \left\|\varepsilon^{(k+1)} + \boldsymbol{h}^{(k)}\right\|_\infty \le K\left(\left\|\varepsilon^{(k+1)}\right\|_\infty + \left\|\boldsymbol{h}^{(k)}\right\|_\infty\right), \quad \text{by triangle inequality}$$

Hence,

$$\left\|\varepsilon^{(k+1)}\right\|_\infty \le \frac{K}{1-K} \left\|\boldsymbol{h}^{(k)}\right\|_\infty \tag{6.72}$$

which is an estimate of error.

6.4.1.2 Algorithm for the Gauss–Jacobi Method

Input: Enter the number of unknown variables n, the coefficient matrix $\boldsymbol{A} = [a_{ij}]_{n\times n}$, the constant column vector $\boldsymbol{b} = [b_i]_{n\times 1}$, and the error tolerance level ε.

Output: Solution $\boldsymbol{x}$ of the given system of equations, that is, x_i, $i = 1,2,\ldots,n$.

Initial step: Choose an initial guess $\boldsymbol{x}^{(0)}$ to the solution $\boldsymbol{x}$.

Step 1: Set $k = 0$

Step 2: for $i = 1(1)n$ do

    sum $= 0$;
    for $j = 1(1)n$ do
        if $j \neq i$ then
            sum $=$ sum $+ a_{ij} * x_j^{(k)}$;
    end
    $x_i^{(k+1)} = \dfrac{(b_i - \text{sum})}{a_{ii}}$;
end

Step 3: If $\left\|\boldsymbol{x}^{(k+1)} - \boldsymbol{x}^{(k)}\right\|_\infty < \varepsilon$ then Go To Step 4
    else
    set $k = k + 1$ and Go To Step 2.

Step 4: Print $\boldsymbol{x}^{(k+1)}$.

Step 5: Stop. ■

MATHEMATICA® Program for Numerical Solution of System of Algebraic Equations by the Gauss–Jacobi Method (Chapter 6, Example 6.9)

```
∈=0.00001;
n=100;
x[1][0]=x[2][0]=x[3][0]=0;

Do[x[1][k]=1/8*(-11.5-2*x[2][k-1]-x[3][k-1]);
  x[2][k]=1/6*(18.5-x[1][k-1]-2*x[3][k-1]);
  x[3][k]=1/5*(12.5-4*x[1][k-1]);

  e[1][k]=Abs[x[1][k]-x[1][k-1]];
  e[2][k]=Abs[x[2][k]-x[2][k-1]];
  e[3][k]=Abs[x[3][k]-x[3][k-1]];

  Print["Step ",k,":"];
  Print["Error=",{e[1][k],e[2][k],e[3][k]}];
  Print["Max Error=",Max[{e[1][k],e[2][k],e[3][k]}]];

  Print["x[1][",k,"]=",x[1][k]];
  Print["x[2][",k,"]=",x[2][k]];
  Print["x[3][",k,"]=",x[3][k]];
  Print["-------------------------------------"];
  If[Max[{e[1][k],e[2][k],e[3][k]}]<∈,Break[]],{k,1,n}]
```

Output:

```
Step 1:
Error={1.4375,3.08333,2.5}
Max Error=3.08333
x[1][1]=-1.4375
```

```
x[2][1]=3.08333
x[3][1]=2.5
--------------------------------------------
Step 2:
Error={1.08333,0.59375,1.15}
Max Error=1.15
x[1][2]=-2.52083
x[2][2]=2.48958
x[3][2]=3.65
--------------------------------------------
Step 3:
Error={0.0046875,0.202778,0.866667}
Max Error=0.866667
x[1][3]=-2.51615
x[2][3]=2.28681
x[3][3]=4.51667
--------------------------------------------
Step 4:
Error={0.0576389,0.28967,0.00375}
Max Error=0.28967
x[1][4]=-2.57378
x[2][4]=1.99714
x[3][4]=4.51292
--------------------------------------------
Step 5:
Error={0.0728863,0.0108565,0.0461111}
Max Error=0.0728863
x[1][5]=-2.5009
x[2][5]=2.00799
x[3][5]=4.55903
--------------------------------------------
Step 6:
Error={0.00847801,0.0275181,0.058309}
Max Error=0.058309
x[1][6]=-2.50938
x[2][6]=1.98047
x[3][6]=4.50072
--------------------------------------------
Step 7:
Error={0.0141681,0.0208493,0.00678241}
Max Error=0.0208493
x[1][7]=-2.49521
x[2][7]=2.00132
x[3][7]=4.5075
--------------------------------------------
Step 8:
Error={0.00606014,0.00462216,0.0113345}
Max Error=0.0113345
x[1][8]=-2.50127
x[2][8]=1.9967
x[3][8]=4.49617
--------------------------------------------
Step 9:
Error={0.00257236,0.0047882,0.00484811}
Max Error=0.00484811
```

```
x[1][9]=-2.4987
x[2][9]=2.00149
x[3][9]=4.50101
-------------------------------------------
Step 10:
Error={0.00180306,0.00204476,0.00205788}
Max Error=0.00205788
x[1][10]=-2.5005
x[2][10]=1.99944
x[3][10]=4.49896
-------------------------------------------
Step 11:
Error={0.000768426,0.000986472,0.00144245}
Max Error=0.00144245
x[1][11]=-2.49973
x[2][11]=2.00043
x[3][11]=4.5004
-------------------------------------------
Step 12:
Error={0.000426924,0.000608888,0.000614741}
Max Error=0.000614741
x[1][12]=-2.50016
x[2][12]=1.99982
x[3][12]=4.49978
-------------------------------------------
Step 13:
Error={0.000229065,0.000276068,0.000341539}
Max Error=0.000341539
x[1][13]=-2.49993
x[2][13]=2.0001
x[3][13]=4.50013
-------------------------------------------
Step 14:
Error={0.000111709,0.000152024,0.000183252}
Max Error=0.000183252
x[1][14]=-2.50004
x[2][14]=1.99995
x[3][14]=4.49994
-------------------------------------------
Step 15:
Error={0.0000609124,0.0000797021,0.0000893675}
Max Error=0.0000893675
x[1][15]=-2.49998
x[2][15]=2.00003
x[3][15]=4.50003
-------------------------------------------
Step 16:
Error={0.0000310965,0.0000399412,0.0000487299}
Max Error=0.0000487299
x[1][16]=-2.50001
x[2][16]=1.99999
x[3][16]=4.49998
-------------------------------------------
Step 17:
Error={0.0000160765,0.0000214261,0.0000248772}
```

```
Max Error=0.0000248772
x[1][17]=-2.49999
x[2][17]=2.00001
x[3][17]=4.50001
--------------------------------------------
Step 18:
Error={8.46616 * 10^-6,0.0000109718,0.0000128612}
Max Error=0.0000128612
x[1][18]=-2.5
x[2][18]=2.
x[3][18]=4.5
--------------------------------------------
Step 19:
Error={4.35061 * 10^-6,5.69811 * 10^-6,6.77293 * 10^-6}
Max Error=6.77293 * 10^-6
x[1][19]=-2.5
x[2][19]=2.
x[3][19]=4.5
--------------------------------------------
```

Example 6.9

Solve the following system of equations by the Gauss–Jacobi method:

$$4x_1+5x_3=12.5$$

$$x_1+6x_2+2x_3=18.5$$

$$8x_1+2x_2+x_3=-11.5$$

Solution:

We first rearrange the given system of equations so that the resulting system is as follows:

$$8x_1+2x_2+x_3=-11.5$$

$$x_1+6x_2+2x_3=18.5$$

$$4x_1+5x_3=12.5$$

Now the system of equations is strictly diagonally dominant. So, the Gauss–Jacobi method will certainly converge.

Again, we rewrite the above equations in the following form:

$$x_1=\frac{1}{8}(-11.5-2x_2-x_3)$$

$$x_2=\frac{1}{6}(18.5-x_1-2x_3)$$

$$x_3=\frac{1}{5}(12.5-4x_1)$$

The successive iterations in the Gauss–Jacobi method will stop if

$\left\|\boldsymbol{x}^{(k+1)}-\boldsymbol{x}^{(k)}\right\|_\infty<\varepsilon$, where $k\geq 0$ and ε is the prescribed error tolerance.

Here, we take $\varepsilon=0.01$

Initial step:

$$x_1^{(0)} = x_2^{(0)} = x_3^{(0)} = 0$$

First iteration:

$$x_1^{(1)} = \frac{1}{8}(-11.5 - 2x_2^{(0)} - x_3^{(0)}) = \frac{-11.5}{8} = -1.4375$$

$$x_2^{(1)} = \frac{1}{6}(18.5 - x_1^{(0)} - 2x_3^{(0)}) = \frac{18.5}{6} = 3.0833$$

$$x_3^{(1)} = \frac{1}{5}(12.5 - 4x_1^{(0)}) = \frac{12.5}{5} = 2.5$$

Here,

$$\left\| \boldsymbol{x}^{(1)} - \boldsymbol{x}^{(0)} \right\|_\infty = \text{Max}\{1.4375,\ 3.0833,\ 2.5\} = 3.0833 > \varepsilon$$

Second iteration:

$$x_1^{(2)} = \frac{1}{8}(-11.5 - 2x_2^{(1)} - x_3^{(1)}) = \frac{1}{8}[-11.5 - 2 \times (3.0833) - 2.5] = -2.5208$$

$$x_2^{(2)} = \frac{1}{6}(18.5 - x_1^{(1)} - 2x_3^{(1)}) = \frac{1}{6}(18.5 + 1.4375 - 5) = 2.4896$$

$$x_3^{(2)} = \frac{1}{5}(12.5 - 4x_1^{(1)}) = \frac{1}{5}[12.5 - 4 \times (-1.4375)] = 3.65$$

$$\left\| \boldsymbol{x}^{(2)} - \boldsymbol{x}^{(1)} \right\|_\infty = \text{Max}\{1.0833,\ 0.5937,\ 1.15\} = 1.15 > \varepsilon$$

Third iteration:

$$x_1^{(3)} = \frac{1}{8}(-11.5 - 2x_2^{(2)} - x_3^{(2)}) = \frac{1}{8}[-11.5 - 2 \times (2.4896) - 3.65] = -2.5162$$

$$x_2^{(3)} = \frac{1}{6}(18.5 - x_1^{(2)} - 2x_3^{(2)}) = \frac{1}{6}[18.5 + 2.5208 - 2 \times (3.65)] = 2.2868$$

$$x_3^{(3)} = \frac{1}{5}(12.5 - 4x_1^{(2)}) = \frac{1}{5}[12.5 + 4 \times (2.5208)] = 4.5167$$

In this case,

$$\left\| \boldsymbol{x}^{(3)} - \boldsymbol{x}^{(2)} \right\|_\infty = \text{Max}\{0.0046,\ 0.2028,\ 0.8667\} = 0.8667 > \varepsilon$$

Fourth iteration:

$$x_1^{(4)} = \frac{1}{8}(-11.5 - 2x_2^{(3)} - x_3^{(3)}) = \frac{1}{8}[-11.5 - 2 \times (2.2868) - 4.5167] = -2.5738$$

$$x_2^{(4)} = \frac{1}{6}(18.5 - x_1^{(3)} - 2x_3^{(3)}) = \frac{1}{6}[18.5 + 2.5162 - 2 \times (4.5167)] = 1.9971$$

$$x_3^{(4)} = \frac{1}{5}(12.5 - 4x_1^{(3)}) = \frac{1}{5}[12.5 - 4 \times (-2.5162)] = 4.5129$$

Also,

$$\left\| \boldsymbol{x}^{(4)} - \boldsymbol{x}^{(3)} \right\|_\infty = \text{Max}\{0.0576, 0.2897, 0.0038\} = 0.2897 > \varepsilon$$

Fifth iteration:

$$x_1^{(5)} = \frac{1}{8}(-11.5 - 2x_2^{(4)} - x_3^{(4)}) = \frac{1}{8}[-11.5 - 2\times(1.9918) - 4.5129] = -2.5009$$

$$x_2^{(5)} = \frac{1}{6}(18.5 - x_1^{(4)} - 2x_3^{(4)}) = \frac{1}{6}[18.5 + 2.5738 - 2\times(4.5129)] = 2.0080$$

$$x_3^{(5)} = \frac{1}{5}(12.5 - 4x_1^{(4)}) = \frac{1}{5}[12.5 - 4\times(-2.5738)] = 4.5590$$

$$\left\| \boldsymbol{x}^{(5)} - \boldsymbol{x}^{(4)} \right\|_\infty = \text{Max}\{0.0729, 0.0109, 0.0461\} = 0.0729 > \varepsilon$$

Sixth iteration:

$$x_1^{(6)} = \frac{1}{8}(-11.5 - 2x_2^{(5)} - x_3^{(5)}) = \frac{1}{8}[-11.5 - 2\times(2.0080) - 4.5590] = -2.5094$$

$$x_2^{(6)} = \frac{1}{6}(18.5 - x_1^{(5)} - 2x_3^{(5)}) = \frac{1}{6}[18.5 + 2.5009 - 2\times(4.5590)] = 1.9805$$

$$x_3^{(6)} = \frac{1}{5}(12.5 - 4x_1^{(5)}) = \frac{1}{5}[12.5 - 4\times(-2.5009)] = 4.5007$$

Here,

$$\left\| \boldsymbol{x}^{(6)} - \boldsymbol{x}^{(5)} \right\|_\infty = \text{Max}\{0.0085, 0.0275, 0.0583\} = 0.0583 > \varepsilon$$

Seventh iteration:

$$x_1^{(7)} = \frac{1}{8}(-11.5 - 2x_2^{(6)} - x_3^{(6)}) = \frac{1}{8}[-11.5 - 2\times(1.9805) - 4.5007] = -2.4952$$

$$x_2^{(7)} = \frac{1}{6}(18.5 - x_1^{(6)} - 2x_3^{(6)}) = \frac{1}{6}[18.5 + 2.5094 - 2\times(4.5007)] = 2.0013$$

$$x_3^{(7)} = \frac{1}{5}(12.5 - 4x_1^{(6)}) = \frac{1}{5}[12.5 - 4\times(-2.5094)] = 4.5075$$

Also,

$$\left\| \boldsymbol{x}^{(7)} - \boldsymbol{x}^{(6)} \right\|_\infty = \text{Max}\{0.0142, 0.0208, 0.0068\} = 0.0208 > \varepsilon$$

Eighth iteration:

$$x_1^{(8)} = \frac{1}{8}(-11.5 - 2x_2^{(7)} - x_3^{(7)}) = \frac{1}{8}[-11.5 - 2\times(2.0013) - 4.5075] = -2.5013$$

$$x_2^{(8)} = \frac{1}{6}(18.5 - x_1^{(7)} - 2x_3^{(7)}) = \frac{1}{6}[18.5 + 2.4952 - 2\times(4.5075)] = 1.9967$$

$$x_3^{(8)} = \frac{1}{5}(12.5 - 4x_1^{(7)}) = \frac{1}{5}[12.5 - 4\times(-2.4952)] = 4.4962$$

$$\left\| \boldsymbol{x}^{(8)} - \boldsymbol{x}^{(7)} \right\|_\infty = \text{Max}\{0.0061,\ 0.0046,\ 0.0113\} = 0.0113 > \varepsilon$$

Ninth iteration:

$$x_1^{(9)} = \frac{1}{8}(-11.5 - 2x_2^{(8)} - x_3^{(8)}) = \frac{1}{8}[-11.5 - 2\times(1.9967) - 4.4962] = -2.4987$$

$$x_2^{(9)} = \frac{1}{6}(18.5 - x_1^{(8)} - 2x_3^{(8)}) = \frac{1}{6}[18.5 + 2.5013 - 2\times(4.4962)] = 2.0015$$

$$x_3^{(9)} = \frac{1}{5}(12.5 - 4x_1^{(8)}) = \frac{1}{5}[12.5 - 4\times(-2.5013)] = 4.5010$$

Finally,

$$\left\| \boldsymbol{x}^{(9)} - \boldsymbol{x}^{(8)} \right\|_\infty = \text{Max}\{0.0026,\ 0.0048,\ 0.0048\} = 0.0048 < \varepsilon$$

Hence, we shall stop here.
So, the sequence of successive approximations terminates at ninth iteration.
Therefore, the required solution of the given system of equations is

$$x_1 = -2.5,\ x_2 = 2 \text{ and } x_3 = 4.5$$

6.4.2 Gauss–Seidel Iteration Method

This is an iterative method of great practical importance. In this method, the iteration is generated by the following formulae

$$\begin{aligned}
x_1^{(k+1)} &= -\frac{1}{a_{11}}\left(a_{12}x_2^{(k)} + a_{13}x_3^{(k)} + \ldots + a_{1n}x_n^{(k)} - b_1\right) \\
x_2^{(k+1)} &= -\frac{1}{a_{22}}\left(a_{21}x_1^{(k+1)} + a_{23}x_3^{(k)} + \ldots + a_{2n}x_n^{(k)} - b_2\right) \\
&\vdots \\
x_n^{(k+1)} &= -\frac{1}{a_{nn}}\left(a_{n1}x_1^{(k+1)} + a_{n2}x_3^{(k+1)} + \ldots + a_{n,n-1}x_{n-1}^{(k+1)} - b_n\right)
\end{aligned} \tag{6.73}$$

where the initial guess $x_i^{(0)}\ (i = 1, 2, \ldots, n)$ can be chosen arbitrarily.

In brief,

$$x_i^{(k+1)} = -\frac{1}{a_{ii}}\left[\sum_{\substack{j=1 \\ j<i}}^{n} a_{ij}x_j^{(k+1)} + \sum_{\substack{j=1 \\ j>i}}^{n} a_{ij}x_j^{(k)} - b_i\right], \quad i = 1, 2, \ldots, n \quad k = 0, 1, 2, \ldots \tag{6.74}$$

which can be rearranged in the following form:

$$a_{11}x_1^{(k+1)} = -\sum_{j=2}^{n} a_{ij}x_j^{(k)} + b_1$$

$$a_{21}x_1^{(k+1)} + a_{22}x_2^{(k+1)} = -\sum_{j=3}^{n} a_{2j}x_j^{(k)} + b_2$$

$$a_{n1}x_1^{(k+1)} + a_{n2}x_2^{(k+1)} + \ldots + a_{nn}x_2^{(k+1)} = b_n$$

In matrix form,

$$(\boldsymbol{L}+\boldsymbol{D})\boldsymbol{x}^{(k+1)} = -\boldsymbol{U}\,\boldsymbol{x}^{(k)} + \boldsymbol{b}$$

$$\boldsymbol{x}^{(k+1)} = -(\boldsymbol{L}+\boldsymbol{D})^{-1}\boldsymbol{U}\,\boldsymbol{x}^{(k)} + (\boldsymbol{L}+\boldsymbol{D})^{-1}\boldsymbol{b}, \quad k = 0,1,2,\ldots \tag{6.75}$$

Therefore, from Equation 6.75, we have

$$\boldsymbol{x}^{(k+1)} = \boldsymbol{Q}_{GS}\,\boldsymbol{x}^{(k)} + \boldsymbol{C}_{GS} \tag{6.76}$$

where $\boldsymbol{Q}_{GS} = -(\boldsymbol{L}+\boldsymbol{D})^{-1}\boldsymbol{U}$ and $\boldsymbol{C}_{GS} = (\boldsymbol{L}+\boldsymbol{D})^{-1}\boldsymbol{b}$.

Now, the Equation 6.75 can also be written as

$$\begin{aligned}\boldsymbol{x}^{(k+1)} &= \boldsymbol{x}^{(k)} - \left[\boldsymbol{I} + (\boldsymbol{L}+\boldsymbol{D})^{-1}\boldsymbol{U}\right]\boldsymbol{x}^{(k)} + (\boldsymbol{L}+\boldsymbol{D})^{-1}\boldsymbol{b} \\ &= \boldsymbol{x}^{(k)} - (\boldsymbol{L}+\boldsymbol{D})^{-1}[\boldsymbol{L}+\boldsymbol{D}+\boldsymbol{U}]\boldsymbol{x}^{(k)} + (\boldsymbol{L}+\boldsymbol{D})^{-1}\boldsymbol{b} \\ &= \boldsymbol{x}^{(k)} - (\boldsymbol{L}+\boldsymbol{D})^{-1}\boldsymbol{A}\,\boldsymbol{x}^{(k)} + (\boldsymbol{L}+\boldsymbol{D})^{-1}\boldsymbol{b} \\ &= \boldsymbol{x}^{(k)} + (\boldsymbol{L}+\boldsymbol{D})^{-1}(\boldsymbol{b} - \boldsymbol{A}\,\boldsymbol{x}^{(k)})\end{aligned} \tag{6.77}$$

Therefore,

$$\boldsymbol{h}^{(k)} = (\boldsymbol{L}+\boldsymbol{D})^{-1}(\boldsymbol{b} - \boldsymbol{A}\,\boldsymbol{x}^{(k)}) = (\boldsymbol{L}+\boldsymbol{D})^{-1}\boldsymbol{r}^{(k)} \tag{6.78}$$

where $\boldsymbol{h}^{(k)} = \boldsymbol{x}^{(k+1)} - \boldsymbol{x}^{(k)}$ is the error in approximation and $\boldsymbol{r}^{(k)} = \boldsymbol{b} - \boldsymbol{A}\boldsymbol{x}^{(k)}$ is the residual vector.

We may solve the following equation

$$(\boldsymbol{L}+\boldsymbol{D})\boldsymbol{h}^{(k)} = \boldsymbol{r}^{(k)} \tag{6.79}$$

for the vector $\boldsymbol{h}^{(k)}$, and then, we determine

$$\boldsymbol{x}^{(k+1)} = \boldsymbol{x}^{(k)} + \boldsymbol{h}^{(k)} \tag{6.80}$$

These equations constitute the Gauss–Seidel method in an error format.

6.4.2.1 Convergence of the Gauss–Seidel Iteration Method

To analyze the convergence of the Gauss–Seidel iteration method, let $\varepsilon^{(k)} = \boldsymbol{x} - \boldsymbol{x}^{(k)}$, $k \geq 0$ be the error at the kth approximation.

Subtracting Equation 6.73 from Equation 6.56, we get

$$\varepsilon_i^{(k+1)} = -\frac{1}{a_{ii}}\left[\sum_{\substack{j=1\\j<i}}^{n} a_{ij}\varepsilon_j^{(k+1)} + \sum_{\substack{j=1\\j>i}}^{n} a_{ij}\varepsilon_j^{(k)}\right] \tag{6.81}$$

$$= -\frac{1}{a_{ii}}\left[\sum_{j=1}^{i-1} a_{ij}\varepsilon_j^{(k+1)} + \sum_{j=i+1}^{n} a_{ij}\varepsilon_j^{(k)}\right], \quad i = 1,2,\ldots,n \text{ and } k \ge 0$$

$$\left|\varepsilon_i^{(k+1)}\right| \le \left[\sum_{j=1}^{i-1}\left|\frac{a_{ij}}{a_{ii}}\right|\left|\varepsilon_j^{(k+1)}\right| + \sum_{j=i+1}^{n}\left|\frac{a_{ij}}{a_{ii}}\right|\left|\varepsilon_j^{(k)}\right|\right] \le K_1\left\|\varepsilon^{(k+1)}\right\|_\infty + (K - K_1)\left\|\varepsilon^{(k)}\right\|_\infty \tag{6.82}$$

where according to Equation 6.68 of the Gauss–Jacobi method

$$K = \underset{1\le i\le n}{\text{Max}} \sum_{\substack{j=1\\j\ne i}}^{n}\left|\frac{a_{ij}}{a_{ii}}\right|,$$

$$K_1 = \underset{1\le i\le n}{\text{Max}}\, \mu_i \text{ and } \mu_i = \sum_{j=1}^{i-1}\left|\frac{a_{ij}}{a_{ii}}\right|, \quad i = 1,2,\ldots,n$$

with $\mu_1 = 0$.

Then, from Equation 6.82, we get

$$\left\|\varepsilon^{(k+1)}\right\|_\infty \le K_1\left\|\varepsilon^{(k+1)}\right\|_\infty + (K - K_1)\left\|\varepsilon^{(k)}\right\|_\infty$$

Consequently,

$$\left\|\varepsilon^{(k+1)}\right\|_\infty \le \frac{(K - K_1)}{(1 - K_1)}\left\|\varepsilon^{(k)}\right\|_\infty \tag{6.83}$$

Since,

$$\frac{(K - K_1)}{(1 - K_1)} \le K \quad \text{as} \quad K < 1$$

Now, from Equation 6.83, we get

$$\left\|\varepsilon^{(k+1)}\right\|_\infty \le K\left\|\varepsilon^{(k)}\right\|_\infty \tag{6.84}$$

This shows that the rate of convergence is linear.

This implies that

$$\left\|\varepsilon^{(k)}\right\|_\infty \le K^k\left\|\varepsilon^{(0)}\right\|_\infty \tag{6.85}$$

If $K < 1$, then $\varepsilon^{(k)} \to \mathbf{0}$ as $k \to \infty$, that is, the Gauss–Seidel iteration method converges.

If $K < 1$, the coefficient matrix $\boldsymbol{A}$ must be diagonally dominant, that is,

$$|a_{ii}| > \sum_{\substack{j=1 \\ j \neq i}}^{n} |a_{ij}|, \quad i = 1, 2, \ldots, n \tag{6.86}$$

Thus, the Gauss–Seidel iteration method also converges if the given system of linear equations is strictly diagonally dominant. Here also the condition of convergence is sufficient but not necessary.

Moreover, if $K < 1$, in a similar way as we derived in case of the Gauss–Jacobi method, an estimate of the error is given by

$$\left\| \varepsilon^{(k+1)} \right\|_{\infty} \leq \frac{K}{1-K} \left\| \boldsymbol{h}^{(k)} \right\|_{\infty} \tag{6.87}$$

Note: It may be noted that the Gauss–Seidel method converges faster than the Gauss–Jacobi method. Since, the improved values of the unknowns are used at each stage of iteration, the Gauss–Seidel method converges twice as fast as the Gauss–Jacobi method.

6.4.2.2 Algorithm for the Gauss–Seidel Method

Input: Enter the number of unknown variables n, the coefficient matrix $\boldsymbol{A} = [a_{ij}]_{n \times n}$, the constant column vector $\boldsymbol{b} = [b_i]_{n \times 1}$, and the error tolerance level ε.

Output: Solution $\boldsymbol{x}$ of the given system of equations, that is, x_i, $i = 1, 2, \ldots, n$.

Initial step: Choose an initial guess $\boldsymbol{x}^{(0)}$ to the solution $\boldsymbol{x}$.

Step 1: Set $k = 0$

Step 2: for $i = 1(1)n$ do

sum $= 0$;

for $j = 1(1)\overline{i-1}$ do

sum $=$ sum $+ a_{ij} * x_j^{(k+1)}$;

end;

for $j = \overline{i+1}(1)n$ do

sum $=$ sum $+ a_{ij} * x_j^{(k)}$;

end;

$$x_i^{(k+1)} = \frac{(b_i - \text{sum})}{a_{ii}};$$

end

Step 3: If $\left\| \boldsymbol{x}^{(k+1)} - \boldsymbol{x}^{(k)} \right\|_{\infty} < \varepsilon$, then Go To Step 4

else

set $k = k + 1$ and Go To Step 2

Step 4: Print $\boldsymbol{x}^{(k+1)}$.

Step 5: Stop. ■

MATHEMATICA® Program for Numerical Solution of System of Algebraic Equations by the Gauss–Seidel Method (Chapter 6, Example 6.10)

```
ϵ=0.00001;
n=100;
x[1][0]=x[2][0]=x[3][0]=1;
```

```
Do[x[1][k]=N[1/5*(19-x[2][k-1]-2*x[3][k-1])];x[2][k]=N[1/4*(-2-x[1]
[k]+2*x[3][k-1])];x[3][k]=N[1/8*(39-2*x[1][k]-3*x[2][k])];

  e[1][k]=Abs[x[1][k]-x[1][k-1]];
  e[2][k]=Abs[x[2][k]-x[2][k-1]];
  e[3][k]=Abs[x[3][k]-x[3][k-1]];


  Print["Step ",k,":"];
  Print["Error=",{e[1][k],e[2][k],e[3][k]}];
  Print["Max Error=",Max[{e[1][k],e[2][k],e[3][k]}]];

Print["x[1][",k,"]=",x[1][k]];Print["x[2][",k,"]=",x[2][k]];Print["x[3]
[",k,"]=",x[3][k]];Print["-------------------------------------
-----"];If[Max[{e[1][k],e[2][k],e[3][k]}]<ϵ,Break[]],{k,1,n}];
```

Output:

```
Step 1:
Error={2.2,1.8,3.375}
Max Error=3.375
x[1][1]=3.2
x[2][1]=-0.8
x[3][1]=4.375
-------------------------------------------
Step 2:
Error={0.99,1.935,0.478125}
Max Error=1.935
x[1][2]=2.21
x[2][2]=1.135
x[3][2]=3.89687
-------------------------------------------
Step 3:
Error={0.19575,0.190125,0.120234}
Max Error=0.19575
x[1][3]=2.01425
x[2][3]=0.944875
x[3][3]=4.01711
-------------------------------------------
Step 4:
Error={0.0100688,0.0626344,0.0209707}
Max Error=0.0626344
x[1][4]=2.00418
x[2][4]=1.00751
x[3][4]=3.99614
-------------------------------------------
Step 5:
Error={0.00413859,0.0094507,0.00457866}
Max Error=0.0094507
x[1][5]=2.00004
x[2][5]=0.998059
x[3][5]=4.00072
-------------------------------------------
Step 6:
Error={0.0000586758,0.00227466,0.000867667}
```

```
Max Error=0.00227466
x[1][6]=2.0001
x[2][6]=1.00033
x[3][6]=3.99985
-------------------------------------------
Step 7:
Error={0.000107866,0.000406867,0.000179542}
Max Error=0.000406867
x[1][7]=1.99999
x[2][7]=0.999926
x[3][7]=4.00003
-------------------------------------------
Step 8:
Error={9.55681 * 10^-6,0.0000873816,0.0000351573}
Max Error=0.0000873816
x[1][8]=2.
x[2][8]=1.00001
x[3][8]=3.99999
-------------------------------------------
Step 9:
Error={3.4134 * 10^-6,0.0000167253,7.12534 * 10^-6}
Max Error=0.0000167253
x[1][9]=2.
x[2][9]=0.999997
x[3][9]=4.
-------------------------------------------
Step 10:
Error={4.94925 * 10^-7,3.43894 * 10^-6,1.41333 * 10^-6}
Max Error=3.43894 * 10^-6
x[1][10]=2.
x[2][10]=1.
x[3][10]=4.
-------------------------------------------
```

Example 6.10

Solve the following system of equations by the Gauss–Seidel method:

$$5x_1 + x_2 + 2x_3 = 19$$
$$x_1 + 4x_2 - 2x_3 = -2$$
$$2x_1 + 3x_2 + 8x_3 = 39$$

Solution:

Clearly, the given system of equations is strictly diagonally dominant. So, the Gauss–Seidel method will certainly converge.

Again, we rewrite the above equations in the following form:

$$x_1 = \frac{1}{5}(19 - x_2 - 2x_3)$$

$$x_2 = \frac{1}{4}(-2 - x_1 + 2x_3)$$

$$x_3 = \frac{1}{8}(39 - 2x_1 - 3x_2)$$

The successive iterations in the Gauss–Seidel method will stop if

$$\left\| \boldsymbol{x}^{(k+1)} - \boldsymbol{x}^{(k)} \right\|_{\infty} < \varepsilon,$$

where $k \geq 0$ and ε is the prescribed error tolerance.

Here, we take $\varepsilon = 0.01$.

Initial step:

$$x_1^{(0)} = x_2^{(0)} = x_3^{(0)} = 1$$

First iteration:

$$x_1^{(1)} = \frac{1}{5}(19 - x_2^{(0)} - 2x_3^{(0)}) = \frac{1}{5}(19 - 1 - 2) = 3.2$$

$$x_2^{(1)} = \frac{1}{4}(-2 - x_1^{(1)} + 2x_3^{(0)}) = \frac{1}{4}(-2 - 3.2 + 2\times 1) = -0.8$$

$$x_3^{(1)} = \frac{1}{8}(39 - 2x_1^{(1)} - 3x_2^{(1)}) = \frac{1}{8}(39 - 6.4 + 2.4) = 4.375$$

Here,

$$\left\| \boldsymbol{x}^{(1)} - \boldsymbol{x}^{(0)} \right\|_{\infty} = \text{Max}\{2.2, 1.8, 3.375\} = 3.375 > \varepsilon$$

Second iteration:

$$x_1^{(2)} = \frac{1}{5}(19 - x_2^{(1)} - 2x_3^{(1)}) = \frac{1}{5}(19 + 0.8 - 2\times 4.375) = 2.21$$

$$x_2^{(2)} = \frac{1}{4}(-2 - x_1^{(2)} + 2x_3^{(1)}) = \frac{1}{4}(-2 - 2.21 + 2\times 4.375) = 1.135$$

$$x_3^{(2)} = \frac{1}{8}(39 - 2x_1^{(2)} - 3x_2^{(2)}) = \frac{1}{8}(39 - 2\times 2.21 - 3\times 1.135) = 3.89687$$

$$\left\| \boldsymbol{x}^{(2)} - \boldsymbol{x}^{(1)} \right\|_{\infty} = \text{Max}\{0.99, 1.935, 0.478125\} = 1.935 > \varepsilon$$

Third iteration:

$$x_1^{(3)} = \frac{1}{5}(19 - x_2^{(2)} - 2x_3^{(2)}) = \frac{1}{5}[19 - 1.135 - 2\times(3.89687)] = 2.01425$$

$$x_2^{(3)} = \frac{1}{4}(-2 - x_1^{(3)} + 2x_3^{(2)}) = \frac{1}{4}[-2 - 2.01425 + 2\times(3.89687)] = 0.94487$$

$$x_3^{(3)} = \frac{1}{8}(39 - 2x_1^{(3)} - 3x_2^{(3)}) = \frac{1}{8}[39 - 2\times(2.01425) - 3\times(0.94487)] = 4.01711$$

In this case,

$$\left\| \boldsymbol{x}^{(3)} - \boldsymbol{x}^{(2)} \right\|_{\infty} = \text{Max}\{0.19575, 0.190125, 0.120234\} = 0.19575 > \varepsilon$$

Fourth iteration:

$$x_1^{(4)} = \frac{1}{5}(19 - x_2^{(3)} - 2x_3^{(3)}) = \frac{1}{5}(19 - 0.944875 - 2\times(4.01711)) = 2.00418$$

$$x_2^{(4)} = \frac{1}{4}(-2 - x_1^{(4)} + 2x_3^{(3)}) = \frac{1}{4}(-2 - 2.00418 + 2\times(4.01711)) = 1.00751$$

$$x_3^{(4)} = \frac{1}{8}(39 - 2x_1^{(4)} - 3x_2^{(4)}) = \frac{1}{8}(39 - 2(2.00418) - 3\times(1.00751)) = 3.99614$$

Also,

$$\left\|\boldsymbol{x}^{(4)} - \boldsymbol{x}^{(3)}\right\|_\infty = \text{Max}\{0.0100688, 0.0626344, 0.0209707\} = 0.0626344 > \varepsilon$$

Fifth iteration:

$$x_1^{(5)} = \frac{1}{5}(19 - x_2^{(4)} - 2x_3^{(4)}) = \frac{1}{5}[19 - 1.00751 - 2\times(3.99614)] = 2.00004$$

$$x_2^{(5)} = \frac{1}{4}(-2 - x_1^{(5)} + 2x_3^{(4)}) = \frac{1}{4}[-2 - 2.00004 + 2\times(3.99614)] = 0.998059$$

$$x_3^{(5)} = \frac{1}{8}(39 - 2x_1^{(5)} - 3x_2^{(5)}) = \frac{1}{8}[39 - 2\times(2.00004) - 3\times(0.998059)] = 4.00072$$

Finally,

$$\left\|\boldsymbol{x}^{(5)} - \boldsymbol{x}^{(4)}\right\|_\infty = \text{Max}\{0.00413859, 0.0094507, 0.00457866\} = 0.0094507 < \varepsilon$$

Hence, we shall stop here.

So, the sequence of successive approximations terminates at fifth iteration.

Therefore, the required solution of the given system of equations is

$$x_1 = 2, x_2 = 1, \text{and } x_3 = 4$$

Example 6.11

Solve the following system of equations

$$2.412x_1 + 9.879x_2 + 1.564x_3 = 4.89$$

$$1.876x_1 + 2.985x_2 - 11.62x_3 = -0.972$$

$$12.214x_1 + 2.367x_2 + 3.672x_3 = 7.814$$

correct to two decimal places by

a. Gauss–Jacobi method
b. Gauss–Seidel method

Solution:

We first rearrange the given system of equations so that the resulting system is as follows:

$$12.214x_1 + 2.367x_2 + 3.672x_3 = 7.814$$

$$2.412x_1 + 9.879x_2 + 1.564x_3 = 4.89$$

$$1.876x_1 + 2.985x_2 - 11.62x_3 = -0.972$$

Now the system of equations is strictly diagonally dominant. So, the Gauss–Jacobi as well as the Gauss–Seidel method will certainly converge.

Again, we rewrite the above equations in the following form:

$$x_1 = \frac{1}{12.214}(7.814 - 2.367x_2 - 3.672x_3)$$

$$x_2 = \frac{1}{9.879}(4.89 - 2.412x_1 - 1.564x_3)$$

$$x_3 = \frac{1}{-11.62}(-0.972 - 1.876x_1 - 2.985x_2)$$

The successive iterations in iterative methods will stop if

$$\left\| \boldsymbol{x}^{(k+1)} - \boldsymbol{x}^{(k)} \right\|_\infty < \varepsilon$$

where $k \geq 0$ and ε is the prescribed error tolerance.
Here, we take $\varepsilon = 0.001$.

a. Gauss–Jacobi method

Initial step:

$$x_1^{(0)} = x_2^{(0)} = x_3^{(0)} = 0$$

First iteration:

$$x_1^{(1)} = \frac{1}{12.214}(7.814 - 2.367x_2^{(0)} - 3.672x_3^{(0)}) = \frac{1}{12.214}(7.814) = 0.639758$$

$$x_2^{(1)} = \frac{1}{9.879}(4.89 - 2.412x_1^{(0)} - 1.564x_3^{(0)}) = \frac{1}{9.879}(4.89) = 0.494989$$

$$x_3^{(1)} = \frac{1}{-11.62}(-0.972 - 1.876x_1^{(0)} - 2.985x_2^{(0)}) = \frac{1}{-11.62}(-0.972) = 0.0836489$$

Here,

$$\left\| \boldsymbol{x}^{(1)} - \boldsymbol{x}^{(0)} \right\|_\infty = \text{Max}\{0.639758, 0.494989, 0.0836489\} = 0.639758 > \varepsilon$$

Second iteration:

$$x_1^{(2)} \frac{1}{12.214}(7.814 - 2.367x_2^{(1)} - 3.672x_3^{(1)})$$

$$= \frac{1}{12.214}[7.814 - 2.367 \times (0.494989) - 3.672 \times (0.0836489)] = 0.518684$$

$$x_2^{(2)} = \frac{1}{9.879}(4.89 - 2.412x_1^{(1)} - 1.564x_3^{(1)})$$

$$= \frac{1}{9.879}[4.89 - 2.412 \times (0.639758) - 1.564 \times (0.0836489)] = 0.325547$$

$$x_3^{(2)} = \frac{1}{-11.62}(-0.972 - 1.876x_1^{(1)} - 2.985x_2^{(1)})$$

$$= \frac{1}{-11.62}[-0.972 - 1.876 \times (0.639758) - 2.985 \times (0.494989)] = 0.31409$$

$$\left\|x^{(2)}-x^{(1)}\right\|_\infty = \text{Max}\{0.121074, 0.169442, 0.230441\} = 0.230441>\varepsilon$$

Third iteration:

$$x_1^{(3)}=\frac{1}{12.214}\left[7.814-2.367x_2^{(2)}-3.672x_3^{(2)}\right]$$

$$=\frac{1}{12.214}[7.814-2.367\times(0.325547)-3.672\times(0.31409)]=0.482241$$

$$x_2^{(3)}=\frac{1}{9.879}\left[4.89-2.412x_1^{(2)}-1.564x_3^{(2)}\right]$$

$$=\frac{1}{9.879}[4.89-2.412\times(0.518684)-1.564\times(0.31409)]=0.318625$$

$$x_3^{(3)}=\frac{1}{-11.62}\left[-0.972-1.876x_1^{(2)}-2.985x_2^{(2)}\right]$$

$$=\frac{1}{-11.62}[-0.972-1.876\times(0.518684)-2.985\times(0.325547)]=0.251016$$

In this case,

$$\left\|x^{(3)}-x^{(2)}\right\|_\infty = \text{Max}\{0.0364426, 0.00692172, 0.0630741\} = 0.0630741>\varepsilon$$

Fourth iteration:

$$x_1^{(4)}=\frac{1}{12.214}\left[7.814-2.367x_2^{(3)}-3.672x_3^{(3)}\right]$$

$$=\frac{1}{12.214}[7.814-2.367\times(0.318625)-3.672\times(0.251016)]=0.502545$$

$$x_2^{(4)}=\frac{1}{9.879}\left[4.89-2.412x_1^{(3)}-1.564x_3^{(3)}\right]$$

$$=\frac{1}{9.879}[4.89-2.412\times(0.482241)-1.564\times(0.251016)]=0.337508$$

$$x_3^{(4)}=\frac{1}{-11.62}\left[-0.972-1.876x_1^{(3)}-2.985x_2^{(3)}\right]$$

$$=\frac{1}{-11.62}[-0.972-1.876\times(0.482241)-2.985\times(0.318625)]=0.243355$$

Also,

$$\left\|x^{(4)}-x^{(3)}\right\|_\infty = \text{Max}\{0.0203039, 0.0188832, 0.00766159\} = 0.0203039>\varepsilon$$

Fifth iteration:

$$x_1^{(5)}=\frac{1}{12.214}\left[7.814-2.367x_2^{(4)}-3.672x_3^{(4)}\right]$$

$$=\frac{1}{12.214}[7.814-2.367\times(0.337508)-3.672\times(0.243355)]=0.501189$$

$$x_2^{(5)} = \frac{1}{9.879}\left[4.89 - 2.412x_1^{(4)} - 1.564x_3^{(4)}\right]$$

$$= \frac{1}{9.879}[4.89 - 2.412\times(0.502545) - 1.564\times(0.243355)] = 0.333764$$

$$x_3^{(5)} = \frac{1}{-11.62}\left[-0.972 - 1.876x_1^{(4)} - 2.985x_2^{(4)}\right]$$

$$= \frac{1}{-11.62}[-0.972 - 1.876\times(0.502545) - 2.985\times(0.337508)] = 0.251483$$

$$\left\|\boldsymbol{x}^{(5)} - \boldsymbol{x}^{(4)}\right\|_\infty = \text{Max}\{0.00135609, 0.00374433, 0.00812879\} = 0.00812879 > \varepsilon$$

Sixth iteration:

$$x_1^{(6)} = \frac{1}{12.214}\left[7.814 - 2.367x_2^{(5)} - 3.672x_3^{(5)}\right]$$

$$= \frac{1}{12.214}[7.814 - 2.367\times(0.333764) - 3.672\times(0.251483)] = 0.499471$$

$$x_2^{(6)} = \frac{1}{9.879}\left[4.89 - 2.412x_1^{(5)} - 1.564x_3^{(5)}\right]$$

$$= \frac{1}{9.879}[4.89 - 2.412\times(0.501189) - 1.564\times(0.251483)] = 0.332808$$

$$x_3^{(6)} = \frac{1}{-11.62}\left[-0.972 - 1.876x_1^{(5)} - 2.985x_2^{(5)}\right]$$

$$= \frac{1}{-11.62}[-0.972 - 1.876\times(0.501189) - 2.985\times(0.333764)] = 0.250303$$

Here,

$$\left\|\boldsymbol{x}^{(6)} - \boldsymbol{x}^{(5)}\right\|_\infty = \text{Max}\{0.0017182, 0.00095582, 0.0011808\} = 0.0017182 > \varepsilon$$

Seventh iteration:

$$x_1^{(7)} = \frac{1}{12.214}\left[7.814 - 2.367x_2^{(6)} - 3.672x_3^{(6)}\right]$$

$$= \frac{1}{12.214}[7.814 - 2.367\times(0.332808) - 3.672\times(0{,}250303)] = 0.500011$$

$$x_2^{(7)} = \frac{1}{9.879}\left[4.89 - 2.412x_1^{(6)} - 1.564x_3^{(6)}\right]$$

$$= \frac{1}{9.879}[4.89 - 2.412\times(0.499471) - 1.564\times(0.250303)] = 0.333415$$

$$x_3^{(7)} = \frac{1}{-11.62}\left[-0.972 - 1.876x_1^{(6)} - 2.985x_2^{(6)}\right]$$

$$= \frac{1}{-11.62}[-0.972 - 1.876\times(0.499471) - 2.985\times(0.332808)] = 0.24978$$

Finally,

$$\left\|x^{(7)}-x^{(6)}\right\|_{\infty}=\text{Max}\{0.000540225, 0.000606444, 0.000522932\}=0.000606444<\varepsilon$$

Hence, we shall stop here.

So, the sequence of successive approximations terminates at the seventh iteration.

Therefore, the required solution of the given system of equations is

$$x_1 = 0.50, x_2 = 0.33 \text{ and } x_3 = 0.25 \text{ correct to two decimal places.}$$

b. Gauss–Seidel method

Initial step:

$$x_1^{(0)} = x_2^{(0)} = x_3^{(0)} = 0$$

First iteration:

$$x_1^{(1)} = \frac{1}{12.214}\left[7.814 - 2.367x_2^{(0)} - 3.672x_3^{(0)}\right]$$

$$= \frac{1}{12.214}\times(7.814) = 0.639758$$

$$x_2^{(1)} = \frac{1}{9.879}\left[4.89 - 2.412x_1^{(1)} - 1.564x_3^{(0)}\right]$$

$$= \frac{1}{9.879}(4.89 - 2.412\times 0.639758 - 1.564\times 0) = 0.33879$$

$$x_3^{(1)} = \frac{1}{-11.62}\left[-0.972 - 1.876x_1^{(1)} - 2.985x_2^{(1)}\right]$$

$$= \frac{1}{-11.62}(-0.972 - 1.876\times 0.639758 - 2.985\times 0.33879) = 0.273965$$

Here,

$$\left\|x^{(1)}-x^{(0)}\right\|_{\infty}=\text{Max}\{0.639758, 0.33879, 0.273965\}=0.639758>\varepsilon$$

Second iteration:

$$x_1^{(2)} = \frac{1}{12.214}\left[7.814 - 2.367x_2^{(1)} - 3.672x_3^{(1)}\right]$$

$$= \frac{1}{12.214}[7.814 - 2.367\times(0.33879) - 3.672\times(0.273965)] = 0.491738$$

$$x_2^{(2)} = \frac{1}{9.879}\left[4.89 - 2.412x_1^{(2)} - 1.564x_3^{(1)}\right]$$

$$= \frac{1}{9.879}[4.89 - 2.412\times(0.491738) - 1.564\times(0.273965)] = 0.331557$$

$$x_3^{(2)} = \frac{1}{-11.62}\left[-0.972 - 1.876x_1^{(2)} - 2.985x_2^{(2)}\right]$$

$$= \frac{1}{-11.62}[-0.972 - 1.876\times(0.491738) - 2.985\times(0.331557)] = 0.24821$$

$$\left\|\boldsymbol{x}^{(2)}-\boldsymbol{x}^{(1)}\right\|_{\infty}=\text{Max}\{0.14802,\ 0.00723325,\ 0.0257553\}=0.14802>\varepsilon$$

Third iteration:

$$x_1^{(3)}=\frac{1}{12.214}\left[7.814-2.367x_2^{(2)}-3.672x_3^{(2)}\right]$$

$$=\frac{1}{12.214}[7.814-2.367\times(0.331557)-3.672\times(0.24821)]=0.500883$$

$$x_2^{(3)}=\frac{1}{9.879}\left[4.89-2.412x_1^{(3)}-1.564x_3^{(2)}\right]$$

$$=\frac{1}{9.879}[4.89-2.412\times(0.500883)-1.564\times(0.24821)]=0.333401$$

$$x_3^{(3)}=\frac{1}{-11.62}\left[-0.972-1.876x_1^{(3)}-2.985x_2^{(3)}\right]$$

$$=\frac{1}{-11.62}[-0.972-1.876\times(0.500883)-2.985\times(0.333401)]=0.25016$$

In this case,

$$\left\|\boldsymbol{x}^{(3)}-\boldsymbol{x}^{(2)}\right\|_{\infty}=\text{Max}\{0.0091448,0.00184472,0.00195027\}=0.0091448>\varepsilon$$

Fourth iteration:

$$x_1^{(4)}=\frac{1}{12.214}\left[7.814-2.367x_2^{(3)}-3.672x_3^{(3)}\right]$$

$$=\frac{1}{12.214}[7.814-2.367\times(0.333401)-3.672\times(0.25016)]=0.499939$$

$$x_2^{(4)}=\frac{1}{9.879}\left[4.89-2.412x_1^{(4)}-1.564x_3^{(3)}\right]$$

$$=\frac{1}{9.879}[4.89-2.412\times(0.499939)-1.564\times(0.25016)]=0.333323$$

$$x_3^{(4)}=\frac{1}{-11.62}\left[-0.972-1.876x_1^{(4)}-2.985x_2^{(4)}\right]$$

$$=\frac{1}{-11.62}[-0.972-1.876\times(0.499939)-2.985\times(0.333323)]=0.249987$$

In this case,

$$\left\|\boldsymbol{x}^{(4)}-\boldsymbol{x}^{(3)}\right\|_{\infty}=\text{Max}\{0.000943823,0.0000783198,0.000172495\}=0.000943823<\varepsilon$$

Hence, we shall stop here.

So, the sequence of successive approximations terminates at third iteration.

Therefore, the required solution of the given system of equations is

$$x_1=0.5, x_2=0.33, \text{and } x_3=0.25$$

6.4.3 SOR Method

In practical purpose for large system of equations, computations need to accelerate convergence of iteration methods. SOR method is a generalization of the Gauss–Seidel method. This method is based on an accelerating parameter ω which accelerates the convergence than that of the Gauss–Seidel method.

Let us recall the Equation 6.74 of the Gauss–Seidel method and introduce an acceleration parameter ω to consider the following modification of Equation 6.74:

$$z_i^{(k+1)} = -\frac{1}{a_{ii}}\left[\sum_{j=1}^{i-1} a_{ij}x_j^{(k+1)} + \sum_{j=i+1}^{n} a_{ij}x_j^{(k)} - b_i\right]$$

$$x_i^{(k+1)} = \omega z_i^{(k+1)} + (1-\omega)x_i^{(k)}, \quad i = 1,2,\ldots,n \text{ and } k = 0,1,2,\ldots \tag{6.88}$$

When $\omega = 1$, we have the regular Gauss–Seidel method. The parameter ω is also known as the relaxation factor, which generally lies in the range $0 < \omega < 2$. If $1 < \omega < 2$, then the scheme is known as an over-relaxation method, and if $0 < \omega < 1$, then it is called an under-relaxation method. The iterative scheme in Equation 6.88 with an optimal value of ω is called the SOR method.

6.4.3.1 Convergence of the SOR Method

Let us rewrite the Equation 6.88 in the following matrix form:

$$\boldsymbol{z}^{(k+1)} = \boldsymbol{D}^{-1}\left[\boldsymbol{b} - \boldsymbol{L}\boldsymbol{x}^{(k+1)} - \boldsymbol{U}\boldsymbol{x}^{(k)}\right] \tag{6.89}$$

$$\boldsymbol{x}^{(k+1)} = \omega\,\boldsymbol{z}^{(k+1)} + (1-\omega)\boldsymbol{x}^{(k)}, \quad k = 0,1,2,\ldots \tag{6.90}$$

Therefore, from Equations 6.89 and 6.90, we obtain

$$\boldsymbol{x}^{(k+1)} = \omega\boldsymbol{D}^{-1}\left[\boldsymbol{b} - \boldsymbol{L}\boldsymbol{x}^{(k+1)} - \boldsymbol{U}\boldsymbol{x}^{(k)}\right] + (1-\omega)\boldsymbol{x}^{(k)}, \quad k = 0,\ 1,2,\ldots \tag{6.91}$$

This implies that

$$\begin{aligned}\boldsymbol{x}^{(k+1)} &= (\boldsymbol{I} + \omega\boldsymbol{D}^{-1}\boldsymbol{L})^{-1}\left[(1-\omega)\boldsymbol{I} - \omega\boldsymbol{D}^{-1}\boldsymbol{U}\right]\boldsymbol{x}^{(k)} + \omega(\boldsymbol{I} + \omega\boldsymbol{D}^{-1}\boldsymbol{L})^{-1}\boldsymbol{D}^{-1}\boldsymbol{b}\\ &= (\boldsymbol{D} + \omega\boldsymbol{L})^{-1}\left[(1-\omega)\boldsymbol{D} - \omega\boldsymbol{U}\right]\boldsymbol{x}^{(k)} + \omega(\boldsymbol{D} + \omega\boldsymbol{L})^{-1}\boldsymbol{b}, \quad k = 0,1,2,\ldots\end{aligned} \tag{6.92}$$

Therefore, from Equation 6.92, we have

$$\boldsymbol{x}^{(k+1)} = \boldsymbol{Q}_{SOR}\,\boldsymbol{x}^{(k)} + \boldsymbol{C}_{SOR} \tag{6.93}$$

where $\boldsymbol{Q}_{SOR} = (\boldsymbol{D} + \omega\boldsymbol{L})^{-1}[(1-\omega)\boldsymbol{D} - \omega\boldsymbol{U}]$ and $\boldsymbol{C}_{SOR} = \omega(\boldsymbol{D} + \omega\boldsymbol{L})^{-1}\boldsymbol{b}$.

It has been established that if $\boldsymbol{A}$ is symmetric and positive-definite, then $\rho(\boldsymbol{Q}_{SOR}) < 1$ for $0 < \omega < 2$. Thus, convergence of the iteration process follows, but we are generally interested in fastest convergence rather than just convergence.

The parameter ω is to be chosen optimally to minimize $\rho(\boldsymbol{Q}_{SOR})$, that is, spectral radius of $\boldsymbol{Q}_{SOR}$ in order to make $\boldsymbol{x}^{(k)}$ converges to $\boldsymbol{x}$ as rapidly as possible.

The optimal value ω^* of the relaxation factor for which fastest convergence takes place is given by

$$\omega^* = \frac{2}{1 + \sqrt{1 - \rho(\boldsymbol{Q}_{GS})}} \tag{6.94}$$

where $Q_{GS} = -(L + D)^{-1} U$ is the iteration matrix of the Gauss–Seidel method.

Now, the Equation 6.92 can also be written as

$$\begin{aligned} x^{(k+1)} &= x^{(k)} - (D + \omega L)^{-1}[(D + \omega L) - (1-\omega) D + \omega U] x^{(k)} + \omega (D + \omega L)^{-1} b \\ &= x^{(k)} - \omega (D + \omega L)^{-1} [L + D + U] x^{(k)} + \omega (D + \omega L)^{-1} b \\ &= x^{(k)} - \omega (D + \omega L)^{-1} A\, x^{(k)} + \omega (D + \omega L)^{-1} b \\ &= x^{(k)} + \omega (D + \omega L)^{-1} r^{(k)}, \quad k = 0,1,2,\ldots \end{aligned} \tag{6.95}$$

where $r^{(k)} = b - A\, x^{(k)}$ is the residual vector.

Therefore, from Equation 6.95, we get

$$h^{(k)} = \omega (D + \omega L)^{-1} r^{(k)}, \quad k = 0,1,2,\ldots \tag{6.96}$$

where $h^{(k)} = x^{(k+1)} - x^{(k)}$ is the error in approximation.

We may solve the following equation

$$(D + \omega L)\, h^{(k)} = \omega r^{(k)} \tag{6.97}$$

for the vector $h^{(k)}$, and then, we determine

$$x^{(k+1)} = x^{(k)} + h^{(k)} \tag{6.98}$$

These equations describe SOR method in its error format.

6.4.3.2 Algorithm for the SOR Method

Input: Enter the number of unknown variables n, the coefficient matrix $A = [a_{i,j}]_{n \times n}$, the constant column vector $b = [b_i]_{n \times 1}$, the error tolerance ε, and the relaxation factor ω.

Output: Solution x of the given system of equations, that is, x_i, $i = 1,2,\ldots,n$.

Initial step: Choose an initial guess $x^{(0)}$ to the solution x.

Step 1: Set $k = 0$

Step 2: for $i = 1(1)n$ do

  sum $= 0$;

  for $j = 1(1)\overline{i-1}$ do

    sum $=$ sum $+ a_{i,j} * x_j^{(k+1)}$;

  end;

  for $j = \overline{i+1}(1)n$ do

    sum $=$ sum $+ a_{i,j} * x_j^{(k)}$;

  end;

  sum $= \dfrac{(b_i - \text{sum})}{a_{ii}}$;

  $x_i^{(k+1)} = \omega * \text{sum} + (1-\omega) * x_i^{(k)}$;

 end

Step 3: If $\left\| x^{(k+1)} - x^{(k)} \right\|_\infty < \varepsilon$, then Go To Step 4

  else

 set $k = k + 1$ and Go To Step 2.

Step 4: Print $x^{(k+1)}$.

Step 5: Stop. ■

MATHEMATICA® Program for Numerical Solution of System of Algebraic Equations by SOR Method (Chapter 6, Example 6.12)

```
A={{6.7,1.1,2.2},{3.1,9.4,-1.5},{2.1,-1.5,8.4}};
Diag=DiagonalMatrix[{6.7,9.4,8.4}];
L=LowerTriangularize[A,-1];
U=UpperTriangularize[A,1];
Q=-Inverse[Diag+L].U;

ro=0.214589;
w=(2/(1+√(1-ro)));

b={{20.5},{22.9},{28.8}};
Print[MatrixForm[Inverse[Diag+w*L]]];
x[0]={{0},{0},{0}};
MatrixForm[x[0]];

ϵ=0.00001;
n=100;

Do[
 Print["Step ",k+1,":"];
 r[k]=b-A.x[k];
 Print["r[",k,"]=",MatrixForm[r[k]]];
    h[k]=w*(Inverse[Diag+w*L].r[k]);
 Print["h[",k,"]=",MatrixForm[h[k]]];
 x[k+1]=x[k]+h[k];

Print["x[",k+1,"]=",MatrixForm[x[k+1]]];If[Max[Abs[
x[k+1]-x[k]]]<ϵ,Break[]],{k,0,n}
 ]
```

Output:

$$\begin{pmatrix} 0.149254 & 0. & 0. \\ -0.0521907 & 0.106383 & 0. \\ -0.0494458 & 0.0201427 & 0.119048 \end{pmatrix}$$

Step 1:

$$r[0] = \begin{pmatrix} 20.5 \\ 22.9 \\ 28.8 \end{pmatrix}$$

$$h[0] = \begin{pmatrix} 3.24424 \\ 1.44866 \\ 3.04968 \end{pmatrix}$$

$$x[1] = \begin{pmatrix} 3.24424 \\ 1.44866 \\ 3.04968 \end{pmatrix}$$

Step 2:

$$r[1] = \begin{pmatrix} -9.53925 \\ 3.79991 \\ -1.45719 \end{pmatrix}$$

$$h[1] = \begin{pmatrix} -1.50964 \\ 0.956516 \\ 0.397344 \end{pmatrix}$$

$$x[2] = \begin{pmatrix} 1.7346 \\ 2.40518 \\ 3.44702 \end{pmatrix}$$

Step 3:

$$r[2] = \begin{pmatrix} -1.35098 \\ 0.0845681 \\ -0.189858 \end{pmatrix}$$

$$h[2] = \begin{pmatrix} -0.2138 \\ 0.0843004 \\ 0.04867 \end{pmatrix}$$

$$x[3] = \begin{pmatrix} 1.5208 \\ 2.48948 \\ 3.49569 \end{pmatrix}$$

Step 4:

$$r[3] = \begin{pmatrix} -0.118322 \\ 0.0279297 \\ -0.0232553 \end{pmatrix}$$

$$h[3] = \begin{pmatrix} -0.0187251 \\ 0.00969821 \\ 0.00386443 \end{pmatrix}$$

$$x[4] = \begin{pmatrix} 1.50208 \\ 2.49918 \\ 3.49955 \end{pmatrix}$$

Step 5:

$$r[4] = \begin{pmatrix} -0.0120334 \\ 0.000611023 \\ -0.00184649 \end{pmatrix}$$

$$h[4] = \begin{pmatrix} -0.00190435 \\ 0.000734831 \\ 0.000410857 \end{pmatrix}$$

$$x[5] = \begin{pmatrix} 1.50017 \\ 2.49991 \\ 3.49996 \end{pmatrix}$$

Step 6:

$$r[5] = \begin{pmatrix} -0.000986424 \\ 0.000223372 \\ -0.000196314 \end{pmatrix}$$

$$h[5] = \begin{pmatrix} -0.000156107 \\ 0.0000797835 \\ 0.0000317067 \end{pmatrix}$$

$$x[6] = \begin{pmatrix} 1.50002 \\ 2.49999 \\ 3.5 \end{pmatrix}$$

Step 7:

$$r[6] = \begin{pmatrix} -0.0000980216 \\ 4.89984 \times 10^{-6} \\ -0.00001515 \end{pmatrix}$$

$$h[6] = \begin{pmatrix} -0.0000155125 \\ 5.97707 \times 10^{-6} \\ 3.33139 \times 10^{-6} \end{pmatrix}$$

$$x[7] = \begin{pmatrix} 1.5 \\ 2.5 \\ 3.5 \end{pmatrix}$$

Step 8:

$$r[7] = \begin{pmatrix} -7.99178 \times 10^{-6} \\ 1.80114 \times 10^{-6} \\ -1.59179 \times 10^{-6} \end{pmatrix}$$

$$h[7] = \begin{pmatrix} -1.26475 \times 10^{-6} \\ 6.45421 \times 10^{-7} \\ 2.56534 \times 10^{-7} \end{pmatrix}$$

$$x[8] = \begin{pmatrix} 1.5 \\ 2.5 \\ 3.5 \end{pmatrix}$$

Example 6.12

Solve the following system of equations

$$3.1x_1 + 9.4x_2 - 1.5x_3 = 22.9$$

$$2.1x_1 - 1.5x_2 + 8.4x_3 = 28.8$$

$$6.7x_1 + 1.1x_2 + 2.2x_3 = 20.5$$

correct to two decimal places by

a. Gauss–Seidel method
b. SOR method

Hence compare the results.

Solution:

We first rearrange the given system of equations so that the resulting system is as follows:

$$6.7x_1 + 1.1x_2 + 2.2x_3 = 20.5$$

$$3.1x_1 + 9.4x_2 - 1.5x_3 = 22.9$$

$$2.1x_1 - 1.5x_2 + 8.4x_3 = 28.8$$

Now the system of equations is strictly diagonally dominant. So, the Gauss–Seidel method will certainly converge.

We have

$$\boldsymbol{L} = \begin{pmatrix} 0 & 0 & 0 \\ 3.1 & 0 & 0 \\ 2.1 & -1.5 & 0 \end{pmatrix}, \ \boldsymbol{D} = \begin{pmatrix} 6.7 & 0 & 0 \\ 0 & 9.4 & 0 \\ 0 & 0 & 8.4 \end{pmatrix}, \text{ and } \boldsymbol{U} = \begin{pmatrix} 0 & 1.1 & 2.2 \\ 0 & 0 & -1.5 \\ 0 & 0 & 0 \end{pmatrix}$$

Here, we take $\varepsilon = 0.0002$.

a. Gauss–Seidel method

Using Equation 3.2.8, we get

$$\boldsymbol{x}^{(k+1)} = \boldsymbol{x}^{(k)} + \boldsymbol{h}^{(k)}, \quad k = 0,1,2,\ldots$$

where

$$\boldsymbol{h}^{(k)} = (\boldsymbol{L} + \boldsymbol{D})^{-1} \boldsymbol{r}^{(k)} \text{ and } \boldsymbol{r}^{(k)} = \boldsymbol{b} - \boldsymbol{A}\,\boldsymbol{x}^{(k)}$$

Initial step:

$$\boldsymbol{x}^{(0)} = \mathbf{0} \quad \text{that is,} \quad x_1^{(0)} = x_2^{(0)} = x_3^{(0)} = 0$$

First iteration:

$$\boldsymbol{r}^{(0)} = \boldsymbol{b} - \boldsymbol{A}\,\boldsymbol{x}^{(0)} = \begin{pmatrix} 20.5 \\ 22.9 \\ 28.8 \end{pmatrix}$$

$$h^{(0)} = (L+D)^{-1} r^{(0)} = \begin{pmatrix} 0.149254 & 0 & 0 \\ -0.049222 & 0.106383 & 0 \\ -0.0461031 & 0.018997 & 0.119048 \end{pmatrix} \begin{pmatrix} 20.5 \\ 22.9 \\ 28.8 \end{pmatrix} = \begin{pmatrix} 3.0597 \\ 1.42712 \\ 2.91849 \end{pmatrix}$$

$$x^{(1)} = x^{(0)} + h^{(0)} = \begin{pmatrix} 3.0597 \\ 1.42712 \\ 2.91849 \end{pmatrix}$$

$$\left\| x^{(1)} - x^{(0)} \right\|_\infty = \text{Max}\{3.0597, 1.42712, 2.91849\} = 3.0597 > \varepsilon$$

Second iteration:

$$r^{(1)} = b - Ax^{(1)} = \begin{pmatrix} -7.99051 \\ 4.37773 \\ 0 \end{pmatrix}$$

$$h^{(1)} = (L+D)^{-1} r^{(1)} = \begin{pmatrix} 0.149254 & 0 & 0 \\ -0.049222 & 0.106383 & 0 \\ -0.0461031 & 0.018997 & 0.119048 \end{pmatrix} \begin{pmatrix} -7.99051 \\ 4.37773 \\ 0 \end{pmatrix} = \begin{pmatrix} -1.19261 \\ 0.859025 \\ 0.451551 \end{pmatrix}$$

$$x^{(2)} = x^{(1)} + h^{(1)} = \begin{pmatrix} 1.86709 \\ 2.28614 \\ 3.37004 \end{pmatrix}$$

$$\left\| x^{(2)} - x^{(1)} \right\|_\infty = \text{Max}\{1.19261, 0.859025, 0.451551\} = 1.19261 > \varepsilon$$

Third iteration:

$$r^{(2)} = b - A\,x^{(2)} = \begin{pmatrix} -1.93834 \\ 0.677326 \\ 0 \end{pmatrix}$$

$$h^{(2)} = (L+D)^{-1} r^{(2)} = \begin{pmatrix} 0.149254 & 0 & 0 \\ -0.049222 & 0.106383 & 0 \\ -0.0461031 & 0.018997 & 0.119048 \end{pmatrix} \begin{pmatrix} -1.93834 \\ 0.677326 \\ 0 \end{pmatrix} = \begin{pmatrix} -0.289304 \\ 0.167465 \\ 0.10223 \end{pmatrix}$$

$$x^{(3)} = x^{(2)} + h^{(2)} = \begin{pmatrix} 1.57778 \\ 2.45361 \\ 3.47227 \end{pmatrix}$$

$$\left\| x^{(3)} - x^{(2)} \right\|_\infty = \text{Max}\{0.289304, 0.167465, 0.10223\} = 0.289304 > \varepsilon$$

Fourth iteration:

$$r^{(3)} = b - Ax^{(3)} = \begin{pmatrix} -0.409118 \\ 0.153346 \\ 0 \end{pmatrix}$$

$$h^{(3)} = (L+D)^{-1} r^{(3)} = \begin{pmatrix} 0.149254 & 0 & 0 \\ -0.049222 & 0.106383 & 0 \\ -0.0461031 & 0.018997 & 0.119048 \end{pmatrix} \begin{pmatrix} -0.409118 \\ 0.153346 \\ 0 \end{pmatrix} = \begin{pmatrix} -0.0610624 \\ 0.036451 \\ 0.0217747 \end{pmatrix}$$

$$x^{(4)} = x^{(3)} + h^{(3)} = \begin{pmatrix} 1.51672 \\ 2.49006 \\ 3.49404 \end{pmatrix}$$

$$\left\| x^{(4)} - x^{(3)} \right\|_\infty = \text{Max}\{0.0610624, 0.036451, 0.0217747\} = 0.0610624 > \varepsilon$$

Fifth iteration:

$$r^{(4)} = b - A\,x^{(4)} = \begin{pmatrix} -0.0880005 \\ 0.0326621 \\ 0 \end{pmatrix}$$

$$h^{(4)} = (L+D)^{-1} r^{(4)} = \begin{pmatrix} 0.149254 & 0 & 0 \\ -0.049222 & 0.106383 & 0 \\ -0.0461031 & 0.018997 & 0.119048 \end{pmatrix} \begin{pmatrix} -0.0880005 \\ 0.0326621 \\ 0 \end{pmatrix} = \begin{pmatrix} -0.0131344 \\ 0.00780625 \\ 0.00467757 \end{pmatrix}$$

$$x^{(5)} = x^{(4)} + h^{(4)} = \begin{pmatrix} 1.50359 \\ 2.49787 \\ 3.49872 \end{pmatrix}$$

$$\left\| x^{(5)} - x^{(4)} \right\|_\infty = \text{Max}\{0.0131344, 0.00780625, 0.00467757\} = 0.0131344 > \varepsilon$$

Sixth iteration:

$$r^{(5)} = b - A x^{(5)} = \begin{pmatrix} -0.0188775 \\ 0.00701636 \\ 0 \end{pmatrix}$$

$$h^{(5)} = (L+D)^{-1} r^{(5)} = \begin{pmatrix} 0.149254 & 0 & 0 \\ -0.049222 & 0.106383 & 0 \\ -0.0461031 & 0.018997 & 0.119048 \end{pmatrix} \begin{pmatrix} -0.0188775 \\ 0.00701636 \\ 0 \end{pmatrix} = \begin{pmatrix} -0.00281754 \\ 0.00167561 \\ 0.0010036 \end{pmatrix}$$

$$x^{(6)} = x^{(5)} + h^{(5)} = \begin{pmatrix} 1.50077 \\ 2.49954 \\ 3.49973 \end{pmatrix}$$

$$\left\| x^{(6)} - x^{(5)} \right\|_\infty = \text{Max}\{0.00281754, 0.00167561, 0.0010036\} = 0.00281754 > \varepsilon$$

Seventh iteration:

$$r^{(6)} = b - A\,x^{(6)} = \begin{pmatrix} -0.00405109 \\ 0.0015054 \\ 0 \end{pmatrix}$$

$$h^{(6)} = (L+D)^{-1} r^{(6)} = \begin{pmatrix} 0.149254 & 0 & 0 \\ -0.049222 & 0.106383 & 0 \\ -0.0461031 & 0.018997 & 0.119048 \end{pmatrix} \begin{pmatrix} -0.00405109 \\ 0.0015054 \\ 0 \end{pmatrix} = \begin{pmatrix} -0.000604641 \\ 0.000359552 \\ 0.000215366 \end{pmatrix}$$

$$x^{(7)} = x^{(6)} + h^{(6)} = \begin{pmatrix} 1.50017 \\ 2.4999 \\ 3.49994 \end{pmatrix}$$

$$\left\| x^{(7)} - x^{(6)} \right\|_\infty = \text{Max}\{0.000604641, 0.000359552, 0.000215366\} = 0.000604641 > \varepsilon$$

Eighth iteration:

$$r^{(7)} = b - A\,x^{(7)} = \begin{pmatrix} -0.000869312 \\ 0.000323049 \\ 0 \end{pmatrix}$$

$$h^{(7)} = (L+D)^{-1} r^{(7)} = \begin{pmatrix} 0.149254 & 0 & 0 \\ -0.049222 & 0.106383 & 0 \\ -0.0461031 & 0.018997 & 0.119048 \end{pmatrix} \begin{pmatrix} -0.000869312 \\ 0.000323049 \\ 0 \end{pmatrix} = \begin{pmatrix} -0.000129748 \\ 0.0000771562 \\ 0.0000462149 \end{pmatrix}$$

$$x^{(8)} = x^{(7)} + h^{(7)} = \begin{pmatrix} 1.50004 \\ 2.49998 \\ 3.49999 \end{pmatrix}$$

$$\left\| x^{(8)} - x^{(7)} \right\|_\infty = \text{Max}\{0.000129748, 0.0000771562, 0.0000462149\} = 0.000129748 < \varepsilon$$

Hence, we shall stop here.
So, the sequence of successive approximations terminates at eighth iteration.
Therefore, the required solution of the given system of equations is

$$x_1 = 1.5, x_2 = 2.5, \text{ and } x_3 = 3.5.$$

b. SOR method
Using Equation 6.98, we get

$$x^{(k+1)} = x^{(k)} + h^{(k)}, \quad k = 0, 1, 2, \ldots$$

where

$$h^{(k)} = \omega(D+\omega L)^{-1} r^{(k)} \text{ and } r^{(k)} = b - Ax^{(k)}$$

Again, from Equation 6.94, the optimal value ω^* of the relaxation factor is given by

$$\omega^* = \frac{2}{1+\sqrt{1-\rho(Q_{GS})}}$$

where $Q_{GS} = -(L+D)^{-1}U$ is the iteration matrix of the Gauss–Seidel method.
Now, $\rho(Q_{GS}) = 0.214589$.

Therefore, the optimal value ω^* of the relaxation factor is

$$\omega^* = \frac{2}{1+\sqrt{1-\rho(\boldsymbol{Q}_{GS})}} = 1.06031$$

Initial step:

$$\boldsymbol{x}^{(0)} = \mathbf{0}, \quad \text{that is,} \quad x_1^{(0)} = x_2^{(0)} = x_3^{(0)} = 0$$

First iteration:

$$\boldsymbol{r}^{(0)} = \boldsymbol{b} - \boldsymbol{A}\,\boldsymbol{x}^{(0)} = \begin{pmatrix} 20.5 \\ 22.9 \\ 28.8 \end{pmatrix}$$

$$\boldsymbol{h}^{(0)} = \omega(\boldsymbol{D}+\omega\boldsymbol{L})^{-1}\,\boldsymbol{r}^{(0)} = (1.06031)\begin{pmatrix} 0.149254 & 0 & 0 \\ -0.0521907 & 0.106383 & 0 \\ -0.0494458 & 0.0201427 & 0.119048 \end{pmatrix}\begin{pmatrix} 20.5 \\ 22.9 \\ 28.8 \end{pmatrix} = \begin{pmatrix} 3.24424 \\ 1.44866 \\ 3.04968 \end{pmatrix}$$

$$\boldsymbol{x}^{(1)} = \boldsymbol{x}^{(0)} + \boldsymbol{h}^{(0)} = \begin{pmatrix} 3.24424 \\ 1.444866 \\ 3.04968 \end{pmatrix}$$

$$\left\| \boldsymbol{x}^{(1)} - \boldsymbol{x}^{(0)} \right\|_\infty = \text{Max}\{3.24424, 1.44866, 3.04968\} = 3.24424 > \varepsilon$$

Second iteration:

$$\boldsymbol{r}^{(1)} = \boldsymbol{b} - \boldsymbol{A}\,\boldsymbol{x}^{(1)} = \begin{pmatrix} -9.53925 \\ 3.79991 \\ -1.45719 \end{pmatrix}$$

$$\boldsymbol{h}^{(1)} = \omega(\boldsymbol{D}+\omega\boldsymbol{L})^{-1}\,\boldsymbol{r}^{(1)} = (1.06031)\begin{pmatrix} 0.149254 & 0 & 0 \\ -0.0521907 & 0.106383 & 0 \\ -0.0494458 & 0.0201427 & 0.119048 \end{pmatrix}\begin{pmatrix} -9.53925 \\ 3.79991 \\ -1.45719 \end{pmatrix} = \begin{pmatrix} -1.50964 \\ 0.956516 \\ 0.397344 \end{pmatrix}$$

$$\boldsymbol{x}^{(2)} = \boldsymbol{x}^{(1)} + \boldsymbol{h}^{(1)} = \begin{pmatrix} 1.7346 \\ 2.40518 \\ 3.44702 \end{pmatrix}$$

$$\left\| \boldsymbol{x}^{(2)} - \boldsymbol{x}^{(1)} \right\|_\infty = \text{Max}\{1.7346, 2.40518, 3.44702\} = 3.44702 > \varepsilon$$

Third iteration:

$$\boldsymbol{r}^{(2)} = \boldsymbol{b} - \boldsymbol{A}\,\boldsymbol{x}^{(2)} = \begin{pmatrix} -1.35098 \\ 0.0845681 \\ -0.189858 \end{pmatrix}$$

$$h^{(2)} = \omega(D+\omega L)^{-1} r^{(2)} = (1.06031)\begin{pmatrix} 0.149254 & 0 & 0 \\ -0.0521907 & 0.106383 & 0 \\ -0.0494458 & 0.0201427 & 0.119048 \end{pmatrix}\begin{pmatrix} -1.35098 \\ 0.0845681 \\ -0.189858 \end{pmatrix} = \begin{pmatrix} -0.2138 \\ 0.0843004 \\ 0.04867 \end{pmatrix}$$

$$x^{(3)} = x^{(2)} + h^{(2)} = \begin{pmatrix} 1.5208 \\ 2.48948 \\ 3.49569 \end{pmatrix}$$

$$\left\| x^{(3)} - x^{(2)} \right\|_\infty = \text{Max}\{0.2138, 0.0843004, 0.04867\} = 0.2138 > \varepsilon$$

Fourth iteration:

$$r^{(3)} = b - A\,x^{(3)} = \begin{pmatrix} -0.118322 \\ 0.0279297 \\ -0.0232553 \end{pmatrix}$$

$$h^{(3)} = \omega(D+\omega L)^{-1} r^{(3)}$$

$$= (1.06031)\begin{pmatrix} 0.149254 & 0 & 0 \\ -0.0521907 & 0.106383 & 0 \\ -0.0494458 & 0.0201427 & 0.119048 \end{pmatrix}\begin{pmatrix} -0.118322 \\ 0.0279297 \\ -0.0232553 \end{pmatrix} = \begin{pmatrix} -0.0187251 \\ 0.00969821 \\ 0.00386443 \end{pmatrix}$$

$$x^{(4)} = x^{(3)} + h^{(3)} = \begin{pmatrix} 1.50208 \\ 2.49918 \\ 3.49955 \end{pmatrix}$$

$$\left\| x^{(4)} - x^{(3)} \right\|_\infty = \text{Max}\{0.0187251, 0.00969821, 0.00386443\} = 0.0187251 > \varepsilon$$

Fifth iteration:

$$r^{(4)} = b - A\,x^{(4)} = \begin{pmatrix} -0.0120334 \\ 0.000611023 \\ 0.00184649 \end{pmatrix}$$

$$h^{(4)} = \omega(D+\omega L)^{-1} r^{(4)}$$

$$= (1.06031)\begin{pmatrix} 0.149254 & 0 & 0 \\ -0.0521907 & 0.106383 & 0 \\ -0.0494458 & 0.0201427 & 0.119048 \end{pmatrix}\begin{pmatrix} -0.0120334 \\ 0.000611023 \\ 0.00184649 \end{pmatrix} = \begin{pmatrix} -0.00190435 \\ 0.000734831 \\ 0.000410857 \end{pmatrix}$$

$$x^{(5)} = x^{(4)} + h^{(4)} = \begin{pmatrix} 1.50017 \\ 2.49991 \\ 3.49996 \end{pmatrix}$$

$$\left\| x^{(5)} - x^{(4)} \right\|_\infty = \text{Max}\{0.00190435, 0.000734831, 0.000410857\} = 0.00190435 > \varepsilon$$

Sixth iteration:

$$\mathbf{r}^{(5)} = \mathbf{b} - \mathbf{A}\,\mathbf{x}^{(5)} = \begin{pmatrix} -0.000986424 \\ 0.000223372 \\ 0.000196314 \end{pmatrix}$$

$$\mathbf{h}^{(5)} = \omega(\mathbf{D}+\omega\mathbf{L})^{-1}\,\mathbf{r}^{(5)}$$

$$= (1.06031)\begin{pmatrix} 0.149254 & 0 & 0 \\ -0.0521907 & 0.106383 & 0 \\ -0.0494458 & 0.0201427 & 0.119048 \end{pmatrix}\begin{pmatrix} -0.000986424 \\ 0.000223372 \\ 0.000196314 \end{pmatrix} = \begin{pmatrix} -0.000156107 \\ 0.0000797835 \\ 0.0000317067 \end{pmatrix}$$

$$\mathbf{x}^{(6)} = \mathbf{x}^{(5)} + \mathbf{h}^{(5)} = \begin{pmatrix} 1.50002 \\ 2.49999 \\ 3.5 \end{pmatrix}$$

$$\left\|\mathbf{x}^{(6)} - \mathbf{x}^{(5)}\right\|_\infty = \text{Max}\{0.000156107, 0.0000797835, 0.0000317067\} = 0.000156107 < \varepsilon$$

Hence, we shall stop here.

So, the sequence of successive approximations terminates at sixth iteration.

Therefore, the required solution of the given system of equations is

$$x_1 = 1.5, x_2 = 2.5, \text{and}\, x_3 = 3.5$$

Comparing the above results, it may be easily observed that the SOR method converges faster than the Gauss–Seidel method.

6.5 CONVERGENT ITERATION MATRICES

According to Theorem 6.2, we know that the matrix $\boldsymbol{M}$ is convergent, that is,

$$\lim_{n\to\infty} \boldsymbol{M}^n = \mathbf{0}$$

if and only if $\rho(\boldsymbol{M}) < 1$, where $\rho(\boldsymbol{M})$ is the spectral radius of $\boldsymbol{M}$ defined by

$$\rho(\boldsymbol{M}) \equiv \underset{i}{\text{Max}}\,|\lambda_i|$$

where λ_i are the eigenvalues of $\boldsymbol{M}$.

Therefore, an iteration matrix $\boldsymbol{M}$ is convergent if and only if $\rho(\boldsymbol{M}) < 1$.

Corollary: The matrix $\boldsymbol{M} = [m_{ij}]$ is convergent if either

$$\text{(Row } sum \text{ matrix norm)} \quad \|\boldsymbol{M}\|_\infty = \max_i \sum_{j=1}^{n} |m_{ij}| < 1 \tag{6.99}$$

$$\text{(Column } sum \text{ matrix norm)} \quad \|\boldsymbol{M}\|_1 = \max_j \sum_{i=1}^{n} |m_{ij}| < 1 \tag{6.100}$$

6.6 CONVERGENCE OF ITERATIVE METHODS

To discuss the convergence of the iterative method in Equation 6.54, we see that the exact solution $\boldsymbol{x}$ will satisfy the following equation:

$$\boldsymbol{x} = \boldsymbol{Q}\,\boldsymbol{x} + \boldsymbol{C} \tag{6.101}$$

Subtracting Equation 6.54 from Equation 6.101, we get

$$\varepsilon^{(k+1)} = \boldsymbol{Q}\,\varepsilon^{(k)}, \quad k = 0,1,2,.. \tag{6.102}$$

from which we obtain

$$\varepsilon^{(k)} = \boldsymbol{Q}^{(k)}\,\varepsilon^{(0)}, \quad k = 0,\ 1,\ 2,\ldots \tag{6.103}$$

In order that $\varepsilon^{(k)} \to \mathbf{0}$ as $k \to \infty$, for arbitrary initial guesses $\boldsymbol{x}^{(0)}$, it is necessary and sufficient that

$$\boldsymbol{Q}^k \to \mathbf{0} \quad \text{as} \quad k \to \infty$$

Or equivalently from Theorem 6.2,

$$\rho(\boldsymbol{Q}) < 1 \tag{6.104}$$

Thus, we see that $\varepsilon^{(k)} \to \mathbf{0}$ as $k \to \infty$, that is, iterative method converges with arbitrary initial guesses $\boldsymbol{x}^{(0)}$, if $\|\boldsymbol{Q}\| < 1$ for any matrix norm.

6.6.1 Rate of Convergence

The rate of convergence of the iterative scheme Equation 6.54 is defined as

$$R \equiv -\log \rho(\boldsymbol{Q})$$

where $\rho(\boldsymbol{Q})$ is the spectral radius of the convergent iteration matrix $\boldsymbol{Q}$.

Theorem 6.4

The number of iteration steps required by a convergent iterative method to reduce the initial error vector $\varepsilon^{(0)}$ by the factor $10^{-m}\,(m > 0)$ is inversely proportional to the rate of convergence R.

Proof:

Without loss of generality, let us assume that the iteration matrix $\boldsymbol{Q}$ possesses a set of n linearly independent eigenvectors $\boldsymbol{v}_1, \boldsymbol{v}_2, \ldots, \boldsymbol{v}_n$. Then, the set of eigenvectors forms a basis for the n-dimensional vector space. The initial error vector $\varepsilon^{(0)}$ can be expressed as

$$\varepsilon^{(0)} = c_1\boldsymbol{v}_1 + c_2\boldsymbol{v}_2 + \ldots + c_n\boldsymbol{v}_n$$

where $c_1, c_2, \ldots, c_n$ are scalars.

Therefore,

$$\begin{aligned} Q\varepsilon^{(0)} &= c_1 Q v_1 + c_2 Q v_2 + \cdots + c_n Q v_n \\ &= c_1\lambda_1 v_1 + c_2\lambda_2 v_2 + \cdots + c_n\lambda_n v_n \\ &= \rho(Q)\left[c_1\frac{\lambda_1}{\rho(Q)}v_1 + c_2\frac{\lambda_2}{\rho(Q)}v_2 + \cdots + c_n\frac{\lambda_n}{\rho(Q)}v_n\right] \end{aligned}$$

It follows that

$$\varepsilon^{(k)} = Q^k\varepsilon^{(0)} = [\rho(Q)]^k\left[c_1\left(\frac{\lambda_1}{\rho(Q)}\right)^k v_1 + c_2\left(\frac{\lambda_2}{\rho(Q)}\right)^k v_2 + \cdots + c_n\left(\frac{\lambda_n}{\rho(Q)}\right)^k v_n\right]$$

Now, in the kth step of iteration, we see that the amplitude of the error is required to be reduced at least by a factor of 10^{-m}, $m > 0$, if

$$[\rho(Q)]^k \le 10^{-m}$$

Since, $0 \le \rho(Q) < 1$, it follows that

$$k \ge -\frac{m}{\log_{10}\rho(Q)} = \frac{m}{R}$$

Hence, the number of iterations required to reduce the initial error by the factor 10^{-m} is inversely proportional to the rate of convergence R. ■

6.7 INVERSION OF A MATRIX BY THE GAUSSIAN METHOD

Usually, the Gauss–Jordan method is used to compute the inverse of a matrix. We start with the augmented matrix A and the identity matrix I of the same order, viz., $[A \mid I]$. When the Gauss–Jordan procedure is completed, we obtain $[I \mid A^{-1}]$. Therefore, using the Gauss–Jordan method, we have

$$[A \mid I] \xrightarrow{\text{Gauss–Jordan}} [I \mid A^{-1}]$$

MATHEMATICA® Program Implementing the Gauss–Jordan Method for Finding Inverse of a Matrix (Chapter 6, Example 6.13)

```
ROWINDEX[M_, k_, n_]:=Module[{s, max, index},
   max=M[[k, k]];
   index=k;
   For[s=k+1,s<=n, s++,
    If[max<M[[k, s]], max=M[[k, s]];index=s]];
   Return [index]];

n=3;
A={{4,-1,-5,1,0,0},{15,1,-5,0,1,0},{5,4,9,0,0,1}};
For[i=1,i<=n, i++,
  For[j=1,j<=2*n, j++,
   a[i, j]=A[[i, j]]]];
Print[
  "..................................................................
   ..........."];
```

```
For[k=1,k<=n, k++,
  Print["Step ",k,":"];
  Print[
   "..................................................................
     ..........."];
  i0=ROWINDEX[A, k,n];
  Print["i0=",i0];
  If[i0>k,
   For[j=1,j<=2*n, j++,
    temp1=a[k, j];
    a[k, j]=a[i0,j];
    a[i0,j]=temp1]];

  For[i=1,i<=n, i++,
   For[j=1,j<=2*n, j++,
    temp[i, j]=a[i, j]]];

  For[i=1,i<=n, i++,
   If[i!=k, For[j=k, j<=2*n, j++,
     a[k, j]=temp[k, j]/temp[k, k];
    a[i, j]=temp[i, j]-temp[i, k]*a[k, j];
    Print["a[",i,",",j,"]=",a[i, j]]]]];
  Print[
   "..................................................................
     ..........."]];

AA=Table[a[i, j],{i,1,n},{j, n+1,2*n}];
MatrixForm[AA]
```

Output:

```
..........................................................................
Step 1:
..........................................................................
i0 = 1
a[2,1]=0
a[2,2]=19/4
a[2,3]=55/ 4
a[2,4]=-(15/4)
a[2,5]=1
a[2,6]=0
a[3,1]=0
a[3,2]=21/4
a[3,3]=61/4
a[3,4]=-(5/4)
a[3,5]=0
a[3,6]=1
..........................................................................
Step 2:
..........................................................................
i0 = 2
a[1,2]=0
a[1,3]=-(10/19)
a[1,4]=1/19
a[1,5]=1/19
a[1,6]=0
a[3,2]=0
a[3,3]=1/19
```

```
a[3,4]=55/19
a[3,5]=-(21/19)
a[3,6]=1
.................................................................
Step 3:
.................................................................
i0 = 3
a[1,3]=0
a[1,4]=29
a[1,5]=-11
a[1,6]=10
a[2,3]=0
a[2,4]=-160
a[2,5]=61
a[2,6]=-55
.................................................................
```

$$\begin{pmatrix} 29 & -11 & 10 \\ -160 & 61 & -55 \\ 55 & -21 & 19 \end{pmatrix}$$

Example 6.13

Using the Gauss–Jordan method, find the inverse of the following matrix:

$$\begin{bmatrix} 4 & -1 & -5 \\ 15 & 1 & -5 \\ 5 & 4 & 9 \end{bmatrix}$$

Solution:

Let

$$\boldsymbol{A} = \begin{bmatrix} 4 & -1 & -5 \\ 15 & 1 & -5 \\ 5 & 4 & 9 \end{bmatrix}$$

We start with the following augmented matrix:

$$\left[\begin{array}{ccc|ccc} 4 & -1 & -5 & 1 & 0 & 0 \\ 15 & 1 & -5 & 0 & 1 & 0 \\ 5 & 4 & 9 & 0 & 0 & 1 \end{array}\right]$$

Step 1:

$$\left[\begin{array}{ccc|ccc} 4 & -1 & -5 & 1 & 0 & 0 \\ 15 & 1 & -5 & 0 & 1 & 0 \\ 5 & 4 & 9 & 0 & 0 & 1 \end{array}\right] \xrightarrow{\frac{1}{4}R_1} \left[\begin{array}{ccc|ccc} 1 & -\frac{1}{4} & -\frac{5}{4} & \frac{1}{4} & 0 & 0 \\ 15 & 1 & -5 & 0 & 1 & 0 \\ 5 & 4 & 9 & 0 & 0 & 1 \end{array}\right]$$

$$\xrightarrow[R_3' \leftarrow R_3 - 5R_1]{R_2' \leftarrow R_2 - 15R_1} \left[\begin{array}{ccc|ccc} 1 & -\frac{1}{4} & -\frac{5}{4} & \frac{1}{4} & 0 & 0 \\ 0 & \frac{19}{4} & \frac{55}{4} & -\frac{15}{4} & 1 & 0 \\ 0 & \frac{21}{4} & \frac{61}{4} & -\frac{5}{4} & 0 & 1 \end{array}\right]$$

Step 2:

$$\left[\begin{array}{ccc|ccc} 1 & -\frac{1}{4} & -\frac{5}{4} & \frac{1}{4} & 0 & 0 \\ 0 & \frac{19}{4} & \frac{55}{4} & -\frac{15}{4} & 1 & 0 \\ 0 & \frac{21}{4} & \frac{61}{4} & -\frac{5}{4} & 0 & 1 \end{array}\right] \xrightarrow{\frac{4}{19}R_2} \left[\begin{array}{ccc|ccc} 1 & -\frac{1}{4} & -\frac{5}{4} & \frac{1}{4} & 0 & 0 \\ 0 & 1 & \frac{55}{19} & -\frac{15}{19} & \frac{4}{19} & 0 \\ 0 & \frac{21}{4} & \frac{61}{4} & -\frac{5}{4} & 0 & 1 \end{array}\right]$$

$$\xrightarrow[R_3' \leftarrow R_3 - \frac{21}{4}R_2]{R_1' \leftarrow R_1 + \frac{1}{4}R_2} \left[\begin{array}{ccc|ccc} 1 & 0 & -\frac{10}{19} & \frac{1}{19} & \frac{1}{19} & 0 \\ 0 & 1 & \frac{55}{19} & -\frac{15}{19} & \frac{4}{19} & 0 \\ 0 & 0 & \frac{1}{19} & \frac{55}{19} & -\frac{21}{19} & 1 \end{array}\right]$$

Step 3:

$$\left[\begin{array}{ccc|ccc} 1 & 0 & -\frac{10}{19} & \frac{1}{19} & \frac{1}{19} & 0 \\ 0 & 1 & \frac{55}{19} & -\frac{15}{19} & \frac{4}{19} & 0 \\ 0 & 0 & \frac{1}{19} & \frac{55}{19} & -\frac{21}{19} & 1 \end{array}\right] \xrightarrow{19R_3} \left[\begin{array}{ccc|ccc} 1 & 0 & -\frac{10}{19} & \frac{1}{19} & \frac{1}{19} & 0 \\ 0 & 1 & \frac{55}{19} & -\frac{15}{19} & \frac{4}{19} & 0 \\ 0 & 0 & 1 & 55 & -21 & 19 \end{array}\right]$$

$$\xrightarrow[R_2' \leftarrow R_2 - \frac{55}{19}R_3]{R_1' \leftarrow R_1 + \frac{10}{19}R_3} \left[\begin{array}{ccc|ccc} 1 & 0 & 0 & 29 & -11 & 10 \\ 0 & 1 & 0 & -160 & 61 & -55 \\ 0 & 0 & 1 & 55 & -21 & 19 \end{array}\right]$$

Hence, the required inverse of the given matrix is

$$\begin{bmatrix} 29 & -11 & 10 \\ -160 & 61 & -55 \\ 55 & -21 & 19 \end{bmatrix}$$

Example 6.14

Find the inverse of the following matrix:

$$\begin{pmatrix} 1 & 1 & 1 \\ 4 & 3 & -1 \\ 3 & 5 & 3 \end{pmatrix}$$

using the Gauss elimination method.

Solution:

Let $\boldsymbol{X} = \begin{pmatrix} x_{11} & x_{12} & x_{13} \\ x_{21} & x_{22} & x_{23} \\ x_{31} & x_{32} & x_{33} \end{pmatrix}$ be the inverse of $\boldsymbol{A} = \begin{pmatrix} 1 & 1 & 1 \\ 4 & 3 & -1 \\ 3 & 5 & 3 \end{pmatrix}$.

So, $\boldsymbol{AX} = \boldsymbol{I}_3$ where $\boldsymbol{I}_3$ is the 3×3 identity matrix.
The augmented system can be written as

$$\bar{\boldsymbol{A}} = \left(\begin{array}{ccc|ccc} 1 & 1 & 1 & 1 & 0 & 0 \\ 4 & 3 & -1 & 0 & 1 & 0 \\ 3 & 5 & 3 & 0 & 0 & 1 \end{array}\right)$$

Applying the Gauss elimination method, we obtain

$$\left(\begin{array}{ccc|ccc} 1 & 1 & 1 & 1 & 0 & 0 \\ 4 & 3 & -1 & 0 & 1 & 0 \\ 3 & 5 & 3 & 0 & 0 & 1 \end{array}\right) \xrightarrow[R_3 \leftarrow R_3 - 3R_1]{R_2 \leftarrow R_2 - 4R_1} \left(\begin{array}{ccc|ccc} 1 & 1 & 1 & 1 & 0 & 0 \\ 0 & -1 & -5 & -4 & 1 & 0 \\ 0 & 2 & 0 & -3 & 0 & 1 \end{array}\right)$$

$$\xrightarrow{R_3 \leftarrow R_3 + 2R_2} \left(\begin{array}{ccc|ccc} 1 & 1 & 1 & 1 & 0 & 0 \\ 0 & -1 & -5 & -4 & 1 & 0 \\ 0 & 0 & -10 & -11 & 2 & 1 \end{array}\right)$$

The equation $\boldsymbol{AX} = \boldsymbol{I}_3$ is equivalent to the following three systems:

$$\begin{pmatrix} 1 & 1 & 1 \\ 0 & -1 & -5 \\ 0 & 0 & -10 \end{pmatrix} \begin{pmatrix} x_{11} \\ x_{21} \\ x_{31} \end{pmatrix} = \begin{pmatrix} 1 \\ -4 \\ -11 \end{pmatrix} \tag{6.105}$$

$$\begin{pmatrix} 1 & 1 & 1 \\ 0 & -1 & -5 \\ 0 & 0 & -10 \end{pmatrix} \begin{pmatrix} x_{12} \\ x_{22} \\ x_{32} \end{pmatrix} = \begin{pmatrix} 0 \\ 1 \\ 2 \end{pmatrix} \tag{6.106}$$

$$\begin{pmatrix} 1 & 1 & 1 \\ 0 & -1 & -5 \\ 0 & 0 & -10 \end{pmatrix} \begin{pmatrix} x_{13} \\ x_{23} \\ x_{33} \end{pmatrix} = \begin{pmatrix} 0 \\ 0 \\ 1 \end{pmatrix} \tag{6.107}$$

From Equation 6.105, we get

$$x_{11} + x_{21} + x_{31} = 1$$
$$-x_{21} - 5x_{31} = -4$$
$$-10x_{31} = -11$$

By the back substitution method, we get

$$x_{31} = \frac{11}{10}; \quad x_{21} = \frac{-3}{2}; \quad x_{11} = \frac{7}{5} \tag{6.108}$$

From Equation 6.106, we get

$$x_{12} + x_{22} + x_{32} = 0$$
$$-x_{22} - 5x_{32} = 1$$
$$-10x_{32} = 2$$

By the back substitution method, we get

$$x_{32} = \frac{-1}{5}; \quad x_{22} = 0; \quad x_{12} = \frac{1}{5} \tag{6.109}$$

From Equation 6.107, we get

$$x_{13} + x_{23} + x_{33} = 0$$

$$-x_{23} - 5x_{33} = 0$$

$$-10x_{33} = 1$$

By the back substitution method, we get

$$x_{33} = \frac{-1}{10}; \quad x_{23} = \frac{1}{2}; \quad x_{13} = \frac{-2}{5} \tag{6.110}$$

From Equations 6.108 through 6.110, we get the inverse of $\boldsymbol{A}$ as

$$\boldsymbol{X} = \begin{pmatrix} \frac{7}{5} & \frac{1}{5} & \frac{-2}{5} \\ \frac{-3}{2} & 0 & \frac{1}{2} \\ \frac{11}{10} & \frac{-1}{5} & \frac{-1}{10} \end{pmatrix}$$

6.8 ILL-CONDITIONED SYSTEMS

A computational problem is called ill-conditioned or ill-posed if small changes or perturbations in data or input cause large changes or perturbations in the solution or the output. On the other hand, a problem is called well-conditioned or well-posed if small changes or perturbations in data or input cause only small changes or perturbations in the solution or the output.

Theorem 6.5

A linear system of equations $\boldsymbol{Ax} = \boldsymbol{b}$ and its coefficient matrix $\boldsymbol{A}$ whose condition number $\kappa(\boldsymbol{A})$ is small are well-conditioned. A large condition number indicates ill-conditioning.

Proof:

We have $\boldsymbol{Ax} = \boldsymbol{b}$, therefore Equation 6.11 yields

$$\|\boldsymbol{b}\| \leq \|\boldsymbol{A}\| \, \|\boldsymbol{x}\| \tag{6.111}$$

Dividing both sides of Equation 6.111 by $\|\boldsymbol{b}\| \, \|\boldsymbol{x}\|$, we get

$$\frac{1}{\|\boldsymbol{x}\|} \leq \frac{\|\boldsymbol{A}\|}{\|\boldsymbol{b}\|} \tag{6.112}$$

The residual $\boldsymbol{r}$ for an approximate solution $\tilde{\boldsymbol{x}}$ of $\boldsymbol{Ax} = \boldsymbol{b}$ is defined as

$$\begin{aligned} \boldsymbol{r} &= \boldsymbol{b} - \boldsymbol{A}\tilde{\boldsymbol{x}} \\ &= \boldsymbol{Ax} - \boldsymbol{A}\tilde{\boldsymbol{x}} \\ &= \boldsymbol{A}(\boldsymbol{x} - \tilde{\boldsymbol{x}}) \end{aligned} \tag{6.113}$$

Multiplying both sides of Equation 6.113 by A^{-1} from the left, we get

$$A^{-1}r = A^{-1}A(x - \tilde{x}) = (x - \tilde{x}) \tag{6.114}$$

Therefore, according Equation 6.11, we obtain

$$\|x - \tilde{x}\| = \|A^{-1}r\| \le \|A^{-1}\|\,\|r\| \tag{6.115}$$

Dividing both sides of Equation 6.115 by $\|x\|\left(\|x\| \neq 0\right)$, we get

$$\frac{\|x - \tilde{x}\|}{\|x\|} = \frac{\|A^{-1}r\|}{\|x\|} \le \frac{\|A^{-1}\|\,\|r\|}{\|x\|} \tag{6.116}$$

Using Equation 6.112, it follows that

$$\frac{\|x - \tilde{x}\|}{\|x\|} \le \frac{\|A\|\|A^{-1}\|\|r\|}{\|b\|} = \frac{\kappa(A)\|r\|}{\|b\|} \tag{6.117}$$

This shows that if $\|r\|/\|b\|$ be small, a large $\kappa(A)$ does not imply a small relative error. Hence, the system is ill-conditioned for large $\kappa(A)$. ■

Corollary: The condition number $\kappa(A)$ will vary with the norm being used, but it is always bounded below by one, since

$$1 \le \|I\| = \|AA^{-1}\| \le \|A\|\,\|A^{-1}\| = \kappa(A).$$

If the condition number $\kappa(A)$ is nearly 1, then we see from Equation 6.117 that small relative perturbations in b will lead to similar small relative perturbations in the solution x. But if $\kappa(A)$ is large, then from Equation 6.117 it may be observed that there may be small relative perturbations of b, though it will lead to large relative error for the solution of x. If $\kappa(A)$ is small, then a small relative perturbation of b implies a small relative error in the solution x, so that the system is well-conditioned.

- Estimate of relative error for perturbations in coefficient matrix and constant vector. Let x be exact solution of the given system of equations

$$Ax = b \tag{6.118}$$

Let us suppose that the data A and b in Equation 6.117 are perturbed by the quantities δA and δb, respectively. If the perturbation in the solution x of Equation 6.117 be δx, then we have

$$(A + \delta A)(x + \delta x) = b + \delta b \tag{6.119}$$

Theorem 6.6

Let A be nonsingular coefficient matrix of the system $Ax = b$. Also, let δA and δb be perturbations of A and b, respectively, and the perturbation δA be so small that

$$\|\delta A\| < \frac{1}{\|A^{-1}\|} \tag{6.120}$$

If $\boldsymbol{x}$ and $\delta\boldsymbol{x}$ satisfy Equations 6.118 and 6.119, then

$$\frac{\|\delta\boldsymbol{x}\|}{\|\boldsymbol{x}\|} \le \frac{\kappa(\boldsymbol{A})}{1-\|\boldsymbol{A}^{-1}\|\|\delta\boldsymbol{A}\|}\left(\frac{\|\delta\boldsymbol{b}\|}{\|\boldsymbol{b}\|}+\frac{\|\delta\boldsymbol{A}\|}{\|\boldsymbol{A}\|}\right) \tag{6.121}$$

where $\kappa(\boldsymbol{A}) \equiv \|\boldsymbol{A}^{-1}\|\|\boldsymbol{A}\|$ is the condition number.

Proof:
Using Equation 3.120, we get

$$\|\boldsymbol{A}^{-1}\delta\boldsymbol{A}\| \le \|\boldsymbol{A}^{-1}\|\;\|\delta\boldsymbol{A}\| < 1 \tag{6.122}$$

Now, using Equation 6.26, we have

$$\left\|\left(\boldsymbol{I}+\boldsymbol{A}^{-1}\delta\boldsymbol{A}\right)^{-1}\right\| \le \frac{1}{1-\|\boldsymbol{A}^{-1}\delta\boldsymbol{A}\|} \le \frac{1}{1-\|\boldsymbol{A}^{-1}\|\|\delta\boldsymbol{A}\|}, \tag{6.123}$$

since by Equation 6.122

$$\|\boldsymbol{A}^{-1}\delta\boldsymbol{A}\| \le \|\boldsymbol{A}^{-1}\|\;\|\delta\boldsymbol{A}\|$$

From Equation 6.119, solving for $\delta\boldsymbol{x}$, we obtain

$$\begin{aligned}\delta\boldsymbol{x} &= \left(\boldsymbol{A}+\delta\boldsymbol{A}\right)^{-1}\left(\delta\boldsymbol{b}-\delta\boldsymbol{A}\boldsymbol{x}\right)\\ &= \left(\boldsymbol{I}+\boldsymbol{A}^{-1}\delta\boldsymbol{A}\right)^{-1}\boldsymbol{A}^{-1}\left(\delta\boldsymbol{b}-\delta\boldsymbol{A}\boldsymbol{x}\right)\end{aligned}$$

Therefore,

$$\|\delta\boldsymbol{x}\| = \left\|\left(\boldsymbol{I}+\boldsymbol{A}^{-1}\delta\boldsymbol{A}\right)^{-1}\boldsymbol{A}^{-1}\left(\delta\boldsymbol{b}-\delta\boldsymbol{A}\boldsymbol{x}\right)\right\| \le \left\|\left(\boldsymbol{I}+\boldsymbol{A}^{-1}\delta\boldsymbol{A}\right)^{-1}\right\|\;\left\|\boldsymbol{A}^{-1}\left(\delta\boldsymbol{b}-\delta\boldsymbol{A}\boldsymbol{x}\right)\right\|$$

So, using Equation 6.123, we have

$$\begin{aligned}\|\delta\boldsymbol{x}\| = \left\|\left(\boldsymbol{I}+\boldsymbol{A}^{-1}\delta\boldsymbol{A}\right)^{-1}\boldsymbol{A}^{-1}\left(\delta\boldsymbol{b}-\delta\boldsymbol{A}\boldsymbol{x}\right)\right\| &\le \frac{1}{1-\|\boldsymbol{A}^{-1}\|\|\delta\boldsymbol{A}\|}\left\|\boldsymbol{A}^{-1}\left(\delta\boldsymbol{b}-\delta\boldsymbol{A}\boldsymbol{x}\right)\right\|\\ &\le \frac{1}{1-\|\boldsymbol{A}^{-1}\|\|\delta\boldsymbol{A}\|}\|\boldsymbol{A}^{-1}\|\|\delta\boldsymbol{b}-\delta\boldsymbol{A}\boldsymbol{x}\|\end{aligned} \tag{6.124}$$

Now,

$$\|\delta\boldsymbol{b}-\delta\boldsymbol{A}\boldsymbol{x}\| \le \|\delta\boldsymbol{b}\| + \|\delta\boldsymbol{A}\boldsymbol{x}\| \le \|\delta\boldsymbol{b}\| + \|\delta\boldsymbol{A}\|\;\|\boldsymbol{x}\| \tag{6.125}$$

Thus from Equation 6.124, we get

$$\|\delta\boldsymbol{x}\| \le \frac{1}{1-\|\boldsymbol{A}^{-1}\|\|\delta\boldsymbol{A}\|}\|\boldsymbol{A}^{-1}\|\left(\|\delta\boldsymbol{b}\| + \|\delta\boldsymbol{A}\|\;\|\boldsymbol{x}\|\right) \tag{6.126}$$

Dividing both sides of Equation 6.126 by $\|x\|$, we obtain

$$\frac{\|\delta x\|}{\|x\|} \le \frac{1}{1-\|A^{-1}\|\|\delta A\|}\|A^{-1}\|\left(\frac{\|\delta b\|}{\|x\|}+\|\delta A\|\right) \tag{6.127}$$

Again by Equation 6.112, we have from Equation 6.127

$$\frac{\|\delta x\|}{\|x\|} \le \frac{1}{1-\|A^{-1}\|\|\delta A\|}\|A^{-1}\|\,\|A\|\left(\frac{\|\delta b\|}{\|b\|}+\frac{\|\delta A\|}{\|A\|}\right) \tag{6.128}$$

Hence,

$$\frac{\|\delta x\|}{\|x\|} \le \frac{\kappa(A)}{1-\|A^{-1}\|\|\delta A\|}\left(\frac{\|\delta b\|}{\|b\|}+\frac{\|\delta A\|}{\|A\|}\right)$$

■

Corollary:

If there is no error in b then $\|\delta b\| = 0$. Then, Equation 6.121 becomes

$$\frac{\|\delta x\|}{\|x\|} \le \frac{\kappa(A)}{1-\|A^{-1}\|\,\|\delta A\|}\frac{\|\delta A\|}{\|A\|}$$

On the other hand, if there is no error in A, then $\|\delta A\| = 0$. Then, Equation 6.121 becomes

$$\frac{\|\delta x\|}{\|x\|} \le \kappa(A)\frac{\|\delta b\|}{\|b\|}$$

So, if $\kappa(A)$ is small, then small perturbation in A or b leads to only small change in the solution x. If $\kappa(A)$ is large, then small perturbation in A or b leads to large change in the solution x, and hence, the system of equations (6.118) is ill-conditioned. If the condition number $\kappa(A)$ is nearly 1, then the system of equations (6.118) is well-conditioned.

Example 6.15

Find the condition number of the following matrix:

$$A = \begin{bmatrix} 4 & -1 & 0 \\ -1 & 4 & -1 \\ 0 & -1 & 4 \end{bmatrix}$$

If $b = [17.4\ \ 23.6\ \ 30.8]^T$, then check whether the system of equations $Ax = b$ is ill-conditioned or not?

Solution:

The inverse of A is

$$A^{-1} = \frac{1}{56}\begin{bmatrix} 15 & 4 & 1 \\ 4 & 16 & 4 \\ 1 & 4 & 15 \end{bmatrix}$$

$$\|A\|_1 = \max\{5, 6, 5\} = 6$$

$$\|A\|_\infty = 6, \quad \text{since } A \text{ is symmetric}$$

Similarly,

$$\|A^{-1}\|_1 = \max\left\{\frac{20}{56}, \frac{24}{56}, \frac{20}{56}\right\} = \frac{3}{7}$$

$$\|A^{-1}\|_\infty = \frac{3}{7}, \quad \text{since } A \text{ is symmetric}$$

Therefore,

$$\kappa(A) = \|A\|_1 \|A^{-1}\|_1 = \|A\|_\infty \|A^{-1}\|_\infty = 2.57143$$

Hence, a linear system of equations $\boldsymbol{Ax} = \boldsymbol{b}$ with this coefficient matrix A is well-conditioned.

For instance, if $\boldsymbol{b} = [17.4 \;\; 23.6 \;\; 30.8]^T$, then the system of equations $\boldsymbol{Ax} = \boldsymbol{b}$ has the following solution:

$$x_1 = 6.89643, \quad x_2 = 10.1857, \quad x_3 = 10.2464$$

Now, if there is a small perturbation in $\boldsymbol{A}$ yielding

$$\tilde{A} = \begin{bmatrix} 4 & -1 & 0 \\ -1.01 & 4 & -1 \\ 0 & -1 & 4 \end{bmatrix},$$

then the system of equations $\tilde{\boldsymbol{A}}\boldsymbol{x} = \boldsymbol{b}$ has the following solution:

$$\tilde{x}_1 = 6.90136, \quad \tilde{x}_2 = 10.2054, \quad \tilde{x}_3 = 10.2514$$

The relative error in the solution for x is given by

$$\frac{\|\boldsymbol{x} - \tilde{\boldsymbol{x}}\|}{\|\boldsymbol{x}\|} = \frac{0.0197}{10.2464} \approx 0.00192263$$

which is obviously desirably small, since the coefficient matrix is well-conditioned and more precisely, the system of equations is well-conditioned.

Therefore, a small perturbation in $\boldsymbol{A}$ causes also small perturbation in the solution $\boldsymbol{x}$.

Example 6.16

Compute the condition number of the coefficient matrix of the following system of equations and hence check the system for ill-condition:

$$\begin{bmatrix} 4.5 & 3.55 \\ 3.55 & 2.8 \end{bmatrix} \begin{bmatrix} x_1 \\ x_2 \end{bmatrix} = \begin{bmatrix} 5.2 \\ 4.1 \end{bmatrix}$$

Solution:

Let the coefficient matrix

$$A = \begin{bmatrix} 4.5 & 3.55 \\ 3.55 & 2.8 \end{bmatrix}$$

The inverse A is

$$A^{-1} = \begin{bmatrix} -1120 & 1420 \\ 1420 & -1800 \end{bmatrix}$$

$$\|A\|_1 = \max\{8.05, 6.35\} = 8.05$$

$$\|A\|_\infty = 8.05, \text{ since } A \text{ is symmetric}$$

Similarly,

$$\|A^{-1}\|_1 = \max\{2540, 3220\} = 3220$$

$$\|A^{-1}\|_\infty = 3220, \text{ since } A \text{ is symmetric}$$

Therefore,

$$\kappa(A) = \|A\|_1 \|A^{-1}\|_1 = \|A\|_\infty \|A^{-1}\|_\infty = 8.05 \times 3220 = 25921$$

Hence, the linear system of equations $Ax = b$ with this coefficient matrix A is ill-conditioned.

For instance, the given system of equations $Ax = b$ has the following solution

$$x_1 = -2, \quad x_2 = 4$$

Now, if there is a small perturbation in A yielding

$$\tilde{A} = \begin{bmatrix} 4.501 & 3.55 \\ 3.55 & 2.8 \end{bmatrix}$$

then the system of equations $\tilde{A}x = b$ has the following solution

$$\tilde{x}_1 = 16.6667, \quad \tilde{x}_2 = -19.6667$$

That means a small perturbation in A causes very large change in the solution x. Thus, the system is ill-conditioned.

On the other hand, if there is a small perturbation in b yielding

$$\tilde{b} = \begin{bmatrix} 5.2 \\ 4.0 \end{bmatrix}$$

then the system of equations $\boldsymbol{Ax} = \tilde{\boldsymbol{b}}$ has the following solution

$$\tilde{x}_1 = -144, \quad \tilde{x}_2 = 184$$

Therefore, a small perturbation in $\boldsymbol{b}$ causes very large change in the solution $\boldsymbol{x}$, since the system is ill-conditioned.

6.9 THOMAS ALGORITHM

We present here an algorithm known as the Thomas algorithm for the solution of linear system of equations. It is a special case of Gaussian elimination or $\boldsymbol{LU}$ decomposition.

Suppose, we wish to solve the system of linear equations

$$\boldsymbol{Mx} = \boldsymbol{F} \tag{6.129}$$

where the tridiagonal matrix

$$\boldsymbol{M} = \begin{pmatrix} B_1 & C_1 & 0 & 0 & & \cdots & & 0 \\ A_2 & B_2 & C_2 & 0 & & \cdots & & 0 \\ \vdots & \vdots & \vdots & \vdots & \vdots & \cdots & \vdots & \vdots \\ 0 & & \cdots & & 0 & A_{N-1} & B_{N-1} & C_{N-1} \\ 0 & 0 & \cdots & \cdots & \cdots & 0 & A_N & B_N \end{pmatrix}$$

$$\boldsymbol{x} = (x_1, x_2, \ldots, x_N)^T \quad \text{and} \quad \boldsymbol{F} = (F_1, F_2, \ldots, F_N)^T$$

Now, the coefficient matrix $\boldsymbol{M}$ can be transformed into a tridiagonal matrix. Thus, without loss of generality, we shall suppose that $\boldsymbol{M}$ is tridiagonal.

Then, we wish to express $\boldsymbol{M}$ as a product $\boldsymbol{LU}$ of a lower triangular matrix

$$\boldsymbol{L} = \begin{bmatrix} 1 & 0 & 0 & 0 & 0 & \cdots & 0 \\ l_2 & 1 & 0 & 0 & 0 & \cdots & 0 \\ 0 & l_3 & 1 & 0 & 0 & \cdots & 0 \\ \vdots & \vdots & \vdots & \vdots & \vdots & \cdots & \vdots \\ 0 & 0 & 0 & \cdots & l_{N-1} & 1 & 0 \\ 0 & 0 & 0 & \cdots & 0 & l_N & 1 \end{bmatrix}$$

and an upper triangular matrix

$$\boldsymbol{U} = \begin{bmatrix} u_1 & v_1 & 0 & 0 & \cdots & 0 \\ 0 & u_2 & v_2 & 0 & \cdots & 0 \\ 0 & 0 & u_3 & v_3 & \cdots & 0 \\ \vdots & \vdots & \vdots & \vdots & \cdots & \vdots \\ 0 & 0 & 0 & \cdots & u_{N-1} & v_{N-1} \\ 0 & 0 & 0 & \cdots & 0 & u_N \end{bmatrix}$$

On multiplying $\boldsymbol{L}$ and $\boldsymbol{U}$ and equating the resulting matrix with $\boldsymbol{M}$, we find that

$$\begin{aligned} &u_1 = B_1, v_1 = C_1 \\ &l_i u_{i-1} = A_i, l_i v_{i-1} + u_i = B_i, v_i = C_i, \quad i = 2,3,\ldots,N-1 \\ &l_N u_{N-1} = A_N, l_N v_{N-1} + u_N = B_N \end{aligned} \tag{6.130}$$

Therefore,

$$\begin{aligned} l_i &= \frac{A_i}{u_{i-1}}, \quad i = 2,3,\ldots,N \\ u_1 = B_1, u_i &= B_i - \frac{A_i C_{i-1}}{u_{i-1}}, \quad i = 2,3,\ldots,N \end{aligned} \tag{6.131}$$

The system of linear equations $\boldsymbol{Mx} = \boldsymbol{F}$ can be written in the following equivalent form

$$\boldsymbol{Lz} = \boldsymbol{F} \tag{6.132}$$

$$\boldsymbol{Ux} = \boldsymbol{z}$$

Thus to obtain the solution of Equation 6.129, we solve two linear systems in succession, at first the lower triangular system $\boldsymbol{Lz} = \boldsymbol{F}$ is solved for $\boldsymbol{z}$, followed by solving the upper triangular system $\boldsymbol{Ux} = \boldsymbol{z}$ for $\boldsymbol{x}$.

Therefore, the lower triangular system $\boldsymbol{Lz} = \boldsymbol{F}$ yields

$$z_1 = F_1,$$

$$z_i = F_i - l_i z_{i-1}, \quad i = 2,3,\ldots,N$$

Next, solving the upper triangular system $\boldsymbol{Ux} = \boldsymbol{z}$, we get

$$x_N = \frac{z_N}{u_N}$$

$$x_j = \frac{z_j - v_j x_{j+1}}{u_j}, \quad j = N-1, N-2,\ldots,1$$

Expressing in the above formulae, the values of u_j and v_j in terms of A_j, B_j, C_j, we obtain

$$x_N = \frac{z_N}{u_N}$$

$$x_j = \alpha_{j+1} x_{j+1} + \beta_{j+1}, \quad j = N-1, N-2,\ldots,1 \tag{6.133}$$

where

$$\alpha_{j+1} = -\frac{v_j}{u_j} = -\frac{C_j}{B_j + \alpha_j A_j}, \quad j = 2,3,\ldots,N; \quad \alpha_2 = -\frac{v_1}{u_1} = -\frac{C_1}{B_1}$$

$$\beta_{j+1} = \frac{z_j}{u_j} = \frac{F_j - \beta_j A_j}{B_j + \alpha_j A_j}, \quad j = 2,3,\ldots,N; \quad \beta_2 = \frac{z_1}{u_1} = \frac{F_1}{B_1}$$

The above formulae in Equation 6.133 are usually referred as the Thomas algorithm. This procedure is widely used in solving tridiagonal systems arising in numerical solution of initial/boundary value problems for ordinary and partial differential equations by finite difference or finite element methods.

The tridiagonal systems belong to the category of sparse systems, in view of the relatively large number of zero elements.

6.9.1 Operational Count for Thomas Algorithm

A small number of arithmetic operations are required for the Thomas algorithm. The ***LU*** decomposition of a tridiagonal matrix requires approximately $3N$ operations. The forward and back substitutions together needs approximately $5N$ operations. Thus, the whole solution procedure requires approximately $8N$ operations.

Therefore, for large N, the total operational count is approximately $O(N)$, which is less compared to the Gauss elimination or the Gauss–Jordan method of $O(N^3)$.

6.9.2 Algorithm

Solve $N \times N$ tridiagonal system $\boldsymbol{Mx} = \boldsymbol{F}$ by Thomas algorithm. $B_i, i = 1,\ldots,N$, are the principal diagonal elements of $\boldsymbol{M}$, subdiagonal elements are $A_i, i = 2,\ldots,N$, and superdiagonal elements are $C_i, i = 1,\ldots,N-1$.

Step 1: $\alpha_2 = -C_1 / B_1$;
Step 2: $\beta_2 = F_1 / B_1$;
Step 3: for $j = 2(1)N$ do

$$\alpha_{j+1} = -\frac{C_j}{(B_j + \alpha_j A_j)};$$

$$\beta_{j+1} = \frac{(F_j - A_j\beta_j)}{(B_j + A_j\alpha_j)};$$

end.

$x_N = \beta_{N+1}$;

Step 4: for $j = N-1(1)1$ do

$$x_j = \alpha_{j+1}x_{j+1} + \beta_{j+1};$$

end

Step 5: Print x_j, $j = 1,2,\ldots,N$;
Step 6: Stop. ■

***MATHEMATICA® Program for Thomas Algorithm in Solving Tridiagonal System of Equations $Mx = b$, Where**

$$M = \begin{pmatrix} 4 & -1 & 0 \\ -1 & 4 & -1 \\ 0 & -1 & 4 \end{pmatrix} \text{ and } b = \begin{pmatrix} 1 \\ 2 \\ 1 \end{pmatrix}*$$

```
n=3;
M={{4,-1,0},{-1,4,-1},{0,-1,4}};
b={{1},{2},{1}};
```

```
For[i=1,i<=n,i++,
  B[i]=M[[i,i]];
 ];

For[i=1,i<=n-1,i++,
  C1[i]=M[[i,i+1]];
 ];

For[i=2,i<=n,i++,
  A[i]=M[[i,i-1]];
 ];

α[2]=-C1[1]/B[1];
β[2]=b[[1]][[1]]/B[1];

For[j=2,j<=n,j++,
  α[j+1]=-C1[j]/(B[j]+α[j]*A[j]);
  β[j+1]=(b[[j]][[1]]-β[j]*A[j])/(B[j]+α[j]*A[j]);
   ];
x[n]=β[n+1];

For[r=n-1,r>=1,r--,
  x[r]=α[r+1]*x[r+1]+β[r+1]];

For[i=1,i<=n,i++,
  Print["x[",i,"]=",N[x[i]]]];
```

Output:

```
x[ 1 ]= 0.428571
x[ 2 ]= 0.714286
x[ 3 ]= 0.428571
```

EXERCISES

1. Solve the following system of equations by the Gauss elimination method:
   a. $x+y+z=6;\quad 3x+3y+4z=20;\quad 2x+y+3z=13$
   b. $10x-2y+3z=23;\quad 2x+10y-5z=-33;\quad 3x-4y+10z=41$
   c. $3x+4y+5z=18;\quad 2x-y+8z=13;\quad 5x-2y+7z=20$
   d. $3.85x-1.96y+3.15z=12.95, -2.89x+5.12y+2.13z=-8.61, 2.15x+3.05y+5.92z=6.88$
   e. $x+y+z=9$

      $2x-3y+4z=13$

      $3x+4y+5z=40$
   f. $x_1+2x_2+x_3-x_4=-2$

      $2x_1+3x_2-x_3+2x_4=7$

      $x_1+x_2+3x_3-2x_4=-6$

      $x_1+x_2+x_3+x_4=-2$
2. Solve the following system of equations using the Gauss–Jordan methods:
   a. $2x-6y+8z=24$

      $5x+4y-3z=2$

      $3x+y+2z=16$

b. $2x + y + 4z = 12$

$8x - 3y + 2z = 20$

$4x + 11y - z = 33$

c. $3.85x - 1.96y + 3.15z = 12.95$

$-2.89x + 5.12y + 2.13z = -8.61$

$2.15x + 3.05y + 5.92z = 6.88$

d. $4x_1 + 3x_2 - x_3 = 6$

$3x_1 + 5x_2 + 3x_3 = 4$

$x_1 + x_2 + x_3 = 1$

3. Solve the following system of equations using the (i) Gauss elimination method and (ii) Gauss–Jordan method:

a. $2x + y + 4z = 12$

$8x - 3y + 2z = 1$

$4x + 11y - z = 33$

b. $x + y - z + w = 2$

$2x + y + z - 3w = 1$

$3x - y - z + w = 2$

$5x + y + 3z - 2w = 7$

c. $2.38x + 1.95y - 3.27z + 1.58w = 2.16$

$3.21x - 0.86y + 2.42z - 3.20w = 13$

$1.44x + 2.95y - 2.14z + 1.86w = 1.42$

$4.17x + 3.62y - 1.68z + 2.26w = 5.21$

d. $1.02x - 0.05y - 0.10z = 0.795$

$-0.11x + 1.03y - 0.05z = 0.849$

$-0.11x - 0.12y + 1.04z = 1.398$

4. Solve the following system of equations by the Gaussian elimination without and with partial pivoting, correct to six decimal places:

a. $10x_1 + 8x_2 + 3x_3 = 21$

$8x_1 + 7x_2 + 5x_3 = 20$

$3x_1 + 5x_2 + 10x_3 = 18$

b. $5x_1 + x_2 - 5x_3 = 2$

$-20x_1 + 3x_2 + 20x_3 = 6$

$5x_1 + 3x_2 + 5x_3 = 3$

5. Solve the following set of simultaneous linear equations using the method of the Gaussian elimination:
   a. $2x+y-3z=11$

      $4x-2y+3z=8$

      $-2x+2y-z=-6$
   b. $x_1+2x_2+3x_3+4x_4=8$

      $2x_1-2x_2-x_3-x_4=-3$

      $x_1-3x_2+4x_3-4x_4=8$

      $2x_1+2x_2-33x_3+4x_4=-2$
   c. $2x_1+x_2+x_3-x_4=10$

      $x_1+5x_2-5x_3+6x_4=25$

      $-7x_1+3x_2-7x_3-5x_4=5$

      $x_1-5x_2+2x_3+7x_4=11$
   d. $3.85x-1.96y+3.15z=12.95$

      $-2.89x+5.12y+2.13z=-8.61$

      $2.15x+3.05y+5.92z=6.88$
   e. $1.2x_1+2.1x_2-1.1x_3=1.8776$

      $-1.1x_1+2.0x_2+3.1x_3=-0.1159$

      $-2.1x_1-2.2x_2+3.7x_3=-4.2882$
6. Using the Gauss–Jordon method, determine the inverse of the following matrices:
   a. $\begin{bmatrix} -1 & 1 & 2 \\ 3 & -1 & 1 \\ -1 & 3 & 4 \end{bmatrix}$
   b. $\begin{bmatrix} 1 & 2 & 0 \\ 3 & -1 & -2 \\ 1 & 0 & -3 \end{bmatrix}$
   c. $\begin{bmatrix} 10 & 3 & 10 \\ 8 & -2 & 9 \\ 8 & 1 & -10 \end{bmatrix}$
   d. $\begin{bmatrix} 1 & 0 & 3 \\ 2 & 1 & -1 \\ 1 & -1 & 1 \end{bmatrix}$
7. Solve by the Gaussian elimination with partial pivoting, correct to four decimal places:
   a. $x_2+8x_3+5x_4=26$

      $x_1+4x_2+13x_3=25$

      $2x_1+8x_2+x_3+x_4=17$

      $6x_1+7x_3+4x_4=18$

b. $0.01x_1 + 5x_2 + 7x_3 = 12$

$3x_1 + 4x_2 - 5x_3 = 6$

$13x_1 - 3x_2 + 2x_3 = 10$

8. Solve the following equations by the Gaussian elimination, without and with pivoting the following systems, correct up to four significant figures:
   a. $0.0002x_1 + 2.303x_2 = 2.31$

   $2x_1 - 15.01x_2 = 4.5$

   b. $0.1x_1 + x_2 = 2.1$

   $2x_1 - 5x_2 = 15$

   c. $0.04x + 0.08y + 4z = 20$

   $0.09x + 3y - 0.15z = 9$

   $4x + 0.24y - 0.08z = 8$

9. Solve the Gauss–Jordan elimination with partial pivotal condensation of the systems, correct to four decimal places:
   a. $8x_2 + 2x_3 = 21$

   $4x_1 + 5x_2 + 9x_3 = 20$

   $7x_1 - x_2 + 5x_3 = 8$

   b. $x_2 + 14x_3 = 18$

   $10x_1 - 8x_2 + x_3 = 4$

   $9x_1 + x_2 + 15x_3 = 25$

10. Calculate the condition numbers of the coefficient matrices of the following systems and test the systems for ill-condition:
   a. $8x_2 + 2x_3 = 21$

   $4x_1 + 5x_2 + 9x_3 = 20$

   $7x_1 - x_2 + 5x_3 = 8$

   b. $x_2 + 14x_3 = 18$

   $10x_1 - 8x_2 + x_3 = 4$

   $9x_1 + x_2 + 15x_3 = 25$

11. Find the inverse of the following matrices by the Gauss elimination method:

   a. $\begin{bmatrix} 4 & 1 & 2 \\ 2 & 3 & -1 \\ 1 & -2 & 2 \end{bmatrix}$

   b. $\begin{bmatrix} 1 & -1 & 1 \\ 1 & -2 & 4 \\ 1 & 2 & 2 \end{bmatrix}$

   c. $\begin{bmatrix} 3 & 7 & 8 & 15 \\ 2 & 5 & 6 & 11 \\ 2 & 6 & 10 & 19 \\ 4 & 11 & 19 & 38 \end{bmatrix}$

12. Find the inverse of the following matrices by the Gauss–Jordan method:

a. $\begin{bmatrix} 2 & -1 & 1 \\ -15 & 6 & -5 \\ 5 & -2 & 2 \end{bmatrix}$

b. $\begin{bmatrix} 50 & 107 & 36 \\ 25 & 54 & 20 \\ 31 & 66 & 21 \end{bmatrix}$

c. $\begin{bmatrix} 4 & -2 & -9 & 5 \\ 11 & 4 & 1 & 3 \\ 5 & 3 & -7 & 10 \\ -4 & -1 & 8 & -6 \end{bmatrix}$

13. Find the inverse of the following matrix by the (i) Gauss elimination method and (ii) Gauss–Jordan method:

a. $\begin{bmatrix} 2 & 1 & 0 \\ 4 & 3 & 1 \\ 1 & 1 & 1 \end{bmatrix}$

b. $\begin{bmatrix} 1 & 2 & 0.5 & 1 \\ 2 & 5 & 0 & -2 \\ 0.5 & 0 & 2.25 & 7.5 \\ 1 & -2 & 7.5 & 27 \end{bmatrix}$

c. $\begin{bmatrix} 13 & 14 & 6 & 4 \\ 8 & -1 & 13 & 9 \\ 6 & 7 & 3 & 2 \\ 9 & 5 & 16 & 11 \end{bmatrix}$

14. Solve the following system of equations by Crout's method:
    a. $2x+y+4z=12;\quad 8x-3y+2z=20;\quad 4x+11y-z=33$
    b. $x+5y+z=21;\quad 2x+y+3z=20;\quad 3x+y+4z=26$
    c. $x+y+z=3;\quad 2x-y+3z=16;\quad 3x+y-z=-3$
    d. $3x+2y+7z=4;\quad 2x+3y+z=5;\quad 3x+4y+z=7$
15. Solve the system of equations by Cholesky's method:

$$\begin{bmatrix} 4 & -1 & 0 \\ -1 & 4 & -1 \\ 0 & -1 & 4 \end{bmatrix}\begin{bmatrix} x_1 \\ x_2 \\ x_3 \end{bmatrix}=\begin{bmatrix} 1 \\ 0 \\ 0 \end{bmatrix}$$

16. Solve the following set of simultaneous linear equations by Cholesky's factorization method:

$$12x_1-6x_2-6x_3+1.5x_4=1$$

$$-6x_1+4x_2+3x_3+0.5x_4=6.5$$

$$-6x_1+3x_2+6x_3+1.5x_4=3$$

$$1.5x_1+0.5x_2+1.5x_3+x_4=4$$

17. Given

$$A = \begin{bmatrix} 5.5 & 0 & 0 & 0 & 0 & 3.5 \\ 0 & 5.5 & 0 & 0 & 0 & 1.5 \\ 0 & 0 & 6.25 & 0 & 3.75 & 0 \\ 0 & 0 & 0 & 5.5 & 0 & 0.5 \\ 0 & 0 & 3.75 & 0 & 6.25 & 0 \\ 3.5 & 1.5 & 0 & 0.5 & 0 & 5.5 \end{bmatrix} \text{ and } \boldsymbol{b} \begin{bmatrix} 1 \\ 1 \\ 1 \\ 1 \\ 1 \\ 1 \end{bmatrix}$$

   a. Find the lower triangular matrix $\boldsymbol{L}$ of Cholesky's factorization (i.e., $\boldsymbol{L}\boldsymbol{L}^T = \boldsymbol{A}$)
   b. Hence, solve the system $\boldsymbol{Ax} = \boldsymbol{b}$

18. Find the necessary and sufficient conditions on k, so that the (i) Jacobi method and (ii) Gauss–Seidal method converge for solving the system of equations $\boldsymbol{Ax} = \boldsymbol{b}$, where

$$\boldsymbol{A} = \begin{bmatrix} 1 & 0 & k \\ 2 & 1 & 3 \\ k & 0 & 1 \end{bmatrix} \text{and } \boldsymbol{b} \text{ arbitrary}$$

19. Check whether the following system of equations is a diagonal system (diagonally dominant). If not make it a diagonal system by rearranging the equations. Also check whether the Gauss–Seidel iteration method can be used to solve the following system of equations in a less number of iterations? If possible solve them.
   a. $2x-3y+20z=25;\quad 20x+y-2z=17;\quad 3x+20y-z=-18$
   b. $x+5y-z=10;\quad x+y+8z=20;\quad 4x+2y+z=14$

20. Solve the following system of equations by the (i) Gauss–Jacobi method and (ii) Gauss–Seidel method:
   a. $x+17y-2z=48; 2x+2y+18z=30; 30x-2y+3z=48$ correct to three decimal places
   b. $6x+15y+2z=72; x+y+54z=110; 27x+6y-z=85$ correct to three decimal places
   c. $8x-y+z=18; 2x+5y-2z=3; x+y-3z=-6$ correct to two decimal places
   d. $10x-2y+z=12; x+9y-z=10; 2x-y+11z=20$ correct to three decimal places
   e. $x+y+54z=110; 27x+6y-z=85; 6x+15y+2z=72$ correct to three decimal places

21. Solve the following set of simultaneous linear equations by the Gauss–Seidel method:
   a. $4x-3y+5z=34$

      $2x-y-z=6$

      $x+y+4z=15$
   b. $2x-y+5z=15$

      $2x+y+z=7$

      $x+3y+z=10$
   c. $15x+3y-2z=85$

      $2x+10y+z=51$

      $x-2y+8z=5$

d. $4x+2y+z=14$

$x+5y-z=10$

$x+y+8z=20$

e. $28x+4y-z=32$

$x+3y+10z=24$

$2x+17y+4z=35,$ correct up to four decimal places

f. $0.89x+4.32y-0.47z+0.95u=3.36$

$1.13x-0.89y+0.61z+5.63u=4.27$

$6.32x-0.73y-0.65z+1.06u=2.95$

$0.74x+1.01y+5.28z-0.88u=1.97,$ correct up to four significant figures

g. $10x_1-2x_2-x_3-x_4=3$

$-2x_1+10x_2-x_3-x_4=15$

$-x_1-x_2+10x_3-2x_4=27$

$-x_1-x_2-2x_3+10x_4=-9$

22. The following system of equations are given:

a. $4x_1+x_2+x_3=4$

$x_1+4x_2-2x_3=4$

$3x_1+2x_2-4x_3=6$

b. $x_1+x_2-x_3=2$

$2x_1+3x_2+5x_3=-3$

$3x_1+2x_2-3x_3=6$

c. $83x+11y-4z=95$

$7x+52y+13z=104$

$3x+8y+29z=71$

d. $10x+7y+8z+7u=32$

$7x+5y+6z+5u=23$

$8x+6y+10z+9u=33$

$7x+5y+9z+10u=31$

Solve the above systems by ***LU*** decomposition method.

23. Solve the following system of equations using the Gauss–Jordan method:

a. $$\begin{bmatrix} 2 & 2 & 1 \\ 4 & 2 & 3 \\ 1 & 1 & 1 \end{bmatrix}\begin{bmatrix} x_1 \\ x_2 \\ x_3 \end{bmatrix}=\begin{bmatrix} 1 \\ 2 \\ 3 \end{bmatrix}$$

b. $\begin{bmatrix} 2 & 1 & 3 \\ 4 & -3 & 5 \\ -3 & 2 & 4 \end{bmatrix}\begin{bmatrix} x_1 \\ x_2 \\ x_3 \end{bmatrix} = \begin{bmatrix} 1 \\ -7 \\ -3 \end{bmatrix}$

24. Solve the following set of simultaneous linear equations by the Gauss–Jordan method:
    a. $4x-3y+5z=34$

       $2x-y-z=6$

       $x+y+4z=15$
    b. $2x-y+z=-1$

       $3x+3y+9z=0$

       $3x+3y+5z=4$
    c. $x+y-z=1$

       $x+2y-2z=0$

       $-2x+y+z=1$
    d. $2x_1+4x_2+x_3=3$

       $3x_1+2x_2-2x_3=-2$

       $x_1-x_2+x_3=6$
    e. $15x_1+3x_2-4x_3+2x_4=49.207$

       $3x_1+27x_2-5x_3-x_4=18.024$

       $-4x_1-5x_2+16x_3+5x_4=-23.871$

       $2x_1-x_2+5x_3+19x_4=54.907$
    f. $4x_1-2x_2-3x_3+6x_4=12$

       $-5x_1+7x_2+6.5x_3-6x_4=6.5$

       $x_1+7.5x_2+6.25x_3+5.5x_4=16$

       $-12x_1+22x_2+15.5x_3-x_4=17$
25. Solve the following systems of equations by triangular decomposition method:
    a. $4x+y+2z=12,\quad 2x-3y+8z=20,\quad -x+11y+4z=33$
    b. $4x+3y+z=1,\quad 3x+4y+z=-2,\quad 8x+3y+z=4$
26. Solve the following systems of equations by Crout's reduction method:
    a. $2x-6y+8z=24,\quad 5x+4y-3z=2,\quad 3x+y+2z=16$
    b. $5x+2y+z=-12,\quad -x+4y+2z=20,\quad 2x-3y+10z=3$
    c. $x_1+3x_2-x_3+2x_4=9, -2x_1+x_2+2x_3-x_4=2,\quad 3x_1-4x_2+x_3+3x_4=3,\quad -6x_1+3x_2-4x_3+5x_4=2$
27. Solve the following system of equations by Crout's factorization correct to four decimal places:
    a. $0.33x_1+0.25x_2=0.583$

       $0.2x_1+0.163x_2=0.33$

b. $12x_2 + 5x_3 = 25$

$8x_1 - 2x_2 + x_3 = 7$

$x_1 + x_2 + 3x_3 = 4$

28. Solve the following system of equations using Cholesky's method:

a. $3x_1 + x_2 + 2x_3 = 3$

$2x_1 - 3x_2 - x_3 = -3$

$x_1 + 2x_2 + x_3 = 4$

b. $2x_1 + x_2 + 4x_3 = 12$

$8x_1 - 3x_2 + 2x_3 = 20$

$4x_1 + 11x_2 - x_3 = 33$

c. $x_1 + x_2 - x_3 + x_4 = 2$

$2x_1 + x_2 + x_3 - 3x_4 = 1$

$3x_1 - x_2 - x_3 + x_4 = 2$

$5x_1 + x_2 + 3x_3 - 2x_4 = 7$

d. $10x_1 - 7x_2 + 3x_3 + 5x_4 = 6$

$-6x_1 + 8x_2 - x_3 - 4x_4 = 5$

$3x_1 + x_2 + 4x_3 + 11x_4 = 2$

$5x_1 - 9x_2 - 2x_3 + 4x_4 = 7$

29. Solve the following system of equations, correct to three decimal places, by the Gauss–Jacobi method:

a. $83x + 11y - 4z = 95$, $7x + 52y + 13z = 104$, $3x + 8y + 29z = 71$

b. $27x + 6y - z = 85$, $6x + 15y + 2z = 72$, $x + y + 54z = 110$

c. $10x_1 + x_2 + x_3 + x_4 = 21.09$, $x_1 + 10x_2 + x_3 + x_4 = 31.08$, $x_1 + x_2 + 10x_3 + x_4 = 41.07$, $x_1 + x_2 + x_3 + 10x_4 = 51.06$

30. Solve the following system of equations, correct to three decimal places, by the Gauss–Seidal iteration method:

a. $x + y + 54z = 110$, $27x + 6y - z = 85$, $6x + 15y + 2z = 72$

b. $x - 2y + 10z = 30.6$, $2x + 5y - z = 10.5$, $3x + y + z = 9.3$

c. $10x_1 - 2x_2 - x_3 - x_4 = 3$, $-2x_1 + 10x_2 - x_3 - x_4 = 15$, $-x_1 - x_2 + 10x_3 - 2x_4 = 27$,

$-x_1 - x_2 - 2x_3 + 10x_4 = -9$

31. Given the matrix $\boldsymbol{A} = \boldsymbol{I} + \boldsymbol{L} + \boldsymbol{U}$,

$$\boldsymbol{A} = \begin{bmatrix} 1 & 2 & -2 \\ 1 & 1 & 1 \\ 2 & 2 & 1 \end{bmatrix}$$

where $\boldsymbol{L}$ and $\boldsymbol{U}$ are strictly lower and upper triangular matrices, respectively, decide whether the (a) Jacobi and (b) Gauss–Seidal methods converge to the solution of $\boldsymbol{Ax}=\boldsymbol{b}$.

32. Is the following system of equations diagonally dominant? If not, make it diagonally dominant and solve by Jacobi's iteration method:
    a. $3x+4y+15z=54.8$

       $x+12y+3z=39.66$

       $10x+y-2z=7.74$
    b. $5x-y-z=3.245$

       $x+4y+z=7.075$

       $x+y+3z=8.870,$ correct up to six significant figures
    c. $10x-2y-z-w=3$

       $-x-y+10z-2w=27$

       $x+y+2z-10w=9$

       $-2x+10y-z-w=15$
33. For the following systems of equations

    i. $4x+y+2z=4$ ii. $10x+4y-2z=12$

    $3x+5y+z=7$ $x-10y-z=-10$

    $x+y+3z=3$ $5x+2y-10z=-3$
    a. Show that the Jacobi iteration scheme converges
    b. Obtain the Jacobi iteration scheme in matrix form
    c. Starting with $\boldsymbol{x}^{(0)}=\boldsymbol{0}$, iterate three times
34. For the following system of equations

    i. $\begin{bmatrix}-3 & 1 & 0\\ 2 & -3 & 1\\ 0 & 2 & -3\end{bmatrix}\boldsymbol{x}=\begin{bmatrix}-2\\ 0\\ -1\end{bmatrix}$,   ii. $\begin{bmatrix}5 & 1 & -2\\ 3 & 4 & -1\\ 2 & -3 & 5\end{bmatrix}\boldsymbol{x}=\begin{bmatrix}2\\ -2\\ 10\end{bmatrix}$
    a. Set up the Gauss–Seidal iteration scheme in matrix form
    b. Show that the iteration scheme is convergent and hence find its rate of convergence
    c. Starting with $\boldsymbol{x}^{(0)}=\boldsymbol{0}$, iterate six times
35. Show that the Gauss–Seidal method for solving the system of equations diverges:

    i. $\begin{bmatrix}1 & 1 & -1\\ 2 & 3 & 5\\ 3 & 2 & -3\end{bmatrix}\begin{bmatrix}x\\ y\\ z\end{bmatrix}=\begin{bmatrix}-1\\ -6\\ 4\end{bmatrix}$

    ii. $\begin{bmatrix}1 & 2 & 4\\ 2 & 1 & 2\\ 4 & 2 & 1\end{bmatrix}\begin{bmatrix}x\\ y\\ z\end{bmatrix}=\begin{bmatrix}-1\\ 5\\ 3\end{bmatrix}$
36. Show that both the (i) Jacobi method and (ii) Gauss–Seidal methods diverge for solving the system of equations:

$$\begin{bmatrix}2 & 3 & 1\\ 3 & 2 & 2\\ 1 & 2 & 2\end{bmatrix}\begin{bmatrix}x_1\\ x_2\\ x_3\end{bmatrix}=\begin{bmatrix}-1\\ 1\\ 6\end{bmatrix}$$

37. For the following systems of given equations

i. $\begin{bmatrix} 3 & 2 & 0 \\ 2 & 3 & -1 \\ 0 & -1 & 2 \end{bmatrix} \boldsymbol{x} = \begin{bmatrix} 5 \\ 4 \\ 1 \end{bmatrix}$, ii. $\begin{bmatrix} 4 & -1 & 0 \\ -1 & 4 & -1 \\ 0 & -1 & 4 \end{bmatrix} \boldsymbol{x} = \begin{bmatrix} 3 \\ 2 \\ 3 \end{bmatrix}$

a. Find the optimal relaxation parameter w^* for the SOR iteration scheme. Also, determine the rate of convergence of this scheme
b. Write the SOR iteration scheme in residual vector form. Starting with $x^{(0)} = [0.5, 0.5, 0.5]^T$, iterate four times

38. Solve the following system of equations using SOR method:

a. $2x + y + 4z = 12$

$8x - 3y + 2z = 20$

$4x + 11y - z = 33$

b. $x + y + z = 3$

$2x - y + 3z = 16$

$3x + y - z = -3$

c. $2x - 3y + 10z = 3$

$-x + 4y + 2z = 20$

$5x + 2y + z = -12$

d. $10x - 7y + 3z + 5w = 6$

$-6x + 8y - z - 4w = 5$

$3x + y + 4z + 11w = 2$

$5x - 9y - 2z + 4w = 7$

39. Solve the following system of equations, correct to three decimal places, by relaxation method:

a. $9x - 2y + z = 50$; $x + 5y - 3z = 18$; $-2x + 2y + 7z = 19$
b. $10x + 2y + z = 9$; $x + 10y - z = 22$; $-2x + 3y + 10z = 22$
c. $50x + 2y - 3z = 196$, $3x + 65y + 2z = 81$, $-x + y + 33z = 63$
d. $10x_1 + 2x_2 - x_4 = 11.0$, $-x_1 + 20x_2 + 2x_3 = 49.5$, $-x_1 + 10x_3 - x_4 = 27.5$, $-x_2 + 2x_3 + 20x_4 = 92.4$
e. $15x_1 + 3x_2 - 4x_3 + 2x_4 = 49.207$, $3x_1 + 7x_2 - 5x_3 - 4x_4 = 18.024$,

$-4x_1 - 5x_2 + 16x_3 + 5x_4 = -23.871$, $2x_1 - x_2 + 5x_3 - 19x_4 = 54.907$

40. Define norm of a matrix if

$$\boldsymbol{A} = \begin{bmatrix} 1 & 3 & 5 \\ 1 & 4 & 3 \\ 1 & 3 & 2 \end{bmatrix}$$

Find $\|\boldsymbol{A}\|_1$, $\|\boldsymbol{A}\|_e$, and $\|\boldsymbol{A}\|_\infty$.

41. Explain what is meant by ill-conditioning of a matrix. Give two examples of ill-conditioned matrices.

Is the matrix given ill-conditioned?

$$A = \begin{bmatrix} 1 & \frac{1}{2} & \frac{1}{3} \\ \frac{1}{2} & \frac{1}{3} & \frac{1}{4} \\ \frac{1}{3} & \frac{1}{4} & \frac{1}{5} \end{bmatrix}$$

42. Find the inverse of the following matrices using the Gauss–Jordan method:

a. $\begin{bmatrix} 2 & 2 & -3 \\ -3 & 2 & 2 \\ 2 & -3 & 2 \end{bmatrix}$

b. $\begin{bmatrix} 1 & 1 & 1 \\ 4 & 3 & -1 \\ 3 & 5 & 3 \end{bmatrix}$

43. Solve the following using Doolittle's decomposition method:

$$x_1 + x_2 + x_3 = 1$$

$$4x_1 + 3x_2 - x_3 = 6$$

$$3x_1 + 5x_2 + 3x_3 = 4$$

44. Solve the following linear system of equations using the Gauss–Seidel method:

a. $x_1 + 9x_2 - 2x_3 = 36$

$2x_1 - x_2 + 8x_3 = 121$

$6x_1 + x_2 + x_3 = 107$

b. $8x_1 - 3x_2 + 2x_3 = 20$

$4x_1 + 11x_2 - x_3 = 33$

$6x_1 + 3x_2 + 12x_3 = 35$

c. $4x_1 + 11x_2 - x_3 = 33$

$6x_1 + 3x_2 + 12x_3 = 35$

$8x_1 - 3x_2 + 2x_3 = 20$

45. Solve the following system of equations using the Gauss elimination method:

$$x_1 - 5x_2 + 3x_3 = -1$$

$$2x_1 - x_2 - x_3 = 5$$

$$5x_1 - 7x_2 + x_3 = 2$$

46. Find the inverse of the following matrices using the Gauss–Jordan method:

$$\begin{bmatrix} -2 & 4 & -1 \\ -2 & 3 & 0 \\ 7 & -12 & 2 \end{bmatrix}$$

47. Solve by Cholesky's method:

$$9x_1 + 6x_2 + 12x_3 = 17.4$$

$$6x_1 + 13x_2 + 11x_3 = 23.6$$

$$12x_1 + 13x_2 + 26x_3 = 30.8$$

48. Solve the following linear system of equations by Crout's method:

$$5x_1 + 4x_2 + x_3 = 3.4$$

$$10x_1 + 9x_2 + 4x_3 = 8.8$$

$$10x_1 + 13x_2 + 15x_3 = 19.2.$$

7 Numerical Solutions of Ordinary Differential Equations

7.1 INTRODUCTION

Differential equations occur in modeling many physical problems and play an important role in modeling real physical problems mathematically. Many science and engineering problems are solved numerically by using differential equations. Differential equations, which are encountered in real physical problems in the fields of science and engineering, are often very difficult and cumbersome to solve analytically. In this chapter, we shall derive and analyze the numerical methods for solving ordinary differential equations.

The general form of an ordinary differential equation is given by the following equation:

$$\mathcal{F}\left(x, y, y', y'', \ldots, y^{(n)}\right) = 0 \tag{7.1}$$

where x is the independent variable, the dependent variable y and its derivatives are functions of x and n is the order of the ordinary differential equation. The degree of the differential equation is the greatest exponent of the highest order derivative after the equation has been rationalized in derivatives. If the dependent variable and its derivatives occur to the first order only and not as higher powers or products, then the equation is said to be linear, otherwise it is nonlinear.

Now, we shall study the numerical solution for the initial value problem. Let us consider the numerical solution of an ordinary differential equation of the first order

$$\frac{dy}{dx} = f(x, y) \tag{7.2}$$

subject to a given initial condition $y(x_0) = y_0$.

The function $f(x, y)$ is a continuous function for all (x, y) in some domain D of the xy-plane, say $D: a = x_0 \le x \le b, -\infty < y < \infty$, and (x_0, y_0) is a given point in D.

The existence and uniqueness of the initial value problem (Equation 7.2) is based on Picard–Lindelöf theorem or Cauchy–Lipschitz–Picard existence theorem. This theorem states that

1. If $f(x, y)$ be a real valued function defined and continuous in the given region.

$$D: a = x_0 \le x \le b, -\infty < y < \infty. \tag{7.3}$$

2. If $f(x, y)$ be Lipschitz continuous function in the given region D, that is, there exists a constant $L > 0$ called *Lipschitz constant* such that for any $x \in [x_0, b]$ and any two values y_1 and y_2 of y.

$$\left|f(x, y_1) - f(x, y_2)\right| \le L\left|y_1 - y_2\right| \tag{7.4}$$

Then for any $y(x_0) = y_0$, the initial value problem Equation 7.2 has a unique solution $y(x)$ for $x \in [x_0, b]$.

The first step in obtaining a numerical solution of the differential equation (Equation 7.2) is to partition the interval $[a, b]$ in which the solution is desired into a finite number of subinterval by the points

$$a = x_0 < x_1 < x_2 < \cdots < x_n = b$$

These points are called the *mesh points* or *grid points*. Usually, the grid points are taken to be equally spaced, that is,

$$x_i = x_0 + ih,\ h = \frac{b-a}{n}(i = 0, 1, 2, \ldots, n)$$

where h is called the *step length* or *step size*.

The methods for the numerical solution of differential equations can be classified into the following two types:

1. Single-step methods, in which the value of y_{i+1} depends on the knowledge of the preceding value y_i only.
2. Multistep methods, in which the value of y_{i+1} depends on the knowledge of the preceding values $y_i, y_{i-1}, y_{i-2}, \ldots, y_{i-k+1}$ only. In this case, it is said to be a k-step method.

7.2 SINGLE-STEP METHODS

7.2.1 Picard's Method of Successive Approximations

Consider the first-order differential equation

$$\frac{dy}{dx} = f(x, y) \tag{7.5}$$

with initial condition $y = y_0$ when $x = x_0$. Integrating (Equation 7.5) between the limits x_0 and x, we get

$$\int_{x_0}^{x} dy = \int_{x_0}^{x} f(x, y)dx$$

Thus,

$$y = y_0 + \int_{x_0}^{x} f(x, y)dx \tag{7.6}$$

This is an integral equation that contains the unknown y under the integral sign. Equation 7.6 is equivalent to Equation 7.5, because any solution of Equation 7.6 is a solution of Equation 7.5 and vice versa.

The first approximation $y^{(1)}$ to y is obtained by putting $y = y_0$ in $f(x, y)$ and so from Equation 7.6 we have

$$y^{(1)} = y_0 + \int_{x_0}^{x} f(x, y^{(0)})dx, \tag{7.7}$$

where initial approximation $y^{(0)} = y_0$. The resulting $y^{(1)}$ is substituted for y in the integrand of Equation 7.6 to obtain the second approximation

$$y^{(2)} = y_0 + \int_{x_0}^{x} f(x, y^{(1)})dx \tag{7.8}$$

Proceeding in this way, we obtain $y^{(1)}$, $y^{(2)}$, …, $y^{(n-1)}$ and $y^{(n)}$, where

$$y^{(n)} = y_0 + \int_{x_0}^{x} f(x, y^{(n-1)})dx, n = 1, 2, \dots, \tag{7.9}$$

with $y^{(0)} = y_0$. This is known as *Picard's iteration formula.*

Picard's method yields a sequence of successive approximations $y^{(1)}$, $y^{(2)}$, …, $y^{(n)}$. Therefore, the method is called *Picard's method of successive approximation.* However, this method can be applied only to those differential equations in which the successive integration can be obtained easily.

The sequence of successive approximations $y^{(1)}$, $y^{(2)}$, …, $y^{(n)}$, … converges to the exact solution $y(x)$ provided that the function $f(x, y)$ is bounded in the neighborhood of the point (x_0, y_0) in some region D and satisfies the Lipschitz condition

$$|f(x,y) - f(x,\tilde{y})| \le L|y - \tilde{y}|, \text{ for all } (x,y),(x,\tilde{y}) \in D,$$

where $L \ge 0$ and the function $f(x, y)$ is continuous in the region D containing the point (x_0, y_0). The process of iteration is concluded when

$$\left|y^{(n)} - y^{(n-1)}\right| < \varepsilon,$$

where ε is the error tolerance for the desired degree of accuracy. The method is not practically much useful as it involves integration at each iteration, and sometimes it may be very complicated for integration.

Example 7.1

Find the value of $y(0.1)$ by Picard's method given that

$$\frac{dy}{dx} = \frac{y - x}{y + x}, y(0) = 1.$$

Solution:

The Picard's iteration formula for the differential equation $dy/dx = f(x, y)$ is

$$y^{(n)} = y_0 + \int_{x_0}^{x} f(x, y^{(n-1)})dx, \quad n = 1, 2, \dots$$

With $y_0 = 1$ for $x_0 = 0$, the first approximation is

$$\begin{aligned} y^{(1)} &= y_0 + \int_{x_0}^{x} f(x, y^{(0)})dx \\ &= 1 + \int_0^x \frac{1 - x}{1 + x} dx \\ &= 1 - x + 2\log(1 + x) \end{aligned} \tag{7.10}$$

The second approximation is

$$\begin{aligned} y^{(2)} &= y_0 + \int_{x_0}^{x} f(x, y^{(1)})dx \\ &= 1 + \int_0^x \left[1 - \frac{2x}{1 + 2\log(1 + x)}\right] dx \end{aligned}$$

which is very difficult to integrate. Hence, we use the first approximation and by substituting $x = 0.1$ in Equation 7.10, we obtain $y(0.1) = 1.09062$.

Example 7.2

Given the differential equation $dy/dx = x^2/(y^2+1)$ with the initial condition $y = 0$ when $x = 0$, use Picard's method to obtain y for $x = 0.25$, 0.5 and 1.0 correct to three decimal places.

Solution:

The Picard's iteration formula is

$$y^{(n)} = y_0 + \int_{x_0}^{x} f(x, y^{(n-1)})dx, \quad n = 1, 2, \ldots,$$

For $y_0 = 0$ and $x_0 = 0$, the first approximation is

$$\begin{aligned} y^{(1)} &= y_0 + \int_{x_0}^{x} f(x, y^{(0)})dx \\ &= \int_0^x x^2 dx \\ &= \frac{x^3}{3} \end{aligned} \tag{7.11}$$

The second approximation is

$$\begin{aligned} y^{(2)} &= y_0 + \int_{x_0}^{x} f(x, y^{(1)})\,dx \\ &= \int_0^x \frac{x^2}{(x^6/9)+1} dx \\ &= \tan^{-1}\left(\frac{x^3}{3}\right) \end{aligned} \tag{7.12}$$

The third approximation is

$$\begin{aligned} y^{(3)} &= y_0 + \int_{x_0}^{x} f(x, y^{(2)})dx \\ &= \int_0^x \frac{x^2}{\left(\tan^{-1}\left(x^3/3\right)\right)^2 + 1} dx \end{aligned} \tag{7.13}$$

which is very difficult to integrate. Hence, we use the second approximation and substituting $x = 0.25$, 0.5, and 1.0, respectively, in Equation 7.12, we obtain $y(0.25) = 0.005$, $y(0.5) = 0.042$ and $y(1) = 0.322$.

Example 7.3

Use Picard's method of successive approximation to find $y(0.1)$ and $y(0.2)$ correct to four decimal places, from the following equation:

$$\frac{dy}{dx} = x^3 + y, \; y(0) = 1$$

Solution:

We start with $y_0 = 1$ and $x_0 = 0$, the first approximation is

$$y^{(1)} = y_0 + \int_{x_0}^{x} f(x, y^{(0)})dx$$

$$= 1 + \int_0^x 1dx$$

$$= 1 + x$$

Therefore,

$$y^{(1)}(0.1) = 1.1$$

The second approximation is

$$y^{(2)} = y_0 + \int_{x_0}^{x} f(x, y^{(1)})dx$$

$$= 1 + \int_0^x (x^3 + x + 1)dx$$

$$= 1 + x + \frac{x^2}{2} + \frac{x^4}{4}$$

Therefore,

$$y^{(2)}(0.1) = 1.105025$$

The third approximation is

$$y^{(3)} = y_0 + \int_0^x f(x, y^{(2)})dx$$

$$= 1 + \int_0^x \left(x^3 + 1 + x + \frac{x^2}{2} + \frac{x^4}{4} \right) dx$$

$$= 1 + x + \frac{x^2}{2} + \frac{x^3}{6} + \frac{x^4}{4} + \frac{x^5}{20}$$

$$y^{(3)}(0.1) = 1.105192$$

The fourth approximation is

$$y^{(4)} = 1 + \int_0^x \left(x^3 + 1 + x + \frac{x^2}{2} + \frac{x^3}{6} + \frac{x^4}{4} + \frac{x^5}{20} \right) dx$$

$$= 1 + x + \frac{x^2}{2} + \frac{x^3}{6} + \frac{7x^4}{24} + \frac{x^5}{20} + \frac{x^6}{120}$$

Therefore,

$$y^{(4)}(0.1) = 1.105196$$

Since

$$\left| y^{(4)}(0.1) - y^{(3)}(0.1) \right| < 0.00001$$

we can stop the successive iterations. Therefore, the required value of $y(0.1)$ is $y(0.1) = 1.1052$, correct to four decimal places.

Now, taking $x_0 = 0.1$, $y_0 = 1.1052$, and $x = 0.2$, from Equation 7.9, we have

$$y^{(1)}(0.2) = 1.1052 + \int_{0.1}^{0.2} (x^3 + 1.1052)dx$$

$$= 1.216095$$

The second approximation is

$$y^{(2)}(0.2) = 1.1052 + \int_{0.1}^{0.2} (x^3 + 1.216095)dx$$

$$= 1.227184$$

The third approximation is

$$y^{(3)}(0.2) = 1.1052 + \int_{0.1}^{0.2} (x^3 + 1.227184)dx$$

$$= 1.228293$$

The fourth approximation is

$$y^{(4)}(0.2) = 1.1052 + \int_{0.1}^{0.2} (x^3 + 1.228293)dx$$

$$= 1.228404$$

The fifth approximation is

$$y^{(5)}(0.2) = 1.1052 + \int_{0.1}^{0.2} (x^3 + 1.228404)dx$$

$$= 1.228415$$

Since

$$\left| y^{(5)}(0.2) - y^{(4)}(0.2) \right| < 0.000012,$$

we can stop the successive iterations. Hence, the required value of $y(0.2)$ is $y(0.2) = 1.2284$, correct to four decimal places.

MATHEMATICA® Program for Solving ODE by Picard's Method of Successive Approximation (Chapter 7, Example 7.3)

```
f[x_,y_]:= x³ + y;
x[0] = 0;
y[0][x] = 1;
For[i=0,i≤4,i++,
  y[i+1][x]=y[0][x] + Integrate[f[x, y[i][x]], {x, x[0], x}];
  Print[y[i+1][x]]];
```

Output:
```
1 + x + x⁴/4
1 + x + x²/2 + x⁴/4 + x⁵/20
1 + x + x²/2 + x³/6 + x⁴/4 + x⁵/20 + x⁶/120
1 + x + x²/2 + x³/6 +(7x⁴)/24 + x⁵/20 + x⁶/120 + x⁷/840
1 + x + x²/2 + x³/6 + (7x⁴)/24 + (7x⁵)/120 + x⁶/120 + x⁷/840 + x⁸/6720
```

7.2.2 Taylor's Series Method

We consider the initial value problem

$$\frac{dy}{dx} = f(x, y) \tag{7.14}$$

with the initial conditions $y(x_0) = y_0$. Let us assume that the given function $f(x, y)$ is continuously differentiable a sufficient number of times in the region D containing the point (x_0, y_0), so that the solution $y(x)$ is also continuously differentiable a sufficient number of times in D.

Now,

$$y' = f(x, y) \tag{7.15}$$

Differentiating Equation 7.15 with respect to x, we get

$$y'' = \frac{\partial f}{\partial x} + \frac{\partial f}{\partial y}\frac{dy}{dx}, \text{ using chain rule}$$

This implies

$$y'' = f_x + f_y y' = f_x + ff_y \tag{7.16}$$

Similarly, again using chain rule, we can obtain

$$\begin{aligned} y''' &= f_{xx} + f_{yx} f + f\left[f_{xy} + f_{yy} f\right] + \left[f_x + ff_y\right] f_y \\ &= f_{xx} + 2ff_{xy} + f^2 f_{yy} + f_x f_y + ff_y^2 \end{aligned} \tag{7.17}$$

and so on. Putting $x = x_0$ and $y = y_0$, we get y_0', y_0'', y_0''', and so on from Equations 7.15 to 7.17, respectively. Now, the Taylor's series expansion of $y(x)$ in the neighborhood of $x = x_0$ is given by

$$y(x) = y_0 + (x - x_0)y_0' + \frac{(x - x_0)^2}{2!} y_0'' + \cdots \tag{7.18}$$

Substituting the values of y_0', y_0'', y_0''', … in Equation 7.18, we can obtain $y(x)$ for all values of x.

7.2.2.1 Error Estimate

If x_i and $y_i = y(x_i)$ are known exactly, then the Taylor series expansion

$$y(x_{i+1}) = y(x_i + h) = y(x_i) + hy'(x_i) + \frac{h^2}{2!}y''(x_i) + \cdots + \frac{h^p}{p!}y^{(p)}(x_i) \tag{7.19}$$

can be used to compute $y_{i+1} = y(x_{i+1})$ with an error

$$T_{i+1} = \frac{h^{p+1}}{(p+1)!}y^{(p+1)}(\xi), x_i < \xi < x_{i+1} \tag{7.20}$$

The number of terms to be included in Equation 7.19 is fixed by the permissible error ε. If the series is truncated at the term $y^{(p)}(x_i)$, then

$$\left|\frac{h^{p+1}}{(p+1)!}y^{(p+1)}(\xi)\right| < \varepsilon$$

This implies

$$h^{p+1}\left|f^{(p)}(\xi, y(\xi))\right| < (p+1)!\varepsilon \tag{7.21}$$

Since, ξ is unknown, we replace $\left|f^{(p)}(\xi, y(\xi))\right|$ by its maximum value in $[x_0, b]$. In order to determine this value, we take one more nonvanishing term in the series than is required and then differentiate this series n times. The maximum value of this quantity in $[x_0, b]$ gives a rough upper bound.

7.2.2.2 Alternatively

The Taylor series method of order p has the property that the final global error is of order $O(h^{p+1})$; hence, p can be chosen as large as necessary to make the error as small as desired.

The truncation error due to the terms neglected in the series is given by

$$E_p = \frac{h^{p+1}}{(p+1)!}y^{(p+1)}(x + \theta h), 0 < \theta < 1 \tag{7.22}$$

Making use of finite differences, the $(p+1)$th derivative of y at $x + \theta h$ can be approximated as

$$E_p = \frac{h^p}{(p+1)!}\left(y^{(p)}(x + \theta h) - y^{(p)}(x)\right), 0 < \theta < 1 \tag{7.23}$$

Equation 7.23 is more useful form and can be incorporated in the algorithm to monitor the error in each step. However, in practice, one usually computes two sets of approximations using step sizes h and $h/2$ and then compares the solutions.

Example 7.4

Using Taylor's series method, obtain $y(0.1)$ and $y(0.2)$, if $y(x)$ satisfies the following differential equation:

$$\frac{dy}{dx} = y^2 - x^2 \text{ with } y(0) = 1$$

Solution:

Here, $x_0 = 0$ and $y_0 = 1$. Now,

$$y' = y^2 - x^2, y_0' = 1$$
$$y'' = 2yy' - 2x, y_0'' = 2$$
$$y''' = 2y'^2 + 2yy'' - 2, y_0''' = 4$$
$$y^{iv} = 6y'y'' + 2yy''', y_0^{iv} = 20$$
$$y^{v} = 6y''^2 + 8y'y''' + 2yy^{iv}, y_0^{v} = 96,$$

and so on.

To compute $y(0.1)$ correct to four decimal places, we consider Taylor's series of order 4, that is, up to the term x^4.

The Taylor's series expansion for $y(x)$ about $x = x_0$ is

$$\begin{aligned} y(x) &= y_0 + xy_0' + \frac{x^2}{2!}y_0'' + \frac{x^3}{3!}y_0''' + \frac{x^4}{4!}y^{iv} \\ &= 1 + x + x^2 + \frac{2}{3}x^3 + \frac{5}{6}x^4 \end{aligned} \tag{7.24}$$

Hence, putting $x = 0.1$ and 0.2 in Equation 7.24, we get

$$y(0.1) = 1.1108 \quad \text{and} \quad y(0.2) = 1.2467$$

Moreover, the error in the approximation is

$$E_4 = \frac{h^4}{5!}\left(y^{(4)}(0.2) - y^{(4)}(0.1)\right) = \frac{(0.1)^4 \times (58.1042 - 33.1419)}{5!} = 0.0000208019 = 2.08019 \times 10^{-5}$$

Example 7.5

Use Taylor's series method to solve the equation

$$\frac{dy}{dx} = \frac{1}{x^2} - \frac{y}{x}$$

for $x = 1.1$, where $y(1) = 0$.

Solution:

Here, $x_0 = 1$ and $y_0 = 0$. Now,

$$y' = \frac{1}{x^2} - \frac{y}{x}, y_0' = 1$$
$$y'' = -\frac{2}{x^3} + \frac{y}{x^2} - \frac{y'}{x}, y_0'' = -3$$
$$y''' = \frac{6}{x^4} + \frac{1}{x} - \frac{2y}{x^3} + \frac{2y'}{x^2}, y_0''' = 9$$
$$y^{iv} = -\frac{24}{x^5} - \frac{3}{x^2} - \frac{3}{x} + \frac{6y}{x^4} - \frac{6y'}{x^3}, y_0^{iv} = -36$$

and so on.

The Taylor's series expansion for $y(x)$ about $x = x_0$ up to fourth order is

$$\begin{aligned} y(x) &= y_0 + (x-1)\, y_0' + \frac{(x-1)^2}{2!} y_0'' + \frac{(x-1)^3}{3!} y_0''' + \frac{(x-1)^4}{4!} y_0^{iv} \\ &= (x-1) - 3\frac{(x-1)^2}{2!} + \frac{(x-1)^3}{3!} \times 9 + \frac{(x-1)^4}{4!} \times (-36) \end{aligned} \tag{7.25}$$

Hence, putting $x = 1.1$ in Equation 7.25, we get

$$y(1.1) = 0.0863628$$

It is easy to see that the exact solution of the given problem is

$$y = \frac{\log x}{x}$$

so that the exact solution is

$$y(1.1) = 0.0866456$$

Therefore, the absolute error is

$$|0.0866456 - 0.0863628| = 0.0002828 = 2.828 \times 10^{-4}$$

7.2.3 General Form of a Single-Step Method

In a single-step method, we determine a function $F(x, y; h)$ depending on $f(x, y)$ and its derivatives such that

$$y(x+h) = y(x) + hF(x, y; h) + O(h^{p+1}) \tag{7.26}$$

where p is a positive integer, called the *order of the method.* Now, putting $x = x_i$, where $i = 0, 1, 2, \ldots$, in Equation 7.26, we get

$$y_{i+1} = y_i + hF(x_i, y_i; h) + O(h^{p+1}) \tag{7.27}$$

where

$$|O(h^{p+1})| \le Kh^{p+1}, \quad K \text{ is a constant} \tag{7.28}$$

Neglecting the last term in Equation 7.26, we get the recursive formula together with the initial condition as

$$\bar{y}_{i+1} = \bar{y}_i + hF(x_i, \bar{y}_i; h), \bar{y}_0 = y(x_0) \tag{7.29}$$

where $\bar{y}_i$ and $\bar{y}_{i+1}$ are the computed values of y_i and y_{i+1}, respectively. From Equation 7.29, we can successively obtain $\bar{y}_1, \bar{y}_2, \bar{y}_3, \ldots, \bar{y}_n$. The term $O(h^{p+1})$ in Equation 7.27 neglected in above approximation is called the *truncation error* or simply *error in the method.*

7.2.3.1 Error Estimate

If the function $f(x, y)$ has sufficiently many continuous and bounded partial derivatives for $(x, y) \in D$, where $D : a \le x \le b, -\infty < y < \infty$, then the function $F(x, y; h)$ and its first-order derivatives are also continuous and hence bounded in D.

Let $|F_y(x,y;h)| \le M$ in D, where M is a constant. So that for any $y, \tilde{y}$, using Lagrange's mean value theorem, we have

$$F(x,y;h) - F(x,\tilde{y};h) = (y - \tilde{y})F_y(x,\eta;h) \tag{7.30}$$

where $\min\{y,\tilde{y}\} < \eta < \max\{y,\tilde{y}\}$. Therefore,

$$|F(x,y;h) - F(x,\tilde{y};h)| \le M\,|y - \tilde{y}| \tag{7.31}$$

From Equations 7.27 and 7.29, we get

$$\begin{aligned}|y_{i+1} - \bar{y}_{i+1}| &\le |y_i - \bar{y}_i| + h|F(x_i, y_i;h) - F(x_i,\bar{y}_i;h)| + |O(h^{p+1})| \\ &\le |y_i - \bar{y}_i| + hM\,|y_i - \bar{y}_i| + Kh^{p+1}, \text{ using Equations 7.31 and 7.28} \\ &= (1+Mh)|y_i - \bar{y}_i| + Kh^{p+1}\end{aligned}$$

Therefore,

$$|\varepsilon_{i+1}| \le (1+Mh)|\varepsilon_i| + Kh^{p+1} \tag{7.32}$$

Now, setting $\lambda = 1 + Mh$ and $\mu = Kh^{p+1}$, from Equation 7.32 we get,

$$|\varepsilon_{i+1}| \le \lambda|\varepsilon_i| + \mu, i = 0,1,2,\ldots \tag{7.33}$$

It, therefore, follows that

$$\begin{aligned}|\varepsilon_{i+1}| &\le \lambda|\varepsilon_i| + \mu \le \lambda(\lambda|\varepsilon_{i-1}| + \mu) + \mu = \lambda^2|\varepsilon_{i-1}| + \mu(1+\lambda) \\ &\le \lambda^2(\lambda|\varepsilon_{i-2}| + \mu) + \mu(1+\lambda) = \lambda^3|\varepsilon_{i-2}| + \mu(1+\lambda+\lambda^2) \\ &\;\vdots \\ &\le \lambda^{i+1}|\varepsilon_0| + \mu(1+\lambda+\lambda^2+\cdots+\lambda^i)\end{aligned} \tag{7.34}$$

If we neglect the initial error, that is, $\varepsilon_0 = 0$, then from Equation 7.34, we obtain

$$|\varepsilon_{i+1}| \le \mu\frac{\lambda^{i+1}-1}{\lambda-1}, \quad i = 0,1,2,\ldots, \tag{7.35}$$

It follows that

$$\begin{aligned}|\varepsilon_i| &\le \mu\frac{\lambda^i - 1}{\lambda - 1}, \quad i = 1,2,\ldots \\ &= \mu\frac{(1+Mh)^i - 1}{Mh}, \quad i = 1,2,\ldots \\ &\le \mu\frac{(1+Mh)^n - 1}{Mh}, \quad \text{since } i \le n \\ &\le \mu\frac{e^{nMh} - 1}{Mh}, \quad \text{since } 1 + x \le e^x\end{aligned}$$

Thus,

$$|\varepsilon_i| \le \frac{Kh^p}{M}\left(e^{(b-a)M} - 1\right) \tag{7.36}$$

Therefore, the grid error or the error on the grid, defined by

$$E = \max_{1 \le i \le n} |\varepsilon_i|$$

is given by

$$E = \frac{Kh^p}{M}\left(e^{(b-a)M} - 1\right) \approx O\left(h^p\right) \tag{7.37}$$

This shows that the grid error E for a single-step method of order p is $O\left(h^p\right)$.

7.2.3.2 Convergence of the Single-Step Method

Equation 7.37 shows that $E \to 0$ as $h \to 0$, that is, as $n \to \infty$. Therefore, the single-step method is invariably convergent. Thus, we can obtain accurate results as we wish by sufficiently reducing the step length h. However, in practical computation, round-off errors at each step and consequently the accumulated round-off error through a large number of steps affect the accurate result. Thus, for practical purpose, it is quite wise to choose the optimal step length, which would keep both the truncation error and the cumulative round-off error reasonably low.

7.2.3.2.1 A General Single-Step Method

A general single-step method of order p can be obtained by expanding $y(x+h)$ by Taylor's theorem as follows:

$$y(x+h) = y(x) + hy'(x) + \frac{h^2}{2!}y''(x) + \cdots + \frac{h^p}{p!}y^{(p)}(x) + \frac{h^{p+1}}{(p+1)!}y^{(p+1)}(x+\theta h), \quad 0<\theta<1 \tag{7.38}$$

Since $y^{(p+1)}(x)$ is continuous in $[a,b]$ and hence bounded therein. Therefore, Equation 7.38 can be written as

$$y(x+h) = y(x) + hF_T(x,y;h) + O\left(h^{p+1}\right) \tag{7.39}$$

where

$$F_T(x,y;h) = y'(x) + \frac{h}{2!}y''(x) + \cdots + \frac{h^{p-1}}{p!}y^{(p)}(x) \tag{7.40}$$

and $y'(x)$, $y''(x)$, …, $y^{(p)}(x)$ are expressed as functions of x, y in terms of the function $f(x,y)$ and its partial derivatives given in Equations 7.16 and 7.17. The recursion is carried out by Equation 7.29. Equation 7.39 represents a general single-step method of order p.

We shall now consider some single-step methods to express the formula (Equation 7.40) in a competent form, which is more useful for practical computations.

7.2.4 Euler Method

Expanding $y(x+h)$ by Taylor's theorem, we get

$$y(x+h) = y(x) + hy'(x) + \frac{h^2}{2!}y''(x) + \cdots + \frac{h^p}{p!}y^{(p)}(x) + \frac{h^{p+1}}{(p+1)!}y^{(p+1)}(x+\theta h), \quad 0<\theta<1$$

Now, we can write

$$y(x+h) = y(x) + hF_T(x,y;h) + O(h^{p+1}) \tag{7.41}$$

where

$$F_T(x,y;h) = y'(x) + \frac{h}{2!}y''(x) + \cdots + \frac{h^{p-1}}{p!}y^{(p)}(x) + \cdots \tag{7.42}$$

The recursive formula of this method is given by

$$\overline{y}_{i+1} = \overline{y}_i + hF_T(x_i, \overline{y}_i; h) \quad (i = 0,1,2,\ldots,n-1) \tag{7.43}$$

with the initial condition $\overline{y}_0 = y(x_0)$. This is called the Taylor's series method of order p. Substituting $p = 1$ in Equation 7.42 we get,

$$F_T(x,y;h) = y' = f(x,y)$$

This single-step method of order 1 is known as *Euler's method*. Therefore, Euler's method can also be called as the Taylor's series method of order 1. In Euler's method, the recursion is given by

$$\overline{y}_{i+1} = \overline{y}_i + hF_T(x_i, \overline{y}_i; h) = \overline{y}_i + hf(x_i, \overline{y}_i), \quad (i = 0,1,2,\ldots,n-1) \tag{7.44}$$

with the initial condition $\overline{y}_0 = y(x_0)$. According to Equation 7.37, the grid error is $E = O(h)$.

Euler's method is the simplest approach to compute a numerical solution of an initial value problem. However, it has about the lowest possible accuracy. If we wish to compute very accurate solutions or solutions that are accurate over a long interval, then Euler's method requires a large number of small steps. For sufficiently small h, the method yields better results.

7.2.4.1 Local Truncation Error

The error in the approximation, that is, the difference between the exact solution at x_{i+1} and the numerical solution $\overline{y}_{i+1}$, is called the *local truncation error* (assuming that $\overline{y}_{i+1}$ is calculated without any round off error). The truncation error is given by

$$\begin{aligned} T_{i+1} &= y_{i+1} - \overline{y}_{i+1} \\ &= y_{i+1} - y_i - hf(x_i, y_i), \text{ using Equation 7.44} \\ &= \frac{h^2}{2!}y''(\xi_i) \end{aligned} \tag{7.45}$$

where $x_i < \xi_i < x_{i+1}$. Hence, the local truncation error for Euler scheme is $O(h^2)$.

At each step of the Euler's method, a truncation error given in Equation 7.45 is introduced. The cumulative effect of these errors is called *global truncation error* or *global error*. From Equation 7.37, it can be easily observed that the global truncation error is $O(h)$. The truncation error can be reduced by using smaller step length h. However, if h becomes too small such that round-off errors become significant, the total error might increase.

7.2.4.2 Geometrical Interpretation

The geometrical interpretation of the Euler method is that for a small interval of length h near (x_n, y_n), the function $y(x)$ has a constant slope equal to the slope at (x_n, y_n). Based on this, the next point (x_{n+1}, y_{n+1}) of the numerical solution is obtained as

$$y_{n+1} = y_n + hf(x_n, y_n)$$

Thus, if (x, y) is a determined point on the solution curve $y = y(x)$, the next point $(x+h, y(x+h))$ lies approximately on the tangent to the solution curve at (x, y). Euler's explicit method has been schematically illustrated in Figure 7.1.

Problem 7.6

Consider the initial value problem $y' = x(x+y)-2$, $y(0) = 2$, use Euler's method with step sizes $h = 0.3$, $h = 0.2$, and $h = 0.15$ to compute approximations to $y(0.6)$ up to five decimal places.

Solution:

Here, $f(x, y) = x(y+x)-2$, $x_0 = 0$, $\bar{y}_0 = y_0 = 2$, and $x_i = x_0 + ih$, where $i = 0,1,2,\ldots,n$. When $h = 0.15$,

$$y(0.15) = \bar{y}_1 = \bar{y}_0 + hf(x_0, \bar{y}_0) = 1.7$$

$$y(0.3) = \bar{y}_2 = \bar{y}_1 + hf(x_1, \bar{y}_1) = 1.441625$$

$$y(0.45) = \bar{y}_3 = \bar{y}_2 + hf(x_2, \bar{y}_2) = 1.2199981$$

$$y(0.6) = \bar{y}_4 = \bar{y}_3 + hf(x_3, \bar{y}_3) = 1.032723$$

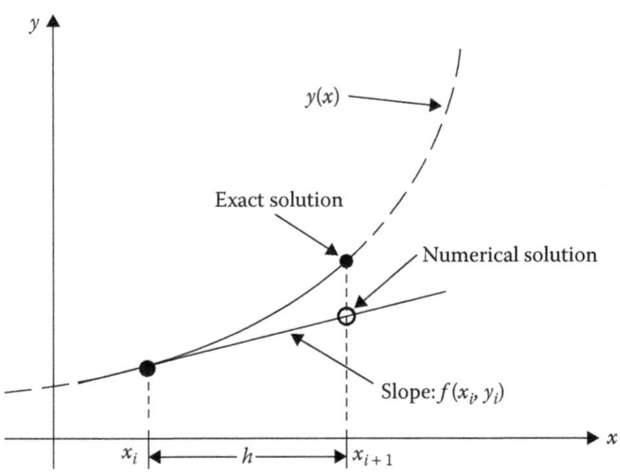

FIGURE 7.1 Euler's explicit method.

Therefore,

$$y(0.6) = 1.03272$$

When $h = 0.2$,

$$y(0.2) = \bar{y}_1 = \bar{y}_0 + hf(x_0, \bar{y}_0) = 1.6$$
$$y(0.4) = \bar{y}_2 = \bar{y}_1 + hf(x_1, \bar{y}_1) = 1.272$$
$$y(0.6) = \bar{y}_4 = \bar{y}_3 + hf(x_3, \bar{y}_3) = 1.00576$$

Therefore,

$$y(0.6) = 1.00576$$

When $h = 0.3$,

$$y(0.3) = \bar{y}_1 = \bar{y}_0 + hf(x_0, \bar{y}_0) = 1.4$$
$$y(0.6) = \bar{y}_4 = \bar{y}_3 + hf(x_3, \bar{y}_3) = 0.953$$

Therefore,

$$y(0.6) = 0.953$$

7.2.4.3 Backward Euler Method

We approximate y' in the differential Equation 7.2 at the mesh point x_{i+1} by

$$\frac{y_{i+1} - y_i}{h} + \frac{h}{2} y''(\xi_{i+1})$$

where

$$x_i \le \xi_{i+1} \le x_{i+1}$$

Then from the differential equation 7.2 at the mesh point x_{i+1}, we get

$$\frac{y_{i+1} - y_i}{h} + \frac{h}{2} y''(\xi_{i+1}) = f(x_{i+1}, y_{i+1}) \tag{7.46}$$

where

$$x_i \le \xi_{i+1} \le x_{i+1}$$

From Equation 7.46, solving for y_{i+1}, we have

$$y_{i+1} = y_i + hf(x_{i+1}, y_{i+1}) - \frac{h^2}{2} y''(\xi_{i+1}) \tag{7.47}$$

Now, dropping the last term of Equation 7.47, we get

$$\bar{y}_{i+1} = \bar{y}_i + hf(x_{i+1}, \bar{y}_{i+1}), \quad i = 0, 1, 2, \ldots, n-1 \tag{7.48}$$

The Equation 7.48 is called the *backward Euler method*. Since the unknown value $\bar{y}_{i+1}$ appears on both sides of Equation 7.48, it defines $\bar{y}_{i+1}$ implicitly. Therefore, this method is also called an *implicit method*. The local truncation error is given by

$$\begin{aligned} T_{i+1} &= y_{i+1} - \bar{y}_{i+1} \\ &= y_{i+1} - y_i - hf(x_{i+1}, y_{i+1}) \\ &= -\frac{h^2}{2} y''(\xi_{i+1}), \text{ using Equation 7.47} \end{aligned} \tag{7.49}$$

Hence, the local truncation error for the backward Euler method is $O(h^2)$. From Equation 7.36, it can be easily observed that the global error is $O(h)$. Moreover, it may be noted that this method is first order, which is same as Euler forward scheme.

In the backward Euler method, at each step, we need to solve a nonlinear algebraic equation 7.48 for $\bar{y}_{i+1}$. Now, any traditional rootfinding methods (e.g., Newton's method, the secant method, and the bisection method) can be applied to Equation 7.48 to find its root $\bar{y}_{i+1}$, but often it is a very time-consuming process. Instead, Equation 7.48 is usually solved by a simple iteration technique. Given an initial guess $\bar{y}_{i+1}^{(0)} \approx \bar{y}_{i+1}$, we then determine $\bar{y}_{i+1}^{(1)}$, $\bar{y}_{i+1}^{(2)}$, and so on by

$$\bar{y}_{i+1}^{(k+1)} = \bar{y}_i + hf(x_{i+1}, \bar{y}_{i+1}^{(k)}), \quad i = 0, 1, 2, \ldots, n-1, \quad k = 0, 1, 2, \ldots \tag{7.50}$$

If h is sufficiently small, then the iterates $\bar{y}_{i+1}^{(k)}$ will converge to $\bar{y}_{i+1}$ as $k \to \infty$. Now, subtracting Equation 7.50 from 7.48 yields

$$\bar{y}_{i+1} - \bar{y}_{i+1}^{(k+1)} = h[f(x_{i+1}, \bar{y}_{i+1}) - f(x_{i+1}, \bar{y}_{i+1}^{(k)})] \tag{7.51}$$

Applying the mean value theorem to Equation 7.51, we obtain

$$\bar{y}_{i+1} - \bar{y}_{i+1}^{(k+1)} \approx h(\bar{y}_{i+1} - \bar{y}_{i+1}^{(k)}) \frac{\partial f(x_{i+1}, \bar{y}_{i+1})}{\partial y} \tag{7.52}$$

Equation 7.52 gives a relation between the errors in the successive iterates. Therefore, if

$$\left| h \frac{\partial f(x_{i+1}, \bar{y}_{i+1})}{\partial y} \right| < 1 \tag{7.53}$$

then the errors will converge to zero, as long as the initial guess $\bar{y}_{i+1}^{(0)}$ is a sufficiently close approximation to $\bar{y}_{i+1}$.

7.2.4.4 Midpoint Method

We approximate y' in the differential equation (Equation 7.2) at the mesh point x_i by

$$\frac{y_{i+1}-y_{i-1}}{2h}-\frac{h^2}{6}y'''(\xi_i)$$

where

$$x_{i-1} \le \xi_i \le x_{i+1}$$

Then from the differential equation (Equation 7.2), we get

$$\frac{y_{i+1}-y_{i-1}}{2h}-\frac{h^2}{6}y'''(\xi_i)=f(x_i,y_i) \tag{7.54}$$

where

$$x_{i-1} \le \xi_i \le x_{i+1}$$

From Equation 7.54, solving for y_{i+1}, we have

$$y_{i+1}=y_{i-1}+2hf(x_i,y_i)+\frac{h^3}{3}y'''(\xi_i) \tag{7.55}$$

Now, dropping the last term in Equation 7.55, we get

$$\bar{y}_{i+1}=\bar{y}_{i-1}+2hf(x_i,\bar{y}_i), \quad i=1,2,\ldots,n-1 \tag{7.56}$$

Equation 7.56 is called the *midpoint method*. It is an explicit two-step method. In this method, the value of y_1 must be obtained by another method.

The local truncation error is given by

$$\begin{aligned} T_{i+1} &= y_{i+1}-\bar{y}_{i+1} \\ &= y_{i+1}-y_{i-1}-2hf(x_i,y_i) \\ &= \frac{h^3}{3}y'''(\xi_i), \text{ using Equation 7.55} \end{aligned} \tag{7.57}$$

Hence, the local truncation error for the midpoint method is $O(h^3)$. From Equation 7.36, it can be easily observed that the global error is $O(h^2)$. Also, the bound on the truncation error is given by

$$|T_{i+1}| \le \frac{h^3}{6}M \tag{7.58}$$

where

$$M=\max_{x\in[x_0,b]}|y'''(x)|$$

The midpoint can also be obtained by integrating Equation 7.2 over $[x_{i-1}, x_{i+1}]$, yielding

$$y_{i+1} = y_{i-1} + \int_{x_{i-1}}^{x_{i+1}} f(x, y)dx \tag{7.59}$$

We can replace the integrand $f(x, y)$ with a constant interpolant, in particular,

$$\int_{x_{i-1}}^{x_{i+1}} f(x, y)dx \approx \int_{x_{i-1}}^{x_{i+1}} f(x_i, y_i)dx = 2hf(x_i, y_i)$$

This leads to the numerical method

$$\begin{aligned} \bar{y}_0 &= y_0 \\ \bar{y}_{i+1} &= \bar{y}_{i-1} + 2hf(x_i, \bar{y}_i), \quad i = 1, 2, \ldots, n-1 \end{aligned} \tag{7.60}$$

This is called the *midpoint method.* Also, the midpoint method in Equation 7.60 can be obtained if we club together Euler's forward and backward schemes, that is,

$$\bar{y}_{i+1} = \bar{y}_i + hf(x_i, \bar{y}_i)$$

$$\bar{y}_i = \bar{y}_{i-1} + hf(x_i, \bar{y}_i)$$

Example 7.6

Consider the initial value problem

$$y' = -y^2, \quad y(0) = 1$$

Determine the value of y at $x = 0.2$ by the backward Euler method.

Solution:
Here, $f(x, y) = -y^2$, $x_0 = 0$, and $y_0 = 1$. Taking $h = 0.1$ and $n = (0.2 - x_0)/h = 2$. By Euler's method,

$$y_1^{(0)} = y_0 + hf(x_0, y_0) = 1 + (0.1) \times (-0 \times 1^2) = 1$$

By the backward Euler method,

$$y_1^{(1)} = y_0 + hf(x_1, y_1^{(0)}) = 1 + (0.1) \times (-0.1 \times 1^2) = 0.919$$

$$y_1^{(2)} = y_0 + hf(x_1, y_1^{(1)}) = 1 + (0.1) \times (-0.1 \times 0.919^2) = 0.915544$$

$$y_1^{(3)} = y_0 + hf(x_1, y_1^{(2)}) = 1 + (0.1) \times (-0.1 \times 0.915544^2) = 0.916178$$

Now, $y_1 = 0.916$. By Euler's method,

$$y_2^{(0)} = y_1 + hf(x_1, y_1) = 0.916 + (0.1) \times (-0.1 \times 0.916^2) = 0.9076$$

Again, by the backward Euler method,

$$y_2^{(1)} = y_1 + hf(x_1, y_1^{(0)}) = 0.916 + (0.1) \times (-0.2 \times 0.9076^2) = 0.846916$$

$$y_2^{(2)} = y_1 + hf(x_1, y_1^{(1)}) = 0.916 + (0.1) \times (-0.2 \times 0.846916^2) = 0.844451$$

$$y_2^{(3)} = y_1 + hf(x_1, y_1^{(2)}) = 0.916 + (0.1) \times (-0.2 \times 0.844451^2) = 0.844868$$

Hence $y_2 = y(0.2) = 0.844$.

Example 7.7

Using the midpoint method, find the value of y at $x = 1.4$, given that

$$y' = -xy^2, \quad y(1) = 1$$

Solution:
Here, $f(x,y) = -xy^2$, $x_0 = 1$, and $y_0 = 1$. Taking $h = 0.2$ and $n = (1.4 - x_0)/h = 2$. By Euler's method,

$$y_1 = y_0 + hf(x_0, y_0) = 1 + (0.2) \times (-1 \times 1^2) = 0.8$$

By the midpoint method,

$$y_2 = y_0 + 2hf(x_1, y_1) = 1 + 2 \times (0.2) \times (-1.2 \times 0.8^2) = 0.6928$$

Hence, $y(1.4) = 0.6928$.

7.2.4.5 Algorithm for Euler's Method

Step 1: Start the program.
Step 2: Define the function $f(x, y)$.
Step 3: Read x_0, y_0, h, x.
Step 4: Compute $n = (x - x_0)/h$.
Step 5: for $i = 0(1)n - 1$ do

$x_i = x_0 + i * h$

$y_{i+1} = y_i + h * f(x_i, y_i)$.

Step 6: Print the value of y_{i+1}.
Step 7: Stop the program. ■

MATHEMATICA® Program for Solving ODE by Euler's Method (Chapter 7, Example 7.6)

```
f[x_, y_]:=x*(x+y)-2;
x[0]=0;
y[0]=2;
h=0.15;
x1 = 0.6;
n=(x1-x[0])/h;
For[i=0,i<=n-1,i++,x[i]=x[0]+i*h;
  y[i+1]=y[i]+h*f[x[i], y[i]];
  Print[x[i]+h," ",y[i+1]]];
```

Output:

```
0.15   1.7
0.3   1.44163
0.45   1.22
0.6   1.03272
```

7.2.5 Improved Euler Method

The Euler's forward scheme may be very easy to implement; however, one main drawback of both Euler's methods is the low convergence order. A very small step size is required for satisfactory result. Next, we present a method that has a higher convergence order. This method is a modification of Euler's method. The improved Euler method, also known as *Heun's method*, is a single-step explicit numerical technique for solving a first-order ordinary differential equation. In Euler's method, it is assumed that in each subinterval, the slope between the points (x_i, y_i) and (x_{i+1}, y_{i+1}) is constant and equal to the slope of $y(x)$ at the point (x_i, y_i). This is usually not the case. In this scheme, the slope used for computing the value of y_{i+1} is modified to include the effect of that the slope changes within the subinterval. Therefore, an improvement is made by taking the arithmetic average of the slopes at the end points x_i and x_{i+1} of each subinterval. It works first by approximating a value to y_{i+1} and then improving it by making use of average slope.

If $y = y_i$ be the value of y corresponding to $x = x_i$, then integrating Equation 7.2 in the subinterval $[x_i, x_{i+1}]$, we get

$$y_{i+1} = y_i + \int_{x_i}^{x_{i+1}} f(x,y)dx \tag{7.61}$$

If we approximate the integral in Equation 7.61 by means of the trapezoidal rule, we obtain

$$y_{i+1} = y_i + \frac{h}{2}\left[f(x_i, y_i) + f(x_{i+1}, y_{i+1})\right] - \frac{h^3}{12}y'''(\xi_i) \tag{7.62}$$

where $x_i < \xi_i < x_{i+1}$. By dropping the error term in Equation 7.62 and then equating both sides, we obtain the trapezoidal method for solving the initial value problem (7.2):

$$\overline{y}_{i+1} = \overline{y}_i + \frac{h}{2}\left[f(x_i, \overline{y}_i) + f(x_{i+1}, \overline{y}_{i+1})\right], \quad i = 0, 1, 2, \ldots \tag{7.63}$$

with $\overline{y}_0 = y_0$. The truncation error for the trapezoidal method is

$$T_{i+1} = -\frac{h^3}{12}y'''(\xi_i) \tag{7.64}$$

It can be shown that the global error is

$$\max_{1 \le i \le n} \left|y(x_i) - \overline{y}_h(x_i)\right| \approx O(h^2) \tag{7.65}$$

where h is sufficiently small. In the improved Euler method or improved Euler–Cauchy method, also known as Heun's method, we first compute an approximate value of the slope y_{i+1} at the point x_{i+1}.

The usual choice of the initial guess or initial approximation $\overline{y}_{i+1}^{(0)}$ is based on Euler's method

$$\overline{y}_{i+1}^{(0)} = y_i + hf(x_i, \overline{y}_i), \quad i = 0, 1, 2, \ldots \tag{7.66}$$

with $\overline{y}_0 = y_0$.

This is called the *predictor formula*. Then this value of y_{i+1} is improved using the average of two slopes in Equation 7.63. Thus, it leads to the following iteration formula:

$$\bar{y}_{i+1}^{(k+1)} = \bar{y}_i + \frac{h}{2}\left[f(x_i, \bar{y}_i) + f(x_{i+1}, \bar{y}_{i+1}^{(k)})\right], \quad k = 0, 1, 2, \ldots; \quad i = 0, 1, 2, \ldots \tag{7.67}$$

This is called the *corrector formula*. Thus, the improved Euler method is a predictor–corrector method, because in each step, we first predict the value $\bar{y}_{i+1}^{(0)}$ by the predictor formula in Equation 7.66 and then correct it by the corrector formula in Equation 7.67. It will thus generate a sequence of successive approximations $\bar{y}_{i+1}^{(k)}$ for $k = 0, 1, 2, \ldots$.

The iteration process generated by Equation 7.67 is terminated at each step if the following condition is satisfied

$$\left|\bar{y}_{i+1}^{(k+1)} - \bar{y}_{i+1}^{(k)}\right| < \varepsilon, \quad k = 0, 1, 2, \ldots \tag{7.68}$$

where ε is the error tolerance depending on the level of accuracy to be accomplished.

- *Geometrical interpretation of the improved Euler method*: Initially, Euler's method is used to predict the slope $\bar{y}_{i+1}^{(0)}$ at the end of the subinterval, that is, at the point x_{i+1}. Then the slope $\bar{y}_{i+1}^{(k+1)}$ at end of the subinterval is estimated by using the average of the two slopes at the points $(x_i, \bar{y}_i)$ and $(x_{i+1}, \bar{y}_{i+1}^{(k)})$. This process is repeated for the next subinterval $[x_{i+1}, x_{i+2}]$, and so on, until x reaches $x_n = b$. The improved Euler method is illustrated in Figure 7.2.
- *Local error in the improved Euler method*: Equation 7.64 shows that the local truncation error for the improved Euler formula is $O(h^3)$ as opposed to $O(h^2)$ for Euler's method. It can also be shown that for a finite interval, the global truncation error for the improved Euler formula is bounded by $O(h^2)$, so this method is a second-order method.
- *Convergence of iterative solution and control of local error*: Let $\bar{y}_{i+1}^{(0)}$ be a suitable initial approximation to the solution y_{i+1}. Now, the iterative formula in the improved Euler method is

$$\bar{y}_{i+1}^{(k+1)} = \bar{y}_i + \frac{h}{2}\left[f(x_i, \bar{y}_i) + f(x_{i+1}, \bar{y}_{i+1}^{(k)})\right], \quad k = 0, 1, 2, \ldots; \quad i = 0, 1, 2, \ldots \tag{7.69}$$

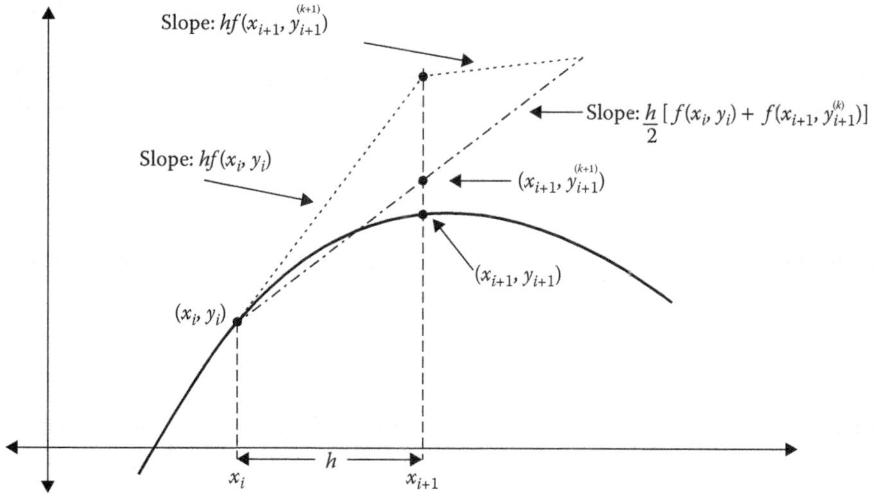

FIGURE 7.2 Improved Euler method.

This corrector formula will generate a sequence of successive approximations $\overline{y}_{i+1}^{(k)}$ for $k = 0, 1, 2, \ldots$. Now, to determine the condition under which the iteration process converges, we subtract Equation 7.69 from 7.63 to obtain

$$\overline{y}_{i+1} - \overline{y}_{i+1}^{(k+1)} = \frac{h}{2}\left[f(x_{i+1}, \overline{y}_{i+1}) - f(x_{i+1}, \overline{y}_{i+1}^{(k)})\right] \tag{7.70}$$

Using Lipchitz's condition (Equation 7.4), we get

$$\left|\overline{y}_{i+1} - \overline{y}_{i+1}^{(k+1)}\right| \le \frac{hL}{2}\left|\overline{y}_{i+1} - \overline{y}_{i+1}^{(k)}\right|, \quad k = 0, 1, 2, \ldots \tag{7.71}$$

Therefore,

$$\begin{aligned}\left|\overline{y}_{i+1} - \overline{y}_{i+1}^{(k)}\right| &\le \frac{hL}{2}\left|\overline{y}_{i+1} - \overline{y}_{i+1}^{(k-1)}\right| \\ &\le \left(\frac{hL}{2}\right)^2 \left|\overline{y}_{i+1} - \overline{y}_{i+1}^{(k-2)}\right| \\ &\;\vdots \\ &\le \left(\frac{hL}{2}\right)^k \left|\overline{y}_{i+1} - \overline{y}_{i+1}^{(0)}\right|\end{aligned} \tag{7.72}$$

Hence, if

$$\frac{hL}{2} < 1 \tag{7.73}$$

then the sequence of iterates $\overline{y}_{i+1}^{(k)}$ will converge to $\overline{y}_{i+1}$ as $k \to \infty$. Practically, sometimes, the step size h and the initial approximation $\overline{y}_{i+1}^{(0)}$ are so chosen to ensure that only one iteration is necessary to compute and then we may accept $\overline{y}_{i+1} \cong \overline{y}_{i+1}^{(1)}$.

Now, we know that the truncation error is $O(h^3)$ in computing $\overline{y}_{i+1}$ from $\overline{y}_i$. In order to maintain this order of accuracy, the iterate $\overline{y}_{i+1}^{(k)}$, which is chosen as solution $\overline{y}_{i+1}$, should at least satisfy

$$\left|\overline{y}_{i+1} - \overline{y}_{i+1}^{(k)}\right| \cong O(h^3)$$

If we want the iteration error to be sufficiently small, then the iterate $\overline{y}_{i+1}^{(k)}$ should be chosen to satisfy

$$\left|\overline{y}_{i+1} - \overline{y}_{i+1}^{(k)}\right| \cong O(h^4) \tag{7.74}$$

Now, let us analyze the local error in computing from $\overline{y}_{i+1}$ from $\overline{y}_i$. Let $u_i(x)$ be the solution of $y' = f(x, y)$ that passes through (x_i, y_i), that is,

$$\overline{u_i'}(x) = f(x, u_i(x)), \quad u_i(x_i) = \overline{y}_i.$$

Now, at step x_i, knowing $\bar{y}_i$, we compute $u_i(x_{i+1})$ using Equation 7.62 yielding

$$u_i(x_{i+1}) = \bar{y}_i + \frac{h}{2}\left[f(x_i,\bar{y}_i) + f(x_{i+1},u_i(x_{i+1}))\right] - \frac{h^3}{12}y'''(\xi_i) \tag{7.75}$$

Subtracting Equation 7.63 from Equation 7.75, we obtain the local error as

$$\text{LE}_{i+1} \equiv u_i(x_{i+1}) - \bar{y}_{i+1} = \bar{y}_i + \frac{h}{2}\left[f(x_{i+1},u_i(x_{i+1})) - f(x_{i+1},\bar{y}_{i+1})\right] - \frac{h^3}{12}y'''(\xi_i) \tag{7.76}$$

It can be easily shown that for all sufficiently small h

$$\text{LE}_{i+1} \equiv u_i(x_{i+1}) - \bar{y}_{i+1} \cong O(h^3) \tag{7.77}$$

Thus, the local error is similar to the truncation error. Now, expanding $u_i(x_{i+1})$ in Taylor series about $x = x_i$, we have

$$\begin{aligned} u_i(x_{i+1}) &= u_i(x_i) + hf(x_i,u_i(x_i)) + \frac{h^2}{2}u_i''(\tilde{x}_i), x_i < \tilde{x}_i < x_{i+1} \\ &= \bar{y}_i + hf(x_i,\bar{y}_i) + \frac{h^2}{2}u_i''(\tilde{x}_i) \end{aligned} \tag{7.78}$$

If we use Euler's method to compute the initial approximation $\bar{y}_{i+1}^{(0)}$, then using the predictor formula, from Equation 7.78, we get

$$u_i(x_{i+1}) - \bar{y}_{i+1}^{(0)} = \frac{h^2}{2}u_i''(\tilde{x}_i), x_i < \tilde{x}_i < x_{i+1} \tag{7.79}$$

Therefore, Equations 7.77 and 7.79 show that

$$\bar{y}_{i+1} - \bar{y}_{i+1}^{(0)} = O(h^2) \tag{7.80}$$

In order to satisfy the condition, Equations 7.74, 7.72 along with Equation 7.80 imply that at least two iterates have to be computed and then we may accept $\bar{y}_{i+1} \cong \bar{y}_{i+1}^{(2)}$ for the solution $\bar{y}_{i+1}$.

7.2.5.1 Algorithm of the Improved Euler Method

Step 1: Start the program.
Step 2: Define the function $f(x, y)$.
Step 3: Read x_0, y_0, h, x and error tolerance level ε.
Step 4: Compute $n = \overline{(x - x_0)/h}$.
Step 5: for $i = 0(1)n-1$ do

$y_{i+1}^{(0)} = y_i + hf(x_i, y_i)$; (Euler formula for predicted value of y_{i+1})
$x_{i+1} = x_0 + (i+1)h$;

Step 6: $k = 0$;

$y_{i+1}^{(k+1)} = y_i + \left(h/2\left[f(x_i, y_i) + f(x_{i+1}, y_{i+1}^{(k)})\right]\right)$; (Heun's formula for corrected value of y_{i+1})

while $\left|y_{i+1}^{(k+1)} - y_{i+1}^{(k)}\right| > \varepsilon$ do
$k = k+1$;
$y_{i+1}^{(k+1)} = y_i + (h/2)\left[f(x_i,y_i) + f(x_{i+1}, y_{i+1}^{(k)})\right]$;
end

Step 7: Set $y_{i+1} = y(x_{i+1}) = y_{i+1}^{(k+1)}$.
Step 8: end.
Step 9: Print y_{i+1}.
Step 10: Stop the program. ■

MATHEMATICA® Program for Solving ODE by Using the Improved Euler Method (Chapter 7, Example 7.8)

```
f[x_,y_]:=x² + y²;
x[0]=0;
y[0]=1;
h = 0.05;
a = 0.1;
n = (a-x[0])/h;
For[i=0,i<=n-1,i++,
  x[i]=x[0]+i*h;
  y[i+1,0]=y[i]+h*f[x[i],y[i]];
  For[j=0,j<=3,j++,
   y[i+1,j+1]=y[i]+h/2*(f[x[i],y[i]]+f[x[i]+h,y[i+1,j]]);
   Print["y[",x[i]+h,"]=",y[i+1,j+1]]];
  y[i+1]= y[i+1,j]];
```

Output:

```
y[0.05]=1.05263
y[0.05]=1.05276
y[0.05]=1.05277
y[0.05]=1.05277
y[0.1]=1.1115
y[0.1]=1.11168
y[0.1]=1.11169
y[0.1]=1.11169
```

Example 7.8

Determine the value of y when $x = 0.1$ given that

$$y' = x^2 + y^2, \quad y(0) = 1$$

Solution:

We take, $h = 0.05$ Here, $x_0 = 0$ and $y_0 = 1$. Therefore, $f(x_0, y_0) = 1$ By Euler's formula, we get

$$y_1^{(0)} = y_0 + hf(x_0,y_0) = 1.05$$

This predicted value of $y_1^{(0)}$ is used in the corrector formula as an initial approximation. Using the corrector formula of the improved Euler formula, we get the first approximation to y_1 as

$$y_1^{(1)} = y_0 + \frac{h}{2}\left[f(x_0, y_0) + f(x_1, y_1^{(0)})\right] = 1.05256$$

Again, the second approximation to y_1 is

$$y_1^{(2)} = y_0 + \frac{h}{2}\left[f(x_0, y_0) + f(x_1, y_1^{(1)})\right] = 1.0527$$

Hence, we take $y_1 = 1.053$, correct to three decimal places, that is, $y(0.05) = 1.053$. Next, again by using Euler's formula, we get

$$y_2^{(0)} = y_1 + hf(x_1, y_1) = 1.10857$$

This predicted value of $y_2^{(0)}$ is used in the corrector formula as an initial approximation. Using the corrector formula of the improved Euler method, we get the second approximation to y_2 as

$$y_2^{(1)} = y_1 + \frac{h}{2}\left[f(x_1, y_1) + f(x_2, y_2^{(0)})\right] = 1.11157$$

$$y_2^{(2)} = y_1 + \frac{h}{2}\left[f(x_1, y_1) + f(x_2, y_2^{(1)})\right] = 1.11173$$

Therefore, we take $y_2 = 1.112$, correct to three decimal places, that is, $y(0.1) = 1.112$. Therefore, the required solution is

$$y(0.1) = 1.112$$

7.2.6 Runge–Kutta Methods

Runge–Kutta (R–K) methods are the most widely used numerical methods for solving ordinary differential equations. It is especially suitable in the case of ordinary differential equations, where the computation of higher derivatives is complicated. As we have already seen that Euler's method is less efficient in practical problems because it requires step length h to be very small for obtaining reasonable accuracy. R–K methods aim to achieve higher accuracy by sacrificing the efficiency of Euler's method through re-evaluating the function $f(x, y)$ at the points intermediate between $(x_n, y(x_n))$ and $(x_{n+1}, y(x_{n+1}))$. In compared to simpler Euler's explicit method, R–K methods provide more accurate solution. Various types of R–K methods are classified according to their order.

The general s-stage R–K family is defined by

$$\bar{y}_{i+1} = \bar{y}_i + hF(x_i, \bar{y}_i; h), \quad i = 0, 1, 2, \ldots \tag{7.81}$$

$$hF(x, y; h) = \sum_{r=1}^{s} \omega_r k_1 \tag{7.82}$$

$$k_1 = hf(x, y) \tag{7.83}$$

$$k_r = hf(x + a_r h, y + \sum_{j=1}^{r-1} b_{rj} k_j), \quad r = 2,3,\ldots, s \tag{7.84}$$

$$a_r = \sum_{j=1}^{r-1} b_{rj}, \quad r = 2,3,\ldots, s \tag{7.85}$$

- *One-stage R–K methods*: Suppose that $s = 1$. Then, the resulting one-stage R–K method is simply Euler's explicit method

$$\bar{y}_{i+1} = \bar{y}_i + hf(x_i, \bar{y}_i), \quad i = 0,1,2,\ldots \tag{7.86}$$

- *Two-stage R–K methods*: Let us consider the case of $s = 2$, corresponding to the following family of methods

$$\bar{y}_{i+1} = \bar{y}_i + hF(x_i, \bar{y}_i; h), \quad i = 0,1,2,\ldots \tag{7.87}$$

where

$$hF(x, y; h) = (\omega_1 k_1 + \omega_2 k_2) \tag{7.88}$$

$$k_1 = hf(x, y) \tag{7.89}$$

$$k_2 = hf(x + a_2 h, y + b_{21} k_1) \tag{7.90}$$

Now, the parameters ω_1, ω_2, a_2, and b_{21} are to be determined, so that when the exact solution $y(x)$ is substituted in Equation 7.87, the truncation error

$$T_{i+1} = y_{i+1} - y_i - hF(x_i, y_i; h) \tag{7.91}$$

will satisfy

$$T_{i+1} \equiv O(h^3) \tag{7.92}$$

Now, from Equations 7.88 through 7.90, we obtain

$$hF(x, y; h) = \omega_1 hf(x, y) + \omega_2 hf(x + a_2 h, y + b_{21} k_1)$$

We expand the function $f(x + a_2 h, y + b_{21} k_1)$ in two variables around the point (x, y) to obtain

$$f(x + a_2 h, y + b_{21} k_1) = f(x, y) + a_2 h f_x + b_{21} k_1 f_y + O(h^2)$$

In addition,

$$y(x+h) = y(x) + hy' + \frac{h^2}{2!} y'' + O(h^3)$$

$$= y(x) + hy' + \frac{h^2}{2!}(f_x + ff_y) + O(h^3), \text{ using Equation 7.16}$$

Then

$$T_{i+1} = \left[y(x+h) - y - hF(x,y;h)\right]_{x=x_i,y=y_i}$$

$$= \left[y + hy' + \frac{h^2}{2!}(f_x + ff_y) - y - \omega_1 hf(x,y) - \omega_2 h\left[f(x,y) + a_2 hf_x + b_{21}k_1 f_y\right] + O(h^3)\right]_{x=x_i,y=y_i}$$

$$= \left[hf(x,y) + \frac{h^2}{2!}(f_x + ff_y) - \omega_1 hf(x,y) - \omega_2 h\left[f(x,y) + a_2 hf_x + b_{21}hff_y\right] + O(h^3)\right]_{x=x_i,y=y_i}$$

Therefore, the requirement Equation 7.92 implies that the coefficients must satisfy the system of following equations:

$$1 - \omega_1 - \omega_2 = 0$$

$$\frac{1}{2} - \omega_2 a_2 = 0$$

$$\frac{1}{2} - \omega_2 b_{21} = 0$$

Therefore,

$$\omega_1 = 1 - \omega_2$$

$$a_2 = b_{21} = \frac{1}{2\omega_2} \tag{7.93}$$

$$\omega_2 \neq 0$$

Thus, there is a family of R–K methods of order 2, depending on the choice of ω_2. Among different values of ω_2, the most suitable and favorable choices are $\omega_2 = 1/2$ and 1.

If $\omega_2 = 1/2$, we obtain the following numerical formula:

$$\bar{y}_{i+1} = \bar{y}_i + \frac{h}{2}\left[f(x_i, \bar{y}_i) + f(x_i + h, \bar{y}_i + hf(x_i, \bar{y}_i))\right], \quad i = 0,1,2,\ldots \tag{7.94}$$

This method is usually known as Heun's method or sometimes the improved Euler method. Again, if $\omega_2 = 1$, the resulting numerical formula is

$$\bar{y}_{i+1} = \bar{y}_i + hf\left(x_i + \frac{h}{2}, \bar{y}_i + \frac{h}{2}f(x_i, \bar{y}_i)\right), \quad i = 0,1,2,\ldots \tag{7.95}$$

This method is also known as the modified Euler's method. These two are well-known examples of second-order R–K methods form Equations 7.87 through 7.90. The family Equations 7.87 through 7.90 is referred to as the class of explicit two-stage R–K methods.

- *Three-stage R–K methods*: Let us consider the case of $s = 3$, corresponding to the following family of methods

$$\bar{y}_{i+1} = \bar{y}_i + hF(x_i, \bar{y}_i; h), \quad i = 0,1,2,\ldots \tag{7.96}$$

where

$$hF(x,y;h) = (\omega_1 k_1 + \omega_2 k_2 + \omega_3 k_3) \tag{7.97}$$

$$k_1 = hf(x,y), \tag{7.98}$$

$$k_2 = hf(x + a_2 h, y + b_{21} k_1) \tag{7.99}$$

$$k_3 = hf(x + a_3 h, y + b_{31} k_1 + b_{32} k_2) \tag{7.100}$$

$$a_2 = b_{21}, a_3 = b_{31} + b_{32} \tag{7.101}$$

Now, the parameters ω_1, ω_2, ω_3, a_2, a_3, b_{31}, and b_{32} are to be determined. Now, expanding k_2 and k_3 into Taylor's series about the point (x, y) yields

$$k_2 = h\left[f + a_2 h f_x + b_{21} k_1 f_y + \frac{1}{2}\left(a_2^2 h^2 f_{xx} + 2a_2 h b_{21} k_1 f_{xy} + b_{21}^2 k_1^2 f_{yy}\right) + O(h^3)\right]$$

$$= hf + a_2 h^2 f_x + a_2 h^2 f f_y + \frac{1}{2} a_2^2 h^3 \left(f_{xx} + 2 f f_{xy} + f^2 f_{yy}\right) + O(h^4)$$

$$= hf + a_2 h^2 F_1 + \frac{1}{2} a_2^2 h^3 F_2 + O(h^4)$$

where

$$F_1 = f_x + f f_y \quad \text{and} \quad F_2 = f_{xx} + 2 f f_{xy} + f^2 f_{yy}$$

and

$$k_3 = h\left[f + a_3 h f_x + (b_{31} k_1 + b_{32} k_2) f_y + \frac{1}{2}\left(a_3^2 h^2 f_{xx} + 2a_3 h (b_{31} k_1 + b_{32} k_2) f_{xy} + (b_{31} k_1 + b_{32} k_2)^2 f_{yy}\right) + O(h^3)\right]$$

$$= hf + a_3 h^2 f_x + h\left[(a_3 - b_{32}) k_1 + b_{32} k_2\right] f_y + \frac{h}{2}\left[a_3^2 h^2 f_{xx} + 2a_3 h[(a_3 - b_{32}) k_1 + b_{32} k_2] f_{xy}\right.$$

$$\left. + [(a_3 - b_{32}) k_1 + b_{32} k_2]^2 f_{yy}\right] + O(h^4)$$

$$k_3 = hf + h^2 a_3 F_1 + h^3\left(a_3 b_{32} F_1 f_y + \frac{1}{2} a_3^2 F_2\right) + O(h^4)$$

Substituting these values of k_1, k_2, and k_3 in Equation 7.97, we get

$$hF(x,y;h) = \omega_1 hf + \omega_2\left(hf + a_2 h^2 F_1 + \frac{1}{2} a_2^2 h^3 F_2\right) + \omega_3\left[hf + h^2 a_3 F_1 + h^3\left(a_3 b_{32} F_1 f_y + \frac{1}{2} a_3^2 F_2\right)\right] + O(h^4)$$

This implies

$$F(x,y;h) = \omega_1 f + \omega_2\left(f + a_2 h F_1 + \frac{1}{2} a_2^2 h^2 F_2\right) + \omega_3\left[f + h a_3 F_1 + h^2\left(a_3 b_{32} F_1 f_y + \frac{1}{2} a_3^2 F_2\right)\right] + O(h^3) \tag{7.102}$$

Now, if the exact solution $y(x)$ satisfies the Equation 7.96, then we have

$$y(x+h)=y(x)+hF(x,y(x);h) \tag{7.103}$$

Again,

$$y(x+h)=y(x)+hy'+\frac{h^2}{2!}y''+\frac{h^3}{3!}y'''+O(h^4)$$

$$=y(x)+hf+\frac{h^2}{2!}(f_x+ff_y)+\frac{h^3}{3!}\left[(f_x+ff_y)f_y+f_{xx}+2ff_{xy}+f^2f_{yy}\right]+O(h^4)$$

Thus,

$$F(x,y;h)=\frac{y(x+h)-y(x)}{h}=f+\frac{h}{2!}F_1+\frac{h^2}{3!}\left[F_1f_y+F_2\right]+O(h^3) \tag{7.104}$$

Now, comparing Equations 7.102 and 7.104 shows that

$$\omega_1+\omega_2+\omega_3=1$$

$$\omega_2a_2+\omega_3a_3=\frac{1}{2}$$

$$\frac{1}{2}\omega_2a_2^2F_2+\omega_3a_3b_{32}F_1f_y+\frac{1}{2}\omega_3a_3^2F_2=\frac{1}{6}\left[F_1f_y+F_2\right]$$

The above identity yields

$$\omega_2a_2^2+\omega_3a_3^2=\frac{1}{3}$$

$$\omega_3a_3b_{32}=\frac{1}{6}$$

Thus, we obtain the system of following equations:

$$\omega_1+\omega_2+\omega_3=1$$

$$\omega_2a_2+\omega_3a_3=\frac{1}{2}$$

$$\omega_2a_2^2+\omega_3a_3^2=\frac{1}{3}$$

$$\omega_3a_2b_{32}=\frac{1}{6}$$

Solving this system of four equations for the six unknowns, that is to say, ω_1, ω_2, ω_3, a_2, a_3, and b_{32}, we can obtain a two-parameter family of the three-stage R–K methods. Now, we shall focus on two significant examples of the family.

1. *Heun's method*: This method is obtained by the following values

$$\omega_1=\frac{1}{4},\ \omega_2=0,\ \omega_3=\frac{3}{4},\ a_2=\frac{1}{3},\ a_3=\frac{2}{3},\ \text{and}\ b_{32}=\frac{2}{3}$$

yielding the following recursive formula

$$\bar{y}_{i+1} = \bar{y}_i + \frac{1}{4}(k_1 + 3k_3), \quad i = 0,1,2,\dots \tag{7.105}$$

where

$$k_1 = hf(x_i, \bar{y}_i) \tag{7.106}$$

$$k_2 = hf\left(x_i + \frac{1}{3}h, \bar{y}_i + \frac{1}{3}k_1\right) \tag{7.107}$$

$$k_3 = hf\left(x_i + \frac{2}{3}h, \bar{y}_i + \frac{2}{3}k_2\right) \tag{7.108}$$

2. *Standard third-order R–K method*: This method is also achieved by the following values:

$$\omega_1 = \frac{1}{6}, \omega_2 = \frac{2}{3}, \omega_3 = \frac{1}{6}, a_2 = \frac{1}{2}, a_3 = 1, \text{and } b_{32} = 2$$

yielding the following recursive formula

$$\bar{y}_{i+1} = \bar{y}_i + \frac{1}{6}(k_1 + 4k_2 + k_3), \quad i = 0,1,2,\dots \tag{7.109}$$

where

$$k_1 = hf(x_i, \bar{y}_i) \tag{7.110}$$

$$k_2 = hf\left(x_i + \frac{1}{2}h, \bar{y}_i + \frac{1}{2}k_1\right) \tag{7.111}$$

$$k_3 = hf(x_i + h, \bar{y}_i - k_1 + 2k_2) \tag{7.112}$$

- *Four-stage R–K methods*: For $s = 4$, an analogous argument leads to a two-parameter family of the four-stage R–K methods of order 4. In case of $s = 4$, we have the following family of methods:

$$\bar{y}_{i+1} = \bar{y}_i + hF(x_i, \bar{y}_i; h), \quad i = 0,1,2,\dots \tag{7.113}$$

where

$$hF(x, y; h) = (\omega_1 k_1 + \omega_2 k_2 + \omega_3 k_3 + \omega_4 k_4) \tag{7.114}$$

$$k_1 = hf(x, y) \tag{7.115}$$

$$k_2 = hf(x + a_2 h, y + b_{21} k_1) \tag{7.116}$$

$$k_3 = hf(x + a_3 h, y + b_{31} k_1 + b_{32} k_2) \tag{7.117}$$

$$k_4 = hf(x + a_4 h, y + b_{41} k_1 + b_{42} k_2 + b_{43} k_3) \tag{7.118}$$

$$a_2 = b_{21}, a_3 = b_{31} + b_{32}, a_4 = b_{41} + b_{42} + b_{43} \tag{7.119}$$

Now, expanding k_2, k_3, and k_4 into Taylor's series about the point (x, y) yields

$$k_2 = h\Big[f + a_2hf_x + b_{21}k_1f_y + \frac{1}{2}\left(a_2^2h^2f_{xx} + 2a_2hb_{21}k_1f_{xy} + b_{21}^2k_1^2f_{yy}\right)$$

$$+\frac{1}{3!}\left(a_2^3h^3f_{xxx} + 3a_2^2h^2b_{21}k_1f_{xxy} + 3a_2hb_{21}^2k_1^2f_{xyy} + b_{21}^3k_1^3f_{yyy}\right) + O(h^4)\Big]$$

$$= hf + a_2h^2f_x + a_2h^2ff_y + \frac{1}{2}a_2^2h^3\left(f_{xx} + 2ff_{xy} + f^2f_{yy}\right)$$

$$+\frac{1}{6}a_2^3h^4\left(f_{xxx} + 3ff_{xxy} + 3f^2f_{xyy} + f^3f_{yyy}\right) + O(h^5)$$

$$= hf + a_2h^2F_1 + \frac{1}{2}a_2^2h^3F_2 + \frac{1}{6}a_2^3h^4F_3 + O(h^5)$$

where

$$F_1 = f_x + ff_y, F_2 = f_{xx} + 2ff_{xy} + f^2f_{yy} \quad \text{and} \quad F_3 = f_{xxx} + 3ff_{xxy} + 3f^2f_{xyy} + f^3f_{yyy}$$

and

$$k_3 = h\Big[f + a_3hf_x + (b_{31}k_1 + b_{32}k_2)f_y + \frac{1}{2}\left(a_3^2h^2f_{xx} + 2a_3h(b_{31}k_1 + b_{32}k_2)f_{xy} + (b_{31}k_1 + b_{32}k_2)^2f_{yy}\right)$$

$$+\frac{1}{6}\left(a_3^3h^3f_{xxx} + 3a_3^2h^2(b_{31}k_1 + b_{32}k_2)f_{xxy} + 3a_3h(b_{31}k_1 + b_{32}k_2)^2f_{xyy} + (b_{31}k_1 + b_{32}k_2)^3f_{yyy}\right) + O(h^4)\Big]$$

$$= hf + a_3h^2f_x + h\big[(a_3 - b_{32})k_1 + b_{32}k_2\big]f_y + \frac{h}{2}\Big[a_3^2h^2f_{xx} + 2a_3h[(a_3 - b_{32})k_1 + b_{32}k_2]f_{xy} + [(a_3 - b_{32})k_1 + b_{32}k_2]^2f_{yy}\Big]$$

$$+\frac{h}{6}\Big[a_3^3h^3f_{xxx} + 3a_3^2h^2[(a_3 - b_{32})k_1 + b_{32}k_2]f_{xxy} + 3a_3h[(a_3 - b_{32})k_1 + b_{32}k_2]^2f_{xyy}$$

$$+[(a_3 - b_{32})k_1 + b_{32}k_2]^3f_{yyy}\Big] + O(h^5)$$

$$k_4 = h\Big[f + a_4hf_x + (b_{41}k_1 + b_{42}k_2 + b_{43}k_3)f_y + \frac{1}{2}\Big(a_4^2h^2f_{xx} + 2a_4h(b_{41}k_1 + b_{42}k_2 + b_{43}k_3)f_{xy}$$

$$+(b_{41}k_1 + b_{42}k_2 + b_{43}k_3)^2f_{yy} + \frac{1}{6}\Big(a_4^3h^3f_{xxx} + 3a_4^2h^2(b_{41}k_1 + b_{42}k_2 + b_{43}k_3)f_{xxy}$$

$$+3a_4h(b_{41}k_1 + b_{42}k_2 + b_{43}k_3)^2f_{xyy}\Big) + (b_{41}k_1 + b_{42}k_2 + b_{43}k_3)^3f_{yyy}\Big) + O(h^4)\Big]$$

$$= hf + a_4h^2f_x + h(b_{41}k_1 + b_{42}k_2 + b_{43}k_3)f_y + \frac{h}{2}\Big(a_4^2h^2f_{xx} + 2a_4h(b_{41}k_1 + b_{42}k_2 + b_{43}k_3)f_{xy}$$

$$+(b_{41}k_1 + b_{42}k_2 + b_{43}k_3)^2f_{yy}\Big) + \frac{h}{6}\Big(a_4^3h^3f_{xxx} + 3a_4^2h^2(b_{41}k_1 + b_{42}k_2 + b_{43}k_3)f_{xxy}$$

$$+3a_4h(b_{41}k_1 + b_{42}k_2 + b_{43}k_3)^2f_{xyy} + (b_{41}k_1 + b_{42}k_2 + b_{43}k_3)^3f_{yyy}\Big) + O(h^5)$$

$$= hf + a_4h^2f_x + h[(a_4 - b_{42} - b_{43})k_1 + b_{42}k_2 + b_{43}k_3]f_y + \frac{h}{2}\Big[a_4^2h^2f_{xx} + 2a_4h[(a_4 - b_{42} - b_{43})k_1 + b_{42}k_2 + b_{43}k_3]f_{xy}$$

$$+[(a_4 - b_{42} - b_{43})k_1 + b_{42}k_2 + b_{43}k_3]^2f_{yy}\Big] + \frac{h}{6}\Big[a_4^3h^3f_{xxx} + 3a_4^2h^2[(a_4 - b_{42} - b_{43})k_1 + b_{42}k_2 + b_{43}k_3]f_{xxy}$$

$$+3a_4h[(a_4 - b_{42} - b_{43})k_1 + b_{42}k_2 + b_{43}k_3]^2f_{xyy} + [(a_4 - b_{42} - b_{43})k_1 + b_{42}k_2 + b_{43}k_3]^3f_{yyy}\Big] + O(h^5)$$

Now, if the exact solution $y(x)$ satisfies the Equation 7.113, then we have

$$y(x+h) = y(x) + hF(x, y(x); h) \tag{7.120}$$

Also,

$$\begin{aligned} y(x+h) &= y(x) + hy' + \frac{h^2}{2!}y'' + \frac{h^3}{3!}y''' + \frac{h^4}{4!}y^{iv} + O(h^5) \\ &= y(x) + hf + \frac{h^2}{2!}F_1 + \frac{h^3}{3!}\left[F_1 f_y + F_2\right] + \frac{h^4}{4!}\left[F_3 + F_2 f_y + 3(f_{xy} + ff_{yy})F_1 + f_y^2 F_1\right] + O(h^5) \end{aligned} \tag{7.121}$$

Now, from Equations 7.114, 7.120, and 7.121, we obtain

$$\begin{aligned} hF(x, y; h) = (\omega_1 k_1 + \omega_2 k_2 + \omega_3 k_3 + \omega_4 k_4) &= hf + \frac{h^2}{2!}F_1 + \frac{h^3}{3!}\left[F_1 f_y + F_2\right] \\ &+ \frac{h^4}{4!}\left[F_3 + F_2 f_y + 3(f_{xy} + ff_{yy})F_1 + f_y^2 F_1\right] + O(h^5) \end{aligned} \tag{7.122}$$

Substituting the values of k_1, k_2, k_3, and k_4 in Equation 7.122 and then equating the coefficients of h, h^2, h^3, and h^4, we obtain the following system of equations:

$$\omega_1 + \omega_2 + \omega_3 + \omega_4 = 1$$

$$\omega_2 a_2 + \omega_3 a_3 + \omega_4 a_4 = \frac{1}{2}$$

$$\omega_2 a_2^2 + \omega_3 a_3^2 + \omega_4 a_4^2 = \frac{1}{3}$$

$$\omega_3 a_2 b_{32} + \omega_4 (a_2 b_{42} + a_3 b_{43}) = \frac{1}{6}$$

$$\omega_2 a_2^3 + \omega_3 a_3^3 + \omega_4 a_4^3 = \frac{1}{4}$$

$$\omega_3 a_2^2 b_{32} + \omega_4 (a_2^2 b_{42} + a_3^2 b_{43}) = \frac{1}{12}$$

$$\omega_3 a_2 a_3 b_{32} + \omega_4 (a_2 b_{42} + a_3 b_{43}) a_4 = \frac{1}{8}$$

$$\omega_4 a_2 b_{32} b_{43} = \frac{1}{24}$$

The above equations along with Equation 7.119 constitute 11 equations in 13 unknowns. Therefore, there are two arbitrary parameters, so that these parameters can be chosen freely. A most popular example from this family can be obtained by the following values:

$$\omega_1 = \omega_4 = \frac{1}{6},\ \omega_2 = \omega_3 = \frac{1}{3},\ a_2 = a_3 = \frac{1}{2},\ a_4 = 1,\ b_{21} = \frac{1}{2},\ b_{31} = 0,\ b_{32} = \frac{1}{2},\ b_{41} = 0,\ b_{42} = 0, \text{ and } b_{43} = 1$$

Thus, we obtain the following four-stage R–K formula of order 4, which has been considered as classical R–K method.

$$\bar{y}_{i+1} = \bar{y}_i + \frac{1}{6}(k_1 + 2k_2 + 2k_3 + k_4), \quad i = 0,1,2,\ldots \tag{7.123}$$

where

$$k_1 = hf(x_i, \bar{y}_i) \tag{7.124}$$

$$k_2 = hf\left(x_i + \frac{h}{2}, \bar{y}_i + \frac{k_1}{2}\right) \tag{7.125}$$

$$k_3 = hf\left(x_i + \frac{h}{2}, \bar{y}_i + \frac{k_2}{2}\right) \tag{7.126}$$

$$k_4 = hf(x_i + h, \bar{y}_i + k_3) \tag{7.127}$$

Clearly, the truncation error of this method is $O(h^5)$. From Equation 7.37, the global error of $O(h^4)$ can be easily obtained.

As a particular case, if $f(x, y)$ does not depends on y, then this formula reduces to Simpson's integration rule. The explicit R–K method of the fourth order is illustrated in Figure 7.3.

7.2.6.1 Algorithm for R–K Method of Order 4

Step 1: Start the program.
Step 2: Define the function $f(x, y)$.
Step 3: Read x_0, y_0, h, x.
Step 4: Compute $n = (x - x_0)/h$.
Step 5: for $i = 0(1)n - 1$ do.

$x_i = x_0 + i * h$
$k_1 = h * f(x_i,\ y_i)$
$k_2 = h * f(x_i + h/2,\ y_i + k_1/2)$
$k_3 = h * f(x_i + h/2,\ y_i + k_2/2)$
$k_4 = h * f(x_i + h,\ y_i + k_3)$
$k = 1/6(k_1 + 2k_2 + 2k_3 + k_4)$
$y_{i+1} = y_i + k.$

Step 6: Print the value of y_{i+1}.
Step 7: Stop the program. ■

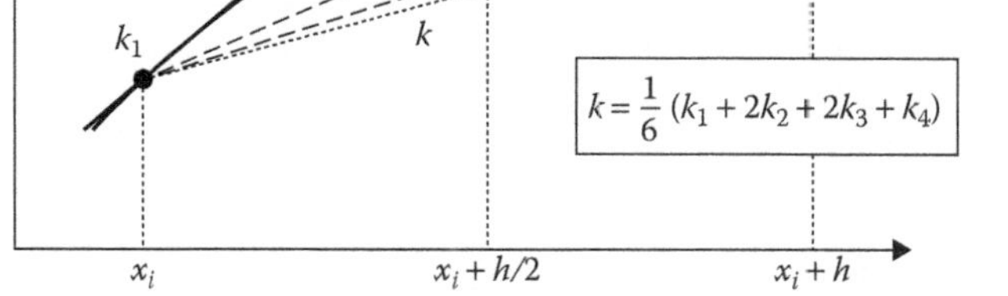

FIGURE 7.3 Fourth-order Runge–Kutta method.

MATHEMATICA® Program for Solving ODE by R–K Method (Chapter 7, Example 7.11)

```
f[x_,y_]:=-y+y²;
x[0]=0;
y[0]=-1;
h=0.1;
For[i=0,i≤4,i++,
  x[i]=x[0]+i*h;
  Print[step[i+1]];
  k[1]=h*f[x[i],y[i]];
  Print[k[1]];
  k[2]=h*f[x[i]+h/2,y[i]+k[1]/2];
  Print[k[2]];
  k[3]=h*f[x[i]+h/2,y[i]+k[2]/2];
  Print[k[3]];
  k[4]=h*f[x[i]+h,y[i]+k[3]];
  Print[k[4]];
  y[i+1]=y[i]+1/6*(k[1]+2*k[2]+2*k[3]+k[4]);
  yexact[i+1]=(1-2*Exp[x])⁻¹/.x->(x[i]+h);
  e[i]=Abs[y[i+1]-yexact[i+1]];

  Print[x[i]+h,"    ",y[i+1],"     ",yexact[i+1],"       ",e[i]]];
```

Output:

```
step[1]
0.2
0.171
0.175081
0.150541
0.1 -0.826216 -0.826213 3.28525×10-6

step[2]
0.150885
0.131443
0.133885
0.117165
0.2 -0.6930908 -0.693094 4.28461×10-6

step[3]
0.117348
0.103692
0.105246
0.0933423
0.3 -0.588337 -0.588333 4.39999×10-6

step[4]
0.0934478
0.0834959
0.084535
0.0757619
0.4 -0.504126 -0.504121 4.17319×10-6

step[5]
0.0758268
0.0683566
0.0690798
0.0624311
0.5 -0.43527 -0.435267 3.82653×10-6
```

7.2.6.2 A General Form for Explicit R–K Methods

An explicit R–K formula with s stages has the following form:

$$u_r = \bar{y}_i + h\sum_{l=1}^{r-1} b_{rl} f(x_i + a_l h, u_l), \quad r = 1,2,\ldots,s \tag{7.128}$$

$$\bar{y}_{i+1} = \bar{y}_i + h\sum_{l=1}^{s} \omega_l f(x_i + a_l h, u_l) \tag{7.129}$$

The coefficients may be determined from Table 7.1, called a *Butcher tableau* (named after J. C. Butcher). The coefficients $\{a_i\}$ and $\{b_{ij}\}$ are usually assumed to satisfy the following conditions:

$$a_i = \sum_{j=1}^{i-1} b_{ij}, \quad i = 2,3,\ldots,s \tag{7.130}$$

7.2.6.3 Estimation of the Truncation Error and Control

In the numerical solution of differential equations, it is desirable to control the size of the local truncation error. For that, we must estimate the local truncation error at each step. The easiest way to estimate the error $y(x) - \bar{y}_h(x)$ in a numerical solution $\bar{y}_h(x)$ is to use Richardson's extrapolation. We shall solve the initial value problem twice on the given interval $[x_0, b]$, with the step sizes $2h$ and h. Then we can use Richardson's extrapolation to estimate $y(x) - \bar{y}_h(x)$ in terms of $y(x) - \bar{y}_{2h}(x)$. The R–K methods have asymptotic error formulae. This asymptotic error formula can be obtained as

$$y(x_i) - \bar{y}_h(x_i) = \varphi(x_i)h^p + O(h^{p+1}) \tag{7.131}$$

where $\varphi(x)$ satisfies a certain initial value problem. The asymptotic result (7.131) justifies the use of Richardson's extrapolation to estimate the error and to accelerate the convergence.

For classical R–K method of order 4, the asymptotic error formula becomes

$$y(x) - \bar{y}_h(x) = \varphi(x)h^4 + O(h^5) \tag{7.132}$$

For a step size of $2h$, the asymptotic error formula leads to

$$y(x) - \bar{y}_{2h}(x) = 16\varphi(x)h^4 + O(h^5) \tag{7.133}$$

Now, using Richardson's extrapolation, we obtain

TABLE 7.1
Butcher Tableau

$0 = a_1$					
a_2	b_{21}				
a_3	b_{31}	b_{32}			
$\vdots$	$\vdots$		$\ddots$		
a_s	b_{s1}	b_{s2}	$\cdots$	$b_{s\,\overline{s-1}}$	
	ω_1	ω_2		ω_{s-1}	ω_s

$$y(x)-\bar{y}_h(x)=\frac{1}{15}\left[\bar{y}_h(x)-\bar{y}_{2h}(x)\right]+O(h^5) \tag{7.134}$$

The first term on the right-hand side of Equation 7.134 is an asymptotic error estimate of the left-hand error.

Therefore, in order to determine whether the values obtained by R–K methods are sufficiently accurate or not, we have to recompute the value obtained by R–K method at the end of each subinterval with the step size halved. If only small change in the value of y_{i+1} occurs, then the results are accepted. Otherwise, the step must be halved again until the results are satisfactory. This procedure is very expensive. The cost of estimating error in this way is an approximately 50% increase in the amount of computation, as compared to the cost of computing just y_{i+1}.

In order to better understand the expense of error estimate in R–K methods, let us consider the fourth-order R–K method in which four evaluations are required per step. Now, in order to move from (x_i, y_i) to (x_i+h, y_i+h), we require eight function evaluations to compute $y_{h/2}(x_i+h)$ and three additional function evaluations to obtain $y_h(x_i+h)$. Thus, in the best case, the variable-step size algorithm would require a total of 11 function evaluations to go from (x_i, y_i) to (x_i+h, y_i+h). Although it is quite expensive, still a variable-step size R–K method is very stable, reliable, and easily programmable.

7.2.6.4 R–K–Fehlberg Method

In the 1970s, a novel technique was devised by E. Fehlberg, in which the local error is computed by comparing the computed value of y_{i+1} with that obtained by an associated higher order R–K formula. It has led to the most popular R–K methods. These methods are often called *Fehlberg methods*. Instead of computing with a method of fixed order, in this case, it simultaneously computes by using two methods of different orders. In the present analysis, we consider only the most popular pair of R–K–Fehlberg formulae of order 4 and 5. These formulae are computed simultaneously and then the higher order formula is used to estimate the local truncation error in the fourth-order formula. Fehlberg discovered two R–K formulae that together need only six function evaluations per step.

Thus, in both the formulae, we need only six different function evaluations altogether, which are as follows:

$$k_1 = hf(x_i, \bar{y}_i) \tag{7.135}$$

$$k_2 = hf\left(x_i+\frac{h}{4}, \bar{y}_i+\frac{k_1}{4}\right) \tag{7.136}$$

$$k_3 = hf\left(x_i+\frac{3h}{8}, \bar{y}_i+\frac{3k_1}{32}+\frac{9k_2}{32}\right) \tag{7.137}$$

$$k_4 = hf\left(x_i+\frac{12}{13}h, \bar{y}_i+\frac{1932k_1}{2197}-\frac{7200k_2}{2197}+\frac{7296k_3}{2197}\right) \tag{7.138}$$

$$k_5 = hf\left(x_i+h, \bar{y}_i+\frac{439k_1}{216}-8k_2+\frac{3680k_3}{513}-\frac{845k_4}{4104}\right) \tag{7.139}$$

$$k_6 = hf\left(x_i+\frac{1}{2}h, \bar{y}_i-\frac{8k_1}{27}+2k_2-\frac{3544k_3}{2565}+\frac{1859k_4}{4104}-\frac{11k_5}{40}\right) \tag{7.140}$$

Then the fourth and fifth-order formulae are given by

$$\bar{y}_{i+1} = \bar{y}_i + \sum_{j=1}^{5} \delta_j k_j, \quad i = 0,1,2,\ldots \tag{7.141}$$

$$\hat{\bar{y}}_{i+1} = \bar{y}_i + \sum_{j=1}^{6} \gamma_j k_j, \quad i = 0,1,2,\ldots \tag{7.142}$$

where the coefficient vector

$$\delta \equiv [\delta_1 \quad \delta_2 \quad \delta_3 \quad \delta_4 \quad \delta_5] = \begin{bmatrix} \frac{25}{216} & 0 & \frac{1408}{2565} & \frac{2197}{4104} & \frac{-1}{5} \end{bmatrix}$$

and the coefficient vector

$$\gamma \equiv [\gamma_1 \quad \gamma_2 \quad \gamma_3 \quad \gamma_4 \quad \gamma_5 \quad \gamma_6] = \begin{bmatrix} \frac{16}{135} & 0 & \frac{6656}{12825} & \frac{28561}{56430} & \frac{-9}{50} & \frac{2}{55} \end{bmatrix}$$

The local error in the fourth-order formula (7.141) is estimated by

$$\varepsilon_{i+1} \approx \hat{\bar{y}}_{i+1} - \bar{y}_{i+1}, \quad i = 0,1,2,\ldots$$

This error estimate requires only 6 function evaluations as compared to 11 function evaluations required in variable-step size R–K method. By using this estimate, one can increase or decrease h depending on the values of the estimated error.

7.2.6.4.1 Algorithm for R–K–Fehlberg Method

Step 1: Start the program.
Step 2: Define the function $f(x, y)$.
Step 3: Read $x_0, \bar{y}_0, h, x, \delta_i, i = 1, \ldots, 5, \gamma_j, j = 1, \ldots, 6$.
Step 4: Compute $n = (x - x_0)/h$.
Step 5: for $i = 0(1)n-1$ do

$x_i = x_0 + i * h;$

$k_1 = h * f(x_i, \bar{y}_i);$

$$k_2 = h * f\left(x_i + \frac{h}{4}, \bar{y}_i + \frac{k_1}{4}\right);$$

$$k_3 = h * f\left(x_i + \frac{3h}{8}, \bar{y}_i + \frac{3k_1}{32} + \frac{9k_2}{32}\right);$$

$$k_4 = h * f\left(x_i + \frac{12h}{13}, \bar{y}_i + \frac{1932k_1}{2197} - \frac{7200k_2}{2197} + \frac{7296k_3}{2197}\right);$$

$$k_5 = h * f\left(x_i + h, \bar{y}_i + \frac{439k_1}{216} - 8k_2 + \frac{3680k_3}{513} - \frac{845k_4}{4104}\right);$$

$$k_6 = h * f\left(x_i + \frac{h}{2}, \bar{y}_i - \frac{8k_1}{27} + 2k_2 - \frac{3544k_3}{2565} + \frac{1859k_4}{4104} - \frac{11k_5}{40}\right);$$

Sum=0;

For $j = 1(1)5$ do

Sum=sum+$\delta_j k_j$;

End

$\bar{y}_{i+1} = \bar{y}_i + \text{Sum}$;
Sum1 = 0;
For $j = 1(1)6$ do
Sum1=sum1+$\gamma_j k_j$;
End
$\hat{\bar{y}}_{i+1} = \bar{y}_i + \text{Sum1}$;
$\varepsilon_{i+1} = \hat{\bar{y}}_{i+1} - \bar{y}_{i+1}$;

Step 6: Print the value of $\hat{\bar{y}}_{i+1}$.

Step 7: Stop the program. ■

MATHEMATICA® Program for Solving ODE by R–K–Fehlberg Method (Chapter 7, Example 7.12)

```
f[x_,y_]:=x*y+y²;
δ={25/216,0,1408/2565,2197/4104,(-1)/5};
γ={16/135,0,6656/12825,28561/56430,(-9)/50,2/55};
x[0]=0;
y[0]=1;
h=0.1;
For[i=0,i<=4,i++,
  x[i]=x[0]+i*h;
  Print[step[i+1]];
  k[1]=h*f[x[i],y[i]];
  Print[k[1]];
  k[2]=h*f[x[i]+h/4,y[i]+k[1]/4];
  Print[k[2]];
  k[3]=h*f[x[i]+(3*h)/8,y[i]+(3*k[1])/32+(9*k[2])/32];
  Print[k[3]];
  k[4]=h*f[x[i]+(12*h)/13,
   y[i]+(1932*k[1])/2197-(7200*k[2])/2197+(7296*k[3])/2197];
  Print[k[4]];
  k[5]=h*f[x[i]+h,
   y[i]+(439*k[1])/216-8*k[2]+(3680*k[3])/513-(845*k[4])/4104];
  Print[k[5]];
  k[6]=h*f[x[i]+h/2,y[i]-(8*k[1])/27+2*k[2]-
   (3544*k[3])/2565 + (1859*k[4])/4104-(11*k[5])/40];
  Print[k[6]];
  y[i+1]=y[i]+ Σ_{j=1}^{5}(δ[[j]]*k[j]);
  y2[i+1]=y[i]+Σ_{j=1}^{6}(γ[[j]]*k[j]);
  ϵ[i+1]=y2[i+1]-y[i+1];
  Print[x[i],"    ",y[i+1],"    ",y2[i+1],"    ",ϵ[i+1]]];
```

Output:

```
step[1]
0.1
0.107625
0.111985
0.132791
0.136224
0.116272
0.    1.11689      1.11689      -1.56753*10^-7

step[2]
0.135913
0.146835
```

```
0.153207
0.184073
0.189321
0.15944
0.1      1.27739      1.27739      -3.97024*10^-7
step[3]
0.188721
0.205253
0.21514
0.263981
0.272608
0.224741
0.2      1.50413      1.50413      -1.06895*10^-6
step[4]
0.271364
0.298198
0.314772
0.398951
0.414585
0.330698
0.3      1.83898      1.83897      -3.19285*10^-6
step[5]
0.411743
0.459634
0.490545
0.654403
0.686973
0.519798
0.4      2.36883      2.36882      -0.000011222
```

Example 7.9

Use the R–K method to find the numerical solution at $x = 0.8$ for

$$\frac{dy}{dx} = \sqrt{x+y}, y(0.4) = 0.41$$

assuming the step length $h = 0.2$.

Solution:

The formula for R–K method of order 4 is given by

$$\bar{y}_{i+1} = \bar{y}_i + \frac{1}{6}(k_1 + 2k_2 + 2k_3 + k_4), \quad i = 0,1,2,\ldots,n \tag{7.143}$$

where

$$k_1 = hf(x_i, \bar{y}_i) \tag{7.144}$$

$$k_2 = hf\left(x_i + \frac{h}{2}, \bar{y}_i + \frac{k_1}{2}\right) \tag{7.145}$$

$$k_3 = hf\left(x_i + \frac{h}{2}, \bar{y}_i + \frac{k_2}{2}\right) \tag{7.146}$$

$$k_4 = hf(x_i + h, \bar{y}_i + k_3) \tag{7.147}$$

Here,

$$n = \frac{b-a}{h} = \frac{0.8-0.4}{0.2} = 2$$

For $i = 0$, we have

$$x_0 = 0.4, \quad y_0 = 0.41$$

$$k_1 = hf(x_0, \bar{y}_0) = 0.2 \times \sqrt{0.4+0.41} = 0.18$$

$$k_2 = hf\left(x_0 + \frac{h}{2}, \bar{y}_0 + \frac{k_1}{2}\right) = 0.2 \times \sqrt{0.5+0.5} = 0.2$$

$$k_3 = hf\left(x_0 + \frac{h}{2}, \bar{y}_0 + \frac{k_2}{2}\right) = 0.2 \times \sqrt{0.5+0.51} = 0.200998$$

$$k_4 = hf\left(x_0 + h, \bar{y}_0 + k_3\right) = 0.2 \times \sqrt{0.6+0.6109975} = 0.220091$$

$$\begin{aligned} y(0.6) = y_1 &= y_0 + \frac{1}{6}(k_1 + 2k_2 + 2k_3 + k_4) \\ &= 0.41 + 0.200348 \\ &= 0.610348 \end{aligned}$$

For $i = 1$, $x_1 = 0.6$, and $y_1 = 0.6103476$, similarly, we obtain

$$k_1 = hf(x_1, \bar{y}_1) = 0.220032$$

$$k_2 = hf\left(x_1 + \frac{h}{2}, \bar{y}_1 + \frac{k_1}{2}\right) = 0.238358$$

$$k_3 = hf\left(x_1 + \frac{h}{2}, \bar{y}_1 + \frac{k_2}{2}\right) = 0.239126$$

$$k_4 = hf\left(x_1 + h, \bar{y}_1 + k_3\right) = 0.256864$$

$$\begin{aligned} y(0.8) = y_2 &= y_1 + \frac{1}{6}(k_1 + 2k_2 + 2k_3 + k_4) \\ &= 0.610348 + 0.238644 \\ &= 0.848991 \end{aligned}$$

Hence, we have

$$y(0.8) \approx 0.84899$$

Example 7.10

Using R–K method of order 3, find the solution $y(1.06)$ for the following equation:

$$2xy\frac{dy}{dx} + \frac{2}{x} + y^2 = 0$$

with initial condition $y(1) = 1$.

Solution:

According to the third-order R–K method formulae, we have

$$\bar{y}_{i+1} = \bar{y}_i + \frac{1}{6}(k_1 + 4k_2 + k_3), \quad i = 0,1,2,\ldots,n-1$$
$$k_1 = hf(x_i, \bar{y}_i)$$
$$k_2 = hf\left(x_i + \frac{h}{2}, \bar{y}_i + \frac{k_1}{2}\right)$$
$$k_3 = hf(x_i + h, \bar{y}_i - k_1 + 2k_2)$$

x	k_1	k_2	k_3	k	Approximate Solution $\bar{y}(x)$	Exact Solution $y(x)$
1.01	−1.5	−1.491337	−1.482759	−0.014914	0.985086	0.985087
1.02	−1.482804	−1.474441	−1.46616	−0.014745	0.970328	0.970342
1.03	−1.466204	−1.458135	−1.450143	−0.014581	0.955733	0.955760
1.04	−1.450188	−1.442404	−1.434696	−0.014424	0.941296	0.941336
1.05	−1.434740	−1.427236	−1.419805	−0.014272	0.927011	0.927064
1.06	−1.419849	−1.412619	−1.405459	−0.014126	0.912873	0.912938

Therefore, the requires solution is $y(1.06) = 0.912873$.

Example 7.11

Consider the initial value problem

$$\frac{dy}{dx} + y = y^2, \quad y(0) = -1$$

1. Use R–K method with step sizes $h = 0.05$ and $h = 0.1$ to compute the approximate value of $y(0.4)$.
2. Estimate the error in the computed value by Richardson extrapolation.

Solution:

Here, $f(x,y) = -y + y^2$. The fourth-order R–K method is given by

$$y_{i+1} = y_i + \frac{1}{6}(k_1 + 2k_2 + 2k_3 + k_4), \quad i = 0,1,2,\ldots$$

where $y_0 = -1$ at $x = 0$

For $h = 0.05$,

Step 1:

$$k_1 = 0.1, k_2 = 0.092625, k_3 = 0.0931604, \text{and } k_4 = 0.0864599$$

$$y_1 = y_0 + \frac{1}{6}(k_1 + 2k_2 + 2k_3 + k_4)$$

$$y_1 = y(0.05) = -1 + \frac{1}{6}(0.1 + 2 \times 0.092625 + 2 \times 0.0931604 + 0.0864599) = -0.90699490$$

Step 2:

$$k_1 = 0.0864817, k_2 = 0.0804913, k_3 = 0.0809002, \text{and } k_4 = 0.0754264$$

$$y_2 = y(0.1) = -0.90699490 + \frac{1}{6}(0.0864817 + 2\times 0.0804913 + 2\times 0.0809002 + 0.0754264)$$
$$= -0.82621307$$

Step 3:

$$k_1 = 0.0754421,\ k_2 = 0.0705106,\ k_3 = 0.0708286, \text{ and } k_4 = 0.0662995$$

$$y_3 = y(0.15) = -0.82621307 + \frac{1}{6}(0.0754421 + 2\times 0.0705106 + 2\times 0.0708286 + 0.0662995)$$
$$= -0.75547641$$

Step 4:

$$k_1 = 0.0663111,\ k_2 = 0.0622034,\ k_3 = 0.0624547, \text{ and } k_4 = 0.058665$$

$$y_4 = y(0.2) = -0.75547641 + \frac{1}{6}(0.0663111 + 2\times 0.0622034 + 2\times 0.0624547 + 0.058665)$$
$$= -0.69309437$$

Step 5:

$$k_1 = 0.0586737,\ k_2 = 0.0552166,\ k_3 = 0.0554179, \text{ and } k_4 = 0.0522154$$

$$y_5 = y(0.25) = -0.69309437 + \frac{1}{6}(0.0586737 + 2\times 0.0552166 + 2\times 0.0554179 + 0.0522154)$$
$$= -0.63773470$$

Step 6:

$$k_1 = 0.052222,\ k_2 = 0.0492854,\ k_3 = 0.0494487, \text{ and } k_4 = 0.0467183$$

$$y_6 = y(0.3) = -0.63773470 + \frac{1}{6}(0.052222 + 2\times 0.0492854 + 2\times 0.0494487 + 0.0467183)$$
$$= -0.58833329$$

Step 7:

$$k_1 = 0.0467235,\ k_2 = 0.0442082,\ k_3 = 0.0443422, \text{ and } k_4 = 0.0419959$$

$$y_7 = y(0.35) = -0.58833329 + \frac{1}{6}(0.0467235 + 2\times 0.0442082 + 2\times 0.0443422 + 0.0419959)$$
$$= -0.54402992$$

Step 8:

$$k_1 = 0.0419999,\ k_2 = 0.0398295,\ k_3 = 0.0399406, \text{ and } k_4 = 0.0379098$$

$$y_8 = y(0.4) = -0.54402992 + \frac{1}{6}(0.0419999 + 2\times 0.0398295 + 2\times 0.0399406 + 0.0379098)$$

$$= -0.50412160$$

For $h = 0.1$,

Step 1:

$$k_1 = 0.2,\ k_2 = 0.171,\ k_3 = 0.175081, \text{ and } k_4 = 0.150541$$

$$y_1 = y(0.1) = -1 + \frac{1}{6}(0.2 + 2\times 0.171 + 2\times 0.175081 + 0.150541) = -0.82621615$$

Step 2:

$$k_1 = 0.150885, k_2 = 0.131443, k_3 = 0.133885, \text{and } k_4 = 0.117165$$

$$y_2 = y(0.2) = -0.82621615 + \frac{1}{6}(0.150885 + 2 \times 0.131443 + 2 \times 0.133885 + 0.117165)$$
$$= -0.69309839$$

Step 3:

$$k_1 = 0.117348, k_2 = 0.103692, k_3 = 0.105246, \text{and } k_4 = 0.0933423$$

$$y_3 = y(0.3) = -0.69309839 + \frac{1}{6}(0.117348 + 2\times 0.103692 + 2\times 0.105246 + 0.0933423) = -0.58833742$$

Step 4:

$$k_1 = 0.0934478, k_2 = 0.0834959, k_3 = 0.084535, \text{and } k_4 = 0.0757619$$

$$y_4 = y(0.4) = -0.58833742 + \frac{1}{6}(0.0934478 + 2\times 0.0834959 + 2\times 0.084535 + 0.0757619) = -0.50412552$$

The exact solution of this problem is $y(x) = 1/(1-2e^x)$. The absolute errors for $h = 0.05$, $h = 0.1$ and the error estimates are presented in Table 7.2. In Table 7.2, the column Ratio indicates the ratio of the errors for corresponding node points as step size h is halved. The last column shows the error estimates using Equation 7.134 obtained by Richardson extrapolation.

Example 7.12

Use R–K–Fehlberg method to find the numerical solution at $x = 0.4$ for

$$\frac{dy}{dx} = xy + y^2, \quad y(0) = 1$$

assuming the step length $h = 0.1$ and also find the local error.

Solution:

Here, $f(x,y)=xy+y^2$. The R–K–Fehlberg method of fourth and fifth-order are given by

$$\bar{y}_{i+1} = \bar{y}_i + \sum_{j=1}^{5} \delta_j k_j, \quad i = 0,1,2,\ldots$$

$$\hat{\bar{y}}_{i+1} = \bar{y}_i + \sum_{j=1}^{6} \gamma_j k_j, \quad i = 0,1,2,\ldots$$

where $\bar{y}_0=1$ at $x_0=0$

TABLE 7.2
Estimated Errors by Richardson Extrapolation

x	$\bar{y}_{h/2}$	$\bar{y}_h$	$y-\bar{y}_{h/2}$	$y-\bar{y}_h$	Ratio	$\frac{1}{15}(\bar{y}_{h/2}-\bar{y}_h)$
0.1	−0.82621307	−0.82621615	2.02E−7	3.28E−6	6.15	2.05E−7
0.2	−0.69309437	−0.69309839	2.62E−7	4.28E−6	6.12	2.68E−7
0.3	−0.58833329	−0.58833742	2.68E−7	4.39E−6	6.10	2.75E−7
0.4	−0.50412160	−0.50412552	2.54E−7	4.17E−6	6.09	2.61E−7

Step 1:

$$k_1 = 0.1,\ k_2 = 0.107625,\ k_3 = 0.111985,\ k_4 = 0.132791,\ k_5 = 0.136224,$$

$$\text{and } k_6 = 0.116272$$

$$\bar{y}_1 = \bar{y}_0 + \delta_1 k_1 + \delta_2 k_2 + \delta_3 k_3 + \delta_4 k_4 + \delta_5 k_5$$

$$= 1 + \left(\frac{25}{216} \times 0.1\right) + (0 \times 0.107625) + \left(\frac{1,408}{2,565} \times 0.111985\right) + \left(\frac{2,197}{4,104} \times 0.132791\right)$$

$$+ \left(-\frac{1}{5} \times 0.136224\right)$$

$$= 1.1168879197$$

$$\hat{\bar{y}}_1 = \bar{y}_0 + \gamma_1 k_1 + \gamma_2 k_2 + \gamma_3 k_3 + \gamma_4 k_4 + \gamma_5 k_5 + \gamma_6 k_6$$

$$= 1 + \left(\frac{16}{135} \times 0.1\right) + (0 \times 0.107625) + \left(\frac{6,656}{12,825} \times 0.111985\right) + \left(\frac{28,561}{56,430} \times 0.132791\right)$$

$$+ \left(-\frac{9}{50} \times 0.136224\right) + \left(\frac{2}{55} \times 0.116272\right)$$

$$= 1.1168877629$$

The local error in the fourth-order formula is

$$\varepsilon_1 = \hat{\bar{y}}_1 - \bar{y}_1 = 1.1168877629 - 1.1168879197 = -1.56753 \times 10^{-7}$$

Step 2:

$$k_1 = 0.135913,\ k_2 = 0.146835,\ k_3 = 0.153207,\ k_4 = 0.184073,\ k_5 = 0.189321,$$

$$\text{and } k_6 = 0.15944$$

$$\bar{y}_2 = \bar{y}_1 + \delta_1 k_1 + \delta_2 k_2 + \delta_3 k_3 + \delta_4 k_4 + \delta_5 k_5$$

$$= 1.1168879197 + \left(\frac{25}{216} \times 0.135913\right) + (0 \times 0.146835) + \left(\frac{1,408}{2,565} \times 0.153207\right)$$

$$+ \left(\frac{2,197}{4,104} \times 0.184073\right) + \left(-\frac{1}{5} \times 0.189321\right)$$

$$= 1.2773942356$$

$$\hat{\bar{y}}_2 = \bar{y}_1+\gamma_1 k_1+\gamma_2 k_2+\gamma_3 k_3+\gamma_4 k_4+\gamma_5 k_5+\gamma_6 k_6$$

$$= 1.1168879197+\left(\frac{16}{135}\times 0.135913\right)+(0\times 0.146835)+\left(\frac{6,656}{12,825}\times 0.153207\right)$$

$$+\left(\frac{28,561}{56,430}\times 0.184073\right)+\left(-\frac{9}{50}\times 0.189321\right)+\left(\frac{2}{55}\times 0.15944\right)$$

$$= 1.2773938386$$

The local error in the fourth-order formula is

$$\varepsilon_2 = \hat{\bar{y}}_2 - \bar{y}_2 = 1.2773938386 - 1.2773942356 = -3.97024\times 10^{-7}$$

Step 3:

$k_1 = 0.188721$, $k_2 = 0.205253$, $k_3 = 0.21514$, $k_4 = 0.263981$, $k_5 = 0.272608$,

and $k_6 = 0.224741$

$$\bar{y}_3 = \bar{y}_2 + \delta_1 k_1 + \delta_2 k_2 + \delta_3 k_3 + \delta_4 k_4 + \delta_5 k_5$$

$$= 1.2773942356+\left(\frac{25}{216}\times 0.188721\right)+(0\times 0.205253)+\left(\frac{1,408}{2,565}\times 0.21514\right)$$

$$+\left(\frac{2,197}{4,104}\times 0.263981\right)+\left(-\frac{1}{5}\times 0.272608\right)$$

$$= 1.5041290867$$

$$\hat{\bar{y}}_3 = \bar{y}_2 + \gamma_1 k_1 + \gamma_2 k_2 + \gamma_3 k_3 + \gamma_4 k_4 + \gamma_5 k_5 + \gamma_6 k_6$$

$$= 1.2773942356 + \left(\frac{16}{135}\times 0.188721\right)+(0\times 0.205253)$$

$$+\left(\frac{6,656}{12,825}\times 0.21514\right)+\left(\frac{28,561}{56,430}\times 0.263981\right)$$

$$+\left(-\frac{9}{50}\times 0.272608\right)+\left(\frac{2}{55}\times 0.224741\right)$$

$$= 1.5041280177$$

The local error in the fourth-order formula is

$$\varepsilon_3 = \hat{\bar{y}}_3 - \bar{y}_3 = 1.5041280177 - 1.5041290867 = -1.06895\times 10^{-6}$$

Step 4:

$$k_1 = 0.271364, k_2 = 0.298198, k_3 = 0.314772, k_4 = 0.398951, k_5 = 0.414585,$$

$$\text{and } k_6 = 0.330698$$

$$\begin{aligned}\bar{y}_4 &= \bar{y}_3 + \delta_1 k_1 + \delta_2 k_2 + \delta_3 k_3 + \delta_4 k_4 + \delta_5 k_5 \\ &= 1.5041290867 + \left(\frac{25}{216} \times 0.271364\right) + (0 \times 0.298198) + \left(\frac{1{,}408}{2{,}565} \times 0.314772\right) \\ &\quad + \left(\frac{2{,}197}{4{,}104} \times 0.398951\right) + \left(-\frac{1}{5} \times 0.414585\right) \\ &= 1.8389779119\end{aligned}$$

$$\begin{aligned}\hat{\bar{y}}_4 &= \bar{y}_3 + \gamma_1 k_1 + \gamma_2 k_2 + \gamma_3 k_3 + \gamma_4 k_4 + \gamma_5 k_5 + \gamma_6 k_6 \\ &= 1.5041290867 + \left(\frac{16}{135} \times 0.271364\right) + (0 \times 0.298198) + \left(\frac{6{,}656}{12{,}825} \times 0.314772\right) \\ &\quad + \left(\frac{28{,}561}{56{,}430} \times 0.398951\right) + \left(-\frac{9}{50} \times 0.414585\right) + \left(\frac{2}{55} \times 0.330698\right) \\ &= 1.8389747190\end{aligned}$$

The local error in the fourth-order formula is

$$\varepsilon_4 = \hat{\bar{y}}_4 - \bar{y}_4 = 1.8389747190 - 1.8389779119 = -3.19285 \times 10^{-6}$$

7.3 MULTISTEP METHODS

We know that in a $(k+1)$-step method, the computation of the value of y_{i+1} depends on the values of the previous $(k+1)$ computed values of $y_i, y_{i-1}, y_{i-2}, \ldots, y_{i-k}$. We shall now consider some multistep methods, which are called *linear multistep methods*.

These methods are classified into two types: (1) explicit linear multistep method and (2) implicit linear multistep method. We shall now discuss two such methods in details:

1. *Explicit linear multistep method*: The general form of an explicit linear $(k+1)$-step method of order p is given by the formula

$$y_{i+1} = \sum_{j=0}^{k} \alpha_j y_{i-j} + h \sum_{j=0}^{k} \beta_j f_{i-j} + O(h^{p+1}), \quad i \geq k \tag{7.148}$$

where the coefficients $\alpha_0, \alpha_1, \ldots, \alpha_k$ and $\beta_0, \beta_1, \ldots, \beta_k$ are constants and $k \geq 0$. The constant coefficients α_j, β_j $(j = 0, 1, 2, \ldots, k)$ are characterizing the method such that α_k and β_k do not vanish simultaneously. If either $\alpha_k \neq 0$ or $\beta_k \neq 0$, the method is called a $(k+1)$-step method, since $(k+1)$ previous values are being used to compute y_{i+1}. The values $y_1, y_2, \ldots, y_k$ are obtained by some other methods.

Here, the truncation error is

$$T_{i+1} = O(h^{p+1}), \quad i \geq k \tag{7.149}$$

Now, neglecting the truncation error and replacing y_i by the corresponding computed value $\bar{y}_i$ in Equation 7.148, we obtain

$$\bar{y}_{i+1} = \sum_{j=0}^{k} \alpha_j \bar{y}_{i-j} + h \sum_{j=0}^{k} \beta_j \bar{f}_{i-j}, \quad i \geq k \tag{7.150}$$

where $\bar{f}_i = f(x_i, \bar{y}_i)$. The Equation 7.150 represents the recursive formula of the explicit linear $(k+1)$-step method and it gives $\bar{y}_{i+1}$ explicitly in terms of $(k+1)$ preceding values of $\bar{y}_i, \bar{y}_{i-1}, \bar{y}_{i-2}, \ldots, \bar{y}_{i-k}$. Moreover, it may be noted that the recursive formula Equation 7.150 can be started only if the initial $(k+1)$ values $\bar{y}_0, \bar{y}_1, \bar{y}_2, \ldots, \bar{y}_k$ are known by some other method.

2. *Implicit linear multistep method*: An implicit linear $(k+1)$-step method of order p is given by the formula

$$y_{i+1} = \sum_{j=0}^{k} \alpha_j y_{i-j} + h \sum_{j=-1}^{k} \beta_j f_{i-j} + O(h^{p+1}), \quad i \geq k, \quad \beta_{-1} \neq 0, \tag{7.151}$$

where the right-hand side contains an addition term $h\beta_{-1} f_{i+1} = h\beta_{-1} f(x_{i+1}, y_{i+1})$ involving y_{i+1}. Neglecting the truncation error and replacing y_i by the corresponding computed value $\bar{y}_i$ in Equation 7.151, we obtain

$$\bar{y}_{i+1} = \sum_{j=0}^{k} \alpha_j \bar{y}_{i-j} + h \sum_{j=0}^{k} \beta_j f(x_{i-j}, \bar{y}_{i-j}) + h\beta_{-1} f(x_{i+1}, \bar{y}_{i+1}), i \geq k \tag{7.152}$$

Since $\beta_{-1} \neq 0$, $\bar{y}_{i+1}$ occurs on both sides of Equation 7.152 and thus the formula leads to an implicit method. The implicit formula in Equation 7.152 is a nonlinear equation with root $\bar{y}_{i+1}$. Therefore, any of the general technique discussed in Chapter 2 can be used to solve it. In this case, the fixed-point iteration method will be most convenient and simple one. Thus, $\bar{y}_{i+1}$ is a root of the following equation:

$$\phi(y) = \sum_{j=0}^{k} \alpha_j \bar{y}_{i-j} + h \sum_{j=0}^{k} \beta_j f(x_{i-j}, \bar{y}_{i-j}) + h\beta_{-1} f(x_{i+1}, y) \tag{7.153}$$

For this, we require an initial guess or approximation $\bar{y}_{i+1}^{(0)}$ for the root $\bar{y}_{i+1}$, which can be obtained by the explicit method of the same order. Thus, the sequence of iterates $\bar{y}_{i+1}^{(k)}$, $k = 0, 1, 2, \ldots$ is generated by the following formula:

$$\bar{y}_{i+1}^{(k+1)} = \sum_{j=0}^{k} \alpha_j \bar{y}_{i-j} + h \sum_{j=0}^{k} \beta_j f(x_{i-j}, \bar{y}_{i-j}) + h\beta_{-1} f\left(x_{i+1}, \bar{y}_{i+1}^{(k)}\right), \quad k = 0, 1, 2, \ldots \tag{7.154}$$

Now, subtracting Equation 7.154 from Equation 7.152, we get

$$\bar{y}_{i+1} - \bar{y}_{i+1}^{(k+1)} = h\beta_{-1} f(x_{i+1}, \bar{y}_{i+1}) - h\beta_{-1} f\left(x_{i+1}, \bar{y}_{i+1}^{(k)}\right) \tag{7.155}$$

Using Lipchitz's condition (Equation 7.4), we get

$$\left|\bar{y}_{i+1} - \bar{y}_{i+1}^{(k+1)}\right| \le hL\left|\beta_{-1}\right|\left|\bar{y}_{i+1} - \bar{y}_{i+1}^{(k)}\right|, k = 0,1,2,\ldots \tag{7.156}$$

Again,

$$\begin{aligned}\left|\bar{y}_{i+1} - \bar{y}_{i+1}^{(k+1)}\right| &\le hL\left|\beta_{-1}\right|\left|\bar{y}_{i+1} - \bar{y}_{i+1}^{(k+1)} + \bar{y}_{i+1}^{(k+1)} - \bar{y}_{i+1}^{(k)}\right| \\ &\le hL\left|\beta_{-1}\right|\left|\bar{y}_{i+1} - \bar{y}_{i+1}^{(k+1)}\right| + hL\left|\beta_{-1}\right|\left|\bar{y}_{i+1}^{(k+1)} - \bar{y}_{i+1}^{(k)}\right|\end{aligned} \tag{7.157}$$

Therefore, the error estimate is given by

$$\left|\bar{y}_{i+1} - \bar{y}_{i+1}^{(k+1)}\right| \le \frac{hL\left|\beta_{-1}\right|}{1 - hL\left|\beta_{-1}\right|}\left|\bar{y}_{i+1}^{(k+1)} - \bar{y}_{i+1}^{(k)}\right| \tag{7.158}$$

Thus, the sequence of iterates generated by Equation 7.154 converges if $hL\left|\beta_{-1}\right| < 1/2$, which is satisfied provided h is sufficiently small. For sufficiently small h so that $hL\left|\beta_{-1}\right| < 1/2$, we have

$$\left|\bar{y}_{i+1} - \bar{y}_{i+1}^{(k+1)}\right| < \left|\bar{y}_{i+1}^{(k+1)} - \bar{y}_{i+1}^{(k)}\right| \tag{7.159}$$

This implies that if two iterates agree up to the desired tolerance level, then the computation may be terminated.

The explicit formula Equation 7.150 that gives the initial approximation $\bar{y}_{i+1}^{(0)}$ for $\bar{y}_{i+1}$ is called the *predictor formula*. On the other hand, the implicit formula Equation 7.154 is known as the corrector formula. Thus, the computation by an implicit formula requires a companion explicit formula for initial approximation or guess $\bar{y}_{i+1}^{(0)}$ apart from the knowledge of the initial set of values $\bar{y}_0$, $\bar{y}_1$, $\bar{y}_2$, …, $\bar{y}_k$.

7.3.1 Adams–Bashforth and Adams–Moulton Predictor–Corrector Method

We assume that the values $\bar{y}_1, \bar{y}_2, \ldots, \bar{y}_{i-1}, \bar{y}_i$ have already been calculated by some other method. Integrating Equation 1.2 from x_i to x_{i+1}, we obtain

$$y_{i+1} = y_i + \int_{x_i}^{x_{i+1}} f(x,y)dx = y_i + \int_{x_i}^{x_{i+1}} y'(x)dx \tag{7.160}$$

In the above integral, the integrand $y' = f(x,y)$ is now approximated by the cubic polynomial passing through the points $(x_i, y_i'), (x_{i-1}, y_{i-1}'), (x_{i-2}, y_{i-2}')$, and (x_{i-3}, y_{i-3}') as given by Newton's backward interpolation formula

$$\begin{aligned}y'(x) &= y_i' + u\nabla y_i' + \frac{u(u+1)}{2!}\nabla^2 y_i' + \frac{u(u+1)(u+2)}{3!}\nabla^3 y_i' + E(x) \\ &= f_i + u\nabla f_i + \frac{u(u+1)}{2!}\nabla^2 f_i + \frac{u(u+1)(u+2)}{3!}\nabla^3 f_i + E(x), \quad f_i = f(x_i, y_i)\end{aligned} \tag{7.161}$$

where $u = (x - x_i)/h$ and interpolation error $E(x)$ is given by

$$E(x) = (x - x_i)(x - x_{i-1})(x - x_{i-2})(x - x_{i-3})y'[x, x_{i-3}, x_{i-2}, x_{i-1}, x_i]$$

Substituting Equation 7.161 in 7.160, we get

$$y_{i+1} = y_i + h\int_0^1\left[f_i + u\nabla f_i + \frac{u(u+1)}{2}\nabla^2 f_i + \frac{u(u+1)(u+2)}{6}\nabla^3 f_i\right]du$$

$$+\int_{x_i}^{x_{i+1}}(x-x_i)(x-x_{i-1})(x-x_{i-2})(x-x_{i-3})\,y'[x, x_{i-3}, x_{i-2}, x_{i-1}, x_i]dx$$

Let us assume $y(x)$ is continuously differentiable sufficient number of times for $x_{i-3} \le x \le x_{i+1}$ and then $y'[x, x_{i-3}, x_{i-2}, x_{i-1}, x_i]$ is also a continuous function of x. Since the polynomial $(x-x_i)(x-x_{i-1})(x-x_{i-2})(x-x_{i-3})$ is nonnegative on $[x_i, x_{i+1}]$, using the integral mean value theorem, we obtain

$$y_{i+1} = y_i + h\left[f_i + \frac{(f_i - f_{i-1})}{2} + \frac{5}{12}\left(f_i - 2f_{i-1} + f_{i-2}\right) + \frac{3}{8}\left(f_i - 3f_{i-1} + 3f_{i-2} - f_{i-3}\right)\right]$$

$$+\,y'[\zeta, x_{i-3}, x_{i-2}, x_{i-1}, x_i]\int_{x_i}^{x_{i+1}}(x-x_i)(x-x_{i-1})(x-x_{i-2})(x-x_{i-3})dx, \quad x_i < \zeta < x_{i+1}$$

$$= y_i + \frac{h}{24}\left[55f_i - 59f_{i-1} + 37f_{i-2} - 9f_{i-3}\right]$$

$$+\,y'[\zeta, x_{i-3}, x_{i-2}, x_{i-1}, x_i]\int_{x_i}^{x_{i+1}}(x-x_i)(x-x_{i-1})(x-x_{i-2})(x-x_{i-3})dx, \quad x_i < \zeta < x_{i+1}$$

Now, using Equation 3.87, we get

$$= y_i + \frac{h}{24}\left[55f_i - 59f_{i-1} + 37f_{i-2} - 9f_{i-3}\right] + \frac{h^5}{4!}y^{(5)}(\xi_i)\int_0^1 u(u+1)(u+2)(u+3)du, \quad x_{i-3} < \xi_i < x_{i+1}$$

$$= y_i + \frac{h}{24}\left[55f_i - 59f_{i-1} + 37f_{i-2} - 9f_{i-3}\right] + \frac{251}{720}h^5 y^{(5)}(\xi_i), \quad x_{i-3} < \xi_i < x_{i+1}$$

Replacing y_i by the corresponding computed value $\overline{y}_i$ and neglecting the truncation error in the last term of above equation, we obtain the following formula, which is called the *Adams–Bashforth formula*, given by

$$\overline{y}_{i+1}^P = \overline{y}_i + \frac{h}{24}\left[55\overline{f}_i - 59\overline{f}_{i-1} + 37\overline{f}_{i-2} - 9\overline{f}_{i-3}\right] \tag{7.162}$$

where $\overline{f}_i = f(x_i, \overline{y}_i)$. This formula is used as a predictor formula. The superscript P indicates that it provides a predicted value.

The corrector formula is obtained by approximating $y' = f(x, y)$ by the cubic Newton's backward interpolation polynomial passing through the points $(x_{i+1}, y'_{i+1}), (x_i, y'_i), (x_{i-1}, y'_{i-1})$, (x_{i-2}, y'_{i-2}) which is given by

$$\begin{aligned} y'(x) &= y'_{i+1} + u\nabla y'_{i+1} + \frac{u(u+1)}{2!}\nabla^2 y'_{i+1} + \frac{u(u+1)(u+2)}{3!}\nabla^3 y'_{i+1} + E(x) \\ &= f_{i+1} + u\nabla f_{i+1} + \frac{u(u+1)}{2!}\nabla^2 f_{i+1} + \frac{u(u+1)(u+2)}{3!}\nabla^3 f_{i+1} + E(x) \end{aligned} \tag{7.163}$$

where $u = (x - x_{i+1})/h$ and interpolation error $E(x)$ is given by

$$E(x) = (x - x_{i-2})(x - x_{i-1})(x - x_i)(x - x_{i+1})y'[x, x_{i-2}, x_{i-1}, x_i, x_{i+1}]$$

Substituting Equation 7.163 in Equation 7.160, we get

$$y_{i+1} = y_i + h\int_{-1}^{0}\left[f_{i+1} + u\nabla f_{i+1} + \frac{u(u+1)}{2!}\nabla^2 f_{i+1} + \frac{u(u+1)(u+2)}{6}\nabla^3 f_{i+1}\right]du$$

$$-\int_{x_i}^{x_{i+1}}(x - x_{i-2})(x - x_{i-1})(x - x_i)(x_{i+1} - x)\,y'[x, x_{i-2}, x_{i-1}, x_i, x_{i+1}]dx$$

Let us assume $y(x)$ is continuously differentiable sufficient number of times for $x_{i-2} \le x \le x_{i+1}$ and then $y'[x, x_{i-2}, x_{i-1}, x_i, x_{i+1}]$ is also a continuous function of x. Since the polynomial $(x - x_{i-2})(x - x_{i-1})(x - x_i)(x_{i+1} - x)$ is nonnegative on $[x_i, x_{i+1}]$, using the integral mean value theorem, we obtain

$$y_{i+1} = y_i + h\left[f_{i+1} - \frac{(f_{i+1} - f_i)}{2} - \frac{1}{12}\left(f_{i+1} - 2f_i + f_{i-1}\right) - \frac{1}{24}\left(f_{i+1} - 3f_i + 3f_{i-1} - f_{i-2}\right)\right]$$

$$+\, y'[\tilde{\zeta}, x_{i-2}, x_{i-1}, x_i, x_{i+1}]\int_{x_i}^{x_{i+1}}(x - x_{i-2})(x - x_{i-1})(x - x_i)(x - x_{i+1})dx, \quad x_i < \tilde{\zeta} < x_{i+1}$$

$$= y_i + \frac{h}{24}\left[9f_{i+1} + 19f_i - 5f_{i-1} + f_{i-2}\right] + \frac{h^5}{4!}y^{(v)}(\tilde{\xi}_i)\int_0^1 (u+2)(u+1)u(u-1)du, \quad x_{i-2} < \tilde{\xi}_i < x_{i+1}$$

$$= y_i + \frac{h}{24}\left[9f_{i+1} + 19f_i - 5f_{i-1} + f_{i-2}\right] - \frac{19}{720}h^5 y^{(v)}(\tilde{\xi}_i), \quad x_{i-2} < \tilde{\xi}_i < x_{i+1}$$

Replacing y_i by the corresponding computed value $\bar{y}_i$ and neglecting the truncation error in the last term of above equation, we obtain the following formula which is called the Adams–Moulton formula, given by

$$\bar{y}^C_{i+1} = \bar{y}_i + \frac{h}{24}\left[9\bar{f}_{i+1} + 19\bar{f}_i - 5\bar{f}_{i-1} + \bar{f}_{i-2}\right] \tag{7.164}$$

where $\bar{f}_i = f(x_i, \bar{y}_i)$. This formula is used as a corrector formula. The superscript C indicates that it provides a corrected value.

In the above predictor–corrector method, the initial approximation to $\bar{y}_{i+1}$ is provided by the predictor formula Equation 7.162, and then a sequence of successive approximations to $\bar{y}_{i+1}$ is generated by the corrector formula Equation 7.164.

7.3.1.1 Error Estimate

Practically, the step size h and the initial approximation $\bar{y}^P_{i+1}$ are such that only one iteration is need to compute and then we may accept $\bar{y}_{i+1} \cong \bar{y}^{(1)}_{i+1} = \bar{y}^C_{i+1}$.

Now, we know that the truncation error in Adams–Bashforth method is $O(h^5)$ in computing $\overline{y}_{i+1}$ from $\overline{y}_i$. In order to maintain this order of accuracy, the iterate $\overline{y}_{i+1}^C$ obtained by Adams–Moulton method, which is chosen as solution $\overline{y}_{i+1}$, should at least satisfy

$$\overline{y}_{i+1} - \overline{y}_{i+1}^C = -\frac{19}{720} h^5 y^{(v)}(\tilde{\xi}_i) \cong O(h^5), \quad x_{i-2} < \tilde{\xi}_i < x_{i+1} \tag{7.165}$$

Similarly, from Adams–Bashforth method, we get

$$\overline{y}_{i+1} - \overline{y}_{i+1}^P = \frac{251}{720} h^5 y^{(v)}(\xi_i) \cong O(h^5), \quad x_{i-3} < \xi_i < x_{i+1} \tag{7.166}$$

Now, assuming $y^{(5)}(x)$ to be approximately same over the interval (x_{i-3}, x_{i+1}), we have

$$\frac{\overline{y}_{i+1} - \overline{y}_{i+1}^P}{251/720} \cong \frac{\overline{y}_{i+1} - \overline{y}_{i+1}^C}{-(19/720)} \cong \frac{\overline{y}_{i+1}^C - \overline{y}_{i+1}^P}{3/8}$$

Thus, we obtain

$$\overline{y}_{i+1} - \overline{y}_{i+1}^C \cong -\frac{19}{270}\left(\overline{y}_{i+1}^C - \overline{y}_{i+1}^P\right) \tag{7.167}$$

This is the required error estimate. It also shows that the error of the corrected value is approximately –(19/270) of the difference between the corrected and the predicted values.

7.3.1.2 Algorithm of Adams Predictor–Corrector Method

Step 1: Start the program.
Step 2: Define the function $f(x, y)$.
Step 3: Read $x_0,\ y_0,\ h,\ x$.
Step 4: Compute $n = (x - x_0)/h$.
Step 5: Compute $y_1,\ y_2,\ y_3$ by R–K method

for $i = 0(1)2$ do

$x_i = x_0 + i * h;$

$k_1 = hf(x_i, \overline{y}_i);$

$k_2 = hf\left(x_i + \frac{h}{2}, \overline{y}_i + \frac{k_1}{2}\right);$

$k_3 = hf\left(x_i + \frac{h}{2}, \overline{y}_i + \frac{k_2}{2}\right);$

$k_4 = hf\left(x_i + h, \overline{y}_i + k_3\right);$

$k = \frac{1}{6}\left[k_1 + 2k_2 + 2k_3 + k_4\right];$

$y_{i+1} = y_i + k;$

End

Step 6: for $i = 3(1)\overline{n-1}$ do

$y_{i+1} = y_i + \frac{h}{24}\left[55f(x_i, y_i) - 59f(x_{i-1}, y_{i-1}) + 37f(x_{i-2}, y_{i-2}) - 9f(x_{i-3}, y_{i-3})\right];$

(Adam–Bashforth formula for predicted value of y_{i+1})

$x_{i+1} = x_0 + (i+1)h;$

Step 7: $k = 0$;

$$y_{i+1}^{(k+1)} = y_i + \frac{h}{24}\left[9f\left(x_{i+1}, y_{i+1}^{(k)}\right) + 19f(x_i, y_i) - 5f(x_{i-1}, y_{i-1}) + f(x_{i-2}, y_{i-2})\right];$$

(Adam–Moulton formula for corrected value of y_{i+1})

while $\left(\left|y_{i+1}^{(k+1)} - y_{i+1}^{(k)}\right| > \varepsilon\right)$ do

$k = k + 1$;

$$y_{i+1}^{(k+1)} = y_i + \frac{h}{24}\left[9f\left(x_{i+1}, y_{i+1}^{(k)}\right) + 19f(x_i, y_i) - 5f(x_{i-1}, y_{i-1}) + f(x_{i-2}, y_{i-2})\right];$$

end

Step 8: Set $y_{i+1} = y(x_{i+1}) = y_{i+1}^{(k+1)}$;

Step 9: end.

Step 10: Print y_{i+1}.

Step 11: Stop the program. ■

MATHEMATICA® Program Implementing Adams–Moulton Method for Solving ODE (Chapter 7, Example 7.13)

```
f[x_,y_]:=x^2*(1+y);
x[0]=1;
y[0]=1;
h=0.1;
a=1.4;
n=(a-x[0])/h;
For[i=0,i<=2,i++,
  x[i]=x[0]+i*h;
  k[1]=h*f[x[i], y[i]];
  k[2]=h*f[x[i]+h/2,y[i]+k[1]/2];
  k[3]=h*f[x[i]+h/2,y[i]+k[2]/2];
  k[4]=h*f[x[i]+h, y[i]+k[3]];
  y[i+1]=y[i]+(1/6)*(k[1]+2*k[2]+2*k[3]+k[4]);
  Print["y[",x[i]+h,"]=",y[i+1]]
];
For[i=3,i<=n-1,i++,
  x[i]=x[0]+i*h;
  y[i+1,0]=y[i]+(h/24)*(55*f[x[i],y[i]]-59*f[x[i-1],y[i-1]]+37*f[x[i-2],
y[i-2]]-9*f[x[i-3],y[i-3]]);
For[j=0,j<=6,j++,
  y[i+1,j+1]=y[i]+(h/24)*(9*f[x[i]+h,y[i+1,j]]+19*f[x[i],y[i]]-5*f[x[i-1],
y[i-1]]+f[x[i-2],y[i-2]]);
  Print["y[",x[i]+h,"]=",y[i+1,j+1]]];
y[i+1]=y[i+1,j]];
```

Output:

```
y[1.1]=1.2333
y[1.2]=1.54929
y[1.3]=1.98066
y[1.4]=2.57694
y[1.4]=2.57714
```

```
y[1.4]=2.57715
y[1.4]=2.57715
y[1.4]=2.57715
y[1.4]=2.57715
y[1.4]=2.57715
```

Example 7.13

Use Adam–Moulton method to determine $y(1.4)$ given that $dy/dx = x^2(1+y)$, $y(1) = 1$.

Solution:

Here, $x_0 = 1$, $y_0 = 1$, and $h = 0.1$, say

$$y' = x^2(1+y), \quad y'_0 = 2$$

$$y'' = (2x+x^4)(1+y), \quad y''_0 = 6$$

$$y''' = (2+6x^3+x^6)(1+y), \quad y'''_0 = 18$$

$$y^{iv} = (20x^2+12x^5+x^8)(1+y), \quad y^{iv}_0 = 66$$

$$y^{v} = (40x+80x^4+20x^7+x^{10})(1+y), \quad y^{v}_0 = 282$$

$$y(1.1) = y_1 = y_0 + hy'_0 + \frac{h^2}{2!}y''_0 + \frac{h^3}{3!}y'''_0 + \frac{h^4}{4!}y^{iv}_0 + \frac{h^5}{5!}y^{v}_0$$

$$= 1+0.1\times 2+\frac{(0.1)^2}{2!}\times 6+\frac{(0.1)^3}{3!}\times 18+\frac{(0.1)^4}{4!}\times 66+\frac{(0.1)^5}{5!}\times 282$$

$$= 1.2333$$

$$y(1.2) = y_2 = y_0 + 2hy'_0 + \frac{(2h)^2}{2!}y''_0 + \frac{(2h)^3}{3!}y'''_0 + \frac{(2h)^4}{4!}y^{iv}_0 + \frac{(2h)^5}{5!}y^{v}_0$$

$$= 1+0.2\times 2+\frac{(0.2)^2}{2!}\times 6+\frac{(0.2)^3}{3!}\times 18+\frac{(0.2)^4}{4!}\times 66+\frac{(0.2)^5}{5!}\times 282$$

$$= 1.54915$$

Similarly,

$$y(1.3) = y_3 = y_0 + 3hy'_0 + \frac{(3h)^2}{2!}y''_0 + \frac{(3h)^3}{3!}y'''_0 + \frac{(3h)^4}{4!}y^{iv}_0 - \frac{(3h)^5}{5!}y^{v}_0$$

$$= 1 + 0.3\times 2 + \frac{(0.3)^2}{2!}\times 6 + \frac{(0.3)^3}{3!}\times 18 + \frac{(0.3)^4}{4!}\times 66 + \frac{(0.3)^5}{5!}\times 282$$

$$= 1.97899$$

Now, let us consider constructing the following table:

x	y	$f(x,y) = x^2(1+y)$
1	1	2
1.1	1.2333	2.70229
1.2	1.54915	3.67078
1.3	1.97899	5.03449

The values in the above table will be used in the Adams predictor–corrector formulae. Using Adams–Bashforth predictor formula in Equation 7.162, we get

$$y^P(1.4) = \bar{y}_4^P = \bar{y}_3 + \frac{h}{24}\left[55\bar{f}_3 - 59\bar{f}_2 + 37\bar{f}_1 - 9\bar{f}_0\right] \quad \text{where } \bar{f}_i = f(x_i, \bar{y}_i)$$

$$= 1.97899 + \frac{0.1}{24}\left[55 \times 5.03449 - 59 \times 3.67078 + 37 \times 2.70229 - 9 \times 2\right]$$

$$= 2.57193$$

$$y'(1.4) = (1.4)^2 \times (1 + 2.57193) = 7.0009828$$

Again, using Adams–Moulton corrector formula in Equation 7.164, we get

First iteration:

$$y^C(1.4) = \bar{y}_4^C = \bar{y}_3 + \frac{h}{24}\left[9\bar{f}_4 + 19\bar{f}_3 - 5\bar{f}_2 + \bar{f}_1\right]$$

$$= 1.97899 + \frac{0.1}{24}\left[9 \times 7.0009828 + 19 \times 5.03449 - 5 \times 3.67078 + 2.70229\right]$$

$$= 2.57488$$

$$y'(1.4) = (1.4)^2 \times (1 + 2.57488) = 7.0067648$$

Second iteration:

$$y^C(1.4) = \bar{y}_4^C = \bar{y}_3 + \frac{h}{24}\left[9\bar{f}_4 + 19\bar{f}_3 - 5\bar{f}_2 + \bar{f}_1\right]$$

$$= 1.97899 + \frac{0.1}{24}\left[9 \times 7.0067648 + 19 \times 5.03449 - 5 \times 3.67078 + 2.70229\right]$$

$$= 2.57509$$

$$y'(1.4) = (1.4)^2 \times (1 + 2.57509) = 7.0071764$$

Third iteration:

$$y^C(1.4) = \bar{y}_4^C = \bar{y}_3 + \frac{h}{24}\left[9\bar{f}_4 + 19\bar{f}_3 - 5\bar{f}_2 + \bar{f}_1\right]$$

$$= 1.97899 + \frac{0.1}{24}\left[9 \times 7.0071764 + 19 \times 5.03449 - 5 \times 3.67078 + 2.70229\right]$$

$$= 2.57511$$

$$y'(1.4) = (1.4)^2 \times (1 + 2.57511) = 7.0072156$$

Fourth iteration:

$$y^C(1.4) = \bar{y}_4^C = \bar{y}_3 + \frac{h}{24}\left[9\bar{f}_4 + 19\bar{f}_3 - 5\bar{f}_2 + \bar{f}_1\right]$$

$$= 1.97899 + \frac{0.1}{24}\left[9 \times 7.0072156 + 19 \times 5.03449 - 5 \times 3.67078 + 2.70229\right]$$

$$= 2.57511$$

The value of $y(1.4)$ in the fourth iteration coincides with that of the third iteration. Therefore, we can stop here, and the required solution is $y(1.4) = 2.5751$, correct to four decimal places.

7.3.2 Milne's Method

Milne's method is a multistep method in which integration is taken over more than one step. The value $\bar{y}_0 = y(x_0)$ being given, we assume that the values $\bar{y}_1, \bar{y}_2, \ldots, \bar{y}_{i-1}, \bar{y}_i$ have already been calculated by some other method, say Picard's or Taylor's series method. Integrating Equation 7.2 from x_{i-3} to x_{i+1}, we obtain

$$y_{i+1} = y_{i-3} + \int_{x_{i-3}}^{x_{i+1}} f(x, y)dx = y_{i-3} + \int_{x_{i-3}}^{x_{i+1}} y'(x)dx \tag{7.168}$$

Now, evaluating the above integral in Equation 7.168 by using Equation 5.54 of the three-point Newton–Cotes quadrature formula discussed in Chapter 5, we get

$$\begin{aligned} y_{i+1} &= y_{i-3} + \int_{x_{i-3}}^{x_{i+1}} y'(x)dx \\ &= y_{i-3} + 4h\sum_{j=0}^{2} K_j^{(2)} y'_{i-j} + \frac{14}{45}h^5 f^{(iv)}(\xi), \quad \text{where} \\ K_j^{(2)} &= \frac{(-1)^{2-j}}{4\,j!\,(2-j)!}\int_{-1}^{3} \frac{t(t-1)(t-2)}{(t-j)}dt, \quad j = 0(1)2 \\ &= y_{i-3} + \frac{4h}{3}\left(2f_i - f_{i-1} + 2f_{i-2}\right) + \frac{14}{45}h^5 f^{(iv)}(\xi_i), \quad x_{i-3} < \xi_i < x_{i+1} \end{aligned} \tag{7.169}$$

Replacing y_i by the corresponding computed value $\bar{y}_i$ and neglecting the truncation error in the last term of above equation, we obtain the following four-step explicit formula of order 4, which is called Milne's predictor formula of order 4, given by

$$\bar{y}_{i+1}^P = \bar{y}_{i-3} + \frac{4h}{3}\left(2\bar{f}_i - \bar{f}_{i-1} + 2\bar{f}_{i-2}\right) \tag{7.170}$$

where $\bar{f}_i = f(x_i, \bar{y}_i)$. This formula is used as a predictor formula. The superscript P indicates that it provides a predicted value.

Now, the corresponding corrector formula is obtained by integrating the differential equation (7.2) from x_{i-1} to x_{i+1} yielding

$$y_{i+1} = y_{i-1} + \int_{x_{i-1}}^{x_{i+1}} y'(x)dx \tag{7.171}$$

Evaluating the integral in Equation 7.171 by Simpson's one-third rule with the node points x_{i-1}, x_i, and x_{i+1}, we get

$$\begin{aligned} y_{i+1} &= y_{i-1} + \int_{x_{i-1}}^{x_{i+1}} y'(x)dx \\ &= y_{i-1} + \frac{h}{3}\left(f_{i-1} + 4f_i + f_{i+1}\right) - \frac{1}{90}h^5 f^{(iv)}(\xi_i), \quad x_{i-1} < \xi_i < x_{i+1} \end{aligned} \tag{7.172}$$

Again, replacing y_i by the corresponding computed value $\overline{y}_i$ and neglecting the truncation error in the last term of above equation, we obtain the following two-step implicit recursive formula of order 4, which is called Milne's corrector formula of order 4, given by

$$\overline{y}_{i+1}^{C} = y_{i-1} + \frac{h}{3}\left(\overline{f}_{i-1} + 4\overline{f}_i + \overline{f}_{i+1}\right), \tag{7.173}$$

This formula is used as a corrector formula. The superscript C indicates that it provides a corrected value. If $f^{(iv)}(x)$ does not vary strongly in $[x_{i-1}, x_{i+1}]$, an estimate of truncation error is given by

$$T_{i+1} \cong -\frac{1}{90}h\Delta^4(x_{i-2}) = -\frac{h}{90}(\overline{f}_{i-2} - 4\overline{f}_{i-1} + 6\overline{f}_i - 4\overline{f}_{i+1} + \overline{f}_{i+2}) \tag{7.174}$$

In order to use this formula, we need additional points at the ends of the grid for computed values of $\overline{f}_{i-2}$ and $\overline{f}_{i+2}$. In the above predictor–corrector method, the initial approximation to $\overline{y}_{i+1}$ is provided by the predictor formula Equation 7.170 and then a sequence of successive approximations to $\overline{y}_{i+1}$ is generated by the corrector formula Equation 7.173.

7.3.2.1 Error Estimate

Practically, the step size h and the initial approximation $\overline{y}_{i+1}^{P}$ are such that only one iteration is needed to compute and then we may accept $\overline{y}_{i+1} \cong \overline{y}_{i+1}^{(1)} = \overline{y}_{i+1}^{C}$.

Again, we know that the truncation error in Milne's predictor formula is $O(h^5)$ in computing $\overline{y}_{i+1}$ from $\overline{y}_i$. In order to maintain this order of accuracy, the iterate $\overline{y}_{i+1}^{C}$ obtained by Milne's corrector formula, which is chosen as solution $\overline{y}_{i+1}$, should at least satisfy

$$\overline{y}_{i+1} - \overline{y}_{i+1}^{C} = -\frac{1}{90}h^5 y^{(iv)}(\tilde{\xi}_i) \cong O(h^5), \quad x_{i-1} < \tilde{\xi}_i < x_{i+1} \tag{7.175}$$

Similarly, from Milne's predictor formula, we get

$$\overline{y}_{i+1} - \overline{y}_{i+1}^{P} = \frac{14}{45}h^5 y^{(iv)}(\xi_i) \cong O(h^5), \quad x_{i-3} < \xi_i < x_{i+1} \tag{7.176}$$

Now, assuming $y^{(iv)}(x)$ does not vary strongly over the interval (x_{i-3}, x_{i+1}), we have

$$\frac{\overline{y}_{i+1} - \overline{y}_{i+1}^{P}}{14/45} \cong \frac{\overline{y}_{i+1} - \overline{y}_{i+1}^{C}}{-(1/90)} \cong \frac{\overline{y}_{i+1}^{C} - \overline{y}_{i+1}^{P}}{29/90}$$

Thus, we obtain

$$\bar{y}_{i+1} - \bar{y}_{i+1}^{C} \cong -\frac{1}{29}\left(\bar{y}_{i+1}^{C} - \bar{y}_{i+1}^{P}\right) \tag{7.177}$$

This is the required error estimate. It also shows that the error of the corrected value is approximately –(1/29) of the difference between the corrected and the predicted values.

Now, comparing the truncation errors in Equations 7.169 and 7.172, it may be observed that the corrector formula is more accurate than the predictor formula.

7.3.2.2 Algorithm of Milne's Method

Step 1: Start the program.

Step 2: Define the function $f(x, y)$.

Step 3: Read x_0, y_0, h, x.

Step 4: Compute $n = \dfrac{x - x_0}{h}$.

Step 5: Compute y_1, y_2, y_3 by R–K method.

For $i = 0(1)2$ do

$$x_i = x_0 + i * h;$$

$$k_1 = hf(x_i, \bar{y}_i);$$

$$k_2 = hf\left(x_i + \frac{h}{2}, \bar{y}_i + \frac{k_1}{2}\right);$$

$$k_3 = hf\left(x_i + \frac{h}{2}, \bar{y}_i + \frac{k_2}{2}\right);$$

$$k_4 = hf\left(x_i + h, \bar{y}_i + k_3\right);$$

$$k = \frac{1}{6}\left[k_1 + 2k_2 + 2k_3 + k_4\right];$$

$$y_{i+1} = y_i + k;$$

End

Step 6: for $i = 3(1)\overline{n-1}$ do

$$y_{i+1}^{(0)} = y_{i-3} + \frac{4h}{3}\left[2f(x_i, y_i) - 2f(x_{i-1}, y_{i-1}) + 2f(x_{i-2}, y_{i-2})\right];$$

(Milne's predictor formula for y_{i+1})

$$x_{i+1} = x_0 + (i+1)h;$$

Step 7: $k = 0$;

$$y_{i+1}^{(k+1)} = y_{i-1} + \frac{h}{3}\left[f(x_{i-1}, y_{i-1}) + 4f(x_i, y_i) + f(x_{i+1}, y_{i+1}^{(k)})\right];$$

(Milne's corrector formula to correct y_{i+1})

while $\left(\left|y_{i+1}^{(k+1)} - y_{i+1}^{(k)}\right| > \varepsilon\right)$ do

$$k = k + 1;$$

$$y_{i+1}^{(k+1)} = y_{i-1} + \frac{h}{3}\left[f(x_{i-1}, y_{i-1}) + 4f(x_i, y_i) + f\left(x_{i+1}, y_{i+1}^{(k)}\right)\right];$$

end

Step 8: Set $y_{i+1} = y(x_{i+1}) = y_{i+1}^{(k+1)}$.

Step 9: end.

Step 10: Print y_{i+1}.

Step 11: Stop the program. ■

MATHEMATICA® Program for Solving ODE by Milne's Method (Chapter 7, Example 7.14)

```
f[x_,y_]:=x+y²;
x[0]=0;
y[0]=0;
h=0.2;
a=1.0;
n=(a-x[0])/h;
For[i=0,i<=2,i++,
  x[i]=x[0]+i*h;
  k[1]=h*f[x[i],y[i]];
  k[2]=h*f[x[i]+h/2,y[i]+k[1]/2];
  k[3]=h*f[x[i]+h/2,y[i]+k[2]/2];
  k[4]=h*f[x[i]+h,y[i]+k[3]];
  y[i+1]=y[i]+1/6*(k[1]+2*k[2]+2*k[3]+k[4]);
  Print["y[",x[i]+h,"]=",y[i+1]]
 ];
For[i=3,i<=n-1,i++,
  x[i]=x[0]+i*h;
  y[i+1,0]=
    y[i-3]+(4*h)/3*(2*f[x[i],y[i]]-f[x[i-1],y[i-1]]+2*f[x[i-2],y[i-2]]);
  For[j=0,j<=6,j++,
  y[i+1,j+1]=
    y[i-1]+h/3*(f[x[i-1],y[i-1]]+4*f[x[i],y[i]]+f[x[i]+h,y[i+1,0]]);
  Print["y[",x[i]+h,"]=",y[i+1,j+1]]];
  y[i+1]=y[i+1,j]];
```

Output:

```
y[0.2]=0.02002
y[0.4]=0.0805244
y[0.6]=0.184009
y[0.8]=0.337537
y[0.8]=0.337537
y[0.8]=0.337537
y[0.8]=0.337537
y[0.8]=0.337537
y[0.8]=0.337537
y[0.8]=0.337537
y[1.]=0.557199
y[1.]=0.557199
y[1.]=0.557199
y[1.]=0.557199
y[1.]=0.557199
y[1.]=0.557199
y[1.]=0.557199
```

Example 7.14

Use Milne's predictor–corrector method to obtain the solution of the equation $dy/dx = x + y^2$, where $y(0) = 0$ at $x = 0.8$ and 1.0.

Solution:

Here, $f(x,y) = x + y^2, x_0 = 0$ and $y_0 = 0$. Let us choose $h = 0.2$. Then, we obtain

$$y' = x + y^2, y_0' = 0$$

$$y'' = 1 + 2yy', y_0'' = 1$$

$$y''' = 2yy'' + 2y'^2, y_0''' = 0$$

$$y^{(iv)} = 2yy''' + 6y'y'', y_0^{(iv)} = 0$$

$$y^{(v)} = 2yy^{(iv)} + 8y'y''' + 6y''^2, y_0^{(v)} = 6$$

and so on.

$$y(0.2) = y_1 = y_0 + hy_0' + \frac{h^2}{2!}y_0'' + \frac{h^3}{3!}y_0''' + \frac{h^4}{4!}y_0^{(iv)} + \frac{h^5}{5!}y_0^{(v)} + \cdots$$

$$= 0 + 0.2\times 0 + \frac{(0.2)^2}{2!}\times 1 + \frac{(0.2)^3}{3!}\times 0 + \frac{(0.2)^4}{4!}\times 0 + \frac{(0.2)^5}{5!}\times 6 + \cdots$$

$$= 0.020016$$

$$y(0.4) = y_2 = y_0 + 2hy_0' + \frac{(2h)^2}{2!}y_0'' + \frac{(2h)^3}{3!}y_0''' + \frac{(2h)^4}{4!}y_0^{(iv)} + \frac{(2h)^5}{5!}y_0^{(v)} + \cdots$$

$$= 0 + 0.4\times 0 + \frac{(0.4)^2}{2!}\times 1 + \frac{(0.4)^3}{3!}\times 0 + \frac{(0.4)^4}{4!}\times 0 + \frac{(0.4)^5}{5!}\times 6 + \cdots$$

$$= 0.080512$$

$$y(0.6) = y_3 = y_0 + 3hy_0' + \frac{(3h)^2}{2!}y_0'' + \frac{(3h)^3}{3!}y_0''' + \frac{(3h)^4}{4!}y_0^{(iv)} + \frac{(3h)^5}{5!}y_0^{(v)} + \cdots$$

$$= 0 + 0.6\times 0 + \frac{(0.6)^2}{2!}\times 1 + \frac{(0.6)^3}{3!}\times 0 + \frac{(0.6)^4}{4!}\times 0 + \frac{(0.6)^5}{5!}\times 6 + \cdots$$

$$= 0.183888$$

Now, let us consider construct the following table:

x	y	$f(x,y) = x+y^2$
0	0	0
0.2	0.020016	0.200401
0.4	0.080512	0.406482
0.6	0.183888	0.633815

The values in the above table will be used in Milne's predictor–corrector formulae. Using Milne's predictor formula in Equation 7.170, we get

$$y^P(0.8) = \bar{y}_4^P = \bar{y}_0 + \frac{4h}{3}\left(2\bar{f}_3 - \bar{f}_2 + 2\bar{f}_1\right), \quad \text{where } \bar{f}_i = f(x_i, \bar{y}_i), \quad i = 0,1,2,3$$

$$= 0 + \frac{4\times 0.2}{3}[2\times 0.633815 - 0.406482 + 2\times 0.200401]$$

$$= 0.33652$$

$$y'(0.8) = 0.8 + (0.33652)^2 = 0.913246$$

Again, using Milne's corrector formula in Equation 7.173, we get

First iteration:

$$y^C(0.8) = \overline{y}_4^C = y_2 + \frac{h}{3}\left(\overline{f}_2 + 4\overline{f}_3 + \overline{f}_4\right)$$

$$= 0.080512 + \frac{0.2}{3}\left[0.406482 + 4 \times 0.633815 + 0.913246\right]$$

$$= 0.337511$$

$$y'(0.8) = 0.8 + (0.337511)^2 = 0.913914$$

Second iteration:

$$y^C(0.8) = \overline{y}_4^C = y_2 + \frac{h}{3}\left(\overline{f}_2 + 4\overline{f}_3 + \overline{f}_4\right)$$

$$= 0.080512 + \frac{0.2}{3}\left[0.406482 + 4 \times 0.633815 + 0.913914\right]$$

$$= 0.337556$$

Since the value of $y(0.8)$ in the second iteration coincides with that of the first iteration, up to four decimal places, we can stop here and the required solution is $y(0.8) = 0.3376$, correct to four decimal places.

Therefore, $y'(0.8) = 0.8 + (0.3376)^2 = 0.913974$. Again, using Milne's predictor formula in Equation 7.170, we get

$$y^P(1.0) = \overline{y}_5^P = \overline{y}_1 + \frac{4h}{3}\left(2\overline{f}_4 - \overline{f}_3 + 2\overline{f}_2\right), \quad \text{where } \overline{f}_i = f(x_i, \overline{y}_i), \quad i = 0,1,2,3$$

$$= 0.20016 + \frac{4 \times 0.2}{3}\left[2 \times 0.913974 - 0.633815 + 2 \times 0.406482\right]$$

$$= 0.735386$$

$$y'(1.0) = 1 + (0.735386)^2 = 1.54079$$

Again, using Milne's corrector formula in Equation 7.173, we get

First iteration:

$$y^C(1.0) = \overline{y}_5^C = y_3 + \frac{h}{3}\left(\overline{f}_3 + 4\overline{f}_4 + \overline{f}_5\right)$$

$$= 0.183888 + \frac{0.2}{3}\left[0.633815 + 4 \times 0.913974 + 1.54079\right]$$

$$= 0.572588$$

$$y'(1.0) = 1 + (0.572588)^2 = 1.32786$$

Second iteration:

$$y^C(1.0) = \overline{y}_5^C = y_3 + \frac{h}{3}\left(\overline{f}_3 + 4\overline{f}_4 + \overline{f}_5\right)$$

$$= 0.183888 + \frac{0.2}{3}\left[0.633815 + 4 \times 0.913974 + 1.32786\right]$$

$$= 0.558393$$

$$y'(1.0) = 1 + (0.558393)^2 = 1.3118$$

Third iteration:

$$y^C(1.0) = \overline{y}_5^C = y_3 + \frac{h}{3}\left(\overline{f}_3 + 4\overline{f}_4 + \overline{f}_5\right)$$

$$= 0.183888 + \frac{0.2}{3}\left[0.633815 + 4 \times 0.913974 + 1.3118\right]$$

$$= 0.557322$$

$$y'(1.0) = 1 + (0.557322)^2 = 1.31061$$

Fourth iteration:

$$y^C(1.0) = \overline{y}_5^C = y_3 + \frac{h}{3}\left(\overline{f}_3 + 4\overline{f}_4 + \overline{f}_5\right)$$

$$= 0.183888 + \frac{0.2}{3}\left[0.633815 + 4 \times 0.913974 + 1.31061\right]$$

$$= 0.557243$$

$$y'(1.0) = 1 + (0.557243)^2 = 1.31052$$

Fifth iteration:

$$y^C(1.0) = \overline{y}_5^C = y_3 + \frac{h}{3}\left(\overline{f}_3 + 4\overline{f}_4 + \overline{f}_5\right)$$

$$= 0.183888 + \frac{0.2}{3}\left[0.633815 + 4 \times 0.913974 + 1.31052\right]$$

$$= 0.557237$$

Since the value of $y(0.8)$ in the fifth iteration coincides with that of the fourth iteration, up to 4 decimal places, we can stop here and the required solution is $y(1.0) = 0.5572$, correct to four decimal places.

7.3.3 Nyström Method

We assume that the values $\overline{y}_1, \overline{y}_2, \ldots, \overline{y}_{i-1}, \overline{y}_i$ have already been calculated by some other method. Integrating Equation 7.2 from x_{i-1} to x_{i+1}, we obtain

$$y_{i+1} = y_{i-1} + \int_{x_{i-1}}^{x_{i+1}} f(x, y)dx = y_{i-1} + \int_{x_{i-1}}^{x_{i+1}} y'(x)dx \tag{7.178}$$

In the above integral, the integrand $y' = f(x, y)$ is now approximated by the polynomial of degree $\leq p$ passing through the interpolating points $x_{i-p}, \ldots, x_i$ as given by Newton's backward interpolation formula

$$\begin{aligned} y'(x) &= y_i' + u\nabla y_i' + \frac{u(u+1)}{2!}\nabla^2 y_i' + \frac{u(u+1)(u+2)}{3!}\nabla^3 y_i' + \cdots + E(x) \\ &= f_i + u\nabla f_i + \frac{u(u+1)}{2!}\nabla^2 f_i + \frac{u(u+1)(u+2)}{3!}\nabla^3 f_i + \cdots + E(x), \quad f_i = f(x_i, y_i) \end{aligned} \tag{7.179}$$

where $u = (x - x_i)/h$ and interpolation error $E(x)$ is given by

$$E(x) = (x - x_{i-p})\ldots(x - x_i)y'[x, x_{i-p}, \ldots, x_i]$$

Substituting Equation 7.179 in 7.178, we get

$$\begin{aligned} y_{i+1} &= y_{i-1} + h\int_{-1}^{1}\left[f_i + u\nabla f_i + \frac{u(u+1)}{2}\nabla^2 f_i + \frac{u(u+1)(u+2)}{6}\nabla^3 f_i + \cdots \right]du \\ &\quad + \int_{x_{i-1}}^{x_{i+1}} (x - x_{i-p})\ldots(x - x_i)y'[x, x_{i-p}, \ldots, x_i]dx \\ &= y_{i-1} + h\sum_{m=0}^{p} \gamma_m^* \nabla^m f_i + T_i(y) \end{aligned} \tag{7.180}$$

where

$$\gamma_m^* = \frac{1}{m!}\int_{-1}^{1} u(u+1)\ldots(u+m-1)du, \quad m \geq 1 \tag{7.181}$$

with $\gamma_0^* = 2$ and the truncation error $T_i(y)$ is given by

$$\begin{aligned} T_i(y) &= \int_{x_{i-1}}^{x_{i+1}} (x - x_{i-p})\ldots(x - x_i)y'[x, x_{i-p}, \ldots, x_i]dx \\ &= h^{p+2}\int_{-1}^{1} \frac{u(u+1)\ldots(u+p-1)}{p!} f^{(p+1)}(\xi_i), \quad x_{i-p} \leq \xi_i \leq x_{i+1} \end{aligned} \tag{7.182}$$

Let, $g(u) = \dfrac{u(u+1)\ldots(u+p-1)}{p!}$, then we have

$$T_i(y) = h^{p+2}\int_{-1}^{1} g(u)f^{(p+1)}(\xi_i)du \tag{7.183}$$

Since $g(u)$ changes sign in $(-1,1)$, the integral mean value theorem cannot be applied to Equation 7.183. Now,

$$|T_i(y)| \le h^{p+2} M_k \int_{-1}^{1} |g(u)| du, \tag{7.184}$$

where

$$M_k = \max_{-1 \le w \le 1} \left| f^{(p+1)}(w) \right|$$

The Equation 7.184 represents the error bound. In particular, if $p = 1$, then from Equation 7.180, neglecting the error term and replacing y_i by the corresponding computed value $\bar{y}_i$, we obtain

$$\bar{y}_{i+1} = \bar{y}_{i-1} + h \sum_{m=0}^{1} \gamma_m^* \nabla^m \bar{f}_i, \quad \bar{f}_i = f(x_i, \bar{y}_i)$$

$$= \bar{y}_{i-1} + 2h\bar{f}_i$$

The truncation error of this formula is $O(h^3)$, and this method is same as the midpoint method.

7.4 SYSTEM OF ORDINARY DIFFERENTIAL EQUATIONS OF FIRST-ORDER

Let us consider the system of first-order ordinary differential equations of the form

$$\frac{d\boldsymbol{y}}{dx} = \boldsymbol{f}(x, y_1, y_2, \ldots, y_n) \tag{7.185}$$

with initial condition $\boldsymbol{y}(x_0) = \alpha$. Here,

$$\boldsymbol{y} = [y_1, y_2, \ldots, y_n]^T, \boldsymbol{f} = [f_1(x,y), f_2(x,y), \ldots, f_n(x,y)]^T \quad \text{and} \quad \alpha = [c_1, c_2, \ldots, c_n]^T$$

The Equation 7.185 can be solved by the methods discussed in the earlier sections.

1. *Taylor's method*: We can write Equation 7.19 in the following vector form as

$$\boldsymbol{y}_{i+1} = \boldsymbol{y}_i + h\boldsymbol{y}_i' + \frac{h^2}{2!}\boldsymbol{y}_i'' + \cdots + \frac{h^p}{p!}\boldsymbol{y}_i^{(p)}, \quad i = 0,1,2,\ldots, n-1 \tag{7.186}$$

where

$$y_i^{(r)} = \begin{bmatrix} y_{1,i}^{(r)} \\ y_{2,i}^{(r)} \\ \vdots \\ y_{n,i}^{(r)} \end{bmatrix} = \begin{bmatrix} \dfrac{d^{r-1}}{dx^{r-1}} f(x_i, y_{1,i}) \\ \dfrac{d^{r-1}}{dx^{r-1}} f(x_i, y_{2,i}) \\ \vdots \\ \dfrac{d^{r-1}}{dx^{r-1}} f(x_i, y_{n,i}) \end{bmatrix}$$

2. *Euler's method*: In vector form, Euler's method can be written as

$$\boldsymbol{y}_{i+1} = \boldsymbol{y}_i + h\boldsymbol{y}'_i, \quad i = 0,1,2,\ldots,n-1 \tag{7.187}$$

3. *Classical R–K method of the fourth order*: The recursion formula can be written in vector form as

$$\boldsymbol{y}_{i+1} = \boldsymbol{y}_i + \boldsymbol{k}, \quad i = 0,1,2,\ldots,n-1 \tag{7.188}$$

where

$$\boldsymbol{k} = \frac{1}{6}\left(\boldsymbol{k}_1 + 2\boldsymbol{k}_2 + 2\boldsymbol{k}_3 + \boldsymbol{k}_4\right)$$

$$\boldsymbol{k}_1 = \begin{bmatrix} k_{11} \\ k_{12} \\ \vdots \\ k_{1n} \end{bmatrix}, \boldsymbol{k}_2 = \begin{bmatrix} k_{21} \\ k_{22} \\ \vdots \\ k_{2n} \end{bmatrix}, \boldsymbol{k}_3 = \begin{bmatrix} k_{31} \\ k_{32} \\ \vdots \\ k_{3n} \end{bmatrix}, \text{ and } \boldsymbol{k}_4 = \begin{bmatrix} k_{41} \\ k_{42} \\ \vdots \\ k_{4n} \end{bmatrix}$$

$$k_{1,j} = hf_j(x_i, \overline{y}_{1,i}, \overline{y}_{2,i}, \ldots, \overline{y}_{n,i}),$$

$$k_{2,j} = hf_j\left(x_i + \frac{h}{2}, \overline{y}_{1,i} + \frac{k_{11}}{2}, \overline{y}_{2,i} + \frac{k_{12}}{2}, \ldots, \overline{y}_{n,i} + \frac{k_{1n}}{2}\right)$$

$$k_{3,j} = hf_j\left(x_i + \frac{h}{2}, \overline{y}_{1,i} + \frac{k_{21}}{2}, \overline{y}_{2,i} + \frac{k_{22}}{2}, \ldots, \overline{y}_{n,i} + \frac{k_{2n}}{2}\right)$$

$$k_{4,j} = hf_j\left(x_i + h, \overline{y}_{1,i} + k_{31}, \overline{y}_{2,i} + k_{32}, \ldots, \overline{y}_{n,i} + k_{3n}\right)$$

$$j = 1(1)n$$

7.4.1 Algorithm of R–K Method of the Fourth Order for Solving System of Ordinary Differential Equations

Step 1: Start the program.
Step 2: Define the function $f_j(x_i, \overline{y}_{1,i}, \overline{y}_{2,i}, \ldots, \overline{y}_{m,i})$.
Step 3: Read $x_0, \overline{y}_{1,0}, \overline{y}_{2,0}, \ldots, \overline{y}_{m,0}, h, x$.
Step 4: Compute $n = (x - x_0)/h$.
Step 5: for $i = 0(1)n-1$ do

$x_i = x_0 + i * h$;
for $j = 1(1)m$ do
$k_{1,j} = hf_j(x_i, \overline{y}_{1,i}, \overline{y}_{2,i}, \ldots, \overline{y}_{m,i})$
end
for $j = 1(1)m$ do
$k_{2,j} = hf_j\left(x_i + \frac{h}{2}, \overline{y}_{1,i} + \frac{k_{11}}{2}, \overline{y}_{2,i} + \frac{k_{12}}{2} \ldots, \overline{y}_{m,i} + \frac{k_{1n}}{2}\right)$
end
for $j = 1(1)m$ do
$k_{3,j} = hf_j\left(x_i + \frac{h}{2}, \overline{y}_{1,i} + \frac{k_{21}}{2}, \overline{y}_{2,i} + \frac{k_{22}}{2} \ldots, \overline{y}_{m,i} + \frac{k_{2n}}{2}\right)$
end

for $j = 1(1)m$ do
$k_{4,j} = hf_j\left(x_i + h, \overline{y}_{1,i} + k_{31}, \overline{y}_{2,i} + k_{32}, \ldots, \overline{y}_{m,i} + k_{3n}\right)$
end
for $j = 1(1)m$ do
$k_j = \frac{1}{6}\left(k_{1,j} + 2k_{2,j} + 2k_{3,j} + k_{4,j}\right)$
$y_{j,i+1} = y_{j,i} + k_j$
end
end
Step 6: Print the value of $y_{j,i+1}$, $i = 0,1 \ldots, n-1$ and $j = 1,2 \ldots, m$.
Step 7: Stop the program. ■

MATHEMATICA® Program for Solving System of ODE by R–K Method (Chapter 7, Example 7.15)

```
f[x_,y_,z_]:=-x*z;
g[x_,y_,z_]:=y²;
x[0]=0;
y[0]=1;
z[0]=1;
h=0.2;
For[i=0,i<=2,i++,
  x[i]=x[0]+i*h;
  Print[step[i+1]];
  k[1]=h*f[x[i],y[i],z[i]];
  l[1]=h*g[x[i],y[i],z[i]];
  Print[k[1]];
  Print[l[1]];
  k[2]=h*f[x[i]+h/2,y[i]+k[1]/2,z[i]+l[1]/2];
  l[2]=h*g[x[i]+h/2,y[i]+k[1]/2,z[i]+l[1]/2];
  Print[k[2]];
  Print[l[2]];
  k[3]=h*f[x[i]+h/2,y[i]+k[2]/2,z[i]+l[2]/2];
  l[3]=h*g[x[i]+h/2,y[i]+k[2]/2,z[i]+l[2]/2];
  Print[k[3]];
  Print[l[3]];
  k[4]=h*f[x[i]+h,y[i]+k[3],z[i]+l[3]];
  l[4]=h*g[x[i]+h,y[i]+k[3],z[i]+l[3]];
  Print[k[4]];
  Print[l[4]];
  y[i+1]=y[i]+1/6*(k[1]+2*k[2]+2*k[3]+k[4]);
  z[i+1]=z[i]+1/6*(l[1]+2*l[2]+2*l[3]+l[4]);
  Print[x[i]+h,"    ",y[i+1],"     ",z[i+1]]];
```

Output:

```
step[1]
0.
0.2
-0.022
0.2
-0.022
```

```
0.195624
-0.047825
0.191297
0.2  0.977363  1.19709
step[2]
-0.0478836
0.191047
-0.0775569
0.181802
-0.0772795
0.176188
-0.109862
0.16203
0.4  0.899459  1.37527
step[3]
-0.110021
0.161805
-0.145617
0.142619
-0.144658
0.13667
-0.181433
0.113945
0.6  0.754126  1.51432
```

Example 7.15

Using R–K method, solve the following system of the equations for $x = 0.2$ and 0.4:

$$\frac{dy}{dx} = -xz, \frac{dz}{dx} = y^2 \text{ with } y(0)=1, \quad z(0)=1$$

Solution:

Given $dy/dx = -xz, dz/dx = y^2$ with initial conditions $y(0) = 1$ and $z(0) = 1$. According to the fourth-order R–K method formulae, we have

$$y_{i+1} = y_i + k,\ z_{i+1} = z_i + l,\ x_{i+1} = x_i + h, \quad i = 0,1,2,\ldots,n-1$$

$$k = \frac{1}{6}(k_1 + 2k_2 + 2k_3 + k_4)$$

$$l = \frac{1}{6}(l_1 + 2l_2 + 2l_3 + l_4)$$

$$k_1 = hf_1(x_i, y_i, z_i)$$

$$l_1 = hf_2(x_i, y_i, z_i)$$

$$k_2 = hf_1\left(x_i + \frac{h}{2}, y_i + \frac{k_1}{2}, z_i + \frac{l_1}{2}\right)$$

$$l_2 = hf_2\left(x_i + \frac{h}{2}, y_i + \frac{k_1}{2}, z_i + \frac{l_1}{2}\right)$$

$$k_3 = hf_1\left(x_i + \frac{h}{2}, y_i + \frac{k_2}{2}, z_i + \frac{l_2}{2}\right)$$

$$l_3 = hf_2\left(x_i + \frac{h}{2}, y_i + \frac{k_2}{2}, z_i + \frac{l_2}{2}\right)$$

$$k_4 = hf_1\left(x_i + h, y_i + k_3, z_i + l_3\right)$$

$$l_4 = hf_2\left(x_i + h, y_i + k_3, z_i + l_3\right)$$

Let $f_1(x,y,z) = -xz, f_2(x,y,z) = y^2$. Here, $x_0 = 0, y_0 = 1, z_0 = 1$. In addition, we take the step size $h = 0.2$, then when $x = 0.2$, we get

$$k_1 = hf_1(x_0, y_0, z_0) = 0$$

$$l_1 = hf_2(x_0, y_0, z_0) = 0.2$$

$$k_2 = hf_1\left(x_0 + \frac{h}{2}, y_0 + \frac{k_1}{2}, z_0 + \frac{l_1}{2}\right) = -0.022$$

$$l_2 = hf_2\left(x_0 + \frac{h}{2}, y_0 + \frac{k_1}{2}, z_0 + \frac{l_1}{2}\right) = 0.2$$

$$k_3 = hf_1\left(x_0 + \frac{h}{2}, y_0 + \frac{k_2}{2}, z_0 + \frac{l_2}{2}\right) = -0.022$$

$$l_3 = hf_2\left(x_0 + \frac{h}{2}, y_0 + \frac{k_2}{2}, z_0 + \frac{l_2}{2}\right) = 0.195624$$

$$k_4 = hf_1\left(x_0 + h, y_0 + k_3, z_0 + l_3\right) = -0.047825$$

$$l_4 = hf_2\left(x_0 + h, y_0 + k_3, z_0 + l_3\right) = 0.191297$$

$$k = \frac{1}{6}\left(k_1 + 2k_2 + 2k_3 + k_4\right) = -0.022637$$

$$l = \frac{1}{6}\left(l_1 + 2l_2 + 2l_3 + l_4\right) = 0.197091$$

$$x_1 = x_0 + h = 0.2$$

Thus,

$$y(0.2) = y_1 = y_0 + k = 0.977363$$

$$z(0.2) = z_1 = z_0 + l = 1.197091$$

To find $y(0.4)$, taking $x_1 = 0.2, y_1 = 0.977363, \text{and}\, z_1 = 1.197091$, we get

$$k_1 = hf_1(x_1, y_1, z_1) = -0.0478836$$

$$l_1 = hf_2(x_1, y_1, z_1) = 0.191047$$

$$k_2 = hf_1\left(x_1 + \frac{h}{2}, y_1 + \frac{k_1}{2}, z_1 + \frac{l_1}{2}\right) = -0.0775569$$

$$l_2 = hf_2\left(x_1 + \frac{h}{2}, y_1 + \frac{k_1}{2}, z_1 + \frac{l_1}{2}\right) = 0.181802$$

$$k_3 = hf_1\left(x_1 + \frac{h}{2}, y_1 + \frac{k_2}{2}, z_1 + \frac{l_2}{2}\right) = -0.0772795$$

$$l_3 = hf_2\left(x_1 + \frac{h}{2}, y_1 + \frac{k_2}{2}, z_1 + \frac{l_2}{2}\right) = 0.176188$$

$$k_4 = hf_1(x_1 + h, y_1 + k_3, z_1 + l_3) = -0.109862$$

$$l_4 = hf_2(x_1 + h, y_1 + k_3, z_1 + l_3) = 0.16203$$

$$k = \frac{1}{6}(k_1 + 2k_2 + 2k_3 + k_4) = -0.0779031$$

$$l = \frac{1}{6}(l_1 + 2l_2 + 2l_3 + l_4) = 0.178176$$

$$x_2 = x_1 + h = 0.4$$

Therefore,

$$y(0.4) = y_2 = y_1 + k = 0.8994599$$

$$z(0.4) = z_2 = z_1 + l = 1.375267$$

Hence, the required solutions are $y(0.2) = 0.980146$, $z(0.2) = 1.197091$ and $y(0.4) = 0.9225571$, $z(0.4) = 1.375267$.

7.5 DIFFERENTIAL EQUATIONS OF HIGHER ORDER

Let us consider the numerical solutions of ordinary differential equations of higher order. The second-order and higher order differential equations can be solved by converting into an equivalent system of first-order differential equations. Let us consider a second-order differential equation

$$y'' = f(x, y, y') \tag{7.189}$$

subject to initial conditions $y(x_0) = y_0$ and $y'(x_0) = y_0'$. Equation 7.189 can be converted into an equivalent system of differential equations of first-order. Substituting $y' = z$, we get an equivalent system

$$\begin{aligned} y' &= z \\ z' &= f(x, y, z) \end{aligned} \tag{7.190}$$

with initial conditions $y(x_0) = y_0, z(x_0) = y_0'$. Similarly, higher order differential equations can also be transformed into an equivalent system of first-order simultaneous differential equations. Then these simultaneous differential equations can be solved by method as discussed in Section 7.4. Next, we discuss this technique by the following illustrative examples.

Example 7.16

Use R–K method to obtain the solution of the equation $d^2y/dx^2 - x(dy/dx)^2 + y^2 = 0$ with $y(0) = 1$ and $y'(0) = 0$ for $x = 0.2$ and 0.4.

Solution:

Substituting $dy/dx = z$, the given second-order differential equation transformed to

$$\frac{dz}{dx} = xz^2 - y^2$$

with initial conditions $y(0)=1$ and $z(0)=0$. According to the fourth-order R–K method formulae, we have

$$y_{i+1}=y_i+k, x_{i+1}=x_i+h \quad i=0,1,2,\ldots,n-1$$

$$k=\frac{1}{6}(k_1+2k_2+2k_3+k_4)$$

$$l=\frac{1}{6}(l_1+2l_2+2l_3+l_4)$$

$$k_1=hf_1(x_i,y_i,z_i)$$
$$l_1=hf_2(x_i,y_i,z_i)$$

$$k_2=hf_1\left(x_i+\frac{h}{2},y_i+\frac{k_1}{2},z_i+\frac{l_1}{2}\right)$$

$$l_2=hf_2\left(x_i+\frac{h}{2},y_i+\frac{k_1}{2},z_i+\frac{l_1}{2}\right)$$

$$k_3=hf_1\left(x_i+\frac{h}{2},y_i+\frac{k_2}{2},z_i+\frac{l_2}{2}\right)$$

$$l_3=hf_2\left(x_i+\frac{h}{2},y_i+\frac{k_2}{2},z_i+\frac{l_2}{2}\right)$$

$$k_4=hf_1(x_i+h,y_i+k_3,z_i+l_3)$$
$$l_4=hf_2(x_i+h,y_i+k_3,z_i+l_3)$$

Let $f_1(x,y,z)=z, f_2(x,y,z)=xz^2-y^2$. Here, $x_0=0$, $y_0=1$, $z_0=0$. In addition, we take the step size $h=0.2$, then when $x=0.2$, we get

$$k_1=hf_1(x_0,y_0,z_0)=0$$
$$l_1=hf_2(x_0,y_0,z_0)=-0.2$$

$$k_2=hf_1\left(x_0+\frac{h}{2},y_0+\frac{k_1}{2},z_0+\frac{l_1}{2}\right)=-0.02$$

$$l_2=hf_2\left(x_0+\frac{h}{2},y_0+\frac{k_1}{2},z_0+\frac{l_1}{2}\right)=-0.1998$$

$$k_3=hf_1\left(x_0+\frac{h}{2},y_0+\frac{k_2}{2},z_0+\frac{l_2}{2}\right)=-0.01998$$

$$l_3=hf_2\left(x_0+\frac{h}{2},y_0+\frac{k_2}{2},z_0+\frac{l_2}{2}\right)=-0.19582$$

$$k_4=hf_1(x_0+h,y_0+k_3,z_0+l_3)=-0.0391641$$
$$l_4=hf_2(x_0+h,y_0+k_3,z_0+l_3)=-0.190554$$

$$k = \frac{1}{6}(k_1 + 2k_2 + 2k_3 + k_4) = -0.019854$$

$$l = \frac{1}{6}(l_1 + 2l_2 + 2l_3 + l_4) = -0.196966$$

$$x_1 = x_0 + h = 0.2,$$

Thus,

$$y(0.2) = y_1 = y_0 + k = 0.980146$$

$$z(0.2) = z_1 = z_0 + l = -0.196966$$

To find $y(0.4)$, taking $x_1 = 0.2$, $y_1 = 0.980146$, and $z_1 = -0.196966$, we get

$$k_1 = hf_1(x_1, y_1, z_1) = -0.0393932$$

$$l_1 = hf_2(x_1, y_1, z_1) = -0.190585$$

$$k_2 = hf_1\left(x_1 + \frac{h}{2}, y_1 + \frac{k_1}{2}, z_1 + \frac{l_1}{2}\right) = -0.0584517$$

$$l_2 = hf_2\left(x_1 + \frac{h}{2}, y_1 + \frac{k_1}{2}, z_1 + \frac{l_1}{2}\right) = -0.179368$$

$$k_3 = hf_1\left(x_1 + \frac{h}{2}, y_1 + \frac{k_2}{2}, z_1 + \frac{l_2}{2}\right) = -0.0573299$$

$$l_3 = hf_2\left(x_1 + \frac{h}{2}, y_1 + \frac{k_2}{2}, z_1 + \frac{l_2}{2}\right) = -0.17592$$

$$k_4 = hf_1(x_1 + h, y_1 + k_3, z_1 + l_3) = -0.0745771$$

$$l_4 = hf_2(x_1 + h, y_1 + k_3, z_1 + l_3) = -0.159194$$

$$k = \frac{1}{6}(k_1 + 2k_2 + 2k_3 + k_4) = -0.0575889$$

$$l = \frac{1}{6}(l_1 + 2l_2 + 2l_3 + l_4) = -0.1767258$$

$$x_2 = x_1 + h = 0.4$$

Therefore,

$$y(0.4) = y_2 = y_1 + k = 0.9225571$$

Hence, the required solutions are $y(0.2) = 0.980146$ and $y(0.4) = 0.9225571$

7.6 BOUNDARY VALUE PROBLEMS

So far, we have considered only initial value problems for numerical solutions of ordinary differential equations. In initial value problems, conditions on the solution of the differential equation are specified at the initial point. On the other hand, in a boundary value problem boundary conditions are prescribed at the end points of the domain.

For simplicity, let us consider a typical boundary value problem for the following second-order linear differential equation:

$$y'' + p(x)y' + q(x)y = r(x), \quad a < x < b \tag{7.191}$$

with the boundary conditions

$$y(a) = \alpha \quad \text{and} \quad y(b) = \beta \tag{7.192}$$

Many numerical methods are available for solving such boundary value problems. Here, we present the finite difference method, where the differential equations are discretized by means of finite difference approximations. As a result of this, a differential equation is transformed into a system of equations, which can be solved by a standard procedure. In addition, we present the shooting method that can be applied to both linear and nonlinear problems. Finally, we introduce collocation method for solving the boundary value problems (Equations 7.191 and 7.192).

7.6.1 Finite Difference Method

In finite difference method, discrete equations are obtained by replacing every derivative occurring in the equation as well as the boundary conditions by means of their finite difference approximations and then solving the resulting system of equations by any standard method.

We first discretize the domain of the problem, that is, the interval $[a,b]$ by dividing the interval $[a,b]$ into n equal subintervals by the grid (or node) points

$$a = x_0 < x_1 < x_2 < \cdots < x_n = b$$

Usually, the grid points are taken to be equally spaced, that is,

$$x_i = x_0 + ih, \; h = \frac{b-a}{n} \quad \left(i = 0,1,2,\ldots,n\right)$$

where h is called the step length or step size. Let us discretize the differential equation at the internal node points $x_1, x_2, x_3, \ldots, x_{n-1}$. For this purpose, let us use central difference approximation formulae

$$y'(x_i) = \frac{y_{i+1} - y_{i-1}}{2h} + O(h^2) \tag{7.193}$$

$$y''(x_i) = \frac{y_{i+1} - 2y_i + y_{i-1}}{h^2} + O(h^2) \tag{7.194}$$

for $i = 1,2,\ldots,n-1$. We use the notation $p_i = p(x_i)$, $q_i = q(x_i)$, $r_i = r(x_i)$, $0 \le i \le n$, and we denote $\bar{y}_i, 0 \le i \le n$ as numerical approximations of the exact solution values $y_i = y(x_i), 0 \le i \le n$.

Now, using these relations, the differential equation (Equation 7.191) at $x = x_i$ becomes

$$\frac{y_{i+1} - 2y_i + y_{i-1}}{h^2} + p_i \frac{y_{i+1} - y_{i-1}}{2h} + q_i y_i = r_i + O(h^2), \quad 1 \le i \le n-1 \tag{7.195}$$

Neglecting the remainder term $O(h^2)$ and replacing y_i by $\bar{y}_i$, we obtain the following difference equations:

$$\frac{\bar{y}_{i+1} - 2\bar{y}_i + \bar{y}_{i-1}}{h^2} + p_i \frac{\bar{y}_{i+1} - \bar{y}_{i-1}}{2h} + q_i \bar{y}_i = r_i, \quad 1 \le i \le n-1 \tag{7.196}$$

which can be written as

$$\left(1 - \frac{1}{2}hp_i\right)\bar{y}_{i-1} + (h^2 q_i - 2)\bar{y}_i + \left(1 + \frac{1}{2}hp_i\right)\bar{y}_{i+1} = h^2 r_i, \quad 1 \le i \le n-1 \tag{7.197}$$

The difference equation (Equation 7.197) consist of $n-1$ equations in $n+1$ unknowns $y_0, y_1, y_2, \ldots, y_n$. Therefore, we need two more equations; these two can be obtained from discretization of the boundary equations. The discretized boundary conditions are as follows:

$$\bar{y}_0 = \alpha \text{ and } \bar{y}_n = \beta \tag{7.198}$$

Equations 7.197 and 7.198 together form a system of linear equations. Since the values of y_0 and y_n are explicitly known by Equation 7.198, we can eliminate y_0 and y_n from the linear system of equations. For $i = 1$, we can rewrite Equation 7.197 as

$$\left(h^2 q_1 - 2\right)\bar{y}_1 + \left(1 + \frac{1}{2}hp_1\right)\bar{y}_2 = h^2 r_1 - \left(1 - \frac{1}{2}hp_1\right)\alpha \tag{7.199}$$

Similarly, from Equation 7.197 for $i = n-1$, we obtain

$$\left(1 - \frac{1}{2}hp_{n-1}\right)\bar{y}_{n-2} + \left(h^2 q_{n-1} - 2\right)\bar{y}_{n-1} = h^2 r_{n-1} - \left(1 + \frac{1}{2}hp_{n-1}\right)\beta \tag{7.200}$$

Now, the finite difference system of $n-1$ equations in $n-1$ unknowns $y_1, y_2, \ldots, y_{n-1}$ can be written in the following matrix form for the unknown solution vector $\boldsymbol{y} = \left[y_1, y_2, \ldots, y_{n-1}\right]^T$ as

$$\boldsymbol{Ay} = \boldsymbol{b} \tag{7.201}$$

where the coefficient matrix

$$\boldsymbol{A} = \begin{bmatrix} h^2 q_1 - 2 & 1 + \frac{1}{2}hp_1 & & \\ 1 - \frac{1}{2}hp_2 & h^2 q_2 - 2 & 1 + \frac{1}{2}hp_2 & \\ & \ddots & & \ddots \\ & & 1 - \frac{1}{2}hp_{n-1} & h^2 q_{n-1} - 2 \end{bmatrix},$$

and

$$\boldsymbol{b} = \begin{bmatrix} h^2 r_1 - \left(1 - \frac{1}{2}hp_1\right)\alpha \\ h^2 r_2 \\ \vdots \\ h^2 r_{n-2} \\ h^2 r_{n-1} - \left(1 + \frac{1}{2}hp_{n-1}\right)\beta \end{bmatrix}$$

Thus, the linear system of equations in Equation 7.201 is tridiagonal system of linear equations. The solution of tridiagonal linear systems is a well-studied problem.

MATHEMATICA® Program for Solving BVP of Second Order by Finite Difference Method (Chapter 7, Example 7.17)

```
a=0;
b=1;
h=1/8;
n=(b-a)/h;
y[0]=0;
y[n]=0;
For[i=1,i<=n-1,i++,
  eqn[i]=Simplify[64*y[i+1]-192*y[i]+64*y[i-1]+10]==0;
  Print["eqn[",i,"]=",eqn[i]]];
eqns=Table[eqn[i],{i,1,n-1}];
unkn=Table[y[i],{i,1,n-1}];
sol=NSolve[eqns, unkn]
```

Output:

```
eqn[1] = 10 - 192 y[1]+64 y[2]==0
eqn[2] = 2 (5 + 32 y[1]-96 y[2]+32 y[3])==0
eqn[3] = 2 (5 + 32 y[2]-96 y[3]+32 y[4])==0
eqn[4] = 2 (5 + 32 y[3]-96 y[4]+32 y[5])==0
eqn[5] = 2 (5 + 32 y[4]-96 y[5]+32 y[6])==0
eqn[6] = 2 (5 + 32 y[5]-96 y[6]+32 y[7])==0
eqn[7] = 2 (5 + 32 y[6]-96 y[7])==0
{{y[1]->0.0964096,y[2]->0.132979,y[3]->0.146277,
  y[4]->0.149601, y[5]->0.146277,y[6]->0.132979, y[7]->0.0964096}}
```

- *Error estimate and convergence*: It can be shown that if the exact solution $y(x)$ is sufficiently differentiable, say with continuous derivatives up to order 4, then the solution of Equation 7.201 satisfies

$$\max_{0\le i\le n}\left|y(x_i)-\bar{y}_i\right| = O\left(h^2\right) \tag{7.202}$$

Moreover, if $y(x)$ has six continuous derivatives, the following asymptotic error expansion formula can be obtained

$$y(x_i)-\bar{y}_i = g(x_i)h^2 + O\left(h^4\right) \tag{7.203}$$

where $g(x)$ is a function independent of h. In order to obtain numerical solution that converges more rapidly, Equation 7.203 can be used to justify Richardson's extrapolation. We can use Equation 7.203 to estimate the solution error and to improve the quality of the numerical solution by applying Richardson's extrapolation.

Now, using Richardson's extrapolation, we obtain

$$y(x)-\bar{y}_h(x) = \frac{1}{3}\left[\bar{y}_h(x)-\bar{y}_{2h}(x)\right]+O\left(h^4\right) \tag{7.204}$$

By dropping the higher order term $O\left(h^4\right)$ in Equation 7.204, we obtain Richardson's extrapolation formula

$$y(x) \approx \tilde{y}_h(x) = \frac{1}{3}\left[4\bar{y}_h(x)-\bar{y}_{2h}(x)\right] \tag{7.205}$$

with Richardson's error estimate

$$y(x) - \overline{y}_h(x) \approx \frac{1}{3}\left[\overline{y}_h(x) - \overline{y}_{2h}(x)\right] \tag{7.206}$$

Example 7.17

Solve the boundary value problem $y'' - 64y + 10 = 0$ with $y(0) = y(1) = 0$ by the finite difference method. Hence, estimate the error in the computed value by Richardson's extrapolation and compare the approximate values with exact solutions.

Solution:

Let $h = 1/4 = 0.25$, then $x_0 = 0, x_1 = 0.25, x_2 = 0.5, x_3 = 0.75, x_4 = 1$. The finite difference approximation of the given differential equation is

$$\frac{\overline{y}_{i+1} - 2\overline{y}_i + \overline{y}_{i-1}}{h^2} - 64\overline{y}_i + 10 = 0, \quad i = 1,2,3 \tag{7.207}$$

This implies that

$$16\overline{y}_{i+1} - 96\overline{y}_i + 16\overline{y}_{i-1} + 10 = 0, \quad i = 1,2,3 \tag{7.208}$$

Using boundary conditions $\overline{y}_0 = 0$ and $\overline{y}_4 = 0$, we get the following system of equations

$$\begin{aligned} 16\overline{y}_2 - 96\overline{y}_1 + 10 &= 0 \\ 16\overline{y}_3 - 96\overline{y}_2 + 16\overline{y}_1 + 10 &= 0 \\ -96\overline{y}_3 + 16\overline{y}_2 + 10 &= 0 \end{aligned} \tag{7.209}$$

In matrix form, Equation 7.209 can be written in the following tridiagonal linear system of the form:

$$\begin{bmatrix} -96 & 16 & 0 \\ 16 & -96 & 16 \\ 0 & 16 & -96 \end{bmatrix} \begin{bmatrix} \overline{y}_1 \\ \overline{y}_2 \\ \overline{y}_3 \end{bmatrix} = \begin{bmatrix} -10 \\ -10 \\ -10 \end{bmatrix} \tag{7.210}$$

Solving the above system of equations, we get

$$y(0.25) = \overline{y}_1 = 0.128676, \; y(0.5) = \overline{y}_2 = 0.147059, \; y(0.75) = \overline{y}_3 = 0.128676$$

Again, taking $h \mapsto h/2 = 1/8$, the finite difference approximation of the given differential equation is

$$\frac{\overline{y}_{i+1} - 2\overline{y}_i + \overline{y}_{i-1}}{(h/2)^2} - 64\overline{y}_i + 10 = 0, \quad i = 1,2,\ldots,7 \tag{7.211}$$

This implies that

$$64\overline{y}_{i+1} - 192\overline{y}_i + 64\overline{y}_{i-1} + 10 = 0, \quad i = 1,2,\ldots,7 \tag{7.212}$$

Using boundary conditions $y_0 = 0$ and $y_8 = 0$, we get the following system of equations:

$$\begin{aligned}
64\bar{y}_2 - 192\bar{y}_1 + 10 &= 0 \\
64\bar{y}_3 - 192\bar{y}_2 + 64\bar{y}_1 + 10 &= 0 \\
64\bar{y}_4 - 192\bar{y}_3 + 64\bar{y}_2 + 10 &= 0 \\
64\bar{y}_5 - 192\bar{y}_4 + 64\bar{y}_3 + 10 &= 0 \\
64\bar{y}_6 - 192\bar{y}_5 + 64\bar{y}_4 + 10 &= 0 \\
64\bar{y}_7 - 192\bar{y}_6 + 64\bar{y}_5 + 10 &= 0 \\
-192\bar{y}_7 + 64\bar{y}_6 + 10 &= 0
\end{aligned} \tag{7.213}$$

In matrix form, Equation 7.213 can be written in the following tridiagonal linear system of the form:

$$\begin{bmatrix} -192 & 64 & 0 & 0 & 0 & 0 & 0 \\ 64 & -192 & 64 & 0 & 0 & 0 & 0 \\ 0 & 64 & -192 & 64 & 0 & 0 & 0 \\ 0 & 0 & 64 & -192 & 64 & 0 & 0 \\ 0 & 0 & 0 & 64 & -192 & 64 & 0 \\ 0 & 0 & 0 & 0 & 64 & -192 & 64 \\ 0 & 0 & 0 & 0 & 0 & 64 & -192 \end{bmatrix} \begin{bmatrix} \bar{y}_1 \\ \bar{y}_2 \\ \bar{y}_3 \\ \bar{y}_4 \\ \bar{y}_5 \\ \bar{y}_6 \\ \bar{y}_7 \end{bmatrix} = \begin{bmatrix} -10 \\ -10 \\ -10 \\ -10 \\ -10 \\ -10 \\ -10 \end{bmatrix} \tag{7.214}$$

Solving the above system of equations, we get

$$y(0.125) = \bar{y}_1 = 0.0964096,\ y(0.25) = \bar{y}_2 = 0.132979,\ y(0.375) = \bar{y}_3 = 0.146277,$$

$$y(0.5) = \bar{y}_4 = 0.149601,\ y(0.625) = \bar{y}_5 = 0.146277,\ y(0.75) = \bar{y}_6 = 0.132979,$$

$$y(0.825) = \bar{y}_7 = 0.0964096$$

Error estimates by Richardson's extrapolation and also comparison between the approximate values with the exact solutions have been presented in the following table.

x	Approximate value $\bar{y}_{h/2}$	Approximate value $\bar{y}_h$	$\frac{1}{15}(\bar{y}_{h/2}-\bar{y}_h)$	Exact solution y	Absolute error $\lvert y-\bar{y}_h \rvert$
0.25	0.132979	0.128676	2.87×10^{-4}	0.134724	0.006048
0.5	0.149601	0.147059	1.69×10^{-4}	0.150528	0.003469
0.75	0.132979	0.128676	2.87×10^{-4}	0.134724	0.006048

7.6.1.1 Boundary Conditions Involving the Derivative

1. Let us consider the boundary condition at $x = b$ is

$$k_1\, y(b) + k_2 y'(b) = \beta \tag{7.215}$$

where k_1, k_2, and β are constants. One obvious discretization is to approximate $y'(b)$ by $(y_n - y_{n-1})/h$. Since

$$y'(b) - \frac{y_n - y_{n-1}}{h} = O(h) \tag{7.216}$$

the accuracy of this approximation is $O(h)$. Consequently, the corresponding difference solution with the following discretize boundary condition

$$k_1\bar{y}_n + k_2\frac{\bar{y}_n - \bar{y}_{n-1}}{h} = \beta \tag{7.217}$$

will have an accuracy of $O(h)$ only. Now, in order to retain the second-order convergence of the difference solution, we required to discretize the boundary condition in Equation 7.215 more accurately. This can be done by using the following formula:

$$y'(b) = \frac{3y_n - 4y_{n-1} + y_{n-2}}{2h} + O(h^2) \tag{7.218}$$

Consequently, the corresponding discretize boundary condition becomes

$$k_1\bar{y}_n + k_2\frac{3\bar{y}_n - 4\bar{y}_{n-1} + \bar{y}_{n-2}}{2h} = \beta \tag{7.219}$$

A similar treatment may be given if the general boundary conditions involve the derivatives $y'(a)$ and $y'(b)$. It can be shown that the resulting difference scheme is second-order accurate.

2. We now consider the boundary conditions of the form

$$k_1 y(a) - k_2 y'(a) = \alpha \tag{7.220}$$

$$k_1 y(b) + k_2 y'(b) = \beta \tag{7.221}$$

The discretization of the differential equation (7.191) at the internal points $i = 1, 2, \ldots, n-1$ is given by Equation 7.197, which has $n-1$ equations in $n+1$ unknowns $y_0, y_1, y_2, \ldots, y_n$. Therefore, we need two more equations, these two can be obtained from discretization of the boundary equations. At $x = a$, we discretize the boundary condition (7.230) as follows:

$$k_1 y_0 - k_2\left(\frac{\bar{y}_1 - \bar{y}_{-1}}{2h}\right) = \alpha$$

Thus,

$$\bar{y}_{-1} = \frac{2hk_1}{k_2}\bar{y}_0 + \bar{y}_1 + \frac{2h\alpha}{k_2} \tag{7.222}$$

At $x = b$, again, we discretize the boundary condition (7.218) as follows:

$$k_1 y_n + k_2\left(\frac{\bar{y}_{n+1} - \bar{y}_{n-1}}{2h}\right) = \beta$$

Therefore,

$$\bar{y}_{n+1} = \bar{y}_{n-1} - \frac{2hk_1}{k_2}\bar{y}_n + \frac{2h\beta}{k_2} \tag{7.223}$$

where y_{-1} and y_{n+1} are function values at x_{-1} and x_{n+1}, respectively. These points x_{-1} and x_{n+1} lie outside the interval $[a, b]$ and are called *fictitious points*.

The values of x_{-1} and x_{n+1} may be eliminated by assuming Equation 7.197 also holds for $i = 0$ and $i = n$, that is, at the boundary points x_{-1} and x_{n+1}. For $i = 0$, we can rewrite Equation 7.197 as

$$\left(1-\frac{1}{2}hp_0\right)\left(-\frac{2hk_1}{k_2}\bar{y}_0+\bar{y}_1+\frac{2h\alpha}{k_2}\right)+\left(h^2q_0-2\right)\bar{y}_0+\left(1+\frac{1}{2}hp_0\right)\bar{y}_1=h^2r_0$$

This implies that

$$\left(-\frac{2hk_1}{k_2}+\frac{h^2k_1p_0}{k_2}+h^2q_0-2\right)\bar{y}_0+2\bar{y}_1=h^2r_0-\frac{2h\alpha}{k_2}+\frac{h^2\alpha p_0}{k_2} \tag{7.224}$$

Similarly, from Equation 7.197 for $i=n$, we obtain

$$\left(1-\frac{1}{2}hp_n\right)\bar{y}_{n+1}+\left(h^2q_n-2\right)\bar{y}_n+\left(1+\frac{1}{2}hp_n\right)\left(\bar{y}_{n-1}-\frac{2hk_1}{k_2}\bar{y}_n+\frac{2h\beta}{k_2}\right)=h^2r_n$$

This implies that

$$2\bar{y}_{n-1}+\left(-\frac{2hk_1}{k_2}-\frac{h^2k_1p_n}{k_2}+h^2q_n-2\right)\bar{y}_n=h^2r_n-\frac{2h\beta}{k_2}-\frac{h^2\beta p_n}{k_2} \tag{7.225}$$

Now, Equations 7.224 and 7.225 along with Equation 7.197 form a system of $n+1$ linear equations in $n+1$ unknowns $y_0, y_1, y_2, \ldots, y_n$. This tridiagonal system of linear equations can be solved by any standard procedure.

Since the difference approximations in Equation 7.197 for the differential equation (Equation 7.191) and the difference approximations in Equations 7.223 and 7.225 are all of second order, this method is also of second order.

Example 7.18

Use a second-order method for the solution of the boundary value problem

$$y''=xy+1,\quad 0<x<1$$

$$y(0)+y'(0)=1,\quad y(1)=1$$

with step length $h=0.25$.

Solution:

Here, $h=0.25$, then $x_0=0$, $x_1=0.25$, $x_2=0.5$, $x_3=0.75$, and $x_4=1$. Using finite difference approximation in the given differential equation, we obtain

$$\frac{\bar{y}_{i+1}-2\bar{y}_i+\bar{y}_{i-1}}{h^2}=x_i\bar{y}_i+1,\quad i=1,2,3 \tag{7.226}$$

This implies that

$$\bar{y}_{i+1}-2\bar{y}_i+\bar{y}_{i-1}=\left(x_i\bar{y}_i+1\right)h^2,\quad i=1,2,3 \tag{7.227}$$

Therefore, the discretization of the given differential equation at the internal points $i=1,2,3$ is given by Equation 7.227, which has three equations in five unknowns $y_0, y_1, \ldots, y_4$. Therefore, we need two more equations, these two can be obtained from discretization of the boundary equations. Now, we discretize the given boundary conditions as follows:

$$\bar{y}_0 + \frac{1}{2h}(\bar{y}_1 - \bar{y}_{-1}) = 1, \quad \bar{y}_4 = 1$$

Thus,

$$\bar{y}_{-1} = \frac{1}{2}\bar{y}_0 + \bar{y}_1 - \frac{1}{2}, \quad \bar{y}_4 = 1 \tag{7.228}$$

Here, the fictitious point x_{-1} lies outside the interval [0,1]. The value of x_{-1} may be eliminated by assuming Equation 7.227 also holds for $i = 0$, that is, at the boundary point x_{-1}. After substituting the value of $\bar{y}_{-1}$ from Equation 7.227 for $i = 0$, we obtain

$$-24\bar{y}_0 + 32\bar{y}_1 = 9 \tag{7.229}$$

Now, Equation 7.229 along with Equation 7.227 form a system of four linear equations in four unknowns y_0, y_1, y_2, and y_3, which can be written in the following tridiagonal linear system of the matrix form:

$$\begin{bmatrix} -24 & 32 & 0 & 0 \\ 1 & -\frac{129}{64} & 1 & 0 \\ 0 & 1 & -\frac{65}{32} & 1 \\ 0 & 0 & 1 & -\frac{131}{64} \end{bmatrix} \begin{bmatrix} \bar{y}_0 \\ \bar{y}_1 \\ \bar{y}_2 \\ \bar{y}_3 \end{bmatrix} = \begin{bmatrix} 9 \\ \frac{1}{16} \\ \frac{1}{16} \\ -\frac{15}{16} \end{bmatrix} \tag{7.230}$$

Solving Equation 7.230, we get

$$y(0) = \bar{y}_0 = -7.46162,\ y(0.25) = \bar{y}_1 = -5.31496,\ y(0.5) = \bar{y}_2 = -3.18885 \text{ and}$$
$$y(0.75) = \bar{y}_3 = -1.0999$$

7.6.1.2 Nonlinear Second-Order Differential Equation

Let us consider the following nonlinear second-order differential equation:

$$y'' = f(x, y, y'), \quad a < x < b \tag{7.231}$$

with boundary conditions

$$y(a) = \alpha, \quad y(b) = \beta \tag{7.232}$$

Now using finite difference approximation Equations 7.193 and 7.194 in differential equation, we have the following nonlinear system:

$$\frac{y_{i+1} - 2y_i + y_{i-1}}{h^2} = f\left(x_i, y_i, \frac{y_{i+1} - y_{i-1}}{2h}\right), \quad i = 1, 2, \ldots, n-1 \tag{7.233}$$

Since the values $\bar{y}_0 = \alpha$ and $\bar{y}_n = \beta$ are explicitly known from the boundary conditions in Equation 7.232; therefore, Equation 7.233 consists of a system of $n-1$ equations in $n-1$ unknowns $y_1, y_2, \ldots, y_{n-1}$. In matrix form, Equation 7.233 can be written as

$$\boldsymbol{F}(\boldsymbol{y}) \equiv \boldsymbol{A}\boldsymbol{y} - h^2\tilde{\boldsymbol{F}}(\boldsymbol{y}) - \boldsymbol{b} = \boldsymbol{0} \tag{7.234}$$

where

$$A = \begin{bmatrix} -2 & 1 & 0 & & & \cdots & 0 \\ 1 & -2 & 1 & & & & \vdots \\ \vdots & & & \ddots & & & \\ & & & & 1 & -2 & 1 \\ 0 & \cdots & & & 0 & 1 & -2 \end{bmatrix}$$

$$\tilde{F}(y) = \begin{bmatrix} f\left(x_1, y_1, \dfrac{y_2 - y_0}{2h}\right) \\ f\left(x_2, y_2, \dfrac{y_3 - y_1}{2h}\right) \\ \vdots \\ f\left(x_{n-1}, y_{n-1}, \dfrac{y_n - y_{n-2}}{2h}\right) \end{bmatrix}$$

$$b = \begin{bmatrix} -\alpha \\ 0 \\ \vdots \\ -\beta \end{bmatrix}$$

The nonlinear system in Equation 7.234 can solved by Newton's method as discussed in Section 2.4.7 of Chapter 2. Equation 7.234 gives a system of $(n-1)$ nonlinear equations with $(n-1)$ unknowns. This system of equations can be written as

$$f_1(y_1, y_2, \ldots, y_{n-1}) = 0$$
$$f_2(y_1, y_2, \ldots, y_{n-1}) = 0$$
$$\vdots$$
$$f_{n-1}(y_1, y_2, \ldots, y_{n-1}) = 0$$

In n-dimensional vector form,

$$F(y) = 0 \tag{7.235}$$

where

$$F = [f_1, f_2, \ldots, f_{n-1}]^T, \quad y = [y_1, y_2, \ldots, y_{n-1}]^T$$

Using Newton's method, the $(k+1)$th approximation is given by

$$y^{(k+1)} = y^{(k)} - J^{-1}\left(y^{(k)}\right)F\left(y^{(k)}\right), \quad k = 0,1,2,\ldots \tag{7.236}$$

with initial guess $y^{(0)} = \left[y_1^{(0)}, y_2^{(0)}, \ldots, y_{n-1}^{(0)}\right]^T = [0,0,\ldots,0]^T$, say and the Jacobian matrix

$$\{J(y)\}_{ij} = \frac{\partial}{\partial x_j} f_i(y)$$

The iteration process will continue until $\left\| y^{(k+1)} - y^{(k)} \right\| < \varepsilon$ where ε is the given tolerance level for error and in this case $\| \;.\; \|$ is the L_∞ norm.

Example 7.19

Solve the boundary value problem $y'' + 3yy' = 0$, $y(0) = 0$, $y(2) = 1$ by using the second-order finite difference method.

Solution:

Here, $x_0 = 0$ and $x_n = 2$, where $n = (x_n - x_0)/h$ take $h = 1/3$.

$$y_0 = y(x_0) = 0, \quad y_n = y(x_n) = 1$$

For the second-order finite difference method, we can approximate y, y', and y'' as

$$y = \frac{1}{2}(y_{i+1} + y_{i-1})$$

$$y' = \frac{1}{2h}(y_{i+1} - y_{i-1})$$

$$y'' = \frac{1}{h^2}(y_{i+1} - 2y_i + y_{i-1})$$

Now, the given problem is reduced to

$$4(y_{i+1} - 2y_i + y_{i-1}) + 3h(y_{i+1} + y_{i-1})(y_{i+1} - y_{i-1}) = 0, \quad i = 1, 2, \ldots, n-1 \tag{7.237}$$

Equation 7.237 gives $(n - 1)$ nonlinear equations with $(n - 1)$ unknowns. This system of equations can be represented as

$$f_1(y_1, y_2, \ldots, y_{n-1}) = 0$$

$$f_2(y_1, y_2, \ldots, y_{n-1}) = 0$$

$$\vdots$$

$$f_{n-1}(y_1, y_2, \ldots, y_{n-1}) = 0$$

This can be written as

$$F(y) = \mathbf{0} \tag{7.238}$$

where:

$F = [f_1, f_2, \ldots, f_{n-1}]^T$

$y = [y_1, y_2, \ldots, y_{n-1}]^T$

The system of equation in Equation 7.238 can be solved by Newton's method using the following formula:

$$y^{(k+1)} = y^{(k)} - J^{-1}(y^{(k)})F(y^{(k)}), \quad k = 0, 1, 2, \ldots \tag{7.239}$$

with initial guess $\boldsymbol{y}^{(0)}=\left[y_1^{(0)},y_2^{(0)},\ldots,y_{n-1}^{(0)}\right]^T=\left[0,0,\ldots,0\right]^T$, where the Jacobian matrix

$$J(\boldsymbol{y})=\begin{bmatrix} -8 & 4+6hy_2 & 0 & 0 & 0 \\ 4-6hy_1 & -8 & 4+6hy_3 & 0 & 0 \\ 0 & 4-6hy_2 & -8 & 4+6hy_4 & 0 \\ 0 & 0 & 4-6hy_3 & -8 & 4+6hy_5 \\ 0 & 0 & 0 & 4-6hy_4 & -8 \end{bmatrix} \tag{7.240}$$

Step 1:

$$\boldsymbol{y}^{(1)}=\boldsymbol{y}^{(0)}-J^{-1}\left(\boldsymbol{y}^{(0)}\right)F\left(\boldsymbol{y}^{(0)}\right)=\begin{bmatrix}0\\0\\0\\0\\0\end{bmatrix}-\begin{bmatrix} -8 & 4 & 0 & 0 & 0 \\ 4 & -8 & 4 & 0 & 0 \\ 0 & 4 & -8 & 4 & 0 \\ 0 & 0 & 4 & -8 & 4 \\ 0 & 0 & 0 & 4 & -8 \end{bmatrix}^{-1}\begin{bmatrix}0\\0\\0\\0\\5\end{bmatrix}=\begin{bmatrix}0.208333\\0.416667\\0.625\\0.833333\\1.04167\end{bmatrix}$$

Step 2:

$$\boldsymbol{y}^{(2)}=\boldsymbol{y}^{(1)}-J^{-1}\left(\boldsymbol{y}^{(1)}\right)F\left(\boldsymbol{y}^{(1)}\right)$$

$$=\begin{bmatrix}0.208333\\0.416667\\0.625\\0.833333\\1.04167\end{bmatrix}-\begin{bmatrix} -8 & 4.83333 & 0 & 0 & 0 \\ 3.58333 & -8 & 5.25 & 0 & 0 \\ 0 & 3.16667 & -8 & 5.66667 & 0 \\ 0 & 0 & 2.75 & -8 & 6.08333 \\ 0 & 0 & 0 & 2.33333 & -8 \end{bmatrix}^{-1}\begin{bmatrix}0.173611\\0.347222\\0.520833\\0.694444\\-0.69444\end{bmatrix}$$

$$=\begin{bmatrix}0.459405\\0.796314\\0.966006\\1.01069\\1.00659\end{bmatrix}$$

Step 3:

$$\boldsymbol{y}^{(3)}=\boldsymbol{y}^{(2)}-J^{-1}\left(\boldsymbol{y}^{(2)}\right)F\left(\boldsymbol{y}^{(2)}\right)$$

$$=\begin{bmatrix}0.459405\\0.796314\\0.966006\\1.01069\\1.00659\end{bmatrix}-\begin{bmatrix} -8 & 5.59263 & 0 & 0 & 0 \\ 3.08119 & -8 & 5.93201 & 0 & 0 \\ 0 & 2.40737 & -8 & 6.02137 & 0 \\ 0 & 0 & 2.06799 & -8 & 6.01318 \\ 0 & 0 & 0 & 1.97863 & -8 \end{bmatrix}^{-1}\begin{bmatrix}0.144132\\0.0532482\\-0.112678\\-0.115055\\-0.031454\end{bmatrix}$$

$$=\begin{bmatrix}0.452262\\0.760324\\0.912204\\0.972306\\0.993165\end{bmatrix}$$

Step 4:

$$y^{(4)} = y^{(3)} - J^{-1}\left(y^{(3)}\right)F\left(y^{(3)}\right)$$

$$= \begin{bmatrix} 0.452262 \\ 0.760324 \\ 0.912204 \\ 0.972306 \\ 0.993165 \end{bmatrix} - \begin{bmatrix} -8 & 5.52065 & 0 & 0 & 0 \\ 3.09548 & -8 & 5.82441 & 0 & 0 \\ 0 & 2.47935 & -8 & 5.94461 & 0 \\ 0 & 0 & 2.17559 & -8 & 5.98633 \\ 0 & 0 & 0 & 2.05539 & -8 \end{bmatrix}^{-1} \begin{bmatrix} 0.0012952 \\ 0.0028436 \\ 0.0001778 \\ -0.0027145 \\ -0.001473 \end{bmatrix}$$

$$= \begin{bmatrix} 0.452445 \\ 0.760356 \\ 0.911661 \\ 0.971533 \\ 0.992782 \end{bmatrix}$$

Step 5:

$$y^{(5)} = y^{(4)} - J^{-1}\left(y^{(4)}\right)F\left(y^{(4)}\right)$$

$$= \begin{bmatrix} 0.452445 \\ 0.760356 \\ 0.911661 \\ 0.971533 \\ 0.992782 \end{bmatrix} - \begin{bmatrix} -8 & 5.52071 & 0 & 0 & 0 \\ 3.09511 & -8 & 5.82332 & 0 & 0 \\ 0 & 2.47929 & -8 & 5.94307 & 0 \\ 0 & 0 & 2.17668 & -8 & 5.98556 \\ 0 & 0 & 0 & 2.05693 & -8 \end{bmatrix}^{-1} \begin{bmatrix} 9.80\times10^{-10} \\ 2.60\times10^{-7} \\ 5.97\times10^{-7} \\ -1.48\times10^{-7} \\ -5.98\times10^{-7} \end{bmatrix}$$

$$= \begin{bmatrix} 0.452445 \\ 0.760356 \\ 0.911661 \\ 0.971533 \\ 0.992782 \end{bmatrix}$$

Hence, the required solutions at the node points are as follows:

$$y_0 = y(0) = 0$$

$$y_1 = y(1/3) = 0.452445$$

$$y_2 = y(2/3) = 0.760356$$

$$y_3 = y(1) = 0.911661$$

$$y_4 = y(4/3) = 0.971533$$

$$y_5 = y(5/3) = 0.992782$$

$$y_6 = y(2) = 1$$

7.6.2 Shooting Method

The shooting method is based on converting the boundary value problem into an equivalent initial value problem. Let us consider the following two-point boundary value problem:

$$y'' = f(x, y, y'), \quad a < x < b \tag{7.241}$$

with boundary conditions

$$\alpha_0 y(a) - \alpha_1 y'(a) = \gamma_1, \quad \beta_0 y(b) + \beta_1 y'(b) = \gamma_2 \tag{7.242}$$

where $\alpha_0, \alpha_1, \beta_0, \beta_1, \gamma_1$, and γ_2 are constants. We shall suppose that Equation 7.242 has a unique solution. The motivation behind shooting methods is to convert the two-point boundary value problem into solving a sequence of initial value problems whose solutions converge to that of the boundary value problem, so that one can use existing program developed for the numerical solution of initial value problems. Let us consider the following initial value problem

$$y'' = f(x, y, y'), \quad a < x < b \tag{7.243}$$

with initial conditions

$$y(a) = \alpha_1 s - c_1\gamma_1, \quad y'(a) = \alpha_0 s - c_0\gamma_1 \tag{7.244}$$

depending on the parameter s, where c_0 and c_1 are arbitrary constants so chosen that

$$\alpha_1 c_0 - \alpha_0 c_1 = 1$$

Let $y(x;s)$ be the solution of the initial value problem (Equations 7.243 and 7.244). Now, we introduce the notation

$$u(x;s) = y(x;s), \quad v(x;s) = \frac{\partial}{\partial x} y(x;s) \tag{7.245}$$

Then, we can rewrite Equation 7.245 as a system of first-order ordinary differential equations:

$$\begin{aligned} \frac{\partial u(x;s)}{\partial x} &= v(x;s), \quad u(a;s) = \alpha_1 s - c_1\gamma_1 \\ \frac{\partial v(x;s)}{\partial x} &= f(x,u,v), \quad v(a;s) = \alpha_0 s - c_0\gamma_1 \end{aligned} \tag{7.246}$$

This shows that $u(x;s)$ is a solution of Equation 7.243, and it satisfies the first boundary condition in Equation 7.244. Furthermore, the solution $u(x;s)$ of the initial value problem in Equation 7.247 will be the same as the solution $y(x)$ of the boundary value problem in Equations 7.241 and 7.239 provided that $u(x;s)$ satisfies the remaining boundary condition at $x = b$ so that

$$\varphi(s) \equiv \beta_0 u(b;s) + \beta_1 u'(b;s) - \gamma_2 = 0 \tag{7.247}$$

This is a nonlinear equation in s. If s^* is a root of $\varphi(s)$, then $u(x;s^*)$ will satisfy the boundary value problem in Equations 7.241 and 7.242. Thus, the essence of the shooting method for the numerical solution of the boundary value problem (Equations 7.241 and 7.242) is to find a root to the Equation 7.247. Any standard root-finding technique can be used. Here, we shall consider rapidly convergent method, that is to say, Newton's method, given by

$$s_{m+1} = s_m - \frac{\varphi(s_m)}{\varphi'(s_m)}, \quad m = 0,1,2,\ldots \tag{7.248}$$

with the starting value s_0 chosen arbitrarily in a sufficiently small interval containing the root. If the initial guess s_0 is a sufficiently good approximation to the required root of Equations 7.247, then the Newton–Raphson method ensures that the sequence of successive approximations $\{s_n\}_{n=0}^{\infty}$ converges quadratically to the root s. To calculate $\varphi'(s)$, we differentiate Equation 7.247 with respect to s to obtain

$$\varphi'(s) = \beta_0 \frac{\partial u(b;s)}{\partial s} + \beta_1 \frac{\partial u'(b;s)}{\partial s} \tag{7.249}$$

Now, we introduce the new dependent variables

$$\xi(x;s) = \frac{\partial u(x;s)}{\partial s}, \quad \eta(x;s) = \frac{\partial v(x;s)}{\partial s} \tag{7.250}$$

and differentiate the initial value problem (Equation 7.247) with respect to s to obtain a second initial value problem:

$$\begin{aligned} \frac{\partial \xi(x;s)}{\partial x} &= \eta(x;s), \quad \xi(a;s) = \alpha_1 \\ \frac{\partial \eta(x;s)}{\partial x} &= p(x;s)\xi(x;s) + q(x;s)\eta(x;s), \quad \eta(a;s) = \alpha_0 \end{aligned} \tag{7.251}$$

where

$$p(x;s) = \frac{\partial f(x, u(x;s), v(x;s))}{\partial u} \quad \text{and} \quad q(x;s) = \frac{\partial f(x, u(x;s), v(x;s))}{\partial v}.$$

Now, if we assign the value s_m to s, $m \geq 0$, then the initial value problem (Equations 7.246 and 7.251) can be solved by a predictor–corrector method of order p or a R–K method of order p on the interval $[a,b]$. Thus, we obtain an approximation to $u(b;s_m)$ from which we can calculate $\varphi(s_m) = \beta_0 u(b;s_m) + \beta_1 v(b;s_m)$ $-\gamma_2$. Consequently, we obtain an approximation to $\varphi'(s_m) = \beta_0 \xi(b;s_m) + \beta_1 \eta(b;s_m)$. Having computed $\varphi(s_m)$ and $\varphi'(s_m)$, we obtain the next Newton–Raphson iterate s_{m+1} from Equation 7.248. The process is then repeated until the iterates s_m reach desire degree of accuracy.

A number of problems may arise with the shooting method. The coupled initial value problem (Equations 7.246 and 7.251) may be very sensitive to perturbations of the initial guess s_0; a bad initial guess of s_0 may result in a sequence of Newton–Raphson iterates $\{s_n\}_{n=0}^{\infty}$, which does not converge to the root s^*. Therefore, since there is no general guess s_0 for Newton iteration, the sequence of successive approximations $\{s_n\}_{n=0}^{\infty}$ may also diverge. A second problem is that the approximate solution $\bar{y}(x,s)$ to $y(x)$ may be very sensitive to h, s, and other characteristics of the boundary value problem.

Example 7.20

Using the shooting method, solve the following boundary value problem:

$$y'' = \frac{-2}{x} yy', \quad 1 < x < 2; \quad y(1) = \frac{1}{2}, y(2) = \frac{2}{3}$$

Hence, compare the result with exact solution $y(x) = x/(1+x)$.

Solution:

The shooting technique for the second-order boundary value problem $y'' = (-2/x)yy'$ can be converted to a sequence of initial value problems involving a parameter t. The problem is of the following form:

$$y''=\frac{-2}{x}yy', \quad \text{for } 1< x<2, \text{ with } y(1)=\frac{1}{2} \text{ and } y'(1)=t \tag{7.252}$$

Rewriting the initial value problem (Equation 7.252), emphasizing that the solution depends on both x and the parameter t:

$$y''(x,t)=\frac{-2}{x}y(x,t)y'(x,t), \text{ with } y(1,t)=\frac{1}{2} \text{ and } y'(1,t)=t \tag{7.253}$$

Taking partial derivatives with respect to t, we get

$$\frac{\partial}{\partial t}y''(x,t)=\left(\frac{-2}{x}\right)y'(x,t)\frac{\partial}{\partial t}y(x,t)+\left(\frac{-2}{x}\right)y(x,t)\frac{\partial}{\partial t}y'(x,t) \tag{7.254}$$

with initial condition $\partial/\partial t y(1,t)=0$ and $\partial/\partial t y'(1,t)=1$. Putting

$$\frac{\partial}{\partial t}y(x,t) = z(x,t)$$

in Equation 7.254 we get

$$z''(x,t)=\left(\frac{-2}{x}\right)y'(x,t)z(x,t)+\left(\frac{-2}{x}\right)y(x,t)z'(x,t) \tag{7.255}$$

with initial condition $z(1,t)=0$ and $z'(1,t)=1$. Substituting $y'(x,t)=u(x,t)$ and $z'(x,t)=v(x,t)$ in Equations 7.253 and 7.255, we get system of initial value problem of the form

$$\begin{aligned} y'(x,t)&=u(x,t)\\ u'(x,t)&=\frac{-2}{x}y(x,t)u(x,t)\\ z'(x,t)&=v(x,t)\\ v'(x,t)&=\left(\frac{-2}{x}\right)u(x,t)z(x,t)+\left(\frac{-2}{x}\right)y(x,t)v(x,t)\end{aligned} \tag{7.256}$$

with initial conditions

$$y(1,t)=\frac{1}{2}, u(1,t)=t, z(1,t)=0, \quad \text{and} \quad v(1,t)=1 \tag{7.257}$$

The above system of equations can be solved using the fourth-order R–K technique. The solution $y(x,t)$ to the initial value problem (Equation 7.256) is a nonlinear equation in the variable t, which can be determined from $y(2,t)-(2/3)=0$ using the Newton–Raphson method. The results of the calculations with $h=0.25$ are given in the following table.

x	Y_{approx}	Y_{exact}
1.25	0.555555	0.555556
1.50	0.599999	0.6
1.75	0.636363	0.636363
2.0	0.666667	0.666667

Example 7.21

Using the shooting method, solve the following boundary value problem.

$$y''=\frac{3}{2}y^2, \quad 0<x<1; \quad y(0)=1, y(1)=4$$

Hence, compare the result with the exact solution $y(x)=4/(2-x)^2$.

Solution:

The shooting technique for the second-order boundary value problem $y'' = (3/2)y^2$ can be converted to a sequence of initial value problems involving a parameter t. The problem is of the following form:

$$y'' = \frac{3}{2}y^2, \quad \text{for } 0 < x < 1, \text{ with } y(0) = 1 \text{ and } y(1) = 4 \tag{7.258}$$

Rewriting the initial value problem (Equation 7.258), emphasizing that the solution depends on both x and the parameter t:

$$y''(x,t) = \frac{3}{2}y^2(x,t), \text{ with } y(0,t) = 1 \text{ and } y'(0,t) = t \tag{7.259}$$

Taking partial derivatives with respect to t, we get

$$\frac{\partial}{\partial t}y''(x,t) = 3y(x,t)\frac{\partial}{\partial t}y(x,t) \tag{7.260}$$

with initial condition

$$\frac{\partial}{\partial t}y(0,t) = 0 \text{ and } \frac{\partial}{\partial t}y'(0,t) = 1$$

Putting $(\partial/\partial t)y(x,t) = z(x,t)$ in Equation 7.260, we get

$$z''(x,t) = 3y(x,t)z(x,t) \tag{7.261}$$

with initial condition $z(0,t) = 0$ and $z'(0,t) = 1$. Substituting $y'(x,t) = u(x,t)$ and $z'(x,t) = v(x,t)$ in Equations 7.259 and 7.261, we get system of initial value problem of the following form:

$$\begin{aligned} y'(x,t) &= u(x,t) \\ u'(x,t) &= \frac{3}{2}y^2(x,t) \\ z'(x,t) &= v(x,t) \\ v'(x,t) &= 3y(x,t)z(x,t) \end{aligned} \tag{7.262}$$

with initial conditions

$$y(0,t) = 1,\ u(0,t) = t,\ z(0,t) = 0 \text{ and } v(0,t) = 1 \tag{7.263}$$

The above system of equations can be solved using the fourth-order R–K technique. The solution $y(x,t)$ to the initial value problem (Equations 7.262) is a nonlinear equation in the variable t, which can be determined from $y(1,t) - 4 = 0$ using the Newton–Raphson method. The results of the calculations with $h = 0.25$ are given in the following table.

x	Y_{approx}	Y_{exact}
0.25	1.30667	1.30612
0.50	1.77884	1.77778
0.75	2.56139	2.56
1.0	4	4

MATHEMATICA® Program for Solving Second-Order BVP by Shooting Method (Chapter 7, Example 7.20)

(*After reducing to the system of first-order IVP *)

```
f[x_,y_,z_,u_,v_]:=u;
g[x_,y_,z_,u_,v_]:=-(2/x)*y*u;
f1[x_,y_,z_,u_,v_]:=v;
g1[x_,y_,z_,u_,v_]:=-(2/x)*u*z-(2/x)*y*v;
a=1.;b=2.;
x[0.]=1.;
y[0.]=1/2;y1[b]=2/3;          (*Boundary conditions*)
z[0.]=0.;u[0.]=t;v[0.]=1.;
h=0.25;
n1=(b-a)/h;
For[s=1.,s≤2.,s++,
  For[i=0.,i≤n1-1,i++,
   x[i]=x[0.]+i*h;

   k[1.]=Simplify[h*f[x[i],y[i],z[i],u[i],v[i]]];
   l[1.]=Simplify[h*f1[x[i],y[i],z[i],u[i],v[i]]];
   m[1.]=Simplify[h*g[x[i],y[i],z[i],u[i],v[i]]];
   n[1.]=Simplify[h*g1[x[i],y[i],z[i],u[i],v[i]]];

   k[2.]=Simplify[h*f[x[i]+h/2,y[i]+k[1.]/2,z[i]+l[1.]/2,u[i]+m[1.]/2,
   v[i]+ n[1.]/2]];
   l[2.]=Simplify[h*f1[x[i]+h/2,y[i]+k[1.]/2,z[i]+l[1.]/2,u[i]+m[1.]/2,
   v[i]+ n[1.]/2]];
   m[2.]=Simplify[h*g[x[i]+h/2,y[i]+k[1.]/2,z[i]+l[1.]/2,u[i]+m[1.]/2,
   v[i]+n[1.]/2]];
   n[2.]=Simplify[h*g1[x[i]+h/2,y[i]+k[1.]/2,z[i]+l[1.]/2,u[i]+m[1.]/2,
   v[i]+n[1.]/2]];

   k[3.]=Simplify[h*f[x[i]+h/2,y[i]+k[2.]/2,z[i]+l[2.]/2,u[i]+m[2.]/2,
   v[i]+n[2.]/2]];
   l[3.]=Simplify[h*f1[x[i]+h/2,y[i]+k[2.]/2,z[i]+l[2.]/2,u[i]+m[2.]/2,
   v[i]+n[2.]/2]];m[3.]=Simplify[h*g[x[i]+h/2,y[i]+k[2.]/2,z[i]+l[2.]/2,u
   [i]+m[2.]/2,v[i]+n[2.]/2]];
   n[3.]=Simplify[h*g1[x[i]+h/2,y[i]+k[2.]/2,z[i]+l[2.]/2,u[i]+m[2.]/2,
   v[i]+n[2.]/2]];

   k[4.]=Simplify[h*f[x[i]+h,y[i]+k[3.],z[i]+l[3.],u[i]+m[3.],v[i]+
   n[3.]]];
   l[4.]=Simplify[h*f1[x[i]+h,y[i]+k[3.],z[i]+l[3.],u[i]+m[3.],
   v[i]+n[3.]]];
   m[4.]=Simplify[h*g[x[i]+h,y[i]+k[3.],z[i]+l[3.],u[i]+m[3.],v[i]+n[3.]]];
   n[4.]=Simplify[h*g1[x[i]+h,y[i]+k[3.],z[i]+l[3.],u[i]+m[3.],v[i]+n[3.]]];

   y[i+1]=Simplify[y[i]+1/6*(k[1.]+2*k[2.]+2*k[3.]+k[4.])];
   z[i+1]=Simplify[z[i]+1/6*(l[1.]+2*l[2.]+2*l[3.]+l[4.])];
   u[i+1]=Simplify[u[i]+1/6*(m[1.]+2*m[2.]+2*m[3.]+m[4.])];
   v[i+1]=Simplify[v[i]+1/6*(n[1.]+2*n[2.]+2*n[3.]+n[4.])];
   If[s==1.,True,
    Print[x[i]+h,"                    ",y[i+1]]]];
  If[s==1.,
   sol=FindRoot[{y[n1]-y1[b]==0.},{t,0.5}];
   u[0.]=sol[[1,2]]]];
```

Output:

```
1.25    0.555555
1.5     0.599999
1.75    0.636363
2.      0.666667
```

7.6.3 Collocation Method

For simplicity, let us consider a typical boundary value problem for the following second-order linear differential equation:

$$y'' = f(x, y, y'), a < x < b \tag{7.264}$$

with the homogeneous boundary conditions

$$y(a) = 0 \quad \text{and} \quad y(b) = 0 \tag{7.265}$$

Suppose that $\{\phi_i(x)\}_{i=1}^{n}$ be a set of linearly independent functions defined on the interval $[a,b]$ satisfying the boundary conditions in Equation 7.265. We shall also suppose that each function $\phi_i(x)$ is twice continuously differentiable on $[a,b]$.

We assume that the solution $y(x)$ of Equations 7.264 and 7.265 is approximated by a linear combination of n given functions $\phi_i(x), \ldots, \phi_n(x)$, given by

$$y(x) = \sum_{i=1}^{n} c_i\phi_i(x) \tag{7.266}$$

Since the functions are all assumed to satisfy the boundary conditions

$$\phi_i(a) = \phi_i(b) = 0, \quad i = 1, 2, \ldots, n \tag{7.267}$$

and hence any linear combination (Equation 7.266) also satisfy the boundary conditions. Thus, the essence of the collocation method is to seek an approximate solution $\overline{y}_n(x)$ to the boundary value problem (Equations 7.264 and 7.265) in the following form:

$$\overline{y}_n(x) = \sum_{i=1}^{n} c_i\phi_i(x) \tag{7.268}$$

which satisfies Equation 7.264 at n distinct points $\xi_i, i = 1, 2, \ldots, n$, referred to as the *collocation points*.

Now we shall determine the coefficients $c_1, c_2, \ldots, c_n$ such that $\overline{y}_n(x)$ satisfies the differential equation (7.264) exactly at n collocation points in (a,b),

$$\overline{y}_n''(\xi_i) = f(\xi_i, \overline{y}_n(\xi_i), \overline{y}_n'(\xi_i)), \quad i = 1, 2, \ldots, n \tag{7.269}$$

with collocation points

$$a < \xi_1 < \xi_2 < \cdots < \xi_n < b \tag{7.270}$$

The procedure of finding $y_n(x)$ implicitly through Equation 7.269 is known as the *collocation method.* Substituting Equation 7.268 into Equation 7.269, we obtain

$$\sum_{i=1}^{n} c_i\phi_i''(\xi_j) = f\left(\xi_j, \sum_{i=1}^{n} c_i\phi_i(\xi_j), \sum_{i=1}^{n} c_i\phi_i'(\xi_j)\right), \quad j = 1,2,\ldots,n \tag{7.271}$$

This is a system of n nonlinear equations in the n unknowns $c_1, c_2, \ldots, c_n$. Solving this system of equations by Newton's method, we can determine the values of $c_1, c_2, \ldots, c_n$ and hence, the approximate solution $\bar{y}_n(x)$ is obtained. The specific properties of the collocation method depend on the choice of the basis functions $\phi_i(x)$ and the collocation points ξ_i. Moreover, if each of the basis functions $\phi_i(x)$ has compact support contained in (a,b), for example, suppose the basis functions are B-spline functions. Then, the resulting sparse discretized system of equations can be solved effectively. As an illustrative example, let us consider the following simple boundary value problem:

$$y''(x) + p(x)y(x) = f(x), \quad 0 < x < \pi. \tag{7.272}$$

with boundary conditions $y(0) = 0$ and $y(\pi) = 0$, where $p(x) \geq 0$ for all $x \in [0,\pi]$. Let us suppose that the basis functions be $\phi_i(x) = \sin ix$, $i = 1,2,\ldots,n$, which satisfy the boundary conditions, and these basis functions are linearly independent on $[0,\pi]$. According to the collocation method, we seek an approximate solution $\bar{y}_n(x)$ in the following form:

$$\bar{y}_n(x) = \sum_{i=1}^{n} c_i \sin ix \tag{7.273}$$

Substituting Equation 7.273 into the differential equation (7.272) yields the following system of equations:

$$-\sum_{i=1}^{n} c_i i^2 \sin i\xi_j + p(\xi_j)\sum_{i=1}^{n} c_i \sin i\xi_j = f(\xi_j), \quad j = 1,2,\ldots,n \tag{7.274}$$

The collocation points ξ_j are usually chosen as

$$\xi_j = \frac{j\pi}{n+1}, \quad j = 1,2,\ldots,n \tag{7.275}$$

Using these collocation points ξ_j defined in Equation 7.275, the system of equations (7.274) can be solved for the unknown coefficients $c_1, c_2, \ldots, c_n$ using any standard method.

Example 7.22

Solve the boundary value problem $y'' + 3yy' = 0$, $y(0) = 0$, $y(2) = 1$ by using the collocation method in terms of cubic Hermite basis functions.

Solution:

Let, $\mathcal{L}y \equiv y'' + 3yy' = 0$ with $y(0) = 0$ at $x = 0$ and $y(2) = 1$ at $x = 2$.

Taking, $h = 1/3$, we have $n = (2-0)/h = 6$.

We approximate the unknown function $y(x)$ by cubic Hermite basis functions as

$$\bar{y}(x) = \sum_{i=0}^{n} \left[c_i\varphi_i^3(x) + d_i w_i^3(x)\right] \tag{7.276}$$

where

$$\varphi_i^3(x) = \begin{cases} 1-3\left(\dfrac{x-x_i}{h_i}\right)^2 - 2\left(\dfrac{x-x_i}{h_i}\right)^3, & x_{i-1} \le x < x_i \\ 1-3\left(\dfrac{x-x_i}{h_{i+1}}\right)^2 + 2\left(\dfrac{x-x_i}{h_{i+1}}\right)^3, & x_i \le x < x_{i+1} \\ 0, & \text{otherwise} \end{cases}$$

$$w_i^3(x) = \begin{cases} (x-x_i)\left(1+\dfrac{x-x_i}{h_i}\right)^2, & x_{i-1} \le x < x_i \\ (x-x_i)\left(1-\dfrac{x-x_i}{h_{i+1}}\right)^2, & x_i \le x < x_{i+1} \\ 0, & \text{otherwise} \end{cases}$$

with $0 = x_0 < x_1 < \cdots < x_n = 2$. Now,

$$\mathcal{L}\bar{y} = \mathcal{L}\left[\sum_{i=0}^{n}\left[c_i\varphi_i^3(x) + d_i w_i^3(x)\right]\right] = 0 \tag{7.277}$$

Putting the collocation points

$$\xi_{i,j} = \frac{1}{2}(x_{i-1} + x_i) + \frac{1}{2}h\rho_j, \quad j = 1,2, \quad i = 1,2,\ldots,n$$

with $\rho_1 = -1/\sqrt{3}$, $\rho_2 = 1/\sqrt{3}$, in Equation 7.277, we can get a system of $2n$ number of nonlinear algebraic equations.

$$\text{i.e., } \mathcal{L}\bar{y} = \mathcal{L}\left[\sum_{i=0}^{n}\left[c_i\varphi_i^3(\xi_{i,j}) + d_i w_i^3(\xi_{i,j})\right]\right] = 0, \quad j = 1,2, \quad i = 1,2,\ldots,n \tag{7.278}$$

Again from boundary conditions, we get

$$\bar{y}(0) = \sum_{i=0}^{n}\left[c_i\varphi_i^3(0) + d_i w_i^3(0)\right] = 0 \tag{7.279}$$

$$\bar{y}(2) = \sum_{i=0}^{n}\left[c_i\varphi_i^3(2) + d_i w_i^3(2)\right] = 1 \tag{7.280}$$

Equations 7.278 through 7.280 give $2n+2$ number of nonlinear algebraic equations with same number of unknowns for c_i and $d_i, i = 0,1,2,\ldots,n$.

Solving this system by Newton's method, we can get the approximate value of $2n+2$ coefficients c_i and d_i, $i = 0,1,2,\ldots,n$ as

$$c_0 = 0, c_1 = 0.4666, c_2 = 0.767528, c_3 = 0.910875, c_4 = 0.969352, c_5 = 0.991686, c_6 = 1,$$

$$d_0 = 1.51455, d_1 = 1.18812, d_2 = 0.631012, d_3 = 0.270051, d_4 = 0.105109, d_5 = 0.0394089,$$

$$d_6 = 0.0145691$$

Therefore, the approximate solution is

$$\bar{y}(x)=\sum_{i=0}^{n}\left[c_i\varphi_i^3(x)+d_iw_i^3(x)\right]$$

which yields

$$\bar{y}(x)=\begin{cases}1.51455\,x-0.0534633\,x^2-0.872353\,x^3, & 0\le x<1/3\\ -0.0335983+1.82663\,x-1.01882\,x^2+0.122102\,x^3, & 1/3\le x<2/3\\ -0.0850209+2.09055\,x-1.46346\,x^2+0.368807\,x^3, & 2/3\le x<1\\ 0.0653621+1.63967\,x-1.01286\,x^2+0.218698\,x^3, & 1\le x<4/3\\ 0.345604+0.998733\,x-0.524356\,x^2+0.0946237\,x^3, & 4/3\le x<5/3\\ 0.600856+0.531886\,x-0.239813\,x^2+0.0368279\,x^3, & 5/3\le x<2\end{cases}$$

MATHEMATICA® Program for Solving ODE by Collocation Method (Chapter 7, Example 7.22)

```
a=0;
b=2;
h=1/3;
n=(b-a)/h;
x[0]=0;
Phi[i_,x_]:=Piecewise[{{1-3*((x-(a+i*h))/h)²-2*((x-(a+i*h))/h)³,
(a+(i-1)*h)<=x<(a+i*h)},{1-3*((x-(a+i*h))/h)²+2*((x-(a+i*h))/h)³,
(a+i*h)<=x<(a+(i+1)*h)}}];
w[i_,x_]:=Piecewise[{{(x-(a+i*h))*(1+(x-(a+i*h))/h)², (a+(i-1)*h)<=
x<(a+i*h)},{(x-(a+i*h))*(1-(x-(a+i*h))/h)², (a+i*h)<=x<(a+(i+1)*h)}}];
Y[x]=Σ_{j=0}^{n}(c[j]*Phi[j,x]+d[j]*w[j,x]);
Y1[x]=D[Y[x],{x,1}];
Y2[x]=D[Y[x],{x,2}];
For[i=1,i<=n,i++,
  x[i]=x[0]+i*h;
  ξ1[i]=(1/2)*(x[i-1]+x[i])+(1/2)*h*(-1/√3);
  ξ2[i]=(1/2)*(x[i-1]+x[i])+(1/2)*h*(1/√3);
  eqn[i]=Simplify[(Y2[x]/.x->ξ1[i])+3*(Y[x]/.x->ξ1[i])*
  (Y1[x]/.x->ξ1[i])]==0;
  eqn[n+i]=Simplify[(Y2[x]/.x->ξ2[i])+3*(Y[x]/.x->ξ2[i])*
  (Y1[x]/.x->ξ2[i])]==0;
  Print["eqn[",i,"]=",eqn[i]];
  Print["eqn[",n+i,"]=",eqn[n+i]]];
eqn[2*n+1]=Simplify[Y[x]/.x->0]==0;
eqn[2*n+2]=Simplify[Y[x]/.x->2]==1;
eqns=Table[eqn[i],{i,1,2*n+2}];
A1=Table[c[i],{i,0,n}];
A2=Table[d[i],{i,0,n}];
A=Join[A1,A2];
AA=Table[{A[[i]],0},{i,1,2*n+2}];
sol=FindRoot[eqns,AA]
AA1=Table[sol[[i]][[2]],{i,1,n+1}];
AA2=Table[sol[[i]][[2]],{i,n+2,2*(n+1)}];
```

```
y[x]=Σ_{j=0}^n(AA1[[j+1]]*Phi[j,x]+AA2[[j+1]]*w[j,x]);
Simplify[y[x]]
For[i=1,i<n,i++,
  x[i]=x[0]+i*h;
  y[i]=y[x]/.x->x[i];
  Print[N[x[i]],"       ",N[y[i]]]];
```

Output:

$$\begin{cases} 0. & x \le \frac{7}{3} \,\|\, x < -\frac{1}{3} \\ 1.53114\times 10^{-25}+1.51455x-0.0534633x^2-0.872353x^3 & 0 \le x < \frac{1}{3} \\ 0.600856+0.531886x-0.239813x^2+0.0368279x^3 & \frac{5}{3} \le x < 2 \\ 0.345604+0.998733x-0.524356x^2+0.0946237x^3 & \frac{4}{3} \le x < \frac{5}{3} \\ -0.0335983+1.82663x-1.01882x^2+0.122102x^3 & \frac{1}{3} \le x < \frac{2}{3} \\ 0.0653621+1.63967x-1.01286x^2+0.218698x^3 & 1 \le x < \frac{4}{3} \\ -0.0850209+2.09055x-1.46346x^2+0.368807x^3 & \frac{2}{3} \le x < 1 \\ 1.53114\times 10^{-25}+1.51455x+9.08729x^2+13.6309x^3 & -\frac{1}{3} \le x < 0 \\ -540.428+757.938x-351.874x^2+54.1311x^3 & \text{True} \end{cases}$$

7.6.4 Galerkin Method

This method is also known as Weighted residual method. For solving the boundary value problem,

$$\mathcal{L}y \equiv f(x), \quad a \le x \le b, \quad y(a)=\alpha,\; y(b)=\beta, \tag{7.281}$$

where $\mathcal{L}$ is the linear operator, by Galerkin method, first we have to choose trial solutions in such a way that it satisfy all the essential boundary conditions. Approximate $y(x)$ as the linear combination of trial solutions

$$\bar{y}(x)=\sum_{i=1}^{n} c_i\varphi_i(x), \quad a \le x \le b$$

where c_j's are unknown coefficients. The residual is given by

$$R(\bar{y})=\mathcal{L}\bar{y}-f(x)=0 \tag{7.282}$$

We choose the weight functions $w_j(x)$ such that it is orthogonal to trial solutions $\varphi_i(x)$ as $\langle w_j(x),\varphi_i(x)\rangle=\delta_{ij}$ (where δ_{ij} denotes the kronecker delta function).

Now, multiplying weight functions $w_j(x)$ both sides of Equation 7.282 and taking inner product, we have

$$\langle w_j(x),R(\bar{y})\rangle=\langle w_j(x),\mathcal{L}\bar{y}-f(x)\rangle=\int_0^1 w_j(x)(\mathcal{L}\bar{y}-f(x))dx=0 \tag{7.283}$$

For $j=1,2,\ldots,n$, Equation 7.283 gives n number of linear algebraic equations with n number of unknowns c_i, $i=1,2,\ldots,n$. Solving these linear equations, we get the approximate values of c_i, $i=1,2,\ldots,n$ and hence obtain the solution $\bar{y}(x)=\sum_{i=1}^{n} c_i\varphi_i(x)$.

MATHEMATICA® Program for Solving ODE by Galerkin Method (Chapter 7, Example 7.23)

```
n=3;
For[i=1,i≤n,i++,
  phi[i][x]=(x*(x-1))^i];
y[x]= Σ_{i=1}^{n}(c[i]*phi[i][x]);
y1[x]=D[y[x],{x,1}];
y2[x]=D[y[x],{x,2}];
R[x]=y2[x]+y[x]+2*x*(1-x);
For[i=1,i≤n,i++,
  eqn[i]=Simplify[Integrate[R[x]*phi[i][x],{x,0,1}]]==0;
  Print["eqn[",i,"]=",eqn[i]]];
eqns=Table[eqn[i],{i,1,n}];
A=Table[c[i],{i,1,n}];
sol=NSolve[eqns,A]
yy[x]= Σ_{i=1}^{n}sol[[1]][[i]][[2]]*phi[i][x];
Print[Simplify[yy[x]]]
For[i=0,i≤1,i=i+0.2,
  yy[i]=yy[x]/.x->i;
  Print[i,"     ",yy[i]]];
```

Output:

```
eqn[1]=(-84-378 c[1]+75 c[2]-16 c[3])/1260==0
eqn[2]=(198+825 c[1]-242 c[2]+61 c[3])/13860==0
eqn[3]=(-572-2288 c[1]+793 c[2]-219 c[3])/180180==0
{{c[1]->-0.18521,c[2]->0.185203,c[3]->-0.00626989}}
(-1+x) x (-0.18521+0.185203 (-1.+x) x-0.00626989 (-1.+x)^2 x^2)
0          0.
0.2        0.0344005
0.4        0.0552048
0.6        0.0552048
0.8        0.0344005
1.         0.
```

Example 7.23

Solve the following boundary value problem by the Galerkin method

$$y''+y+2x(1-x),\ y(0)=0, y(1)=0$$

Solution:

Here,

$$\mathcal{L}\bar{y} \equiv y''+y = 2x(x-1),\ y(0)=0,\ y(1)=0 \tag{7.284}$$

We choose the trial solutions $\varphi_i(x) = x^i(x-1)^i$, which satisfy all the boundary conditions

$$\varphi_i(0)=0,\ \varphi_i(1)=0, \quad i=1,2,\ldots$$

We approximate the unknown function $\bar{y}(x)$ by the linear combination of finite number of trial solutions, that is, for $n=3$,

$$\bar{y}(x)=\sum_{i=1}^{3}c_i\varphi_i(x) \tag{7.285}$$

We have the residual

$$R(\bar{y})=\mathcal{L}\bar{y}-f(x)=\bar{y}''+\bar{y}'-2x(x-1)=0 \tag{7.286}$$

Now, we choose the weight functions same as trial solutions, that is, $w_j(x)=\varphi_j(x),\ j=1,2,3$. Multiplying $w_j(x)$ both sides of Equation 7.286 and taking inner product both sides, we have

$$\begin{aligned}\langle w_j(x),R(\bar{y})\rangle &= \langle w_j(x),\mathcal{L}\bar{y}-f(x)\rangle\\ &=\int_0^1\varphi_j(x)\left(\sum_{i=1}^{3}c_i\varphi_i''(x)+\sum_{i=1}^{3}c_i\varphi_i(x)-2x(x-1)\right)dx=0\end{aligned} \tag{7.287}$$

For $j=1,2,3$, Equation 7.287 gives three equations with three unknowns c_1, c_2, and c_3.

$$378c_1-75c_2+16c_3=-84$$

$$825c_1-242c_2+61c_3=-198$$

$$2288c_1-793c_2+219c_3=-552$$

Solving the above system for unknowns, we have

$$c_1=-0.18521,\ c_2=0.185203,\ \text{and}\ c_3=-0.0062698$$

Hence, we get the following approximate solution:

$$\bar{y}(x)=\sum_{i=1}^{3}c_i\varphi_i(x)=0.18521x-0.364136x^3+0.166393x^4+0.0188097x^5-0.00626989x^6$$

7.7 STABILITY OF AN INITIAL VALUE PROBLEM

The stability for the solution $y(x)$ of an initial value problem is examined when there is a small change in the initial value problem. We consider the perturbed initial value problem

$$y'=f(x,y)+\delta(x),\quad y(x_0)=\bar{y}_0+\varepsilon \tag{7.288}$$

Here, $\delta(x)$ is assumed to be continuous for all x such that $(x,y)\in D$ for some y. Under the same hypotheses in Equation 7.3, it can be shown that the initial value problem in Equation 7.288 has a unique solution $y(x;\delta,\varepsilon)$.

In order to simplify our discussion, we consider only a small perturbation in the initial value for the initial value problem. Perturbing the initial value $\bar{y}_0$, let $y_\varepsilon(x)$ denote the perturbed solution. Then,

$$y_\varepsilon'(x)=f(x,y_\varepsilon(x)),\quad y_\varepsilon(x_0)=\bar{y}_0+\varepsilon \tag{7.289}$$

Now, under the same hypotheses in Equation 7.3, it can be shown that for all small values of ε, $y(x)$, and $y_\varepsilon(x)$ exist in the interval $[x_0,b]$ and moreover,

$$\max_{x_0\le x\le b}|y(x)-y_\varepsilon(x)|\le k\varepsilon \tag{7.290}$$

for some $k > 0$, which is independent of ε. Thus, small change in the initial value $\bar{y}_0$ leads to small change in the solution $y(x)$ of the initial value problem. This is a desirable property for practical problems. Although all initial value problems (Equation 7.2) are stable virtually according to Equation 7.290. But sometimes, these initial value problems may be well-conditioned or ill-conditioned. If the condition in Equation 7.290 is satisfied, then the corresponding initial value problem is well-conditioned. Otherwise, it is considered to be ill-conditioned.

To better understand, the conditioning of initial value problem, we estimate the perturbation in $y(x)$ due to small change in the initial value of the initial value problem. Subtracting Equation 7.2 from 7.289, we get

$$y'_\varepsilon(x) - y'(x) = f(x, y_\varepsilon(x)) - f(x, y(x)), \quad y_\varepsilon(x_0) - y(x_0) = \varepsilon \tag{7.291}$$

$$\begin{aligned} y'_\varepsilon(x) - y'(x) &= f(x, y_\varepsilon(x)) - f(x, y(x)) \\ &= (y_\varepsilon(x) - y(x))\frac{\partial f(x, y(x))}{\partial y} \end{aligned} \tag{7.292}$$

applying Lagrange's mean value theorem. Solving Equation 7.292, we get

$$y_\varepsilon(x) - y(x) = \varepsilon \exp\left(\int_{x_0}^{x} \frac{\partial f(x, y(x))}{\partial y} dx\right) \tag{7.293}$$

If $[\partial f(x, y(x))]/\partial y \le 0$, for x sufficiently close to x_0, then we have that $y_\varepsilon(x) - y(x)$ probably remains bounded by ε as x increases. In this case, we can say conveniently the initial value problem is well-conditioned.

Thus, a stable numerical method is one for which the numerical solution is well posed when considering small perturbations, provided that the step size h is sufficiently small. In actual computations, however, the step size h cannot be too small because a very small step size decreases the efficiency of the numerical method. Therefore, we need to further analyze the stability of numerical methods when h is not assumed to be small.

Now, we consider the following model problem that is generally used to investigate the performance of various numerical methods

$$\begin{aligned} y' &= \lambda y, \quad x > 0 \\ y(0) &= 1 \end{aligned} \tag{7.294}$$

In the model problem (or test equation) (7.294), we always assume that the real constant $\lambda < 0$ or λ is complex with $\mathrm{Re}(\lambda) < 0$. The exact solution of Equation 7.294 is

$$y(x) = e^{\lambda x} \tag{7.295}$$

which decays exponentially in x since the parameter λ has a negative real part.

As a kind of stability property, we seek when a numerical method is applied to Equation 7.294, the numerical solution satisfies

$$y(x_i) \to 0 \text{ as } x_i \to \infty \tag{7.296}$$

for any choice of the step size h. The set of values λh is called the *region of absolute stability* of the numerical method. This set is considered as a subset of the complex plane, for which $y_i \to 0$ as $i \to \infty$.

Example 7.24

Show that $y' = -xy^2, y(0) = 1$ is well-posed differential equation.

Solution:
The initial value problem $y' = -xy^2, y(0) = 1$ has the solution

$$y(x) = \frac{2}{x^2 + 2}$$

For perturbed problem,

$$y'_\varepsilon = -xy_\varepsilon^2, \quad y_\varepsilon(0) = 1 + \varepsilon$$

Now,

$$y - y_\varepsilon = -\varepsilon \exp\left(\int_0^x \frac{\partial f(x,y)}{\partial y} dx\right)$$

Here,

$$f(x,y) = -xy^2$$

$$g(x) = \frac{\partial f(x,y)}{\partial y} = -2xy = \frac{-4x}{x^2 + 2}$$

$$\int_0^x g(s)\,ds = -2\log\left(\frac{x^2 + 2}{2}\right)$$

$$\exp\left(\int_0^x g(s)\,ds\right) = e^{\log\left(\left(x^2+2\right)/2\right)^{-2}} = \frac{4}{\left(x^2 + 2\right)^2}$$

Now,

$$y - y_\varepsilon \approx \frac{-4\varepsilon}{\left(x^2 + 2\right)^2}$$

which indicates that the above problem is a well-posed problem. Otherwise, we directly show that

$$\frac{\partial f(x,y)}{\partial y} = -2xy = \frac{-4x}{x^2 + 2} \le 0, \quad 0 \le x \le x_n = b$$

which indicates that the above problem is a well-posed problem.

7.7.1 Stability Analysis of Single Step Methods

We shall now discuss the stability analysis of some single-step methods.

7.7.1.1 Stability of Euler's Method

Let us consider the following perturbed initial value problem

$$y' = f(x, y), \quad y(x_0) = \bar{y}_0 + \varepsilon \tag{7.297}$$

To perform a stability analysis for Euler's method, we consider the following numerical method:

$$u_{i+1} = u_i + hf(x_i, u_i), \quad i = 0, 1, \ldots, n-1 \tag{7.298}$$

with $u_0 = \bar{y}_0 + \varepsilon$. This is analogous to finding the solution $y_\varepsilon(x)$ to the perturbed initial value problem in Equation 7.289.

Let, $e_i = u_i - y_i, i \geq 0$. Then $e_0 = \varepsilon$, since $y_0 = \bar{y}_0$. Now subtracting $y_{i+1} = y_i + hf(x_i, y_i)$ from Equation 7.298, we get

$$e_{i+1} = e_i + h\left[f(x_i, u_i) - f(x_i, y_i)\right] \tag{7.299}$$

Taking bounds, using Equation 7.4, we obtain

$$|e_{i+1}| \leq |e_i| + hL|u_i - y_i| = |e_i| + hL|e_i| = (1 + hL)|e_i|, \quad i = 0, 1, 2, \ldots, n-1 \tag{7.300}$$

Applying Equations 7.300 recursively, we obtain

$$|e_{i+1}| \leq (1 + hL)^{i+1}|e_0|, \quad i = 0, 1, 2, \ldots$$

Thus,

$$|e_{i+1}| \leq (1 + hL)^i|e_0|, \quad i = 1, 2, \ldots$$

$$\leq (1 + hL)^n|e_0|, \quad \text{since } i \leq n$$

$$\leq e^{nhL}|e_0|, \quad \text{since } 1 + x \leq \mathrm{e}^x$$

Therefore,

$$\max_{1 \leq i < n}|e_i| = \max_{1 \leq i \leq n}|u_i - y_i| \leq \mathrm{e}^{(b - x_0)L}|\varepsilon|$$

Consequently, there is a constant $k > 0$, independent of h, such that

$$\max_{1 \leq i \leq n}|u_i - y_i| \leq k|\varepsilon| \tag{7.301}$$

This is the analog to the result (Equation 7.290) for the original initial value problem. This shows that Euler's method is a stable numerical method for the solution of the initial value problem (7.2).

Now, let us examine the performance of Euler's method on the model problem (7.291). We have

$$\bar{y}_{i+1} = \bar{y}_i + h\lambda\bar{y}_i = (1 + h\lambda)\bar{y}_i, \quad i \geq 0, \quad \bar{y}_0 = 1 \tag{7.302}$$

Applying Equations 7.302 recursively, we obtain

$$\bar{y}_i = (1 + h\lambda)^i\,\bar{y}_0 = (1 + h\lambda)^i\,\bar{y}_0, \quad i \geq 0 \tag{7.303}$$

In particular, for a fixed-node point $x_i = ih \equiv \tilde{x}$, as $i \to \infty$, we obtain

$$\bar{y}_i = \left(1 + \frac{\lambda\tilde{x}}{i}\right)^i \bar{y}_0 \text{ which tends to } \mathrm{e}^{\lambda\tilde{x}}.$$

This establishes the convergence of Euler's method. This is an asymptotic property that is valid in the limit as $h \to 0$. From Equation 7.303, we see that $\bar{y}_i \to 0$ as $i \to \infty$ if and only if

$$|1 + \lambda h| < 1 \tag{7.304}$$

If λ is real and negative, the condition in Equation 7.304, yielding

$$-2 < \lambda h < 0 \tag{7.305}$$

This implies that

$$0 < h < -\frac{2}{\lambda}, \quad \text{since } \lambda < 0 \tag{7.306}$$

This restriction on the range h can be taken into account if we apply Euler's method. For such value of h, the Euler's explicit method is said to be absolutely stable and the interval $(-2,0)$ is referred to as the *interval of absolute stability* of the method. Thus, Euler's explicit method is stable for $|1+\lambda h| < 1$ and unstable for $|1+\lambda h| \geq 1$. The stability region of Euler's method is a disc of unit radius with center on the negative real (λh)-axis at -1 and the imaginary (λh)-axis touching it at the origin, as shown in Figure 7.4.

7.7.1.2 Stability of the Backward Euler Method

Let us apply backward Euler method to the model problem (Equation 7.294). We obtain

$$\overline{y}_{i+1} = \overline{y}_i + h\lambda \overline{y}_{i+1}, \quad i \geq 0, \quad \overline{y}_0 = 1 \tag{7.307}$$

Thus,

$$\overline{y}_{i+1} = (1-\lambda h)^{-1}\,\overline{y}_i, \quad i \geq 0 \tag{7.308}$$

Applying Equations 7.307 recursively, we obtain

$$\overline{y}_i = (1-\lambda h)^{-1}, \quad i \geq 0 \tag{7.309}$$

For any step size $h > 0$, we have $|1-\lambda h| > 1$ and so $\overline{y}_i \to 0$ as $i \to \infty$. Therefore, the backward Euler method satisfies the property (7.293) for any step size $h > 0$ when it is applied to the model problem (7.291). Such numerical method is said to be unconditionally stable. The stability region of the backward Euler method has been shown in Figure 7.5.

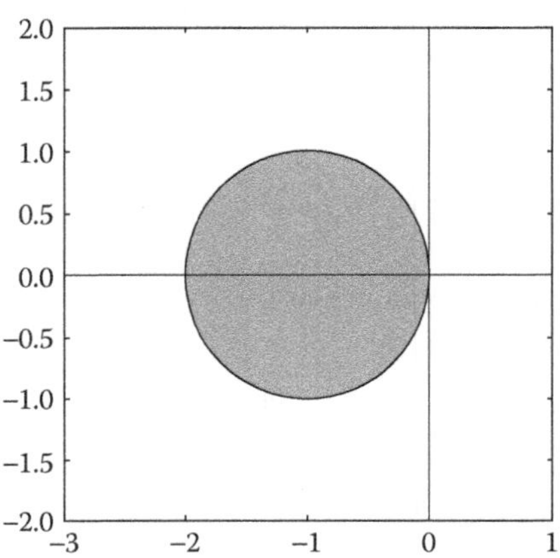

FIGURE 7.4 Stability region of Euler's explicit method.

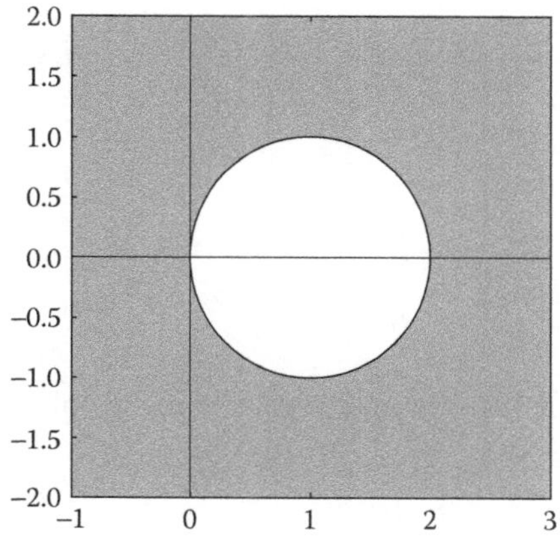

FIGURE 7.5 Stability region of the backward Euler method.

7.7.1.3 Stability of R–K Methods

For simplicity, we shall restrict our attention to the case of s-stage methods with $1 \leq s \leq 4$. For $s = 1$, one-stage R–K method is simply Euler's explicit method. The stability of this method has been already discussed earlier.

Now let us consider $s = 2$ corresponding to two-stage second-order R–K methods:

$$\bar{y}_{i+1} = \bar{y}_i + (\omega_1 k_1 + \omega_2 k_2), \quad i = 0,1,2,\ldots \tag{7.310}$$

where

$$k_1 = hf(x, y) \tag{7.311}$$

$$k_2 = hf(x + a_2 h, y + b_{21} k_1) \tag{7.312}$$

with

$$\omega_1 + \omega_2 = 1, \quad \omega_2 a_2 = \omega_2 b_{21} = \frac{1}{2} \tag{7.313}$$

Applying the two-stage R–K method to Equation 7.294, we get

$$\bar{y}_{i+1} = \left(1 + \lambda h + \frac{\lambda^2 h^2}{2}\right)\bar{y}_i, \quad i \geq 0 \tag{7.314}$$

This implies that

$$\bar{y}_i = \left(1 + \lambda h + \frac{\lambda^2 h^2}{2}\right)^i \bar{y}_0 = \left(1 + \lambda h + \frac{\lambda^2 h^2}{2}\right)^i, \quad i \geq 0 \tag{7.315}$$

Hence, the method is absolutely stable if and only if

$$\left|1 + \lambda h + \frac{\lambda^2 h^2}{2}\right| < 1 \tag{7.316}$$

which is valid when $-2 < \lambda h < 0$. Here also, $(-2,0)$ is the interval of absolute stability of the two-stage R–K method. In the case of $s = 3$ corresponding to the three-stage R–K methods, a similar argument shows that

$$\bar{y}_{i+1} = \left(1 + \lambda h + \frac{\lambda^2 h^2}{2} + \frac{\lambda^3 h^3}{6}\right)\bar{y}_i, \quad i \ge 0 \tag{7.317}$$

This implies that

$$\bar{y}_i = \left(1 + \lambda h + \frac{\lambda^2 h^2}{2} + \frac{\lambda^3 h^3}{6}\right)^i \bar{y}_0 = \left(1 + \lambda h + \frac{\lambda^2 h^2}{2} + \frac{\lambda^3 h^3}{6}\right)^i, \quad i \ge 0 \tag{7.318}$$

Hence, the three-stage R–K method is absolutely stable if and only if

$$\left|1 + \lambda h + \frac{\lambda^2 h^2}{2} + \frac{\lambda^3 h^3}{6}\right| < 1, \tag{7.319}$$

which is valid when $-2.51 < \lambda h < 0$. Therefore, $(-2.51, 0)$ is the interval of absolute stability of the three-stage R–K method.

When $s = 4$, we have the four-stage R–K method of order 4. In this case also, analogous to previous argument, we may obtain

$$\bar{y}_{i+1} = \left(1 + \lambda h + \frac{\lambda^2 h^2}{2} + \frac{\lambda^3 h^3}{6} + \frac{\lambda^4 h^4}{24}\right)\bar{y}_i, \quad i \ge 0 \tag{7.320}$$

This implies that

$$\bar{y}_i = \left(1 + \lambda h + \frac{\lambda^2 h^2}{2} + \frac{\lambda^3 h^3}{6} + \frac{\lambda^4 h^4}{24}\right)^i \bar{y}_0 = \left(1 + \lambda h + \frac{\lambda^2 h^2}{2} + \frac{\lambda^3 h^3}{6} + \frac{\lambda^4 h^4}{24}\right)^i, \quad i \ge 0 \tag{7.321}$$

Hence the four-stage R–K method is absolutely stable if and only if

$$\left|1 + \lambda h + \frac{\lambda^2 h^2}{2} + \frac{\lambda^3 h^3}{6} + \frac{\lambda^4 h^4}{24}\right| < 1 \tag{7.322}$$

which is valid when $-2.78 < \lambda h < 0$. Therefore, $(-2.78, 0)$ is the interval of absolute stability of the four-stage R–K method. Figure 7.6 shows the stability region of the fourth-order R–K method.

7.7.2 Stability Analysis of General Multistep Methods

The general form of the multistep methods can be considered as follows:

$$\bar{y}_{i+1} = \sum_{j=0}^{k} \alpha_j \bar{y}_{i-j} + h \sum_{j=-1}^{k} \beta_j f\left(x_{i-j}, \bar{y}_{i-j}\right), i \ge k \tag{7.323}$$

The coefficients $\alpha_0, \alpha_1, \ldots, \alpha_k$ and $\beta_0, \beta_1, \ldots, \beta_k$ are constants and $k \ge 0$. Assuming that $|\alpha_k| + |\beta_k| \ne 0$, we consider this method a $(k+1)$-step method, because $(k+1)$ previous values are being used to compute y_{i+1}. The values $y_1, y_2, \ldots, y_k$ are obtained by some other methods.

To investigate the stability of Equation 7.323, we consider only the model problem (Equation 7.294). The results obtained will transfer to the study of stability for a general differential equation. Applying Equation 7.323 to the model problem (Equation 7.294), we obtain

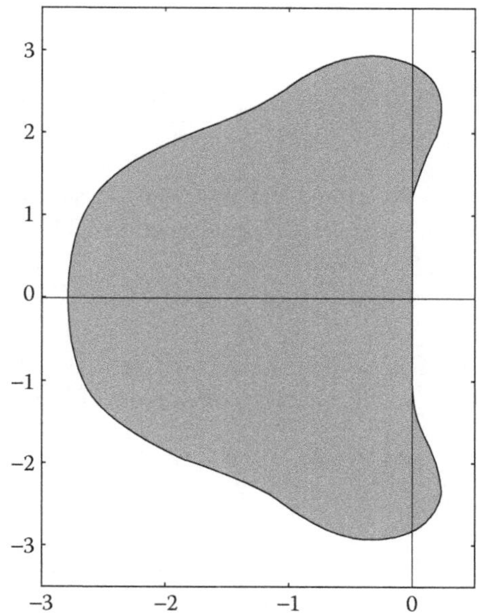

FIGURE 7.6 Stability region of the Runge–Kutta method of order 4.

$$\bar{y}_{i+1} = \sum_{j=0}^{k} \alpha_j \bar{y}_{i-j} + h\lambda \sum_{j=-1}^{k} \beta_j \bar{y}_{i-j}, \quad i \geq k \tag{7.324}$$

This implies that

$$\left(1 - h\lambda\beta_{-1}\right)\bar{y}_{i+1} - \sum_{j=0}^{k}\left(\alpha_j + h\lambda\beta_j\right)\bar{y}_{i-j} = 0, \quad i \geq k \tag{7.325}$$

This is a homogeneous linear difference equation of order $(k+1)$. Now, to find a general solution of Equation 7.328, let us consider the solution of the following form:

$$\bar{y}_i = r^i, \quad i \geq 0 \tag{7.326}$$

Substituting Equation 7.326 in 7.325, we get

$$\left(1 - h\lambda\beta_{-1}\right)r^{i+1} - \sum_{j=0}^{k}\left(\alpha_j + h\lambda\beta_j\right)r^{i-j} = 0, \quad i \geq k \tag{7.327}$$

This implies that

$$\left(1 - h\lambda\beta_{-1}\right)r^{k+1} - \sum_{j=0}^{k}\left(\alpha_j + h\lambda\beta_j\right)r^{k-j} = 0 \tag{7.328}$$

Equation 7.328 represents the characteristic equation. The stability of Equation 7.323 is linked to the roots of the polynomial.

$$\rho(r) = r^{k+1} - \sum_{j=0}^{k} \alpha_j r^{k-j} \tag{7.329}$$

Now, we define

$$\sigma(r) = \beta_{-1} r^{k+1} + \sum_{j=0}^{k} \beta_j r^{k-j} \tag{7.330}$$

The polynomials $\rho(r)$ and $\sigma(r)$ are called the first and the second characteristic polynomials, respectively. Using Equations 7.329 and 7.330, from Equation 7.328, we obtain

$$\pi(r;\lambda h) \equiv \rho(r) - h\lambda\sigma(r) = 0 \tag{7.331}$$

The $(k+1)$th degree polynomial $\pi(r;\lambda h)$ defined in Equation 7.331 is called the stability polynomial of the linear $(k+1)$-step method. Let the characteristics roots of Equation 7.328 be

$$r_0(\lambda h), r_1(\lambda h), \ldots, r_k(\lambda h) \tag{7.332}$$

When $\lambda h = 0$, Equation 7.331 becomes $\rho(r) = 0$ and consequently we have $r_j(0) = r_j$, $j = 0,1,\ldots,k$, which are the roots of $\rho(r) = 0$. Let $r_0 = 1$ is a root of $\rho(r) = 0$ and also let $r_0(\lambda h)$ be the root of Equation 7.331 for which $r_0(0) = 1$. The root $r_0(\lambda h)$ is called the *principal root*. If the roots $r_j(\lambda h)$, $j = 0,1,\ldots,k$ are all distinct, then the general solution of Equation 7.325 is

$$\bar{y}_i = \sum_{j=0}^{k} c_j r_j(\lambda h)^i, \quad i \geq 0 \tag{7.333}$$

Now, if $r_j(\lambda h)$ is a root of multiplicity $v > 1$, then there are v linearly independent solutions of Equation 7.325, that is to say,

$$\left\{\left[r_j(\lambda h)\right]^i\right\}, \left\{i\left[r_j(\lambda h)\right]^i\right\}, \ldots, \left\{i^{v-1}\left[r_j(\lambda h)\right]^i\right\} \tag{7.334}$$

Thus, the solution in Equation 7.333 contains the following part:

$$\left(c_j + c_{j+1} i + \cdots + c_{j+v-1} i^{v-1}\right)\left[r_j(\lambda h)\right]^i \tag{7.335}$$

The Equation 7.335 together with the remaining part arising from the other roots gives the general solution of Equation 7.325. In particular, for consistent methods, it can be shown that

$$\left[r_0(\lambda h)\right]^i = e^{\lambda x_i} + O(h) \tag{7.336}$$

as $h \to 0$. The remaining roots $r_j(\lambda h)$, $j = 1,\ldots,k$ are called *parasitic roots* of the numerical method. The term

$$\sum_{j=1}^{k} c_j r_j(\lambda h)^i \tag{7.337}$$

is called a *parasitic solution*. In view of the general solution of Equation 7.325, it shows that $\bar{y}_i$ will converge to zero with fixed $h > 0$ as $i \to \infty$ if and only if all the roots of the stability polynomial $\pi(r;\lambda h)$ have modulus <1. This motivates the following definition.

Absolute stability: The linear $(k+1)$ multistep method (Equation 7.323) is called *absolutely stable* for a given value of λh if and only if for λh all the roots $r_j(\lambda h)$ of the stability polynomial $\pi(r;\lambda h)$ defined in Equation 7.331 satisfy $|r_j(\lambda h)|<1$, $j=0,1,\ldots,k$. Otherwise, the method is said to be absolutely unstable. An interval (α,β) on the real line is called the *interval of absolute stability*, provided the method is absolutely stable for $\alpha<\lambda h<\beta$. If the method is absolutely unstable for all λh, then it has no interval of absolute stability.

The convergence and stability of Equation 7.323 are linked to the roots of the first characteristic polynomial $\rho(r)$, given in Equation 7.329.

Root condition: The linear multistep method (Equation 7.323) is said to satisfy the root condition if the roots of the equation $\rho(r)=0$ lie inside the unit circle in the complex plane, and are simple if they lie on the unit circle.

Thus, the linear multistep method (Equation 7.323) satisfies the root condition if

i. $|r_j|\le 1, \quad j=0,1,\ldots,k$

ii. $|r_j|=1 \Rightarrow \rho'(r_j)\ne 0$

where $r_0, r_1, r_2, \ldots, r_k$ are the roots of the equation $\rho(r)=0$.

Zero stability: **Definition 1:** The linear multistep method (Equation 7.323) is said to be zero stable if all solutions of the homogeneous linear difference equation

$$\sum_{j=0}^{k} \alpha_j \bar{y}_{i-j} = 0$$

are bounded for all i.

Remark. A linear multistep method (Equation 7.323) is zero-stable if and only if the root condition is satisfied.

Definition 2: A linear $(k+1)$-step method is said to be consistent if and only if it has order

$$|p|\ge 1, \quad \text{that is, } \sum_{j=0}^{k}\alpha_j = 1, -\sum_{j=0}^{k} j\alpha_j + \sum_{j=-1}^{k}\beta_j = 1$$

Remark. The linear multistep method (Equation 7.323) is consistent if and only if

$$\rho(1)=0 \quad \text{and} \quad \rho'(1)-\sigma(1)=0$$

Usually, a numerical method is consistent if it has an order greater than 0. The (forward) Euler method and the backward Euler method both have order 1, so they are consistent. Most methods being used in practice attain higher order. Consistency is a necessary condition for convergence, but not sufficient; for a method to be convergent, it must be both consistent and zero-stable. Analogically, the necessary and sufficient conditions for a linear multistep method (Equation 7.323) to converge are that it be consistent and zero-stability.

Now suppose that a consistent linear $(k+1)$ multistep method is applied to a sufficiently smooth differential equation and that the starting values $\bar{y}_1, \bar{y}_2, \ldots, \bar{y}_k$ all converge to the initial value $\bar{y}_0 = y(x_0)$ as $h\to 0$. Then, the numerical solution converges to the exact solution as $h\to 0$ if and only if the method is zero-stable. This result is known as the *Dahlquist equivalence theorem*, named after Germund Dahlquist.

***Relative stability and weak stability*:** The linear multistep method (Equation 7.323) is said to be relatively stable if the characteristic roots $r_j(\lambda h),\ j = 1, \ldots, k$ satisfy

$$\left|r_j(\lambda h)\right| \le r_0(\lambda h), \quad j = 1, \ldots, k \tag{7.338}$$

for all sufficiently small nonzero values of $|\lambda h|$. Furthermore, the method is said to satisfy the strong root condition if

$$\left|r_j(0)\right| < 1, \quad j = 1, \ldots, k \tag{7.339}$$

This condition can be easily verified, and it implies relative stability. If a multistep method is stable but not relatively stable, then it will be called *weakly stable*.

Example 7.25:

Let us consider the following two-step midpoint method:

$$\bar{y}_{i+1} = \bar{y}_{i-1} + 2hf(x_i, \bar{y}_i), \quad i = 1, 2, \ldots, n-1$$

Applying it to the test equation (7.294) gives rise to the following difference equation:

$$\bar{y}_{i+1} = \bar{y}_{i-1} + 2h\lambda\bar{y}_i, \quad i \ge 1 \tag{7.340}$$

To find a general solution of Equation 7.340, let us consider the solution of the following form:

$$\bar{y}_i = r^i, \quad i \ge 0. \tag{7.341}$$

Substituting Equation 7.341 in 7.340, we get

$$r^{i+1} = r^{i-1} + 2h\lambda r^i, \quad i \ge 1 \tag{7.342}$$

This implies that

$$r^2 = 1 + 2h\lambda r \tag{7.343}$$

Equation 7.343 represents the characteristic equation for the midpoint method. Its characteristic roots are

$$r_0(\lambda h) = \lambda h + \sqrt{1 + \lambda^2 h^2} \text{ and } r_1(\lambda h) = \lambda h - \sqrt{1 + \lambda^2 h^2} \tag{7.344}$$

Alternatively, using Equation 7.331, we can obtain the stability polynomial

$$\pi(r; \lambda\, h) = r^2 - 1 - 2h\lambda r$$

whose roots are

$$r_0(\lambda h) = \lambda h + \sqrt{1 + \lambda^2 h^2}, \quad r_1(\lambda h) = \lambda h - \sqrt{1 + \lambda^2 h^2}$$

Therefore, the general solution of Equation 7.340 is

$$\bar{y}_i = c_0\left[r_0(\lambda h)\right]^i + c_1\left[r_1(\lambda h)\right]^i, \quad i \ge 0 \tag{7.345}$$

The coefficients c_0 and c_1 are determined from the following equations:

$$c_0 + c_1 = \bar{y}_0 = 1$$

$$c_0 r_0(\lambda h) + c_1 r_1(\lambda h) = \bar{y}_1 = e^{\lambda h},$$

since the initial condition $\bar{y}_0 = y_0 = 1$ and $\bar{y}_1 = e^{\lambda x_1} = e^{\lambda h}$ from Equations 7.295. Solving the above equations yields

$$c_0 = \frac{e^{\lambda h} - r_1(\lambda h)}{2\sqrt{1+\lambda^2 h^2}} = \frac{e^{\lambda h} - \lambda h + \sqrt{1+\lambda^2 h^2}}{2\sqrt{1+\lambda^2 h^2}} = 1 + O(\lambda^3 h^3),$$

$$c_1 = \frac{r_0(\lambda h) - e^{\lambda h}}{2\sqrt{1+\lambda^2 h^2}} = \frac{\lambda h + \sqrt{1+\lambda^2 h^2} - e^{\lambda h}}{2\sqrt{1+\lambda^2 h^2}} = O(\lambda^3 h^3)$$

From these values, $c_0 \to 1$ and $c_1 \to 0$ as $h \to 0$. Therefore, in the general solution given by Equation 7.345, the term $c_0\left[r_0(\lambda h)\right]^i$ should corresponds to the exact solution $e^{\lambda x_i}$, since the term $c_1\left[r_1(\lambda h)\right]^i \to 0$ as $h \to 0$. Moreover, for $\lambda < 0$, we have

$$|r_1(\lambda h)| > r_0(\lambda h)$$

for all small values of $h > 0$ and thus Equation 7.338 is not satisfied. Therefore, the midpoint method is not relatively stable, but it is only weakly stable.

It can be shown that if λh is a pure imaginary number of the form $\lambda h = i\gamma$ with $|\gamma| < 1$, then $|r_0(\lambda h)| = |r_1(\lambda h)| = 1$ and $r_0(\lambda h) \neq r_1(\lambda h)$, and hence the root condition is satisfied. For any other λh, the root condition is not satisfied. In particular, if $\lambda h = \pm i$, then $r_0(\lambda h) = r_1(\lambda h) = \pm i$ is a repeated root of modulus 1. Therefore, the stability region consists only of the open interval from $-i$ to i on the imaginary axis, as shown in Figure 7.7.

7.7.2.1 General Methods for Finding the Interval of Absolute Stability

In this section, we shall discuss two methods for locating the interval of absolute stability: Schur criterion and Routh–Hurwitz criterion.

7.7.2.1.1 Schur Criterion

Let us consider the polynomial

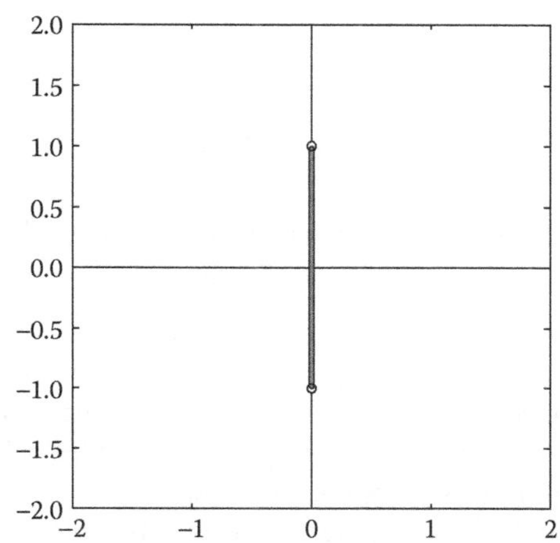

FIGURE 7.7 Stability region of the midpoint method.

$$p(r) = c_{k+1}r^{k+1} + c_k r^k + \cdots + c_1 r + c_0, \quad c_{k+1} \neq 0, \quad c_0 \neq 0$$

where c_j, $j = 0,1,\ldots,k+1$ are the complex coefficients. The polynomial $p(r)$ is said to be a Schur polynomial if all the roots r_l satisfies $|r_l| < 1$, $l = 0,1,\ldots,k$. Let us consider another polynomial

$$\bar{p}(r) = \bar{c}_0 r^{k+1} + \bar{c}_1 r^k + \cdots + \bar{c}_k r + \bar{c}_{k+1}$$

where $\bar{c}_j$ are the complex conjugates of c_j, $j = 0,1,\ldots,k+1$. Furthermore, let us define

$$p_1(r) = \frac{1}{r}\left[\bar{p}(0)p(r) - p(0)\bar{p}(r)\right]$$

Obviously $p_1(r)$ is a polynomial of degree k.

Now, in the following equation, the Schur's criterion has been stated. The polynomial $p(r)$ is a Schur polynomial if and only if $|\bar{p}(0)| > |p(0)|$ and $p_1(r)$ is a Schur polynomial.

Repeating the above procedure, we only determine whether a polynomial of low degree is a Schur polynomial.

7.7.2.1.2 Routh–Hurwitz Criterion

Let us consider the mapping

$$z = \frac{r-1}{r+1}$$

which maps the interior of the open unit disc $|r| < 1$ onto the open left half-plane $\operatorname{Re}(z) < 0$ of the complex z-plane. The inverse of this mapping is given by

$$r = \frac{1+z}{1-z}$$

Under this transformation, the function

$$\pi(r;\lambda h) = \rho(r) - h\lambda\sigma(r)$$

becomes

$$\rho\left(\frac{1+z}{1-z}\right) - h\lambda\sigma\left(\frac{1+z}{1-z}\right) \tag{7.346}$$

Multiplying Equation 7.346 by $(1-z)^{k+1}$, we obtain a polynomial of the following form:

$$a_0 z^{k+1} + a_1 z^k + \cdots + a_{k+1} \tag{7.347}$$

where the coefficients a_i are real constants, $i = 0,1,\ldots,k+1$. The roots of the stability polynomial $\pi(r;\lambda h)$ lie inside the open unit disc $|r| < 1$ if and only if the roots of the polynomial in Equation 7.347 lie in the open left half-plane $\operatorname{Re}(z) < 0$, as shown in Figure 7.8.

We can now apply the Routh–Hurwitz criterion, which gives the necessary and sufficient conditions for the roots Equation 7.347 to have negative real parts, that is, the roots will lie in the open left half-plane of the complex z-plane.

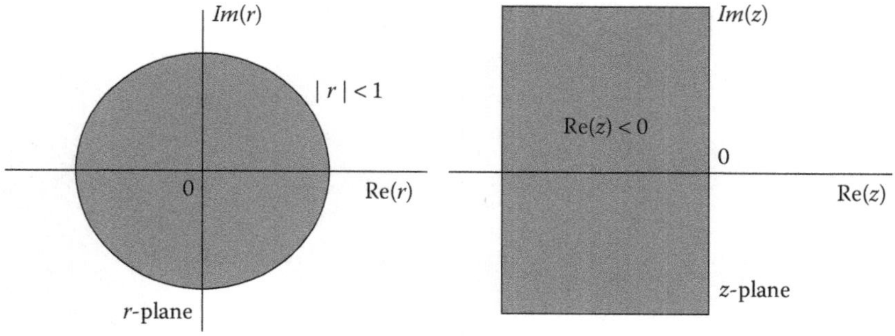

FIGURE 7.8 Mapping of the interior of open unit circle onto the left half plane.

Routh–Hurwitz stability criterion: The roots of Equation 7.347 lie in the open left half-plane of the complex z-plane if and only if all the leading principal minors of the $(k+1)\times(k+1)$ matrix

$$M=\begin{bmatrix} a_1 & a_3 & a_5 & \cdots & a_{2k+1} \\ a_0 & a_2 & a_4 & \cdots & a_{2k} \\ 0 & a_1 & a_3 & \cdots & a_{2k-1} \\ 0 & a_0 & a_2 & \cdots & a_{2k-2} \\ \cdots & \cdots & \cdots & \cdots & \cdots \\ 0 & 0 & 0 & \cdots & a_{k+1} \end{bmatrix}$$

are positive, where $a_0 > 0$ and $a_j = 0$ for $j > k+1$. In particular, for polynomials of degree 2, 3 and 4, the Routh–Hurwitz's criteria are summarized as follows:

The Routh–Hurwitz criteria for $k = 1,2,3$.

1. For $k=1$: $a_0 > 0, a_1 > 0, a_2 > 0$.
2. For $k=2$: $a_0 > 0, a_1 > 0, a_2 > 0, a_3 > 0, a_1a_2 > a_0a_3$.
3. For $k=3$: $a_0 > 0, a_1 > 0, a_2 > 0, a_3 > 0, a_4 > 0, a_1a_2a_3 - a_1^2a_4 - a_0a_3^2 > 0$.

The above criteria represent the necessary and sufficient conditions for guaranteeing that all the roots of Equation 7.347 lie in the open left half-plane of the complex z-plane.

To illustrate, the intervals of absolute stability of the four-step Adams–Bashforth method and the three-step Adams–Moulton method are as follows:

1. For the four-step Adams–Bashforth method

$$\bar{y}_4 = \bar{y}_3 + \frac{h}{24}\left[55\bar{f}_3 - 59\bar{f}_2 + 37\bar{f}_1 - 9\bar{f}_0\right]$$

the interval of absolute stability is $(-(3/10),0)$.

2. For the three-step Adams–Bashforth method

$$\bar{y}_3 = \bar{y}_2 + \frac{h}{24}\left[9\bar{f}_3 + 19\bar{f}_2 - 5\bar{f}_1 + \bar{f}_0\right]$$

the interval of absolute stability is $(-3,0)$.

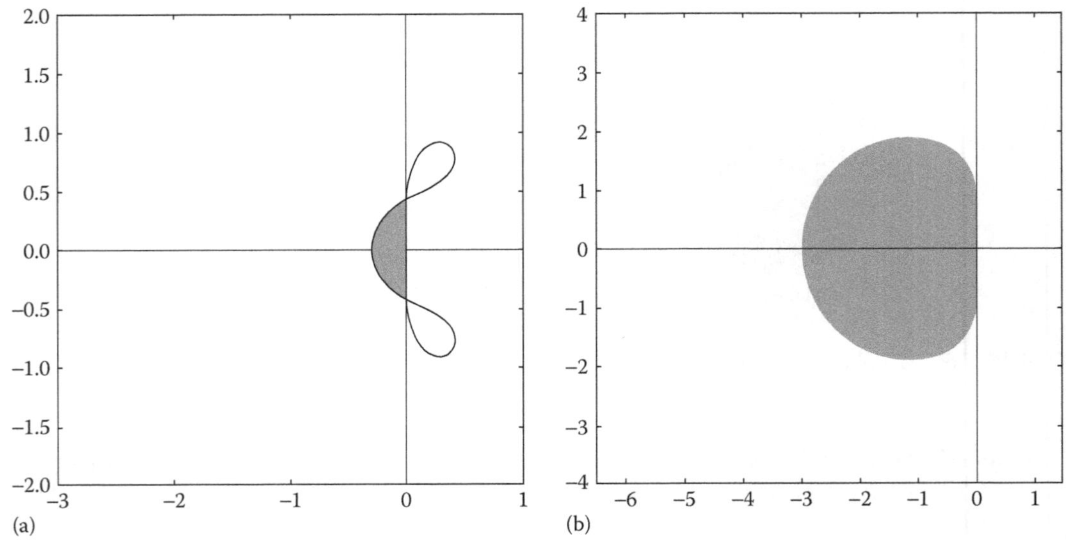

FIGURE 7.9 Stability region of (a) Adams–Bashforth four-step method and (b) Adams–Moulton three-step method.

The intervals of absolute stability of the four-step Adams–Bashforth method and the three-step Adams–Moulton method have been shown in Figure 7.9, respectively.

7.8 STIFF DIFFERENTIAL EQUATIONS

A stiff equation is a *differential equation* for which certain *numerical methods* for solving the equation are numerically unstable, unless the step size is taken to be extremely small. There is no unique definition of stiffness in the literature, but the main idea is that the equation includes some terms that can lead to rapid variation in the solution.

Usually we assess the accuracy of numerical methods for initial value problems in terms of the rate at which the error approaches zero, when the step size h approaches zero. However, this characterization of accuracy is not always informative, because it neglects the fact that the local truncation error of any single-step or multistep method also depends on higher order derivatives of the solution. When these higher derivatives are bounded, then the method gives predictable error bound. In some cases, these derivatives can be quite large in magnitude, even when the solution itself is relatively small, which requires that h be chosen particularly very small in order to achieve even reasonable accuracy. This leads to the concept of a stiff differential equation. A differential equation of the form (7.2) is said to be stiff if its exact solution $y(x)$ includes a term that decays exponentially to zero as x increases, but whose derivatives are much greater in magnitude than the term itself. An example of such a term is e^{-cx}, where c is a large, positive constant, because its nth derivative is $(-1)^n c^n e^{-cx}$. As a result of the factor of c^n, this derivative decays to zero much more slowly than e^{-cx} as x increases. In such cases the step size h should be drastically decreased to maintain stability. Furthermore, the larger c is, the smaller h must be to maintain accuracy.

Thus, the stiff equations are characterized by those differential equations having a term of the form e^{-cx} in the exact solution with large c. Such equations are very common in physical and biological systems. One typical example is mass spring system with a large spring constant (stiffness).

Example 7.26

Consider the initial value problem

$$y' = -100y, x>0, y(0) = 1$$

The exact solution is

$$y(x) = e^{-100x}$$

which rapidly decays to zero as x increases. If we solve this problem using Euler's method, with step size $h = 0.1$, then we have

$$\bar{y}_{i+1} = \bar{y}_i - 100h\bar{y}_i = -9\bar{y}_i, \quad i = 0,1,\ldots,n$$

which yields the exponentially growing solution $\bar{y}_i = (-9)^i$. On the other hand, if we choose $h = 10^{-3}$, we obtain the computed approximate solution $\bar{y}_i = (0.9)^i$, which is much more accurate, and correctly adhere to the qualitative behavior of the exact solution, in that it rapidly decays to zero.

7.9 A-STABILITY AND L-STABILITY

7.9.1 A-Stability

The behavior of numerical methods on stiff problems can be analyzed by applying these methods to the test equation $y' = \lambda y$ subject to the initial condition $y(0) = 1$ with $\lambda \in C$. The solution of this equation is $y(x) = e^{\lambda x}$. This solution approaches zero as $x \to \infty$ when $\text{Re}(\lambda) < 0$. If the numerical method also exhibits this behavior (for a fixed step size), then the method is said to be A-stable. A larger region of absolute stability allows a larger step size h to be chosen for a given value of λ, it is preferable to use a method that has as large a region of absolute stability as possible. The ideal situation is when a method is A-stable, which means that its region of absolute stability contains the entire left half plane, because then, the solution will decay to zero regardless of the choice of h. A-stable methods do not exhibit the instability problems as described in the following motivating examples.

An example of an A-stable single-step method is the backward Euler method, which is an implicit method. Applying this method to the test equation yields

$$\bar{y}_{i+1} = \bar{y}_i + h\lambda\bar{y}_{i+1}$$

Similar to previous argument for this method, from Equation 7.309 we see that $\bar{y}_i \to 0$ as $i \to \infty$ if and only if the stability function $\pi(\lambda h) = 1/(1-\lambda h)$ satisfies

$$|\pi(\lambda h)| = \left|\frac{1}{1-\lambda h}\right| < 1$$

This is true for all λh with $\text{Re}(\lambda) < 0$, regardless of the value of h. Therefore, the backward Euler method has the region of absolute stability that contains the set $\{\lambda \in \boldsymbol{C} \mid \text{Re}(\lambda) < 0\}$, that is, the left half plane. Thus, the backward Euler method solution better reflects the behavior of the exact solution of the model equation. Another example of A-stable method is the trapezoidal method

$$\bar{y}_{i+1} = \bar{y}_i + \frac{h}{2}\left[f\left(x_i, \bar{y}_i\right) + f\left(x_{i+1}, \bar{y}_{i+1}\right)\right]$$

Applying this method to test equation $y' = \lambda y$, $y(0) = 1$, we get

$$\bar{y}_{i+1} = \bar{y}_i + \frac{h\lambda}{2}\left(\bar{y}_i + \bar{y}_{i+1}\right) \tag{7.348}$$

Solving Equation 7.348 yields

$$\bar{y}_{i+1} = \left[\frac{1+(\lambda h/2)}{1-(\lambda h/2)}\right]\bar{y}_i, \quad i \geq 0 \tag{7.349}$$

This implies that

$$\bar{y}_i = \left[\frac{1+(\lambda h/2)}{1-(\lambda h/2)}\right]^i \bar{y}_0, \quad i \geq 0 \tag{7.350}$$

provided $\lambda h \neq 2$. For $\text{Re}(\lambda) < 0$, we see that the stability function $\pi(\lambda h) = [1+(\lambda h/2)]/[1-(\lambda h/2)]$ satisfies

$$\left|\pi(\lambda h)\right| = \left|\frac{1+\dfrac{\lambda h}{2}}{1-\dfrac{\lambda h}{2}}\right| < 1 \tag{7.351}$$

for all values of $h > 0$. Thus, $\bar{y}_i \to 0$ as $i \to \infty$ hold for all $h > 0$ and all complex λ with $\text{Re}(\lambda) < 0$. Therefore, the trapezoidal method has the region of absolute stability that contains the set $\{\lambda \in \boldsymbol{C} \mid \text{Re}(\lambda) < 0\}$, that is, the left half plane. Hence, the trapezoidal method is A-stable. Figure 7.10 shows the stability region of the trapezoidal method.

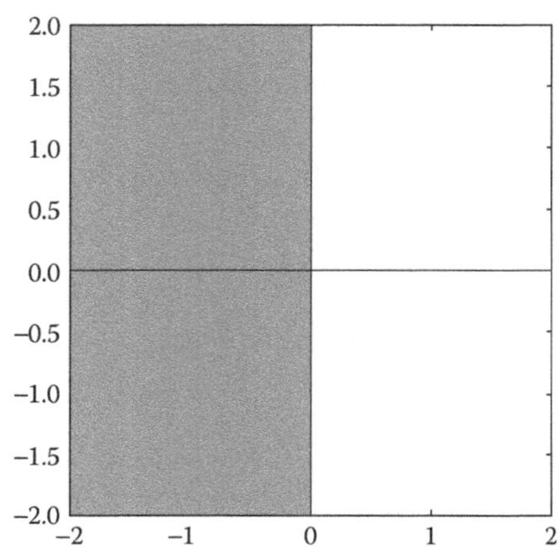

FIGURE 7.10 Stability region of the trapezoidal method.

7.9.2 L-Stability

Definition: A numerical method is L-stable if it is A-stable and $\pi(\mu) \to 0$ as $\mu \to \infty$, where $\mu = \lambda h$. The trapezoidal method is A-stable but not L-stable.

The *backward Euler method* is an example of an L-stable method, because

$$\pi(\mu) = \frac{1}{1-\mu} \to 0 \text{ as } \mu \to \infty$$

Thus, we see that a numerical method that is L-stable has the stronger property that the solution approaches zero in a single step as the step size goes to infinity.

EXERCISES

1. Use Picard's method to obtain $y(0.1)$ and $y(0.2)$ of the problem defined by

$$\frac{dy}{dx} = x + yx^4, \quad y(0) = 3$$

2. Using Picard's method, find $y(0.1)$, given that

$$\frac{dy}{dx} = \frac{y-x}{y+x}, \quad \text{and} \quad y(0) = 1$$

3. Use Picard's method to obtain the second approximation to the solution of $(d^2y/dx^2) - x^3(dy/dx) - x^3 y = 0$ with the initial conditions $y(0) = 1$ and $y'(0) = 1/2$.
4. Solve the equation $dy/dx = x^2/(1+y^2)$ with the initial condition $y(0) = 0$ by Picard's method to obtain y for $x = 0.25$ and $x = 0.5$, correct to four decimal places.
5. Find a series solution of the equation

$$\frac{dy}{dx} = \sqrt{x}\left(2 + y^2\right), y(0) = 0$$

by using Picard's method and hence find $y(0.9)$.

6. Given $(dy/dx) = 1 + xy$, $y(0) = 1$, obtain Taylor's series for $y(x)$ and compute $y(0.1)$, correct to four decimal places.
7. Find by Taylor's series method, the value of $y(0.1)$ given that

$$y'' - xy' - y = 0, y(0) = 1, \quad \text{and} \quad y'(0) = 0$$

8. If $(dy/dx) = 1/(x^2 + y)$ with $y(4) = 4$, compute the values of $y(4.1)$ and $y(4.2)$ by Taylor's series method.
9. Using Taylor's series, find $y(0.1)$, $y(0.2)$, and $y(0.3)$ given that

$$\frac{dy}{dx} = xy + y^2, \quad y(0) = 1$$

10. Using Taylor's series method, produce a fourth-order method to solve $y' = x - y^2$, $y(0) = 0$. Use fixed step sizes, $h = 0.5,\ 0.25,$ and 0.125 in succession, and solve for $0 \le x \le 10$. Estimate the global error using the error estimate based on the Richardson extrapolation.
11. Find the value of $y(1.1)$ and $y(1.2)$, correct to three decimal places, given that $dy/dx = xy^{1/3}$, $y(1) = 1$ using the first three terms of Taylor's series expansions.
12. Solve $y'' + 0.1y' + x = 0$, $y(0) = 0$, and $y'(0) = 1$ from $x = 0$ to 2 with Taylor's series method of order four using $h = 0.25$.

13. Find $y(0.1)$ and $y(0.2)$ given $dy/dx = y + xy^2$, $y(0) = 1$ by Taylor's series method.
14. Using Taylor's series method, obtain the values of y at $x = 0.1(0.1)0.3$ to four significant figures if y satisfies the equation $y'' = -xy$ given that $y' = 0.5$ and $y = 1$ when $x = 0$.
15. The system

$$y' = tz + 1, \quad y(0) = 0$$

$$z' = -ty, \quad z(0) = 1$$

is solved using the second-order Taylor's series method with step-length h. Show that

$$\begin{bmatrix} y_{j+1} \\ z_{j+1} \end{bmatrix} = \mathbf{A} \begin{bmatrix} y_j \\ z_j \end{bmatrix} + \mathbf{b}$$

where $\mathbf{A}$ is a 2×2 matrix and $\mathbf{b}$ is a column vector.

a. Determine the form of $\mathbf{A}$ and $\mathbf{b}$.

b. Compute the solution $y(0.2)$ and $z(0.2)$ with $h = 0.1$.

16. Use Taylor's series method of the fourth-order to solve the following differential equation and find the required values as indicated:
Find

$$y(0.1) \text{and}\, y(0.2), \quad \text{if } \frac{dy}{dx} = \sqrt{x + y};\ y(0) = 1$$

17. Use Taylor series method of fourth-order to solve the following simultaneous differential equations and find the values of the dependent variables at the required points:

a. Find the values of

$$x(\pm 0.1) \text{ and } y(\pm 0.1), \text{if } \frac{dx}{dt} = x - 5y; \frac{dy}{dt} = 2x - y; x(0) = 2; y(0) = 3$$

b. Find

$$y(0.1) \text{ and } z(0.1), \text{given that } \frac{dy}{dx} = 2x + yz; \frac{dz}{dx} = 2xz + y; y(0) = 1; z(0) = -1$$

c. Find

$$x(0.1) \text{ and } y(0.1), \text{if } \frac{dx}{dt} = y; \frac{dy}{dt} = \frac{y^2}{x}; x(0) = 1;\ y(0) = -1$$

18. Use Taylor series method of the forth-order to solve the following higher order differential equations and find the required values of y at the required points:

a. Find

$$y(0.1) \text{ and } y(0.2), \quad \text{if } y'' - x(y')^2 + y^2 = 0;\ y(0) = 1;\ y'(0) = -1$$

b. Find

$$y(0.1) \text{ and } y(0.2), \quad \text{if } y'' - y' + xy^2 = 0; y(0) = 1; y'(0) = -1$$

c. Find

$$y(0.1) \text{ and } y(0.2), \quad \text{if } y''' + 2y'' + y' - y = \cos x;\ y(0) = 0;\ y'(0) = 1;\ y''(0) = 2$$

19. Use Euler's method with $h = 0.025$ to find the solution of the equation $y' = (y-x)/(y+x)$ with $y(0) = 1$ in the range $0 \le x \le 0.1$.
20. Find $y(-0.1)$ and $y(-0.2)$ by using improved Euler's method with $h = -0.1$, given that

$$\frac{dy}{dx} = \frac{y^2 - x^2}{y^2 + x^2}, y(0) = 1$$

21. Using improved Euler's method solve numerically the equation $(dy/dx) = 2 + \sqrt{xy}$ with $y(1) = 1$ for $x = 1(0.2)1.4$.
22. Compute $y(0.3)$ from $(dy/dx) = 1 + xy$ by modified Euler's method given that $y(0) = 2$.
23. Given the differential equation

$$\frac{dy}{dx} = x^2 + y$$

with $y(0) = 1$, compute $y(0.02)$ by using Euler's modified method.
24. Find $y(-0.1)$ and $y(-0.2)$, by using improved Euler's method with $h = -0.1$, given that

$$\frac{dy}{dx} = \frac{y^2 - x^2}{y^2 + x^2}, \; y(0) = 1$$

25. Compute, by improved Euler method, the first two steps of the solution of the following initial value problems, taking $h = 0.2$. Also, compare with the corresponding exact solutions:
    a. $xy' = x^3 - y$, $y(1) = 1$.
    b. $x^2 y' = e^y - x, x > 0, y(1) = 0$.
    c. $y' = y - y^2, y(0) = 0.5$.
26. Solve $(dy/dx) = \sqrt{xy+1}$ with $y(0) = 1$ for finding $y(0.075)$ by using improved Euler's method taking $h = 0.025$.
27. Solve the following first-order ordinary differential equation $y' - yx + x^2 = 0$ from $x = 0$ to $x = 1.8$ with $y(0) = 1$ by applying
    a. Euler's method.
    b. Fourth-order R–K method with $h = 0.6$.

    Hence, compare the results in both cases with the exact solution: $y = x^2 - e^{-0.5x^2} + 2$ by computing the error between the computed and exact solution.
28. Given the initial value problem defined by

$$\frac{dy}{dx} = y\left(1 + x^2\right), y(0) = 1$$

Find the values of y for $x = 0.2, 0.4, 0.6, 0.8$, and 1 using the Euler, the modified Euler, and the fourth-order R–K methods. Compare the computed values with the exact values.
29. Given $y' = -y^2 + 3x/(1+x^2)$ and $y(0) = -0.4122$, compute the value of $y(0.6)$, correct to four significant figures, by (a) Euler's method, (b) Heun's method, and (c) modified Euler's method using step-length 0.1.
30. Use Euler's method to solve $(dy/dx) = -1.2y + 7e^{-0.3x}$, from $x = 0$ to $x = 2$ with the initial condition $y = 3$ at $x = 0$. Take $h = 0.5$.
31. From the differential equation

$$\frac{dy}{dx} = \frac{1 - x^2 - y^2}{1 - xy}, y(0) = 1$$

compute the value of $y(0.6)$ correct to five decimal places by the fourth-order R–K method by using step length 0.1. Estimate the error in the computed value by using the Richardson extrapolation by doubling the step length.

32. Find $y(-0.2)$ and $y(-0.4)$, by using the improved Euler's method with $h=-0.2$, given that

$$\frac{dy}{dx}=x^2+y^2, y(0)=1.$$

33. Solve numerically the differential equation $(dy/dx)=(1/2)\left(y^3-(y/x)\right)$ using Euler's method at $x=1.6$, given that $y=1$ when $x=1$. Also compare with the exact solution at $x=1.6$.
34. Compute by the four-step Adams–Bashforth method (ordinate form) the values of $y(0.4)$ and $y(0.5)$, correct to five decimal places, from the differential equation $y'=(\cos x/1+y^2)$, $y(0)=0$, by taking step length 0.1 and starting the solution by the fourth-order R–K method.
35. For the differential equation

$$\frac{dy}{dx}=\frac{\cos x}{1+x}+\frac{y^2}{10}\left(y(0)=0\right)$$

start the solution by Taylor's method and continue the solution till $x=0.8$ by using the fourth-order Adams–Bashforth method with step length 0.1. Give results correct to five decimal places.

36. Given $(dy/dx)=x+\left|\sqrt{y}\right|$ and $y=1$ at $x=0$. Find approximate value of y for the range $0\le x\le 0.4$ in steps of 0.2 by the improved Euler's method.
37. Find $y(0.1)$ and then $y(0.2)$, by using the improved Euler's method, given that

$$\frac{dy}{dx}=\log(x+y);y(0)=1$$

38. Find $y(1.9)$, $y(1.8)$ and $y(1.7)$ by using simple, improved, and modified Euler's methods, respectively, if $(dy/dx)=x+\sqrt{y}$, $y(2)=4$. Take $h=-0.1$.
39. Solve the following problem by using Euler's predictor–corrector method. Find $y(1.1)$ and (1.2), given that $(dy/dx)+(1/x)y^2=0$, $y(1)=1$.
40. Use Euler's method to solve $y''-2y'+2y-e^{2t}\sin t=0$ with $0\le t\le 1$, $y(0)=-0.4$, and $y'(0)=0.6$. Take the time step of $h=0.1$ and determine $y(0.2)$.
41. Consider the following system of two ordinary differential equations:

$$x'-xt+y=0$$

$$y'-yt-x=0$$

from $t=0$ to $t=1.2$ with $x(0)=1$, and $y(0)=0.5$. Solve the equation by

a. Euler's method with $h=0.4$.
b. Fourth-order R–K method using $h=0.4$.

42. Give $y'=(1-x)y^2-y, (x=0, y=1)$, compute the value of y when $x=1$ correct to five significant figures by taking step length $1/8$ and using (a) the three-step Adams–Moulton method and (b) Milne's method. Start the solution by Taylor's method.
43. Consider the initial value problem $y'=x(y+x)-2, y(0)=2$.
    a. Use Euler method with step sizes $h=0.3$, $h=0.2$, and $h=0.15$ to compute approximations to $y(0.6)$, correct up to five decimal places.
    b. Improve the approximations in (a) to $O(h^3)$ by using the Richardson extrapolation.
44. Solve the following initial value problems
    i. $y'=t+y, y(1)=0$
    ii. $y'=-y^2, y(1)=1$

using (a) Euler method, (b) backward Euler method, and (c) midpoint method. Compute $y(1.2)$, using $h=0.1$ in each case.

45. Use the fourth-order R–K method to find the value of y when $x=1$ given that $(dy/dx) = [(y-x)/(y+x)]$, $y(0)=1$.
46. Use the classical fourth-order R–K method to solve $(dy/dx)=-1.2y+7e^{-0.3x}$ from $x=0$ to $x=1.5$ with the initial condition $y=3$ at $x=0$. Take $h=0.5$.
47. Solve $y'=\sin x+\cos y$ for $x=3(0.5)4$ with initial value $y=2.5$ by the classical R–K method.
48. Use the fourth-order R–K formula to find $y(0.2)$ and $y(0.4)$, given that

$$y'=\frac{y^2-x^2}{y^2+x^2}, y(0)=1$$

49. Use the fourth-order R–K method to integrate $y'=3y-4e^{-x}$, $y(0)=1$ from $x=0$ to 1 in steps of $h=0.2$. Compare the result with the analytical solution $y=e^{-x}$.
50. Using the fourth-order R–K method, compute the value of $y(0.2)$, given that

$$\frac{d^2y}{dx^2}+y=0$$

with $y(0)=1$ and $y'(0)=0$.

51. Given that

$$y''-xy'+4y=0,\ y(0)=3,\ y'(0)=0$$

compute the value of $y(0.2)$ using the fourth-order R–K formula.

52. Solve $y''=-0.1y'-x$, $y(0)=0$, $y'(0)=1$ from $x=0$ to 2 in increments of $h=0.25$ with the fourth-order R–K method.
53. Solve using the R–K method $y''-x(y')^2-y^2=0$ for $x=0.2$, correct to four decimal places, with the initial conditions $x=0, y=1,$ and $y'=0$.
54. Solve using the fourth-order R–K method for systems, the initial value problem

$$\frac{d^2u}{dx^2}+0.8\sin u=0, x=0, u=0.2, \frac{du}{dx}=0.$$

Taking step length $h=0.25$, compute the first two steps of the solution.

55. Solve the initial value problem

$$y'=t^2-y^2, y(0)=1, t\in[0,0.6]$$

Use the third-order Adams–Bashforth method with $h=0.1$. Obtain the starting values using the third-order Taylor series method.

56. Find the values of $y(0.1)$, $y(0.2)$, and $y(0.3)$ using the fourth-order R–K method, given that $(dy/dx)=xy+y^2; y(0)=1$. Also find the value of $y(0.4)$ by using the Adams–Bashfourth method.
57. Find $y(0.8)$ by using the Adams–Moulton method for the equation $y'=y-x^2$, $y(0)=1$, obtaining the staring values by Taylor's series method.
58. Find the solution at $t=1.2$ for the initial value problem

$$u'=t^2+u^2, u(1)=2$$

using the third-order Adams–Moulton method with $h=0.1$. Use the Newton–Raphson method to solve the nonlinear differential equations.

59. Evaluate $y(1.4)$, given that $y'+(y/x)=(1/x^2)$ and $y(1)=1$, by using Adam's predictor–corrector formula.
60. Use Milne's method to compute the solution at $x=0.4$ given that $(dy/dx)=xy+y^2, y(0)=1$. Take $h=0.1$ and obtain the starting value for Milne's method using the fourth-order R–K method.
61. Solve the following problems by using Milne's predictor–corrector method. Find $y(-0.1)$ and $y(-0.2)$, given that $(dy/dx)=[(y^2-x^2)/(y^2+x^2)]$, $y(0)=1$.
62. Using Milne's predictor–corrector method determine $y(0.3)$, given that $y'=x^2+y^2-2$, $y(0)=1$. Obtain the starting values by Taylor series method.
63. Use Milne's predictor–corrector method to find $y(0.8)$ taking $h=0.2$. Given that $(dy/dx)=y+x^2$ with $y(0)=1$.
64. Using Picard's method, find approximate values of y and z corresponding to $x=0.1$, given that $y(0)=2$, $z(0)=1$ and $(dy/dx)=x+z$, $(dy/dx)=x-y^2$.
65. Solve the following system of equations by Euler's method when $t=0.1$

$$\frac{dx}{dt}=y-t,\quad \frac{dy}{dt}=x+t$$

given $x=1$ and $y=1$ when $t=0$.
66. Solve the boundary value problem defined by

$$y''-y=0, y(0)=0, y(1)=1$$

by using the finite difference method. Compare the solution obtained at $y(0.5)$ with the exact value, taking $h=0.5$ and $h=0.25$.
67. Solve by finite difference the boundary value problem

$$\frac{d^2u}{dx^2}=x\frac{du}{dx}-4u, u(0)=3, u(1)=-2$$

Take step length $= 0.25$.
68. Using Galerkin's method, compute the value of $y(0.5)$ given that

$$y''+y=x^2, 0<x<1, y(0)=0, y(1)=0$$

69. Solve the boundary value problem $y''-y+x=0$, $y(0)=y(1)=0$. Apply the shooting method.
70. Using the shooting method, solve the following boundary value problems.
    a. $y''=(-2/x)yy', 1<x<2; y(1)=(1/2), y(2)=2/3$.
    b. $y''=2yy', 1<x<(\pi/4); y(0)=0, y(\pi/4)=1$.
71. Solve the boundary value problem $y''+xy'+1=0, y(0)=y(1)=0$ by using the shooting method.
72. Discretize the initial value problem

$$y'=y,\ y(0)=1$$

using the midpoint method with step length h.
    a. Solve the initial value problem to estimate $y(0.3)$ using $h=0.1$. The value of $y(0.1)$ may be obtained from the analytical solution $y(t)=e^t$. Determine the percentage and relative error at $t=0.3$.
    b. Find the stability condition.
    c. Show that $y_j-e^{jh}=O(h^2)$ as $h\to 0$, $h=$ constant.

73. Consider the problem

$$y' = Ay$$

where

$$y(0) = \begin{bmatrix} 1 \\ 0 \end{bmatrix}, A = \begin{bmatrix} -2 & 1 \\ 1 & -20 \end{bmatrix}$$

   a. Show that the system is asymptotically stable.
   b. Examine the method

$$y_{j+1} = y_j + \frac{h}{2}(3f_{j+1} - f_j)$$

   for the equation $y' = f(x, y)$. What is its order of approximation? Is it stable? Is it A-stable?
   c. Choose step size $h = 0.2$ and $h = 0.1$ and compute approximations to $y(0.2)$ using the method in (b). Finally, make a suitable extrapolation. The exact solution is $y(0.2) = [0.68, 0.036]^T$ with two significant digits.

74. Given that $(dy/dx) - 2y = 3e^x; y(0) = 0$, find $y(-0.1)$ and $y(0.1)$ by Taylor's series method, $y(0.2)$ by the R–K method, $y(0.3)$ by Milne's method, and hence $y(0.4)$ by the Adams–Bashforth method.

75. Solve the following system of equation by Taylor's series method:

$$\frac{dy}{dx} = x + y + t, \frac{d^2y}{dt^2} = x - t \text{ for } t = 0.4, \text{given } x = 0, y = 1, \frac{dy}{dx} = -1 \text{ at } t = 0$$

76. Use Picard's method with three iterations to find $y(0.3)$ and $z(0.3)$, given that

$$\frac{dy}{dx} = xz + 1, \frac{dz}{dx} = -xy;\ y(0) = 0;\ z(0) = 1$$

77. Use the fourth-order R–K method to solve the equations in the following problems and find the required values.
   a. Find

$$y(0.1) \text{ and } z(0.1), \text{given that } \frac{dy}{dx} = \sin x - z + 1; \frac{dz}{dx} = \cos x - y; y(0) = 1; z(0) = 2$$

   b. Find

$$y(0.1) \text{ and } z(0.1), \text{given that } \frac{dy}{dx} = z + x; \frac{dz}{dx} = y - x; y(0) = 1; z(0) = 1$$

78. Solve the following boundary value problems, by taking the indicated step size:
   a. $y''(x) + (1 + x^2)y(x) + 1 = 0; y(\pm 1) = 0; h = 1/4.$
   b. $y''(x) + xy(x) + y(x) = 2x; y(0) = 1, y(1) = 0; h = 1/4.$
   c. $y''(x) + (1 - x)y(x) + xy(x) = x; y(0) = 0, y(1) = 0; h = 1/4.$

79. Check the conditioning of the initial value problem

$$y'(t) = y^2(t) - 5\sin t - 25\cos^2 t, y(0) = 6$$

80. Check the conditioning of the initial value problem

$$y'(t) = -y(t) + \sin t + \cos t,\ y(0) = 0$$

81. Solve the differential equation

$$y'(t) = -y(t) + \frac{1}{1+t^2} + \tan^{-1} t,\ y(0) = 0$$

on the interval $[0,1]$, using the backward Euler method and trapezoidal method taking step size $h = 0.2$. Hence discuss the results.

82. Using the midpoint method, solve the initial value problem

$$y'(t) = -y(t) + 2\cos t,\ y(0) = 1$$

on the interval $[0,1]$ taking step size $h = 0.25$.

8 Matrix Eigenvalue Problem

8.1 INTRODUCTION

In this chapter, we address the fundamental problems of numerical linear algebra (techniques of finding the eigenvalues and eigenvectors of a matrix). Eigenvalues and eigenvectors are important in many areas of science and engineering, solving differential equations and finding physical characteristics, such as the principal stress and moment of inertia. Eigenvalues play an important role in determining the convergence of many iterative methods.

This chapter starts with the Gerschgorin's circle theorem which finds the location of the eigenvalues of a given matrix **A**. According to the Gerschgorin's circle theorem, the eigenvalues of **A** must lie within the Gerschgorin discs whose radius is equal to the sum of the absolute values of the off-diagonal entries along the columns of **A**.

In this chapter, Householder's method has been discussed. In this method, a matrix **A** is reduced to the tridiagonal form by orthogonal transformations.

A matrix **B** is called Hessenberg if

$$b_{ij} = 0, \quad \text{for all} \quad i > j + 1$$

It is an upper triangular matrix except for a single nonzero subdiagonal.

The *QR* method, which uses *QR* factorization, has been presented in this chapter to find all the eigenvalues of a matrix. Since this method requires repeated factorization, a preliminary similarity transformation is usually applied to transform the matrix to Hessenberg matrix. If **A** is symmetric, then this Hessenberg form will be tridiagonal. The *QR* method produces a sequence of matrices that converge to a similar matrix for which the eigenvalues are easy to determine.

Next we consider an iterative method, known as the power method, for finding the dominant eigenvalue, that is, the eigenvalue of largest magnitude and its associated eigenvector for a given matrix **A**. Using inverse power method, we can find any eigenvalue of a matrix **A**.

The Jacobi method has been also described in this chapter. In Jacobi method, the symmetric matrix **A** is reduced to a diagonal matrix **D**. Also Givens method has been presented. In this method, the symmetric matrix **A** is reduced to a tridiagonal matrix **T**. Then applying Sturm sequence property, all the eigenvalues and the corresponding eigenvectors are obtained.

8.1.1 Characteristic Equation, Eigenvalue, and Eigenvector of a Square Matrix

Let **A** be an $n \times n$ matrix. Then, $\det(\boldsymbol{A} - \lambda \boldsymbol{I})$ is called the characteristic polynomial of **A** and $\det(\boldsymbol{A} - \lambda \boldsymbol{I}) = 0$ is called the characteristic equation of **A**.

A root of the characteristic equation of a square matrix **A** is said to be a characteristic value or an eigenvalue of **A**. The set of all eigenvalues of **A** is called the spectrum of **A**. The spectral radius of **A**, denoted by $\rho(\boldsymbol{A})$, is defined by

$$\rho(\boldsymbol{A}) = \max_{1 \le i \le n} |\lambda_i|$$

A non-null vector $\boldsymbol{X}(\boldsymbol{X} \neq \boldsymbol{0})$ is said to be an eigenvector of **A** if there exists a scalar λ such that $\boldsymbol{AX} = \lambda \boldsymbol{X}$ holds.

If λ be a root of the characteristic equation $\det(\boldsymbol{A}-\lambda \boldsymbol{I})=0$ of multiplicity r, then r is called the algebraic multiplicity of λ.

The number of linearly independent eigenvectors corresponding to an eigenvalue λ is called geometric multiplicity of λ.

It can be shown that, for an eigenvalue λ, algebraic multiplicity $\geq$ geometric multiplicity ≥ 1. If the algebraic multiplicity and geometric multiplicity of an eigenvalue are equal, then the corresponding eigenvalue is called regular.

Let $\boldsymbol{A}$ has an eigenvalue λ and an eigenvector $\boldsymbol{X}$. Then, we have

$$\boldsymbol{AX}=\lambda \boldsymbol{X} \tag{8.1}$$

Premultiplying Equation 8.1 $m-1$ times by $\boldsymbol{A}$, we obtain

$$\boldsymbol{A}^m \boldsymbol{X}=\lambda^m \boldsymbol{X} \tag{8.2}$$

which shows that $\boldsymbol{A}^m$ has an eigenvalue λ^m and the same eigenvector $\boldsymbol{X}$.

Again, if $\boldsymbol{A}$ has an inverse $\boldsymbol{A}^{-1}$, then Equation 8.1 can be written as

$$\boldsymbol{A}^{-1}\boldsymbol{X}=\frac{1}{\lambda}\boldsymbol{X} \tag{8.3}$$

Therefore, the inverse matrix $\boldsymbol{A}^{-1}$ has the same eigenvector as $\boldsymbol{A}$ but has the eigenvalue $1/\lambda$.

8.1.2 Similar Matrices and Diagonalizable Matrix

A square matrix $\hat{\boldsymbol{A}}$ of order n is said to be similar to a square matrix $\boldsymbol{A}$ of order n if there exists a nonsingular matrix $\boldsymbol{P}$ of order n such that $\hat{\boldsymbol{A}}=\boldsymbol{P}^{-1}\boldsymbol{AP}$, that is, $\boldsymbol{A}=\boldsymbol{P}\hat{\boldsymbol{A}}\boldsymbol{P}^{-1}$.

This transformation which gives $\hat{\boldsymbol{A}}$ from $\boldsymbol{A}$ is called a similarity transformation.

Multiplying both sides of Equation 8.1 from the left by $\boldsymbol{P}^{-1}$, we get

$$\boldsymbol{P}^{-1}\boldsymbol{AX}=\lambda \boldsymbol{P}^{-1}\boldsymbol{X} \tag{8.4}$$

Substituting $\boldsymbol{X}=\boldsymbol{PY}$ in Equation 8.4, we obtain

$$\boldsymbol{P}^{-1}\boldsymbol{APY}=\lambda \boldsymbol{P}^{-1}\boldsymbol{PY}=\lambda \boldsymbol{IY}=\lambda \boldsymbol{Y} \tag{8.5}$$

Again, substituting $\hat{\boldsymbol{A}}=\boldsymbol{P}^{-1}\boldsymbol{AP}$ in Equation 8.5, we get

$$\hat{\boldsymbol{A}}\boldsymbol{Y}=\lambda \boldsymbol{Y} \tag{8.6}$$

Therefore, $\hat{\boldsymbol{A}}=\boldsymbol{P}^{-1}\boldsymbol{AP}$ has the same eigenvalue as $\boldsymbol{A}$ and its eigenvector is given by

$$\boldsymbol{Y}=\boldsymbol{P}^{-1}\boldsymbol{X}$$

If a square matrix $\boldsymbol{A}$ of order n has n linearly independent eigenvectors, then $\boldsymbol{A}$ is diagonalizable. In this case, there exists a diagonal matrix $\boldsymbol{D}$ which is similar to $\boldsymbol{A}$. This implies that

$$\boldsymbol{D}=\boldsymbol{P}^{-1}\boldsymbol{AP}, \quad \text{that is,} \quad \boldsymbol{A}=\boldsymbol{PDP}^{-1}$$

where $\boldsymbol{D}=\begin{bmatrix} \lambda_1 & 0 & \cdots & 0 \\ 0 & \lambda_2 & \cdots & 0 \\ \vdots & \vdots & \cdots & \vdots \\ 0 & 0 & \cdots & \lambda_n \end{bmatrix}$ and $\lambda_1, \lambda_2, \ldots, \lambda_n$ are the eigenvalues of $\boldsymbol{A}$.

Thus, a similarity transformation, where $\boldsymbol{P}$ is the matrix of eigenvectors, reduces the matrix $\boldsymbol{A}$ into a diagonal matrix $\boldsymbol{D}$. The ith column of $\boldsymbol{P}$ is an eigenvector of $\boldsymbol{A}$ corresponding to the eigenvalue λ_i of $\boldsymbol{A}$. This matrix $\boldsymbol{P}$ is called the diagonalizing matrix, which diagonalize $\boldsymbol{A}$ into a diagonal matrix $\boldsymbol{D}$.

Now, the solution of the characteristic equation $\det(\boldsymbol{A}-\lambda\boldsymbol{I})=0$ becomes difficult and laborious if $n>3$. Moreover, for a highly accurate approximation to an eigenvalue λ, the eigenvector obtained from Equation 8.1 may be different from the exact eigenvector.

In order to overcome the above difficulties, we may use several numerical methods for determining eigenvalues and the corresponding eigenvectors of a given matrix $\boldsymbol{A}=(a_{ij})_{n\times n}$. If the given matrix is symmetric, we generally use the methods of Jacobi, Givens, and Householder in which the matrix $\boldsymbol{A}$ is transformed to a special form and consequently the eigenvalues are readily obtained. On the other hand, if $\boldsymbol{A}$ is nonsymmetric, then QR method may be applied.

In all these methods, a series of similarity transformations are used and so these methods are referred to as direct or transformation methods.

Another techniques, called iterative methods, are usually used to find some eigenvalues and the corresponding eigenvectors of a given matrix $\boldsymbol{A}$. The power method and inverse power method belong to this category.

8.2 INCLUSION OF EIGENVALUES

The Gerschgorin's theorem provides a region that contains all the eigenvalues of a complex square matrix. This theorem gives a region consisting of closed circular discs in the complex plane, including all the eigenvalues of a given matrix.

8.2.1 GERSCHGORIN'S DISCS

Definition 8.1

Let $n\geq 2$ and $\boldsymbol{A}\in\boldsymbol{C}^{n\times n}$. The Gerschgorin's discs D_i, $i=1,2,\ldots,n$, of the matrix $\boldsymbol{A}$ are defined as the closed circular regions

$$D_i=\left\{z\in\boldsymbol{C}\,\middle|\,|z-a_{ii}|\leq r_i\right\} \tag{8.7}$$

in the complex plane, where

$$r_i=\sum_{\substack{j=1\\ j\neq i}}^{n}|a_{ij}| \tag{8.8}$$

is the radius of D_i.

8.2.2 GERSCHGORIN'S THEOREM

Let λ be a given eigenvalue of an arbitrary matrix $A=(a_{ij})_{n\times n}$, then for some integer i $(1\leq i\leq n)$, we have

$$|a_{ii}-\lambda|\leq|a_{i1}|+|a_{i2}|+\cdots+|a_{i,i-1}|+|a_{i,i+1}|+\cdots+|a_{in}| \tag{8.9}$$

Proof:

Let $\boldsymbol{x}$ be an eigenvector corresponding to eigenvalue λ of matrix $\boldsymbol{A}$. Then,

$$\boldsymbol{A}\boldsymbol{x}=\lambda\boldsymbol{x} \tag{8.10}$$

where

$$x = \begin{bmatrix} x_1 \\ x_2 \\ \vdots \\ x_n \end{bmatrix}$$

The x_i be a component of the eigenvector $\boldsymbol{x}$ that is maximum in magnitude. Then, we have

$$\left|\frac{x_r}{x_i}\right| \leq 1, \quad r = 1,2,3,\ldots,n \tag{8.11}$$

The i-th equation of the system of equations in Equation 8.10 is

$$a_{i1}x_1 + a_{i2}x_2 + \cdots + a_{i,i-1}x_{i-1} + (a_{ii} - \lambda)x_i + a_{i,i+1}x_{i+1} + \cdots + a_{in}x_n = 0$$

This implies that

$$(a_{ii} - \lambda)x_i = -a_{i1}x_1 - a_{i2}x_2 - \cdots - a_{i,i-1}x_{i-1} - a_{i,i+1}x_{i+1} - \cdots - a_{in}x_n \tag{8.12}$$

Dividing both sides of Equation 8.12 by x_i and taking modulus, we get

$$|a_{ii} - \lambda| = \left| -a_{i1}\frac{x_1}{x_i} - a_{i2}\frac{x_2}{x_i} - a_{i,i-1}\frac{x_{i-1}}{x_i} - a_{i,i+1}\frac{x_{i+1}}{x_i} \cdots - a_{in}\frac{x_n}{x_i} \right| \leq |a_{i1}|\left|\frac{x_1}{x_i}\right| + |a_{i2}|\left|\frac{x_2}{x_i}\right| + \cdots + |a_{in}|\left|\frac{x_n}{x_i}\right|$$

Using Equation 8.11, we obtain

$$|a_{ii} - \lambda| \leq |a_{i1}| + |a_{i2}| + \cdots + |a_{in}| \tag{8.13}$$

That is,

$$|a_{ii} - \lambda| \leq \sum_{\substack{j=1 \\ j \neq i}}^{n} |a_{ij}| \tag{8.14}$$

■

For each i, $i = 1,2,\ldots,n$, the Equation 8.7 of the Gerschgorin's theorem determines a closed circular discs in the complex λ-plane with center a_{ii} and radius r_i given by the right side of Equation 8.7. The Gerschgorin's theorem states that each of the eigenvalues of $\boldsymbol{A}$ lies in one of these n discs.

Example 8.1

Use the Gerschgorin circle theorem to estimate the eigenvalues of the following matrix:

$$\boldsymbol{A} = \begin{bmatrix} 5 & -2 & 2 \\ 2 & 0 & 4 \\ 4 & 2 & 7 \end{bmatrix}$$

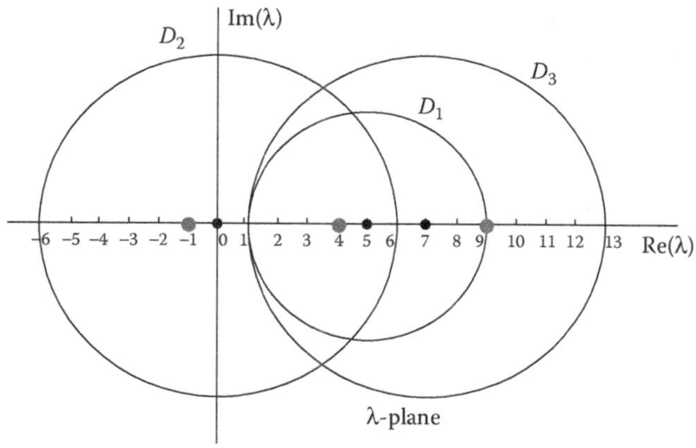

FIGURE 8.1 **(See color insert.)** Gerschgorin's discs containing eigenvalues.

Solution:

We have the following Gerschgorin's discs that bound the eigenvalues:

1. D_1: Center at (5,0) with radius 4
2. D_2: Center at (0,0) with radius 6
3. D_3: Center at (7,0) with radius 6

The red dots in Figure 8.1 indicate the actual location of the eigenvalues. Thus, we see that all the eigenvalues lie inside the union of all the discs, that is, $D_1 \cup D_2 \cup D_3$. The exact eigenvalues are $\lambda_1 = 9$, $\lambda_2 = 4$, and $\lambda_3 = -1$, which lie in $D_1 \cup D_2 \cup D_3$.

8.3 HOUSEHOLDER'S METHOD

This procedure can be used to find the eigenvalues of a real symmetric matrix. In the first stage, an orthogonal transformation is applied to reduce a real symmetric matrix to a tridiagonal matrix. This can be done in a finite number of steps by using Householder matrices.

Let $\boldsymbol{A} = (a_{ij})_{n\times n}$ be a real symmetric matrix of order n. In this method, a real symmetric orthogonal matrix $\boldsymbol{P}$ is obtained in such a way that $\boldsymbol{PAP}$ is in tridiagonal form. In this method, $\boldsymbol{A}$ is reduced to the tridiagonal form by orthogonal transformations. These orthogonal transformations are carried out by using the Householder matrix of the form

$$\boldsymbol{P} = \boldsymbol{I} - 2\boldsymbol{w}\boldsymbol{w}^T \tag{8.15}$$

where $\boldsymbol{w} \in \boldsymbol{R}^n$ is column vector, such that

$$\boldsymbol{w}^T\boldsymbol{w} = w_1^2 + w_2^2 + \cdots + w_n^2 = 1 \tag{8.16}$$

Then,

$$\boldsymbol{P}^T = (\boldsymbol{I} - 2\boldsymbol{w}\boldsymbol{w}^T)^T = \boldsymbol{I} - 2\boldsymbol{w}\boldsymbol{w}^T = \boldsymbol{P} \tag{8.17}$$

Also,

$$\begin{aligned}\boldsymbol{P}^T\boldsymbol{P} &= (\boldsymbol{I} - 2\boldsymbol{w}\boldsymbol{w}^T)(\boldsymbol{I} - 2\boldsymbol{w}\boldsymbol{w}^T)\\ &= \boldsymbol{I} - 4\boldsymbol{w}\boldsymbol{w}^T + 4\boldsymbol{w}\boldsymbol{w}^T\boldsymbol{w}\boldsymbol{w}^T = \boldsymbol{I}\end{aligned} \tag{8.18}$$

Therefore, $\boldsymbol{P}$ is a symmetric and orthogonal matrix.

The vectors are constructed with the first $(m-1)$ components as zeros, that is,

$$\boldsymbol{w}^{(m)} = (0,0,\ldots,0,w_m,w_{m+1},\ldots,w_n)^T \tag{8.19}$$

Since, $\left(\boldsymbol{w}^{(m)}\right)^T \boldsymbol{w}^{(m)} = 1$, we have

$$w_m^2 + w_{m+1}^2 + \cdots + w_n^2 = 1$$

Consequently,

$$\boldsymbol{P}_m = \boldsymbol{I} - 2\boldsymbol{w}^{(m)}\left(\boldsymbol{w}^{(m)}\right)^T \tag{8.20}$$

Using similarity transformation, we get

$$\boldsymbol{A}^{(m)} = \boldsymbol{P}_m^{-1}\boldsymbol{A}^{(m-1)}\boldsymbol{P}_m = \boldsymbol{P}_m^T\boldsymbol{A}^{(m-1)}\boldsymbol{P}_m = \boldsymbol{P}_m\boldsymbol{A}^{(m-1)}\boldsymbol{P}_m \tag{8.21}$$

since $\boldsymbol{P}_m$ is a symmetric and orthogonal matrix.

Thus, we successively form the matrices

$$\boldsymbol{A}^{(m)} = \boldsymbol{P}_m\boldsymbol{A}^{(m-1)}\boldsymbol{P}_m, \quad m = 2,3,\ldots,n-1 \tag{8.22}$$

with $\boldsymbol{A}^{(1)} = \boldsymbol{A}$. At the first transformation, we find w_m's so that the zeros are obtained in the positions $(1,3),(1,4),\ldots,(1,n)$ and $(3,1),(4,1),\ldots,(n,1)$.

Proceeding in this manner, we obtain a tridiagonal matrix after $n-1$ such transformations. Thus, the tridiagonalization is completed with exactly $n-2$ Householder transformations.

- *Determination of unit vector* $\boldsymbol{w}$: In the first step, we determine the unit vector $\boldsymbol{w}^{(2)} = (w_1,w_2,\ldots,w_n)^T$, where $\left(\boldsymbol{w}^{(2)}\right)^T \boldsymbol{w}^{(2)} = w_1^2 + w_2^2 + \cdots + w_n^2 = 1$, so that in the matrix

$$\boldsymbol{A}^{(2)} = \boldsymbol{P}_2\boldsymbol{A}^{(1)}\boldsymbol{P}_2 = \left[\boldsymbol{I} - 2\boldsymbol{w}^{(2)}\left(\boldsymbol{w}^{(2)}\right)^T\right]\boldsymbol{A}\left[\boldsymbol{I} - 2\boldsymbol{w}^{(2)}\left(\boldsymbol{w}^{(2)}\right)^T\right], \tag{8.23}$$

  we have $a_{11}^{(2)} = a_{11}^{(1)} = a_{11}$ and $a_{i1}^{(2)} = 0, \quad i = 3,4,\ldots,n.$

  Next, setting $w_1 = 0$ which ensures $a_{11}^{(2)} = a_{11}$, we determine

$$\boldsymbol{P}_2 = \boldsymbol{I} - 2\boldsymbol{w}^{(2)}\left(\boldsymbol{w}^{(2)}\right)^T$$

  so that

$$\boldsymbol{P}_2(a_{11},a_{21},a_{31},\ldots,a_{n1})^T = (a_{11},a_{21}^{(2)},0,\ldots,0)^T \tag{8.24}$$

  Let $\tilde{\boldsymbol{w}}^{(2)} = (w_2,w_3,\ldots,w_n)^T$ and $\boldsymbol{v} = (a_{21},a_{31},\ldots,a_{n1})^T$ and $\hat{\boldsymbol{P}}_2$ be the $(n-1)\times(n-1)$ Householder transformation

$$\hat{\boldsymbol{P}}_2 = \boldsymbol{I}_{n-1} - 2\tilde{\boldsymbol{w}}_2\tilde{\boldsymbol{w}}_2^T$$

From Equation 8.24, we get

$$P_2\begin{bmatrix} a_{11} \\ \\ a_{21} \\ \vdots \\ \\ a_{n1} \end{bmatrix} = \begin{bmatrix} 1 & \vdots & 0 & \cdots & 0 \\ \cdots & \cdots & \cdots & \cdots & \cdots \\ 0 & \vdots & & & \\ \vdots & \vdots & & \hat{P}_2 & \\ 0 & \vdots & & & \end{bmatrix}\begin{bmatrix} a_{11} \\ \cdots \\ v \\ \\ \end{bmatrix} = \begin{bmatrix} a_{11} \\ \cdots \\ \hat{P}_2 v \\ \\ \end{bmatrix} = \begin{bmatrix} a_{11} \\ \cdots \\ a_{21}^{(2)} \\ 0 \\ \vdots \\ 0 \end{bmatrix} \tag{8.25}$$

where

$$\hat{P}_2 v = (I_{n-1} - 2\tilde{w}_2\tilde{w}_2^T)v = v - 2(\tilde{w}_2^T v)\tilde{w}_2 = (a_{21}^{(2)}, 0, \ldots, 0)^T \tag{8.26}$$

Let $r = \tilde{w}_2^T v$. Then,

$$(a_{21}^{(2)}, 0, \ldots, 0)^T = (a_{21} - 2rw_2, a_{31} - 2rw_3, \ldots, a_{n1} - 2rw_n)^T \tag{8.27}$$

Once we know $a_{21}^{(2)}$ and r, all values of w_i, $i = 2, 3, \ldots, n$, can be determined.
Now, from Equation 8.27, we get

$$a_{21}^{(2)} = a_{21} - 2rw_2$$

and

$$a_{i1} - 2rw_i = 0, \quad i = 3, \ldots, n$$

Thus,

$$2rw_2 = a_{21} - a_{21}^{(2)} \tag{8.28}$$

and

$$2rw_i = a_{i1}, \quad i = 3, \ldots, n \tag{8.29}$$

Squaring both sides of Equations 8.28 and 8.29 and then adding the corresponding terms, we obtain

$$4r^2 \sum_{i=2}^{n} w_i^2 = \left(a_{21} - a_{21}^{(2)}\right)^2 + \sum_{i=3}^{n} a_{i1}^2 \tag{8.30}$$

This implies that

$$4r^2 = \sum_{i=2}^{n} a_{i1}^2 - 2a_{21}a_{21}^{(2)} + \left(a_{21}^{(2)}\right)^2 \tag{8.31}$$

Using Equation 8.26, we have

$$\left(a_{21}^{(2)}\right)^2 = \begin{bmatrix} a_{21}^{(2)} & 0 & 0 & 0 \end{bmatrix} \begin{bmatrix} a_{21}^{(2)} \\ 0 \\ \vdots \\ 0 \end{bmatrix} = \left(\hat{\boldsymbol{P}}_2 \boldsymbol{v}\right)^T \hat{\boldsymbol{P}}_2 \boldsymbol{v} = \boldsymbol{v}^T \hat{\boldsymbol{P}}_2^T \hat{\boldsymbol{P}}_2 \boldsymbol{v} = \boldsymbol{v}^T \boldsymbol{v}$$

since $\hat{\boldsymbol{P}}_2$ is an orthogonal matrix.

Therefore,

$$\left(a_{21}^{(2)}\right)^2 = \sum_{i=2}^{n} a_{i1}^2 \tag{8.32}$$

From Equations 8.31 and 8.32, we obtain

$$2r^2 = \sum_{i=2}^{n} a_{i1}^2 - a_{21} a_{21}^{(2)} \tag{8.33}$$

Now, taking into account Equations 8.32 and 8.33, we choose

$$a_{21}^{(2)} = -\text{sgn}(a_{21}) \left(\sum_{i=2}^{n} a_{i1}^2 \right)^{1/2} \tag{8.34}$$

which implies that

$$2r^2 = \sum_{i=2}^{n} a_{i1}^2 + |a_{21}| \left(\sum_{i=2}^{n} a_{i1}^2 \right)^{1/2} \tag{8.35}$$

Next, using this values of $a_{21}^{(2)}$ and $2r^2$, from Equations 8.28 and 8.29 we obtain

$$w_2 = \frac{a_{21} - a_{21}^{(2)}}{2r} \tag{8.36}$$

$$w_i = \frac{a_{i1}}{2r}, \quad i = 3, \ldots, n \tag{8.37}$$

with $w_1 = 0$.

In this way, having obtained $\boldsymbol{P}_2$ and computed $\boldsymbol{A}^{(2)} = \boldsymbol{P}_2 \boldsymbol{A}^{(1)} \boldsymbol{P}_2$, this process is repeated for $k = 3, \ldots, n-1$ as follows:

$$a_{k,k-1}^{(k)} = -\text{sgn}(a_{k,k-1}) \left(\sum_{i=k}^{n} a_{i,k-1}^2 \right)^{1/2}$$

$$r = \left(\frac{1}{2} \sum_{i=k}^{n} a_{i,k-1}^2 - \frac{1}{2} a_{k,k-1} a_{k,k-1}^{(k)} \right)^{1/2}$$

$$w_1^{(k)} = w_2^{(k)} = w_3^{(k)} = \cdots = w_{k-1}^{(k)} = 0$$

$$w_k^{(k)} = \frac{a_{k,k-1}^{(k-1)} - a_{k,k-1}^{(k)}}{2r}$$

$$w_i^{(k)} = \frac{a_{k,k-1}^{(k-1)}}{2r}, \quad i = k+1,\ldots,n$$

$$\boldsymbol{P}_k = \boldsymbol{I} - 2\boldsymbol{w}^{(k)}\left(\boldsymbol{w}^{(k)}\right)^T$$

and

$$\boldsymbol{A}^{(k)} = \boldsymbol{P}_k \boldsymbol{A}^{(k-1)} \boldsymbol{P}_k$$

Continuing in this manner, the tridiagonal and symmetric matrix $\boldsymbol{A}^{(n-1)}$ is formed, where

$$\boldsymbol{A}^{(n-1)} = \boldsymbol{P}_{n-1}\boldsymbol{P}_{n-2}\ldots\boldsymbol{P}_2\boldsymbol{A}\boldsymbol{P}_2\ldots\boldsymbol{P}_{n-2}\boldsymbol{P}_{n-1}$$

8.3.1 Algorithm for Householder's Method

Input: Enter the order of the matrix n and matrix $\boldsymbol{A} = \boldsymbol{A}^{(1)} = [a_{i,j}^{(1)}]_{n\times n}$.
Output: Tridiagonal and symmetric matrix $\boldsymbol{A}^{(n-1)}$.
Step 1: for $k = 2(1)n-1$ do
  for $j = 1(1)k-1$ do
    $w_j^{(k)} = 0;$
  end.
  $sum = 0;$
  for $i = k(1)n$ do
    $\text{sum} = \text{sum} + \left[a_{i,k-1}^{(k-1)}\right]^2;$
  end.
  $a_{k,k-1}^{(k)} = -\text{sgn}\left(a_{k,k-1}^{(k-1)}\right) * \text{sum}^{1/2};$
  $r = \left[\frac{1}{2}\text{sum} - \frac{1}{2}a_{k,k-1}^{(k-1)}\ a_{k,k-1}^{(k)}\right]^{1/2};$
  $w_k^{(k)} = \frac{a_{k,k-1}^{(k-1)} - a_{k,k-1}^{(k)}}{2r};$
  for $i = k+1(1)n$ do
    $w_i^{(k)} = \frac{a_{k,k-1}^{(k-1)}}{2r};$
  end.
  Construct the column vector
  $w^{(k)} = [w_1^{(k)}, w_2^{(k)}, \ldots, w_n^{(k)}]^T;$
  $P_k = I - 2w^{(k)}\left(w^{(k)}\right)^T;$
  $A^{(k)} = P_k A^{(k-1)} P_k;$
end.
Step 2: Print $A^{(n-1)}$;
Step 3: Stop. ■

MATHEMATICA® Program for Householder Method (Chapter 8, Example 8.2)

```
n=3;
A[1]={{2,-1,-1},{-1,2,-1},{-1,-1,2}};
(*A[1]={{4,-1,-1,0},{-1,4,0,-1},{-1,0,4,-1},{0,-1,-1,4}};*)
For[k=2,k≤n-1,k++,

  For[j=1,j≤k-1,j++,
   w[j,k]=0];
  a[k,k-1]=-Sign[A[k-1][[k,k-1]]]*(Σ_{i=k}^n(A[k-1][[i,k-1]])^2)^1/2;
  r=(1/2*Σ_{i=k}^n(A[k-1][[i,k-1]])^2-1/2*A[k-1][[k,k-1]]*a[k,k-1])^1/2;
  w[k,k]=N[(A[k-1][[k,k-1]]-a[k,k-1])/(2*r)];
  For[i=k+1,i≤n,i++,
   w[i,k]=N[A[k-1][[k,k-1]]/(2*r)]];
  w[k]=Table[w[i,k],{i,1,n}];
  p[k]=IdentityMatrix[n]-2*Transpose[{w[k]}].{w[k]};
  Print[MatrixForm[p[k]]];
  A[k]=p[k].A[k-1].p[k];
  Print[MatrixForm[A[k]]]];
```

Output:

$$\begin{pmatrix} 1. & 0. & 0. \\ 0. & -0.707107 & -0.707107 \\ 0. & -0.707107 & 0.707107 \end{pmatrix}$$

$$\begin{pmatrix} 2. & 1.41421 & 1.11022\times10^{-16} \\ 1.41421 & 1. & -1.5702\times10^{-16} \\ 1.11022\times10^{-16} & 7.84962\times10^{-17} & 3. \end{pmatrix}$$

Example 8.2

Use the Householder's transformation to reduce the following matrix

$$A=\begin{bmatrix} 2 & -1 & -1 \\ -1 & 2 & -1 \\ -1 & -1 & 2 \end{bmatrix}$$

into a tridiagonal matrix form and hence find all the eigenvalues.

Solution:

According to Householder's method, let us assume that

$$w_2^T=(0,w_2,w_3),\quad w_2^2+w_3^2=1$$

so that

$$P_2 = I - 2\mathbf{w}_2\mathbf{w}_2^T = \begin{bmatrix} 1 & 0 & 0 \\ 0 & 1-2w_2^2 & -2w_2w_3 \\ 0 & -2w_2w_3 & 1-2w_3^2 \end{bmatrix}$$

For the first iteration of Householder transformation, we have

$$a_{21}^{(2)} = -\text{sgn}(a_{21})\left(\sum_{i=2}^{n} a_{i1}^2\right)^{1/2} = \sqrt{2} \tag{8.38}$$

$$2r^2 = \sum_{i=2}^{n} a_{i1}^2 + |a_{21}|\left(\sum_{i=2}^{n} a_{i1}^2\right)^{1/2} = 2+\sqrt{2} \tag{8.39}$$

Next, using this values of $a_{21}^{(2)}$ and $2r^2$, we obtain

$$w_2 = \frac{a_{21}-a_{21}^{(2)}}{2r} = \frac{-1-\sqrt{2}}{2\left(\dfrac{2+\sqrt{2}}{2}\right)^{1/2}} = -\frac{1}{2}\sqrt{2-\sqrt{2}}\,(1+\sqrt{2}) \tag{8.40}$$

$$w_i = \frac{a_{i1}}{2r} = -\frac{1}{2}\sqrt{2-\sqrt{2}}, \quad i = 3,\ldots,n \tag{8.41}$$

Thus, we obtain

$$P_2AP_2 = \begin{bmatrix} 2 & -1 & -1 \\ 1.41421 & -0.707107 & -0.707107 \\ 0 & -2.12132 & 2.12132 \end{bmatrix}\begin{bmatrix} 1 & 0 & 0 \\ 0 & -0.707107 & -0.707107 \\ 0 & -0.707107 & 0.707107 \end{bmatrix}$$

$$= \begin{bmatrix} 2 & 1.41421 & 0 \\ 1.41421 & 1 & 0 \\ 0 & 0 & 3 \end{bmatrix}$$

which is the required tridiagonal matrix.

The characteristic equation for P_2AP_2 is

$$\begin{vmatrix} 2-\lambda & 1.41421 & 0 \\ 1.41421 & 1-\lambda & 0 \\ 0 & 0 & 3-\lambda \end{vmatrix} = 0$$

giving the solutions $\lambda = 3, 3, 0$ which are the required eigenvalues of the given matrix.

8.4 THE *QR* METHOD

We now investigate a well known and efficient method for finding all the eigenvalues of a general $n \times n$ real matrix. The *QR* method can be used for an arbitrary real matrix, but for a general matrix it takes many iterations and becomes time consuming. To apply the *QR* method, we begin with a symmetric matrix in tridiagonal form. If this is not the form of the symmetric matrix, the first step is to apply Householder's method to compute a symmetric, tridiagonal matrix similar to the given matrix.

Let $\boldsymbol{A}$ be a real symmetric, tridiagonal matrix. We compute $\boldsymbol{A} = \boldsymbol{A}^{(1)}$, $\boldsymbol{A}^{(2)}$, …, $\boldsymbol{A}^{(n)}$ stepwise according to the following iterative method:

Step 1: $\boldsymbol{A}^{(1)} = \boldsymbol{A}$ is factorized as a product $\boldsymbol{A}^{(1)} = \boldsymbol{Q}^{(1)}\boldsymbol{R}^{(1)}$, where $\boldsymbol{Q}^{(1)}$ is an orthogonal matrix and $\boldsymbol{R}^{(1)}$ is an upper triangular matrix. Then, we compute $\boldsymbol{A}^{(2)} = \boldsymbol{R}^{(1)}\boldsymbol{Q}^{(1)}$.

Step 2: $\boldsymbol{A}^{(2)}$ is factorized as a product $\boldsymbol{A}^{(2)} = \boldsymbol{Q}^{(2)}\boldsymbol{R}^{(2)}$. Then, we compute $\boldsymbol{A}^{(3)} = \boldsymbol{R}^{(2)}\boldsymbol{Q}^{(2)}$

General step k:

1. $\boldsymbol{A}^{(k)}$ is factorized as a product

$$\boldsymbol{A}^{(k)} = \boldsymbol{Q}^{(k)}\boldsymbol{R}^{(k)} \tag{8.42}$$

where $\boldsymbol{Q}^{(k)}$ is an orthogonal matrix and $\boldsymbol{R}^{(k)}$ is an upper triangular matrix.

2. We compute

$$\boldsymbol{A}^{(k+1)} = \boldsymbol{R}^{(k)}\boldsymbol{Q}^{(k)} \tag{8.43}$$

Since, $\boldsymbol{Q}^{(k)}$ is an orthogonal matrix, we have $\boldsymbol{R}^{(k)} = \boldsymbol{Q}^{(k)^T}\boldsymbol{A}^{(k)}$, and therefore,

$$\boldsymbol{A}^{(k+1)} = \boldsymbol{R}^{(k)}\boldsymbol{Q}^{(k)} = \boldsymbol{Q}^{(k)^T}\boldsymbol{A}^{(k)}\boldsymbol{Q}^{(k)} \tag{8.44}$$

Thus, $\boldsymbol{A}^{(k+1)}$ is similar to $\boldsymbol{A}^{(k)}$. Hence, by induction $\boldsymbol{A}^{(k+1)}$ is similar to $\boldsymbol{A}^{(1)} = \boldsymbol{A}$, for all k. Therefore, $\boldsymbol{A}^{(k+1)}$ has the same eigenvalues as the original matrix $\boldsymbol{A}$, and $\boldsymbol{A}^{(k+1)}$ tends to a diagonal matrix with the eigenvalues of $\boldsymbol{A}$ along the diagonal.

More precisely, if the eigenvalues of $\boldsymbol{A}$ are different in absolute value, say, $|\lambda_1| > |\lambda_2| > \cdots > |\lambda_n|$, then

$$\lim_{k\to\infty} \boldsymbol{A}^{(k)} = \boldsymbol{D}$$

where $\boldsymbol{D}$ is diagonal, with main diagonal entries $\lambda_1, \lambda_2, \ldots, \lambda_n$.

- *Rotation matrices*: To describe the construction of the matrices $\boldsymbol{Q}^{(k)}$ and $\boldsymbol{R}^{(k)}$, we need the notion of a rotation matrix.

Definition:

A rotation matrix $\boldsymbol{P}$ differs from the identity matrix in at most four entries. These four entries are of the form

$$p_{ii} = p_{jj} = \cos\theta \quad \text{and} \quad p_{ij} = p_{ji} = -\sin\theta, \tag{8.45}$$

for some θ and some $i \neq j$.

Every rotation matrix $\boldsymbol{P}$ is an orthogonal matrix, because the definition implies that $\boldsymbol{P}\boldsymbol{P}^T = \boldsymbol{I}$.

Now, to construct the matrices $\boldsymbol{Q}^{(k)}$ and $\boldsymbol{R}^{(k)}$ at each stage, we proceed as follows:

Let $\boldsymbol{A} = \boldsymbol{A}^{(1)} = (a_{ij})_{n\times n} = \boldsymbol{Q}^{(1)}\boldsymbol{R}^{(1)}$. The tridiagonal matrix $\boldsymbol{A}$ has $n-1$ nonzero entries below the main diagonal. These are $a_{21}, a_{32}, \ldots, a_{n,n-1}$. We multiply $\boldsymbol{A}$ from the left by a rotation matrix $\boldsymbol{P}_2$ such that $\boldsymbol{P}_2\boldsymbol{A} = (a_{ij}^{(2)})_{n\times n}$ has $a_{21}^{(2)} = 0$. We again multiply $\boldsymbol{P}_2\boldsymbol{A}$ from the left by a rotation matrix $\boldsymbol{P}_3$ such that $\boldsymbol{P}_3\boldsymbol{P}_2\boldsymbol{A} = (a_{ij}^{(3)})_{n\times n}$ has $a_{32}^{(3)} = 0$. Continuing this process, after $n-1$ such multiplications we are left with an upper triangular matrix $\boldsymbol{R}^{(1)}$, that is,

$$\boldsymbol{P}_n\boldsymbol{P}_{n-1}\ldots\boldsymbol{P}_3\boldsymbol{P}_2\boldsymbol{A}^{(1)} = \boldsymbol{R}^{(1)} \tag{8.46}$$

where $\boldsymbol{P}_k$, $k = 2,3,\ldots,n$, is of the form

$$\boldsymbol{P}_k = \left[\begin{array}{c|ccc|c} I_{k-2} & & \mathbf{0} & & \mathbf{0} \\ \hline & c_k & & s_k & \\ \mathbf{0} & & & & \mathbf{0} \\ & -s_k & & c_k & \\ \hline \mathbf{0} & & \mathbf{0} & & I_{n-k} \end{array}\right] \begin{array}{l} \\ \leftarrow \text{row } k-1 \\ \\ \leftarrow \text{row } k \\ \\ \end{array} \tag{8.47}$$

$$\qquad\qquad \uparrow \qquad\qquad\qquad \uparrow$$

$$\qquad \text{column } k-1 \qquad \text{column } k$$

where $c_k = \cos\theta_k$ and $s_k = \sin\theta_k$ are constants.

To find the angle of rotation, the constants $\cos\theta_2$ and $\sin\theta_2$ in $\boldsymbol{P}_2$ are chosen such that $a_{21}^{(2)} = 0$ in the product

$$\boldsymbol{P}_2\boldsymbol{A}^{(1)} = \boldsymbol{P}_2\boldsymbol{A} = \begin{bmatrix} c_2 & s_2 & 0 & \ldots \\ -s_2 & c_2 & 0 & \ldots \\ 0 & 0 & 1 & \ldots \\ \vdots & \vdots & \vdots & \ldots \end{bmatrix}\begin{bmatrix} a_{11} & a_{12} & a_{13} & \ldots \\ a_{21} & a_{22} & a_{23} & \ldots \\ \ldots & \ldots & \ldots & \ldots \\ \vdots & \vdots & \vdots & \ldots \end{bmatrix} \tag{8.48}$$

Therefore, from Equation 8.47, we have

$$a_{21}^{(2)} = -s_2a_{11} + c_2a_{21} = -\sin\theta_2 a_{11} + \cos\theta_2 a_{21} = 0 \tag{8.49}$$

Thus,

$$\tan\theta_2 = \frac{s_2}{c_2} = \frac{a_{21}}{a_{11}} \tag{8.50}$$

So,

$$\cos\theta_2 = \frac{1}{\sqrt{1+\tan^2\theta_2}} = \frac{1}{\sqrt{1+\left(\dfrac{a_{21}}{a_{11}}\right)^2}} \tag{8.51}$$

and

$$\sin\theta_2 = \frac{\tan\theta_2}{\sqrt{1+\tan^2\theta_2}} = \frac{a_{21}/a_{11}}{\sqrt{1+\left(\dfrac{a_{21}}{a_{11}}\right)^2}} \tag{8.52}$$

Using similar arguments, we can successively determine θ_3, θ_4, and so on.

The other part of the QR factorization is the matrix

$$\boldsymbol{Q}^{(1)} = (\boldsymbol{P}_n\boldsymbol{P}_{n-1}\ldots\boldsymbol{P}_3\boldsymbol{P}_2)^{-1} = \boldsymbol{P}_2^T\boldsymbol{P}_3^T\ldots\boldsymbol{P}_{n-1}^T\boldsymbol{P}_n^T \tag{8.53}$$

since the rotational matrices $\boldsymbol{P}_k$, $k = 2,3,\ldots,n$, are orthogonal matrices.

Moreover, the matrix $\boldsymbol{Q}^{(1)}$ is orthogonal because

$$\left(\boldsymbol{Q}^{(1)}\right)^T\boldsymbol{Q}^{(1)} = \left(\boldsymbol{P}_2^T\boldsymbol{P}_3^T\ldots\boldsymbol{P}_{n-1}^T\boldsymbol{P}_n^T\right)^T\left(\boldsymbol{P}_2^T\boldsymbol{P}_3^T\ldots\boldsymbol{P}_{n-1}^T\boldsymbol{P}_n^T\right) = (\boldsymbol{P}_n\boldsymbol{P}_{n-1}\ldots\boldsymbol{P}_3\boldsymbol{P}_2)\left(\boldsymbol{P}_2^T\boldsymbol{P}_3^T\ldots\boldsymbol{P}_{n-1}^T\boldsymbol{P}_n^T\right) = \boldsymbol{I} \tag{8.54}$$

This is QR factorization of $\boldsymbol{A}^{(1)}$. Then using Equation 8.43, we get

$$\boldsymbol{A}^{(2)} = \boldsymbol{R}^{(1)}\boldsymbol{Q}^{(1)} = \boldsymbol{R}^{(1)}\boldsymbol{P}_2^T\boldsymbol{P}_3^T\ldots\boldsymbol{P}_{n-1}^T\boldsymbol{P}_n^T \tag{8.55}$$

We do not need $\boldsymbol{Q}^{(1)}$ explicitly, but to get $\boldsymbol{A}^{(2)}$ from Equation 8.55, we first compute $\boldsymbol{R}^{(1)}\boldsymbol{P}_2^T$, then $\left(\boldsymbol{R}^{(1)}\boldsymbol{P}_2^T\right)\boldsymbol{P}_3^T$, and so on. This process is repeated to construct $\boldsymbol{A}^{(3)}$, $\boldsymbol{A}^{(4)}$, … until satisfactory convergence is obtained.

8.4.1 Algorithm for the *QR* Method

Input: Enter the order of the matrix n, matrix $\boldsymbol{A} = \boldsymbol{A}_1^{(1)} = [a_{i,j}^{(1)}]_{n\times n}$, and the number of iterations m.

Output: The eigenvalues of matrix $\boldsymbol{A}$.

Step 1: for $i = 1(1)m$ do

for $k = 2(1)n$ do

$$c_k = \frac{1}{\sqrt{1+\left(\dfrac{a_{k,k-1}^{(k-1)}}{a_{k-1,k-1}^{(k-1)}}\right)^2}};$$

$$s_k = \frac{a_{k,k-1}^{(k-1)}}{a_{k-1,k-1}^{(k-1)}}c_k;$$

Assign

$$a_{k-1,k-1}^{(k)} = c_k;$$

$$a_{k-1,k}^{(k)} = s_k;$$

$$a_{k,k-1}^{(k)} = -s_k;$$

$$a_{k,k}^{(k)} = c_k;$$

for $i = 1,2,\ldots,k-2,k+1,\ldots,n$

for $j = 1,2,\ldots,k-2,k+1,\ldots,n$

If $(i == j)$, then $a_{i,j}^{(k)} = 1$, else $a_{i,j}^{(k)} = 0$;

end.

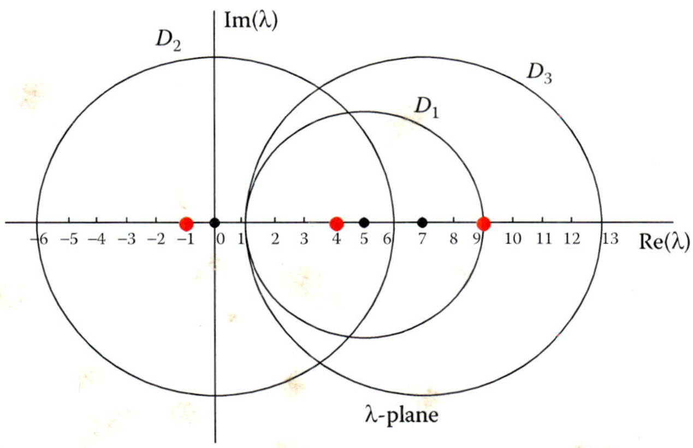

FIGURE 8.1 Gerschgorin's discs containing eigenvalues.

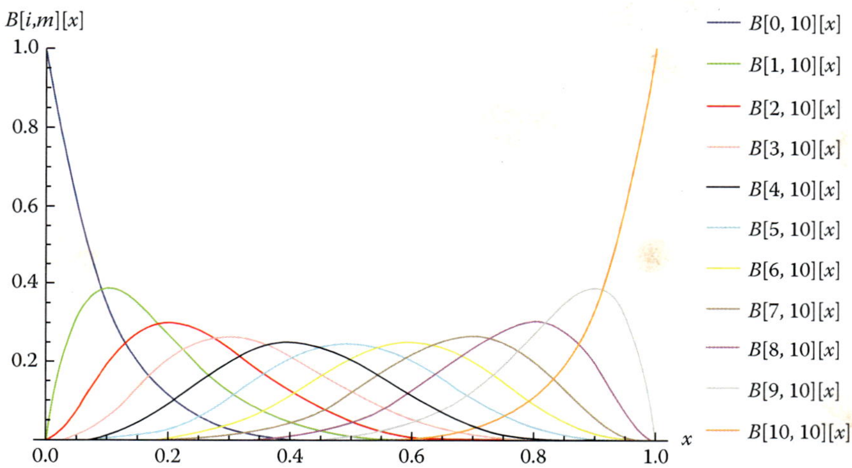

FIGURE 9.1 Graph of 10-degree Bernstein polynomials over [0, 1].

end.

$p_k = \left[a_{i,j}^{(k)}\right]_{n\times n}, \quad i,j = 1,2,\ldots,n;$

$A_i^{(k)} = p_k.A_i^{(k-1)};$

end.

$R^{(i)} = A_i^{(n)};$

$Q^{(i)} = I;$

for $j = 2(1)n$ do

$Q^{(i)} = Q^{(i)}.p_j^{\;T};$

end.

$A_{i+1} = R^{(i)}.Q^{(i)};$

$A_{i+1}^{(1)} = \left[a_{i,j}^{(1)}\right]_{n\times n} = A_{i+1};$

end.

Step 2: Print A_{i+1};

Step 3: Stop. ■

*MATHEMATICA® Program for Finding Eigenvalue by *QR* Method (Chapter 8, Example 8.3)*

```
A[1]={{7,0.1,0},{0.1,4,0.1},{0,0.1,1}};
For[i=1,i<=5,i++,
  Print[Step[i]];
  c2 = 1/Sqrt[1 + (A[i][[2,1]]/A[i][[1,1]])^2];Print["c2=",c2];
  s2=A[i][[2,1]]/A[i][[1,1]]*c2;
  Print["s2=",s2];
  p2={{c2,s2,0},{-s2,c2,0},{0,0,1}};
  A1=p2.A[i];
  Print[MatrixForm[A1]];
  c3 = 1/Sqrt[1 + (A1[[3,2]]/A1[[2,2]])^2];
  Print["c3=",c3];
  s3=A1[[3,2]]/A1[[2,2]]*c3;
  Print["s3=",s3];
  p3={{1,0,0},{0,c3,s3},{0,-s3,c3}};
  R[i]=p3.A1;
  Print[MatrixForm[R[i]]];
  Q[i]=Transpose[p3.p2];
  Print[MatrixForm[Q[i]]];
  A[i+1]=R[i].Q[i];
  Print[MatrixForm[A[i+1]]]];
```

Output:

```
Step [1]
c2 = 0.999898
s2 = 0.0142843
```

$$\begin{pmatrix} 7.00071 & 0.157127 & 0.00142843 \\ 1.38778\times 10^{-17} & 3.99816 & 0.0999898 \\ 0. & 0.1 & 1. \end{pmatrix}$$

```
c3 = 0.999687
s3 = 0.0250037
```

$$\begin{pmatrix} 7.00071 & 0.157127 & 0.00142843 \\ 1.38734\times 10^{-17} & 3.99941 & 0.124962 \\ -3.46996\times 10^{-19} & -1.38778\times 10^{-17} & 0.997187 \end{pmatrix}$$

$$\begin{pmatrix} 0.999898 & -0.0142798 & 0.000357159 \\ 0.0142843 & 0.999585 & -0.0250011 \\ 0. & 0.0250037 & 0.999687 \end{pmatrix}$$

$$\begin{pmatrix} 7.00224 & 0.0571287 & 0. \\ 0.0571287 & 4.00088 & 0.024933 \\ -5.45194\times 10^{-19} & 0.0249333 & 0.996875 \end{pmatrix}$$

```
Step [2]
c2 = 0.999967
s2 = 0.00815835
```

$$\begin{pmatrix} 7.00248 & 0.0897673 & 0.000203415 \\ 0. & 4.00028 & 0.0249325 \\ -5.45194\times 10^{-19} & 0.0249333 & 0.996875 \end{pmatrix}$$

```
c3 = 0.999981
s3 = 0.00623278
```

$$\begin{pmatrix} 7.00248 & 0.0897673 & 0.000203415 \\ -3.39807\times 10^{-21} & 4.00036 & 0.0311453 \\ -5.45183\times 10^{-19} & 0. & 0.996701 \end{pmatrix}$$

$$\begin{pmatrix} 0.999967 & -0.00815819 & 0.0000508492 \\ 0.00815835 & 0.999947 & -0.00623257 \\ 0. & 0.00623278 & 0.999981 \end{pmatrix}$$

$$\begin{pmatrix} 7.00298 & 0.0326363 & 3.25261\times 10^{-19} \\ 0.0326363 & 4.00034 & 0.00621221 \\ -5.45165\times 10^{-19} & 0.00621221 & 0.996681 \end{pmatrix}$$

```
Step [3]
c2 = 0.999989
s2 = 0.0046603
```

$$\begin{pmatrix} 7.00305 & 0.0512788 & 0.0000289508 \\ 0. & 4.00015 & 0.00621214 \\ -5.45165\times 10^{-19} & 0.00621221 & 0.996681 \end{pmatrix}$$

```
c3 = 0.999999
s3 = 0.00155299
```

$$\begin{pmatrix} 7.00305 & 0.0512788 & 0.0000289508 \\ -8.46639\times 10^{-22} & 4.00015 & 0.00775998 \\ -5.45165\times 10^{-19} & 0. & 0.996671 \end{pmatrix}$$

$$\begin{pmatrix} 0.999989 & -0.00466029 & 7.23742 \times 10^{-6} \\ 0.0046603 & 0.999988 & -0.00155298 \\ 0. & 0.00155299 & 0.999999 \end{pmatrix}$$

$$\begin{pmatrix} 7.00322 & 0.0186419 & 4.64174 \times 10^{-19} \\ 0.0186419 & 4.00011 & 0.00154782 \\ -5.45159 \times 10^{-19} & 0.00154782 & 0.996669 \end{pmatrix}$$

```
Step [4]
c2 = 0.999996
s2 = 0.0026619
```

$$\begin{pmatrix} 7.00324 & 0.0292897 & 4.12015 \times 10^{-6} \\ 0. & 4.00005 & 0.00154782 \\ -5.45159 \times 10^{-19} & 0.00154782 & 0.996669 \end{pmatrix}$$

```
c3 = 1.
s3 = 0.000386951
```

$$\begin{pmatrix} 7.00324 & 0.0292897 & 4.12015 \times 10^{-6} \\ -2.1095 \times 10^{-22} & 4.00005 & 0.00193348 \\ -5.45159 \times 10^{-19} & 0. & 0.996669 \end{pmatrix}$$

$$\begin{pmatrix} 0.999996 & -0.0026619 & 1.03002 \times 10^{-6} \\ 0.0026619 & 0.999996 & -0.00038695 \\ 0. & 0.000386951 & 1. \end{pmatrix}$$

$$\begin{pmatrix} 7.00329 & 0.0106477 & 5.31937 \times 10^{-19} \\ 0.0106477 & 4.00004 & 0.000385662 \\ -5.45157 \times 10^{-19} & 0.000385662 & 0.996669 \end{pmatrix}$$

```
Step [5]
c2 = 0.999999
s2 = 0.00152039
```

$$\begin{pmatrix} 7.0033 & 0.0167293 & 5.86355 \times 10^{-7} \\ -1.73472 \times 10^{-18} & 4.00002 & 0.000385661 \\ -5.45157 \times 10^{-19} & 0.000385662 & 0.996669 \end{pmatrix}$$

```
c3 = 1.
s3 = 0.000096415
```

$$\begin{pmatrix} 7.0033 & 0.0167293 & 5.86355 \times 10^{-7} \\ -1.73478 \times 10^{-18} & 4.00002 & 0.000481755 \\ -5.44989 \times 10^{-19} & 0. & 0.996669 \end{pmatrix}$$

$$\begin{pmatrix} 0.999999 & -0.00152039 & 1.46588\times 10^{-7} \\ 0.00152039 & 0.999999 & -0.0000964149 \\ 0. & 0.000096415 & 1. \end{pmatrix}$$

$$\begin{pmatrix} 7.00332 & 0.00608157 & 5.67936\times 10^{-19} \\ 0.00608157 & 4.00001 & 0.0000960938 \\ -5.44989\times 10^{-19} & 0.0000960938 & 0.996669 \end{pmatrix}$$

Example 8.3

Apply the QR method to find all the eigenvalues of the following matrix

$$\boldsymbol{A} = \begin{bmatrix} 7 & 0.1 & 0 \\ 0.1 & 4 & 0.1 \\ 0 & 0.1 & 1 \end{bmatrix}.$$

Solution:

First iteration:
Let $\boldsymbol{A}^{(1)} = \boldsymbol{A}$ be the given matrix and $\boldsymbol{P}_2$ be the rotation matrix.
To create a zero in (2,1) position of $\boldsymbol{P}_2\boldsymbol{A}$, we take

$$\boldsymbol{P}_2 = \begin{bmatrix} \boldsymbol{c}_2 & \boldsymbol{s}_2 & 0 \\ -\boldsymbol{s}_2 & \boldsymbol{c}_2 & 0 \\ 0 & 0 & 1 \end{bmatrix} \text{with } \boldsymbol{c}_2 = \cos\theta_2 = 0.999898 \text{ and } \boldsymbol{s}_2 = \sin\theta_2 = 0.0142843$$

yielding

$$\boldsymbol{P}_2\boldsymbol{A} = \begin{bmatrix} 7.00071 & 0.157127 & 0.00142843 \\ 0 & 3.99816 & 0.0999898 \\ 0 & 0.1 & 1 \end{bmatrix}$$

To create a zero in (3,2) position of $\boldsymbol{P}_3\boldsymbol{P}_2\boldsymbol{A}$, we take

$$\boldsymbol{P}_3 = \begin{bmatrix} 1 & 0 & 0 \\ 0 & \boldsymbol{c}_3 & \boldsymbol{s}_3 \\ 0 & -\boldsymbol{s}_3 & \boldsymbol{c}_3 \end{bmatrix} \text{with } \boldsymbol{c}_3 = \cos\theta_3 = 0.999687 \text{ and } \boldsymbol{s}_3 = \sin\theta_3 = 0.0250037$$

yielding

$$\boldsymbol{R}^{(1)} = \boldsymbol{P}_3\boldsymbol{P}_2\boldsymbol{A} = \begin{bmatrix} 7.00071 & 0.157127 & 0.00142843 \\ 0 & 3.99941 & 0.124962 \\ 0 & 0 & 0.997187 \end{bmatrix}$$

Again,

$$Q^{(1)} = P_2^T P_3^T$$

Therefore,

$$A^{(2)} = R^{(1)} P_2^T P_3^T = \begin{bmatrix} 7.00224 & 0.0571287 & 0 \\ 0.0571286 & 4.00088 & 0.0249332 \\ 0 & 0.0249332 & 0.996875 \end{bmatrix}$$

Second iteration:
To create a zero in (2,1) position of $P_2 A^{(2)}$, we take

$$P_2 = \begin{bmatrix} c_2 & s_2 & 0 \\ -s_2 & c_2 & 0 \\ 0 & 0 & 1 \end{bmatrix} \text{with } c_2 = \cos\theta_2 = 0.999967 \text{ and } s_2 = \sin\theta_2 = 0.00815833$$

yielding

$$P_2 A^{(2)} = \begin{bmatrix} 7.00248 & 0.0897673 & 0.000203413 \\ 0 & 4.00028 & 0.0249324 \\ 0 & 0.0249332 & 0.996875 \end{bmatrix}$$

To create a zero in (3,2) position of $P_3 P_2 A^{(2)}$, we take

$$P_3 = \begin{bmatrix} 1 & 0 & 0 \\ 0 & c_3 & s_3 \\ 0 & -s_3 & c_3 \end{bmatrix} \text{with } c_3 = \cos\theta_3 = 0.999981 \text{ and } s_3 = \sin\theta_3 = 0.00623274$$

yielding

$$R^{(2)} = P_3 P_2 A^{(2)} = \begin{bmatrix} 7.00248 & 0.0897673 & 0.000203413 \\ 0 & 4.00036 & 0.0311452 \\ 0 & 0 & 0.996701 \end{bmatrix}$$

Again,

$$Q^{(2)} = P_2^T P_3^T$$

Therefore,

$$A^{(3)} = R^{(2)} P_2^T P_3^T = \begin{bmatrix} 7.00297 & 0.0326365 & 0 \\ 0.0326364 & 4.00034 & 0.00621219 \\ 0 & 0.00621218 & 0.996681 \end{bmatrix}$$

Similarly, we obtain the subsequent matrices in the next successive iterations

$$A^{(4)} = \begin{bmatrix} 7.00321 & 0.018642 & 0 \\ 0.0186419 & 4.00011 & 0.00154782 \\ 0 & 0.00154781 & 0.996669 \end{bmatrix}$$

$$A^{(5)} = \begin{bmatrix} 7.00328 & 0.0106478 & 0 \\ 0.0106477 & 4.00003 & 0.000385667 \\ 0 & 0.000385657 & 0.996668 \end{bmatrix}$$

So, the element in the (3,3) position converges to real eigenvalue 0.9967 correct to four decimal places. Also the 2×2 submatrices

$$\begin{bmatrix} 7.00321 & 0.018642 \\ 0.0186419 & 4.00011 \end{bmatrix}$$

and

$$\begin{bmatrix} 7.00328 & 0.0106478 \\ 0.0106477 & 4.00003 \end{bmatrix}$$

have eigenvalues given by 7.00332, 3.99999, and 7.00333, 3.99999, respectively. Thus, the eigenvalues converge to 7.0033, 4, respectively.
Therefore, the required eigenvalues of the given matrix is 7.0033, 4 and 0.9967 correct to four decimal places.

Note: We can thus realize the purpose of applying Householder's triangularization before the *QR*-factorization method. It substantially reduces the computational overhead in each *QR*-factorization, in particular if $\boldsymbol{A}$ is large.

The acceleration of the convergence and further reduction of the computational overhead can be achieved by spectral shift in which $\boldsymbol{A}^{(i)} - \sigma_i \boldsymbol{I}$, $i = 2, 3, \ldots, n$, is taken at each iteration instead of $\boldsymbol{A}^{(i)}$, where the constant σ_i is selected close to an eigenvalue of A.

8.4.2 The *QR* Method with Shift

The *QR* algorithm is generally applied with a shift of origin for the eigenvalues in order to accelerate the speed of convergence. A constant σ is selected close to an eigenvalue of A. Some modification of *QR* method modifies the factorization similar to *QR* method by choosing Q_i and R_i such that

$$A^{(i)} - \sigma I = Q^{(i)} R^{(i)}$$

and, correspondingly, the matrix $A^{(i+1)}$ can be obtained as

$$A^{(i+1)} = R^{(i)} Q^{(i)} + \sigma I$$

To be more specific on the choice of shifts σ, we consider only a symmetric tridiagonal matrix A. There are two methods by which σ is chosen:

1. First choose the submatrix at ith step as

$$\begin{bmatrix} a_{n-1,n-1}^{(i)} & a_{n-1,n}^{(i)} \\ a_{n,n-1}^{(i)} & a_{n,n}^{(i)} \end{bmatrix}$$

   and choose σ same as the eigenvalues of this submatrix which is more nearer to $a_{n,n}^{(i)}$.

2. Second, directly choose σ as $a_{n,n}^{(i)}$.

The first strategy is more preferred, but in either case the matrices $A^{(i)}$ converge to a block diagonal matrix at a much more rapid rate than with the original QR method.

MATHEMATICA® Program for Finding Eigenvalues by Shifted *QR* Method (Chapter 8, Example 8.4)

```
AA[1]={{7,0.1,0},{0.1,4,0.1},{0,0.1,1}};
For[i=1,i<=5,i++,
  Print[Step[i]];
  e=Eigenvalues[{{AA[i][[2,2]],AA[i][[2,3]]},{AA[i][[3,2]],AA[i]
[[3,3]]}}];
  Print[e];
  mu=e[[2]];
  A[i]=AA[i]-mu*IdentityMatrix[3];
  Print[MatrixForm[A[i]]];
  c2=1/Sqrt[1+(A[i][[2,1]]/A[i][[1,1]])^2];Print["c2=",c2];
  s2=A[i][[2,1]]/A[i][[1,1]]*c2;
  Print["s2=",s2];
  p2={{c2,s2,0},{-s2,c2,0},{0,0,1}};
  A1=p2.A[i];
  Print[MatrixForm[A1]];
  c3 = 1/Sqrt[1+(A1[[3,2]]/A1[[2,2]])^2];
  Print["c3=",c3];
  s3=A1[[3,2]]/A1[[2,2]]*c3;
  Print["s3=",s3];
  p3={{1,0,0},{0,c3,s3},{0,-s3,c3}};
  R[i]=p3.A1;
  Print[MatrixForm[R[i]]];
  Q[i]=Transpose[p3.p2];
  Print[MatrixForm[Q[i]]];
  AA[i+1]=R[i].Q[i]+mu*IdentityMatrix[3];
  Print[MatrixForm[AA[i+1]]];
  Print["......................................................"]];
```

Output:

Step[1]
{4.00333,0.99667}

$$\begin{pmatrix} 6.00333 & 0.1 & 0. \\ 0.1 & 3.00333 & 0.1 \\ 0. & 0.1 & 0.00332964 \end{pmatrix}$$

c2 = 0.999861
s2 = 0.0166551

$$\begin{pmatrix} 6.00416 & 0.150007 & 0.00166551 \\ -1.38778\times10^{-17} & 3.00125 & 0.0999861 \\ 0. & 0.1 & 0.00332964 \end{pmatrix}$$

c3 = 0.999445
s3 = 0.033301

$$\begin{pmatrix} 6.00416 & 0.150007 & 0.00166551 \\ -1.38701\times10^{-17} & 3.00291 & 0.100042 \\ 4.62144\times10^{-19} & -1.38778\times10^{-17} & -1.84672\times10^{-6} \end{pmatrix}$$

$$\begin{pmatrix} 0.999861 & -0.0166459 & 0.000554632 \\ 0.0166551 & 0.999307 & -0.0332964 \\ 0. & 0.033301 & 0.999445 \end{pmatrix}$$

$$\begin{pmatrix} 7.0025 & 0.0500139 & 0. \\ 0.0500139 & 4.00083 & -6.14977\times10^{-8} \\ 2.30944\times10^{-19} & -6.14977\times10^{-8} & 0.996669 \end{pmatrix}$$

....

Step[2]
{4.00083,0.996669}

$$\begin{pmatrix} 6.00583 & 0.0500139 & 0. \\ 0.0500139 & 3.00416 & -6.14977\times10^{-8} \\ 2.30944\times10^{-19} & -6.14977\times10^{-8} & 1.22125\times10^{-15} \end{pmatrix}$$

c2 = 0.999965
s2 = 0.00832726

$$\begin{pmatrix} 6.00604 & 0.0750286 & -5.12108\times10^{-10} \\ 0. & 3.00364 & -6.14956\times10^{-8} \\ 2.30944\times10^{-19} & -6.14977\times10^{-8} & 1.22125\times10^{-15} \end{pmatrix}$$

c3 = 1.
s3 = -2.04744×10^{-8}

$$\begin{pmatrix} 6.00604 & 0.0750286 & -5.12108\times10^{-10} \\ -4.72843\times10^{-27} & 3.00364 & -6.14956\times10^{-8} \\ 2.30944\times10^{-19} & 1.32349\times10^{-23} & -3.78383\times10^{-17} \end{pmatrix}$$

$$\begin{pmatrix} 0.999965 & -0.00832726 & -1.70495\times10^{-10} \\ 0.00832726 & 0.999965 & 2.04737\times10^{-8} \\ 0. & -2.04744\times10^{-8} & 1. \end{pmatrix}$$

$$\begin{pmatrix} 7.00312 & 0.0250121 & 1.52312\times10^{-19} \\ 0.0250121 & 4.00021 & 1.82901\times10^{-17} \\ 2.30936\times10^{-19} & -1.90912\times10^{-21} & 0.996669 \end{pmatrix}$$

....

Step[3]
{4.00021,0.996669}

$$\begin{pmatrix} 6.00645 & 0.0250121 & 1.52312\times10^{-19} \\ 0.0250121 & 3.00354 & 1.82901\times10^{-17} \\ 2.30936\times10^{-19} & -1.90912\times10^{-21} & 0. \end{pmatrix}$$

c2 = 0.999991
s2 = 0.00416417

$$\begin{pmatrix} 6.00651 & 0.0375192 & 2.28474\times10^{-19} \\ 0. & 3.00341 & 1.82893\times10^{-17} \\ 2.30936\times10^{-19} & -1.90915\times10^{-21} & 0. \end{pmatrix}$$

c3 = 1.
s3 = -6.35651×10^{-22}

$$\begin{pmatrix} 6.0065 & 0.0375192 & 2.28474\times10^{-19} \\ -1.49795\times10^{-40} & 3.00341 & 1.82893\times10^{-17} \\ 2.30936\times10^{-19} & 0. & 1.16257\times10^{-28} \end{pmatrix}$$

$$\begin{pmatrix} 0.999991 & -0.00416417 & -2.64696\times10^{-24} \\ 0.00416417 & 0.999991 & 6.35646\times10^{-22} \\ 0. & -6.35651\times10^{-22} & 1. \end{pmatrix}$$

$$\begin{pmatrix} 7.00328 & 0.0125067 & 2.28482\times10^{-19} \\ 0.0125067 & 4.00005 & 1.82913\times10^{-17} \\ 2.30934\times10^{-19} & -9.61658\times10^{-22} & 0.996669 \end{pmatrix}$$

....

Step[4]
{4.00005,0.996669}

$$\begin{pmatrix} 6.00661 & 0.0125067 & 2.28482\times10^{-19} \\ 0.0125067 & 3.00338 & 1.82913\times10^{-17} \\ 2.30934\times10^{-19} & -9.61658\times10^{-22} & 0. \end{pmatrix}$$

c2 = 0.999998
s2 = 0.00208215

$$\begin{pmatrix} 6.00662 & 0.0187602 & 2.66567\times10^{-19} \\ 1.73472\times10^{-18} & 3.00335 & 1.82907\times10^{-17} \\ 2.3093410^{-19} & -9.61658\times10^{-22} & 0. \end{pmatrix}$$

c3 = 1.
s3 = -3.20195×10^{-22}

$$\begin{pmatrix} 6.00662 & 0.0187602 & 2.66567\times10^{-19} \\ 1.73472\times10^{-18} & 3.00335 & 1.82907\times10^{-17} \\ 2.30934\times10^{-19} & 0. & 5.8566\times10^{-39} \end{pmatrix}$$

$$\begin{pmatrix} 0.999998 & -0.00208215 & -6.66695\times10^{-25} \\ 0.00208215 & 0.999998 & 3.20194\times10^{-22} \\ 0. & -3.20195\times10^{-22} & 1. \end{pmatrix}$$

$$\begin{pmatrix} 7.00332 & 0.00625344 & 2.66569\times10^{-19} \\ 0.00625344 & 4.00001 & 1.82917\times10^{-17} \\ 2.30934\times10^{-19} & -4.8084\times10^{-22} & 0.996669 \end{pmatrix}$$

....

Step [5]
{4.00001,0.996669}

$$\begin{pmatrix} 6.00665 & 0.00625344 & 2.66569\times10^{-19} \\ 0.00625344 & 3.00334 & 1.82917\times10^{-17} \\ 2.30934\times10^{-19} & -4.8084\times10^{-22} & 0. \end{pmatrix}$$

c2 = 0.999999
s2 = 0.00104109

$$\begin{pmatrix} 6.00665 & 0.00938017 & 2.85612\times10^{-19} \\ -8.67362\times10^{-19} & 3.00334 & 1.82914\times10^{-17} \\ 2.30934\times10^{-19} & -4.8084\times10^{-22} & 0. \end{pmatrix}$$

c3 = 1.
s3 = -1.60102×10^{-22}

$$\begin{pmatrix} 6.00665 & 0.00938017 & 2.85612\times10^{-19} \\ -8.67362\times10^{-19} & 3.00334 & 1.82914\times10^{-17} \\ 2.30934\times10^{-19} & 0. & 2.92849\times10^{-39} \end{pmatrix}$$

$$\begin{pmatrix} 0.999999 & -0.00104109 & -1.6668\times10^{-25} \\ 0.00104109 & 0.999999 & 1.60102\times10^{-22} \\ 0. & -1.60102\times10^{-22} & 1. \end{pmatrix}$$

$$\begin{pmatrix} 7.00333 & 0.00312673 & 2.85612\times10^{-19} \\ 0.00312673 & 4. & 1.82919\times10^{-17} \\ 2.30933\times10^{-19} & -2.40422\times10^{-22} & 0.996669 \end{pmatrix}$$

....

Example 8.4

Apply the QR method with shifting to find all the eigenvalues of the following matrix

$$A = \begin{bmatrix} 7 & 0.1 & 0 \\ 0.1 & 4 & 0.1 \\ 0 & 0.1 & 1 \end{bmatrix}$$

Solution:

First iteration:
We determine the eigenvalues of submatrix

$$\begin{bmatrix} 4 & 0.1 \\ 0.1 & 1 \end{bmatrix} \text{ which are } \mu_1 = 4.00333 \text{ and } \mu_2 = 0.99667.$$

Here μ_2 is very nearer to 1. Take $\sigma = 0.99667$.

Let $A^{(1)} = A - \sigma I$ be the required matrix for QR decomposition and let P_2 be the rotation matrix. To create a zero in (2,1) position of $P_2A^{(1)}$, we take

$$P_2 = \begin{bmatrix} c_2 & s_2 & 0 \\ -s_2 & c_2 & 0 \\ 0 & 0 & 1 \end{bmatrix} \text{ with } c_2 = \cos\theta_2 = 0.999861 \text{ and } s_2 = \sin\theta_2 = 0.0166551$$

yielding

$$P_2A^{(1)} = \begin{bmatrix} 6.00416 & 0.150007 & 0.00166551 \\ 0 & 3.00125 & 0.0999861 \\ 0 & 0.1 & 0.00332964 \end{bmatrix}$$

To create a zero in (3,2) position of $P_3P_2A^{(1)}$, we take

$$P_3 = \begin{bmatrix} 1 & 0 & 0 \\ 0 & c_3 & s_3 \\ 0 & -s_3 & c_3 \end{bmatrix} \text{ with } c_3 = \cos\theta_3 = 0.999445 \text{ and } s_3 = \sin\theta_3 = 0.033301$$

yielding

$$R^{(1)} = P_3P_2A^{(1)} = \begin{bmatrix} 6.00416 & 0.150007 & 0.00166551 \\ 0 & 3.00291 & 0.100042 \\ 0 & 0 & 0 \end{bmatrix}$$

Again

$$Q^{(1)} = P_2^T P_3^T$$

Therefore,

$$A^{(2)} = R^{(1)}Q^{(1)} + \sigma I = \begin{bmatrix} 7.0025 & 0.0500139 & 0 \\ 0.0500139 & 4.00083 & 0 \\ 0 & 0 & 0.996669 \end{bmatrix}$$

Second iteration:
We determine the eigenvalues of submatrix

$$\begin{bmatrix} 4.00083 & 0 \\ 0 & 0.996669 \end{bmatrix} \text{ which are } \mu_1 = 4.00083 \text{ and } \mu_2 = 0.996669.$$

Here μ_2 is very nearer to 1. Take $\sigma = 0.996669$.

Let $A^{(1)} - \sigma I$ be the required matrix for QR decomposition, and let P_2 be the rotation matrix. To create a zero in (2,1) position of $P_2A^{(1)}$, we take

$$P_2 = \begin{bmatrix} c_2 & s_2 & 0 \\ -s_2 & c_2 & 0 \\ 0 & 0 & 1 \end{bmatrix} \text{ with } c_2 = \cos\theta_2 = 0.999965 \text{ and } s_2 = \sin\theta_2 = 0.00832726$$

yielding

$$P_2A^{(1)} = \begin{bmatrix} 6.00604 & 0.0750286 & 0 \\ 0 & 3.00364 & 0 \\ 0 & 0 & 0 \end{bmatrix}$$

To create a zero in (3,2) position of $P_3P_2A^{(1)}$, we take

$$P_3 = \begin{bmatrix} 1 & 0 & 0 \\ 0 & c_3 & s_3 \\ 0 & -s_3 & c_3 \end{bmatrix} \text{ with } c_3 = \cos\theta_3 = 1 \text{ and } s_3 = \sin\theta_3 = 0$$

yielding

$$R^{(2)} = P_3P_2A^{(2)} = \begin{bmatrix} 6.00604 & 0.0750286 & 0 \\ 0 & 3.00364 & 0 \\ 0 & 0 & 0 \end{bmatrix}$$

Again

$$Q^{(2)} = P_2^T P_3^T$$

Therefore,

$$A^{(3)} = R^{(2)}Q^{(2)} + \sigma I = \begin{bmatrix} 7.00312 & 0.0250121 & 0 \\ 0.0250121 & 4.00021 & 0 \\ 0 & 0 & 0.996669 \end{bmatrix}$$

Similarly,

$$A^{(4)} = \begin{bmatrix} 7.00328 & 0.0125067 & 0 \\ 0.0125067 & 4.00005 & 0 \\ 0 & 0 & 0.996669 \end{bmatrix}$$

Therefore, the required eigenvalues of the given matrix are 7.0033, 4 and 0.9967 correct to four decimal places. The acceleration of the convergence increases by shifted *QR* method.

8.5 POWER METHOD

The power method is an iterative technique for approximating the dominant eigenvalue of a matrix together with an associated eigenvector. This method applies to any $n \times n$ matrix $\boldsymbol{A}$ that has a dominant eigenvalue $\lambda = \max_{1 \le i \le n} |\lambda_i|$.

To apply the power method, we assume that the $n \times n$ matrix $\boldsymbol{A}$ has n eigenvalues $\lambda_1, \lambda_2, \ldots, \lambda_n$ with an associated linearly independent eigenvectors $\{\boldsymbol{v}_1, \boldsymbol{v}_2, \ldots, \boldsymbol{v}_n\}$.

Let $\boldsymbol{A}$ be an $n \times n$ matrix with eigenvalues satisfying

$$|\lambda_1| > |\lambda_2| > \cdots > |\lambda_n| > 0 \tag{8.56}$$

The largest eigenvalue λ_1 is called the dominant eigenvalue.

If $\boldsymbol{x}^{(0)}$ be any vector in $\boldsymbol{R}^n$, the fact that $\{\boldsymbol{v}_1, \boldsymbol{v}_2, \ldots, \boldsymbol{v}_n\}$ is linearly independent implies that there exists scalars $\alpha_1, \alpha_2, \ldots, \alpha_n$ such that $\boldsymbol{x}^{(0)}$ can be expressed as a linear combination of $\{\boldsymbol{v}_1, \boldsymbol{v}_2, \ldots, \boldsymbol{v}_n\}$. Then, we can write

$$\boldsymbol{x}^{(0)} = \alpha_1 \boldsymbol{v}_1 + \alpha_2 \boldsymbol{v}_2 + \cdots + \alpha_n \boldsymbol{v}_n \tag{8.57}$$

Now construct the sequence $\boldsymbol{x}^{(m)} = \boldsymbol{A}\boldsymbol{x}^{(m-1)}$, for $m \ge 1$.

Multiplying both sides of Equation 8.57 by $\boldsymbol{A}, \boldsymbol{A}^2, \ldots, \boldsymbol{A}^m, \ldots$ yields

$$\begin{aligned}
\boldsymbol{x}^{(1)} &= \boldsymbol{A}\boldsymbol{x}^{(0)} = \alpha_1\lambda_1\boldsymbol{v}_1 + \alpha_2\lambda_2\boldsymbol{v}_2 + \cdots + \alpha_n\lambda_n\boldsymbol{v}_n \\
\boldsymbol{x}^{(2)} &= \boldsymbol{A}^2\boldsymbol{x}^{(0)} = \boldsymbol{A}\boldsymbol{x}^{(1)} = \alpha_1\lambda_1^2\boldsymbol{v}_1 + \alpha_2\lambda_2^2\boldsymbol{v}_2 + \cdots + \alpha_n\lambda_n^2\boldsymbol{v}_n \\
&\vdots \\
\boldsymbol{x}^{(m)} &= \boldsymbol{A}^m\boldsymbol{x}^{(0)} = \boldsymbol{A}\boldsymbol{x}^{(m-1)} = \alpha_1\lambda_1^m\boldsymbol{v}_1 + \alpha_2\lambda_2^m\boldsymbol{v}_2 + \cdots + \alpha_n\lambda_n^m\boldsymbol{v}_n
\end{aligned} \tag{8.58}$$

Since, $|\lambda_i / \lambda_1| < 1, i = 2, 3, \ldots, n$.

$$\frac{\boldsymbol{x}^{(m)}}{\lambda_1^m} = \alpha_1\boldsymbol{v}_1 + \alpha_2\left(\frac{\lambda_2}{\lambda_1}\right)^m \boldsymbol{v}_2 + \cdots + \alpha_n\left(\frac{\lambda_n}{\lambda_1}\right)^m \boldsymbol{v}_n = \alpha_1\boldsymbol{v}_1 + O\left(\left(\frac{\lambda_2}{\lambda_1}\right)^m\right) \tag{8.59}$$

which gives

$$\lim_{m\to\infty} \frac{\boldsymbol{A}^m\boldsymbol{x}^{(0)}}{\lambda_1^m} = \lim_{m\to\infty} \frac{\boldsymbol{x}^{(m)}}{\lambda_1^m} = \alpha_1\boldsymbol{v}_1 \tag{8.60}$$

Hence, the sequence $\{\boldsymbol{x}^{(m)}/\lambda_1^m\}$ converges to an eigenvector associated with the dominant eigenvalue.

Moreover,

$$\frac{\boldsymbol{x}^{(0)^T}\boldsymbol{A}^m\boldsymbol{x}^{(0)}}{\boldsymbol{x}^{(0)^T}\boldsymbol{A}^{m-1}\boldsymbol{x}^{(0)}}=\lambda_1\frac{\eta_1+O\left((\lambda_2/\lambda_1)^m\right)}{\eta_1+O\left((\lambda_2/\lambda_1)^{m-1}\right)},\quad \text{where } \boldsymbol{x}^{(0)^T}\boldsymbol{v}_i=\eta_i, i=1,2,\ldots,n \tag{8.61}$$

This implies that

$$\lim_{m\to\infty}\frac{\boldsymbol{x}^{(0)^T}\boldsymbol{A}^m\boldsymbol{x}^{(0)}}{\boldsymbol{x}^{(0)^T}\boldsymbol{A}^{m-1}\boldsymbol{x}^{(0)}}=\lambda_1 \tag{8.62}$$

- *The power method implementation*: Let us choose the initial guess $\boldsymbol{x}^{(0)}$ such that $\left\|\boldsymbol{x}^{(0)}\right\|_\infty=1=x_{p_0}^{(0)}$.

  Let $\boldsymbol{y}^{(1)}=A\boldsymbol{x}^{(0)}$. We define $\lambda^{(1)}=y_{p_0}^{(1)}$. Then, we can write

$$\lambda^{(1)}=y_{p_0}^{(1)}=\frac{y_{p_0}^{(1)}}{x_{p_0}^{(0)}}=\frac{\alpha_1\lambda_1 v_{p_0}^{(1)}+\sum_{j=2}^{n}\alpha_j\lambda_j v_{p_0}^{(j)}}{\alpha_1 v_{p_0}^{(1)}+\sum_{j=2}^{n}\alpha_j v_{p_0}^{(j)}}=\lambda_1\frac{\alpha_1 v_{p_0}^{(1)}+\sum_{j=2}^{n}\alpha_j(\lambda_j/\lambda_1)v_{p_0}^{(j)}}{\alpha_1 v_{p_0}^{(1)}+\sum_{j=2}^{n}\alpha_j v_{p_0}^{(j)}}$$

  where $y_{p_0}^{(1)}$, $x_{p_0}^{(0)}$, and $v_{p_0}^{(j)}$, $j=1,2,\ldots,n$, are the p_0 components of $\boldsymbol{y}^{(1)}$, $\boldsymbol{x}^{(0)}$, and $\boldsymbol{v}_j$, $j=1,2,\ldots,n$, respectively.

  Let p_1 be the smallest integer index such that

$$\left|y_{p_1}^{(1)}\right|=\left\|\boldsymbol{y}^{(1)}\right\|_\infty$$

  Next, we define

$$\boldsymbol{x}^{(1)}=\frac{\boldsymbol{y}^{(1)}}{y_{p_1}^{(1)}}=\frac{A\boldsymbol{x}^{(0)}}{y_{p_1}^{(1)}}$$

  Let

$$x_{p_1}^{(1)}=1=\left\|\boldsymbol{x}^{(1)}\right\|_\infty$$

  and

$$\boldsymbol{y}^{(2)}=A\boldsymbol{x}^{(1)}=\frac{A^2\boldsymbol{x}^{(0)}}{y_{p_1}^{(1)}}$$

  Now, we define

$$\lambda^{(2)}=y_{p_1}^{(2)}=\frac{y_{p_1}^{(2)}}{x_{p_1}^{(1)}}=\frac{\left(\alpha_1\lambda_1^2 v_{p_1}^{(1)}+\sum_{j=2}^{n}\alpha_j\lambda_j^2 v_{p_1}^{(j)}\right)/y_{p_1}^{(1)}}{\left(\alpha_1\lambda_1 v_{p_1}^{(1)}+\sum_{j=2}^{n}\alpha_j\lambda_j v_{p_1}^{(j)}\right)/y_{p_1}^{(1)}}=\lambda_1\frac{\alpha_1 v_{p_1}^{(1)}+\sum_{j=2}^{n}\alpha_j(\lambda_j/\lambda_1)^2 v_{p_1}^{(j)}}{\alpha_1 v_{p_1}^{(1)}+\sum_{j=2}^{n}\alpha_j(\lambda_j/\lambda_1)v_{p_1}^{(j)}}$$

  where $y_{p_1}^{(2)}$, $x_{p_1}^{(1)}$, and $v_{p_1}^{(j)}$, $j=1,2,\ldots,n$, are the p_1 components of $\boldsymbol{y}^{(2)}$, $\boldsymbol{x}^{(1)}$, and $\boldsymbol{v}_j$, $j=1,2,\ldots,n$, respectively.

Let p_2 be the smallest integer index such that

$$\left|y_{p_2}^{(2)}\right| = \left\|\boldsymbol{y}^{(2)}\right\|_\infty$$

Next, we define

$$\boldsymbol{x}^{(2)} = \frac{\boldsymbol{y}^{(2)}}{y_{p_2}^{(2)}} = \frac{A\boldsymbol{x}^{(1)}}{y_{p_2}^{(2)}} = \frac{A^2\boldsymbol{x}^{(0)}}{y_{p_2}^{(2)} y_{p_1}^{(1)}}$$

Proceeding in this manner, by the method of induction we can obtain the sequences of vectors $\{\boldsymbol{x}^{(m)}\}_0^\infty$, $\{\boldsymbol{y}^{(m)}\}_0^\infty$ and a sequence of scalars $\{\lambda_m\}_0^\infty$ given by

$$\lambda^{(m)} = y_{p_{m-1}}^{(m)} = \lambda_1 \frac{\alpha_1 v_{p_{m-1}}^{(1)} + \sum_{j=2}^{n} \alpha_j \left(\lambda_j/\lambda_1\right)^m v_{p_{m-1}}^{(j)}}{\alpha_1 v_{p_{m-1}}^{(1)} + \sum_{j=2}^{n} \alpha_j \left(\lambda_j/\lambda_1\right)^{m-1} v_{p_{m-1}}^{(j)}} \tag{8.63}$$

and

$$\boldsymbol{x}^{(m)} = \frac{\boldsymbol{y}^{(m)}}{y_{p_m}^{(m)}} = \frac{A^m \boldsymbol{x}^{(0)}}{\prod_{i=1}^{m} y_{p_i}^{(i)}} \tag{8.64}$$

where at the mth step, p_m is the smallest integer index for which

$$\left|y_{p_m}^{(m)}\right| = \left\|\boldsymbol{y}^{(m)}\right\|_\infty$$

From Equation 8.63, we see that

$$\lim_{m\to\infty} \lambda^{(m)} = \lambda_1$$

since $|\lambda_i/\lambda_1| < 1, i = 2,3,\ldots,n$, provided that initial guess vector $\boldsymbol{x}^{(0)}$ is chosen such that $\alpha_1 \neq 0$. Also, from Equation 8.64, it manifests that the sequence of vectors $\{\boldsymbol{x}^{(m)}\}_0^\infty$ converges to an eigenvector associated with the dominant eigenvalue λ_1 with l_∞ norm equal to one.

Note: Since the power method is an iterative scheme, a stopping condition is given by $\left|\lambda^{(m)} - \lambda^{(m-1)}\right| < \varepsilon, \varepsilon$ is the prescribed error tolerance, where $\lambda^{(m)}$ is the dominant eigenvalue approximation during the mth iteration, or $\left\|\boldsymbol{x}^{(m)} - \boldsymbol{x}^{(m-1)}\right\|_\infty < \varepsilon$, ε is the prescribed error tolerance.

Example 8.5

Determine the largest eigenvalue and the corresponding eigenvector of the following matrix:

$$A = \begin{bmatrix} 1 & 3 & -1 \\ 3 & 2 & 4 \\ -1 & 4 & 10 \end{bmatrix}$$

Solution:

Let us choose the initial vector $x^{(0)} = \begin{bmatrix} -1 \\ 0 \\ 1 \end{bmatrix}$ such that $\|x^{(0)}\|_\infty = 1$.

Then,

$$y^{(1)} = Ax^{(0)} = \begin{bmatrix} 1 & 3 & -1 \\ 3 & 2 & 4 \\ -1 & 4 & 10 \end{bmatrix}\begin{bmatrix} -1 \\ 0 \\ 1 \end{bmatrix} = \begin{bmatrix} -2 \\ 1 \\ 11 \end{bmatrix} = 11\begin{bmatrix} -0.181818 \\ 0.0909091 \\ 1 \end{bmatrix} = 11x^{(1)}$$

Next,

$$y^{(2)} = Ax^{(1)} = \begin{bmatrix} 1 & 3 & -1 \\ 3 & 2 & 4 \\ -1 & 4 & 10 \end{bmatrix}\begin{bmatrix} -0.181818 \\ 0.0909091 \\ 1 \end{bmatrix} = \begin{bmatrix} -0.909091 \\ 3.63636 \\ 10.5455 \end{bmatrix} = 10.5455\begin{bmatrix} -0.0862069 \\ 0.344828 \\ 1 \end{bmatrix} = 10.5455x^{(2)}$$

In this similar manner, we can obtain

$$y^{(3)} = Ax^{(2)} = \begin{bmatrix} -0.0517241 \\ 4.43103 \\ 11.4655 \end{bmatrix} = 11.4655\begin{bmatrix} -0.00451128 \\ 0.386466 \\ 1 \end{bmatrix} = 11.4655x^{(3)}$$

$$y^{(4)} = Ax^{(3)} = \begin{bmatrix} 0.154887 \\ 4.7594 \\ 11.5504 \end{bmatrix} = 11.5504\begin{bmatrix} 0.0134097 \\ 0.412056 \\ 1 \end{bmatrix} = 11.5504x^{(4)}$$

$$y^{(5)} = Ax^{(4)} = \begin{bmatrix} 0.249577 \\ 4.86434 \\ 11.6348 \end{bmatrix} = 11.6348\begin{bmatrix} 0.0214509 \\ 0.418085 \\ 1 \end{bmatrix} = 11.6348x^{(5)}$$

$$y^{(6)} = Ax^{(5)} = \begin{bmatrix} 0.275706 \\ 4.90052 \\ 11.6509 \end{bmatrix} = 11.6509\begin{bmatrix} 0.0236639 \\ 0.420614 \\ 1 \end{bmatrix} = 11.6509x^{(6)}$$

$$y^{(7)} = Ax^{(6)} = \begin{bmatrix} 0.285505 \\ 4.91222 \\ 11.6588 \end{bmatrix} = 11.6588\begin{bmatrix} 0.0244884 \\ 0.421332 \\ 1 \end{bmatrix} = 11.6588x^{(7)}$$

$$y^{(8)} = Ax^{(7)} = \begin{bmatrix} 0.288484 \\ 4.91613 \\ 11.6608 \end{bmatrix} = 11.6608\begin{bmatrix} 0.0247395 \\ 0.421593 \\ 1 \end{bmatrix} = 11.6608x^{(8)}$$

$$y^{(9)} = Ax^{(8)} = \begin{bmatrix} 0.289519 \\ 4.9174 \\ 11.6616 \end{bmatrix} = 11.6616\begin{bmatrix} 0.0248266 \\ 0.421674 \\ 1 \end{bmatrix} = 11.6616x^{(9)}$$

$$y^{(10)} = Ax^{(9)} = \begin{bmatrix} 0.289848 \\ 4.91783 \\ 11.6619 \end{bmatrix} = 11.6619\begin{bmatrix} 0.0248543 \\ 0.421701 \\ 1 \end{bmatrix} = 11.6619x^{(10)}$$

$$y^{(11)} = Ax^{(10)} = \begin{bmatrix} 0.289959 \\ 4.91797 \\ 11.662 \end{bmatrix} = 11.662 \begin{bmatrix} 0.0248637 \\ 0.42171 \\ 1 \end{bmatrix} = 11.662x^{(11)}$$

$$y^{(12)} = Ax^{(11)} = \begin{bmatrix} 0.289995 \\ 4.91801 \\ 11.662 \end{bmatrix} = 11.662 \begin{bmatrix} 0.0248667 \\ 0.421713 \\ 1 \end{bmatrix} = 11.662x^{(12)}$$

and so on.

The above results can be summarized in the following computation table:

Step m	$x^{(m-1)^T}$	$y^{(m)^T}$	$\lVert y^{(m)^T} \rVert_\infty$
1	$(-1, 0, 1)$	$(-2, 1, 11)$	11
2	$(-0.0862069, 0.344828, 1)$	$(-0.909091, 3.63636, 10.5455)$	10.5455
3	$(-0.00451128, 0.386466, 1)$	$(-0.0517241, 4.43103, 11.4655)$	11.4655
4	$(0.0134097, 0.412056, 1)$	$(0.154887, 4.7594, 11.5504)$	11.5504
5	$(0.0214509, 0.418085, 1)$	$(0.249577, 4.86434, 11.6348)$	11.6348
6	$(0.0236639, 0.420614, 1)$	$(0.275706, 4.90052, 11.6509)$	11.6509
7	$(0.0244884, 0.421332, 1)$	$(0.285505, 4.91222, 11.6588)$	11.6588
8	$(0.0247395, 0.421593, 1)$	$(0.288484, 4.91613, 11.6608)$	11.6608
9	$(0.0248266, 0.421674, 1)$	$(0.289519, 4.9174, 11.6616)$	11.6616
10	$(0.0248543, 0.421701, 1)$	$(0.289848, 4.91783, 11.6619)$	11.6619
11	$(0.0248637, 0.42171, 1)$	$(0.289959, 4.91797, 11.662)$	11.662
12	$(0.0248667, 0.421713, 1)$	$(0.289995, 4.91801, 11.662)$	11.662

Since, $x^{(10)} \approx x^{(11)}$ correct to five decimal places, the approximate largest eigenvalue is 11.66 correct to four significant figures, and the corresponding approximate l_∞-unit eigenvector for the eigenvalue 11.66 is $(0.025, 0.422, 1)^T$.

- *Symmetric power method*: To approximate the dominant eigenvalue and an associated eigenvector of the $n \times n$ symmetric matrix A, given a nonzero vector x, we have the following theorem.

Theorem 8.1

Let, A be an $n \times n$ real symmetric matrix. Let, $x \neq 0$ be any real vector with n components. Then the quotient

$$q = \frac{x^T A x}{x^T x} \quad \text{(Rayleigh quotient)} \tag{8.65}$$

is an approximation for an eigenvalue λ of A, and if we set $q = \lambda - \varepsilon$, so that ε is the error of q, then the error bound is given by

$$|\varepsilon| \le \delta = \sqrt{\frac{(Ax)^T Ax}{x^T x} - q^2} \tag{8.66}$$

Proof:

$(Ax - qx)^T (Ax - qx) = (Ax)^T Ax - qx^T Ax - q(Ax)^T x + q^2 x^T x = (Ax)^T Ax - 2qx^T Ax + q^2 x^T x$, since A is a symmetric matrix.

Therefore,

$(\boldsymbol{Ax}-q\boldsymbol{x})^T(\boldsymbol{Ax}-q\boldsymbol{x})=(\boldsymbol{Ax})^T\boldsymbol{Ax}-2q\boldsymbol{x}^T\boldsymbol{Ax}+q^2\boldsymbol{x}^T\boldsymbol{x}=(\boldsymbol{Ax})^T\boldsymbol{Ax}-q^2\boldsymbol{x}^T\boldsymbol{x}=\delta^2\boldsymbol{x}^T\boldsymbol{x}$, using Equations 8.65 and 8.66.

Again, since $\boldsymbol{A}$ is symmetric matrix, it has an orthogonal set $\{\boldsymbol{v}_1, \boldsymbol{v}_2,\ldots,\boldsymbol{v}_n\}$ of n linearly independent unit eigenvectors corresponding to the eigenvalues λ_1, λ_2, ..., λ_n, respectively. Then, x can be expressed as linear combination of $\{\boldsymbol{v}_1, \boldsymbol{v}_2,\ldots,\boldsymbol{v}_n\}$ in the following form:

$$\boldsymbol{x}=\alpha_1\boldsymbol{v}_1+\alpha_2\boldsymbol{v}_2+\cdots+\alpha_n\boldsymbol{v}_n \tag{8.67}$$

This implies that

$$\boldsymbol{x}^T\boldsymbol{x}=\alpha_1^2+\alpha_2^2+\cdots+\alpha_n^2$$

since $\boldsymbol{v}_1, \boldsymbol{v}_2,\ldots,\boldsymbol{v}_n$ are orthogonal unit vectors.

Therefore,

$$\boldsymbol{Ax}=\alpha_1\lambda_1\boldsymbol{v}_1+\alpha_2\lambda_2\boldsymbol{v}_2+\cdots+\alpha_n\lambda_n\boldsymbol{v}_n \tag{8.68}$$

Now,

$$(\boldsymbol{Ax}-q\boldsymbol{x})=\alpha_1(\lambda_1-q)\boldsymbol{v}_1+\alpha_2(\lambda_2-q)\boldsymbol{v}_2+\cdots+\alpha_n(\lambda_n-q)\boldsymbol{v}_n$$

Thus,

$$\delta^2\boldsymbol{x}^T\boldsymbol{x}=(\boldsymbol{Ax}-q\boldsymbol{x})^T(\boldsymbol{Ax}-q\boldsymbol{x})=\alpha_1^2(\lambda_1-q)^2+\alpha_2^2(\lambda_2-q)^2+\cdots+\alpha_n^2(\lambda_n-q)^2$$

since $\boldsymbol{v}_1, \boldsymbol{v}_2,\ldots,\boldsymbol{v}_n$ are orthogonal unit vectors.

Now, let $\hat{\lambda}$ be an eigenvalue of $\boldsymbol{A}$, closest to q. Then, $(\hat{\lambda}-q)^2\le(\lambda_j-q)^2$, for $j=1,2,\ldots,n$. Thus, we obtain

$$\delta^2\boldsymbol{x}^T\boldsymbol{x}\ge(\hat{\lambda}-q)^2(\alpha_1^2+\alpha_2^2+\cdots+\alpha_n^2)=(\hat{\lambda}-q)^2\boldsymbol{x}^T\boldsymbol{x}$$

Hence,

$$\delta^2\ge(\hat{\lambda}-q)^2$$

This implies that

$$|\hat{\lambda}-q|=|\varepsilon|\le\delta=\sqrt{\left[(\boldsymbol{Ax})^T\boldsymbol{Ax}/\boldsymbol{x}^T\boldsymbol{x}\right]-q^2}\text{, using Equation 8.66}$$

This shows that the error ε for the approximation of an eigenvalue of $\boldsymbol{A}$ is bounded by δ.

Example 8.6

Determine the largest eigenvalue and the corresponding eigenvector of the following symmetric matrix:

$$\begin{bmatrix} 9 & 4 \\ 4 & 3 \end{bmatrix}$$

Solution:

Let $\boldsymbol{A} = \begin{bmatrix} 9 & 4 \\ 4 & 3 \end{bmatrix}$ and initial guess $x_0 = \begin{bmatrix} 1 \\ 1 \end{bmatrix}$.

Step 1:

$$\boldsymbol{y}^{(1)} = \boldsymbol{A}\boldsymbol{x}^{(0)} = \begin{bmatrix} 13 \\ 7 \end{bmatrix} = 13\begin{bmatrix} 1 \\ 0.538462 \end{bmatrix} = 13\boldsymbol{x}^{(1)}$$

$$q = \frac{\boldsymbol{x}^{(0)^T}\boldsymbol{A}\boldsymbol{x}^{(0)}}{\boldsymbol{x}^{(0)^T}\boldsymbol{x}^{(0)}} = \frac{\begin{bmatrix} 1 & 1 \end{bmatrix}\begin{bmatrix} 9 & 4 \\ 4 & 3 \end{bmatrix}\begin{bmatrix} 1 \\ 1 \end{bmatrix}}{\begin{bmatrix} 1 & 1 \end{bmatrix}\begin{bmatrix} 1 \\ 1 \end{bmatrix}} = \frac{\begin{bmatrix} 1 & 1 \end{bmatrix}\begin{bmatrix} 13 \\ 7 \end{bmatrix}}{2} = 10$$

$$\delta = \sqrt{\frac{\begin{bmatrix} 13 & 7 \end{bmatrix}\begin{bmatrix} 13 \\ 7 \end{bmatrix}}{2} - q^2} = \sqrt{\frac{169+49}{2} - 10^2} = 3$$

Step 2:

$$\boldsymbol{y}^{(2)} = \boldsymbol{A}\boldsymbol{x}^{(1)} = \begin{bmatrix} 9 & 4 \\ 4 & 3 \end{bmatrix}\begin{bmatrix} 1 \\ 0.538462 \end{bmatrix} = \begin{bmatrix} 11.153848 \\ 5.615386 \end{bmatrix} = 11.153848\begin{bmatrix} 1 \\ 0.503448 \end{bmatrix} = 11.153848\boldsymbol{x}^{(2)}$$

$$q = \frac{\boldsymbol{x}^{(1)^T}\boldsymbol{A}\boldsymbol{x}^{(1)}}{\boldsymbol{x}^{(1)^T}\boldsymbol{x}^{(1)}} = 10.990827$$

$$\delta = \sqrt{\frac{\left(\boldsymbol{A}\boldsymbol{x}^{(1)}\right)^T\boldsymbol{A}\boldsymbol{x}^{(1)}}{\boldsymbol{x}^{(1)^T}\boldsymbol{x}^{(1)}} - q^2} = 0.3027$$

Step 3:

$$\boldsymbol{y}^{(3)} = \boldsymbol{A}\boldsymbol{x}^{(2)} = \begin{bmatrix} 9 & 4 \\ 4 & 3 \end{bmatrix}\begin{bmatrix} 1 \\ 0.503448 \end{bmatrix} = \begin{bmatrix} 11.013792 \\ 5.510344 \end{bmatrix} = 11.013792\begin{bmatrix} 1 \\ 0.500313 \end{bmatrix} = 11.013792\boldsymbol{x}^{(3)}$$

$$q = \frac{\boldsymbol{x}^{(2)^T}\boldsymbol{A}\boldsymbol{x}^{(2)}}{\boldsymbol{x}^{(2)^T}\boldsymbol{x}^{(2)}} = 10.999923$$

$$\delta = \sqrt{\frac{\left(\boldsymbol{A}\boldsymbol{x}^{(2)}\right)^T\boldsymbol{A}\boldsymbol{x}^{(2)}}{\boldsymbol{x}^{(2)^T}\boldsymbol{x}^{(2)}} - q^2} = 0.027548$$

Step 4:

$$\boldsymbol{y}^{(4)} = \boldsymbol{A}\boldsymbol{x}^{(3)} = \begin{bmatrix} 9 & 4 \\ 4 & 3 \end{bmatrix}\begin{bmatrix} 1 \\ 0.500313 \end{bmatrix} = \begin{bmatrix} 11.0013 \\ 5.50094 \end{bmatrix} = 11.0013\begin{bmatrix} 1 \\ 0.500028 \end{bmatrix} = 11.0013\boldsymbol{x}^{(4)}$$

$$q = \frac{x^{(3)^T} A x^{(3)}}{x^{(3)^T} x^{(3)}} = 11$$

$$\delta = \sqrt{\frac{\left(Ax^{(3)}\right)^T Ax^{(3)}}{x^{(3)^T} x^{(3)}} - q^2} = 0.00250438$$

So, the required dominant eigenvalue is 11 and the corresponding eigenvector is (1, 0.503448).

The above results can be summarized in the following computation table.

Step m	$x^{(m-1)^T}$	$y^{(m)^T}$	q	δ
1	(1, 1)	(13, 7)	10	3
2	(1, 0.538462)	(11.153848, 5.615386)	10.990827	0.302752
3	(1, 0.503448)	(11.013792, 5.510344)	10.999923	0.027548
4	(1, 0.500313)	(11.0013, 5.50094)	11	0.00250438

8.5.1 Algorithm of Power Method

Input: Enter the order of the matrix n, matrix $A = [a_{ij}]_{n\times n}$, and initial guess $x^{(0)}$ such that $\|x^{(0)}\|_\infty = 1$.

Output: The dominant eigenvalue of matrix A and the corresponding eigenvector.

Step 1: Set $i = 1$;

Step 2: $y^{(i)} = Ax^{(i-1)}$;

Step 3: Choose λ_i such that $|\lambda_i| = \|y^{(i)}\|_\infty$;

Step 4: $x^{(i+1)} = (1/\lambda_i)\, y^{(i)}$;

Step 5: If $\|x^{(i+1)} - x^{(i)}\|_\infty < \varepsilon$, ε being the prescribed error tolerance
then go to Step 7
else
go to Step 6

Step 6: Set $i = i + 1$ and go to Step 2

Step 7: Print the dominant eigenvalue λ_i and the corresponding eigenvector $x^{(i+1)}$

Step 8: Stop. ■

MATHEMATICA® Program for Finding Greatest Eigenvalue by Power Method (Chapter 8, Example 8.5)

```
A={{1,3,-1},{3,2,4},{-1,4,10}};
ϵ=0.00001;
xnew={{-1},{0},{1}};
Print["----------------------------------------"];

Do[
  Print["Step ",i,":"];
  xold=xnew;
  Print[N[MatrixForm[A.xold]]];
  Print[N[Max[A.xold]]];
```

```
  xnew=A.xold/Max[A.xold];
  Print[N[MatrixForm[xnew]]];
    Print["Max Error=",Max[N[Abs[xnew-xold]]]];
  Print["--------------------------------------"];
If[Max[Abs[xnew-xold]]<ϵ,Break[]],
  {i,1,100}];
Print["The Required Eigenvalue is ",N[Max[A.xold]]];
---------------------------------------
```

Output:

Step 1:

$$\begin{pmatrix} -2. \\ 1. \\ 11. \end{pmatrix}$$

11.

$$\begin{pmatrix} -0.181818 \\ 0.0909091 \\ 1. \end{pmatrix}$$

Max Error = 0.818182

Step 2:

$$\begin{pmatrix} -0.909091 \\ 3.63636 \\ 10.5455 \end{pmatrix}$$

10.5455

$$\begin{pmatrix} -0.0862069 \\ 0.344828 \\ 1. \end{pmatrix}$$

Max Error = 0.253918

Step 3:

$$\begin{pmatrix} -0.0517241 \\ 4.43103 \\ 11.4655 \end{pmatrix}$$

11.4655

$$\begin{pmatrix} -0.00451128 \\ 0.386466 \\ 1. \end{pmatrix}$$

Max Error = 0.0816956

```
Step 4:
```

$$\begin{pmatrix} 0.154887 \\ 4.7594 \\ 11.5504 \end{pmatrix}$$

```
11.5504
```

$$\begin{pmatrix} 0.0134097 \\ 0.412056 \\ 1. \end{pmatrix}$$

```
Max Error = 0.0255896
----------------------------------------
Step 5:
```

$$\begin{pmatrix} 0.249577 \\ 4.86434 \\ 11.6348 \end{pmatrix}$$

```
11.6348
```

$$\begin{pmatrix} 0.0214509 \\ 0.418085 \\ 1. \end{pmatrix}$$

```
Max Error = 0.00804116
----------------------------------------
Step 6:
```

$$\begin{pmatrix} 0.275706 \\ 4.90052 \\ 11.6509 \end{pmatrix}$$

```
11.6509
```

$$\begin{pmatrix} 0.0236639 \\ 0.420614 \\ 1. \end{pmatrix}$$

```
Max Error = 0.00252864
----------------------------------------
Step 7:
```

$$\begin{pmatrix} 0.285505 \\ 4.91222 \\ 11.6588 \end{pmatrix}$$

```
11.6588
```

$$\begin{pmatrix} 0.244884 \\ 0.421332 \\ 1. \end{pmatrix}$$

```
Max Error = 0.000824442
----------------------------------------
```

```
Step 8:
```

$$\begin{pmatrix} 0.288484 \\ 4.91613 \\ 11.6608 \end{pmatrix}$$

```
11.6608
```

$$\begin{pmatrix} 0.0247395 \\ 0.421593 \\ 1. \end{pmatrix}$$

```
Max Error = 0.000261274
----------------------------------------
Step 9:
```

$$\begin{pmatrix} 0.289519 \\ 4.9174 \\ 11.6616 \end{pmatrix}$$

```
11.6616
```

$$\begin{pmatrix} 0.0.248266 \\ 0.421674 \\ 1. \end{pmatrix}$$

```
Max Error = 0.0000870672
----------------------------------------
Step 10:
```

$$\begin{pmatrix} 0.289848 \\ 4.91783 \\ 11.6619 \end{pmatrix}$$

```
11.6619
```

$$\begin{pmatrix} 0.0248543 \\ 0.421701 \\ 1. \end{pmatrix}$$

```
Max Error = 0.000027729
----------------------------------------
Step 11:
```

$$\begin{pmatrix} 0.289959 \\ 4.91797 \\ 11.662 \end{pmatrix}$$

```
11.662
```

$$\begin{pmatrix} 0.0248637 \\ 0.42171 \\ 1. \end{pmatrix}$$

Max Error = 9.33009×10^{-6}

```
----------------------------------------
The Required Eigenvalue is  11.662
```

8.6 INVERSE POWER METHOD

The inverse power method is usually more powerful than the power method. The inverse power method is a modification of the power method that provides faster convergence. It has the advantage that it can be used to determine any eigenvalue of $\boldsymbol{A}$ that is closest to a specified number μ.

Let us suppose that the matrix $\boldsymbol{A}$ has eigenvalues $\lambda_1, \lambda_2, \ldots, \lambda_n$ with the linearly independent eigenvectors $\boldsymbol{v}_1, \boldsymbol{v}_2, \ldots, \boldsymbol{v}_n$. For μ not an eigenvalue of $\boldsymbol{A}$, the eigenvalues of $(\boldsymbol{A}-\mu\boldsymbol{I})^{-1}$ are

$$\hat{\lambda}_1 = \frac{1}{\lambda_1-\mu}, \hat{\lambda}_2 = \frac{1}{\lambda_2-\mu}, \ldots, \hat{\lambda}_n = \frac{1}{\lambda_n-\mu}, \quad \mu = \lambda_i, i = 1,2,\ldots,n$$

with the same eigenvectors $\boldsymbol{v}_1, \boldsymbol{v}_2, \ldots, \boldsymbol{v}_n$, respectively.

For some positive integer k, the largest eigenvalue in magnitude of $(\boldsymbol{A}-\mu\boldsymbol{I})^{-1}$ is $\hat{\lambda}_k$ corresponding to λ_k that is closest eigenvalue to μ. Thus, the power method applied to $(\boldsymbol{A}-\mu\boldsymbol{I})^{-1}$ converges to $\hat{\lambda}_k$ and therefore yields $\lambda_k = \mu + (1/\hat{\lambda}_k)$.

Now, applying power method to $(\boldsymbol{A}-\mu\boldsymbol{I})^{-1}$, according to Equations 8.63 and 8.64 we have

$$\boldsymbol{y}^{(m)} = (\boldsymbol{A}-\mu\boldsymbol{I})^{-1}\boldsymbol{x}^{(m-1)} \tag{8.69}$$

$$\hat{\lambda}^{(m)} = y_{p_{m-1}}^{(m)} = \frac{y_{p_{m-1}}^{(m)}}{x_{p_{m-1}}^{(m-1)}} = \frac{\sum_{j=1}^{n} \alpha_j \left[1/(\lambda_j-\mu)\right]^m v_{p_{m-1}}^{(j)}}{\sum_{j=1}^{n} \alpha_j \left[1/(\lambda_j-\mu)\right]^{m-1} v_{p_{m-1}}^{(j)}} \tag{8.70}$$

and

$$\boldsymbol{x}^{(m)} = \frac{\boldsymbol{y}^{(m)}}{y_{p_m}^{(m)}} \tag{8.71}$$

where at the mth step, p_m is the smallest integer index for which

$$\left|y_{p_m}^{(m)}\right| = \left\|\boldsymbol{y}^{(m)}\right\|_\infty$$

From Equation 8.70, we see that

$$\lim_{m\to\infty} \lambda^{(m)} = \frac{1}{(\lambda_k-\mu)}$$

where

$$\frac{1}{|\lambda_k-\mu|} = \max_{1\le i\le n} \frac{1}{|\lambda_i-\mu|}$$

Therefore, $\lambda_k \approx \mu + \left(1/\hat{\lambda}^{(m)}\right)$ is the approximate eigenvalue of $\boldsymbol{A}$ closest to μ.

Now, with regard to the dominant eigenvalue $1/(\lambda_k - \mu)$ of $(\boldsymbol{A}-\mu\boldsymbol{I})^{-1}$, from Equation 8.70, we obtain

$$\hat{\lambda}^{(m)} = \frac{1}{(\lambda_k - \mu)} \frac{\alpha_k v_{p_{m-1}}^{(k)} + \sum_{\substack{j=2\\ j\neq k}}^{n} \alpha_j \left[(\lambda_k - \mu)/(\lambda_j - \mu)\right]^m v_{p_{m-1}}^{(j)}}{\alpha_k v_{p_{m-1}}^{(k)} + \sum_{\substack{j=2\\ j\neq k}}^{n} \alpha_j \left[(\lambda_k - \mu)/(\lambda_j - \mu)\right]^{m-1} v_{p_{m-1}}^{(j)}}$$

$$= \frac{1}{(\lambda_k - \mu)} \frac{\alpha_k v_{p_{m-1}}^{(k)} + O\left(\left|\dfrac{\lambda_k - \mu}{\lambda_j - \mu}\right|^m\right)}{\alpha_k v_{p_{m-1}}^{(k)} + O\left(\left|\dfrac{\lambda_k - \mu}{\lambda_j - \mu}\right|^{m-1}\right)} \tag{8.72}$$

Thus, if $1/(\lambda_k - \mu)$ is the unique largest eigenvalue, then the convergence depends upon suitable choice of μ. Moreover, from Equation 8.72 it can be observed that if μ is closer to an eigenvalue λ_k, then the convergence will be faster.

The system of linear equations in Equation 8.69 can be written as

$$(\boldsymbol{A}-\mu\boldsymbol{I})\boldsymbol{y}^{(m)} = \boldsymbol{x}^{(m-1)} \tag{8.73}$$

The vector $\boldsymbol{y}^{(m)}$ is obtained by solving the above linear system. The suitable value of μ can be chosen based on Gerschgorin's circle theorem.

8.6.1 Algorithm of Inverse Power Method

Input: Enter the order of the matrix n, matrix $\boldsymbol{A} = [a_{ij}]_{n\times n}$ and initial guess $\boldsymbol{x}^{(0)}$ such that $\left\|\boldsymbol{x}^{(0)}\right\|_\infty = 1$

Output: The eigenvalue of matrix $\boldsymbol{A}$ closest to μ and the corresponding eigenvector.

Step 1: Compute $\mu = (x^{(0)})^T Ax^{(0)}/(x^{(0)})^T x^{(0)}$;

Step 2: Set $i = 1$;

Step 3: Solve the linear system $(\boldsymbol{A}-\mu\boldsymbol{I})\boldsymbol{y}^{(i)} = \boldsymbol{x}^{(i-1)}$ for determining $\boldsymbol{y}^{(i)}$.

Step 4: Choose λ_i such that $|\lambda_i| = \left\|\boldsymbol{y}^{(i)}\right\|_\infty$;

Step 5: $\hat{\lambda}_i = (1/\lambda_i) + \mu$;

Step 6: $\boldsymbol{x}^{(i+1)} = (1/\lambda_i)\boldsymbol{y}^{(i)}$;

Step 7: If $\left\|\boldsymbol{x}^{(i+1)} - \boldsymbol{x}^{(i)}\right\|_\infty < \varepsilon$, ε being the prescribed error tolerance.
then go to Step 9.
else
go to Step 8.

Step 8: Set $i = i + 1$ and go to Step 3.

Step 9: Print the eigenvalue $\hat{\lambda}_i$ that is closest to μ and the corresponding eigenvector $\boldsymbol{x}^{(i+1)}$.

Step 10: Stop. ■

MATHEMATICA® Program for Finding Eigenvalue Nearer to μ by Inverse Power Method (Chapter 8, Example 8.7)

```
A={{4,2,1},{0,3,2},{1,1,4}};
n=3;
є=0.00001;
Print["Eigenvalues of A:",N[Eigenvalues[A]]];
xnew={{1},{0.5},{1}};
Y=Table[y[i],{i,1,n}];
μ=Transpose[xnew].A.xnew/Transpose[xnew].xnew;
Print["μ=",N[μ[[1]][[1]]]];
Print["----------------------------------------"];
Do[
  Print["Step ",i,":"];
  xold=xnew;
  slv=Solve[(A-N[μ[[1]][[1]]]*IdentityMatrix[n]).Transpose[{Y}]==xold,Y];
  Print[slv];

  Y1=Transpose[{Table[slv[[1,i,2]],{i,1,n}]}];

  xnew=N[Y1/Max[Abs[Y1]]];
  Print[N[MatrixForm[Y1/Max[Abs[Y1]]]]];
  q=1/Max[Abs[Y1]]+μ;
  Print[«q=»,N[q[[1]][[1]]]];

  Print["Max Error=",Max[Abs[xnew-xold]]];
  Print["----------------------------------------"];
  If[Max[Abs[xnew-xold]]<є,Break[]],
  {i,1,10}];
Print["The Eigenvalue closest to μ is ",N[q[[1]][[1]]]];
```

Output:

```
Eigenvalues of A: {5.91964,2.54018 +0.688173 I,2.54018 -0.688173 I}
 μ=5.88889
 ----------------------------------------
 Step 1:
 {{y[1]->35.4332,y[2]->19.6489,y[3]->28.6317}}
 (    1.    )
 ( 0.554532 )
 ( 0.808047 )

 q=5.91711
 Max Error=0.191953
 ----------------------------------------
 Step 2:
 {{y[1]->32.4788,y[2]->18.0208,y[3]->26.3073}}
 (    1.    )
 ( 0.554847 )
 ( 0.809983 )

 q=5.91968
 Max Error=0.00193668
 ----------------------------------------
```

```
Step 3:
{{y[1]->32.5201,y[2]->18.0434,y[3]->26.3401}}
```

$$\begin{pmatrix} 1. \\ 0.554838 \\ 0.809963 \end{pmatrix}$$

```
q=5.91964
Max Error=0.0000198855
----------------------------------------
Step 4 :
{{y[1]->32.5196,y[2]->18.0431,y[3]->26.3397}}
```

$$\begin{pmatrix} 1. \\ 0.554838 \\ 0.809964 \end{pmatrix}$$

```
q=5.91964
Max Error=1.93522*10^-7
----------------------------------------
The Eigenvalue closest to μ is  5.91964
```

MATHEMATICA® Program for Finding Least Eigenvalue by Inverse Power Method (Chapter 8, Example 8.8)

```
A1={{4,-1,1},{-1,3,-2},{1,-2,3}};
Print["Eigenvalues of A:",N[Eigenvalues[A1]]];
ϵ=0.00001;
A=Inverse[A1];
xnew={{1},{1},{1}};
Print["----------------------------------------"];
Do[
  Print["Step ",i,":"];
  xold=xnew;
  Print[N[MatrixForm[A.xold]]];
  Print[N[Max[A.xold]]];
  xnew=A.xold/Max[A.xold];
  Print[N[MatrixForm[xnew]]];
  Print["Max Error=",Max[N[Abs[xnew-xold]]]];
  Print["----------------------------------------"];
  If[Max[Abs[xnew-xold]]<ϵ,Break[]],
  {i,1,100}];
Print["The Required Eigenvalue is ",N[Max[A.xold]]];
```

Output:

```
Eigenvalues of A: {6.,3.,1.}
----------------------------------------
Step 1:
```

$$\begin{pmatrix} 0.277778 \\ 1.05556 \\ 0.944444 \end{pmatrix}$$

```
1.05556
```

$$\begin{pmatrix} 0.263158 \\ 1. \\ 0.894737 \end{pmatrix}$$

```
Max Error = 0.736842
----------------------------------------
Step 2:
```

$$\begin{pmatrix} 0.0789474 \\ 0.973684 \\ 0.921053 \end{pmatrix}$$

```
0.973684
```

$$\begin{pmatrix} 0.0810811 \\ 1. \\ 0.945946 \end{pmatrix}$$

```
Max Error = 0.182077
----------------------------------------
Step 3:
```

$$\begin{pmatrix} 0.0255255 \\ 0.983483 \\ 0.962462 \end{pmatrix}$$

```
0.983483
```

$$\begin{pmatrix} 0.0259542 \\ 1. \\ 0.978626 \end{pmatrix}$$

```
Max Error = 0.0551269
----------------------------------------
Step 4:
```

$$\begin{pmatrix} 0.00839695 \\ 0.99313 \\ 0.985496 \end{pmatrix}$$

```
0.99313
```

$$\begin{pmatrix} 0.00845503 \\ 1. \\ 0.992314 \end{pmatrix}$$

```
Max Error = 0.0174992
----------------------------------------
Step 5:
```

$$\begin{pmatrix} 0.00277564 \\ 0.997481 \\ 0.994833 \end{pmatrix}$$

```
0.997481
```

$$\begin{pmatrix} 0.00278265 \\ 1. \\ 0.997346 \end{pmatrix}$$

```
Max Error = 0.00567238
----------------------------------------
Step 6:
```

$$\begin{pmatrix} 0.000920416 \\ 0.999122 \\ 0.998223 \end{pmatrix}$$

0.999122

$$\begin{pmatrix} 0.000921225 \\ 1. \\ 0.9991 \end{pmatrix}$$

```
Max Error = 0.00186143
----------------------------------------
Step 7:
```

$$\begin{pmatrix} 0.000305885 \\ 0.999701 \\ 0.999399 \end{pmatrix}$$

0.999701

$$\begin{pmatrix} 0.000305976 \\ 1. \\ 0.999698 \end{pmatrix}$$

```
Max Error = 0.000615249
----------------------------------------
Step 8:
```

$$\begin{pmatrix} 0.000101794 \\ 0.999899 \\ 0.999798 \end{pmatrix}$$

0.999899

$$\begin{pmatrix} 0.000101804 \\ 1. \\ 0.999899 \end{pmatrix}$$

```
Max Error = 0.000204172
----------------------------------------
Step 9:
```

$$\begin{pmatrix} 0.0000339015 \\ 0.999966 \\ 0.999932 \end{pmatrix}$$

0.999966

$$\begin{pmatrix} 0.0000339027 \\ 1. \\ 0.999966 \end{pmatrix}$$

```
Max Error = 0.0000679012
----------------------------------------
Step 10:
```

$$\begin{pmatrix} 0.0000112954 \\ 0.999989 \\ 0.999977 \end{pmatrix}$$

```
0.999989
```

$$\begin{pmatrix} 0.0000112955 \\ 1. \\ 0.999989 \end{pmatrix}$$

```
Max Error = 0.0000226072
----------------------------------------
Step 11:
```

$$\begin{pmatrix} 3.76425*10^{-6} \\ 0.999996 \\ 0.999992 \end{pmatrix}$$

```
0.999996
```

$$\begin{pmatrix} 3.76426*10^{-6} \\ 1. \\ 0.99996 \end{pmatrix}$$

```
Max Error = 7.53124*10^-6
----------------------------------------
The Required Eigenvalue is 0.999996
```

Example 8.7

Apply inverse power method to determine the eigenvalue correct to four decimal places and the corresponding eigenvector of the following matrix:

$$\begin{bmatrix} 4 & 2 & 1 \\ 0 & 3 & 2 \\ 1 & 1 & 4 \end{bmatrix}$$

Solution:

Let us choose the initial guess $\boldsymbol{x}^{(0)} = \begin{bmatrix} 1 \\ 0.5 \\ 1 \end{bmatrix}$ such that $\left\| \boldsymbol{x}^{(0)} \right\|_\infty = 1.$

Next, we compute

$$\mu = \frac{\boldsymbol{x}^{(0)^T} \boldsymbol{A}\boldsymbol{x}^{(0)}}{\boldsymbol{x}^{(0)^T} \boldsymbol{x}^{(0)}} = 5.88889$$

To apply the inverse power method, we consider

$$(A-\mu I)=\begin{bmatrix} -1.88889 & 2 & 1 \\ 0 & -2.88889 & 2 \\ 1 & 1 & -1.88889 \end{bmatrix}.$$

Step 1:
We solve the linear system

$$(A-\mu I)y^{(1)} = x^{(0)}$$

yielding

$$y^{(1)} = \begin{bmatrix} 35.4332 \\ 19.6489 \\ 28.6317 \end{bmatrix}$$

Now,

$$\left\| y^{(1)} \right\|_\infty = 35.4332, \quad x^{(1)} = \frac{1}{35.4332} y^{(1)} = \begin{bmatrix} 1 \\ 0.554532 \\ 0.808047 \end{bmatrix}$$

Also,

$$\hat{\lambda}^{(1)} = \frac{1}{35.4332} + 5.88889 = 5.91711$$

Step 2:
Next, we solve the linear system

$$(A-\mu I)y^{(2)} = x^{(1)}$$

yielding

$$y^{(2)} = \begin{bmatrix} 32.4788 \\ 18.0208 \\ 26.3073 \end{bmatrix}$$

Now,

$$\left\| y^{(2)} \right\|_\infty = 32.4788, \quad x^{(2)} = \frac{1}{32.4788} y^{(2)} = \begin{bmatrix} 1 \\ 0.554847 \\ 0.809983 \end{bmatrix}$$

Also,

$$\hat{\lambda}^{(2)} = \frac{1}{32.4788} + 5.88889 = 5.91968$$

Step 3:
Again, we solve the linear system

$$(A-\mu I)y^{(3)} = x^{(2)}$$

yielding

$$y^{(3)} = \begin{bmatrix} 32.5201 \\ 18.0434 \\ 26.3401 \end{bmatrix}$$

Now,

$$\|y^{(3)}\|_\infty = 32.5201, \quad x^{(3)} = \frac{1}{32.5201}y^{(3)} = \begin{bmatrix} 1 \\ 0.554838 \\ 0.809963 \end{bmatrix}$$

Also,

$$\hat{\lambda}^{(3)} = \frac{1}{32.5201} + 5.88889 = 5.91964$$

Step 4:
Now, we solve the linear system

$$(A - \mu I)y^{(4)} = x^{(3)}$$

yielding

$$y^{(4)} = \begin{bmatrix} 32.5196 \\ 18.0431 \\ 26.3397 \end{bmatrix}$$

Now,

$$\|y^{(4)}\|_\infty = 32.5196, \quad x^{(4)} = \frac{1}{32.5196}y^{(4)} = \begin{bmatrix} 1 \\ 0.554838 \\ 0.809964 \end{bmatrix}$$

Also,

$$\hat{\lambda}^{(4)} = \frac{1}{32.5196} + 5.88889 = 5.91964$$

The above results can be summarized in the following computation table.

Step m	$x^{(m-1)^T}$	$y^{(m)^T}$	$\hat{\lambda}^{(m)}$
1	(1, 0.5, 1)	(35.4332, 19.6489, 28.6317)	5.91711
2	(1, 0.554532, 0.808047)	(32.4788, 18.0208, 26.3073)	5.91968
3	(1, 0.554847, 0.809983)	(32.5201, 18.0434, 26.3401)	5.91964
4	(1, 0.554838, 0.809963)	(32.5196, 18.0431, 26.3397)	5.91964

So, the required eigenvalue is 5.9196 correct to four decimal places and the corresponding eigenvector is $(1, 0.554838, 0.809963)$. Moreover, the eigenvalue 5.9196 is closest to 5.88889.

Example 8.8

Apply inverse power method to determine the smallest eigenvalue correct to three decimal places and the corresponding eigenvector of the following matrix:

$$A = \begin{bmatrix} 4 & -1 & 1 \\ -1 & 3 & -2 \\ 1 & -2 & 3 \end{bmatrix}$$

Solution:

Let us choose the initial guess $x^{(0)} = \begin{bmatrix} 1 \\ 1 \\ 1 \end{bmatrix}$ with $\|x^{(0)}\|_\infty = 1$

Step 1: We solve the linear system $Ay^{(1)} = x^{(0)}$ yielding

$$y^{(1)} = \begin{bmatrix} 0.277778 \\ 1.05556 \\ 0.944444 \end{bmatrix}$$

$$\|y^{(1)}\|_\infty = 1.05556 \quad \text{and} \quad x^{(1)} = \frac{y^{(1)}}{\|y^{(1)}\|_\infty} = \begin{bmatrix} 0.263158 \\ 1 \\ 0.894737 \end{bmatrix}$$

$$\hat{\lambda}^{(1)} = \frac{1}{1.05556} = 0.94736$$

Step 2: Solving the linear system $Ay^{(2)} = x^{(1)}$, we get

$$y^{(2)} = \begin{bmatrix} 0.0789474 \\ 0.973684 \\ 0.921053 \end{bmatrix}$$

$$\|y^{(2)}\|_\infty = 0.973684 \quad \text{and} \quad x^{(2)} = \frac{y^{(2)}}{\|y^{(2)}\|_\infty} = \begin{bmatrix} 0.0810811 \\ 1 \\ 0.945946 \end{bmatrix}$$

$$\hat{\lambda}^{(2)} = \frac{1}{0.973684} = 1.02703.$$

Step 3: Next, we solve the linear system $Ay^{(3)} = x^{(2)}$ yielding

$$y^{(3)} = \begin{bmatrix} 0.0255255 \\ 0.983483 \\ 0.962462 \end{bmatrix}$$

$$\left\|y^{(3)}\right\|_\infty = 0.983483 \quad \text{and} \quad x^{(3)} = \frac{y^{(3)}}{\left\|y^{(3)}\right\|_\infty} = \begin{bmatrix} 0.0259542 \\ 1 \\ 0.978626 \end{bmatrix}$$

$$\hat{\lambda}^{(3)} = \frac{1}{0.983483} = 1.01679$$

Step 4: Again solving the linear system $\boldsymbol{Ay}^{(4)} = \boldsymbol{x}^{(3)}$, we obtain

$$y^{(4)} = \begin{bmatrix} 0.00839695 \\ 0.99313 \\ 0.985496 \end{bmatrix}$$

$$\left\|y^{(4)}\right\|_\infty = 0.99313 \quad \text{and} \quad x^{(4)} = \frac{y^{(4)}}{\left\|y^{(4)}\right\|_\infty} = \begin{bmatrix} 0.00845503 \\ 1 \\ 0.992314 \end{bmatrix}$$

$$\hat{\lambda}^{(4)} = \frac{1}{0.99313} = 1.00692$$

Step 5: Solving the linear system $\boldsymbol{A.y}^{(5)} = \boldsymbol{x}^{(4)}$ yields

$$y^{(5)} = \begin{bmatrix} 0.00277564 \\ 0.997481 \\ 0.994833 \end{bmatrix}$$

$$\left\|y^{(5)}\right\|_\infty = 0.997481 \quad \text{and} \quad x^{(5)} = \frac{y^{(5)}}{\left\|y^{(5)}\right\|_\infty} = \begin{bmatrix} 0.00278265 \\ 1 \\ 0.997346 \end{bmatrix}$$

$$\hat{\lambda}^{(5)} = \frac{1}{0.997481} = 1.00252$$

Step 6: Again we solve the linear system $\boldsymbol{A.y}^{(6)} = \boldsymbol{x}^{(6)}$ yielding

$$y^{(6)} = \begin{bmatrix} 0.000920416 \\ 0.999122 \\ 0.998223 \end{bmatrix}$$

$$\left\|y^{(6)}\right\|_\infty = 0.999122 \quad \text{and} \quad x^{(6)} = \frac{y^{(6)}}{\left\|y^{(6)}\right\|_\infty} = \begin{bmatrix} 0.000921225 \\ 1 \\ 0.9991 \end{bmatrix}$$

$$\hat{\lambda}^{(6)} = \frac{1}{0.999122} = 1.00088.$$

Step 7: Solving the linear system $\boldsymbol{A}.\boldsymbol{y}^{(7)} = \boldsymbol{x}^{(7)}$, we have

$$y^{(7)} = \begin{bmatrix} 0.000305885 \\ 0.999701 \\ 0.999399 \end{bmatrix}$$

$$\left\| y^{(7)} \right\|_\infty = 0.999701 \quad \text{and} \quad \boldsymbol{x}^{(7)} = \frac{y^{(7)}}{\left\| y^{(7)} \right\|_\infty} = \begin{bmatrix} 0.000305976 \\ 1 \\ 0.999698 \end{bmatrix}$$

$$\hat{\lambda}^{(7)} = \frac{1}{0.999701} = 1.00030$$

The above results can be summarized in the following computation table.

Step m	$(x^{(m-1)})^T$	$(y^{(m)})^T$	$\hat{\lambda}^{(m)}$
1	(1, 1, 1)	(0.277778, 1.05556, 0.944444)	0.94736
2	(0.263158, 1, 0.894737)	(0.0789474, 0.973684, 0.921053)	1.02703
3	(0.0810811, 1, 0.945946)	(0.0255255, 0.983483, 0.962462)	1.01679
4	(0.0259542, 1, 0.978626)	(0.00839695, 0.99313, 0.985496)	1.00692
5	(0.00845503, 1, 0.992314)	(0.00277564, 0.997481, 0.994833)	1.00252
6	(0.00278265, 1, 0.997346)	(0.000920416, 0.999122, 0.998223)	1.00088
7	(0.000921225, 1, 0.9991)	(0.000305885, 0.999701, 0.999399)	1.00030

So, the required smallest eigenvalue is 1 correct to three decimal places and the corresponding eigenvector is (0.000921225, 1, 0.9991).

8.7 JACOBI'S METHOD

One of oldest methods for computing eigenvalues is Jacobi's method, which uses similarity transformation based on plane rotations. This method can be used to find the eigenvalues of a symmetric matrix. Let $\boldsymbol{A}$ be a real symmetric matrix. We know that the eigenvalues of real symmetric matrix are real and there exists an orthogonal matrix $\boldsymbol{P}$ such that $\boldsymbol{P}^{-1}\boldsymbol{AP}$ is a diagonal matrix $\boldsymbol{D}$, whose diagonal elements are the eigenvalues of $\boldsymbol{A}$. Also the columns of $\boldsymbol{P}$ are eigenvectors of $\boldsymbol{A}$ corresponding to the eigenvalues of $\boldsymbol{A}$.

In this method, the diagonalization is done by a series of orthogonal transformations $\boldsymbol{P}_1, \boldsymbol{P}_2, \ldots, \boldsymbol{P}_n, \ldots$ as follows.

Let a_{ik} be the numerically largest in magnitude off-diagonal element of $\boldsymbol{A}$. Then, the 2×2 submatrix

$$\boldsymbol{A}_1 = \begin{bmatrix} a_{ii} & a_{ik} \\ a_{ki} & a_{kk} \end{bmatrix} \tag{8.74}$$

can be transferred to a diagonal form by the orthogonal transformation $\left(\boldsymbol{P}_1^*\right)^T \boldsymbol{A}_1 \boldsymbol{P}_1^*$, where

$$P_1^* = \begin{bmatrix} \cos\theta & -\sin\theta \\ \sin\theta & \cos\theta \end{bmatrix} \tag{8.75}$$

and θ is chosen in such a way that the 2×2 submatrix A_1 is diagonalized. Then, we have

$$\begin{aligned}(P_1^*)^T A_1 P_1^* &= \begin{bmatrix} \cos\theta & \sin\theta \\ -\sin\theta & \cos\theta \end{bmatrix}\begin{bmatrix} a_{ii} & a_{ik} \\ a_{ki} & a_{kk} \end{bmatrix}\begin{bmatrix} \cos\theta & -\sin\theta \\ \sin\theta & \cos\theta \end{bmatrix} \\ &= \begin{bmatrix} a_{ii}\cos^2\theta + a_{ik}\sin 2\theta + a_{kk}\sin^2\theta & a_{ik}\cos 2\theta + (a_{kk}-a_{ii})\sin\theta\cos\theta \\ a_{ik}\cos 2\theta + (a_{kk}-a_{ii})\sin\theta\cos\theta & a_{kk}\cos^2\theta - a_{ik}\sin 2\theta + a_{ii}\sin^2\theta \end{bmatrix}\end{aligned} \tag{8.76}$$

Now, the matrix $(P_1^*)^T A_1 P_1^*$ in Equation 8.76 reduces to diagonal matrix, if

$$a_{ik}\cos 2\theta + (a_{kk}-a_{ii})\sin\theta\cos\theta = 0$$

that is, if

$$\tan 2\theta = \frac{2a_{ik}}{a_{ii}-a_{kk}} \tag{8.77}$$

From Equation 8.77, we get

$$\theta = \frac{1}{2}\tan^{-1}\frac{2a_{ik}}{a_{ii}-a_{kk}} \tag{8.78}$$

When $a_{ii} \neq a_{kk}$, we may choose the principal value of θ such that

$$-\frac{\pi}{4} \le \theta \le \frac{\pi}{4} \tag{8.79}$$

$$\text{When } a_{ii} = a_{kk}, \text{ then } \theta = \begin{cases} \dfrac{\pi}{4}, & \text{if } a_{ik} > 0 \\ -\dfrac{\pi}{4}, & \text{if } a_{ik} < 0 \end{cases} \tag{8.80}$$

With the value of θ given in Equations 8.79 and 8.80, after simplification $(P_1^*)^T A_1 P_1^*$ becomes a diagonal matrix. Thus, the first step is now completed by performing the transformation (rotation) $P_1^{-1}AP_1$. In the second step, again the largest off-diagonal element in magnitude in the new transformed (rotated) matrix is found and the same procedure is repeated. After performing m transformations, we get

$$\begin{aligned} Q_m &= P_m^T P_{m-1}^T \ldots P_1^T A P_1 \ldots P_{m-1} P_m \\ &= P_m^{-1} P_{m-1}^{-1} \ldots P_1^{-1} A P_1 \ldots P_{m-1} P_m \\ &= (P_1 P_2 \ldots P_{m-1} P_m)^{-1} A P_1 \ldots P_{m-1} P_m \\ &= P^{-1} A P, \end{aligned} \tag{8.81}$$

where $P = P_1 \ldots P_{m-1} P_m$.

In general, P_r, $r = 1, 2, \ldots, m$ is of the following form:

$$
\begin{array}{cc}
 & \begin{array}{cccccccc}
 & & \text{column } i & & & \text{column } k & & \\
 & & \downarrow & & & \downarrow & &
\end{array} \\
P_r = & \begin{bmatrix}
1 & & & & & & & 0 \\
 & \ddots & & & & & \iddots & \\
 & & \cos\theta & & & -\sin\theta & & \\
 & & & 1 & & & & \\
 & \cdots & \vdots & \ddots & & \vdots & \cdots & \\
 & & & & 1 & & & \\
 & & \sin\theta & & & \cos\theta & & \\
 & \iddots & & & & & \ddots & \\
0 & & & & & & & 1
\end{bmatrix}
\begin{array}{l}
\\ \\ \leftarrow \text{row } i \\ \\ \\ \\ \leftarrow \text{row } k \\ \\ \\
\end{array}
\end{array}
$$

These matrices P_r are orthogonal and are called plane rotation matrix or Jacobi's rotation matrix.

In Equation 8.81, the sequence of plane rotations is chosen to annihilate symmetric pairs of matrix entries, eventually converging to diagonal form. As $m \to \infty$, Q_m tends to a diagonal matrix whose diagonal elements are the eigenvalues of A and the columns of P are the corresponding eigenvectors of A. Jacobi's method has the disadvantage that the elements annihilated by a plane rotation may not necessarily remain zero during subsequent transformations.

Example 8.9

Using Jacobi's method determine the eigenvalues of matrix

$$A = \begin{bmatrix} 2 & -1 & 0 \\ -1 & 2 & -1 \\ 0 & -1 & 2 \end{bmatrix}$$

Solution:

Step 1:

The numerically largest in magnitude off-diagonal element in A is $a_{12} = -1$ (we may take $a_{21} = -1$ or $a_{23} = -1$ or $a_{32} = -1$ also).

We construct the submatrix

$$A_1 = \begin{bmatrix} 2 & -1 \\ -1 & 2 \end{bmatrix}$$

$$P_1^* = \begin{bmatrix} \cos\theta & -\sin\theta \\ \sin\theta & \cos\theta \end{bmatrix} = \begin{bmatrix} \dfrac{1}{\sqrt{2}} & \dfrac{1}{\sqrt{2}} \\ -\dfrac{1}{\sqrt{2}} & \dfrac{1}{\sqrt{2}} \end{bmatrix} \text{ with } \theta = -\frac{\pi}{4}$$

$$P_1 = \begin{bmatrix} 0.707107 & 0.707107 & 0 \\ -0.707107 & 0.707107 & 0 \\ 0 & 0 & 1 \end{bmatrix} \text{ and } A^{(2)} = P_1^T A P_1 = \begin{bmatrix} 3 & 0 & 0.707107 \\ 0 & 1 & -0.707107 \\ 0.707107 & -0.707107 & 2 \end{bmatrix}$$

Step 2:
The numerically largest in magnitude off-diagonal element in $A^{(2)}$ is $a_{13} = 0.707107$, say. Again we construct the submatrix

$$A_2 = \begin{bmatrix} 3 & 0.707107 \\ 0.707107 & 2 \end{bmatrix}$$

$$P_2^* = \begin{bmatrix} \cos\theta & -\sin\theta \\ \sin\theta & \cos\theta \end{bmatrix} = \begin{bmatrix} 0.888074 & -0.459701 \\ 0.459701 & 0.888074 \end{bmatrix} \text{with}\,\theta = 0.477658$$

$$P_2 = \begin{bmatrix} 0.888074 & 0 & -0.459701 \\ 0 & 1 & 0 \\ 0.459701 & 0 & 0.888074 \end{bmatrix}$$

and

$$A^{(3)} = P_2^T A^{(2)} P_2 = \begin{bmatrix} 3.36603 & -0.325058 & 0 \\ -0.325058 & 1 & -0.627963 \\ 0 & -0.627963 & 1.63398 \end{bmatrix}$$

Step 3:
The numerically largest in magnitude off-diagonal element in $A^{(3)}$ is $a_{23} = -0.627963$, say. Now, we construct the submatrix

$$A_3 = \begin{bmatrix} 1 & -0.627963 \\ -0.627963 & 1.63398 \end{bmatrix}$$

$$P_3^* = \begin{bmatrix} \cos\theta & -\sin\theta \\ \sin\theta & \cos\theta \end{bmatrix} = \begin{bmatrix} 0.851655 & -0.524103 \\ 0.524103 & 0.851655 \end{bmatrix} \text{with}\,\theta = 0.551662$$

$$P_3 = \begin{bmatrix} 1 & 0 & 0 \\ 0 & 0.851655 & -0.524103 \\ 0 & 0.524103 & 0.851655 \end{bmatrix}$$

and

$$A^{(4)} = P_3^{-1} A^{(3)} P_3 = \begin{bmatrix} 3.36603 & -0.276837 & 0.170364 \\ -0.276837 & 0.613556 & 0 \\ 0.170364 & 0 & 2.02042 \end{bmatrix}$$

Similarly, in the next steps we obtain the following plane rotation matrices and the similarity transformations as follows:
Step 4:

$$P_4 = \begin{bmatrix} 0.995078 & 0.09909 & 0 \\ -0.09909 & 0.995078 & 0 \\ 0 & 0 & 1 \end{bmatrix}$$

and

$$A^{(5)} = P_4^{-1} A^{(4)} P_4 = \begin{bmatrix} 3.3936 & 0 & 0.169526 \\ 0 & 0.585989 & 0.0168814 \\ 0.169526 & 0.0168814 & 2.02042 \end{bmatrix}$$

Step 5:

$$P_5 = \begin{bmatrix} 0.992684 & 0 & -0.120739 \\ 0 & 1 & 0 \\ 0.120739 & 0 & 0.992684 \end{bmatrix}$$

and

$$A^{(6)} = P_5^{-1} A^{(5)} P_5 = \begin{bmatrix} 3.41422 & 0.00203824 & 0 \\ 0.00203824 & 0.585989 & 0.0167579 \\ 0 & 0.0167579 & 1.9998 \end{bmatrix}$$

Step 6:

$$P_6 = \begin{bmatrix} 1 & 0 & 0 \\ 0 & 0.99993 & 0.0118505 \\ 0 & -0.0118505 & 0.99993 \end{bmatrix}$$

and

$$A^{(7)} = P_6^{-1} A^{(6)} P_6 = \begin{bmatrix} 3.41422 & 0.0020381 & 0 \\ 0.0020381 & 0.58579 & 0 \\ 0 & 0 & 2 \end{bmatrix}$$

Hence, the eigenvalues are 3.4142, 0.5858, and 2.
The corresponding eigenvectors are

$$\begin{bmatrix} 0.499652 \\ -0.707616 \\ 0.499628 \end{bmatrix}, \begin{bmatrix} 0.500362 \\ 0.706597 \\ 0.500359 \end{bmatrix} \text{ and } \begin{bmatrix} -0.707097 \\ 0 \\ 0.707117 \end{bmatrix}$$

for respective eigenvalues.

8.8 GIVENS METHOD

We already know that in Jacobi's method, the nondiagonal elements, which were annihilated by a plane rotation, that is, by an orthogonal transformation, may not remain zero during the subsequent transformations. Givens method preserves the zeros in the off-diagonal positions once they are obtained. In this method, the given symmetric matrix is reduced to a tridiagonal matrix.

Let $\boldsymbol{A}$ be a symmetric matrix, and let us consider the orthogonal matrix

$$P_1 = \begin{bmatrix} 1 & 0 & 0 & 0 & \dots & 0 \\ 0 & \cos\theta & -\sin\theta & 0 & \dots & 0 \\ 0 & \sin\theta & \cos\theta & 0 & \dots & 0 \\ 0 & 0 & 0 & 1 & \dots & 0 \\ \vdots & \vdots & \vdots & \vdots & \dots & \vdots \\ 0 & 0 & 0 & 0 & \dots & 1 \end{bmatrix} \tag{8.82}$$

Then, $\boldsymbol{P}_1^T \boldsymbol{A} \boldsymbol{P}_1$ becomes a tridiagonal matrix if

$$-a_{12}\sin\theta + a_{13}\cos\theta = 0$$

that is,

$$\theta = \tan^{-1}\frac{a_{13}}{a_{12}} \tag{8.83}$$

With this value of θ, the orthogonal transformation $P_1^T A P_1$ gives zeros in the (3,1) and (1,3) positions. Further, let similar orthogonal transformations are performed in order to make the elements (1,4), (1,5), …, (1,n) and (4,1), (5,1), …, (n,1) zero. This would not affect the zeros that have been obtained earlier. Proceeding in this manner, treating all the rows of the given matrix, we get after $(n-2)+(n-3)+\cdots+2+1=[(n-1)(n-2)/2]$ rotations, the following symmetric tridiagonal matrix:

$$T = \begin{bmatrix} a_1 & b_1 & 0 & 0 & \cdots & 0 \\ b_1 & a_2 & b_2 & 0 & \cdots & 0 \\ 0 & b_2 & a_3 & b_3 & \cdots & 0 \\ \vdots & \vdots & \vdots & \ddots & \ddots & \vdots \\ \cdots & \cdots & \cdots & \cdots & a_{n-1} & b_{n-1} \\ 0 & 0 & 0 & 0 & b_{n-1} & a_n \end{bmatrix} \tag{8.84}$$

8.8.1 Eigenvalues of a Symmetric Tridiagonal Matrix

The determinants of the successive principal minors of a matrix of this form can easily be calculated by recurrence relation. Let $p_k(\lambda)$ be the determinant of the leading principal minor of order k of $T-\lambda I$. Then,

$$p_1(\lambda) = a_1 - \lambda$$

$$p_2(\lambda) = (a_1-\lambda)(a_2-\lambda) - b_1^2 = (a_2-\lambda)p_1(\lambda) - b_1^2$$

In general, $p_k(\lambda)$ can be obtained from the recurrence relation

$$p_k(\lambda) = (a_k-\lambda)p_{k-1}(\lambda) - b_{k-1}^2 p_{k-2}(\lambda), \quad k = 2,3,\ldots,n \tag{8.85}$$

with $p_0(\lambda) = 1$.

We assume that $b_1, b_2, \ldots, b_{n-1}$ are not all zero in the matrix T. Under this assumption, all the eigenvalues of T are the simple roots of $p_n(\lambda)$ and the sequence of functions $\{p_k(\lambda)\}_{k=0}^n$ is called the Strum's sequence.

The eigenvalues of T are found, in general, by using Strum's sequence $\{p_k(\lambda)\}_{k=0}^n$, which has the following property:

- *Strum sequence property*: For any $\mu \in \mathbf{R}$, the number of changes in sign of successive terms of the sequence $\{p_k(\mu)\}_{k=0}^n$ is equal to the number of eigenvalues of the matrix T and this number is strictly greater than μ. If the value of some member $p_i(\mu) = 0$, its sign be chosen opposite to that of $p_{i-1}(\mu)$.

Let $s(x)$ be the number of changes in sign in the Sturm's sequence for a given value of x. Now, for $a < b$, the number of roots of the characteristics equation $p_n(\lambda) = \det(T - \lambda I) = 0$ in (a,b) is given by $|s(a) - s(b)|$, provided $p_n(a) \neq 0$ and $p_n(b) \neq 0$.

In this way, we can compute approximately the eigenvalues of T and then improve them by using any of the methods discussed in Chapter 2. After finding out the eigenvalues of T, the corresponding eigenvectors can be obtained by the usual procedure. If Y be an eigenvector of T, then the corresponding eigenvector X of A is given by

$$X = PY$$
$$= P_1P_2 \dots P_rY$$

where $P = P_1P_2 \dots P_r$ is the orthogonal matrix employed for transforming A into T.

Example 8.10

Use the Givens method to find the approximate eigenvalues of following symmetric matrix:

$$A = \begin{bmatrix} 1 & \sqrt{2} & \sqrt{2} & 2 \\ \sqrt{2} & -\sqrt{2} & -1 & \sqrt{2} \\ \sqrt{2} & -1 & \sqrt{2} & \sqrt{2} \\ 2 & \sqrt{2} & \sqrt{2} & -3 \end{bmatrix}$$

Solution:

Here $n = 4$, we have to determine tridiagonal matrix similar to A after three similarity transformations.

Step 1: To obtain zeros in (1,3) and (3,1) positions of $P_1^{-1}AP_1$, we determine the rotation matrix P_1 as

$$P_1 = \begin{bmatrix} 1 & 0 & 0 & 0 \\ 0 & \cos\theta & -\sin\theta & 0 \\ 0 & \sin\theta & \cos\theta & 0 \\ 0 & 0 & 0 & 1 \end{bmatrix} \text{with}\,\theta = \frac{\pi}{4}$$

Therefore,

$$A_1 = P_1^{-1}AP_1 = \begin{bmatrix} 1 & 2 & 0 & 2 \\ 2 & -1 & \sqrt{2} & 2 \\ 0 & \sqrt{2} & 1 & 0 \\ 2 & 2 & 0 & -3 \end{bmatrix}$$

Step 2: To obtain zeros in (1,4) and (4,1) positions of $P_2^{-1}A_1P_2$, we determine the rotation matrix P_2 as

$$P_2 = \begin{bmatrix} 1 & 0 & 0 & 0 \\ 0 & \cos\theta & 0 & -\sin\theta \\ 0 & 0 & 1 & 0 \\ 0 & \sin\theta & 0 & \cos\theta \end{bmatrix} \text{with } \theta = \frac{\pi}{4}$$

Therefore,

$$A_2 = P_2^{-1}A_1P_2 = \begin{bmatrix} 1 & 2\sqrt{2} & 0 & 0 \\ 2\sqrt{2} & 0 & 1 & -1 \\ 0 & 1 & 1 & -1 \\ 0 & -1 & -1 & -4 \end{bmatrix}$$

Step 3: To obtain zeros in (2,4) and (4,2) positions of $P_3^{-1}A_2P_3$, we determine the rotation matrix P_3 as

$$P_3 = \begin{bmatrix} 1 & 0 & 0 & 0 \\ 0 & 1 & 0 & 0 \\ 0 & 0 & \cos\theta & -\sin\theta \\ 0 & 0 & \sin\theta & \cos\theta \end{bmatrix} \text{with}\,\theta = -\frac{\pi}{4}$$

Thus, we have

$$A_3 = P_3^{-1} A_2 P_3 = \begin{bmatrix} 1 & 2\sqrt{2} & 0 & 0 \\ 2\sqrt{2} & 0 & \sqrt{2} & -1 \\ 0 & \sqrt{2} & -1/2 & 5/2 \\ 0 & 0 & 5/2 & -5/2 \end{bmatrix}$$

which is the tridiagonal form original matrix $\boldsymbol{A}$.
Now, we determine the Sturm's sequence $\{p_k(\lambda)\}_{k=0}^{4}$ as follows:

$$p_0(\lambda) = 1$$

$$p_1(\lambda) = 1 - \lambda$$

$$p_2(\lambda) = (1-\lambda)(0-\lambda) - (2\sqrt{2})^2 = \lambda^2 - \lambda - 8$$

$$p_3(\lambda) = (-1/2-\lambda)(\lambda^2-\lambda-8) - (\sqrt{2})^2(1-\lambda) = -\lambda^3 + \frac{1}{2}\lambda^2 + \frac{21}{2}\lambda + 2$$

$$p_4(\lambda) = (-5/2-\lambda)(-\lambda^3 + \frac{1}{2}\lambda^2 + \frac{21}{2}\lambda + 2) - (5/2)^2(\lambda^2-\lambda-8) = \lambda^4 + 2\lambda^3 - 18\lambda^2 - 22\lambda + 45$$

λ	$p_0(\lambda)$	$p_1(\lambda)$	$p_2(\lambda)$	$p_3(\lambda)$	$p_4(\lambda)$	$s(\lambda)$
−5	+1	+6	+22	−63	+80	2 }
−4	+1	+5	+12	−48	−27	1 }
−3	+1	+4	+4	−34	−24	1 }
−2	+1	+3	−2	−21	+17	2 }
−1	+1	+2	−6	−9	+48	2
0	+1	+1	−8	+2	+45	2
1	+1	+0	−8	+12	+8	2 }
2	+1	−1	−6	+21	−39	3 }
3	+1	−2	−2	+29	−48	3 }
4	+1	−3	+4	+36	+53	2 }

The above table shows that the eigenvalues lie in the interval (−5,−4), (−3,−2), (1,2), and (3,4).

Then, we find the approximate root of the characteristic polynomial $p_4(\lambda) = 0$ in the above four intervals, yielding the four eigenvalues of the original matrix.

For the approximation of root, we use Newton–Raphson's method and the results are as follows:

In (−5,−4), the root $\lambda_1 = -4.42488$
In (−3,−2), the root $\lambda_2 = -2.39198$
In (1,2), the root $\lambda_3 = 1.16387$
In (3,4), the root $\lambda_4 = 3.65298$.

EXERCISES

1. Use the Gerschgorin's theorem to locate the region of the eigenvalues of the following matrix:

$$A = \begin{bmatrix} -4 & 8 & 0 \\ -5 & 13 & 0 \\ -1 & 0 & 2 \end{bmatrix}$$

2. Use the Gerschgorin's theorem to determine the approximate location of the eigenvalues of

a. $\begin{bmatrix} 1 & -1 & 0 \\ 1 & 5 & 1 \\ -2 & -1 & 9 \end{bmatrix}$

b. $\begin{bmatrix} -2 & 1 & 1 \\ 1 & 3 & 1 \\ 1 & -1 & 3 \end{bmatrix}$

3. Use the Gerschgorin's theorem to estimate $|\lambda_i - \bar{\lambda}_i|$, $i = 1,2,3$, where λ_i are the eigenvalues of

$$A = \begin{bmatrix} 2 & 3/2 & 0 \\ 1/2 & 1 & 0 \\ 0 & 0 & -1 \end{bmatrix}$$

and $\bar{\lambda}_i$ are the eigenvalues of $\tilde{A} = A + 10^{-2}\begin{bmatrix} 1 & -1 & 1 \\ -1 & 1 & -1 \\ 1 & -1 & 1 \end{bmatrix}$

4. Locate the region of the eigenvalues of the following matrices using the Gerschgorin's theorem:

a. $A = \begin{bmatrix} 1 & 24 & 0 \\ -3 & 8 & 0 \\ -1 & 0 & 2 \end{bmatrix}$

b. $A = \begin{bmatrix} -4 & 14 & 0 \\ -5 & 13 & 0 \\ -1 & 0 & 2 \end{bmatrix}$

5. Give a good estimate of the eigenvalues of the matrix A in the complex plane. Also give an upper estimate of the matrix norm of A, which corresponds to the Euclidean vector norm.

$$A = \begin{bmatrix} -1 & 0 & 1+2i \\ 0 & 2 & 1-i \\ 1-2i & 1+i & 0 \end{bmatrix}$$

6. Using the Gerchgorin's circle theorem, find the intervals which contain all the eigenvalues of the following symmetric matrices:

   a. $\begin{bmatrix} 1 & 2 & -3 \\ 2 & 1 & -1 \\ -3 & -1 & 2 \end{bmatrix}$

   b. $\begin{bmatrix} 2 & 3 & 1 \\ 3 & 2 & 2 \\ 1 & 2 & 1 \end{bmatrix}$

   c. $\begin{bmatrix} -1 & 1 & 1 & 1 \\ 1 & -1 & 1 & 1 \\ 1 & 1 & -1 & 1 \\ 1 & 1 & 1 & -1 \end{bmatrix}$

7. Find all the eigenvalues of the following matrix using the Householder method:

$$\begin{bmatrix} 1 & 2 & -1 \\ 2 & 1 & 2 \\ -1 & 2 & 1 \end{bmatrix}$$

8. Using the Householder method to reduce the following matrices to tridiagonal form:

   a. $\begin{bmatrix} 1 & 2 & 3 \\ 2 & 3 & 5 \\ 3 & 5 & 8 \end{bmatrix}$

   b. $\begin{bmatrix} 5 & 4 & 1 & 1 \\ 4 & 5 & 1 & 1 \\ 1 & 1 & 4 & 2 \\ 1 & 1 & 2 & 4 \end{bmatrix}$

   c. $\begin{bmatrix} 4 & 6 & 242 & 12 \\ 6 & 225 & 3 & 18 \\ 242 & 3 & 25 & 6 \\ 12 & 18 & 6 & 0 \end{bmatrix}$

   d. $\begin{bmatrix} 1 & 3 & 4 \\ 3 & 1 & 2 \\ 4 & 2 & 1 \end{bmatrix}$

   e. $\begin{bmatrix} 4 & -1 & 2 \\ -1 & 3 & 3 \\ -2 & 3 & 1 \end{bmatrix}$

f. $\begin{bmatrix} 7 & 2 & 3 & -1 \\ 2 & 8 & 5 & 1 \\ 3 & 5 & 12 & 9 \\ -1 & 1 & 9 & 7 \end{bmatrix}$

9. Compute eigenvalues and the corresponding eigenvectors of the following matrices, using the Householder method:

a. $\begin{bmatrix} \frac{1}{\sqrt{2}} & 1 & \sqrt{2} \\ 1 & \frac{3}{\sqrt{2}} & 1 \\ \sqrt{2} & 1 & \frac{1}{\sqrt{2}} \end{bmatrix}$

b. $\begin{bmatrix} 2 & -1 & 0 \\ -1 & 2 & -1 \\ 0 & -1 & 2 \end{bmatrix}$

c. $\begin{bmatrix} 5 & 0 & 1 \\ 0 & -2 & 0 \\ 1 & 0 & 5 \end{bmatrix}$

d. $\begin{bmatrix} 2 & 3 & 1 \\ 3 & 2 & 2 \\ 1 & 2 & 1 \end{bmatrix}$

10. Determine the largest eigenvalue and the corresponding eigenvector of the following matrices using power method:

a. $\begin{bmatrix} 1 & 6 & 1 \\ 1 & 2 & 0 \\ 0 & 0 & 3 \end{bmatrix}$

b. $\begin{bmatrix} 5 & 0 & 1 \\ 0 & -2 & 0 \\ 1 & 0 & 5 \end{bmatrix}$

c. $\begin{bmatrix} -15 & 4 & 3 \\ 10 & -12 & 6 \\ 20 & -4 & 2 \end{bmatrix}$

d. $\begin{bmatrix} 2 & -1 & 0 \\ -1 & 2 & -1 \\ 0 & -1 & 2 \end{bmatrix}$

e. $\begin{bmatrix} 3 & 2 & 4 \\ -1 & 4 & 10 \\ 1 & 3 & -1 \end{bmatrix}$

11. Find the smallest eigenvalue of the following matrix using inverse power method:

$$\begin{bmatrix} 1 & 6 & 1 \\ 1 & 2 & 0 \\ 0 & 0 & 3 \end{bmatrix}$$

12. Determine the largest/dominant eigenvalue and the corresponding eigenvector of the matrices:

a. $\begin{bmatrix} 1 & 3 & -1 \\ 3 & 2 & 4 \\ -1 & 4 & 10 \end{bmatrix}$

b. $\begin{bmatrix} 10 & -2 & 1 \\ -2 & 10 & -2 \\ 1 & -2 & 10 \end{bmatrix}$

c. $\begin{bmatrix} 10 & 7 & 8 & 7 \\ 7 & 5 & 6 & 5 \\ 8 & 6 & 10 & 9 \\ 7 & 5 & 9 & 10 \end{bmatrix}$

d. $\begin{bmatrix} 25 & 1 & 2 \\ 1 & 3 & 0 \\ 2 & 0 & -4 \end{bmatrix}$

e. $\begin{bmatrix} 1 & -3 & 2 \\ 4 & 4 & -1 \\ 6 & 3 & 5 \end{bmatrix}$

13. Evaluate the smallest eigenvalue of the following system using the inverse power method:

$$A = \begin{bmatrix} -4 & 14 & 0 \\ -5 & 13 & 0 \\ -1 & 0 & 2 \end{bmatrix}$$

14. Using Householder's method to convert the following matrix A to upper-Hessenberg form:

$$A = \begin{bmatrix} 1 & -2 & 3 \\ -2 & 4 & 1 \\ 3 & 1 & 2 \end{bmatrix}$$

15. Find the numerically largest eigenvalues, the corresponding eigenvectors of the following matrices, using the power method:

a. $\begin{bmatrix} 5 & 2 & 0 & 0 \\ 2 & 2 & 0 & 0 \\ 0 & 0 & 5 & -2 \\ 0 & 0 & -2 & 2 \end{bmatrix}$

b. $\begin{bmatrix} 10 & 4 & -1 \\ 4 & 2 & 3 \\ -1 & 3 & 1 \end{bmatrix}$

c. $\begin{bmatrix} 1 & 3 & 2 \\ -1 & 0 & 2 \\ 3 & 4 & 5 \end{bmatrix}$

16. Use the QR method (a) without shift, and (b) with shift, to calculate the eigenvalues of

a. $\begin{bmatrix} 3 & 1 & 0 \\ 1 & 2 & 1 \\ 0 & 1 & 1 \end{bmatrix}$

b. $\begin{bmatrix} 2 & 1 & 0 & 0 \\ 1 & 2 & 1 & 0 \\ 0 & 1 & 2 & 1 \\ 0 & 0 & 1 & 2 \end{bmatrix}$

c. $\begin{bmatrix} 0 & 1 & 0 & 0 & 0 \\ 1 & 1 & 1 & 0 & 0 \\ 0 & 1 & 1 & 1 & 0 \\ 0 & 0 & 1 & 1 & 1 \\ 0 & 0 & 0 & 1 & 2 \end{bmatrix}$

17. Find all the eigenvalues and the eigenvectors of the following matrices using the Jacobi's method (perform only three iterations):

a. $\begin{bmatrix} 1/\sqrt{2} & 1 & \sqrt{2} \\ 1 & 3/\sqrt{2} & 1 \\ \sqrt{2} & 1 & 1/\sqrt{2} \end{bmatrix}$

b. $\begin{bmatrix} 5 & 0 & 1 \\ 0 & -2 & 0 \\ 1 & 0 & 5 \end{bmatrix}$

c. $\begin{bmatrix} 2 & -1 & 0 \\ -1 & 2 & -1 \\ 0 & -1 & 2 \end{bmatrix}$

18. Use the Householder matrices to produce the QR factorization of

a. $\begin{bmatrix} 1 & 1 & 1 \\ 2 & -1 & -1 \\ 2 & -4 & 5 \end{bmatrix}$

b. $\begin{bmatrix} 1 & 3 & -2 \\ -1 & -2 & 3 \\ 1 & 1 & 2 \end{bmatrix}$

19. Find the eigenvalue correct to two decimal places which is nearest to five for the following matrix using inverse power method. Obtain the corresponding eigenvector. Take the initial approximate vector as $[1 \quad 1 \quad 1]^T$:

$$\begin{bmatrix} 4 & 1 & 0 \\ 1 & 4 & 1 \\ 0 & 1 & 4 \end{bmatrix}$$

20. Find approximately the eigenvalues of the following matrices using the QR method:

a. $\begin{bmatrix} 3 & 1 \\ 1 & 1 \end{bmatrix}$

b. $\begin{bmatrix} -15 & 4 & 3 \\ 10 & -12 & 6 \\ 20 & -4 & 2 \end{bmatrix}$

21. Using Jacobi's method, find all the eigenvalues and eigenvectors of the following matrices:

a. $\begin{bmatrix} 1 & 2 & 3 \\ 0 & -4 & 2 \\ 0 & 0 & 7 \end{bmatrix}$

b. $\begin{bmatrix} 5 & 0 & 1 \\ 0 & -2 & 0 \\ -1 & 0 & 5 \end{bmatrix}$

c. $\begin{bmatrix} 1 & 1 & 0.5 \\ 1 & 1 & 0.25 \\ 0.5 & 0.25 & 2 \end{bmatrix}$

d. $\begin{bmatrix} 2 & -1 & 0 \\ -1 & 2 & -1 \\ 0 & -1 & 2 \end{bmatrix}$

22. Using Jacobi's method to find the eigenvalues and eigenvectors of the following matrices:

a. $\begin{bmatrix} 1 & 2 \\ 2 & 1 \end{bmatrix}$

b. $\begin{bmatrix} 0 & 1 \\ 1 & 0 \end{bmatrix}$

c. $\begin{bmatrix} -2 & -2 & 6 \\ -2 & 3 & 4 \\ 6 & 4 & -1 \end{bmatrix}$

d. $\begin{bmatrix} 3 & 2 & 2 \\ 2 & 5 & 2 \\ 2 & 2 & 3 \end{bmatrix}$

e. $\begin{bmatrix} 1 & -2 & 4 \\ -2 & 5 & -2 \\ 4 & -2 & 1 \end{bmatrix}$

f. $\begin{bmatrix} 1 & \sqrt{3} & 4 \\ \sqrt{3} & 5 & \sqrt{3} \\ 4 & \sqrt{3} & 1 \end{bmatrix}$

g. $\begin{bmatrix} 2 & \sqrt{2} & 4 \\ \sqrt{2} & 6 & \sqrt{2} \\ 4 & \sqrt{2} & 2 \end{bmatrix}$

23. Use the *QR* method to compute the eigenvalues of the following matrices:

a. $\begin{bmatrix} 337 & 304 & 176 \\ 304 & 361 & 128 \\ 176 & 128 & 121 \end{bmatrix}$

b. $\begin{bmatrix} 12 & 3 & 1 \\ -9 & -2 & -3 \\ 14 & 6 & 2 \end{bmatrix}$

c. $\begin{bmatrix} 2 & \frac{1}{3} & 1 \\ 3 & \frac{-5}{3} & 1 \\ 0 & \frac{13}{9} & \frac{5}{3} \end{bmatrix}$

24. Find all eigenvalues and the corresponding eigenvector using the Givens method:

a. $\begin{bmatrix} 1 & 2 & 3 \\ 0 & 2 & 3 \\ 0 & 0 & 2 \end{bmatrix}$

b. $\begin{bmatrix} 1 & \sqrt{2} & 2 \\ \sqrt{2} & 3 & \sqrt{2} \\ 2 & \sqrt{2} & 1 \end{bmatrix}$

c. $\begin{bmatrix} 2 & 3 & 1 \\ 3 & 2 & 2 \\ 1 & 2 & 1 \end{bmatrix}$

d. $\begin{bmatrix} 1.5 & -0.5 & 0 \\ -0.5 & 1 & -0.5 \\ 0 & -0.5 & 1.5 \end{bmatrix}$

25. Transform the following matrices to diagonal form using the Givens method. Using the Strum's sequence obtain exact eigenvalues:

a. $\begin{bmatrix} 3 & 1 & 1 \\ 1 & 3 & 2 \\ 1 & 2 & 3 \end{bmatrix}$

b. $\begin{bmatrix} 2 & 3 & 1 \\ 3 & 2 & 2 \\ 1 & 2 & 1 \end{bmatrix}$

c. $\begin{bmatrix} 1 & 2 & 2 \\ 2 & 1 & -1 \\ 2 & -1 & 1 \end{bmatrix}$

d. $\begin{bmatrix} 2 & 1 & \sqrt{3} \\ 1 & 2 & \sqrt{3} \\ \sqrt{3} & \sqrt{3} & 3 \end{bmatrix}$

26. Show that the matrix A has one eigenvalue in the interval (2, 4) using the Strum's sequence:

$$A = \begin{bmatrix} 5 & -2 & 0 & 0 \\ -2 & 4 & -1 & 0 \\ 0 & -1 & 4 & -2 \\ 0 & 0 & -2 & 5 \end{bmatrix}$$

27. Reduce the following matrices into the tridiagonal form using the Givens method:

a. $\begin{bmatrix} 2 & -1 & -1 \\ -1 & 2 & -1 \\ -1 & -1 & 2 \end{bmatrix}$

b. $\begin{bmatrix} 4 & 2 & 2 \\ 2 & 5 & 1 \\ 2 & 1 & 6 \end{bmatrix}$

c. $\begin{bmatrix} 2 & 1 & 3 \\ 1 & 4 & 2 \\ 3 & 2 & 3 \end{bmatrix}$

d. $\begin{bmatrix} 4 & 1 & 2 \\ 1 & 4 & 1 \\ 2 & 1 & 0 \end{bmatrix}$

9 Approximation of Functions

9.1 INTRODUCTION

In this chapter, we are interested in approximation problems. For evaluating a function $f(x)$ on a computer, in the context of space and time complexity, it is obviously better to have an analytic approximation to $f(x)$ rather than to store a table and use interpolation because the computational overhead in function evaluation through interpolation techniques over stored tabular values is much higher in comparison to the use of efficient function approximations. Generally speaking, starting from a function $f(x)$, we would like to find a different function $\phi(x)$ that belongs to a given class of functions and is *close* to $f(x)$ in some sense. As far as the class of functions that $\phi(x)$ belongs to, we will typically assume that $\phi(x)$ is a polynomial of a given degree (though it can be a trigonometric function or any other function). It is desirable to use the lowest possible degree of polynomial that will give the desired accuracy in approximating $f(x)$. A typical approximation problem will be therefore to find the *closest* polynomial of degree less than equal to $f(x)$.

Other than polynomials, there are different forms of approximating functions such as the Padé approximation. In this approximation, rational functions are quotients of polynomials and they are usually a more efficient form of approximations.

There are different ways of measuring the *distance* between two functions. Among them, we will focus on two such measurements the L_∞ norm and the L_2 norm.

We recall that a norm on a vector space V over $\boldsymbol{R}$ is a function $\|\,.\,\|: V \to \boldsymbol{R}$ with the following properties:

1. $\|f\|$ is a nonnegative real number, that is, $\|f\| \geq 0$ for all $f \in V$ (nonnegativity).
2. $\|f\| = 0$ if and only if f is the zero element of V (positive-definiteness).
3. $\|kf\| = |k|\,\|f\|$, for all $k \in \boldsymbol{R}$ and for all $f \in V$ (absolute homogeneity or absolute scalability).
4. $\|f + g\| \leq \|f\| + \|g\|$, for all $f, g \in V$ (triangle inequality or subadditivity).

We assume that the function $f(x) \in C[a,b]$ be a continuous function on a closed interval $[a,b]$, which attains its maximum in the same interval $[a,b]$. We can therefore define the L_∞-norm (also known as the maximum norm) of such a function $f(x)$ by

$$\|f\|_\infty = \max_{x \in [a,b]} |f(x)| \tag{9.1}$$

The L_∞ norm between two functions $f(x), g(x) \in C[a,b]$ is given by

$$\|f - g\|_\infty = \max_{x \in [a,b]} |f(x) - g(x)| \tag{9.2}$$

We now proceed by defining the L_2 norm of a continuous function $f(x)$ as

$$\|f\|_2 = \sqrt{\int_a^b |f(x)|^2\, dx} \tag{9.3}$$

The L_2 function space is the collection of functions $f(x)$ for which $\|f\|_2 < \infty$. The L_2 norm between two functions $f(x)$ and $g(x)$ is given by

$$\|f-g\|_2 = \sqrt{\int_a^b |f(x)-g(x)|^2\, dx} \tag{9.4}$$

To justify using polynomials to approximate continuous functions, the following theorem, the Weierstrass approximation theorem, which plays a central role in any discussion of approximations of functions, is discussed.

Theorem 9.1: Weierstrass Approximation Theorem

Suppose $f(x)$ be a continuous, real-valued function defined on the real interval $[a, b]$, then for any $\varepsilon > 0$, there exists a polynomial $p(x)$ such that for all x in $[a,b]$,

$$\|f(x)-p(x)\| < \varepsilon. \tag{9.5}$$

This is an important theorem in classical analysis, and there are many proofs of this result. We will provide a constructive proof of the Weierstrass approximation theorem. It is evidently sufficient to consider only the interval [0, 1], an appropriate change of variable will then extend the proof to any bounded closed interval $[a, b]$. For a real-valued function f, defined and continuous on the interval [0, 1], the following proof uses the Bernstein polynomials. First, we will define a family of polynomials, known as the Bernstein polynomials, and then we will show that they uniformly converge to $f(x)$.

We start with the definition and properties of the Bernstein polynomials as follows.

9.1.1 Bernstein Polynomials and Its Properties

The general form of the Bernstein polynomials of nth degree over the interval $[a, b]$ is given by

$$B_{i,n}(x) = \binom{n}{i}\frac{(x-a)^i(b-x)^{n-i}}{(b-a)^n} \quad i = 0,1,\ldots,n \tag{9.6}$$

where $\binom{n}{i} = n!/(i!\cdot(n-i)!)$.

Note that each of these $(n+1)$ polynomials having degree n satisfies the following properties:

1. $B_{i,n}(x) = 0, \quad \text{if } i < 0 \text{ or } i > n$
2. $B_{i,n}(a) = B_{i,n}(b) = 0, \quad \text{for } 1 \le i \le n-1$
3. $\sum_{i=0}^{n} B_{i,n}(x) = 1$

For $n = 10$, the Bernstein polynomial basis functions over the interval [0, 1] are given as follows, and the graph of all its basis functions are shown in Figure 9.1.

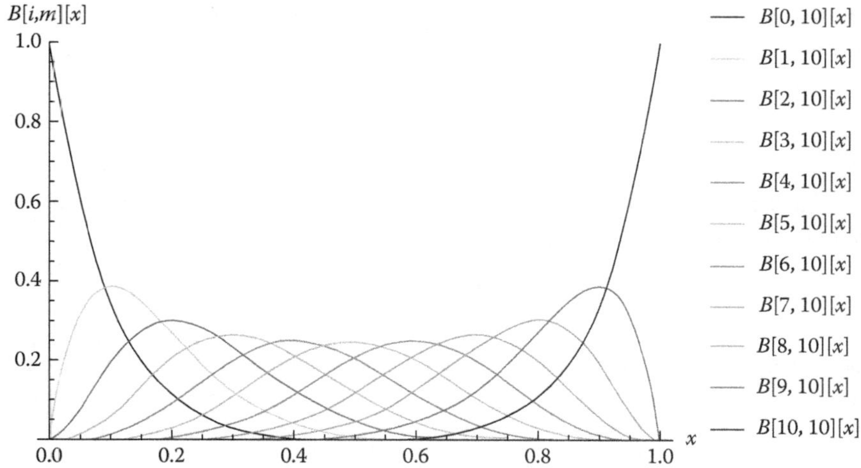

FIGURE 9.1 **(See color insert.)** Graph of 10-degree Bernstein polynomials over [0, 1].

$$B_{0,10}(x) = (1-x)^{10}$$

$$B_{1,10}(x) = 10(1-x)^9 x$$

$$B_{2,10}(x) = 45(1-x)^8 x^2$$

$$B_{3,10}(x) = 120(1-x)^7 x^3$$

$$B_{4,10}(x) = 210(1-x)^6 x^4$$

$$B_{5,10}(x) = 252(1-x)^5 x^5$$

$$B_{6,10}(x) = 210(1-x)^4 x^6$$

$$B_{7,10}(x) = 120(1-x)^3 x^7$$

$$B_{8,10}(x) = 45(1-x)^2 x^8$$

$$B_{9,10}(x) = 10(1-x)x^9$$

$$B_{10,10}(x) = x^{10}$$

A function $f(x)$ defined over $[a, b]$ can be approximated by Bernstein polynomials basis functions of degree n as

$$f(x) \approx \sum_{i=0}^{n} c_i B_{i,n}(x) \tag{9.7}$$

Explicitly, if f be a continuous function on [0, 1], then $B_n(f)$ is called nth Bernstein polynomial for f and defined as

$$B_n(f)(x) = \sum_{i=0}^{n} f\left(\frac{i}{n}\right) B_{i,n}(x) \tag{9.8}$$

Bernstein polynomials defined earlier form a complete basis over the interval [0, 1].

The error bound is given by

$$\left|(B_n f)(x) - f(x)\right| \le \frac{1}{2n} x(1-x)\left\|f''\right\|_\infty, \quad \text{for all } x \in [a,b] \tag{9.9}$$

which shows that the rate of convergence is at least $1/n$ for $f \in C^2[a,b]$. For each function $f:[a,b] \to R$ and a point of continuity x of f, we have

$$\lim_{n\to\infty} (B_n f)(x) = f(x) \tag{9.10}$$

and the convergence is uniform if f is continuous. If the second derivative $f''(x)$ of the function f exists, then

$$\lim_{n\to\infty} n((B_n f)(x) - f(x)) = \frac{1}{2} x(1-x) f''(x) \tag{9.11}$$

Therefore, $(B_n f)(x) = f(x) + O(1/n)$.

We are now ready to prove the Weierstrass approximation theorem, Theorem 9.1.

Proof:

We will prove the theorem in the interval [0, 1]. The proof can be extended to $[a, b]$. Since $f(x)$ is continuous on a closed interval [0, 1], it is uniformly continuous therein.

Then for all $x, y \in [0,1]$, such that $|x - y| < \delta$,

$$|f(x) - f(y)| < \varepsilon \tag{9.12}$$

Since $f(x)$ is continuous on a closed interval [0, 1], it is also bounded in [0, 1]. Let

$$M = \max_{x\in[0,1]} |f(x)|$$

We would now like to estimate the difference between $B_n f$ and f.

$$\begin{aligned} B_n(f)(x) - f(x) &= \sum_{i=0}^{n} \binom{n}{i} x^i (1-x)^{n-i} f\left(\frac{i}{n}\right) - f(x) \sum_{i=0}^{n} \binom{n}{i} x^i (1-x)^{n-i} \\ &= \sum_{i=0}^{n} \binom{n}{i} x^i (1-x)^{n-i} \left(f\left(\frac{i}{n}\right) - f(x) \right) \end{aligned} \tag{9.13}$$

If follows that for all $x \in [0,1]$,

$$|B_n(f)(x) - f(x)| \le \sum_{i=0}^{n} \binom{n}{i} x^i (1-x)^{n-i} \left| f\left(\frac{i}{n}\right) - f(x) \right| \tag{9.14}$$

Since, $f(x)$ is uniformly continuous on [0,1]. So, if $|(i/n) - x| < \delta$, then from Equation 9.12

$$\left| f\left(\frac{i}{n}\right) - f(x) \right| < \varepsilon \tag{9.15}$$

Also, if $|(i/n)-x| \geq \delta$, then

$$\left|f\left(\frac{i}{n}\right)-f(x)\right| \leq 2M \leq 2M\left(\frac{(i/n)-x}{\delta}\right)^2 \tag{9.16}$$

Combining Equations 9.15 and 9.16, we obtain

$$\left|f\left(\frac{i}{n}\right)-f(x)\right| \leq \frac{\varepsilon}{2}+M\left(\frac{(i/n)-x}{\delta}\right)^2 \tag{9.17}$$

Now, from Equation 9.14, we obtain

$$\begin{aligned}|B_n(f)(x)-f(x)| &\leq \frac{\varepsilon}{2}\sum_{i=0}^{n}\binom{n}{i}x^i(1-x)^{n-i}+\frac{M}{\delta^2}\sum_{i=0}^{n}\binom{n}{i}x^i(1-x)^{n-i}\left(\frac{i}{n}-x\right)^2 \\ &= \frac{\varepsilon}{2}+\frac{M}{\delta^2}\frac{x(1-x)}{n} \leq \frac{\varepsilon}{2}+\frac{M}{n\delta^2}\end{aligned} \tag{9.18}$$

For this δ, we can choose N large enough, so that when $n \geq N, N > (2M/\delta^2\varepsilon)$.
Therefore, we have for all $n \geq N$,

$$\|B_n(f)-f\|_\infty < \frac{\varepsilon}{2}+\frac{\varepsilon}{2}=\varepsilon \tag{9.19}$$

and thus $B_n(f)$ converges uniformly to f as $n \to \infty$. ■

9.2 LEAST SQUARE CURVE FITTING

9.2.1 Straight Line Fitting

Let, (x_i, y_i), $i = 1,2,\ldots,n$ be a given set of n pairs of observations.
Let us consider the fitting of a straight line

$$y = a + bx \tag{9.20}$$

Using Legendre's principle of least squares, we have to determine the coefficients a and b, so that

$$S = \sum_{i=1}^{n}(y_i - a - bx_i)^2 \tag{9.21}$$

is minimum.

From the principle of maxima and minima, the partial derivatives of S with respect to a and b should vanish separately.

$$\frac{\partial S}{\partial a} = -2\sum_{i=1}^{n}(y_i - a - bx_i) = 0, \quad \frac{\partial S}{\partial b} = -2\sum_{i=1}^{n}x_i(y_i - a - bx_i) = 0 \tag{9.22}$$

Thus we have

$$\sum_{i=1}^{n} y_i = na + b\sum_{i=1}^{n} x_i, \quad \sum_{i=1}^{n} x_i y_i = a\sum_{i=1}^{n} x_i + b\sum_{i=1}^{n} x_i^2 \tag{9.23}$$

The equations in (9.23) are called normal equations. These equations in (9.23) are solved for constants a and b. With these values of a and b thus obtained, Equation 9.20 is the line of best fit to the given set of points (x_i, y_i), $i = 1, 2, \ldots, n$.

Example 9.1

Using the method of least squares, fit a curve of the form $y = ab^x$ to the following data:

x	1	2	3	4
y	4	11	35	100

Solution:

Taking logarithm on both sides of $y = ab^x$, we get

$$\ln y = \ln a + x \ln b$$

This implies that

$$Y = A + Bx$$

where

$$A = \ln a, \quad B = \ln b, \quad \text{and } Y = \ln y$$

x_i	y_i	$Y_i = \ln y_i$	x_i^2	x_iY_i
1	4	1.3863	1	1.3863
2	11	2.3979	4	4.7958
3	35	3.5553	9	10.6659
4	100	4.6052	16	18.4208
$\sum x_i = 10$	—	$\sum Y_i = 11.9447$	$\sum x_i^2 = 30$	$\sum x_iY_i = 35.2688$

The normal equations are

$$\sum_{i=1}^{4} Y_i = 4A + B\sum_{i=1}^{4} x_i$$

$$\sum_{i=1}^{4} x_i Y_i = A\sum_{i=1}^{4} x_i + B\sum_{i=1}^{4} x_i^2$$

Therefore, we have

$$11.9447 = 4A + 10B$$

$$35.2688 = 10A + 30B$$

Solving the above equations for A and B, we get

$$A = 0.2827, \quad B = 1.0814$$

Thus, we obtain

$$a = 1.3267, \quad b = 2.9488$$

Hence, the required best fitting curve is

$$y = 1.3267(2.9488)^x$$

9.2.2 Fitting of kth Degree Polynomial

Let

$$y = a_0 + a_1 x + a_2 x^2 + \cdots + a_k x^k \tag{9.24}$$

be the kth degree polynomial that best fits to the set of points $(x_i, y_i), i = 1, 2, \ldots, n$.

Using Legendre's principle of least squares, we have to determine the coefficients $a_0, a_1, \ldots, a_k$, so that

$$S = \sum_{i=1}^{n} \left(y - a_0 - a_1 x_i - a_2 x_i^2 - \cdots - a_k x_i^k\right)^2 \tag{9.25}$$

is minimum.

The $(k+1)$ normal equations are

$$\begin{aligned}
\frac{\partial S}{\partial a_1} &= -2\sum_{i=1}^{n}\left(y_i - a_0 - a_1 x_i - a_2 x_i^2 - \cdots - a_k x_i^k\right) = 0 \\
\frac{\partial S}{\partial a_2} &= -2\sum_{i=1}^{n} x_i\left(y_i - a_0 - a_1 x_i - a_2 x_i^2 - \cdots - a_k x_i^k\right) = 0 \\
&\vdots \\
\frac{\partial S}{\partial a_n} &= -2\sum_{i=1}^{n} x_i^k\left(y_i - a_0 - a_1 x_i - a_2 x_i^2 - \cdots - a_k x_i^k\right) = 0
\end{aligned} \tag{9.26}$$

Thus, we obtain

$$\begin{aligned}
&\sum_{i=1}^{n} y_i = n a_0 + a_1 \sum_{i=1}^{n} x_i + a_2 \sum_{i=1}^{n} x_i^2 + \cdots + a_k \sum_{i=1}^{n} x_i^k \\
&\sum_{i=1}^{n} x_i y_i = a_0 \sum_{i=1}^{n} x_i + a_1 \sum_{i=1}^{n} x_i^2 + a_2 \sum_{i=1}^{n} x_i^3 + \cdots + a_k \sum_{i=1}^{n} x_i^{k+1} \\
&\vdots \\
&\sum_{i=1}^{n} x_i^k y_i = a_0 \sum_{i=1}^{n} x_i^k + a_1 \sum_{i=1}^{n} x_i^{k+1} + a_2 \sum_{i=1}^{n} x_i^{k+2} + \cdots + a_k \sum_{i=1}^{n} x_i^{2k}
\end{aligned} \tag{9.27}$$

These are $(k+1)$ linear normal equations in $(k+1)$ unknowns $a_0, a_1, \ldots, a_k$. Solving these normal equations for $a_0, a_1, \ldots, a_k$, we can determine the kth degree polynomial of best fit to the given set of points $(x_i, y_i), i = 1, 2, \ldots, n$.

Example 9.2

Fit a parabola of second degree to the following data:

x	0	1	2	3	4
y	1	1.8	1.3	2.5	6.3

Solution:

Let $y = a + bx + cx^2$ be the second-degree parabola.

x_i	y_i	x_i^2	x_i^3	x_i^4	$x_i y_i$	$x_i^2 y_i$
0	1.0	0	0	0	0	0
1	1.8	1	1	1	1.8	1.8
2	1.3	4	8	16	2.6	5.2
3	2.5	9	27	81	7.5	22.5
4	6.3	16	64	256	25.2	100.8
$\sum x_i = 10$	$\sum y_i = 12.9$	$\sum x_i^2 = 30$	$\sum x_i^3 = 100$	$\sum x_i^4 = 354$	$\sum x_i y_i = 37.1$	$\sum x_i^2 y_i = 130.3$

Using Normal equations, we get

$$12.9 = 5a + 10b + 30c; \quad 37.1 = 10a + 30b + 100c; \quad 130.3 = 30a + 100b + 354c$$

Solving these equations, we get

$$a = 1.42, \quad b = -1.07, \quad \text{and} \quad c = 0.55$$

Therefore, the required equation of the second-degree parabola is

$$y = 1.42 - 1.07x + 0.55x^2$$

Example 9.3

The pressure and volume of a gas are related by the equation $pv^{\lambda} = k$ (λ and k are constants). Fit this equation for the following data using the principle of least squares

p	0.5	1	1.5	2	2.5	3
v	1.62	1.00	0.75	0.62	0.52	0.46

Solution:

Taking logarithm on both sides of $pv^{\lambda} = k$, we get

$$\ln p + \lambda \ln v = \ln k$$

This implies that

$$y = a + bx$$

where $y = \ln p$, $x = \ln v$, $a = \ln k$, and $b = -\lambda$.

v_i	p_i	$x_i = \ln v_i$	$y_i = \ln p_i$	$x_i y_i$	x_i^2
1.62	0.5	0.4824	−0.6931	−0.3344	0.2327
1.00	1	0	0	0	0
0.75	1.5	−0.2877	0.4055	−0.1167	0.0829
0.62	2	−0.4780	0.6931	−0.3313	0.2285
0.52	2.5	−0.6539	0.9163	−0.5992	0.4276
0.46	3	−0.7765	1.0986	−0.8531	0.6030
—	—	$\sum x_i = -1.7137$	$\sum y_i = 2.4204$	$\sum x_i y_i = -2.2347$	$\sum x_i^2 = 1.5745$

The normal equations are

$$2.4204 = 6a + (-1.7137)b$$

$$-2.2347 = (-1.7137)a + (1.5745)b$$

Solving the above equations, we get

$$a = -0.0029 \quad \text{and} \quad b = -1.4224$$

This implies

$$\ln k = -0.0029 \quad \text{and} \quad -\lambda = -1.4224$$

yielding

$$k = 0.9971 \quad \text{and} \quad \lambda = 1.4224$$

Hence, the required best fitting curve is

$$pv^{1.4224} = 0.9971$$

9.3 LEAST SQUARES APPROXIMATION

Let us suppose that $f(x) \in C[a,b]$. We find a polynomial $P_n(x)$ of degree less than equal to n such that the error

$$\|f(x) - P_n(x)\|_2^2 = \int_a^b [f(x) - P_n(x)]^2 dx \tag{9.28}$$

is minimized.

To determine the least squares approximating polynomial, let

$$P_n(x) = a_0 + a_1 x + \cdots + a_{n-1}x^{n-1} + a_n x^n = \sum_{i=0}^{n} a_i x^i \tag{9.29}$$

and we define

$$E = \int_a^b [f(x) - \sum_{i=0}^{n} a_i x^i]^2 dx \tag{9.30}$$

Now, we shall determine the coefficients $a_0, a_1, \ldots, a_n$ that will minimize E. A necessary condition for the numbers $a_0, a_1, \ldots, a_n$ to minimize E is that

$$\frac{\partial E}{\partial a_k} = 0, \quad k = 0,1,2,\ldots,n \tag{9.31}$$

These are called normal equations. So, there are $(n+1)$ normal equations in $(n+1)$ unknowns, viz., $a_0, a_1, \ldots, a_n$.

From Equation 9.30, we have

$$E = \int_a^b [f(x)]^2 dx - 2\sum_{i=0}^{n} a_i \int_a^b x^i f(x) dx + \int_a^b \left(\sum_{i=0}^{n} a_i x^i \right)^2 dx \tag{9.32}$$

Therefore, the $(n+1)$ linear normal equations are

$$\frac{\partial E}{\partial a_k} = -2\int_a^b x^k f(x) dx + 2\sum_{i=0}^{n} a_i \int_a^b x^{i+k} dx = 0, \quad k = 0,1,2,\ldots,n \tag{9.33}$$

Now, to determine the polynomial $P_n(x)$, the $(n+1)$ linear normal equations are solved for $(n+1)$ unknowns $a_0, a_1, \ldots, a_n$. Since the coefficients $a_0, a_1, \ldots, a_n$ are determined uniquely by solving the normal equations in Equation 9.33, the polynomial $P_n(x)$ exists and is unique, provided that $f(x) \in C[a,b]$.

9.4 ORTHOGONAL POLYNOMIALS

The method of least squares approximation discussed earlier has the disadvantage that sometimes it needs to solve a large system of linear equations which may be ill-conditioned. It causes large errors in unknown coefficients. This difficulty can be avoided by using the orthogonal polynomials. It has a great advantage that it does not require solving a system of linear equations.

The orthogonal polynomials play a central role in the solution of least-squares problems. In this section, we will focus on the construction of orthogonal polynomials.

9.4.1 Weight Function

Definition 9.1

An integrable function $w(x)$ is called a weight function on the interval I, if $w(x) \geq 0$ for all x in I, but $w(x) \neq 0$ on any subinterval of I.

We start by defining the weighted inner product between two functions $f(x)$ and $g(x)$ (with respect to the weight $w(x)$):

$$< f, g >_w = \int_a^b w(x) f(x) g(x) dx$$

To simplify the notations, even in the weighted case, we will typically write $< f,g >$ instead of $< f,g >_w$. Some properties of the weighted inner product are

1. $< \alpha f_1 + \beta f_2, g > = \alpha < f_1, g > + \beta < f_2, g >$, for all $\alpha, \beta \in \boldsymbol{R}$
2. $< f,g > = < g,f >$.
3. $< f,f > \geq 0$ and $< f,f > = 0$ if and only if $f \equiv 0$. Here we must assume that $f(x) \in C[a,b]$. If $f(x)$ is not continuous, we can have $< f,f > = 0$ and $f(x)$ can still be nonzero.

The weighted L_2 norm can be obtained from the weighted inner product by

$$\|f\|_{2,w} = \sqrt{< f,f >_w}$$

Let $\{\phi_0, \phi_1, \ldots, \phi_n\}$ be a set of linearly independent functions on $[a,b]$ and $w(x)$ be a weight function on $[a,b]$. Let us suppose that $f(x) \in C[a,b]$, then we consider an approximation

$$f(x) = \sum_{i=0}^{n} c_i \phi_i(x) \tag{9.34}$$

so that the error

$$E = \int_a^b w(x)[f(x) - \sum_{i=0}^{n} c_i \phi_i(x)]^2 dx$$

is minimum.

The normal equations associated with this problem are

$$\frac{\partial E}{\partial c_k} = -2\int_a^b w(x)[f(x) - \sum_{i=0}^{n} c_i \phi_i(x)]\,\phi_k(x)dx = 0, \quad k = 0,1,2,\ldots,n \tag{9.35}$$

Now the system of normal equations can be written as

$$\int_a^b w(x)f(x)\,\phi_k(x)\,dx = \sum_{i=0}^{n} c_i \int_a^b w(x)\,\phi_i(x)\,\phi_k(x)\,dx, \quad k = 0,1,2,\ldots,n \tag{9.36}$$

If the functions are chosen such that

$$\int_a^b w(x)\phi_i(x)\,\phi_k(x)dx = \begin{cases} \alpha_i(>0), & \text{if } i = k \\ 0, & \text{if } i \neq k \end{cases} \tag{9.37}$$

then Equation 9.36 becomes

$$\int_a^b w(x)f(x)\,\phi_k(x)dx = c_k\alpha_k, \quad k = 0,1,2,\ldots,n \tag{9.38}$$

Solving these equations, we obtain

$$c_k = \frac{1}{\alpha_k}\int_a^b w(x)f(x)\phi_k(x)dx, \quad k = 0,1,2,\ldots,n \tag{9.39}$$

Substituting the values of c_i, $i = 0,1,2,\ldots,n$, in Equation 9.34, we obtain the least squares approximation. Thus the least squares approximation problem is greatly simplified if the functions $\phi_i(x)$, $i = 0,1,2,\ldots,n$ are chosen as orthogonal functions, as shown in Equation 9.37. In this way, ill-conditioning in normal equations is avoided, and the coefficients c_i, $i = 0,1,2,\ldots,n$ are all determined.

Definition 9.2

A set $\{\phi_0, \phi_1, \ldots, \phi_n\}$ is said to be an orthogonal set of functions on $[a,b]$ with respect to the weight function $w(x)$ if

$$\int_a^b w(x)\phi_i(x)\phi_k(x)dx = \begin{cases} \alpha_i(>0), & \text{if } i = k \\ 0, & \text{if } i \neq k \end{cases} \tag{9.40}$$

If the set $\{\phi_0, \phi_1, \ldots, \phi_n\}$ satisfies the following condition

$$\int_a^b w(x)\phi_i(x)\phi_k(x)dx = \begin{cases} 1, & \text{if } i = k \\ 0, & \text{if } i \neq k \end{cases} \tag{9.41}$$

then the set is said to be orthonormal.

Thus we have the following theorem.

Theorem 9.2

If $\{\phi_0, \phi_1, \ldots, \phi_n\}$ be an orthogonal set of functions on $[a,b]$ with respect to the weight function $w(x)$, then the least squares approximation to $f(x)$ on $[a,b]$ with respect to $w(x)$ is

$$f(x) = \sum_{i=0}^{n} c_i\phi_i(x) \tag{9.42}$$

where

$$c_k = \frac{1}{\alpha_k}\int_a^b w(x)f(x)\phi_k(x)dx = \frac{\int_a^b w(x)f(x)\phi_k(x)dx}{\int_a^b w(x)[\phi_k(x)]^2 dx}, \quad k = 0,1,2,\ldots,n \tag{9.43}$$

A family of functions is an orthogonal family if each member is orthogonal to every other member of that family. It is called an orthonormal family if it is an orthogonal family and $\|\phi_k\|_2 = 1$, for each $k = 0,1,2,\ldots,n$.

A few examples of orthogonal polynomials are list as follows:

Type	Interval	Orthogonal Polynomial $\phi(x)$	Weight Function
Chebyshev (1st kind)	$[-1,1]$	$T_n(x)$	$\frac{1}{\sqrt{1-x^2}}$
Chebyshev (2nd kind)	$[-1,1]$	$U_n(x)$	$\sqrt{1-x^2}$
Legendre	$[-1,1]$	$P_n(x)$	1
Laguerre	$[0, \infty)$	$L_n(x)$	e^{-x}
Hermite	$(-\infty,\infty)$	$H_n(x)$	e^{-x^2}

There exists a procedure to construct orthogonal polynomials $\phi_i(x)$, $i = 0,1,2,\ldots,n$ on an interval $[a,b]$, with respect to any weight function $w(x)$. This is known as Gram–Schmidt orthogonalization process. It describes how to construct orthogonal polynomials on $[a,b]$ with respect to a weight function $w(x)$.

9.4.2 Gram–Schmidt Orthogonalization Process

In the context of linear algebra, this process is being used to convert one set of linearly independent vectors to an orthogonal set of vectors that spans the same vector space. In this context, we shall consider this process as converting one set of polynomials that spans the space of polynomials of degree $\le n$ to an orthogonal set of polynomials that spans the same space $\boldsymbol{P}_n$ (i.e., the space of all polynomials of degree $\le n$). Typically, the initial set of polynomials will be $\{1, x, x^2, \ldots, x^n\}$, which we would like to convert to orthogonal polynomials with respect to the weight $w(x)$.

We suppose that $\phi_0(x) = 1$. Now we write a linear polynomial with leading x as $\phi_1(x) = x - k_{1,0}\phi_0(x)$, for some constant $k_{1,0}$.

We choose the constant $k_{1,0}$ in such a way that $\phi_0(x)$ and $\phi_1(x)$ are orthogonal. Then,

$$\int_a^b w(x)\phi_1(x)\phi_0(x)dx = \int_a^b w(x)[x - k_{1,0}\phi_0(x)]\phi_0(x)dx = 0 \tag{9.44}$$

It follows that

$$k_{1,0} = \frac{\int_a^b w(x)x\phi_0(x)dx}{\int_a^b w(x)[\phi_0(x)]^2 dx} \tag{9.45}$$

Again, for any quadratic polynomial $\phi_2(x)$ with leading term x^2, we may write

$\phi_2(x) = x^2 - k_{2,0}\phi_0(x) - k_{2,1}\phi_1(x)$, for some constants $k_{2,0}$ and $k_{2,1}$.

We choose the constants $k_{2,0}$ and $k_{2,1}$ in such a way that $\phi_2(x)$ is orthogonal to both $\phi_0(x)$ and $\phi_1(x)$. Then,

$$\int_a^b w(x)\phi_2(x)\phi_0(x)dx = \int_a^b w(x)[x^2 - k_{2,0}\phi_0(x) - k_{2,1}\phi_1(x)]\phi_0(x)dx = 0 \tag{9.46}$$

From Equation 9.46, we obtain

$$k_{2,0} = \frac{\int_a^b w(x)x^2\phi_0(x)dx}{\int_a^b w(x)[\phi_0(x)]^2 dx} \tag{9.47}$$

because $\phi_0(x)$ and $\phi_1(x)$ are orthogonal, that is,

$$\int_a^b w(x)\phi_1(x)\phi_0(x)dx = 0$$

Similarly, the orthogonality of $\phi_1(x)$ and $\phi_2(x)$ implies that

$$k_{2,1} = \frac{\int_a^b w(x)x^2\phi_1(x)dx}{\int_a^b w(x)[\phi_1(x)]^2 dx} \tag{9.48}$$

Proceeding in this way, by the mathematical induction, we can prove that if

$$\phi_i(x) = x^i - k_{i,0}\phi_0(x) - k_{i,1}\phi_1(x) - \cdots - k_{i,n}\phi_n(x)$$

where the constants $k_{i,j}$ are chosen such that $\phi_i(x)$ is orthogonal to each member $\phi_j(x)$, $j = 0,1,2,\ldots,i-1$ of the sequence $\{\phi_i(x)\}_{i=0}^n$, then,

$$k_{i,j} = \frac{\int_a^b w(x)x^i\phi_j(x)dx}{\int_a^b w(x)[\phi_j(x)]^2 dx} \tag{9.49}$$

Thus, we construct a sequence $\{\phi_i(x) | i \geq 0\}$ of orthogonal polynomials on the interval (a,b) with respect to the weight function $w(x)$.

However, to keep the discussion slightly more general, we start with $(n+1)$ linearly independent functions $\{g_i(x)\}_{i=0}^n$ (all $g_i(x) \in L_w^2[a,b]$, $i = 0,1,2,\ldots,n$, that is, $\int_a^b w(x)[g_j(x)]^2 dx < \infty$). The functions $\{g_i(x)\}_{i=0}^n$ will be converted into orthonormal functions $\{f_i(x)\}_{i=0}^n$.

Now we consider

$$f_0(x) = d_0 g_0(x),$$

$$f_1(x) = d_1(g_1(x) - c_{1,0}f_0(x)),$$

$$f_2(x) = d_2(g_2(x) - c_{2,0}f_0(x) - c_{2,1}f_1(x))$$

$$\vdots$$

$$f_n(x) = d_n(g_n(x) - c_{n,0}f_0(x) - c_{n,1}f_1(x) - \cdots - c_{n,n-1}f_{n-1}(x))$$

We shall determine the coefficients d_k and $c_{k,j}$ such that $\{f_i\}_{i=0}^{n}$ is orthonormal with respect to the weighted L^2 norm over $[a,b]$, that is,

$$< f_i, f_j >_w = \int_a^b w(x) f_i(x) f_j(x) dx = \delta_{ij} = \begin{cases} 1, & \text{if } i = j \\ 0, & \text{if } i \neq j \end{cases}$$

We start with $f_0(x)$ to obtain

$$< f_0, f_0 >_w = d_0^2 < g_0, g_0 >_w$$

Therefore,

$$d_0 = \frac{1}{\sqrt{< g_0, g_0 >_w}}$$

For $f_1(x)$, we require that it is orthogonal to $f_0(x)$, that is, $< f_0, f_1 >_w = 0$. Therefore, we obtain

$$0 = d_1(< f_0, g_1 >_w - c_{1,0} < f_0, f_0 >_w)$$

that is,

$$c_{1,0} = < f_0, g_1 >_w$$

Also, the normalization condition $< f_1, f_1 >_w = 1$ implies that

$$1 = d_1^2 < g_1 - c_{1,0} f_0, g_1 - c_{1,0} f_0 >_w$$

Thus,

$$d_1 = \frac{1}{\sqrt{< g_1 - c_{1,0} f_0, g_1 - c_{1,0} f_0 >_w}}$$

The denominator cannot be zero because of the assumption that $\{g_i(x)\}_{i=0}^{n}$ are linearly independent.

In general, for $1 \leq k \leq n$

$$f_k(x) = d_k(g_k(x) - c_{k,0} f_0(x) - c_{k,1} f_1(x) - \cdots - c_{k,k-1} f_{k-1}(x))$$

For $i = 0, 1, 2, \ldots, k-1$, we require the orthogonality conditions

$$< f_k, f_i >_w = 0$$

Using this condition, we obtain

$$0 = d_k(< f_i, g_k >_w - c_{k,i} < f_i, f_i >_w) = d_k(< f_i, g_k >_w - c_{k,i})$$

that is,

$$c_{k,i} = < f_i, g_k >_w, \quad i = 0, 1, 2, \ldots, k-1$$

Also, the coefficient d_k can be obtained from the normalization condition $< f_k, f_k >_w = 1$.

Although the Gram–Schmidt orthogonalization process is available to find a set of orthogonal polynomials, some well-known orthogonal polynomials such as Legendre polynomials, Chebyshev polynomials, and Laguerre polynomials are used in many areas of applied mathematics and engineering. We are now going to present several important examples of orthogonal polynomials.

1. *Legendre polynomials:* We start with the Legendre polynomials. This represents a family of polynomials that are orthogonal with respect to the weight function $w(x) \equiv 1$, on the interval $[-1,1]$.

   The Legendre polynomials $P_n(x)$ defined on $[-1,1]$ are given by

$$P_n(x) = \sum_{i=0}^{M} (-1)^i \frac{(2n-2i)!x^{n-2i}}{2^n i!(n-i)!(n-2i)!} \tag{9.50}$$

   where $M = n/2$ or $(n-1)/2$ whichever is an integer (Figure 9.2).

   The Legendre polynomials satisfy the differential equation

$$(1-x^2)y'' - 2xy' + n(n+1)y = 0 \tag{9.51}$$

   where n is a real number. This differential equation is called Legendre differential equation.

   The Legendre polynomials can be obtained from the recurrence relation

$$(n+1)P_{n+1}(x) - (2n+1)xP_n(x) + nP_{n-1}(x) = 0, \quad n \geq 1 \tag{9.52}$$

   starting with

$$P_0(x) = 1, P_1(x) = x$$

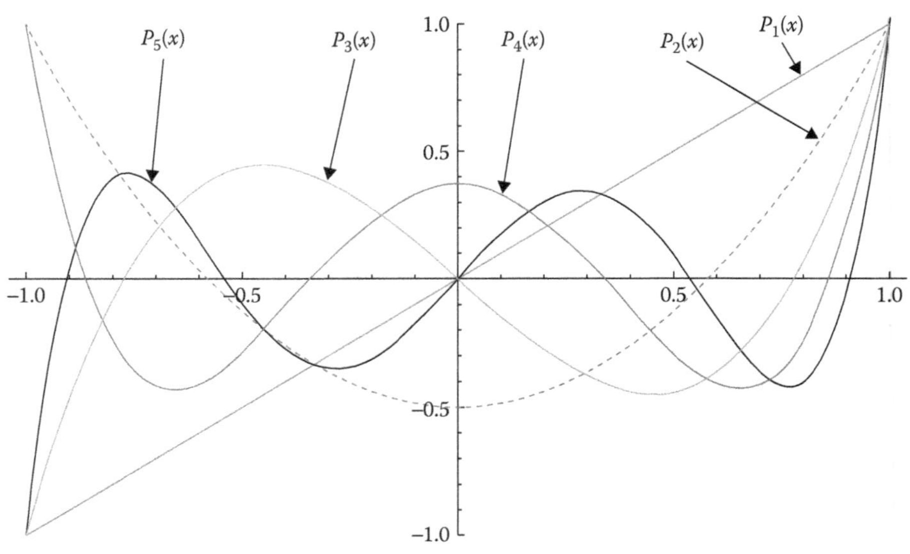

FIGURE 9.2 Legendre polynomials.

It is also possible to calculate Legendre polynomials directly by Rodrigues' formula

$$P_n(x) = \frac{1}{2^n n!}\frac{d^n}{dx^n}(x^2-1)^n, \quad n \geq 0 \tag{9.53}$$

The Legendre polynomials satisfy the orthogonality condition

$$< P_n, P_m > \; = \; \frac{2}{2n+1}\delta_{nm} \tag{9.54}$$

that is

$$\int_{-1}^{1} P_n(x)P_m(x)dx = \begin{cases} 0, & m \neq n \\ \dfrac{2}{2n+1}, & m = n \end{cases}$$

2. *Chebyshev polynomials:* The Chebyshev polynomials satisfy the differential equation

$$(1-x^2)y'' - xy' + n^2 y = 0 \tag{9.55}$$

where n is a real number. This differential equation is called Chebyshev equation. This differential equation has two independent solutions $T_n(x)$ and $U_n(x)$ defined on $[-1,1]$.

The Chebyshev polynomials of the first kind $T_n(x)$ are explicitly given by (Figure 9.3)

$$T_n(x) = \cos(n\cos^{-1} x), \quad n \geq 0 \tag{9.56}$$

On the other hand, the function $U_n(x)$ is known as Chebyshev polynomials of the second kind and is given by (Figure 9.4)

$$U_n(x) = \sin[(n+1)\cos^{-1} x], \quad n \geq 0 \tag{9.57}$$

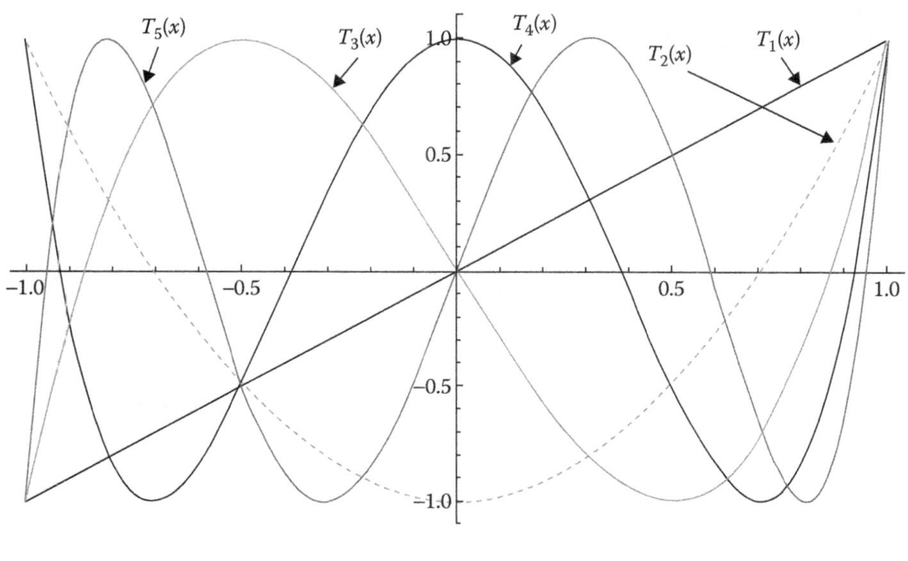

FIGURE 9.3 Chebyshev polynomials of the first kind.

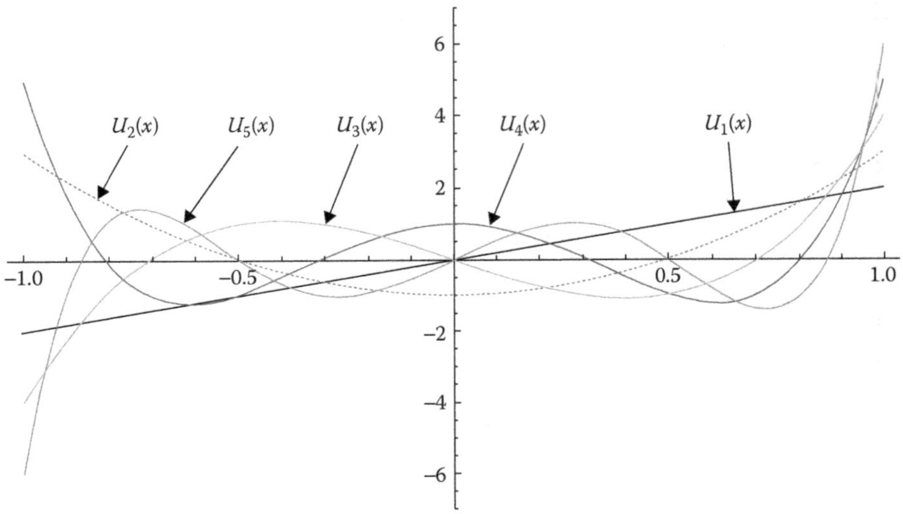

FIGURE 9.4 Chebyshev polynomials of the second kind.

The Chebyshev polynomials $T_n(x)$ satisfy the recurrence relation

$$T_{n+1}(x) = 2xT_n(x) - T_{n-1}(x), \quad n \geq 1 \tag{9.58}$$

with $T_0(x) = 1$ and $T_1(x) = x$.

The Chebyshev polynomials $T_n(x)$ possess the following properties:

a. $T_n(x)$ of degree $n \geq 1$ has n simple zeros at the points

$$x_j = \cos\left(\frac{2j-1}{2n}\pi\right), \quad j = 1, 2, \ldots, n$$

on the interval $[-1, 1]$.

b. $T_n(x)$ assumes its absolute extrema at

$$\bar{x}_j = \cos\left(\frac{j\pi}{n}\right), \quad j = 0, 1, 2, \ldots, n$$

and the extreme value at $\bar{x}_j$ is

$$T_n(\bar{x}_j) = (-1)^j, \quad j = 0, 1, 2, \ldots, n$$

c. $|T_n(x)| \leq 1, x \in [-1, 1]$.

d. $T_n(x)$ polynomials are orthogonal with respect to the weight function

$$w(x) = \frac{1}{\sqrt{1 - x^2}}$$

on the interval $[-1, 1]$.

The orthogonality property that Chebyshev polynomials satisfy is

$$<T_m, T_n> = \begin{cases} 0, & m \neq n \\ \frac{\pi}{2}, & m = n \neq 0 \\ \pi, & m = n = 0 \end{cases}$$

e. $T_n(x)$ satisfies the minimax property, that is, if $P_n(x)$ is a monic polynomial of degree n ($P_n(x)$ is a polynomial of degree n with leading coefficient 1) and $\hat{T}_n(x) = T_n(x)/2^{n-1}$, $n \geq 1$, then the $\hat{T}_n(x)$ polynomials have the property that

$$\frac{1}{2^{n-1}} = \max_{x\in[-1,1]} |\hat{T}_n(x)| \leq \max_{x\in[-1,1]} |P_n(x)|$$

Moreover, equality occurs only if $P_n \equiv \hat{T}_n$.

Proof:
Suppose that, $P_n(x)$ is a monic polynomial of degree n and

$$\max_{x\in[-1,1]} |P_n(x)| \leq \frac{1}{2^{n-1}} = \max_{x\in[-1,1]} |\hat{T}_n(x)|$$

The monic (polynomials with leading coefficient 1) Chebyshev polynomials $\hat{T}_n(x)$ are derived from the Chebyshev polynomials $T_n(x)$ by dividing the leading coefficient 2^{n-1}. Hence,

$$\hat{T}_0(x) = 1 \quad \text{and} \quad \hat{T}_n(x) = \frac{T_n(x)}{2^{n-1}}, \quad \text{for each } n \geq 1$$

Let, $Q \equiv \hat{T}_n - P_n$. Then since, $\hat{T}_n(x)$ and $P_n(x)$ are both monic polynomials of degree n, $Q(x)$ is a polynomial of degree at most $(n-1)$. Also, at the $n+1$ extrema points $\bar{x}_j$ of $\hat{T}_n(x)$, we have

$$Q(\bar{x}_j) = \hat{T}_n(\bar{x}_j) - P_n(\bar{x}_j) = \frac{(-1)^j}{2^{n-1}} - P_n(\bar{x}_j)$$

Since,

$$|P_n(\bar{x}_j)| \leq \frac{1}{2^{n-1}}, \quad j = 0, 1, 2, \ldots, n$$

Therefore,

$Q(\bar{x}_j) \leq 0$, when j is odd
and
$Q(\bar{x}_j) \geq 0$, when j is even.

Since $Q(x)$ is a continuous function, applying the intermediate value theorem, it follows that $Q(x)$ has at least one zero between $\bar{x}_i$ and $\bar{x}_{i+1}$, for $i = 0, 1, 2, \ldots$

Hence $Q(x)$ has at least n zeros in the interval $[-1, 1]$. But the degree of $Q(x)$ is less than n. Therefore, $Q(x) \equiv 0$. Thus, it follows that $P_n \equiv \hat{T}_n$. ■

3. *Laguerre polynomials:* We next proceed with the Laguerre polynomials. Here the interval is given by $[0, \infty)$ with the weight function $w(x) = e^{-x}$.

   The Laguerre polynomials are given by

$$L_n(x) = \frac{e^x}{n!}\frac{d^n}{dx^n}(x^n e^{-x}), \quad n \geq 0$$

   The normalization condition is

$$\|L_n\|_2 = 1, \quad \text{for all } n$$

   Also, $\{L_n\}$ are orthogonal on $[0, \infty)$ with regard to the weight function $w(x) = e^{-x}$.

   A more general form of the Laguerre polynomials is obtained, when the weight function $w(x)$ is taken as

$$w(x) = e^{-x}x^{\alpha}$$

   for an arbitrary real number $\alpha > -1$, on the interval $[0,\infty)$

Example 9.4

Using the Chebyshev polynomials, obtain the least squares approximation of second degree for $f(x) = x^6$ on $[-1, 1]$.

Solution:

Let $f(x) = c_0T_0(x) + c_1T_1(x) + c_2T_2(x)$.

Now according to the least squares approximation

$$E(c_0,c_1,c_2) = \int_{-1}^{1} \frac{1}{\sqrt{1-x^2}}[x^6 - (c_0T_0(x) + c_1T_1(x) + c_2T_2(x))]^2 dx$$

must be minimum.

Therefore,

$$\frac{\partial E}{\partial c_0} = -2\int_{-1}^{1} \frac{1}{\sqrt{1-x^2}}[x^6 - (c_0T_0(x) + c_1T_1(x) + c_2T_2(x))]\,T_0(x)dx = 0$$

$$\frac{\partial E}{\partial c_1} = -2\int_{-1}^{1} \frac{1}{\sqrt{1-x^2}}[x^6 - (c_0T_0(x) + c_1T_1(x) + c_2T_2(x))]\,T_1(x)dx = 0$$

$$\frac{\partial E}{\partial c_2} = -2\int_{-1}^{1} \frac{1}{\sqrt{1-x^2}}[x^6 - (c_0T_0(x) + c_1T_1(x) + c_2T_2(x))]\,T_2(x)dx = 0$$

Thus, we obtain

$$c_0 = \frac{1}{\pi}\int_{-1}^{1} \frac{x^6T_0(x)}{\sqrt{1-x^2}}dx = \frac{5}{16}$$

$$c_1 = \frac{2}{\pi}\int_{-1}^{1} \frac{x^6T_1(x)}{\sqrt{1-x^2}}dx = 0$$

$$c_2 = \frac{2}{\pi}\int_{-1}^{1} \frac{x^6 T_2(x)}{\sqrt{1-x^2}}dx = \frac{15}{32}$$

Hence, the required least squares approximation is

$$f(x) = \frac{5}{16}T_0(x) + \frac{15}{32}T_2(x)$$

4. *Some properties of orthogonal polynomials:* We shall now establish some results which will be useful in the following.

Theorem 9.3

Let $\{\phi_i(x) | i \geq 0\}$ be an orthogonal family of polynomials on $[a, b]$ with weight function $w(x)$, such that degree of $\phi_i(x) = i$, $i \geq 0$. If $f(x)$ be a polynomial of degree n, then

$$f(x) = \sum_{i=0}^{n} \frac{< f, \phi_i >_w}{< \phi_i, \phi_i >_w}\phi_i(x) \tag{9.59}$$

Proof:

We first show that every polynomial can be written as a linear combination of orthogonal polynomials of less degrees. Since the degree of $\phi_0(x)$ is 0, we have $\phi_0(x) = c$, a constant. Thus

$$1 \equiv \frac{1}{c}\phi_0(x)$$

Again, because the degree of $\phi_1(x)$ is 1, the Gram–Schmidt construction of orthogonal polynomials leads to

$$\phi_1(x) = k_{1,1}x + k_{1,0}\phi_0(x), \quad k_{1,1} \neq 0$$

This implies that

$$x = \frac{\phi_1(x) - k_{1,0}\phi_0(x)}{k_{1,1}}$$

Proceeding in this manner, by induction, the Gram–Schmidt process yields

$$\phi_r(x) = k_{r,r}x^r + k_{r,r-1}\phi_{r-1}(x) + \cdots + k_{r,0}\phi_0(x), \quad k_{r,r} \neq 0 \tag{9.60}$$

and

$$x^r = \frac{\phi_r(x) - k_{r,r-1}\phi_{r-1}(x) - \cdots - k_{r,0}\phi_0(x)}{k_{r,r}} \tag{9.61}$$

Thus every monomial can be expressed as a combination of orthogonal polynomials of less degrees. Consequently, it follows that any arbitrary polynomial $f(x)$ of degree n can be expressed as

$$f(x) = \hat{c}_0\,\phi_0(x) + \hat{c}_1\,\phi_1(x) + \cdots + \hat{c}_n\,\phi_n(x) \tag{9.62}$$

for some constants $\hat{c}_0, \hat{c}_1, \ldots, \hat{c}_n$.

Now to calculate each $\hat{c}_i$, $i = 0,1,2,\ldots,n$, we multiply both sides of Equation 9.62 by $w(x)\phi_i(x)$ and then integrate over $[a, b]$ to obtain

$$\int_a^b f(x)w(x)\phi_i(x)dx = \hat{c}_i \int_a^b w(x)[\phi_i(x)]^2 dx$$

Therefore,

$$\hat{c}_i = \frac{\int_a^b f(x)w(x)\phi_i(x)dx}{\int_a^b w(x)[\phi_i(x)]^2 dx} = \frac{< f,\phi_i >_w}{< \phi_i,\phi_i >_w} \tag{9.63}$$

Hence, from Equation 9.62, we derive

$$f(x) = \sum_{i=0}^{n} \frac{< f,\phi_i >_w}{< \phi_i,\phi_i >_w}\phi_i(x)$$

■

Corollary: If $f(x)$ is a polynomial of degree $k < n$, then $f(x)$ can be expressed as

$$f(x) = \hat{c}_0\,\phi_0(x) + \hat{c}_1\,\phi_1(x) + \cdots + \hat{c}_k\,\phi_k(x) = \sum_{i=0}^{k} \hat{c}_i\,\phi_i(x) \tag{9.64}$$

for some constants $\hat{c}_0, \hat{c}_1, \ldots, \hat{c}_k$.

Since ϕ_n is orthogonal to ϕ_i, for each $i = 0,1,2,\ldots,k$ with respect to the weight function $w(x)$ on $[a, b]$, we have

$$\int_a^b w(x)f(x)\phi_n(x)dx = \sum_{i=0}^{k} \hat{c}_i \int_a^b w(x)\phi_i(x)\phi_n(x)dx = \sum_{i=0}^{k} \hat{c}_i .0 = 0$$

Thus $\phi_n(x)$ is orthogonal to $f(x)$.

Theorem 9.4

Let $\{\phi_i \mid i \geq 0\}$ be an orthogonal family of polynomials on $[a, b]$ with respect to the weight $w(x)$, then the roots x_k, $k = 1,2,\ldots,i$ of the polynomial $\phi_i(x)$ are all real, simple, and are in the open interval (a, b).

Proof:

Let $x_1, x_2, \ldots, x_r$ be the roots of $\phi_i(x)$ in (a, b) for which

i. $a < x_j < b$
ii. $\phi_i(x)$ changes sign at x_j.

Since the degree of $\phi_i(x) = i$, we must have $r \leq i$. We assume that $r < i$ and let us define

$$P_r(x) = (x - x_1)(x - x_2)\cdots(x - x_r)$$

Then the polynomial $\phi_i(x)P_r(x) = (x - x_1)(x - x_2)\cdots(x - x_r)\phi_i(x)$ does not change sign in (a, b). Therefore, $\phi_i(x)P_r(x)$ is a polynomial with one sign in (a, b). This implies that

$$\int_a^b w(x)\phi_i(x)P_r(x)dx \neq 0$$

and hence $r = i$, because $\phi_i(x)$ is orthogonal to polynomials of degree less than i.

Now $\phi_i(x)$ can have at most i roots and the assumptions on $x_1, x_2, \ldots, x_i$ imply that they must all be simple. ■

Another important property of orthogonal polynomials is that they can all be written in terms of recursion relations.

Theorem 9.5: Triple Recursion Relation

Let $\{\phi_i | i \geq 0\}$ be an orthogonal family of polynomials on $[a, b]$ with weight function $w(x) \geq 0$. Then any three consecutive orthogonal polynomials are related by a recursion formula of the form

$$\phi_{n+1}(x) = (A_n x + B_n)\phi_n(x) - C_n\phi_{n-1}(x), \quad \text{for } n \geq 1 \tag{9.65}$$

If a_k and b_k are the coefficients of the terms of degree k and degree $k-1$ in $\phi_k(x)$, then

$$A_n = \frac{a_{n+1}}{a_n}, \quad B_n = \frac{a_{n+1}}{a_n}\left(\frac{b_{n+1}}{a_{n+1}} - \frac{b_n}{a_n}\right), \quad C_n = \frac{a_{n+1}a_{n-1}}{a_n^2}\cdot\frac{\gamma_n}{\gamma_{n-1}} \tag{9.66}$$

where

$$\gamma_n = <\phi_n, \phi_n>_w \quad \text{and} \quad \gamma_n > 0$$

Proof:

For $A_n = \dfrac{a_{n+1}}{a_n}$,

let

$$P_n(x) = \phi_{n+1}(x) - A_n x\phi_n(x)$$

$$= a_{n+1}x^{n+1} + b_{n+1}x^n + \cdots - \frac{a_{n+1}}{a_n}x(a_n x^n + b_n x^{n-1} + \cdots) \tag{9.67}$$

$$= \left(b_{n+1} - \frac{a_{n+1}}{a_n}b_n\right)x^n + \cdots$$

Therefore, the degree of $P_n(x) \leq n$, which means that there exists $\alpha_0, \alpha_1, \ldots, \alpha_n$ such that

$$P_n(x) = \alpha_0\phi_0(x) + \alpha_1\phi_1(x) + \cdots + \alpha_n\phi_n(x) \tag{9.68}$$

For $0 \leq i \leq n-2$,

$$\alpha_i = \frac{<P_n, \phi_i>_w}{<\phi_i, \phi_i>_w} = \frac{<\phi_{n+1} - A_n x\phi_n, \phi_i>_w}{<\phi_i, \phi_i>_w} = \frac{<\phi_{n+1}, \phi_i>_w - A_n <x\phi_n, \phi_i>_w}{<\phi_i, \phi_i>_w} = 0$$

because the degree of $x\phi_n = n+1, <x\phi_n, \phi_i>_w = 0$ and also $<\phi_{n+1}, \phi_i>_w = 0$.

Therefore, from Equation 9.68, we get

$$P_n(x) = \alpha_{n-1}\phi_{n-1}(x) + \alpha_n\phi_n(x) \tag{9.69}$$

Now we set $B_n = \alpha_n$ and $C_n = -\alpha_{n-1}$. Then from Equations 9.69 and 9.67, it follows that

$$\phi_{n+1}(x) = (A_n x + B_n)\phi_n(x) - C_n\phi_{n-1}(x) \tag{9.70}$$

Now from Equation 9.70, we have

$$<\phi_{n+1},\phi_{n-1}>_w = A_n <x\phi_n,\phi_{n-1}>_w + B_n <\phi_n,\phi_{n-1}>_w - C_n <\phi_{n-1},\phi_{n-1}>_w$$

This implies that

$$0 = A_n <\phi_n, x\phi_{n-1}>_w - C_n <\phi_{n-1},\phi_{n-1}>_w$$

Thus

$$C_n = \frac{A_n <\phi_n, x\phi_{n-1}>_w}{<\phi_{n-1},\phi_{n-1}>_w} = \frac{a_{n-1}}{a_n}\frac{A_n <\phi_n,\phi_n>_w}{<\phi_{n-1},\phi_{n-1}>_w} = \frac{a_{n+1}a_{n-1}}{a_n^2}\cdot\frac{\gamma_n}{\gamma_{n-1}}$$

Since

$$x\phi_{n-1} = a_{n-1}x^n + b_{n-1}x^{n-1} + \cdots = \frac{a_{n-1}}{a_n}\phi_n(x) + Q_{n-1}(x) \quad \text{and} \quad <\phi_n, x\phi_{n-1}>_w = \frac{a_{n-1}}{a_n} <\phi_n,\phi_n>_w$$

where the degree of the polynomial $Q_{n-1}(x) = n-1$.

Again, Equation 9.70 can be explicitly written as

$$a_{n+1}x^{n+1} + b_{n+1}x^n + \cdots = (A_n x + B_n)\left(a_n x^n + b_n x^{n-1} + \cdots\right) - C_n\left(a_{n-1}x^{n-1} + b_{n-1}x^{n-2} + \cdots\right)$$

Equating coefficients of x^n both sides of the above equation, we get

$$b_{n+1} = a_n B_n + A_n b_n$$

This implies that

$$B_n = \frac{b_{n+1} - A_n b_n}{a_n} = \frac{a_{n+1}}{a_n}\left(\frac{b_{n+1}}{a_{n+1}} - \frac{b_n}{a_n}\right) \qquad ■$$

9.5 THE MINIMAX POLYNOMIAL APPROXIMATION

This approximation is also known as uniform polynomial approximation.

We assume that the function $f(x)$ is continuous on $[a, b]$, and also assume that $P_n(x)$ is a polynomial of degree $\leq n$. The L_∞ norm of the two functions $f(x)$ and $P_n(x)$ on the interval $[a, b]$ is given by

$$\|f - P_n\|_\infty = \max_{a \leq x \leq b}|f(x) - P_n(x)| \tag{9.71}$$

Now the question we would like to address is how close can we get to $f(x)$ (in the L_∞-norm sense) with polynomials of a given degree. This question leads us to the following approximation problem.

Given that $f(x) \in C[a, b]$ and fixed $n \geq 0$, we can find $P_n \in \boldsymbol{P}_n$ such that

$$d_n(f) = \inf_{q \in \boldsymbol{P}_n} \|f - q\|_\infty \tag{9.72}$$

Our goal is to find a polynomial $P_n^*(x)$ for which the infimum in Equation 9.72 is actually obtained, that is,

$$d_n(f) = \|f - P_n^*\|_\infty \tag{9.73}$$

Such a polynomial $P_n^*(x)$ that satisfies Equation 9.72 is called a polynomial of best approximation or the minimax polynomial of degree n to the function $f(x)$ in the L_∞ norm. The minimal distance in Equation 9.73 will be referred to as the minimax error.

In the following discussion, we will explore that the minimax polynomial always exists and is unique. We will also provide a characterization of the minimax polynomial. Let us consider a simple example to illustrate some important properties.

Example 9.5

Let $f(x)$ be strictly monotonic increasing on $[a, b]$ and $f(x) \in C[a, b]$. We wish to find the minimax polynomial of degree 0 for $f(x)$ on $[a, b]$.

We denote this minimax polynomial by

$$P_0^*(x) = c$$

We need to determine $c \in \boldsymbol{R}$ so that

$$\|f - P_0^*\|_\infty = \max_{x \in [a,b]} |f(x) - c|$$

is minimal. Since $f(x)$ is monotonic increasing, $f(x) - c$ attains its minimum when $x = a$ and its maximum when $x = b$. Consequently, $|f(x) - c|$ is maximum at one of the endpoints of $[a, b]$, that is,

$$E(c) = \max_{x \in [a,b]} |f(x) - c| = \max\{|f(a) - c|, |f(b) - c|\}$$

Therefore,

$$E(c) = \begin{cases} f(b) - c, & \text{if } c < \frac{1}{2}(f(a) + f(b)) \\ c - f(a), & \text{if } c \geq \frac{1}{2}(f(a) + f(b)) \end{cases}$$

Figure 9.5 shows that the minimum is attained when $c = (1/2)(f(a) + f(b))$. Consequently, the desired minimax polynomial of degree 0 for the function $f(x)$ is

$$P_0^*(x) = \frac{1}{2}(f(a) + f(b)), \quad x \in [a, b]$$

More generally, if $f(x) \in C[a, b]$ (not necessarily monotonic), and α and β denote two points in $[a, b]$ where $f(x)$ attains its minimum and maximum values, respectively, then it can be shown that the minimax polynomial of degree 0 to $f(x)$ on $[a, b]$ is

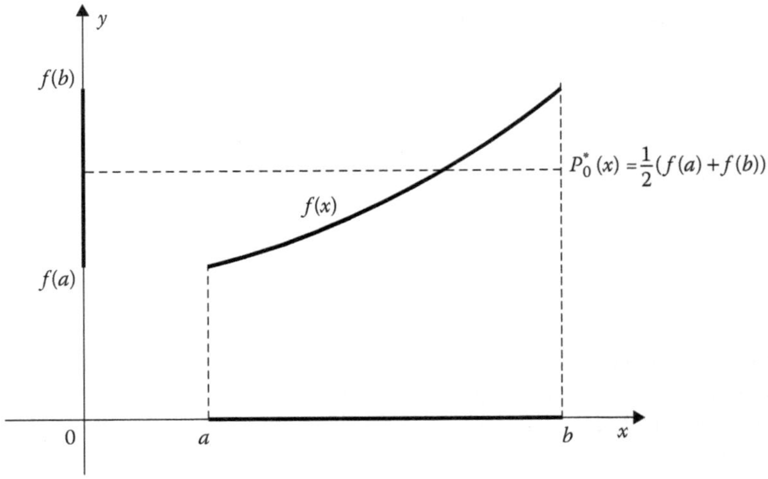

FIGURE 9.5 Minimax approximation $P_0^*(x)$ of strictly monotonic increasing continuous function $f(x)$ on $[a,b]$.

$$P_0^*(x) = \frac{1}{2}(f(\alpha) + f(\beta)), \quad x \in [a, b]$$

Furthermore, the earlier example shows that the minimax polynomial $P_0^*(x)$ of degree 0 for $f(x) \in C[a, b]$ has the property that the approximation error $f(x) - P_0^*(x)$ attains its extrema at two points, say, $x = \alpha$ and $x = \beta$, with the error

$$f(x) - P_0^*(x) = \frac{1}{2}(f(x) - f(\alpha)) + \frac{1}{2}(f(x) - f(\beta))$$

which is negative at the point $x = \alpha$, and positive at the point, $x = \beta$. We shall prove that a property of this kind holds in general because of the oscillating nature of the approximation error, which is usually referred to as the oscillation theorem. This property gives a complete characterization of the minimax polynomial and provides a method for its construction. Next, we discuss a theorem due to de la Vallee-Poussin, which provides the lower bound of the minimax error.

Theorem 9.6: The de la Vallee-Poussin Theorem

Let, $a \le x_0 < x_1 < \cdots < x_n < x_{n+1} \le b$. Let $P_n(x)$ be a polynomial of degree $\le n$, that is, $P_n(x) \in \boldsymbol{P}_n$. Suppose that

$$f(x_j) - P_n(x_j) = (-1)^j e_j, \quad j = 0, 1, 2, \ldots, n+1 \tag{9.74}$$

where all e_j are nonzero and of the same sign. Then

$$\min_{0 \le j \le n+1} |e_j| \le d_n(f)$$

Proof:

We shall prove this by contradiction. We assume that for some $Q_n(x)$

$$\|f - Q_n\|_\infty < \min_j |e_j| \tag{9.75}$$

Therefore,

$$|f(x_j)-Q_n(x_j)|<|f(x_j)-P_n(x_j)|, \quad j=0,1,2,\ldots,n+1$$

Now

$$P_n(x_j)-Q_n(x_j)=[P_n(x_j)-f(x_j)]-[Q_n(x_j)-f(x_j)], \quad j=0,1,2,\ldots,n+1$$

Since the first term on the right always exceeds the second term in absolute value, it follows that $P_n(x_j)-Q_n(x_j)$ and $P_n(x_j)-f(x_j)$ have the same sign for $j=0,1,2,\ldots,n+1$. Therefore, $P_n(x_j)-Q_n(x_j)$ is a polynomial of degree $\le n$, which changes sign $(n+2)$ times, hence it has $(n+1)$ zeros. $P_n(x_j)-Q_n(x_j)$ is a polynomial of degree $\le n$; this is possible only if it is identically zero, that is, if

$P_n(x)\equiv Q_n(x)$, which contradicts Equations 9.74 and 9.75. Hence, it is proved. ■

9.5.1 Characterization of the Minimax Polynomial

The following theorem provides a characterization of the minimax polynomial in terms of its oscillations property.

Theorem 9.7: Chebyshev Equioscillation Theorem

Suppose that $f(x)\in C[a,b]$. A polynomial $P_n^*(x)\in \boldsymbol{P}_n$ is the minimax polynomial of degree n to $f(x)$ on $[a,b]$ if and only if $f(x)-P_n^*(x)$ assumes the values $\pm\|f-P_n^*\|_\infty$ with an alternating change of sign at least $(n+2)$ times in $[a,b]$, that is, a polynomial $P_n^*(x)\in \boldsymbol{P}_n$ is a minimax polynomial for $f(x)$ on $[a,b]$ if and only if there exists a sequence of $(n+2)$ points (referred to as critical points) x_i, $i=0,1,\ldots,n+1$, such that $a\le x_0<\cdots<x_{n+1}\le b$,

1. $|f(x_i)-P_n^*(x_i)|=\|f-P_n^*\|_\infty, \quad i=0,1,\ldots,n+1$
2. $f(x_i)-P_n^*(x_i)=-[f(x_{i+1})-P_n^*(x_{i+1})], \quad i=0,1,\ldots,n$

Proof:

We present here the proof of only the sufficient part of the theorem. For the necessary part of the theorem, the reader may refer to Süli and Mayers (2003).

Without loss of generality, we suppose that

$$f(x_i)-P_n^*(x_i)=(-1)^i\|f-P_n^*\|_\infty, \quad i=0,1,2,\ldots,n+1$$

Let

$$D^*=\|f-P_n^*\|_\infty$$

and let

$$d_n(f)=\min_{P_n\in\boldsymbol{P}_n}\|f-P_n\|_\infty$$

Here, we replace the infimum in the definition of $d_n(f)$ by a minimum because we already know that a minimum exists. Now, according to the de la Vallee-Poussin theorem, we have $D^*\le d_n(f)$.

On the other hand, the definition of $d_n(f)$ implies that $d_n(f) \le D^*$. Hence $d_n(f) = D^*$ and consequently $P_n^*(x)$ is the minimax polynomial. ■

Note: In view of Theorems 9.7, it is quite obvious that the Taylor expansion is a poor uniform approximation because the sum is nonoscillatory.

Example 9.6

Find the polynomial of best approximation of degree at most 2 for $|x|$ on the interval $[-1,1]$.

Solution:

Let $f(x) = |x|$, $P_2(x) = a_0 + a_1x + a_2x^2$ and $x_0 = -1$, $x_1 = -\alpha$, $x_2 = \alpha$, $x_3 = 1$, where $\alpha > 0$. Now

$$\varepsilon(x) = f(x) - P_2(x) = |x| - a_0 - a_1x - a_2x^2$$

Using the Chebyshev equioscillation theorem, we get

$$\varepsilon(-1) = -\varepsilon(-\alpha), \text{ i.e., } \varepsilon(-1) + \varepsilon(-\alpha) = 0$$

$$\varepsilon(-\alpha) = -\varepsilon(\alpha), \text{ i.e., } \varepsilon(-\alpha) + \varepsilon(\alpha) = 0$$

$$\varepsilon(\alpha) = -\varepsilon(1), \text{ i.e., } \varepsilon(\alpha) + \varepsilon(1) = 0$$

and

$$\varepsilon'(-\alpha) = \varepsilon'(\alpha) = 0$$

Therefore

$\varepsilon'(-\alpha) = 0$ implies that

$-1 - a_1 + 2a_2\alpha = 0$

$\varepsilon'(\alpha) = 0$

implies that

$1 - a_1 - 2a_2\alpha = 0$

$\varepsilon(-\alpha) + \varepsilon(\alpha) = 0$

yields

$\alpha - a_0 - a_2\alpha^2 = 0$

$\varepsilon(-1) + \varepsilon(-\alpha) = \varepsilon(\alpha) + \varepsilon(1) = 0$

yields

$1 + \alpha - 2a_0 - (1 + \alpha^2)a_2 = 0$

Solving the above equations, we get

$$a_0 = \frac{1}{2}, \quad a_1 = 0, \quad a_2 = \frac{1}{2}, \quad \text{and } \alpha = 1$$

Hence the best approximation to $|x|$ on the interval $[-1,1]$ is given by

$$P_2(x) = \frac{x^2}{2} + \frac{1}{2}$$

9.5.2 Existence of the Minimax Polynomial

The existence of the minimax polynomial is provided by the following theorem.

Theorem 9.8

Let $f(x) \in C[a,b]$. Then for any $n \in \boldsymbol{N}$ there exists $P_n^*(x) \in \boldsymbol{P}_n$ such that $\|f - P_n^*\|_\infty = \min_{P_n \in \mathrm{P}_n} \|f - P_n\|_\infty$.

Proof:

Let $\xi = (\xi_0, \xi_1, \ldots, \xi_n)$ be an arbitrary point in $\boldsymbol{R}^{n+1}$ and let

$$P_n(x) = \sum_{i=0}^{n} \xi_i x^i \in \boldsymbol{P}_n$$

We also let

$$E(\xi) = E(\xi_0, \xi_1, \ldots, \xi_n) = \|f - P_n\|_\infty$$

We shall first show that E is continuous, which will imply that E attains its bounds on any bounded closed set in $\boldsymbol{R}^{n+1}$, that is, there exists a point $\xi^* = (\xi_0^*, \xi_1^*, \ldots, \xi_n^*)$ such that

$$E(\xi^*) = \min_{\xi \in \boldsymbol{R}^{n+1}} E(\xi_0, \xi_1, \ldots, \xi_n)$$

Step 1: We first show that $E(\xi) = E(\xi_0, \xi_1, \ldots, \xi_n)$ is a continuous function on $\boldsymbol{R}^{n+1}$. For an arbitrary $\eta = (\eta_0, \eta_1, \ldots, \eta_n) \in \boldsymbol{R}^{n+1}$, we define

$$Q_n(x) = \sum_{i=0}^{n} \eta_i x^i$$

Then

$$E(\xi + \eta) = \|f - (P_n + Q_n)\|_\infty \le \|f - P_n\|_\infty + \|Q_n\|_\infty = E(\xi) + \|Q_n\|_\infty$$

Thus

$$E(\xi + \eta) - E(\xi) \le \|Q_n\|_\infty \le \max_{x \in [a,b]} \left(|\eta_0| + |\eta_1||x| + \cdots + |\eta_n||x^n| \right)$$

For any $\varepsilon > 0$, let $\delta = \varepsilon/(1 + K + \cdots + K^n)$, where $K = \max(|a|, |b|)$. Then for any $\eta = (\eta_0, \eta_1, \ldots, \eta_n)$ such that $\max |\eta_i| \le \delta$, $i = 0, 1, 2, \ldots, n$,

$$E(\xi + \eta) - E(\xi) \le \varepsilon \tag{9.76}$$

Similarly,

$$E(\xi)=E(\xi_0,\xi_1,\ldots,\xi_n)=\|f-P_n\|_\infty=\|f-(P_n+Q_n)+Q_n\|_\infty\leq\|f-(P_n+Q_n)\|_\infty+\|Q_n\|_\infty$$
$$=E(\xi+\eta)+\|Q_n\|_\infty$$

which implies that under the same conditions as in Equation 9.76, we get

$$E(\xi)-E(\xi+\eta)\leq\varepsilon \tag{9.77}$$

Combining Equations 9.76 and 9.77, we get

$$|E(\xi+\eta)-E(\xi)|\leq\varepsilon$$

which implies that E is continuous at ξ. Since ξ was an arbitrary point in $\boldsymbol{R}^{n+1}$, E is continuous in the entire $\boldsymbol{R}^{n+1}$.

Step 2: We now construct a compact set in $\boldsymbol{R}^{n+1}$ on which E attains its minimum. We let,

$$S=\left\{\zeta\in\boldsymbol{R}^{n+1}\middle|E(\zeta)\leq\|f\|_\infty\right\}$$

Now we have

$$E(0)=\|f\|_\infty$$

therefore $0\in S$ and the set S is nonempty. We also note that the set S is closed and bounded. Since E is continuous in the entire $\boldsymbol{R}^{n+1}$, it is also continuous on S and hence it must attain its minimum on S, say at $\zeta^*\in\boldsymbol{R}^{n+1}$, that is,

$$\min_{\zeta\in S}E(\zeta)=E(\zeta^*)$$

Step 3:

Since $0\in S$, we have

$$\min_{\zeta\in S}E(\zeta)\leq E(0)=\|f\|_\infty$$

Therefore, if $\zeta\in\boldsymbol{R}^{n+1}$ but $\zeta\notin S$, then

$$E(\zeta)>\|f\|_\infty\geq\min_{\zeta\in S}E(\zeta)=d$$

Thus, the lower bound d of the function E over the set S is the same as the lower bound of E over all values of $\zeta\in\boldsymbol{R}^{n+1}$. The lower bound d is attained at a point $\zeta^*\in S$; letting $P_n^*(x)=\sum_{i=0}^{n}\zeta_i x^i$, we find that $d=\|f-P_n^*\|_\infty$ and therefore P_n^* is the required polynomial of best approximation of degree n to the function f in the ∞ norm on $[a,b]$, that is, it is the minimax polynomial and hence the minimax polynomial exists. ■

9.5.3 Uniqueness of the Minimax Polynomial

Theorem 9.9: Uniqueness Theorem

Let $f(x)\in C[a,b]$. Then its minimax polynomial $P_n^*(x)\in\boldsymbol{P}_n$ is unique.

Proof:

Let

$$E_n(f) = \min_{P_n \in \boldsymbol{P}_n} \|f - P_n\|_\infty$$

We assume that $Q_n(x)$ is also a minimax polynomial for $f(x)$ and that $P_n^*(x)$ and $Q_n(x)$ are distinct. Then

$$\|f - P_n^*\|_\infty = \|f - Q_n\|_\infty = E_n(f)$$

Now the triangle inequality implies that

$$\left\|f - \frac{1}{2}(P_n^* + Q_n)\right\|_\infty = \left\|\frac{1}{2}(f - P_n^*) + \frac{1}{2}(f - Q_n)\right\|_\infty \leq \frac{1}{2}\|f - P_n^*\|_\infty + \frac{1}{2}\|f - Q_n\|_\infty = E_n(f)$$

Therefore, $(1/2)(P_n^* + Q_n) \in \boldsymbol{P}_n$ is also a minimax polynomial approximation to $f(x)$ on $[a,b]$. By the oscillating theorem, there exist $x_0, x_1, \ldots, x_{n+1} \in [a,b]$ such that

$$\left|f(x_i) - \frac{1}{2}(P_n^*(x_i) + Q_n(x_i))\right| = E_n(f), \quad i = 0,1,2,\ldots,n+1 \tag{9.78}$$

Equation 9.78 can be written as

$$\left|f(x_i) - P_n^*(x_i) + f(x_i) - Q_n(x_i)\right| = 2E_n(f), \quad i = 0,1,2,\ldots,n+1$$

Now

$$\left|f(x_i) - P_n^*(x_i)\right| \leq \max_{x \in [a,b]} \left|f(x) - P_n^*(x)\right| = \|f - P_n^*\|_\infty = E_n(f)$$

Similarly,

$$\left|f(x_i) - Q_n(x_i)\right| \leq \|f - Q_n\|_\infty = E_n(f)$$

Therefore, it follows that

$$f(x_i) - P_n^*(x_i) = f(x_i) - Q_n(x_i), \quad i = 0,1,2,\ldots,n+1$$

Thus, the difference $P_n^* - Q_n \in \boldsymbol{P}_n$ vanishes at $(n+2)$ distinct points, which is possible for a polynomial of degree $\leq n$ only if it is identically zero. Therefore,

$$P_n^* \equiv Q_n$$

This contradicts our initial assumption that $P_n^*(x)$ and $Q_n(x)$ are distinct. Hence, it establishes the uniqueness of the minimax polynomial $P_n^* \in \boldsymbol{P}_n$ for $f \in C[a,b]$. ■

9.5.4 The Near-Minimax Polynomial

The de la Vallee-Poussin theorem suggests the notion of a near-minimax polynomial, which is a polynomial $P_n \in \boldsymbol{P}_n$ such that the difference $f(x) - P_n(x)$ changes sign at $(n+2)$ points x_i, $i = 0,1,\ldots,n+1$, with $a \leq x_0 < x_1 < \cdots < x_n < x_{n+1} \leq b$.

We now associate the minimax approximation problem with polynomial interpolation. In order for $f(x)-P_n(x)$ to change its sign $(n+2)$ times, there should be $(n+1)$ points on which $f(x)$ and $P_n(x)$ agree with each other. In other words, we can consider $P_n(x)$ as a polynomial that interpolates $f(x)$ at $(n+1)$ points, say, $x_0, x_1, \ldots, x_n$.

We recall that the interpolation error given in Equation 3.7 of Chapter 3,

$$f(x)-P_n(x)=\Omega(x)\frac{f^{(n+1)}(\xi)}{(n+1)!} \tag{9.79}$$

where $\min\{x, x_0, x_1, \ldots, x_n\} < \xi < \max\{x, x_0, x_1, \ldots, x_n\}$ and $\Omega(x)=(x-x_0)(x-x_1)\cdots(x-x_n)$ is a monic polynomial of degree $(n+1)$.

Theorem 9.10

Suppose that $f(x)$ be a real valued continuous function on the interval $[a, b]$ and the derivative of order $(n+1)$ is continuous on $[a,b]$. Let, $P_n \in \boldsymbol{P}_n$ be the Lagrange interpolation polynomial of $f(x)$ with the interpolation points

$$x_i = \frac{1}{2}(b-a)\cos\left[\frac{\big(i+(1/2)\big)\pi}{n+1}\right]+\frac{1}{2}(b+a), \quad i=0,1,2,\ldots,n$$

then

$$\|f-P_n\|_\infty \le \frac{(b-a)^{n+1}}{2^{2n+1}(n+1)!}M_{n+1} \tag{9.80}$$

where $M_{n+1}=\max\limits_{\xi\in[a,b]}\left|f^{(n+1)}(\xi)\right|$

Proof:

Let $t_i=\cos\big[[(i+(1/2))\pi]/(n+1)\big], i=0,1,2,\ldots,n$, denote the zeros of the polynomial $T_{n+1}(t)$ in the interval $(-1, 1)$.

Let

$$\prod_{i=0}^{n}(t-t_i)=t^{n+1}-q(t)$$

where $q(t)\in \boldsymbol{P}_n$.

Then

$$\begin{aligned}\min_{t\in[-1,1]}\left\|\prod_{i=0}^{n}(t-t_i)\right\|_\infty &= \min_{t\in[-1,1]}\left\|t^{n+1}-q(t)\right\|_\infty \\ &= \left\|t^{n+1}-(t^{n+1}-2^{-n}T_{n+1}(t))\right\|_\infty\end{aligned}$$

because it can easily be shown that $q(t) \in \boldsymbol{P}_n$ is the unique minimax approximation to the function t^{n+1} on the interval $[-1, 1]$.

Thus, we obtain

$$\min_{t\in[-1,1]} \left\| \prod_{i=0}^{n} (t - t_i) \right\|_\infty = \left\| 2^{-n} T_{n+1}(t) \right\|_\infty$$

Therefore, we may consider

$$\prod_{i=0}^{n} (t - t_i) = 2^{-n} T_{n+1}(t), t \in [-1,1] \tag{9.81}$$

Clearly, $x_i \in (a,b)$ is the image of $t_i \in (-1,1)$ under the linear mapping $t \mapsto x = (1/2)(b-a)t + (1/2)(b+a)$. The inverse of this mapping is $x \mapsto t = (2x - a - b)/(b - a)$.

Now

$$\prod_{i=0}^{n} (x - x_i) = \left(\frac{b-a}{2}\right)^{n+1} \prod_{i=0}^{n} (t - t_i) = \left(\frac{b-a}{2}\right)^{n+1} 2^{-n} T_{n+1}(t(x))$$

Since $\left|T_{n+1}(t(x))\right| < 1$ for all $x \in [a,b]$, we have

$$\left\| f - P_n \right\|_\infty = \left\| \Omega(x) \frac{f^{(n+1)}(\xi)}{(n+1)!} \right\|_\infty = \left\| \prod_{i=0}^{n} (x - x_i) \frac{f^{(n+1)}(\xi)}{(n+1)!} \right\|_\infty \leq \frac{(b-a)^{n+1}}{2^{2n+1}(n+1)!} M_{n+1}$$

where $M_{n+1} = \max_{\xi\in[a,b]} \left| f^{(n+1)}(\xi) \right|$. ■

Thus the relation in Equation 9.80 motivates us to refer to the interpolant at the Chebyshev points as the near-minimax polynomial. We have therefore just shown that if $f^{(n+1)}(x)$ exists and is continuous on the interval $[a,b]$ and has the same sign on the open interval (a,b), then the polynomial $P_n \in \boldsymbol{P}_n$ constructed by interpolating at the points x_i, $i = 0, 1, \ldots, n$, obtained by linearly mapping the $(n+1)$ zeros of the Chebyshev polynomial $T_{n+1}(t)$ from $(-1,1)$ to (a,b), is a near-minimax approximation for the function $f(x) \in C[a,b]$.

9.6 B-SPLINES

A B-spline is a piecewise polynomial function of degree $< m$ in a variable x. It is defined over a domain $x_0 \leq x \leq x_m$. The points where $x = x_j$ are known as knots or break points. The number of internal knots is equal to the degree of the polynomial if there are no knot multiplicities. The knots must be in ascending order. The number of knots is the minimum for the degree of the B-spline, which has a nonzero value only in the range between the first and last knot. The B-spline is a continuous function at the knots.

The zero-order B-spline function can be defined as

$$B_{1,i}(x) = \begin{cases} 1, & x_i \leq x < x_{i+1} \\ 0, & \text{otherwise} \end{cases} \tag{9.82}$$

Higher order B-spline function can be defined by the following recursive formula:

$$B_{m,i}(x) = \frac{x - x_i}{x_{i+m-1} - x_i} B_{m-1,i}(x) + \frac{x_{i+m} - x}{x_{i+m} - x_{i+1}} B_{m-1,i+1}(x) \tag{9.83}$$

where, x_i's are the node points on the interval $[a,b]$ and defined as

$$a = x_0 < x_1 < x_2 < \cdots < x_{n-1} < x_n = b, \quad n \le m$$

and it need not be equispaced.

For $m = 2$, the linear B-spline functions are

$$\begin{aligned}
B_{2,i}(x) &= \frac{x - x_i}{x_{i+1} - x_i} B_{1,i}(x) + \frac{x_{i+2} - x}{x_{i+2} - x_{i+1}} B_{1,i+1}(x) \\
&= \frac{x - x_i}{x_{i+1} - x_i} \times \begin{cases} 1, & x_i \le x < x_{i+1}, \\ 0, & \text{otherwise} \end{cases} + \frac{x_{i+2} - x}{x_{i+2} - x_{i+1}} \times \begin{cases} 1, & x_{i+1} \le x < x_{i+2} \\ 0, & \text{otherwise} \end{cases} \\
&= \begin{cases} \dfrac{x - x_i}{x_{i+1} - x_i}, & x_i \le x < x_{i+1} \\ \dfrac{x_{i+2} - x}{x_{i+2} - x_{i+1}}, & x_{i+1} \le x < x_{i+2} \\ 0, & \text{otherwise} \end{cases}
\end{aligned}$$

For $m = 3$, the quadratic B-spline functions are

$$\begin{aligned}
B_{3,i}(x) &= \frac{x - x_i}{x_{i+2} - x_i} B_{2,i}(x) + \frac{x_{i+3} - x}{x_{i+3} - x_{i+1}} B_{2,i+1}(x) \\
&= \frac{x - x_i}{x_{i+2} - x_i} \times \begin{cases} \dfrac{x - x_i}{x_{i+1} - x_i}, & x_i \le x < x_{i+1}, \\ \dfrac{x_{i+2} - x}{x_{i+2} - x_{i+1}}, & x_{i+1} \le x < x_{i+2}, \\ 0, & \text{otherwise} \end{cases} + \frac{x_{i+3} - x}{x_{i+3} - x_{i+1}} \times \begin{cases} \dfrac{x - x_{i+1}}{x_{i+2} - x_{i+1}}, & x_{i+1} \le x < x_{i+2} \\ \dfrac{x_{i+3} - x}{x_{i+3} - x_{i+2}}, & x_{i+2} \le x < x_{i+3} \\ 0, & \text{otherwise} \end{cases} \\
&= \begin{cases} \dfrac{(x - x_i)^2}{(x_{i+1} - x_i)(x_{i+2} - x_i)}, & x_i \le x < x_{i+1} \\ \dfrac{(x - x_i)(x_{i+2} - x)}{(x_{i+2} - x_i)(x_{i+2} - x_{i+1})} + \dfrac{(x_{i+3} - x)(x - x_{i+1})}{(x_{i+3} - x_{i+1})(x_{i+2} - x_{i+1})}, & x_{i+1} \le x < x_{i+2} \\ \dfrac{(x_{i+3} - x)^2}{(x_{i+3} - x_{i+1})(x_{i+3} - x_{i+2})}, & x_{i+2} \le x < x_{i+3} \\ 0, & \text{otherwise} \end{cases}
\end{aligned}$$

For $m = 4$, the cubic B-spline functions are

$$B_{4,i}(x) = \frac{x - x_i}{x_{i+3} - x_i} B_{3,i}(x) + \frac{x_{i+4} - x}{x_{i+4} - x_{i+1}} B_{3,i+1}(x)$$

$$= \frac{x - x_i}{x_{i+3} - x_i} \times \begin{cases} \dfrac{(x - x_i)^2}{(x_{i+1} - x_i)(x_{i+2} - x_i)}, & x_i \le x < x_{i+1} \\[2ex] \dfrac{(x - x_i)(x_{i+2} - x)}{(x_{i+2} - x_i)(x_{i+2} - x_{i+1})} + \dfrac{(x_{i+3} - x)(x - x_{i+1})}{(x_{i+3} - x_{i+1})(x_{i+2} - x_{i+1})}, & x_{i+1} \le x < x_{i+2} \\[2ex] \dfrac{(x_{i+3} - x)^2}{(x_{i+3} - x_{i+1})(x_{i+3} - x_{i+2})}, & x_{i+2} \le x < x_{i+3} \\[2ex] 0, & \text{otherwise} \end{cases}$$

$$+ \frac{x_{i+4} - x}{x_{i+4} - x_{i+1}} \times \begin{cases} \dfrac{(x - x_{i+1})^2}{(x_{i+2} - x_{i+1})(x_{i+3} - x_{i+1})}, & x_{i+1} \le x < x_{i+2} \\[2ex] \dfrac{(x - x_{i+1})(x_{i+3} - x)}{(x_{i+3} - x_{i+1})(x_{i+3} - x_{i+2})} + \dfrac{(x_{i+4} - x)(x - x_{i+2})}{(x_{i+4} - x_{i+2})(x_{i+3} - x_{i+2})}, & x_{i+2} \le x < x_{i+3} \\[2ex] \dfrac{(x_{i+4} - x)^2}{(x_{i+4} - x_{i+2})(x_{i+4} - x_{i+3})}, & x_{i+3} \le x < x_{i+4} \\[2ex] 0, & \text{otherwise} \end{cases}$$

$$= \begin{cases} \dfrac{(x - x_i)^3}{(x_{i+1} - x_i)(x_{i+2} - x_i)(x_{i+3} - x_i)}, & x_i \le x < x_{i+1} \\[2ex] \left(\dfrac{x - x_i}{x_{i+3} - x_i}\right)\left(\dfrac{(x - x_i)(x_{i+2} - x)}{(x_{i+2} - x_i)(x_{i+2} - x_{i+1})} + \dfrac{(x_{i+3} - x)(x - x_{i+1})}{(x_{i+3} - x_{i+1})(x_{i+2} - x_{i+1})}\right) \\ \quad + \left(\dfrac{x_{i+4} - x}{x_{i+4} - x_{i+1}}\right)\left(\dfrac{(x - x_{i+1})^2}{(x_{i+2} - x_{i+1})(x_{i+3} - x_{i+1})}\right), & x_{i+1} \le x < x_{i+2} \\[2ex] \left(\dfrac{x - x_i}{x_{i+3} - x_i}\right)\left(\dfrac{(x_{i+3} - x)^2}{(x_{i+3} - x_{i+1})(x_{i+3} - x_{i+2})}\right) \\ \quad + \left(\dfrac{x_{i+4} - x}{x_{i+4} - x_{i+1}}\right)\left(\dfrac{(x - x_{i+1})(x_{i+3} - x)}{(x_{i+3} - x_{i+1})(x_{i+3} - x_{i+2})} + \dfrac{(x_{i+4} - x)(x - x_{i+2})}{(x_{i+4} - x_{i+2})(x_{i+3} - x_{i+2})}\right), & x_{i+2} \le x < x_{i+3} \\[2ex] \left(\dfrac{x_{i+4} - x}{x_{i+4} - x_{i+1}}\right)\left(\dfrac{(x_{i+4} - x)^2}{(x_{i+4} - x_{i+2})(x_{i+4} - x_{i+3})}\right), & x_{i+3} \le x < x_{i+4} \\[2ex] 0, & \text{otherwise} \end{cases}$$

Let $a = x_{-m+1} = \cdots = x_0 < x_1 < \cdots < x_n = x_{n+1} = \cdots = x_{n+m-1} = b$ be an equally spaced knots sequence such that $x_k - x_{k-1} = h, \quad k = 1, 2, \ldots, n$. The B-spline functions are as follows:

Linear B-spline:

$$B_{2,i}(x) = \begin{cases} 1 + (x - ih)/h, & x_{i-1} \le x < x_i \\ 1 - (x - ih)/h, & x_i \le x < x_{i+1} \\ 0, & \text{otherwise.} \end{cases} \tag{9.84}$$

Quadratic B-spline:

$$B_{3,i}(x)=\frac{1}{h^2}\begin{cases}(x-x_{i-1})^2, & x_{i-1}\le x<x_i\\ 2h^2-(x_{i+1}-x)^2-(x-x_i)^2, & x_i\le x<x_{i+1}\\ (x_{i+2}-x)^2, & x_{i+1}\le x<x_{i+2}\\ 0, & \text{otherwise.}\end{cases} \tag{9.85}$$

Cubic B-spline:

$$B_{4,i}(x)=\frac{1}{h^3}\begin{cases}(x-x_{i-2})^3, & x_{i-2}\le x<x_{i-1}\\ h^3+3h^2(x-x_{i-1})+3h(x-x_{i-1})^2-3(x-x_{i-1})^3, & x_{i-1}\le x<x_i\\ h^3+3h^2(x_{i+1}-x)+3h(x_{i+1}-x)^2-3(x_{i+1}-x)^3, & x_i\le x<x_{i+1}\\ (x_{i+2}-x)^3, & x_{i+1}\le x<x_{i+2}\\ 0, & \text{otherwise.}\end{cases} \tag{9.86}$$

Properties

1. $B_{m,i}(x)$ is an $(m-1)$th degree polynomial in x.
2. $B_{m,i}(x)$ satisfies the recurrence relation

$$B_{m,i}(x)=w_{m,i}(x)B_{m-1,i}(x)+(1-w_{m,i+1})B_{m-1,i+1}(x)$$

where

$$w_{m,i}(x)=\frac{x-x_i}{x_{m+i-1}-x_i}$$

3. For all $m,\ i,\ x,\ B_{m,i}(x)$ is nonnegative.
4. $\sum_{i=0}^{m}B_{m,i}(x)=1$; this is called partition of unity.
5. $B_{m,i}(x)$ has compact support $[x_i,x_{i+1}]$.
6. $B_{m,i}(x)$ is zero on boundary points. that is, $B_{m,i}(a)=0=B_{m,i}(b)$.

9.6.1 Function Approximation by Cubic B-Spline

Any function $f(x)\in L^2[a,b]$ can be approximated by the cubic B-spline functions as

$$f(x)=\sum_{i=-1}^{n+1}\alpha_iB_{4,i}(x) \tag{9.87}$$

where α_i is the coefficients of cubic B-spline functions that can be determined as

$$\alpha_i=\int_a^b f(x)B_{4,i}(x)dx \tag{9.88}$$

MATHEMATICA® Program for Finding Approximate Solution by Cubic B-Spline Method (Chapter 9, Example 9.7)

```
a=0;x[0]=0;
b=1;
n=10;
h=(b-a)/n;
For[i=0,i<=n,i++,
   x[i]=x[0]+i*h];
B[i_][x]:=1/(6*h^3)*Piecewise[{{(x-i*h)^3,i*h<=x<=(i+1)*h},{h^3+3*h^2*
  (x-(i+1)*h)+3*h*(x-(i+1)*h)^2-3*(x-(i+1)*h)^3,(i+1)*h<=x<=(i+2)*h},
  {h^3+3*h^2*((i+3)*h-x)+3*h*((i+3)*h-x)^2-3*((i+3)*h-x)^3,
  (i+2)*h<=x<=(i+3)*h},{((i+4)*h-x)^3,(i+3)*h<=x<=(i+4)*h}}];
B1[i_][x]:=1/(6*h^3)*Piecewise[{{6*(x-i*h),i*h<=x<=(i+1)*h},
  {-3*6*(x-(i+1)*h),(i+1)*h<=x<=(i+2)*h},{-3*6*((i+3)*h-x),
  (i+2)*h<=x<=(i+3)*h},{6*((i+4)*h-x),(i+3)*h<=x<=(i+4)*h}}];
y[x]=Sum[c[i]*B[i][x], {i,-3,n-1}];
y1[x]=D[y[x],{x,2}];
For[i=0,i<=n,i++,
   eqn[i]=N[Simplify[-(y1[x]/.x->i*h)+Pi^2*(y[x]/.x->i*h)-
    2*Pi^2*Sin[Pi*i*h]]]==0;
   Print["eqn[",i,"]=",eqn[i]]];
A=Table[eqn[i],{i,0,n}];
A1=N[Simplify[y[x]/.x->0]]==0
A2=N[Simplify[y[x]/.x->1]]==0
eqns=Join[A,{A1},{A2}];
const=Table[c[j],{j,-3,n-1}];
sol=NSolve[eqns,const]
AA=Table[sol[[1]][[i]][[2]],{i,1,n+3}];
For[i=0,i<=1,i=i+0.1,
  yy[i]=Sum[sol[[1]][[j+4]][[2]]*B[j][x]/.x->i, {j,-3,n-1}];
  yexact[i]=Sin[Pi*i];
  Print[i,"     ",N[yy[i]],"      ",
  N[yexact[i]],"      ",Abs[N[yy[i]]-N[yexact[i]]]]];
```

(In the printed program the sums appear as $\sum_{i=-3}^{n-1}$ c[i]*B[i][x] and $\sum_{j=-3}^{n-1}$ sol[[1]][[j+4]][[2]]*B[j][x]/.x->i; powers are printed as superscripts.)

Output:

```
eqn[ 0 ]= -100. c[-3.]+200. c[-2.]-100. c[-1.]+1.64493 (c[-3.]+4.
  c[-2.]+c[-1.])==0
eqn[ 1 ]= -1.64493 (3.7082 -1. c[-2.]-4. c[-1.]-1. c[0.])-100. (c[-2.]-2.
  c[-1.]+c[0.])==0
eqn[ 2 ]= 0.166667 (-600. (c[-1.]-2. c[0.]+c[1.])+9.8696 (-7.05342+c[-
  1.]+4. c[0.]+c[1.]))==0
eqn[ 3 ]= -1.64493 (9.7082 -1. c[0.]-4. c[1.]-1. c[2.])-100. (c[0.]-2.
  c[1.]+c[2.])==0
eqn[ 4 ]= 0.166667 (-600. (c[1.]-2. c[2.]+c[3.])+9.8696
  (-11.4127+c[1.]+4. c[2.]+c[3.]))==0
eqn[ 5 ]= -100. (c[2.]-2. c[3.]+c[4.])+1.64493 (-12.+c[2.]+4.
  c[3.]+c[4.])==0
```

```
eqn[ 6 ]= 0.166667 (-600. (c[3.]-2. c[4.]+c[5.])+9.8696
  (-11.4127+c[3.]+4. c[4.]+c[5.]))==0
eqn[ 7 ]= -1.64493 (9.7082 -1. c[4.]-4. c[5.]-1. c[6.])-100. (c[4.]-2.
  c[5.]+c[6.])==0
eqn[ 8 ]= 0.166667 (-600. (c[5.]-2. c[6.]+c[7.])+9.8696
  (-7.05342+c[5.]+4. c[6.]+c[7.]))==0
eqn[ 9 ]= -1.64493 (3.7082 -1. c[6.]-4. c[7.]-1. c[8.])-100. (c[6.]-2.
  c[7.]+c[8.])==0
eqn[ 10 ]= -100. c[7.]+200. c[8.]-100. c[9.]+1.64493 (c[7.]+4.
  c[8.]+c[9.])==0
0.166667 (c[-3.]+4. c[-2.]+c[-1.])==0
0.166667 (c[7.]+4. c[8.]+c[9.])==0

{{c[-3]->-0.312851,c[-2]->0.,c[-1]->0.312851,c[0]->0.595079,
  c[1]->0.819055,c[2]->0.962857,c[3]->1.01241,c[4]->0.962857,c[5]-
  >0.819055,c[6]->0.595079,c[7]->0.312851,c[8]->0.,c[9]->-0.312851}}
 0        0.              0.              0.
 0.1      0.307747        0.309017        0.00126968
 0.2      0.58537         0.587785        0.00241508
 0.3      0.805693        0.809017        0.00332407
 0.4      0.947149        0.951057        0.00390768
 0.5      0.995891        1.              0.00410877
 0.6      0.947149        0.951057        0.00390768
 0.7      0.805693        0.809017        0.00332407
 0.8      0.58537         0.587785        0.00241508
 0.9      0.307747        0.309017        0.00126968
 1.       3.88578*10^-16  5.66554*10^-16  1.77976*10^-16
```

Example 9.7

Solve the second-order boundary value problem

$$y''(x) + \frac{\pi^2}{4} y(x) = \frac{\pi^2}{16} \cos\left(\frac{\pi}{4} x\right), \quad y(0) = 0 = y(1)$$

by cubic B-spline polynomials with $h = 0.1$. The analytical solution of this problem is

$$y^*(x) = \frac{-1}{3} \cos\left(\frac{\pi}{2} x\right) - \frac{\sqrt{2}}{6} \sin\left(\frac{\pi}{2} x\right) + \frac{1}{3} \cos\left(\frac{\pi}{4} x\right)$$

Solution:

Here $a = 0,\ b = 1$ and $h = 0.1,\ n = (b - a)/h = 10$

Approximate the unknown function $y(x)$ by cubic B-spline polynomials as

$$y(x) = \sum_{i=-1}^{n+1} \alpha_i B_{4,i}(x) \tag{9.89}$$

and for higher derivatives, we can approximate as follows:

$$y'(x) = \sum_{i=-1}^{n+1} \alpha_i B'_{4,i}(x)$$

TABLE 9.1
Numerical Results for Example 9.7

x	$B_{4,i}(x)$	$y(x)$	$\lvert y(x)-B_{4,i}(x)\rvert$
0.0	0	0	0
0.1	−0.0336885	−0.0337956	0.000107168
0.2	−0.0604311	−0.0606254	0.000194314
0.3	−0.0796274	−0.0798855	0.00025809
0.4	−0.0908996	−0.0911958	0.000296202
0.5	−0.0941016	−0.0944091	0.000307477
0.6	−0.0893215	−0.0896134	0.000291905
0.7	−0.0768784	−0.077129	0.000250628
0.8	−0.0573136	−0.0574995	0.000185891
0.9	−0.0313756	−0.0314765	0.000100961
1.0	0	0	0

$$y''(x)=\sum_{i=-1}^{n+1}\alpha_i B''_{4,i}(x)$$

Now the original equation can be reduced as

$$\sum_{i=-1}^{n+1}\alpha_i B''_{4,i}(x)+\frac{\pi^2}{4}\sum_{i=-1}^{n+1}\alpha_i B_{4,i}(x)=\frac{\pi^2}{16}\cos\left(\frac{\pi}{4}x\right) \tag{9.90}$$

Collocating at the node points $x_j = x_0 + jh$ for $j= 0,1,\ldots, n$ in Equation 9.90, we have

$$\sum_{i=-1}^{n+1}\alpha_i B''_{4,i}(x_j)+\frac{\pi^2}{4}\sum_{i=-1}^{n+1}\alpha_i B_{4,i}(x_j)=\frac{\pi^2}{16}\cos\left(\frac{\pi}{4}x_j\right),\quad j=0,1,\ldots,n \tag{9.91}$$

Now, from the boundary conditions, we have

$$\left.\sum_{i=-1}^{n+1}\alpha_i B_{4,i}(x)\right|_{x=0}=0 \tag{9.92}$$

$$\left.\sum_{i=-1}^{n+1}\alpha_i B_{4,i}(x)\right|_{x=1}=0 \tag{9.93}$$

Equations 9.91 through 9.93 give a system of algebraic equations with $(n + 3)$ number of equations along with same number of unknowns as α_i, $i=-1,0,\ldots,n+1$. Solving this system by the Newton method with appropriate initial guess, we obtain the value of α_i, $i=-1,0,\ldots,n+1$, and hence we obtain the approximate solution from Equation 9.89 (Table 9.1).

9.7 PADÉ APPROXIMATION

In the previous sections, we have discussed the polynomial approximation of a continuous function. There are a sufficient number of polynomials to approximate any continuous function on a closed interval. But using polynomials for approximation, there is a disadvantage because of oscillatory behavior of the polynomials, causing error bounds in polynomial approximation to significantly exceed the average approximation error.

We now discuss a technique that disseminates the approximation error more evenly over the interval of approximation. The procedure is to seek a rational function for the approximation. A rational function $r(x)$ of degree n has the following form:

$$r(x) = \frac{p(x)}{q(x)}$$

where $p(x)$ and $q(x)$ are the polynomials whose degrees add to n.

Given a function $f(x)$ expanded in a Maclaurin series $f(x) = \sum_{n=0}^{\infty} c_n x^n$, we can use the coefficients of the series to represent the function by a ratio of two polynomials

$$\frac{a_0 + a_1 x + \cdots + a_L x^L}{b_0 + b_1 x + \cdots + b_M x^M} \tag{9.94}$$

denoted by [L/M] and called the Padé approximant. The basic idea is to match the series coefficients as far as possible. Even though the series has a finite region of convergence, we can obtain the limit of the function as $x \to \infty$ if $L = M$.

Note that if we wish to obtain [1/1], we will have

$$(a_0 + a_1 x) = (b_0 + b_1 x)(c_0 + c_1 x + c_2 x^2 + \cdots)$$

so that coefficients of x^2 are zero, that is,

$$b_1 c_1 + b_0 c_2 = 0$$

Taking $b_0 = 1$, we have

$$b_1 c_1 + c_2 = 0$$

Now we consider [2/2], yielding

$$(a_0 + a_1 x + a_2 x^2) = (b_0 + b_1 x + b_2 x^2)(c_0 + c_1 x + c_2 x^2 + c_3 x^3 + \cdots)$$

Clearly the coefficients of x^3 are zero, so that we can write

$$b_2 c_1 + b_1 c_2 + b_0 c_3 = 0$$

In general, we note that there are $(L + 1)$ independent coefficients in the numerator and $(M + 1)$ coefficients in the denominator. To make the system determinable, it is customary to let $b_0 = 1$. We then have M independent coefficients in the denominator, and thus we have $(L + M + 1)$ independent coefficients in total. Now the [L/M] approximant can fit the power series through orders $1, x, x^2, \ldots, x^{L+M}$ with an error of $O(x^{L+M+1})$. For example, let us consider

$$f(x) = 1 - \frac{1}{2}x + \frac{1}{3}x^2 + \cdots$$

We have

$$[1/1] = \frac{1 + (1/6)x}{1 + (2/3)x} = f(x) + O(x^3)$$

Consequently,

$$(a_0 + a_1 x + \cdots + a_L x^L) = (b_0 + b_1 x + \cdots + b_M x^M)(c_0 + c_1 x + \cdots)$$

Equating coefficients of $x^{L+1}, x^{L+2}, \ldots, x^{L+M}$ successively to zero, we obtain

$$b_M c_{L-M+1} + b_{M-1} c_{L-M+2} + \cdots + b_0 c_{L+1} = 0$$

$$b_M c_{L-M+2} + b_{M-1} c_{L-M+3} + \cdots + b_0 c_{L+2} = 0$$

$$\vdots$$

$$b_M c_L + b_{M-1} c_{L+1} + \cdots + b_0 c_{L+M} = 0$$

Now setting $b_0 = 1$, we have M linear equations for the M coefficients in the denominator.

$$\begin{bmatrix} c_{L-M+1} & c_{L-M+2} & \cdot & \cdot & \cdot & c_L \\ c_{L-M+2} & c_{L-M+3} & \cdot & \cdot & \cdot & c_{L+1} \\ \cdot & & & & & \\ \cdot & & & & & \\ \cdot & & & & & \\ c_L & c_{L+1} & \cdot & \cdot & \cdot & c_{L-M-1} \end{bmatrix} \begin{bmatrix} b_M \\ b_{M-1} \\ \cdot \\ \cdot \\ \cdot \\ b_1 \end{bmatrix} = \begin{bmatrix} c_{L+1} \\ c_{L+2} \\ \cdot \\ \cdot \\ \cdot \\ c_{L+M} \end{bmatrix}$$

Solving the above system of equations, we determine the coefficients b_i for $i = 1, 2, \ldots, M$. Since we know the coefficients $c_0, c_1, c_2, \ldots$, we can equate the coefficients of $1, x, x^2, \ldots, x^L$ to obtain the remaining coefficients $a_0, a_1, \ldots, a_L$. Thus

$$a_0 = c_0$$

$$a_1 = c_1 + b_1 c_0$$

$$a_2 = c_2 + b_1 c_1 + b_2 c_0$$

$$\vdots$$

$$a_L = c_L + \sum_{i=1}^{N} b_i c_{L-i}, N = \min(L, M)$$

Thus the numerator and denominator of the Padé approximant are all determined and we have agreement with the original series through order x^{L+M}.

The same or higher order accuracy can be achieved by using lower order polynomials in [L/M] than direct polynomial approximation. For higher orders approximants, one can use symbolic programs.

For $f(x) = c_0 + c_1 x + c_2 x^2 + \cdots$, we have

$$[1/1] = \frac{a_0 + a_1 x}{b_0 + b_1 x} \qquad \lim_{x\to\infty}[1/1] = a_1/b_1$$

$$[2/2] = \frac{a_0 + a_1 x + a_2 x^2}{b_0 + b_1 x + b_2 x^2} \qquad \lim_{x\to\infty}[2/2] = a_2/b_2$$

$$[3/3] = \frac{a_0 + a_1 x + a_2 x^2 + a_3 x^3}{b_0 + b_1 x + b_2 x^2 + b_3 x^3} \qquad \lim_{x\to\infty}[3/3] = a_3/b_3$$

When the function $f(x)$ is such that its limiting value is finite as $x \to \infty$, polynomial approximations give very poor results. On the other hand, the Padé approximations give much better results.

Example 9.8

Find the Padé approximation to e^x and hence find the limit for e^x.

Solution:
We have

$$[1/1] = \frac{2+x}{2-x}$$

$$[2/2] = \frac{12+6x+x^2}{12-6x+x^2}$$

$$[3/3] = \frac{120+60x+12x^2+x^3}{120-60x+12x^2-x^3}$$

Note that if we let $x = 1$ to consider the series for e, we get the correct limit.

$$[1/1] = 3$$

$$[2/2] = 2.714$$

$$[3/3] = 2.718$$

$$[4/4] = 2.718$$

Note: The limit is correct for $x = 1$ but the limit at ∞ for [1/1], [2/2], [3/3] fluctuates between ± 1 because we know $e^x \to \infty$ as $x \to \infty$.

For cases where the Padé approximant appears inapplicable, we can sometimes use transformations (Shanks [or Wynn] transformation) of the series. These transformations are effective in accelerating convergence of many slowly convergent series.

Example 9.9

Find the Padé approximations to the following function

$$f(x) = \left(\frac{1+(1/2)x}{1+2x}\right)^{1/2} = 1 - \frac{3}{4}x + \frac{39}{32}x^2 - \cdots$$

Solution:
To approximate $f(x)$ by Padé approximants, we have

$$[1/1] = \frac{1+(7/8)x}{1+(13/8)x}, \lim_{x\to\infty}[1/1] = 0.53846$$

$$[2/2] = \frac{1+(17/8)x+(61/64)x^2}{1+(23/8)x+(121/64)x^2}, \quad \lim_{x\to\infty}[2/2] = 0.504132$$

$$[3/3] = \frac{1+(27/8)x+(111/32)x^2+(547/512)x^3}{1+(33/8)x+(171/32)x^2+(1093/512)x^3}, \quad \lim_{x\to\infty}[3/3] = 0.500457$$

$$[4/4] = \frac{1+(37/8)x+(483/64)x^2+(1307/256)x^3+(4921/4096)x^4}{1+(43/8)x+(663/64)x^2+(2153/256)x^3+(9841/4096)x^4}, \quad \lim_{x\to\infty}[4/4] = 0.500051$$

This shows that the limit approaches to the correct limit of 0.5 for the given function $f(x)$, more and more closely as we go to the higher order [L/M] Padé approximants.

EXERCISES

1. In the following problems, obtain the linear and quadratic least square polynomial approximation to the given data.

   a.

x	0.2	0.4	0.6	0.8	1
$f(x)$	0.108	0.164	0.316	0.612	1.1

   b.

x	−1.0	−0.5	−0.25	0	0.25	0.5	1
$f(x)$	1.0000	0.7500	0.8125	1.0000	1.3125	1.7500	3.0000

2. Fit a curve of the form $y = ab^x$ for the following data:

x	4	8	14	20	28	36	44	56
y	36.4	32.4	27.3	23.0	18.1	14.4	11.4	8.0

3. Find the linear least squares polynomial approximation to $f(x)$ on the indicated interval, if
   a. $f(x) = x^2 + 3x + 2, [0,1]$
   b. $f(x) = x^3, [0,2]$
   c. $f(x) = 1/x, [1,3]$
   d. $f(x) = e^x, [0,2]$
   e. $f(x) = (1/2)\cos x + (1/3)\sin 2x, [0,1]$
   f. $f(x) = x \ln x, [1,3]$
4. Fit the curve $y = ax^b$ to the following data:

   a.

x_i	1	2	3	4	5	6	7	8
y_i	15.3	20.5	27.4	36.6	49.1	65.6	87.8	117.6

   b.

x_i	2	4	7	10	20	40	60	80
y_i	43	25	18	13	8	5	3	2

5. Fit the curve $y = ae^{bx}$ to the following data:

   a.

x_i	1	2	3	4	5	6
y_i	1.6	4.5	13.8	40.2	125.0	300.0

   b.

x_i	0	0.5	1.0	1.5	2.0	2.5
y_i	0.10	0.45	2.15	9.15	40.35	180.75

6. A car is travelling along a straight road with a constant speed $v = b$ (m/s), its position y (m) at time t s is $y = a + bt$. By actual measurement, suppose the following data were found:

t	0	3	6	8	10
Y	100	140	190	210	240

Find a least square straight line fit to these data and estimate from it the speed of the car.
7. Determine the best minimax approximation to $2x^3 + 5x^2 + 1$ with a polynomial of degree 0 and 1 for $x \in [0,1]$.
8. Derive the minimax polynomial approximation to $2x^3 + x^2$, on $[-1, 1]$.
9. Obtain the Chebyshev linear and quadratic polynomial approximations to the function $f(x) = x^3$ on [0, 1].
10. Find the polynomial $P(x)$ of degree 3 minimizing $\|q(x) - P(x)\|_2$ where the norm is defined by

$$< g,h >= \int_0^\infty g(x)h(x)e^{-x}dx$$

and $q(x) = x^5 - 3x^2 + x$.
11. Using Gram–Schmidt process, construct orthogonal polynomials ϕ_0, ϕ_1, ϕ_2, and ϕ_3 on [0, 1] with respect to the weight function $W(x) = 1$ and hence find the cubic polynomial least squares approximation to e^x.
12. Develop the function $f(x) = (1/2)\ln[(1+x)/(1-x)]$ in a series of Chebyshev polynomials.
13. Using the Chebyshev polynomials $T_n(x)$, obtain the least square approximation of second degree for
    a. $3x^4 + 2x^3 + x + 2$ on $[-1, 1]$
    b. $5x^3 + 6x^2 - 5x + 3$ on $[-1, 1]$
14. Fit a curve of the form $y = a.e^{bx}$ to the following data:

x	1.5	3.1	4.7	6.3	7.9
y	2.9	4.9	5.7	8.9	12.4

15. Fit a second-degree parabola of the form $y = a + bx + cx^2$ to the following data:

x	0	10	20	30	40	50
y	115	160	215	270	335	400

16. For the data given, find the equation to the best fitting exponential curve of the form $y = ab^x$.

x	2	4	6	8	10
y	3	13	32	57	91

17. The following table gives the data collected in an experiment to study the relationship between the stopping distance d (m) of an automobile travelling at speeds v (km/h) at the instant the danger is sighted. Fit a least squares parabola of the form $d = a + bv + cv^2$ to the following data:

Speed v (km/h)	32	48	64	80	96	112
Stopping distance d (m)	16.5	27.5	19.5	24.5	29.3	34.2

18. Fit the exponential function of the form $y = ae^{bx}$ to the following data:

x_i	0	0.3	0.6	0.8	1.0
y_i	2.0	1.8	1.65	1.55	1.45

19. Use cubic B-spline basis function to approximate the solution of the boundary value problem:

$$x^2y'' + 2xy' - 2y = 4x^2, \quad 0 \le x \le 1, \; y(0) = y(1) = 0$$

Hence, compare the result with the actual solution $y(x) = x^2 - x$.

20. Determine the Padé approximation of degree 6, that is, [3/3] Padé approximant for $f(x) = \sin x$. Hence, compare the results at $x_i = 0.1i$, for $i = 0,1,2,\ldots,5$, with the exact results and with the results of the sixth Maclaurin polynomial.

10 Numerical Solutions of Partial Differential Equations

10.1 INTRODUCTION

Partial differential equations (PDEs) play an important role in numerous branches of science and engineering. They describe many types of physical phenomena in science and engineering. PDEs have become a useful tool for describing the nature of science and engineering models. Nowadays, most of the phenomena that arise in mathematical physics and engineering fields can be described by PDEs. Many engineering applications are simulated mathematically as PDEs with initial and boundary conditions. Most physical phenomena in fluid dynamics, quantum mechanics, electricity, and many other fields are described using PDEs. Exact analytical solutions can be obtained in only a few cases, and these analytical methods are rather complicated. However, there are efficient numerical methods to obtain good approximate solutions. Therefore, it becomes increasingly important to be familiar with the numerical methods for solving PDEs and implementing these methods. Among the numerical methods, the finite difference method (FDM) has the attractive feature that it is applicable to linear as well to nonlinear PDEs. In this chapter, we shall discuss FDMs for solving PDEs.

10.2 CLASSIFICATION OF PDEs OF SECOND ORDER

The most general form of PDEs of second order in two independent variables is as follows:

$$A\frac{\partial^2 u}{\partial x^2}+B\frac{\partial^2 u}{\partial x\partial y}+C\frac{\partial^2 u}{\partial y^2}+D\frac{\partial u}{\partial x}+E\frac{\partial u}{\partial y}+Fu=G \tag{10.1}$$

where A, B, C, D, E, F, and G are functions of x and y only.

The above equation is said to be elliptic or parabolic or hyperbolic at a point of the domain in the plane according as

$$\begin{aligned}
&1.\ B^2-4AC<0 \quad (\text{elliptic type})\\
&2.\ B^2-4AC=0 \quad (\text{parabolic type})\\
&3.\ B^2-4AC>0 \quad (\text{hyperbolic type})
\end{aligned} \tag{10.2}$$

Some well-known examples of the three types are as follows:

1. Parabolic equation:

$$\frac{\partial u}{\partial t}=c^2\frac{\partial^2 u}{\partial x^2} \quad (\text{one-dimensional heat equation})$$

$$\frac{\partial u}{\partial t}=c^2\left(\frac{\partial^2 u}{\partial x^2}+\frac{\partial^2 u}{\partial y^2}\right) \quad (\text{two-dimensional heat equation})$$

2. Elliptic equation:

$$\frac{\partial^2 u}{\partial x^2}+\frac{\partial^2 u}{\partial y^2}=0 \quad \text{(two-dimensional Laplace's equation)}$$

3. Hyperbolic equation:

$$\frac{\partial^2 u}{\partial t^2}=c^2\frac{\partial^2 u}{\partial x^2} \quad \text{(one-dimensional wave equation)}$$

$$\frac{\partial^2 u}{\partial t^2}=c^2\left(\frac{\partial^2 u}{\partial x^2}+\frac{\partial^2 u}{\partial y^2}\right) \quad \text{(two-dimensional wave equation)}$$

10.3 TYPES OF BOUNDARY CONDITIONS AND PROBLEMS

The parabolic and hyperbolic types of equations are either initial value problems or initial boundary value problems, whereas the elliptic type equation is always a boundary value problem.

The boundary conditions associated with a linear second-order PDE

$$Lu=G(x,y), \quad \text{for } x, y \in R$$

where

$$Lu=A\frac{\partial^2 u}{\partial x^2}+B\frac{\partial^2 u}{\partial x\partial y}+C\frac{\partial^2 u}{\partial y^2}+D\frac{\partial u}{\partial x}+E\frac{\partial u}{\partial y}+Fu$$

can be written in the operator form

$$Bu=f(x,y), \quad \text{for } x, y \in \partial R$$

where ∂R denotes the boundary of the region R and $f(x,y)$ is a given function of x and y. There are three types of boundary conditions encountered in PDEs:

1. *Dirichlet boundary condition*: If the boundary operator $Bu=u$ the boundary condition represents the dependent variable being specified on the boundary. These types of boundary conditions are called Dirichlet conditions.
2. *Neumann boundary condition*: If the boundary operator $Bu=\partial u/\partial n = grad\ u.\hat{n}$ denotes a normal derivative, then the boundary condition is that the normal derivative at each point of the boundary is being specified. These types of boundary conditions are called Neumann conditions. Neumann conditions require the boundary to be such that one can calculate the normal derivative $\partial u/\partial n$ at each point of the boundary of the given region R. This requires that the unit outwards normal vector $\hat{n}$ be known at each point of the boundary.
3. *Mixed boundary condition*: If the boundary operator is a linear combination of the Dirichlet and Neumann boundary conditions, then the boundary operator has the form

$$Bu=\alpha\frac{\partial u}{\partial n}+\beta u$$

where α and β are constants. These types of boundary conditions are said to be of the Robin type.

a. The PDE together with a Dirichlet boundary condition is sometimes referred to as a boundary value problem of the first kind.
b. A PDE with a Neumann boundary condition is sometimes referred to as a boundary value problem of the second kind.
c. A boundary value problem of the third kind is a PDE with a Robin type boundary condition.

A PDE with a boundary condition of the form

$$Bu = \begin{cases} u, & \text{for all} \quad x, y \in \partial R_1 \\ \dfrac{\partial u}{\partial n}, & \text{for all} \quad x, y \in \partial R_2 \end{cases} \qquad \partial R_1 \cap \partial R_2 = \phi, \partial R_1 \cup \partial R_2 = \partial R$$

is called a mixed boundary value problem.

If time t is one of the independent variables in a PDE, then a given condition to be satisfied at the time $t = 0$ is referred to as an initial condition. A PDE subject to both boundary and initial conditions is called a boundary initial value problem.

10.4 FINITE DIFFERENCE APPROXIMATIONS TO PARTIAL DERIVATIVES

We divide the xy plane into sets of equal rectangles of sides $\Delta x = h$ and $\Delta y = k$ by having the equally spaced grid lines parallel to the coordinate axes, defined by $x_i = ih$, $y_j = jk$, $i, j = 0, 1, 2, \ldots$. The points of intersections of these families of lines are called mesh points or grid points or lattice points.

The value of $u(x, y)$ at a mesh point $P(x_i, y_j)$ is denoted by u_{ij}, that is, $u_{ij} = u(x_i, y_j) = u(ih, jk)$.

Now

$$u_x(ih, jk) = \frac{u_{i+1,j} - u_{i,j}}{h} + O(h) \quad \text{(forward-difference approximation)}$$

$$= \frac{u_{i,j} - u_{i-1,j}}{h} + O(h) \quad \text{(backward-difference approximation)}$$

$$= \frac{u_{i+1,j} - u_{i-1,j}}{2h} + O(h^2) \quad \text{(central difference approximation)}$$

Similarly,

$$u_y(ih, jk) = \frac{u_{i,j+1} - u_{i,j}}{k} + O(k) \quad \text{(forward-difference approximation)}$$

$$= \frac{u_{i,j} - u_{i,j-1}}{k} + O(k) \quad \text{(backward-difference approximation)}$$

$$= \frac{u_{i,j+1} - u_{i,j-1}}{2k} + O(k^2) \quad \text{(central difference approximation)}$$

Again,

$$u_{xx}(ih, jk) = \frac{u_{i+1,j} - 2u_{i,j} + u_{i-1,j}}{h^2} + O(h^2)$$

$$u_{yy}(ih, jk) = \frac{u_{i,j+1} - 2u_{i,j} + u_{i,j-1}}{k^2} + O(k^2)$$

10.5 PARABOLIC PDEs

We consider the parabolic PDE, the one-dimensional heat equation given by

$$\frac{\partial u}{\partial t} = c^2 \frac{\partial^2 u}{\partial x^2}, \quad 0 < x < L, t \ge 0 \tag{10.3}$$

where the coefficient $c^2 = k/s\rho$ is called diffusivity, s is the specific heat capacity, ρ is the mass density of the material, and k is the thermal conductivity.

The parabolic equation in Equation 10.3 is also called the diffusion equation. Now we shall develop two methods to determine the numerical solution of Equation 10.3 subject to the following initial conditions:

$$u(x,0) = f(x), \quad 0 \le x \le L$$

and boundary conditions

$$u(0,t) = g_1(t) \quad \text{and} \quad u(L,t) = g_2(t)$$

10.5.1 Explicit FDM

We divide the $x-t$ plane into small rectangles by means of the sets of lines

$$x_i = ih, \quad i = 0,1,2,\ldots,N$$

$$t_n = nk, \quad n = 0,1,2,\ldots$$

Let us denote the mesh points $(x_i, t_n) = (ih, nk)$ by (i,n). Let u_i^n denote the computed value of $u(x_i, t_n)$. To convert the heat equation (10.3) into a difference equation, we apply explicit finite difference approximations for the derivatives terms. Thus the first-order time derivative is approximated with a two-point forward difference, and the second-order space derivative is approximated with a three-point central difference. Then we have

$$u_t(x_i, t_n) = \frac{u_i^{n+1} - u_i^n}{k} + O(k)$$

$$u_{xx}(x_i, t_n) = \frac{u_{i+1}^n - 2u_i^n + u_{i-1}^n}{h^2} + O(h^2)$$

Substituting these differences in Equation 10.3 yields

$$\frac{u_i^{n+1} - u_i^n}{k} = c^2 \frac{u_{i+1}^n - 2u_i^n + u_{i-1}^n}{h^2}$$

which simplifies to

$$u_i^{n+1} = ru_{i-1}^n + (1-2r)u_i^n + ru_{i+1}^n, \quad \text{for } i = 1,2,\ldots,N-1 \text{ and } n = 1,2,\ldots \tag{10.4}$$

where $r = kc^2/h^2$. Equation 10.4 is the finite difference equation corresponding to Equation 10.3. This equation is called an explicit formula because Equation 10.4 gives a formula for the unknown *temperature* u_i^{n+1} at the $(i,n+1)$th mesh point in terms of known *temperatures* along the nth time row (Figure 10.1). The above scheme is also known as forward time central space scheme, abbreviated as Forward Time Central Space (FTCS) scheme. It can be shown that Equation 10.4 is valid only for $0 < r \le 1/2$, which is called the stability condition of the explicit formula.

Now we have

$$u_0^0 = f(x_0), \quad u_1^0 = f(x_1),\ldots,u_N^0 = f(x_N)$$

For $n = 0$, we can generate the next time row by

$$u_0^1 = u(0,t_1) = g_1(t_1)$$

$$u_1^1 = ru_0^0 + (1-2r)u_1^0 + ru_2^0$$

$$u_2^1 = ru_1^0 + (1-2r)u_2^0 + ru_3^0$$

$$\vdots$$

$$u_{N-1}^1 = ru_{N-2}^0 + (1-2r)u_{N-1}^0 + ru_N^0$$

$$u_N^1 = u(L,t_1) = g_2(t_1)$$

Then we can use the values of u_i^1 to generate all the values u_i^2 and so on.

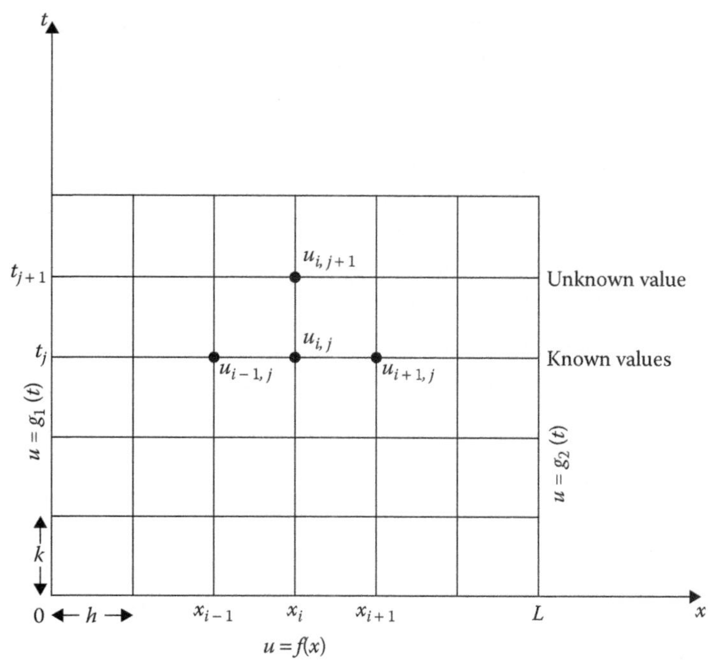

FIGURE 10.1 Forward-difference method.

The explicit nature of this difference method implies the tridiagonal system with the following associated $(N-1)\times(N-1)$ matrix as

$$A = \begin{bmatrix} (1-2r) & r & & & & & \\ r & (1-2r) & r & & & & \\ \cdots & \cdots & \cdots & \cdots & \cdots & \cdots & \cdots \\ & & & & r & (1-2r) & r \\ & & & & & r & (1-2r) \end{bmatrix}$$

Thus, if we let

$$\boldsymbol{u}^{(0)} = (f(x_1), f(x_2), \ldots, f(x_{N-1}))^T$$

and

$$\boldsymbol{u}^{(j)} = (u_{1,j}, u_{2,j}, \ldots, u_{N-1,j})^T, \quad \text{for } j = 1,2,\ldots$$

then the approximate solution is given by

$$\boldsymbol{u}^{(j)} = \boldsymbol{A}\boldsymbol{u}^{(j-1)} + r\boldsymbol{B}, \quad \text{for } j = 1,2,\ldots$$

where $\boldsymbol{B} = (f(x_0), 0, \ldots, 0, f(x_N))^T$. This is known as forward-difference method. If the solution of the PDE has four continuous partial derivatives in x and two in t, then this method has the local truncation error of order $O(h^2 + k)$.

Equation 10.4 is also known as the Bender–Schmidt recurrence formula. The simplest form of the Bender–Schmidt difference scheme is obtained by taking $r = 1/2$. In this case, Equation 10.4 reduces to

$$u_i^{n+1} = \frac{1}{2}\left(u_{i-1}^n + u_{i+1}^n\right) \tag{10.5}$$

In practical computational work in solving Equation 10.3, we shall choose h and k in such a way that $r = kc^2/h^2 = 1/2$. If $r \neq 1/2$, we shall use Equation 10.4. We shall not consider the case when $r > 1/2$, giving an unstable solution of Equation 10.3. Thus it may be observed that Equations 10.4 and 10.5 have a limited application because of the restriction on the values of r. To get rid of this restriction, we require a formula that does not have any restriction on r.

10.5.1.1 Algorithm for Solving Parabolic PDE by FDM

PDE: $\partial u/\partial t = c^2\left(\partial^2 u/\partial x^2\right), \quad 0 < x < L, t \geq 0$
Initial conditions: $u(x,0) = f(x), \quad 0 < x < L, \quad t \geq 0$
Boundary conditions: $u(0,t) = g_1(t), \quad u(L,t) = g_2(t)$.
Input: Read L, h, k, define $f(x)$, $g_1(t)$, $g_2(t)$, the required value of t
Output: Approximate values of u_i^M for $1 \leq i \leq N-1$.
Step 1: Compute $N = L/h; \quad M = t/k; \quad r = c^2 k/h^2;$
Step 2: for $i = 0(1)N$ do

$x_i = ih;$

$u(x_i, 0) = f(x_i)$

end.
Step 3: for $j = 0(1)M$ do

$t_j = jk;$

$u(0, t_j) = g_1(t_j);$

$u(L, t_j) = g_2(t_j);$

end.

Step 4: for $j = 0(1)\overline{M-1}$ do
  for $i = 1(1)\overline{N-1}$ do
    $u(x_i, t_{j+1}) = ru(x_{i-1}, t_j) + (1-2r)u(x_i, t_j) + ru(x_{i+1}, t_j)$;
  end.
end.

Step 5: Print $u_i^M = u(x_i, t_M)$ for $i = 1, \ldots, N-1$

Step 6: Stop. ■

MATHEMATICA® Program for Explicit FDM in Solving Parabolic PDE (Chapter 10, Example 10.1)

```
u1[x_,t_]:= 8/pi^3 * Sum[1/(2*l+1)^3 * Sin[(2*l+1)*Pi*x]*Exp[-(2*l+1)^2 * Pi^2 * t], {l,0,∞}];

s=0.5;
h=0.1;
k=0.005;
r=k/h²;
a=0;
b=1;
m=(b-a)/h;
n=s/k;
For[i=0,i<=m,i++,
  x[i]=a+i*h;
  u[x[i],0.]=x[i]*(1-x[i])];
For[j=0,j<=n,j++,
  t[j]=j*k;
  u[0.,t[j]]=0;
  u[1.,t[j]]=0];
For[j=0,j<n,j++,
    For[i=1,i<=m-1,i++,
      u[x[i],t[j+1]]=
        r*u[x[i-1],t[j]]+(1-2*r)*u[x[i],t[j]]+r*u[x[i+1],t[j]];
      Print[x[i],"      ",t[j+1],"      ",u[x[i],t[j+1]]]
    ]];
```

Output:

```
0.1     0.005     0.08
0.2     0.005     0.15
0.3     0.005     0.2
0.4     0.005     0.23
0.5     0.005     0.24
0.6     0.005     0.23
0.7     0.005     0.2
0.8     0.005     0.15
0.9     0.005     0.08
⋮
⋮
⋮
0.1     0.5     0.00052785
```

```
0.2    0.5    0.00100277
0.3    0.5    0.00138193
0.4    0.5    0.00162251
0.5    0.5    0.00170816
0.6    0.5    0.00162251
0.7    0.5    0.00138193
0.8    0.5    0.00100277
0.9    0.5    0.00052785
```

10.5.2 Crank–Nicolson Implicit Method

Crank and Nicolson (1947) proposed and used a method that reduces the total amount of calculation, and the formula is valid (i.e., convergent and stable) for all finite values of r. In the Crank–Nicolson implicit method, Equation 10.3 is approximated by replacing the space derivative by the average of its finite difference approximations on the nth and $(n+1)$th time levels. Thus Equation 10.3 can be written as

$$\frac{u_i^{n+1}-u_i^n}{k}=c^2\frac{u_{i+1}^n-2u_i^n+u_{i-1}^n+u_{i+1}^{n+1}-2u_i^{n+1}+u_{i-1}^{n+1}}{2h^2}$$

which simplifies to

$$-ru_{i-1}^{n+1}+(2+2r)u_i^{n+1}-ru_{i+1}^{n+1}=ru_{i-1}^n+2(1-r)u_i^n+ru_{i+1}^n, \tag{10.6}$$

where $r=kc^2/h^2$.

This equation is called Crank–Nicolson difference scheme or Crank–Nicolson difference method and it is an implicit formula. In general, the left-hand side of Equation 10.6 contains three unknowns and the right-hand side has three known values of u. The Crank–Nicolson method is unconditionally stable.

For $n=0$ and $i=1,2,\ldots,N-1$, Equation 10.6 gives $(N-1)$ simultaneous equations for $(N-1)$ unknowns $u_{1,1}, u_{2,1},\ldots, u_{N-1,1}$ in terms of known initial and boundary values $u_{0,0}, u_{1,0}, u_{2,0}, \ldots, u_{N,0}$; $u_{0,0}$ and $u_{N,0}$ are the boundary values and $u_{1,0}, u_{2,0}, \ldots, u_{N-1,0}$ are the initial values.

Similarly, for $n=1$ and $i=1,2,\ldots,N-1$, Equation 10.6 gives another set of unknown values $u_{1,2}$, $u_{2,2}, \ldots, u_{N-1,2}$ in terms of calculated values for $n=0$ and so on.

The system of equations (10.6) can be written in the following matrix form:

$$\begin{bmatrix} 2+2r & -r & & & & & \\ & -r & 2+2r & -r & & & \\ \cdots & \cdots & \cdots & \cdots & \cdots & \cdots & \cdots \\ & & & & -r & 2+2r & -r \\ & & & & & -r & 2+2r \end{bmatrix}\begin{bmatrix} u_{1,n+1} \\ u_{2,n+1} \\ u_{3,n+1} \\ \vdots \\ u_{N-1,n+1} \end{bmatrix}=\begin{bmatrix} b_{1,n} \\ b_{2,n} \\ b_{3,n} \\ \vdots \\ b_{N-1,n} \end{bmatrix} \tag{10.7}$$

where

$$b_{1,n} = ru_0^n + 2(1-r)u_1^n + ru_2^n + ru_0^{n+1}$$

$$b_{j,n} = ru_{j-1}^n + 2(1-r)u_j^n + ru_{j+1}^n, \quad j = 2,\ldots,N-2$$

$$b_{N-1,n} = ru_{N-2}^n + 2(1-r)u_{N-1}^n + ru_N^n + ru_N^{n+1}$$

The right-hand side of Equation 10.7 is known. This tridiagonal system of equations can be solved easily.

10.5.2.1 Algorithm for Solving Parabolic PDE by the Crank–Nicolson Method

PDE: $\partial u/\partial t = c^2\left(\partial^2 u/\partial x^2\right), \quad 0 < x < L, t \geq 0$
Initial conditions: $u(x,0) = f(x), \quad 0 < x < L, t \geq 0$
Boundary conditions: $u(0,t) = g_1(t), \quad u(L,t) = g_2(t)$.
Input: Read L, h, k, define $f(x), g_1(t), g_2(t)$, the required value of t
Output: Approximate values of u_i^M for $1 \leq i \leq N-1$.
Step 1: Compute $N = L/h; M = t/k; r = c^2k/h^2$;
Step 2: (Constructing the tridiagonal matrix **A**)

for $i = 1(1)\overline{N-1}$ do
  $c_{i,i} = 2 + 2r$;
end.
for $i = 1(1)\overline{N-2}$ do
  $c_{i,i+1} = -r$;
end.
for $i = 2(1)\overline{N-1}$ do
  $c_{i,i-1} = -r$;
end.
Construct the matrix $\boldsymbol{A} = [c_{i,j}]$, for $i = 1(1)\overline{N-1}$ and $j = 1(1)\overline{N-1}$;

Step 3: for $i = 0(1)N$ do
  $x_i = ih$;
  $u(x_i,0) = f(x_i)$
end.
Step 4: for $j = 0(1)M$ do
  $t_j = jk$;
  $u(0,t_j) = g_1(t_j)$;
  $u(L,t_j) = g_2(t_j)$;
end.
Step 5: Construct the vector $\boldsymbol{u}_0 = [u_1^0, u_2^0, \ldots, u_{N-1}^0]^T$;
Step 6: for $j = 0(1)\overline{M-1}$ do
  $b_{1,j} = ru_0^j + (2-2r)u_1^j + ru_2^j + ru_0^{j+1}$;

for $i = 2(1)\overline{N-2}$ do

$b_{i,j} = ru_{i-1}^j + (2-2r)u_i^j + ru_{i+1}^j$;

end.

$b_{N-1,j} = ru_{N-2}^j + (2-2r)u_{N-1}^j + ru_N^j + ru_N^{j+1}$;

Construct $\boldsymbol{b}_j = [b_{1,j}, b_{2,j}, \ldots, b_{N-1,j}]^T$;

Using the Thomas algorithm, solve the system $\boldsymbol{A}\boldsymbol{u}_{j+1} = \boldsymbol{b}_j$; where $\boldsymbol{u}_j = [u_1^j, u_2^j, \ldots, u_{N-1}^j]^T$;

end.

Step 7: Print $\boldsymbol{u}_M = [u_1^M, u_2^M, \ldots, u_{N-1}^M]^T$;

Step 8: Stop. ■

MATHEMATICA® Program for Solving PDEs by the Crank–Nicolson Method (Chapter 10, Example 10.1)

```
s=0.5;
h=0.1;
k=0.005;
r=k/h²;
a=0;
b1=1;
m=(b1-a)/h;
n=s/k;
For[i=0,i<=m,i++,
  x[i]=a+i*h;
  u[x[i],0]=x[i]*(1-x[i])];
For[j=0,j<=n,j++,
  t[j]=j*k;
  u[0,t[j]]=0;
  u[1,t[j]]=0];
A={{2+2*r,-r,0,0,0,0,0,0,0},
  {-r,2+2*r,-r,0,0,0,0,0,0},{0,-r,2+2*r,-r,0,0,0,0,0},
  {0,0,-r,2+2*r,-r,0,0,0,0},{0,0,0,-r,2+2*r,-r,0,0,0},
  {0,0,0,0,-r,2+2*r,-r,0,0},{0,0,0,0,0,-r,2+2*r,-r,0},
  {0,0,0,0,0,0,-r,2+2*r,-r},{0,0,0,0,0,0,0,-r,2+2*r}};
u[0]=Table[u[x[i],0],{i,1,m-1}];
For[j=0,j<n,j++,

  b[1,j]=r*u[0,t[j]]+(2-2*r)*u[j][[1]]+r*u[j][[2]]+r*u[0,t[j+1]];
  For[i=2,i<=m-2,i++,
   b[i,j]=r*u[j][[i-1]]+(2-2*r)*u[j][[i]]+r*u[j][[i+1]]];

  b[9,j]=r*u[j][[m-2]]+(2-2*r)*u[j][[m-1]]+r*u[1,t[j]]+r*u[1,t[j+1]];
  b[j]=Table[b[l,j],{l,1,m-1}];
  u[j+1]=Inverse[A].Transpose[{b[j]}];
  u[j+1]=Transpose[u[j+1]][[1]];
  Print[u[j+1]];
  Print["............................."]
];
```

Output:

```
{0.0817157,0.150294,0.200051,0.230009,0.240003,0.230009,0.200051,0.150294,
 0.0817157}
.............................
```

```
{0.0758579,0.141421,0.190316,0.220069,0.230027,0.220069,0.190316,0.141421,
 0.0758579}
...............................
{0.0711093,0.133518,0.180984,0.210264,0.22012,0.210264,0.180984,0.133518,
 0.0711093}
...............................
{0.0670196,0.126381,0.172137,0.200688,0.210357,0.200688,0.172137,0.126381,
 0.0670196}
...............................
{0.0633773,0.119844,0.163766,0.191409,0.200818,0.191409,0.163766,0.119844,
 0.0633773}
...............................
{0.0600646,0.113789,0.15584,0.182468,0.191565,0.182468,0.15584,0.113789,
 0.0600646}
...............................
{0.0570091,0.108136,0.148326,0.173884,0.182639,0.173884,0.148326,0.108136,
 0.0570091}
...............................
⋮
⋮
⋮
{0.000690832,0.00131404,0.00180862,0.00212616,0.00223558,0.00212616,
 0.001 80862,0.00131404,0.000690832}
...............................
{0.000657828,0.00125126,0.00172222,0.00202459,0.00212878,0.00202459,
 0.001 72222,0.00125126,0.000657828}
...............................
{0.000626401,0.00119148,0.00163994,0.00192786,0.00202707,0.00192786,
 0.001 63994,0.00119148,0.000626401}
...............................
{0.000596475,0.00113456,0.00156159,0.00183576,0.00193023,0.00183576,
 0.001 56159,0.00113456,0.000596475}
...............................
```

Example 10.1

Solve the PDE $\partial u/\partial t = \partial^2 u/\partial x^2$, with initial conditions $u(x,0) = x(1-x)$, $0 < x < 1$, $t \geq 0$ and boundary conditions $u(0,t) = u(1,t) = 0$, by the explicit FDM and the Crank–Nicolson method. The exact solution is given by

$$u(x,t) = \frac{8}{\pi^3}\sum_{n=0}^{\infty}\frac{1}{(2n+1)^3}\sin[(2n+1)\pi x]\exp(-(2n+1)^2\pi^2 t)$$

Solution:

Here, we take $h = 0.1$ and $k = 0.005$, such that $r = k/h^2 < 0.5$.

Method 1: Using the explicit FDM, the PDE reduces to

$$\frac{u_i^{n+1}-u_i^n}{k} = \frac{u_{i+1}^n-2u_i^n+u_{i-1}^n}{h^2}$$

or $\quad u_i^{n+1} = ru_{i-1}^n+(1-2r)u_i^n+ru_{i+1}^n, \quad i = 1,2,\ldots,N-1$ where $N = 1/h$ and $r = \dfrac{k}{h^2}$

The approximate results at $t = 0.5$ for different values of x have been provided in Table 10.1.

Method 2: (the Crank–Nicolson method)

From Equation 10.6, the given PDE reduces to

$$-ru_{i-1}^{n+1} + (2 + 2r)u_i^{n+1} - ru_{i+1}^{n+1} = ru_{i-1}^{n} + (2 - 2r)u_i^{n} + ru_{i+1}^{n}$$

where $r = k/h^2,\ i = 1,2,\ldots,N-1$.

For each $n,\ n = 0,1,\ldots,$ we have to solve a system of $(N-1)$ equations and obtain the values of next iteration.

Now for $n = 0$, solving the following system

$$-ru_{i-1}^{1} + (2 + 2r)u_i^{1} - ru_{i+1}^{1} = ru_{i-1}^{0} + (2 - 2r)u_i^{0} + ru_{i+1}^{0},\quad i = 1,2,\ldots,N-1$$

we get the values for $u_i^1,\quad i = 1,2,\ldots,N-1$.

Now for $n = 1$, solving the following system

$$-ru_{i-1}^{2} + (2 + 2r)u_i^{2} - ru_{i+1}^{2} = ru_{i-1}^{1} + (2 - 2r)u_i^{1} + ru_{i+1}^{1},\quad i = 1,2,\ldots,N-1$$

we get the values for $u_i^2,\ i = 1,2,\ldots,N-1$, and so on.

Table 10.2 provides the approximate results of u_i at $t = 0.5$.

TABLE 10.1
Results Obtained by the Explicit FDM at $t = 0.5$

x	Exact	EFDM	Absolute Error	Percentage Error
0	0	0	0	0
0.1	0.00057341	0.00052785	0.0000455599	7.94543
0.2	0.00109069	0.00100277	0.0000879229	8.06121
0.3	0.00150121	0.00138193	0.000119277	7.94543
0.4	0.00176477	0.00162251	0.000142262	8.06121
0.5	0.00185559	0.00170816	0.000147435	7.94543
0.6	0.00176477	0.00162251	0.000142262	8.06121
0.7	0.00150121	0.00138193	0.000119277	7.94543
0.8	0.00109069	0.00100277	0.0000879229	8.06121
0.9	0.00057341	0.00052785	0.0000455599	7.94543
1.0	0	0	0	0

TABLE 10.2
Results Obtained by the Crank–Nicolson Method at $t = 0.5$

x	Exact	CNM	Absolute Error	Percentage Error
0	0	0	0	0
0.1	0.00057341	0.000596475	0.000023065	4.02243
0.2	0.00109069	0.00113456	0.00004387	4.02223
0.3	0.00150121	0.00156159	0.00006038	4.02209
0.4	0.00176477	0.00183576	0.00007099	4.02262
0.5	0.00185559	0.00193023	0.00007464	4.02244
0.6	0.00176477	0.00183576	0.00007099	4.02262
0.7	0.00150121	0.00156159	0.00006038	4.02209
0.8	0.00109069	0.00113456	0.00004387	4.02223
0.9	0.00057341	0.000596475	0.000023065	4.02243
1.0	0	0	0	0

10.6 HYPERBOLIC PDEs

We now consider the hyperbolic PDE, the one-dimensional wave equation given by

$$\frac{\partial^2 u}{\partial t^2} = c^2 \frac{\partial^2 u}{\partial x^2}, \quad 0 < x < L, t \geq 0 \tag{10.8}$$

Here, we shall proceed to develop a method to determine the numerical solution of Equation 10.8 subject to the initial conditions

$$u(x,0) = f(x) \quad \text{and} \quad u_t(x,0) = g(x), \quad \text{for } 0 \leq x \leq L$$

and boundary conditions

$$u(0,t) = \varphi(t) \quad \text{and} \quad u(L,t) = \psi(t), \quad \text{for } 0 \leq t \leq T$$

Equation 10.8 is of hyperbolic type, and it models the transverse vibrations of a stretched string.

10.6.1 Explicit Central Difference Method

Replacing the partial derivatives in Equation 10.8 by the finite difference approximations, we obtain

$$\frac{u_i^{n-1} - 2u_i^n + u_i^{n+1}}{k^2} = c^2 \frac{u_{i-1}^n - 2u_i^n + u_{i+1}^n}{h^2}, \tag{10.9}$$

where

$$x_i = ih, \quad (i = 0,1,2,\ldots,N)$$

and

$$t_n = nk, \quad (n = 0,1,2,\ldots).$$

Let u_i^n denote the computed value of $u(x_i, t_n)$. Setting $\lambda = ck/h$ in Equation 10.9 and rearranging the terms, we get

$$u_i^{n+1} = \lambda^2 (u_{i-1}^n + u_{i+1}^n) + 2(1-\lambda^2)u_i^n - u_i^{n-1}, \quad \text{for} \quad i = 1,2,\ldots,N-1 \quad \text{and} \quad n = 1,2,\ldots \tag{10.10}$$

This scheme is called an explicit scheme for the numerical solution of the wave equation (10.8). Further, Equation 10.10 shows that the function values at the nth and $(n-1)$th time levels are required to determine those at the $(n+1)$th time level. This explicit scheme of Equation 10.10 is valid for $\lambda < 1$, which is the condition for stability.

Now the boundary conditions give

$$u_0^n = \varphi(t_n) \quad \text{and} \quad u_N^n = \psi(t_n), \quad n = 1,2,\ldots$$

and the initial condition implies that

$$u_i^0 = f(x_i), \quad \text{for } i = 0,1,2,\ldots,N$$

Also, from the initial condition $u_t(x,0) = g(x)$, we obtain the following forward-difference approximation:

$$\frac{u_i^1 - u_i^0}{k} = g_i, \quad \text{for } i = 0,1,2,\ldots,N$$

where $g_i = g(x_i)$. This implies that

$$u_i^1 = u_i^0 + kg_i, \quad \text{for } i = 0,1,2,\ldots,N$$

This approximation has the truncation error of order $O(k)$. However, the local truncation error of Equation 10.10 is of order $O(k^2)$.

Equation 10.10 can be written in the following matrix form:

$$\begin{bmatrix} u_{1,n+1} \\ u_{2,n+1} \\ u_{3,n+1} \\ \vdots \\ u_{N-1,n+1} \end{bmatrix} = \begin{bmatrix} 2(1-\lambda^2) & \lambda^2 & & & & \\ \lambda^2 & 2(1-\lambda^2) & \lambda^2 & & & \\ \cdots & \cdots & \cdots & \cdots & \cdots & \cdots \\ & & \lambda^2 & 2(1-\lambda^2) & \lambda^2 & \\ & & & \lambda^2 & 2(1-\lambda^2) & \end{bmatrix} \begin{bmatrix} u_{1,n} \\ u_{2,n} \\ u_{3,n} \\ \vdots \\ u_{N-1,n} \end{bmatrix} - \begin{bmatrix} u_{1,n-1} \\ u_{2,n-1} \\ u_{3,n-1} \\ \vdots \\ u_{N-1,n-1} \end{bmatrix} + \lambda^2 \begin{bmatrix} \varphi(t_n) \\ 0 \\ \vdots \\ 0 \\ \psi(t_n) \end{bmatrix} \quad (10.11)$$

Equations 10.10 and 10.11 show that the $(n+1)$th time step requires values from the nth and $(n-1)$ th time steps.

For simplicity, we may choose λ such that $1-\lambda^2 = 0$, that is, $k^2/h^2 = 1/c^2$. Then Equation 10.10 becomes

$$u_i^{n+1} = u_{i-1}^n - u_i^{n-1} + u_{i+1}^n \quad (10.12)$$

Using Equation 10.12, we can determine the values of u at the $(n+1)$th time level, if we know the values of u at the nth and $(n-1)$th time levels.

The boundary conditions $u(0,t) = \varphi(t)$ and $u(L,t) = \psi(t)$ give the value of u along the lines $x = 0$ and $x = L$. From the initial condition $u(x,0) = f(x)$, we have

$$u_i^0 = f_i, \quad \text{where } f_i = f(x_i), i = 0,1,2,\ldots$$

Again the initial condition $u_t(x,0) = g(x)$ implies that

$$\frac{u_i^{n+1} - u_i^{n-1}}{2k} = g_i, \quad \text{where } g_i = g(x_i), i,n = 0,1,2,\ldots$$

Substituting $n = 0$ in Equation 10.12, we get

$$\begin{aligned} u_i^1 &= u_{i-1}^0 - u_i^{-1} + u_{i+1}^0 \\ &= u_{i-1}^0 + 2kg_i - u_i^1 + u_{i+1}^0, \quad \text{since } \frac{u_i^1 - u_i^{-1}}{2k} = g_i \end{aligned}$$

Therefore,

$$u_i^1 = \frac{u_{i-1}^0 + 2kg_i + u_{i+1}^0}{2} \quad (10.13)$$

For $i = 1,2,\ldots$, we get the values of u in the second row. Using Equation 10.12, we can obtain the successive values of u.

10.6.1.1 Algorithm for Solving Hyperbolic PDE by the Explicit Central Difference Method

PDE: $\partial^2 u/\partial t^2 = c^2\left(\partial^2 u/\partial x^2\right), \quad 0 < x < L, t > 0.$
Initial conditions: $u(x,0) = f_1(x), \quad u_t(x,0) = f_2(x) \quad 0 \le x \le L,$
Boundary conditions: $u(0,t) = g_1(t), \quad u(L,t) = g_2(t), \quad t > 0,$
Input: Read $L,\ h,\ k$, define $f_1(x), f_2(x), g_1(t), g_2(t)$, the required value of t
Output: Approximate values of u_i^M for $1 \le i \le N-1$.
Step 1: Compute $N = L/h; \quad M = t/k; \quad \lambda = ck/h$;
Step 2: for $i = 0(1)N$ do
$x_i = ih;$
$u(x_i,0) = f_1(x_i)$
end.
Step 3: for $j = 0(1)M$ do
$t_j = jk;$
$u(0,t_j) = g_1(t_j);$
$u(L,t_j) = g_2(t_j);$
end.
Step 4: for $i = 1(1)\overline{N-1}$ do
$u(x_i,t_{-1}) = f_1(x_i) - kf_2(x_i);$ (using the backward-difference formula to $u_t(x,0) = f_2(x)$)
end.
Step 5: for $j = 0(1)\overline{M-1}$ do
for $i = 1(1)\overline{N-1}$ do
$u(x_i,t_{j+1}) = \lambda^2 u(x_{i-1},t_j) + 2(1-\lambda^2)u(x_i,t_j) + \lambda^2 u(x_{i+1},t_j) - u(x_i,t_{j-1});$
end.
end.
Step 6: Print $u_i^M = u(x_i,t_M)$ for $i = 1,\ldots,N-1$
Step 7: Stop. ■

MATHEMATICA® Program for Solving Hyperbolic PDE by Explicit FDM (Chapter 10, Example 10.2)

```
s=0.5;
h=N[Pi/10];
k=0.05;
r=N[k/h];
a=0;
b=N[Pi];
m=N[(b-a)/h];
n=N[s/k];
For[i=0,i<=m,i++,
  x[i]=N[a+i*h];
  u[x[i],0.]=N[Sin[x[i]]]];
For[j=0,j<=n,j++,
  t[j]=N[j*k];
  u[x[0],t[j]]=0.;
  u[N[Pi],t[j]]=0.];
Print[u[x[m],t[2]]];

For[i=1;j=0,i<=m-1,i++,
```

```
u[x[i],t[j+1]]=N[1/2*(r²*u[x[i-1],t[j]]+(2-2*r²)*u[x[i],
  t[j]]+r²*u[x[i+1],t[j]])];
  Print[N[x[i]],"     ",t[j+1],"      ",u[x[i],t[j+1]]]];
For[j=1,j<=n-1,j++,
  For[i=1,i<=m-1,i++,
   u[x[i],t[j+1]]=r²*u[x[i-1],t[j]]+
    (2-2*r²)*u[x[i],t[j]]+r²*u[x[i+1],t[j]]-u[x[i],t[j-1]];
   u1[x[i],t[j+1]]=N[Sin[x[i]]*Cos[t[j+1]]];
   err[i,j]=N[Abs[u[x[i],t[j+1]]-u1[x[i],t[j+1]]]];
   percerr[i,j]=N[err[i,j]/u1[x[i],t[j+1]]*100];
   Print[N[x[i]],"     ",t[j+1],"      ",u[x[i],t[j+1]],"      ",
    N[u1[x[i],t[j+1]]],"       ",err[i,j],"      ",percerr[i,j]]
  ]];
```

Output:

```
0.314159   0.1    0.307486    0.307473    0.0000123288      0.00400973
0.628319   0.1    0.584872    0.584849    0.0000234509      0.00400973
0.942478   0.1    0.805008    0.804975    0.0000322773      0.00400973
1.25664    0.1    0.946343    0.946305    0.0000379443      0.00400973
1.5708     0.1    0.995044    0.995004    0.000039897       0.00400973
1.88496    0.1    0.946343    0.946305    0.0000379443      0.00400973
2.19911    0.1    0.805008    0.804975    0.0000322773      0.00400973
2.51327    0.1    0.584872    0.584849    0.0000234509      0.00400973
2.82743    0.1    0.307486    0.307473    0.0000123288      0.00400973

⋮
⋮
⋮

0.314159   0.5    0.271484    0.271188    0.00029608        0.109179
0.628319   0.5    0.516393    0.51583     0.000563177       0.109179
0.942478   0.5    0.710754    0.709979    0.000775146       0.109179
1.25664    0.5    0.835542    0.834631    0.000911239       0.109179
1.5708     0.5    0.878541    0.877583    0.000958134       0.109179
1.88496    0.5    0.835542    0.834631    0.000911239       0.109179
2.19911    0.5    0.710754    0.709979    0.000775146       0.109179
2.51327    0.5    0.516393    0.51583     0.000563177       0.109179
2.82743    0.5    0.271484    0.271188    0.00029608        0.109179
```

10.6.2 Implicit FDM

There exist implicit finite difference schemes for Equation 10.8. Two implicit schemes are

$$\frac{u_i^{n-1}-2u_i^n+u_i^{n+1}}{k^2}=\frac{c^2}{2h^2}(u_{i-1}^{n-1}-2u_i^{n-1}+u_{i+1}^{n-1}+u_{i-1}^{n+1}-2u_i^{n+1}+u_{i+1}^{n+1}) \tag{10.14}$$

and

$$\frac{u_i^{n-1}-2u_i^n+u_i^{n+1}}{k^2}=\frac{c^2}{4h^2}\left[(u_{i-1}^{n-1}-2u_i^{n-1}+u_{i+1}^{n-1})+2(u_{i-1}^{n}-2u_i^{n}+u_{i+1}^{n})+(u_{i-1}^{n+1}-2u_i^{n+1}+u_{i+1}^{n+1})\right] \tag{10.15}$$

The above implicit schemes of Equations 10.14 and 10.15 are valid for all values of $\lambda = ck/h$.

Example 10.2

Solve the hyperbolic PDE $\partial^2 u/\partial t^2 = \partial^2 u/\partial x^2$, $0 < x < \pi, t > 0$,
Initial conditions: $u(x,0) = \sin x$, $u_t(x,0) = 0$, $0 \le x \le \pi$,
Boundary conditions: $u(0,t) = 0$, $u(\pi,t) = 0$, $t > 0$,
by the explicit central difference method by taking (i) $h = \pi/10$, $k = 0.05$, (ii) $h = \pi/20$, $k = 0.1$, and (iii) $h = \pi/20$, $k = 0.05$. Compare the numerical results at $= 0.5$ with the exact result $u(x,t) = \cos t \sin x$.

Solution:

Applying the finite difference approximation, the given PDE reduces to

$$\frac{u_i^{n-1} - 2u_i^n + u_i^{n+1}}{k^2} = \frac{u_{i-1}^n - 2u_i^n + u_{i+1}^n}{h^2}$$

or

$$u_i^{n+1} = \lambda^2(u_{i-1}^n + u_{i+1}^n) + 2(1-\lambda^2)u_i^n - u_i^{n-1}$$

Now from the initial condition $u_t(x,0) = 0$, that is, $\dfrac{u_i^1 - u_i^{-1}}{2k} = 0$ or $u_i^1 = u_i^{-1}$.
For $n = 0$,

$$u_i^1 = \lambda^2(u_{i-1}^0 + u_{i+1}^0) + 2(1-\lambda^2)u_i^0 - u_i^{-1}$$

or

$$u_i^1 = \frac{1}{2}\left(\lambda^2(u_{i-1}^0 + u_{i+1}^0) + 2(1-\lambda^2)u_i^0\right), \quad i = 1,2,\ldots,N-1\left(\text{use } u_i^1 = u_i^{-1}\right)$$

For $n = 1$,

$$u_i^2 = \lambda^2(u_{i-1}^1 + u_{i+1}^1) + 2(1-\lambda^2)u_i^1 - u_i^0, \quad i = 1,2,\ldots,N-1$$

and so on.

The comparison of numerical results with exact results for different values of h and k is presented in Table 10.3.

TABLE 10.3
Results Obtained by Explicit FDM at $t = 0.5$

x	Exact	FDM $h = \pi/10, k = 0.05$	FDM $h = \pi/20, k = 0.1$	FDM $h = \pi/20, k = 0.05$
0	0	0	0	0
$\pi/10$	0.271188	0.271484	0.271233	0.271256
$\pi/5$	0.51583	0.516393	0.515916	0.51596
$3\pi/10$	0.709979	0.710754	0.710098	0.710158
$2\pi/5$	0.834631	0.835542	0.83477	0.834841
$\pi/2$	0.877583	0.878541	0.877729	0.877804
$3\pi/5$	0.834631	0.835542	0.83477	0.834841
$7\pi/10$	0.709979	0.710754	0.710098	0.710158
$4\pi/5$	0.51583	0.516393	0.515916	0.51596
$9\pi/10$	0.271188	0.271484	0.271233	0.271256
π	0	0	0	0

10.7 ELLIPTIC PDEs

The numerical solution of elliptic PDE that we consider here is the Poisson equation

$$\nabla^2 u \equiv \frac{\partial^2 u}{\partial x^2} + \frac{\partial^2 u}{\partial y^2} = f(x,y) \tag{10.16}$$

on the rectangular region $R = \{(x,y) | a < x < b, c < y < d\}$, with the following Dirichlet condition:

$$u(x,y) = g(x,y), \quad \text{for}\,(x,y) \in \partial R$$

where ∂R denotes the boundary of the region R.

In order to apply the FDM, the rectangular domain is subdivided into a network by drawing straight lines $x = a + ih, i = 0,1,2,\ldots,m$ and $y = c + jk,\ j = 0,1,2,\ldots,n$, parallel to the coordinate axes, as shown in Figure 10.2. Here, the step sizes are defined by

$$h = \frac{(b-a)}{m} \quad \text{and} \quad k = \frac{(d-c)}{n}$$

Let x_i and y_j be the discrete values of x and y at the grid points. Therefore,

$$x_i = a + ih, \quad i = 0,1,2,\ldots,m$$

$$y_j = c + jk, \quad j = 0,1,2,\ldots,n$$

and $u_{i,j} = u(x_i, y_j)$ be the computed value of $u(x,y)$ at the point (x_i, y_j).

Now we use the finite difference approximations for the second-derivative terms in Equation 10.16. The three-point central difference approximations for the second derivatives are given by

$$u_{xx} = \frac{u_{i+1,j} - 2u_{i,j} + u_{i-1,j}}{h^2} + O(h^2) \tag{10.17}$$

$$u_{yy} = \frac{u_{i,j+1} - 2u_{i,j} + u_{i,j-1}}{k^2} + O(k^2) \tag{10.18}$$

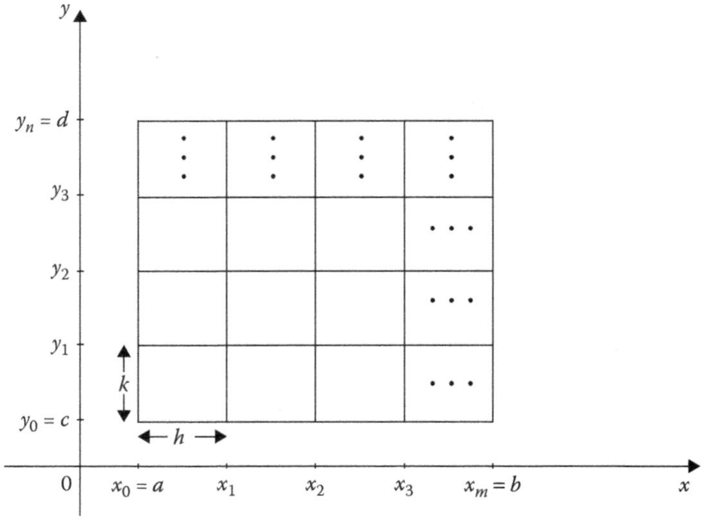

FIGURE 10.2 Finite difference computational domain.

Substituting Equations 10.17 and 10.18 into Equation 10.16 and ignoring the higher order terms, we obtain the system of difference equations for the interior mesh points or grid points (x_i, y_j), for $i = 1,2,\ldots,m-1$ and $j = 1,2,\ldots,n-1$. Thus we obtain

$$\frac{u_{i+1,j} - 2u_{i,j} + u_{i-1,j}}{h^2} + \frac{u_{i,j+1} - 2u_{i,j} + u_{i,j-1}}{k^2} = f_{i,j} \tag{10.19}$$

for $i = 1,2,\ldots,m-1$ and $j = 1,2,\ldots,n-1$; where $f_{i,j} = f(x_i, y_j)$. The boundary conditions are

$$u_{0,j} = g_{0,j} \quad \text{and} \quad u_{m,j} = g_{m,j}; \quad \text{for } j = 0,1,2,\ldots,n$$

$$u_{i,0} = g_{i,0} \quad \text{and} \quad u_{i,n} = g_{i,n}; \quad \text{for } i = 1,2,\ldots,m-1$$

Equation 10.19 is known as the central difference method. This method has the local truncation error of order $O(h^2 + k^2)$. The five-point formula in Equation 10.19 involves the approximations to $u(x, y)$ at the following grid points:

$$(x_{i-1}, y_j), \quad (x_i, y_j), \quad (x_{i+1}, y_j), \quad (x_i, y_{j-1}), \quad \text{and} \quad (x_i, y_{j+1})$$

The system of difference equations in Equation 10.19 produces an $(m-1)(n-1) \times (m-1)(n-1)$ linear system with the unknowns being the approximations $u_{i,j}$ to $u(x, y)$ at the interior mesh points (x_i, y_j). Instead of matrix calculations for the linear system, it is more efficient if a relabelling of the interior mesh points is introduced. A recommended labelling of these points is to let

$$u_l = u_{i,j}$$

where $l = i + (n-1-j)(m-1)$, for $i = 1,2,\ldots,m-1$ and $j = 1,2,\ldots,n-1$. It enables the labelling of the mesh points consecutively form left to right and top to bottom. Thus labelling the mesh points in this manner ensures that the system required for determining the $u_{i,j}$ is a band matrix with the band width at most $(2m-1)$. For example, with $m = 4$ and $n = 4$, the relabelling results in a grid whose points are shown in Figure 10.3.

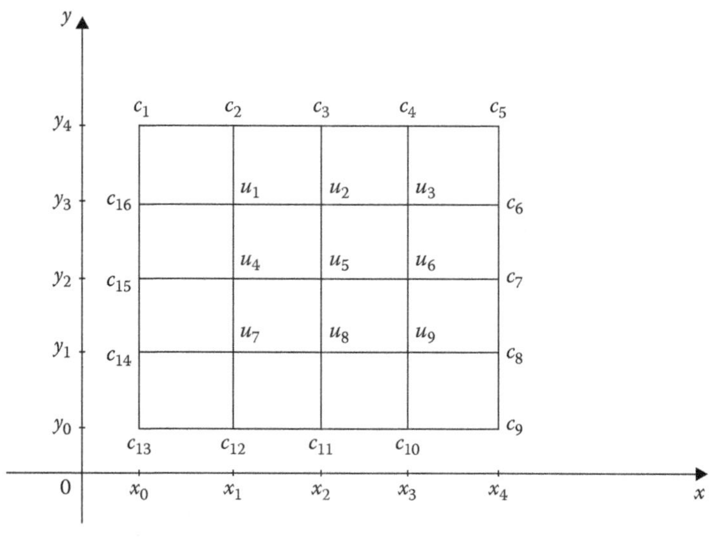

FIGURE 10.3 Interior mesh points and boundary points.

The grid element can often be taken to be a square with sides $h = k$. This simplifies the central difference equation in (10.19) as follows:

$$u_{i+1,j} + u_{i-1,j} - 4u_{i,j} + u_{i,j+1} + u_{i,j-1} = h^2 f_{i,j} \tag{10.20}$$

The iterative methods are ideally suited for the solution of the above finite difference equations.

1. *Gauss–Jacobi method:* Let $u_{i,j}^{(s)}$ denote the sth iterative value of $u_{i,j}$. Then an iterative scheme to solve Equation 10.20 for the interior mesh points is given by

$$u_{i,j}^{(s+1)} = \frac{1}{4}\left[u_{i-1,j}^{(s)} + u_{i+1,j}^{(s)} + u_{i,j-1}^{(s)} + u_{i,j+1}^{(s)} - h^2 f_{i,j}\right] \tag{10.21}$$

   for the interior mesh points. This is called the point Jacobi method.
2. *Gauss–Seidel method:* In this method, the iterative formula for solving Equation 10.20 is given by

$$u_{i,j}^{(s+1)} = \frac{1}{4}\left[u_{i-1,j}^{(s+1)} + u_{i+1,j}^{(s)} + u_{i,j-1}^{(s+1)} + u_{i,j+1}^{(s)} - h^2 f_{i,j}\right] \tag{10.22}$$

   Here we use the latest available iterative values and hence the rate of convergence will be twice as fast as the Jacobi method.
3. *Successive over relaxation method:* In this successive over relaxation (SOR) method, the rate of convergence of an iterative method is accelerated. With regard to Gauss–Seidel over relaxation method, the iteration formula is given by

$$u_{i,j}^{(s+1)} = \frac{1}{4}\omega\left[u_{i-1,j}^{(s+1)} + u_{i+1,j}^{(s)} + u_{i,j-1}^{(s+1)} + u_{i,j+1}^{(s)} - h^2 f_{i,j}\right] + (1-\omega)u_{i,j}^{(s)}, \quad 1 < \omega < 2 \tag{10.23}$$

   The rate of convergence of Equation 10.23 depends on the value of ω, which is called the relaxation factor.

MATHEMATICA® Program for Solving the Poisson Equation by the Gauss–Seidel Method (Chapter 10, Example 10.3)

```
tol=0.001;
K=100;
h=0.1;
a=1.;
b=2.;
n=N[(b-a)/h];
For[i=0.,i<=n,i++,
  x[i]=N[a+i*h];
  For[k=0.,k<=K,k++,
   u[i,0.][k]=x[i]*Log[x[i]];
   u[i,n][k]=2*x[i]*Log[2*x[i]]]];
For[j=0.,j<=n,j++,
  t[j]=N[a+j*h];
  For[k=0.,k<=K,k++,
   u[0.,j][k]=t[j]*Log[t[j]];
   u[n,j][k]=2*t[j]*Log[2*t[j]]]];
For[i=1.,i<=n-1,i++,
  For[j=1.,j<=n-1,j++,
   u[i,j][0.]=0
  ]];
Do[
  For[i=1.,i<=n-1,i++,
```

```
  For[j=1.,j<=n-1,j++,
   u[i,j][k+1]=1/4*(u[i-1,j][k+1]+u[i+1,j][k]+u[i,j-1][k+1]+u[i,j+1]
    [k]-h²*(x[i]/t[j]+t[j]/x[i]));
   u1[i,j]=N[x[i]*t[j]*Log[x[i]*t[j]]];
   e[i,j][k+1]=Abs[u[i,j][k+1]-u1[i,j]]
 ]];
 AA=Table[e[i,j][k+1],{i,1.,n-1},{j,1.,n-1}];
 If[Max[AA]<tol,
  Print[Step[k]];
  Print["..............................................................."];
  Print["x[i]   t[j]    Approximate     Exact      Abs. Error"];
  Print["..............................................................."];
  For[i=1.,i<=n-1,i++,
   For[j=1.,j<=n-1,j++,
    Print[x[i],"    ",t[j],"     ",
     u[i,j][k+1],"      ",u1[i,j],"      ",e[i,j][k+1]]]];
 Break[]],{k,0.,K}];
```

Output:

```
Step[80.]
 .......................................................................
 x[i]      t[j]      Approximate        Exact          Abs. Error
 .......................................................................
 1.1       1.1       0.230524         0.230651         0.000126737
 1.1       1.2       0.366236         0.366474         0.000237527
 1.1       1.3       0.511158         0.511474         0.000316568
 1.1       1.4       0.664589         0.664945         0.000356391
 1.1       1.5       0.825924         0.826279         0.000355712
 1.1       1.6       0.994634         0.994952         0.000318259
 1.1       1.7       1.17025          1.1705           0.000251774
 1.1       1.8       1.35236          1.35253          0.00016709
 1.1       1.9       1.5406           1.54067          0.0000775907
 1.2       1.1       0.366236         0.366474         0.000237527
 1.2       1.2       0.524646         0.525086         0.000439703
 1.2       1.3       0.693127         0.69371          0.000582993
 1.2       1.4       0.870918         0.871574         0.00065519
 1.2       1.5       1.05736          1.05802          0.000654267
 1.2       1.6       1.25188          1.25246          0.000586969
 1.2       1.7       1.45395          1.45442          0.000467139
 1.2       1.8       1.66312          1.66343          0.000313893
 1.2       1.9       1.87897          1.87912          0.000149894
⋮
⋮
⋮
 1.9       1.1       1.5406           1.54067          0.0000775907
 1.9       1.2       1.87897          1.87912          0.000149894
 1.9       1.3       2.23322          2.23342          0.000203303
 1.9       1.4       2.60212          2.60235          0.000231566
 1.9       1.5       2.98463          2.98486          0.000233252
 1.9       1.6       3.37984          3.38005          0.000210551
 1.9       1.7       3.78695          3.78712          0.000168402
 1.9       1.8       4.20526          4.20537          0.000113685
 1.9       1.9       4.63413          4.63419          0.0000545315
```

MATHEMATICA® Program for Solving the Poisson Equation by the SOR Method (Chapter 10, Example 10.3)

```
tol=0.001;
K=100.;
w=1.5;
h=0.1;
a=1.;
b=2.;
n=N[(b-a)/h];
For[i=0.,i<=n,i++,
  x[i]=N[a+i*h];
  For[k=0.,k<=K,k++,
   u[i,0.][k]=x[i]*Log[x[i]];
   u[i,n][k]=2*x[i]*Log[2*x[i]]]];
For[j=0.,j<=n,j++,
  t[j]=N[a+j*h];
  For[k=0.,k<=K,k++,
   u[0.,j][k]=t[j]*Log[t[j]];
   u[n,j][k]=2*t[j]*Log[2*t[j]]]];
For[i=1.,i<=n-1,i++,
  For[j=1.,j<=n-1,j++,
   u[i,j][0.]=0.
  ]];
Do[
  For[i=1.,i<=n-1,i++,
   For[j=1.,j<=n-1,j++,
    z[i,j][k+1]=1/4*(u[i-1,j][k+1]+u[i+1,j][k]+u[i,j-1][k+1]+u[i,j+1]
    [k]-h²*(x[i]/t[j]+t[j]/x[i]));
    u[i,j][k+1]=w*z[i,j][k+1]+(1-w)*u[i,j][k];
    u1[i,j]=N[x[i]*t[j]*Log[x[i]*t[j]]];
    e[i,j][k+1]=Abs[u[i,j][k+1]-u1[i,j]]
   ]];
  AA=Table[e[i,j][k+1],{i,1.,n-1},{j,1.,n-1}];
  If[Max[AA]<tol,
   Print[Step[k]];
   Print["..........................................................."];
   Print["x[i]    t[j]     Approximate      Exact       Abs. Error"];
   Print["..........................................................."];
   For[i=1.,i<=n-1,i++,
    For[j=1.,j<=n-1,j++,
     Print[x[i],"     ",t[j],"     ",
      u[i,j][k+1],"      ",u1[i,j],"     ", e[i,j][k+1]]]];
Break[]],{k,0.,K}];
```

Output:

```
Step[23.]
 ...........................................................
 x[i]     t[j]      Approximate      Exact          Abs. Error
 ...........................................................
 1.1      1.1       0.230424         0.230651       0.00022635
 1.1      1.2       0.366129         0.366474       0.000344712
 1.1      1.3       0.511095         0.511474       0.000379386
 1.1      1.4       0.66459          0.664945       0.000355258
```

1.1	1.5	0.825987	0.826279	0.00029244
1.1	1.6	0.994739	0.994952	0.000213158
1.1	1.7	1.17037	1.1705	0.000134916
1.1	1.8	1.35246	1.35253	0.0000671659
1.1	1.9	1.54065	1.54067	0.0000195743
1.2	1.1	0.366129	0.366474	0.000344712
1.2	1.2	0.524559	0.525086	0.000527251
1.2	1.3	0.693129	0.69371	0.000581155
1.2	1.4	0.871031	0.871574	0.00054246
1.2	1.5	1.05757	1.05802	0.000445283
1.2	1.6	1.25214	1.25246	0.000325088
1.2	1.7	1.45421	1.45442	0.000205831
1.2	1.8	1.66333	1.66343	0.000103658
1.2	1.9	1.87909	1.87912	0.0000324051
⋮				
⋮				
⋮				
1.9	1.1	1.54065	1.54067	0.0000195743
1.9	1.2	1.87909	1.87912	0.0000324051
1.9	1.3	2.23338	2.23342	0.0000351047
1.9	1.4	2.60232	2.60235	0.0000295417
1.9	1.5	2.98484	2.98486	0.0000203347
1.9	1.6	3.38004	3.38005	$9.14884*10^{-6}$
1.9	1.7	3.78712	3.78712	$4.19598*10^{-7}$
1.9	1.8	4.20538	4.20537	$6.45079*10^{-6}$
1.9	1.9	4.63419	4.63419	$6.98949*10^{-6}$

Example 10.3

Solve the Poisson equation

$$\frac{\partial^2 u}{\partial x^2}+\frac{\partial^2 u}{\partial y^2}=\frac{x}{y}+\frac{y}{x}, \quad 1<x, \quad y<2$$

with the boundary conditions

$$u(x,1)=x\ln x, \quad u(x,2)=2x\ln(2x), \quad 1\le x\le 2$$

$$u(1,y)=y\ln y, \quad u(2,y)=2y\ln(2y), \quad 1\le y\le 2$$

using the Gauss–Seidel method and the SOR method with $h = k = 0.1$. The exact solution is given by $u(x,y) = xy\ln(xy)$.

Solution:

Here, we take $h = k = 0.1$.

Using the Gauss–Seidel method from Equation 10.22, the approximation of PDE is given by

$$u_{i,j}^{(s+1)}=\frac{1}{4}\left[u_{i-1,j}^{(s+1)}+u_{i+1,j}^{(s)}+u_{i,j-1}^{(s+1)}+u_{i,j+1}^{(s)}-h^2\left(\frac{x_i}{y_j}+\frac{y_j}{x_i}\right)\right], \quad s=0,1,\ldots \quad \text{and} \quad i,j=1,2,\ldots,N-1$$

The initial guess $u_{i,j}^{(0)}=0, \quad i,j=1,2,\ldots,N-1.$

For

$$s=0, u_{i,j}^{(1)} = \frac{1}{4}\left[u_{i-1,j}^{(1)} + u_{i+1,j}^{(0)} + u_{i,j-1}^{(1)} + u_{i,j+1}^{(0)} - h^2\left(\frac{x_i}{y_j} + \frac{y_j}{x_i}\right)\right], \quad i, j = 1,2,\ldots,N-1$$

which means, it is the first approximation of 81 internal nodes. Some of them are

$$u_{1,1}^{(1)} = 0.0474206, \quad u_{1,2}^{(1)} = 0.0615327, \quad u_{1,3}^{(1)} = 0.0955816,\ldots$$

Similarly, after 60 iterations, we obtain the approximate results of $u_{i,j}, i,j=1,2,\ldots,N-1$.

Using the SOR method from Equation 10.23, the finite difference scheme is given by

$$u_{i,j}^{(s+1)} = \frac{1}{4}\omega\left[u_{i-1,j}^{(s+1)} + u_{i+1,j}^{(s)} + u_{i,j-1}^{(s+1)} + u_{i,j+1}^{(s)} - h^2\left(\frac{x_i}{y_j} + \frac{y_j}{x_i}\right)\right] + (1-\omega)u_{i,j}^{(s)}$$

$$s = 0,1,\ldots \quad \text{and} \quad i,j = 1,2,\ldots,N-1$$

$$\text{or} \quad u_{i,j}^{(s+1)} = \omega z_{i,j}^{(s+1)} + (1-\omega)u_{i,j}^{(s)}$$

where $z_{i,j}^{(s+1)}$ can be obtained initially by the Gauss–Seidel method and choose ω such that $1 < \omega < 2$. Here, we take $\omega = 1.5$. The approximate results obtained by the Gauss–Seidel and SOR methods along with exact results are presented in Table 10.4.

10.7.1 Laplace Equation

We now develop the numerical solution of the two-dimensional Laplace equation given by

$$\frac{\partial^2 u}{\partial x^2} + \frac{\partial^2 u}{\partial y^2} = 0 \tag{10.24}$$

in a bounded rectangular region R with boundary C for which the values of $u(x, y)$ are known at the boundary. For simplicity, we take R to be a square region and divide the region R into small squares of side h. Let the values c_i of $u(x, y)$ are specified on the boundary C and let the interior mesh points and the boundary points be as shown in Figure 10.3.

Let x_i and y_j be the discrete values of x and y at the grid points. Therefore,

$$x_i = x_0 + ih, \quad i = 0,1,2,\ldots$$
$$y_j = y_0 + jh, \quad j = 0,1,2,\ldots$$

and $u_{i,j} = u(x_i, y_j)$ be the computed value of $u(x, y)$ at the point (x_i, y_j).

TABLE 10.4
Approximate Results Obtained by the Gauss–Seidel Method and SOR Method at t = 1.5

x	Exact	GSM	Absolute Error	SOR	Absolute Error
1.1	0.826279	0.822941	3.33×10^{-3}	0.826336	5.69×10^{-5}
1.2	1.05802	1.05197	6.05×10^{-3}	1.05811	9.22×10^{-5}
1.3	1.30227	1.29434	7.92×10^{-3}	1.30238	1.12×10^{-4}
1.4	1.55807	1.5492	8.87×10^{-3}	1.55819	1.20×10^{-4}
1.5	1.82459	1.81572	8.87×10^{-3}	1.82471	1.19×10^{-4}
1.6	2.10112	2.09311	8.02×10^{-3}	2.10124	1.12×10^{-4}
1.7	2.38704	2.38056	6.48×10^{-3}	2.38714	9.75×10^{-5}
1.8	2.68178	2.67731	4.47×10^{-3}	2.68178	7.49×10^{-5}
1.9	2.98486	2.98263	2.23×10^{-3}	2.9849	4.29×10^{-5}

Now we replace the partial derivatives in Equation 10.24 by the finite difference approximations

$$u_{xx} = \frac{u_{i+1,j} - 2u_{i,j} + u_{i-1,j}}{h^2} + O(h^2)$$

$$u_{yy} = \frac{u_{i,j+1} - 2u_{i,j} + u_{i,j-1}}{h^2} + O(h^2)$$

Therefore, Equation 10.24 becomes

$$\frac{u_{i+1,j} - 2u_{i,j} + u_{i-1,j}}{h^2} + \frac{u_{i,j+1} - 2u_{i,j} + u_{i,j-1}}{h^2} = 0$$

This implies that

$$u_{i,j} = \frac{1}{4}\left[u_{i-1,j} + u_{i+1,j} + u_{i,j-1} + u_{i,j+1}\right] \tag{10.25}$$

This shows that the value of u at any interior mesh point is the average of its values at four neighboring points to the left, right above and below as shown in Figure 10.4. Equation 10.25 is called the standard five-point formula (SFPF).

Since the Laplace equation remains invariant when the co-ordinate axes are rotated through 45°, we can also use the following formula instead of Equation 10.25

$$u_{i,j} = \frac{1}{4}\left[u_{i-1,j-1} + u_{i+1,j-1} + u_{i+1,j+1} + u_{i-1,j+1}\right] \tag{10.26}$$

Since Equation 10.26 uses the values of u at the nearest diagonal points as shown in Figure 10.5, it is called the diagonal five-point formula (DFPF).

Now the approximate function values at the interior mesh points can be computed according to the following scheme:

- We first use the DFPF (10.26) and compute

$$u_5 = \frac{1}{4}(c_1 + c_5 + c_9 + c_{13})$$

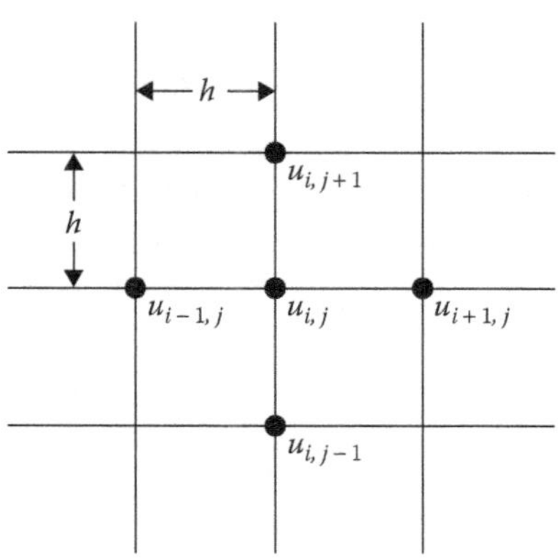

FIGURE 10.4 Standard five-point formula.

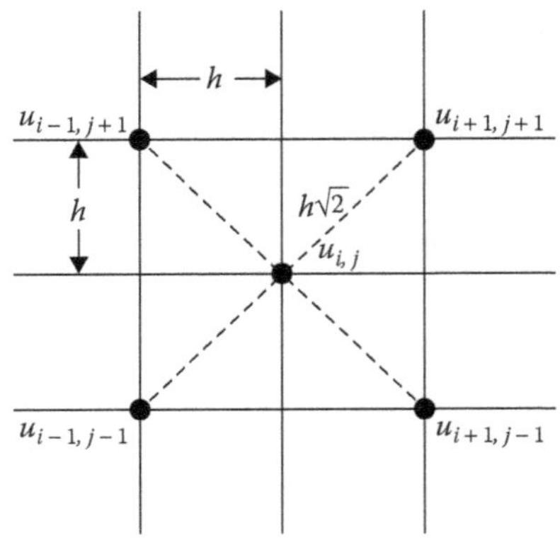

FIGURE 10.5 Diagonal five-point formula.

Next we compute u_1, u_3, u_7, u_9 in this order. Thus, we have

$$u_1 = \frac{1}{4}(c_1 + c_3 + u_5 + c_{15}), u_3 = \frac{1}{4}(c_3 + c_5 + c_7 + u_5)$$

$$u_7 = \frac{1}{4}(u_5 + c_{11} + c_{13} + c_{15}), u_9 = \frac{1}{4}(u_5 + c_7 + c_9 + c_{11})$$

- We then compute the values at the remaining mesh points u_2, u_4, u_6, u_8 by using the SFPF (10.25). Thus, we have

$$u_2 = \frac{1}{4}(u_1 + c_3 + u_3 + u_5), u_4 = \frac{1}{4}(u_1 + u_5 + u_7 + c_{15})$$

$$u_6 = \frac{1}{4}(u_3 + u_5 + c_7 + u_9), u_8 = \frac{1}{4}(u_5 + u_7 + u_9 + c_{11})$$

Having found all the values of u_i $(i = 1,2,\ldots,9)$, the accuracy can be improved by any one of the following iterative methods

1. *Gauss–Jacobi method:* Let $u_{i,j}^{(s)}$ denote the sth iterative value of $u_{i,j}$. Then an iterative procedure for solving Equation 10.24 is given by

$$u_{i,j}^{(s+1)} = \frac{1}{4}\left[u_{i-1,j}^{(s)} + u_{i+1,j}^{(s)} + u_{i,j-1}^{(s)} + u_{i,j+1}^{(s)}\right] \tag{10.27}$$

for the interior mesh points. This is called the point Jacobi method.

2. *Gauss–Seidel method:* In this case, the iterative formula for solving Equation 10.24 is given by

$$u_{i,j}^{(s+1)} = \frac{1}{4}\left[u_{i-1,j}^{(s+1)} + u_{i+1,j}^{(s)} + u_{i,j-1}^{(s+1)} + u_{i,j+1}^{(s)}\right] \tag{10.28}$$

Here we use the latest available iterative values and hence the rate of convergence will be twice as fast as the Jacobi method. This iterative process is also referred to as Leibmann's iterative method.

3. *SOR method:* Equation 10.28 can be written as

$$u_{i,j}^{(s+1)} = u_{i,j}^{(s)} + \frac{1}{4}\left[u_{i-1,j}^{(s+1)} + u_{i+1,j}^{(s)} + u_{i,j-1}^{(s+1)} + u_{i,j+1}^{(s)} - 4u_{i,j}^{(s)}\right]$$

In this SOR method, the iteration formula is given by

$$u_{i,j}^{(s+1)} = u_{i,j}^{(s)} + \frac{1}{4}\omega\left[u_{i-1,j}^{(s+1)} + u_{i+1,j}^{(s)} + u_{i,j-1}^{(s+1)} + u_{i,j+1}^{(s)} - 4u_{i,j}^{(s)}\right]$$

$$= \frac{1}{4}\omega\left[u_{i-1,j}^{(s+1)} + u_{i+1,j}^{(s)} + u_{i,j-1}^{(s+1)} + u_{i,j+1}^{(s)}\right] + (1-\omega)u_{i,j}^{(s)}, \quad 1 < \omega < 2 \qquad (10.29)$$

$$= \omega u_{i,j}^{(s+1)} + (1-\omega)u_{i,j}^{(s)}$$

The rate of convergence of Equation 10.29 depends on the value of ω, which is called the relaxation factor. In general, the choice of the best value of ω is a difficult one.

Further, it may be noted that for $\omega = 1$, Equation 10.29 reduces to the basic iteration formula (10.28).

10.7.2 Algorithm for Solving Laplace Equation by SOR Method

Elliptic PDE: $\partial^2 u/\partial x^2 + \partial^2 u/\partial y^2 = 0, \quad 0 < x < L_1, 0 < y < L_2$
Boundary conditions:

$$u(x,0) = f_1(x), u(x,L_2) = f_2(x)$$

$$u(0,y) = g_1(y), u(L_1,y) = g_2(y)$$

Input: Read L_1, L_2, h, k, ω, define $f_1(x)$, $f_2(x)$, $g_1(y)$, $g_2(y)$, number of iteration K
Output: Approximate values of $u_{i,j}^{(K)}$ for $1 \le i \le N-1, 1 \le j \le M-1$
Step 1: Compute $N = L_1/h$; $M = L_2/k$; $\lambda = h^2/k^2$;
Step 2: for $i = 0(1)N$ do
        $x_i = ih$;
        for $s = 0(1)K$ do
            $u_{i,0}^{(s)} = f_1(x_i)$;
            $u_{i,N}^{(s)} = f_2(x_i)$;
        end.
    end.
Step 3: for $j = 0(1)M$ do
        $t_j = jk$;
        for $s = 0(1)K$ do
            $u_{0,j}^{(s)} = g_1(t_j)$;

            $u_{0,M}^{(s)} = g_2(t_j)$;
        end.
    end.
Step 4: for $i = 1(1)\overline{N-1}$ do
        for $j = 1(1)\overline{M-1}$ do

            $u_{i,j}^{(0)} = 0$;
        end.
    end.

Step 5: for $s = 0(1)\overline{K-1}$ do
for $i = 1(1)\overline{N-1}$ do
for $j = 1(1)\overline{M-1}$ do

$$z_{i,j}^{(s+1)} = \frac{1}{2+2\lambda}\left(u_{i-1,j}^{(s+1)} + u_{i+1,j}^{(s)} + \lambda u_{i,j-1}^{(s+1)} + \lambda u_{i,j+1}^{(s)}\right);$$

$$u_{i,j}^{(s+1)} = \omega z_{i,j}^{(s+1)} + (1-\omega) u_{i,j}^{(s)}$$

end.
end.
end.
Step 6: Print $u_{i,j}^{(K)}$ for $i = 1,\ldots,N-1,\ j = 1,\ldots,M-1$;
Step 7: Stop. ■

MATHEMATICA® Program for Solving the Laplace Equation by the Gauss–Seidel Method (Chapter 10, Example 10.4)

```
tol=0.001;
K=100.;
h=0.2;
a=0.;
b=1.;
n=N[(b-a)/h];
f1[x_]:=0;
f2[x_]:=x;
g1[y_]:=0;
g2[y_]:=y;
For[i=0.,i<=n,i++,
 x[i]=N[a+i*h];
 For[k=0.,k<=K,k++,
     u[i,0.][k]=f1[x[i]];
     u[i,n][k]=f2[x[i]]]];
   For[j=0.,j<=n,j++,
    y[j]=N[a+j*h];
    For[k=0.,k<=K,k++,
     u[0.,j][k]=g1[y[j]];
     u[n,j][k]=g2[y[j]]]];
   For[i=1.,i<=n-1,i++,
    For[j=1.,j<=n-1,j++,
     u[i,j][0.]=0
    ]];
   Do[
    For[i=1.,i<=n-1,i++,
     For[j=1.,j<=n-1,j++,
      u[i,j][k+1]=1/4*(u[i-1,j][k+1]+u[i+1,j][k]+u[i,j-1][k+1]+u[i,j+1][k]);
      u1[i,j]=N[x[i]*y[j]];
      e[i,j][k+1]=Abs[u[i,j][k+1]-u1[i,j]]
     ]];
    AA=Table[e[i,j][k+1],{i,1.,n-1},{j,1.,n-1}];
    If[Max[AA]<tol,
     Print[Step[k]];
     Print["...................................................."];
     Print["x[i]    y[j]    Approximate      Exact      Abs. Error"];
     Print["...................................................."];
     For[i=1.,i<=n-1,i++,
```

```
    For[j=1.,j<=n-1,j++,
     Print[x[i],"     ",y[j],"      ",
      u[i,j][k+1],"        ",u1[i,j],"        ",e[i,j][k+1]]]];
   Break[]],{k,0.,K}];
```

Output:

```
Step[15.]
..............................................................
 x[i]      y[j]       Approximate       Exact        Abs. Error
..............................................................
 0.2       0.2        0.0395272         0.04         0.00047283
 0.2       0.4        0.079381          0.08         0.000618976
 0.2       0.6        0.119499          0.12         0.000500779
 0.2       0.8        0.15975           0.16         0.000250393
 0.4       0.2        0.079381          0.08         0.000618976
 0.4       0.4        0.15919           0.16         0.000810275
 0.4       0.6        0.239344          0.24         0.000655538
 0.4       0.8        0.319672          0.32         0.000327771
 0.6       0.2        0.119499          0.12         0.000500779
 0.6       0.4        0.239344          0.24         0.000655538
 0.6       0.6        0.35947           0.36         0.000530346
 0.6       0.8        0.479735          0.48         0.000265174
 0.8       0.2        0.15975           0.16         0.000250393
 0.8       0.4        0.319672          0.32         0.000327771
 0.8       0.6        0.479735          0.48         0.000265174
 0.8       0.8        0.639867          0.64         0.000132587
```

MATHEMATICA® Program for Solving the Laplace Equation by the SOR Method (Chapter 10, Example 10.4)

```
tol=0.001;
K=100.;
w=1.02;
h=0.2;
a=0.;
b=1.;
n=N[(b-a)/h];
f1[x_]:=0;
f2[x_]:=x;
g1[y_]:=0;
g2[y_]:=y;
For[i=0.,i<=n,i++,
  x[i]=N[a+i*h];
  For[k=0.,k<=K,k++,
   u[i,0.][k]=f1[x[i]];
   u[i,n][k]=f2[x[i]]]];
For[j=0.,j<=n,j++,
  y[j]=N[a+j*h];
  For[k=0.,k<=K,k++,
   u[0.,j][k]=g1[y[j]];
   u[n,j][k]=g2[y[j]]]];
For[i=1.,i<=n-1,i++,
  For[j=1.,j<=n-1,j++,
   u[i,j][0.]=0
  ]];
```

```
Do[
 For[i=1.,i<=n-1,i++,
  For[j=1.,j<=n-1,j++,
   z[i,j][k+1]=1/4*(u[i-1,j][k+1]+u[i+1,j][k]+u[i,j-1][k+1]+u[i,j+1][k]);
   u[i,j][k+1]=w*z[i,j][k+1]+(1-w)*u[i,j][k];
   u1[i,j]=N[x[i]*y[j]];
   e[i,j][k+1]=Abs[u[i,j][k+1]-u1[i,j]]
  ]];
 AA=Table[e[i,j][k+1],{i,1.,n-1},{j,1.,n-1}];
 If[Max[AA]<tol,
  Print[Step[k]];
  Print["............................................................"];
  Print["x[i]    y[j]     Approximate      Exact        Abs. Error"];
  Print["............................................................"];
  For[i=1.,i<=n-1,i++,
   For[j=1.,j<=n-1,j++,
    Print[x[i],"      ",y[j],"       ",
     u[i,j][k+1],"        ",u1[i,j],"         ",e[i,j]   [k+1]]]];
   Break[]],{k,0.,K}];
```

Output:

```
Step[14.]
```

x[i]	y[j]	Approximate	Exact	Abs. Error
0.2	0.2	0.0394439	0.04	0.000556119
0.2	0.4	0.0792799	0.08	0.000720074
0.2	0.6	0.119424	0.12	0.000576223
0.2	0.8	0.159715	0.16	0.000284976
0.4	0.2	0.0792799	0.08	0.000720074
0.4	0.4	0.159068	0.16	0.000932345
0.4	0.6	0.239254	0.24	0.000746078
0.4	0.8	0.319631	0.32	0.000368977
0.6	0.2	0.119424	0.12	0.000576223
0.6	0.4	0.239254	0.24	0.000746078
0.6	0.6	0.359403	0.36	0.000597018
0.6	0.8	0.479705	0.48	0.000295258
0.8	0.2	0.159715	0.16	0.000284976
0.8	0.4	0.319631	0.32	0.000368977
0.8	0.6	0.479705	0.48	0.000295258
0.8	0.8	0.639854	0.64	0.000146021

Example 10.4

Solve the elliptic PDE, that is, the Laplace equation

$$\frac{\partial^2 u}{\partial x^2}+\frac{\partial^2 u}{\partial y^2}=0, \quad 0<x,y<1.$$

with initial conditions $u(x,0)=0$, $u(x,1)=x \quad 0\le x\le 1$, and boundary conditions $u(0,y)=0$, $u(1,y)=y \quad 0\le y\le 1$, using the Gauss–Seidel method and SOR method. The exact solution is given by $u(x,y)=xy$.

TABLE 10.5
Results by the Gauss–Seidel Method

y	x	Exact	GSM	Absolute Error	% Error
0.2	0.2	0.04	0.0399132	0.0000868	0.217
	0.4	0.08	0.0798864	0.0001136	0.142
	0.6	0.12	0.119908	0.000092	0.076
	0.8	0.16	0.159954	0.000046	0.02875
0.6	0.2	0.12	0.119908	0.000092	0.076
	0.4	0.24	0.23988	0.00012	0.05
	0.6	0.36	0.359903	0.000097	0.027
	0.8	0.48	0.479951	0.000049	0.0102

Solution:

Here, we take $h = k = 0.2$, that is, there are 16 number of internal node points and 20 number of boundary node points. By applying the boundary conditions, we obtain the following 16 number of algebraic equations with the same number of unknowns, that is, u_i, $i = 1,2,\ldots,16$.

$$
\begin{aligned}
-4u_1 + u_2 + u_5 &= 0 & u_5 - 4u_9 + u_{10} + u_{13} &= 0\\
u_1 - 4u_2 + u_3 + u_6 &= 0 & u_6 + u_9 - 4u_{10} + u_{11} + u_{14} &= 0\\
u_2 - 4u_3 + u_4 + u_7 &= 0 & u_7 + u_{10} - 4u_{11} + u_{12} + u_{15} &= 0\\
u_3 - 4u_4 + u_8 &= -0.2 & u_8 + u_{11} - 4u_{12} + u_{16} &= -0.6\\
u_1 - 4u_5 + u_6 + u_9 &= 0 & u_9 - 4u_{13} + u_{14} &= -0.2\\
u_2 + u_5 - 4u_6 + u_7 + u_{10} &= 0 & u_{10} + u_{13} - 4u_{14} + u_{15} &= -0.4\\
u_3 + u_6 - 4u_7 + u_8 + u_{11} &= 0 & u_{11} + u_{14} - 4u_{15} + u_{16} &= -0.6\\
u_4 + u_7 - 4u_8 + u_{12} &= -0.4 & u_{12} + u_{15} - 4u_{16} &= -1.6
\end{aligned}
$$

Using the Gauss–Seidel method to the above system up to 20 iterations, we obtain the approximate values of $u(x, y)$ at $y = 0.2$ and 0.6, and these numerical results with absolute errors and percentage errors are given in Table 10.5. (We take initial guess: $u_i^{(0)} = 0$, $i = 1,2,\ldots,16$.)

10.8 ALTERNATING DIRECTION IMPLICIT METHOD

We subdivide the $(x - t)$ plane into a network by drawing the straight lines

$$
\begin{aligned}
x &= ih, \quad i = 0,1,2,\ldots,N\\
t &= nk, \quad n = 0,1,2,\ldots
\end{aligned}
$$

parallel to the co-ordinate axes.

Let us denote the mesh points $(x_i,t_n) = (ih,nk)$ by (i,n). Let u_i^n denote the computed value of $u(x_i,t_n)$.

Now we define the central difference of u_i^n in the spatial x-direction as

$$\delta_x u_i^n = u_{i+1}^n - u_{i-1}^n$$

and also the second-order central difference of u_i^n in the spatial x-direction as

$$\delta_x^2 u_i^n = u_{i+1}^n - 2u_i^n + u_{i-1}^n$$

In this way, we may define higher order central differences of u_i^n in the spatial x-direction.

- *ADI applied to two-dimensional heat equation*

We consider the two-dimensional heat equation

$$\frac{\partial u}{\partial t} = c^2\left(\frac{\partial^2 u}{\partial x^2} + \frac{\partial^2 u}{\partial y^2}\right) \tag{10.30}$$

where $c > 0$ is a constant.

Now we shall discuss the alternating direction implicit (ADI) method applied to the two-dimensional heat equation of the form Equation 10.30.

The Crank–Nicolson scheme for this Equation 10.30) may be written in operator notation as

$$\frac{u_{i,j}^{n+1} - u_{i,j}^n}{\Delta t} = \frac{c^2}{2}\left(\frac{\delta_x^2 u_{i,j}^n + \delta_x^2 u_{i,j}^{n+1}}{\Delta x^2} + \frac{\delta_y^2 u_{i,j}^n + \delta_y^2 u_{i,j}^{n+1}}{\Delta y^2}\right) \tag{10.31}$$

where

$$u_{i,j}^n \equiv u(i\Delta x, j\Delta y, n\Delta t),\quad \delta_x^2 u_{i,j}^n = u_{i+1,j}^n - 2u_{i,j}^n + u_{i-1,j}^n,\quad \delta_x^2 u_i^{n+1} = u_{i+1,j}^{n+1} - 2u_{i,j}^{n+1} + u_{i-1,j}^{n+1},$$

$$\delta_y^2 u_{i,j}^n = u_{i,j+1}^n - 2u_{i,j}^n + u_{i,j-1}^n \quad\text{and}\quad \delta_y^2 u_i^{n+1} = u_{i,j+1}^{n+1} - 2u_{i,j}^{n+1} + u_{i,j-1}^{n+1}$$

Let us take

$$r_1 = \frac{c^2\Delta t}{\Delta x^2} \quad\text{and}\quad r_2 = \frac{c^2\Delta t}{\Delta y^2}$$

Then Equation 10.31 may be simplified as

$$u_i^{n+1} - u_i^n = \frac{r_1}{2}\left(\delta_x^2 u_{i,j}^n + \delta_x^2 u_{i,j}^{n+1}\right) + \frac{r_2}{2}\left(\delta_y^2 u_{i,j}^n + \delta_y^2 u_{i,j}^{n+1}\right)$$

which is rewritten in operator notation as

$$\left(I - \frac{r_1}{2}\delta_x^2 - \frac{r_2}{2}\delta_y^2\right)u_{i,j}^{n+1} = \left(I + \frac{r_1}{2}\delta_x^2 + \frac{r_2}{2}\delta_y^2\right)u_{i,j}^n \tag{10.32}$$

To obtain tridiagonal systems in the numerical solution of heat equation, Peaceman and Rachford (1955) proposed a method called the ADI method. In the ADI scheme, each time step is subdivided into two half-steps. In the first half time step, the spatial x-direction is treated implicitly, whereas the other space direction y is treated explicitly. In the next half-step, the spatial x-direction is treated explicitly, whereas the spatial direction y is treated implicitly. In the operator notation, the steps from nth time level to $(n+1)$th time level may be expressed, assuming that the values at the level n have been already computed as

$$\left(I - \frac{r_1}{2}\delta_x^2\right)u_{i,j}^{n+\frac{1}{2}} = \left(I + \frac{r_2}{2}\delta_y^2\right)u_{i,j}^n \tag{10.33}$$

$$\left(I-\frac{r_2}{2}\delta_y^2\right)u_{i,j}^{n+1}=\left(I+\frac{r_1}{2}\delta_x^2\right)u_{i,j}^{n+\frac{1}{2}} \tag{10.34}$$

It may be noted that each of the half time step Equations 10.33 and 10.34 constitutes a tridiagonal system. This implies that one full time step requires the solution of the two tridiagonal systems.

The ADI scheme has all the desirable features that we would expect from a numerical scheme:

1. It is an implicit method.
2. It is unconditionally stable.
3. It is second-order accurate in both time and space.
4. It has a low amount of computational overhead that is required in each time step as the number of arithmetic operations required for each time step is proportional to the number of mesh points.

Due to these attractive features, the ADI scheme has wide applications in applied science and engineering problems.

10.8.1 Algorithm for Two-Dimensional Parabolic PDE by ADI Method

The two-dimensional PDE:

$$\frac{\partial u}{\partial t}=c^2\left(\frac{\partial^2 u}{\partial x^2}+\frac{\partial^2 u}{\partial y^2}\right),\quad 0<x<L_1, 0<y<L_2, t>0$$

subject to initial conditions $u(x,y,0)=f(x,y), t\geq 0$ and boundary conditions

$$u(0,y,t)=g_1(y,t),\quad u(L_1,y,t)=g_2(y,t)$$

$$u(x,0,t)=h_1(x,t),\quad u(x,L_2,t)=h_2(x,t)$$

Input: Read L_1, L_2, Δx, Δy, Δt, define $f(x,y)$, $g_1(y,t)$, $g_2(y,t)$, $h_1(x,t)$, $h_2(x,t)$, number of time steps K.

Output: Approximate values of $u_{i,j}^K$ for $1\leq i\leq N-1, 1\leq j\leq M-1$

Step 1: Compute $N=L_1/\Delta x, M=L_2/\Delta y, r_1=c^2\Delta t/\Delta x^2, r_2=c^2\Delta t/\Delta y^2$.

Step 2: for $i=0(1)N$ do
- $x_i=i\Delta x$;
- for $j=0(1)M$ do
  - $y_j=j\Delta y$;
  - $u_{i,j}^0=f(x_i,y_j)$
- end.

end.

Step 3: for $k=0(1/2)n$ do
- $t_k=k\Delta t$;
- for $j=0(1)M$ do
  - $y_j=j\Delta y$;
  - $u_{0,j}^k=g_1(y_j,t_k)$;
  - $u_{N,j}^k=g_2(y_j,t_k)$;
- end.

end.

Step 4: for $k = 0(1/2)n$ do

$t_k = k\Delta t;$

for $i = 0(1)N$ do

$x_i = i\Delta x;$

$u_{i,0}^{k} = h_1(x_i, t_k);$

$u_{i,M}^{k} = h_2(x_i, t_k);$

end.

end.

Step 5: for $k = 0(1)\overline{n-1}$ do;

Step 6: for $j = 1(1)\overline{M-1}$ do

$$b_1 = u_{1,j}^{k} + \frac{r_2}{2}\left(u_{1,j+1}^{k} - 2u_{1,j}^{k} + u_{1,j-1}^{k}\right) + \frac{r_1}{2}u_{0,j}^{k+\frac{1}{2}};$$

for $i = 2(1)\overline{N-2}$ do

$$b_i = u_{i,j}^{k} + \frac{r_2}{2}\left(u_{i,j+1}^{k} - 2u_{i,j}^{k} + u_{i,j-1}^{k}\right);$$

end.

$$b_{N-1} = u_{N-1,j}^{k} + \frac{r_2}{2}\left(u_{N-1,j+1}^{k} - 2u_{N-1,j}^{k} + u_{N-1,j-1}^{k}\right) + \frac{r_1}{2}u_{N,j}^{k+\frac{1}{2}};$$

Step 7: (Solving the tridiagonal system)

$$p_2 = \frac{r_1}{2(1+r_1)};$$

$$q_2 = -\frac{b_1}{1+r_1};$$

for $s = 2(1)\overline{N-1}$ do

$$p_{s+1} = \frac{r_1}{2((1+r_1)+(-r_1/2)p_s)};$$

$$q_{s+1} = \frac{b_s - (-r_1/2)q_s}{(1+r_1)+(-r_1/2)p_s};$$

end.

$u_{N-1,j}^{k+\frac{1}{2}} = q_N;$

for $l = \overline{N-2}(1)1$ do

$$u_{l,j}^{k+\frac{1}{2}} = q_{l+1} + p_{l+1}u_{l+1,j}^{k+\frac{1}{2}};$$

end.

Step 8: end. (For Step 6)

Step 9: for $i = 1(1)\overline{N-1}$ do

$$c_1 = u_{i,1}^{k+\frac{1}{2}} + \frac{r_1}{2}(u_{i+1,1}^{k+\frac{1}{2}} - 2u_{i,1}^{k+\frac{1}{2}} + u_{i-1,1}^{k+\frac{1}{2}}) + \frac{r_2}{2}u_{i,0}^{k+1};$$

for $j = 2(1)\overline{M-2}$ do

$$c_j = u_{i,j}^{k+\frac{1}{2}} + \frac{r_1}{2}(u_{i+1,j}^{k+\frac{1}{2}} - 2u_{i,j}^{k+\frac{1}{2}} + u_{i-1,j}^{k+\frac{1}{2}});$$

end.

$$c_{M-1} = u_{i,M-1}^{k+\frac{1}{2}} + \frac{r_1}{2}(u_{i+1,M-1}^{k+\frac{1}{2}} - 2u_{i,M-1}^{k+\frac{1}{2}} + u_{i-1,M-1}^{k+\frac{1}{2}}) + \frac{r_2}{2}u_{i,M}^{k+1};$$

Step 10: (Solving the tridiagonal system)

$$p_2^* = \frac{r_2}{2(1+r_2)};$$

$$q_2^* = \frac{c_1}{1+r_2};$$

for $s = 2(1)\overline{M-1}$ do

$$p^*_{s+1} = \frac{r_2}{2((1+r_2)+(-r_2/2)p^*_s)};$$

$$q^*_{s+1} = \frac{c_s-(-r_2/2)q^*_s}{(1+r_2)+(-r_2/2)p^*_s};$$

end.

$u^{k+1}_{i,M-1} = q^*_M;$

for $l = \overline{M-2}(1)1$ do

$$u^{k+1}_{i,l} = q^*_{l+1} + p^*_{l+1}u^{k+1}_{i,l+1};$$

end.

Step 11: end. (For Step 9)

Step 12: end. (For Step 5)

Step 13: Print $u^K_{i,j}$, for $i = 1,2,\ldots,N-1$, $j = 1,2,\ldots,M-1$.

Step 14: Stop. ■

MATHEMATICA® Program for the Two-Dimensional Parabolic PDE by the ADI Method (Chapter 10, Example 10.5)

```
L1=1.;
L2=1.;
h=0.125;
k=0.125;
kappa=0.5;
delt=0.015625;
t1=0.125;
K=t1/delt;
n=L1/h;
m=L2/k;
r1=kappa*delt/h²;r2=kappa*delt/k²;x[0]=0.;y[0]=0.;
f[x_,y_]:=Sin[Pi*x]*Sin[Pi*y];
g1[y_,t_]:=0.;g2[y_,t_]:=0.;h1[x_,t_]:=0.;h2[x_,t_]:=0.;

For[i=0.,i<=n,i++,
  x[i]=x[0]+i*h;
  For[j=0.,j<=m,j++,
   y[j]=y[0]+j*k;
   u[i,j][0.]=f[x[i],y[j]]
  ]
 ];
For[k1=0.,k1<=K,k1=k1+0.5,
  t[k1]=k1*delt;
  For[j=0.,j<=m,j++,y[j]=y[0]+j*k;
   u[0.,j][k1]=g1[y[j],t[k1]];
   u[n,j][k1]=g2[y[j],t[k1]]
  ]
 ];

For[k1=0.,k1<=K,k1=k1+0.5,
  t[k1]=k1*delt;
  For[i=0.,i<=n,i++,x[i]=x[0]+i*h;
   u[i,0.][k1]=h1[x[i],t[k1]];
   u[i,m][k1]=h2[x[i],t[k1]]
  ]
 ];
```

```
For[k1=0.,k1<=K-1,k1++,
  For[j=1.,j<=m-1,j++,
   For[i=1.,i<=n-1,i++,
    eqn[i]=Simplify[u[i,j][k1+0.5]-u[i,j][k1]-
       r1/2*(u[i+1,j][k1+0.5]-2*u[i,j][k1+0.5]+u[i-1,j][k1+0.5])-
       r2/2*(u[i,j+1][k1]-2*u[i,j][k1]+u[i,j-1][k1])]==0];
   eqns=Table[eqn[i],{i,1.,n-1}];
   uu=Table[u[i,j][k1+0.5],{i,1.,n-1}];
   sol=NSolve[eqns,uu];
   For[i=1.,i<=n-1,i++,
    u[i,j][k1+0.5]=sol[[1,i,2]]]];

  For[i=1.,i<=n-1.,i++,
   For[j=1.,j<=m-1,j++,
    eqn1[j]=Simplify[u[i,j][k1+1.]-u[i,j][k1+0.5]-
         r2/2*(u[i,j+1][k1+1]-2*u[i,j][k1+1]+u[i,j-1][k1+1])-
         r1/2*(u[i+1,j][k1+0.5]-2*u[i,j][k1+0.5]+u[i-1,j][k1+0.5])]==0];
   eqns1=Table[eqn1[j],{j,1.,m-1}];
   uu1=Table[u[i,j][k1+1],{j,1.,m-1}];
   sol1=NSolve[eqns1,uu1];
   For[j=1.,j<=m-1,j++,
    u[i,j][k1+1]=sol1[[1,j,2]]
   ]
  ]
 ];
U[x_,y_,t_]:=Exp[-Pi2*t]*Sin[Pi*x]*Sin[Pi*y];
For[i=1.,i<=n-1.,i=i+2,
  For[j=1.,j<=m-1.,j=j+2,
   Print[i,"   ",j,"     ",N[u[i,j][K]],"       ",
    N[U[x[i],y[j],t1]],"   ",Abs[u[i,j][K]-U[x[i],y[j],t1]]]]];
```

Output:

```
1. 1. 0.0432997 0.0426471 0.000652516
1. 3. 0.104535  0.102959  0.00157531
1. 5. 0.104535  0.102959  0.00157531
1. 7. 0.0432997 0.0426471 0.000652516
3. 1. 0.104535  0.102959  0.00157531
3. 3. 0.252369  0.248566  0.00380314
3. 5. 0.252369  0.248566  0.00380314
3. 7. 0.104535  0.102959  0.00157531
5. 1. 0.104535  0.102959  0.00157531
5. 3. 0.252369  0.248566  0.00380314
5. 5. 0.252369  0.248566  0.00380314
5. 7. 0.104535  0.102959  0.00157531
7. 1. 0.0432997 0.0426471 0.000652516
7. 3. 0.104535  0.102959  0.00157531
7. 5. 0.104535  0.102959  0.00157531
7. 7. 0.0432997 0.0426471 0.000652516
```

Example 10.5

Solve the two-dimensional heat equation

$$\frac{\partial u}{\partial t} = \kappa\left(\frac{\partial^2 u}{\partial x^2} + \frac{\partial^2 u}{\partial y^2}\right), \quad 0<x,y<1, t>0$$

TABLE 10.6
Results of 2D Heat Equation Obtained by the ADI Method

i	j	$u_{i,j}$	$u(x_i, y_j, 0.125)$	Absolute Error
1	1	0.0432997	0.0426471	0.000652516
	3	0.104535	0.102959	0.00157531
	5	0.104535	0.102959	0.00157531
	7	0.0432997	0.0426471	0.000652516
3	1	0.104535	0.102959	0.00157531
	3	0.252369	0.248566	0.00380314
	5	0.252369	0.248566	0.00380314
	7	0.104535	0.102959	0.00157531
5	1	0.104535	0.102959	0.00157531
	3	0.252369	0.248566	0.00380314
	5	0.252369	0.248566	0.00380314
	7	0.104535	0.102959	0.00157531
7	1	0.0432997	0.0426471	0.000652516
	3	0.104535	0.102959	0.00157531
	5	0.104535	0.102959	0.00157531
	7	0.0432997	0.0426471	0.000652516

subject to the initial condition $u(x,y,0) = \sin(\pi x)\sin(\pi y), \quad 0 \le x,y \le 1, t\ 0$
and the boundary conditions: $\partial u = 0$ (for all boundary points).
Exact solution of this problem is given by $u(x,y,t) = e^{-\pi^2 t}\sin(\pi x)\sin(\pi y)$.

Solution:

Here, we take $\Delta x = \Delta y = 0.125, \quad \kappa = 0.5, \quad \Delta t = 0.015625$.

According to the ADI method, the given problem reduces to

$$u_{i,j}^{n+\frac{1}{2}} = u_{i,j}^n + \frac{r_1}{2}\left(u_{i+1,j}^{n+\frac{1}{2}} - 2u_{i,j}^{n+\frac{1}{2}} + u_{i-1,j}^{n+\frac{1}{2}}\right) + \frac{r_2}{2}\left(u_{i,j+1}^n - 2u_{i,j}^n + u_{i,j-1}^n\right) \tag{10.35}$$

$$u_{i,j}^{n+1} = u_{i,j}^{n+\frac{1}{2}} + \frac{r_2}{2}\left(u_{i,j+1}^{n+1} - 2u_{i,j}^{n+1} + u_{i,j-1}^{n+1}\right) + \frac{r_1}{2}\left(u_{i+1,j}^{n+\frac{1}{2}} - 2u_{i,j}^{n+\frac{1}{2}} + u_{i-1,j}^{n+\frac{1}{2}}\right) \tag{10.36}$$

For $n = 0$, first we have to find values of $u_{i,j}$ at the half time step by using Equation 10.35, and then using the values at the half time step, we can obtain the values of $u_{i,j}$ at the first step by using Equation 10.36, and so on.

For $t = 0.125$, that is, at the eighth time step the following results are provided for different values of x_i and y_j, where $x_i = i\Delta x$ and $y_j = j\Delta y$ (Table 10.6).

10.9 STABILITY ANALYSIS OF THE NUMERICAL SCHEMES

Let $\tilde{u}_i^n$ be the numerical solution of the difference equation Equation 10.4. Then we define the error ε_i^n as

$$\varepsilon_i^n = u_i^n - \tilde{u}_i^n$$

where ε_i^n is the error because of the computational round-off error. Since $\tilde{u}_i^n$ must also satisfy Equation 10.4, we can see that ε_i^n satisfies the equation

$$\varepsilon_i^{n+1} = r\varepsilon_{i-1}^n + (1-2r)\varepsilon_i^n + r\varepsilon_{i+1}^n, \quad \text{for } i = 1,2,\ldots,N-1 \text{ and } n = 1,2,\ldots \tag{10.37}$$

This shows that the propagation of errors ε_i^n is governed by the same difference equation (10.4) that is satisfied by the unknown function u_i^n.

We shall now derive the stability of a finite difference scheme Equation 10.4 by means of finding a condition under which the error

$$\varepsilon_i^n = u_i^n - \tilde{u}_i^n \tag{10.38}$$

remains bounded as n tends to infinity.

For the stability analysis, a harmonic decomposition is made of the approximate solution at grid points at a given time level. Then by following the von Neumann idea for stability analysis, a distribution of errors is introduced at the initial time level $t = 0$, which propagates with increase of time t in accordance with the governing Equation 10.4. For difference equations with constant coefficients, the error may be expanded in a finite Fourier series. Then the error can be written as

$$\varepsilon_i^0 = \sum_j A_j \exp\left(\sqrt{-1}\beta_j ih\right) \tag{10.39}$$

where the wave number $\beta_j = j\pi/Nh$ and the number of terms in Equation10.39 is equal to the number of mesh points on the line $t = 0$.

To investigate the error propagation as t increases, it is necessary to find a solution of the finite difference equation (10.37), which reduces to Equation 10.39 at $t = 0$.

Let us assume that

$$\varepsilon_i^n = \sum_j A_j \xi^n(j) \exp\left(\sqrt{-1}\beta_j ih\right) \tag{10.40}$$

where ξ is an arbitrary real or complex number.

Now the difference equation is linear and homogeneous. Superposition of solutions is also a solution and it is enough to consider the growth of error of a typical term

$$\varepsilon_i^n \sim A\xi^n \exp\left(\sqrt{-1}\beta ih\right) \tag{10.41}$$

where A is an arbitrary constant. Here $\xi = \exp(\alpha k)$ and α, in general, is a complex number. The quantity ξ is often called the amplification factor. Equation 10.41 reduces to $\varepsilon_i^0 \sim A\exp\left(\sqrt{-1}\beta ih\right)$, when $t = 0$.

In order that the original error component $\exp\left(\sqrt{-1}\beta ih\right)$ will not grow with time, it is necessary and sufficient that

$$|\xi| \le 1, \quad \text{that is, } |\exp(\alpha k)| \le 1 \text{ for all } \alpha \tag{10.42}$$

This is known as von Neumann's criterion for stability. Equation 10.42 gives the required condition for the stability of the corresponding difference scheme.

This method of stability analysis is known as the von Neumann method or the finite Fourier series method.

- *Stability of the FTCS scheme*: Substituting Equation 10.41 into Equation 10.37 and simplifying, we obtain

$$\begin{aligned}\xi &= 1 + r(e^{i\beta h} + e^{-i\beta h} - 2)\\ &= 1 + r(2\cos\beta h - 2)\\ &= 1 - 4r\sin^2\frac{\beta h}{2}\end{aligned}$$

Therefore, the condition in Equation 10.42 implies that

$$-1 \le 1 - 4r\sin^2\frac{\beta h}{2} \le 1$$

The right-hand side of this inequality is trivially satisfied if $r > 0$ and the left-hand side gives

$$r \le \frac{1}{2\sin^2\beta h/2}, \quad \text{which yields the stability condition } 0 < r \le 1/2 \tag{10.43}$$

The following important points concerning the von Neumann method of stability may be noted.

1. This method applies only to the linear difference equation with constant coefficients
2. Boundary conditions that apply only to initial value problems with periodic initial data are neglected by the von Neumann stability method
3. For two-level difference schemes with one dependent variable and any number of independent variables, the von Neumann condition in Equation 10.42 is sufficient as well as necessary. Otherwise, for difference equations with three or more time levels or two or more dependent variables, the von Neumann condition of stability is always necessary but may not be sufficient.

- *Stability of the Crank–Nicolson scheme*: We shall now investigate the stability of Crank–Nicolson scheme Equation 10.6 by the von Neumann method. Since the difference equation is linear, the error satisfies the following difference equation:

$$-r\varepsilon_{i-1}^{n+1} + (2+2r)\varepsilon_i^{n+1} - r\varepsilon_{i+1}^{n+1} = r\varepsilon_{i-1}^{n} + 2(1-r)\varepsilon_i^{n} + r\varepsilon_{i+1}^{n}. \tag{10.44}$$

Substituting Equation 10.41 into Equation 10.44 and simplifying, we obtain

$$-r\xi e^{-i\beta h} + (2+2r)\xi - r\xi e^{i\beta h} = re^{-i\beta h} + 2(1-r) + re^{i\beta h}$$

This implies that

$$\xi = \frac{r\cos\beta h + (1-r)}{1 + r - r\cos\beta h} = \frac{1 - 2r\sin^2\beta h/2}{1 + 2r\sin^2\beta h/2}$$

It follows that $|\xi| < 1$ for all values of r, $r > 0$. Hence the Crank–Nicolson scheme is unconditionally stable.

- *Stability of the ADI scheme*: In the ADI scheme of the Peaceman–Rachford method, the difference equations are as follows:

$$\left(I - \frac{r_1}{2}\delta_x^2\right)u_{i,j}^{n+\frac{1}{2}} = \left(I + \frac{r_2}{2}\delta_y^2\right)u_{i,j}^{n} \tag{10.45}$$

$$\left(I - \frac{r_2}{2}\delta_y^2\right)u_{i,j}^{n+1} = \left(I + \frac{r_1}{2}\delta_x^2\right)u_{i,j}^{n+\frac{1}{2}} \tag{10.46}$$

Since the difference equation is linear, the error satisfies the following difference equations:

$$\left(I - \frac{r_1}{2}\delta_x^2\right)\varepsilon_{i,j}^{n+\frac{1}{2}} = \left(I + \frac{r_2}{2}\delta_y^2\right)\varepsilon_{i,j}^{n} \tag{10.47}$$

$$\left(I - \frac{r_2}{2}\delta_y^2\right)\varepsilon_{i,j}^{n+1} = \left(I + \frac{r_1}{2}\delta_x^2\right)\varepsilon_{i,j}^{n+\frac{1}{2}} \tag{10.48}$$

Equations 10.47 and 10.48 have constant coefficients and hence their solutions can be sought in the form:

$$\varepsilon_{i,j}^{n} = \xi^n \exp\left(\sqrt{-1}k_1 i\Delta x\right)\exp\left(\sqrt{-1}k_2 j\Delta y\right) \tag{10.49}$$

$$\varepsilon_{i,j}^{n+\frac{1}{2}} = \rho\xi^n \exp\left(\sqrt{-1}k_1 i\Delta x\right)\exp\left(\sqrt{-1}k_2 j\Delta y\right) \tag{10.50}$$

where ρ is a constant, and $\xi = \exp(\alpha\Delta t)$ is the amplification factor. We may rewrite Equations 10.49 and 10.50 as follows:

$$\varepsilon_{i,j}^{n} = \xi^n \exp\left(\sqrt{-1}i\beta_1\right)\exp\left(\sqrt{-1}j\beta_2\right) \tag{10.51}$$

$$\varepsilon_{i,j}^{n+\frac{1}{2}} = \rho\xi^n \exp\left(\sqrt{-1}i\beta_1\right)\exp\left(\sqrt{-1}j\beta_2\right) \tag{10.52}$$

where $\beta_1 = k_1\Delta x$ and $\beta_2 = k_2\Delta x$ are real.

Now, we note that

$$\delta_x^2\varepsilon_{i,j}^{n+\frac{1}{2}} = \varepsilon_{i+1,j}^{n+\frac{1}{2}} - 2\varepsilon_{i,j}^{n+\frac{1}{2}} + \varepsilon_{i-1,j}^{n+\frac{1}{2}}$$

yielding

$$\begin{aligned}\delta_x^2\varepsilon_{i,j}^{n+\frac{1}{2}} &= \rho\xi^n \exp(\sqrt{-1}j\beta_2)(\exp[\sqrt{-1}(i+1)\beta_1] - 2\exp(\sqrt{-1}i\beta_1) + \exp[\sqrt{-1}(i-1)\beta_1]) \\ &= \rho\xi^n \exp(\sqrt{-1}j\beta_2)\exp(\sqrt{-1}i\beta_1)(2\cos\beta_1 - 2)\end{aligned} \tag{10.53}$$

Similarly,

$$\delta_y^2\varepsilon_{i,j}^{n} = \xi^n \exp(\sqrt{-1}i\beta_1)\exp(\sqrt{-1}j\beta_2)(2\cos\beta_2 - 2) \tag{10.54}$$

and

$$\delta_y^2 \varepsilon_{i,j}^{n+1} = \xi^{n+1} \exp(\sqrt{-1} i\beta_1) \exp(\sqrt{-1} j\beta_2)(2\cos\beta_2 - 2) \quad (10.55)$$

Substituting Equations 10.53 through 10.55 into Equations 10.47 and 10.48 and then simplifying, we obtain

$$\rho - r_1\rho(\cos\beta_1 - 1) = 1 + r_2(\cos\beta_2 - 1)$$

$$\xi - r_2\xi(\cos\beta_2 - 1) = \rho + r_1\rho(\cos\beta_1 - 1)$$

We can rewrite the above equations as follows

$$\rho - \rho\lambda_1 = 1 + \lambda_2 \quad (10.56)$$

$$\xi - \xi\lambda_2 = \rho + \rho\lambda_1 \quad (10.57)$$

where

$$\lambda_1 = r_1(\cos\beta_1 - 1) \quad \text{and} \quad \lambda_2 = r_2(\cos\beta_2 - 1)$$

Solving Equations 10.56 and 10.57 simultaneously, we get

$$\rho = \frac{1+\lambda_2}{1-\lambda_1} \quad \text{and} \quad \xi = \rho\frac{1+\lambda_1}{1-\lambda_2}$$

Therefore, the amplification factor ξ is given by

$$\xi = \frac{(1+\lambda_1)(1+\lambda_2)}{(1-\lambda_1)(1-\lambda_2)}$$

Since β_1 and β_2 are real and $r_1, r_2 > 0$, it follows that both $\lambda_1 < 0$ and $\lambda_2 < 0$.

Hence, it follows that $|\xi| \le 1$ and consequently the ADI scheme is unconditionally stable.

EXERCISES

1. Use the forward-difference method to approximate the solution of the following parabolic PDE:

$$\frac{\partial u}{\partial t} - \frac{\partial^2 u}{\partial x^2} = 0, \quad 0 < x < \pi, t > 0$$

$$u(0,t) = u(\pi,t) = 0, \quad t > 0$$

$$u(x,0) = \sin(x), \quad 0 \le x \le \pi$$

Use $h = \pi/10$ and $k = 0.05$ and compare your results at $t = 0.5$ to the analytical solution $u(x,t) = \mathrm{e}^{-t}\sin x$.

2. Solve the PDE

$$\frac{\partial u}{\partial t} = \frac{\partial^2 u}{\partial x^2}$$

subject to the condition $u(x,0) = \sin \pi x, 0 \le x \le 1; u(0,t) = u(1,t) = 0$, using the Crank–Nicolson method.

3. Use the forward-difference method to approximate the solution of the following parabolic differential equation:

$$\frac{\partial u}{\partial t} - \frac{\partial^2 u}{\partial x^2} = 0, \quad 0 < x < 2, t > 0$$

$$u(0,t) = u(2,t) = 0, \quad t > 0$$

$$u(x,0) = \sin(2\pi x), \quad 0 \le x \le 2$$

Use $h = 0.4$ and $k = 0.1$, and compare your results at $t = 0.5$ to the analytical solution $u(x,t) = e^{-4\pi^2 t} \sin 2\pi x$. Then use $h = 0.4$ and $k = 0.05$ and compare the answers.

4. Find the approximate solutions of the wave equation $(\partial^2 u/\partial t^2) - (\partial^2 u/\partial x^2) = 0, 0 < x < 1, t > 0$

$$u(x,0) = \sin \pi x, u_t(x,0) = 0, \quad 0 \le x \le 1$$

$$u(0,t) = 0, u(1,t) = 0, \quad t > 0$$

using the FDM. Hence, compare your results at $t = 1.0$ to the analytical solution $u(x,t) = \sin \pi x \cos \pi t$.

5. Find the solution of the following PDE using the Crank–Nicolson method

$$\frac{\partial u}{\partial t} - \frac{\partial^2 u}{\partial x^2} = 0, \quad 0 < x < 2, 0 < t$$

$$u(0,t) = u(2,t) = 0, \quad t > 0$$

$$u(x,0) = \sin \frac{\pi}{2} x, \quad 0 \le x \le 2$$

Hence, compare your results to the actual solution

$$u(x,t) = e^{\left(\frac{-\pi^2}{4}\right)t} \sin \frac{\pi x}{2}$$

6. Solve the hyperbolic problem

$$\frac{\partial^2 u}{\partial t^2}(x,t) - 4\frac{\partial^2 u}{\partial x^2}(x,t) = 0, \quad 0 < x < 1, t > 0$$

with the boundary condition:

$$u(0,t) = u(1,t) = 0, \quad \text{for } t > 0$$

and initial conditions

$$u(x,0) = \sin \pi x, u_t(x,0) = 0 \quad 0 \le x \le 1$$

7. Solve the following elliptic PDE

$$\frac{\partial^2 u}{\partial x^2} + \frac{\partial^2 u}{\partial y^2} = 4, \quad 0 < x < 1, 0 < y < 2$$

$$u(x,0) = x^2, u(x,2) = (x-2)^2, \quad 0 \le x \le 1$$

$$u(0,y) = y^2, u(1,y) = (y-1)^2, \quad 0 \le y \le 2$$

Use $h = k = 1/2$, and compare the results to the actual solution $u(x,y) = (x-y)^2$.

8. Find the solution of the following elliptic PDE

$$\frac{\partial^2 u}{\partial x^2} + \frac{\partial^2 u}{\partial y^2} = 0, \quad 1 < x < 2, 0 < y < 1$$

$$u(x,0) = 2\ln x, u(x,1) = \ln(x^2+1), \quad 1 \le x \le 2$$

$$u(1,y) = \ln(y^2+1), u(2,y) = \ln(y^2+4), \quad 0 \le y \le 1$$

Use $h = k = 1/3$, and compare the results to the actual solution $u(x,y) = \ln(x^2+y^2)$.

9. Solve the PDE $\partial^2 u/\partial x^2 + \partial^2 u/\partial y^2 = 0$ in the domain of the below figure by the (a) Jacobi method, (b) Gauss–Seidel method, and (c) SOR method.

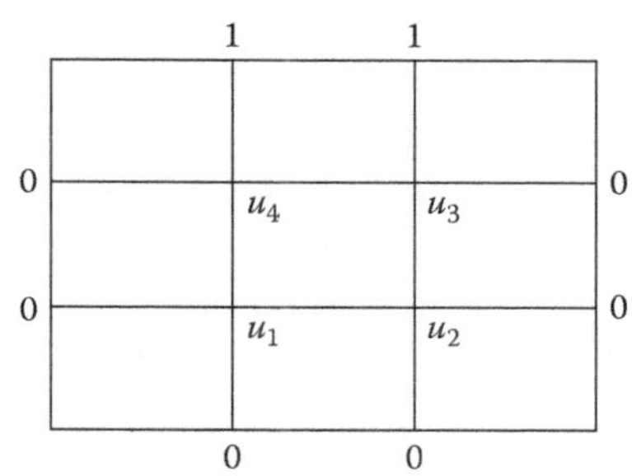

10. Solve the Laplace equation $u_{xx} + u_{yy} = 0$ in the domain shown in the below figure by the (a) Gauss–Seidel method, (b) Jacobi method, and (c) Gauss–Seidel SOR method.

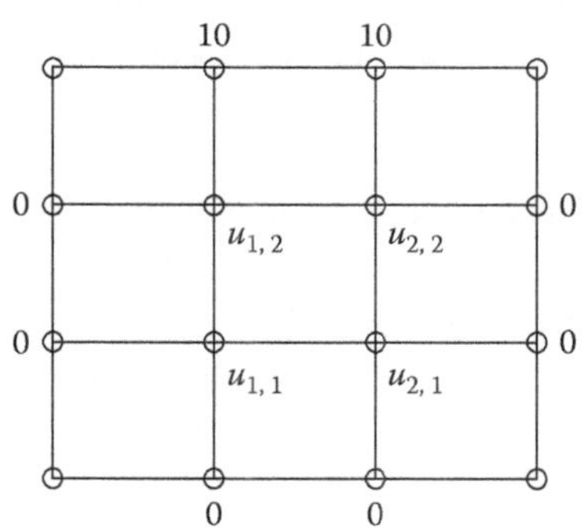

11. Find the solution of the following elliptic PDE:

a. $$\frac{\partial^2 u}{\partial x^2} + \frac{\partial^2 u}{\partial y^2} = 0, \quad 0 < x < 1, 0 < y < 1$$

$$u(x,0) = 0, u(x,1) = x, \quad 0 \le x \le 1$$

$$u(0,y) = 0, u(1,y) = y, \quad 0 \le y \le 1$$

Use $h = k = 0.2$, and compare the results to the actual solution $u(x,y) = xy$.

b. $\frac{\partial^2 u}{\partial x^2}+\frac{\partial^2 u}{\partial y^2}=-(\cos(x+y)+\cos(x-y)),\quad 0<x<\pi, 0<y<\frac{\pi}{2}$

$$u(0,y)=\cos y, u(\pi,y)=-\cos y,\quad 0\le y\le\frac{\pi}{2}$$

$$u(x,0)=\cos x, u\left(x,\frac{\pi}{2}\right)=0,\quad 0\le x\le\pi$$

Use $h=\pi/5$ and $k=\pi/10$ and compare the results to the actual solution $u(x,y)=\cos x\cos y$.

c. $\frac{\partial^2 u}{\partial x^2}+\frac{\partial^2 u}{\partial y^2}=\left(x^2+y^2\right)e^{xy},\quad 0<x<2, 0<y<1$

$$u(0,y)=1,\qquad u(2,y)=e^{2y},\quad 0\le y\le 1$$

$$u(x,0)=1,\qquad u(x,1)=e^{x},\quad 0\le x\le 2$$

Use $h=0.2$ and $k=0.1$, and compare the results to the actual solution $u(x,y)=e^{xy}$.

12. Solve the Poisson's equation $\partial^2u/\partial x^2(x,y)+\partial^2u/\partial y^2(x,y)=xe^y$, $0<x<2, 0<y<1$ with the boundary conditions

$$u(0,y)=0, u(2,y)=2e^y,\quad 0\le y\le 1$$

$$u(x,0)=x, u(x,1)=ex,\quad 0\le x\le 2$$

Compare the solution $u(x,y)=xe^y$. Use the termination criterion for the Gauss–Seidel method with error tolerance of 10^{-10}.

13. Solve the Poisson equation $u_{xx}+u_{yy}=8x^2y^2$ for the square region $0\le x\le 1.0, 0\le y\le 1$ with $h=1/3$ and the values of u on the boundary are everywhere zero. Use the
   a. Gauss–Seidel method
   b. Gauss–Seidel SOR method.

14. Solve using the Crank–Nicolson method, the equation $\partial u/\partial t=\partial^2u/\partial x^2$, $0<x<1, t>0$, satisfying the conditions $u(0,t)=0$, $u(1,t)=0$, and $u(x,0)=100x(1-x)$. Compute u for two time steps with $h=0.25$ and a suitable k.

15. Solve the Laplace equation $\nabla^2u=0$ with the boundary conditions shown in the below figure.

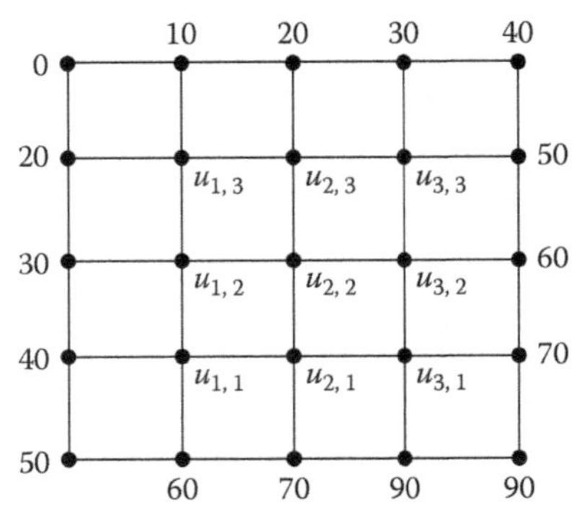

16. Use the Gauss–Jacobi iteration method to solve the Laplace equation $\partial^2 u/\partial x^2 + \partial^2 u/\partial y^2 = 0$ at the nodal points of the following square grids (below figures) satisfying the boundary conditions prescribed therein.

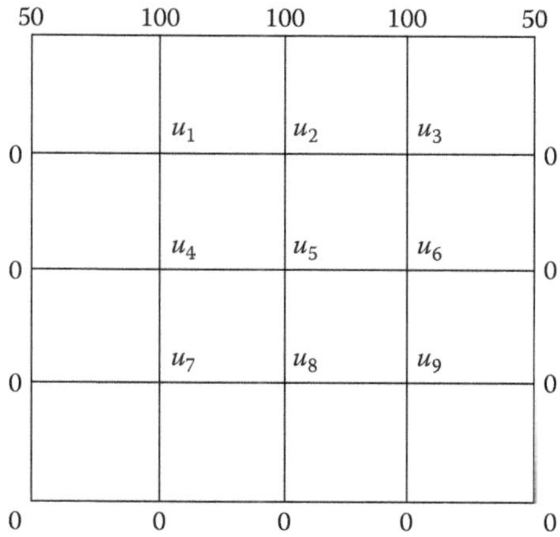

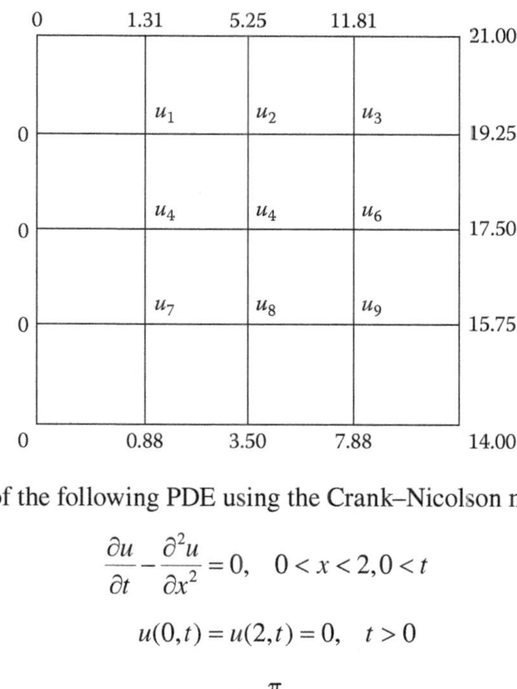

17. Find the solution of the following PDE using the Crank–Nicolson method

$$\frac{\partial u}{\partial t} - \frac{\partial^2 u}{\partial x^2} = 0, \quad 0 < x < 2, 0 < t$$

$$u(0,t) = u(2,t) = 0, \quad t > 0$$

$$u(x,0) = \sin\frac{\pi}{2}x, \quad 0 \le x \le 2$$

Hence compare your results to the actual solution $u(x,t) = e^{\left(\frac{-\pi^2}{4}\right)t}\sin\left(\frac{\pi x}{2}\right)$.

18. Find the pivotal values of the equation $y_{tt} = y_{xx}$ for $x = 0(1)10$ and $t = 0(1)5$, given that

$$y(0,t) = 0, y(10,t) = 0, y(x,0) = 0 \text{ and } \frac{\partial y}{\partial t}(x,0) = \begin{cases} \dfrac{x}{10}, & 0 \le x \le 5 \\ \dfrac{1}{10}(10 - x), & 5 \le x \le 10 \end{cases}$$

19. Solve the equation $\partial u/\partial t = \partial^2 u/\partial x^2$, satisfying the conditions $u(0,t)=0,\ u(6,t)=0; t \geq 0$ and $u(x,0)=100,\quad 0<x<6$. Compute u for one time step by the Crank–Nicolson method with $h=1$ and $k=1$.
20. Solve the equation $y_{tt} = 25y_{xx}$, taking $h=1$, given that $y_t(x,0)=0,\ y(0,t)=0,\ y(5,t)=0$ and

$$y(x,0) = \begin{cases} 20x, & \text{for } 0 \leq x \leq 1 \\ 5(5-x), & \text{for } 1 \leq x \leq 5 \end{cases}$$

Give the solution upto five time steps.
21. Find the values of $u(x,t)$ for two time steps, given that $u_t = u_{xx},\ u(0,t)=0,\ u(1,t)=0$ and $u(x,0)=\sin \pi x$. Use the Crank–Nicolson method with $h=0.2$.
22. Solve the equation $u_{tt} = u_{xx}$ upto $t=1$ with step size of 0.2, given that
   a. $u(0,t)=0,\quad u(1,t)=0,\quad u_t(x,0)=0,\quad$ and $\quad u(x,0)=10+x(1-x);\quad 0<x<1.$
   b. $u(0,t)=0,\quad u(1,t)=0,\quad u_t(x,0)=0,\quad$ and $\quad u(x,0)=\sin \pi x;\quad 0<x<1.$
23. Solve the equation $\partial u/\partial t = (1/2)\left(\partial^2 u/\partial x^2\right),\ 0 \leq x \leq 12; 0 \leq t \leq 12,\ 0 \leq x \leq 12;$ $0 \leq t \leq 12$, with boundary and initial conditions $u(0,t)=0, u(12,t)=9, 0 \leq t \leq 12$ and $u(x,0)=(1/4)x(15-x);\quad 0 \leq x \leq 12$. Use the (i) Bender–Schmidt difference equation and (ii) Crank–Nicolson difference equation, taking $h=3$ and $k=3$.
24. Solve, using the Crank–Nicolson method, the equation $\partial u/\partial t = 4\left(\partial^2 u/\partial x^2\right),\ 0<x<4, t>0,$ satisfying the conditions $u(0,t)=0,\ u(4,t)=0$ and

$$u(x,0) = \begin{cases} 20x, & \text{for } 0 \leq x \leq 2 \\ 20(4-x), & \text{for } 2 \leq x \leq 4 \end{cases}$$

Compute u for two time steps with $h=1$ and a convenient value of k.
25. Solve the poisson equation $\nabla^2 u = -4x^2y^2$ over the square mesh with sides $x=0,\ y=0,$ $x=3$, and $y=3$ with $u=0$ on the boundary and mesh length 1 unit, correct to two places of decimals, using the method of relaxation.
26. Solve the poisson equation $\nabla^2 u = -4xy$ over the square mesh with sides $x=0,\ y=0,\ x=3,$ and $y=3$ with $u=0$ on the boundary and mesh length 1 unit, correct to two places of decimals, using the method of iteration.
27. Given that $u(x,y)$ satisfies the equation $\nabla^2 u = 0$ and the boundary conditions $u(0,y)=0,$ $u(3,y)=3y+9,\ u(x,0)=x^2$, and $u(x,3)=2x^2$, find the values of $u(i,j),\ i=1,2;\ j=1,2$, correct to two places of decimals using the relaxation method.
28. Determine the first approximate value of the interior mesh points of the following Dirchlet's problem:

$$u_{xx} + u_{yy} = 0$$

$$u(x,0)=0,\quad u(0,y)=0$$

$$u(x,1)=10x,\quad u(1,y)=20x$$

29. Solve $25u_{xx} = u_{tt}$ given $u_t(x,0)=0;\ u(0,t)=0;\ u(5,t)=0;$ and

$$u(x,0) = \begin{cases} 20x, & 0 \leq x \leq 1 \\ 5(5-x), & 1 \leq x \leq 5 \end{cases}$$

30. Solve by the Crank–Nicolson method $16\left(\partial u/\partial t\right) = \partial^2 u/\partial x^2$ in $0<x<1$ and $t>0$, given $u(x,0)=0;\ u(0,t)=0;$ and $u(1,t)=100\ t$. Compute u for one step with $h=0.25$.

31. Using the Schmidt process, solve $u_{xx} = 2u_t$ with the conditions $u(x,0) = (1/4)x(15 - x)$ for $0 \le x \le 12$; $u(0,t) = 0$; $u(12,t) = 9$ for $0 < t < 12$, take $h = k = 3$.
32. Solve using the Crank–Nicolson scheme the heat-conduction equation, respectively, with initial conditions

    a. $u(x,0) = \begin{cases} 2x, & 0 \le x \le 0.5 \\ 2(1-x), & 0.5 \le x \le 1 \end{cases}$

    b. $u(x,0) = \sin \pi x, \quad 0 \le x \le 1$

    and boundary conditions $u(0,t) = 0 = u(1,t)$, for all time.
    Taking $c = 1$, $\Delta x = 1/4$, compute the solution for the first two steps.
33. Compute the solution of the Dirichlet problem for the Laplace equation $u_{xx} + u_{yy} = 0$, in the unit square in the first quadrant with the boundary condition

$$u(x,y) = e^{0.2\pi x} \sin(0.2\pi y)$$

    on the boundary of the square. Take $\Delta x = \Delta y = 1/4$ and use the SOR scheme with the optimum relaxation factor. Compute correct to 2D.
34. Compute the solution of the boundary value problem for the Poisson equation

$$u_{xx} + u_{yy} = x^2 + y^2, \quad 0 < x < 1, 0 < y < 1$$

    with the Dirichlet boundary condition $u(x,y) = 0$ on the boundary of the square. Take $\Delta x = \Delta y = 0.25$, and compute the solution, correct to 2D.
35. Using three-point central difference representation for the second derivatives, compute the solution of the boundary value problem for the Poisson equation

$$u_{xx} + 4u_{yy} = -1, \quad |x| \le 1, |y| \le 1$$

    with the Dirichlet boundary condition $u(x,y) = 0$ on the boundary of the square $|x| = 1$, $|y| = 1$. Take $\Delta x = \Delta y = 0.5$, compute the first three steps by the Gauss–Seidel iteration scheme and continue till convergence, correct to 2D, by the SOR.
36. Compute the solution of the boundary value problem for the Poisson equation

$$u_{xx} + u_{yy} = 2xy, \quad 0 < x < 1, 0 < y < 1$$

    with the Dirichlet boundary condition $u(x,y) = 0$ on the boundary of the unit square in the first quadrant. Take $\Delta x = \Delta y = 0.25$, and compute the solution, correct to 2D.
37. Write down the finite difference scheme for the equation $u_{xx} + u_{yy} = 0$ and solve it for the square region given below

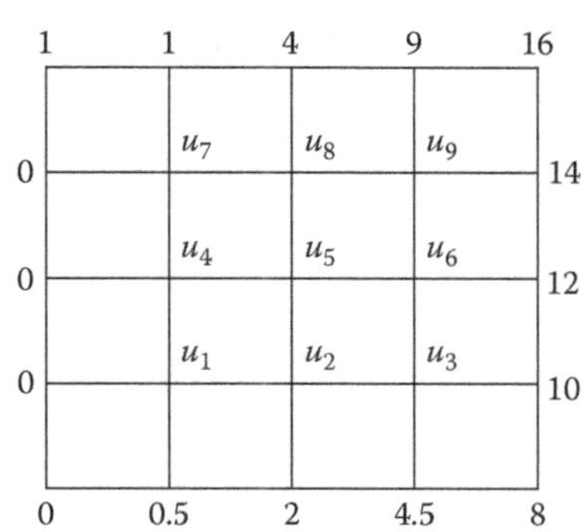

    with $h = k = 1.0$, use the Gauss–Seidel method to compute, correct to two decimal places, values of u at internal mesh points.

38. The function $u(x, y)$ satisfies the Laplace equation at all points within the squares given below and has the boundary values as indicated. Compute a solution correct to two decimal places by the FDM.

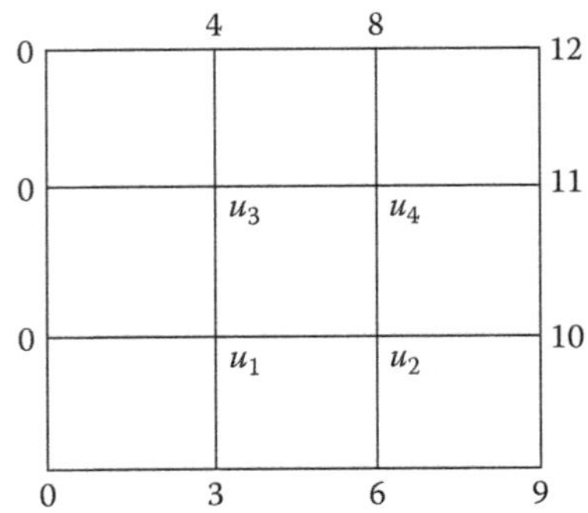

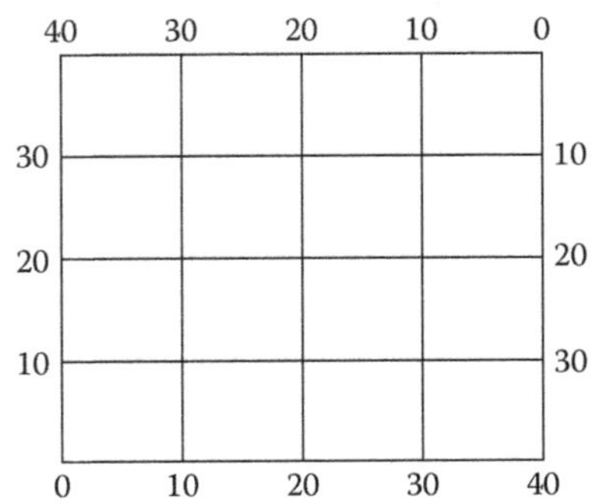

39. Solve the heat conduction equation $\partial u/\partial t = \partial^2 u/\partial x^2$ with the conditions $u(x,0) = \sin x,\ 0 \le x \le \pi,\ \partial u/\partial t\,(0,t) = \mathrm{e}^{-t}$, and $\partial u/\partial t\,(\pi,t) = -\mathrm{e}^{-t}$. Using the Crank–Nicolson formula, taking $h = \pi/2$ and $k = \pi^2/4\sqrt{20}$, compute $u\left(\pi/2, \pi^2/4\sqrt{20}\right)$.
40. Use the explicit formula to solve the equation $\partial u/\partial t = \partial^2 u/\partial x^2$ with the conditions $u(0,t) = u(5,t) = 0,\ u(x,0) = x^2(25 - x^2)$. With $h = 1$ and $k = 0.5$, tabulate the values of u_i^n for $i = 0, 1, 2, 3, 4, 5$ and $n = 0, 1, 2$.
41. Find the approximate solutions of the wave equation

$$\frac{\partial^2 u}{\partial t^2} - \frac{\partial^2 u}{\partial x^2} = 0, \quad 0 < x < 1, 0 < t$$

$$u(0,t) = u(1,t) = 0, \quad t > 0$$

$$u(x,0) = \sin 2\pi\, x, \quad 0 \le x \le 1$$

$$\frac{\partial u}{\partial t}(x,0) = 2\pi \sin 2\pi\, x, \quad 0 \le x \le 1$$

with $h = 0.1$ and $k = 0.1$. Hence compare your results at $t = 0.3$ with the actual solution $u(x,t) = \sin 2\pi\, x(\cos 2\pi t + \sin 2\pi t)$.
42. Solve the wave equation

$$\frac{\partial^2 u}{\partial t^2} - \frac{\partial^2 u}{\partial x^2} = 0, \quad 0 < x < 1, 0 < t$$

$$u(0,t) = u(1,t) = 0, \quad t > 0$$

$$u(x,0) = \begin{cases} 1, & 0 \le x \le \frac{1}{2} \\ -1, & \frac{1}{2} \le x \le 1 \end{cases}$$

$$\frac{\partial u}{\partial t}(x,0) = 0, \quad 0 \le x \le 1$$

with $h = 0.1$ and $k = 0.1$.

43. Use the forward-difference method to approximate the solution of the following parabolic PDEs:

a. $$\frac{\partial u}{\partial t} - \frac{\partial^2 u}{\partial x^2} = 0, \quad 0 < x < 2, 0 < t$$

$$u(0,t) = u(2,t) = 0, \quad t > 0$$

$$u(x,0) = \sin 2\pi x, \quad 0 \le x \le 2$$

use $h = 0.4$ and $k = 0.1$, and compare your results at $t = 0.5$ to the actual solution $u(x,t) = e^{-4\pi^2 t} \sin 2\pi x$. Then use $h = 0.4$ and $k = 0.05$, and compare the answers.

b. $$\frac{\partial u}{\partial t} - \frac{\partial^2 u}{\partial x^2} = 0, \quad 0 < x < \pi, 0 < t$$

$$u(0,t) = u(\pi,t) = 0, \quad t > 0$$

$$u(x,0) = \sin x, \quad 0 \le x \le \pi$$

use $h = \pi/10$ and $k = 0.05$, and compare your results at $t = 0.5$ with the actual solution $u(x,t) = e^{-t} \sin x$.

44. Use the forward difference method to approximate the solution of the following parabolic PDEs:

a. $$\frac{\partial u}{\partial t} - \frac{4}{\pi^2}\frac{\partial^2 u}{\partial x^2} = 0, \quad 0 < x < 4, 0 < t$$

$$u(0,t) = u(4,t) = 0, \quad t > 0$$

$$u(x,0) = \sin\frac{\pi x}{4}\left(1 + 2\cos\frac{\pi x}{4}\right), \quad 0 \le x \le 4$$

Use $h = 0.2$ and $k = 0.04$, and compare your results at $t = 0.4$ with the actual solution $u(x,t) = e^{-t}\sin(\pi x/2) + e^{-t/4}\sin(\pi x/4)$.

b. $$\frac{\partial u}{\partial t} - \frac{1}{\pi^2}\frac{\partial^2 u}{\partial x^2} = 0, \quad 0 < x < 1, 0 < t$$

$$u(0,t) = u(1,t) = 0, \quad t > 0$$

$$u(x,0) = \cos\pi(x - 0.5), \quad 0 \le x \le 1$$

Use $h = 0.1$ and $k = 0.04$, and compare your results at $t = 0.4$ with the actual solution $u(x,t) = e^{-t}\cos\pi(x - 0.5)$.

45. Solve the wave equation

$$\frac{\partial^2 u}{\partial t^2} - \frac{\partial^2 u}{\partial x^2} = 0, \quad 0 < x < 1, 0 < t$$

$$u(0,t) = u(1,t) = 0, \quad t > 0$$

$$u(x,0) = \sin \pi x, \quad 0 \le x \le 1$$

$$\frac{\partial u}{\partial t}(x,0) = 0, \quad 0 \le x \le 1$$

compare your results at $t = 1.0$ to the actual solution $u(x,t) = \cos \pi t \sin \pi x$.

46. Solve the wave equation

$$\frac{\partial^2 u}{\partial t^2} - \frac{1}{16\pi^2}\frac{\partial^2 u}{\partial x^2} = 0, \quad 0 < x < 0.5, 0 < t$$

$$u(0,t) = u(0.5,t) = 0, \quad t > 0$$

$$u(x,0) = 0, \quad 0 \le x \le 0.5$$

$$\frac{\partial u}{\partial t}(x,0) = \sin 4\pi x, \quad 0 \le x \le 0.5$$

compare your results at $t = 0.5$ to the actual solution $u(x,t) = \sin t \sin 4\pi x$.

11 An Introduction to the Finite Element Method

11.1 INTRODUCTION

In this chapter, we shall explore a brief derivation and foundation of the finite difference methods to solve boundary value problems in ordinary differential equations.

In the early 1940s, the finite element method (FEM) was proposed by Richard Courant (1943) in his research paper. The historical foundation of the method can be traced back from earlier work by Galerkin in 1915. The FEM has been developed as one of the most powerful techniques for the numerical solution of differential equations, which is widely used in engineering design and analysis.

Unlike finite difference schemes, which are sought to approximate the unknown analytical solution to a differential equation at a finite number of grid points or mesh points in the computational domain, the FEM provides an approximation to the analytical solution in the form of a piecewise polynomial function, which is defined over the entire computational domain. One advantage of the FEM over finite difference methods is that FEM is the most flexible one in terms of dealing with complex geometry and complicated boundary conditions. Many physical problems have boundary conditions involving derivatives and irregular-shaped boundaries. Boundary conditions of this type are difficult to handle using finite difference techniques, and irregular shaping of the boundary causes difficulty for placing the grid points. The FEM includes the boundary conditions as integrals in a functional that is being minimized, thus leading to the construction procedure independent of the typical boundary conditions of the problem.

Now, we shall discuss two techniques for the construction of finite element approximations:

1. The Rayleigh–Ritz method
2. The Galerkin method

11.2 PIECEWISE LINEAR BASIS FUNCTIONS

Suppose that we wish to approximate a real-valued function $f(x)$ over a finite interval $[a,b]$. The simplest choice of basis functions (shape functions) are piecewise linear polynomials. An usual approach is to first partition the computational domain $[a,b]$ into a number of subintervals $[x_i, x_{i+1}]$, referred to as *elements*, hence the name FEM, by the node points x_0, x_1, …, x_{n+1} such that

$$a = x_0 < x_1 < \cdots < x_n < x_{n+1} = b$$

Now, we define the basis function $\phi_i(x)$, $i = 1, 2, \ldots, n$ by

$$\phi_i(x) = \begin{cases} 0, & \text{if} \quad x_0 \le x \le x_{i-1} \\ \dfrac{(x - x_{i-1})}{h_{i-1}}, & \text{if} \quad x_{i-1} \le x \le x_i \\ \dfrac{(x_{i+1} - x)}{h_i}, & \text{if} \quad x_i \le x \le x_{i+1}, \text{ for } i = 1, 2, \ldots, n \\ 0, & \text{if} \quad x_{i+1} \le x \le x_{n+1} \end{cases} \tag{11.1}$$

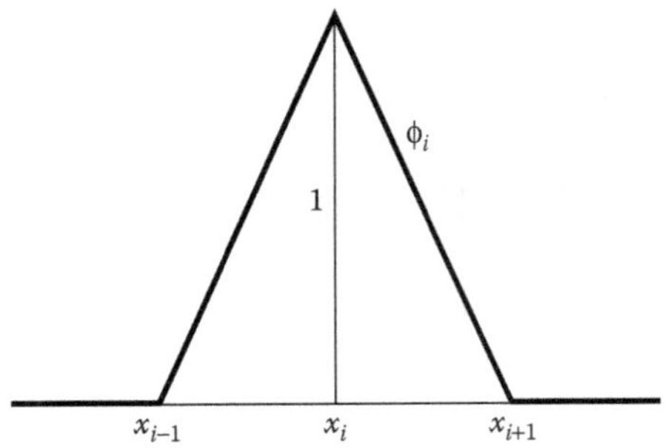

FIGURE 11.1 A piecewise linear basis function ϕ_i, where $i = 1, 2, \ldots, n$.

where $h_i = x_{i+1} - x_i$.

The functions $\phi_i(x)$, $i = 1, 2, \ldots, n$, are called the (piecewise linear) finite element basis functions and the associated approximation $u(x)$ is referred to as the (piecewise linear) finite element approximation of $y(x)$. The closure of the interval (x_{i-1}, x_{i+1}) over which $\phi_i(x)$ is nonzero, called the support of the function $\phi_i(x)$. For FEM, the important property of basis functions $\phi_i(x)$, $i = 1, 2, \ldots, n$, is that they have local support, since the basis functions $\phi_i(x)$, $i = 1, 2, \ldots, n$ are nonzero in only a pair of adjacent intervals, $(x_{i-1}, x_i]$ and $[x_i, x_{i+1})$. It can be easily observed that the basis functions $\phi_i(x)$, $i = 1, 2, \ldots, n$ are identically zero except for the interval $[x_{i-1}, x_{i+1}]$ with $\phi_i(x_i) = 1$. The piecewise linear finite element basis functions $\phi_i(x)$, $i = 1, 2, \ldots, n$ with compact support $[x_{i-1}, x_{i+1}]$ is shown in Figure 11.1.

11.3 THE RAYLEIGH–RITZ METHOD

The Rayleigh–Ritz technique arises from the study of variational principles. This technique is based on converting the boundary value problem into a variational problem, involving the minimization of a certain functional over a region. To describe the Rayleigh–Ritz method, the concept of functional is required. Let us consider the following boundary value problem:

$$-\frac{d}{dx}\left(p(x)\frac{dy}{dx}\right) + q(x)y = f(x), \quad \text{for } a \le x \le b \tag{11.2}$$

with the boundary conditions

$$y(a) = \alpha, \quad y(b) = \beta \tag{11.3}$$

In this boundary vale problem, we assume that $p \in C^1[a,b]$, $q \in C[a,b]$, and $f \in L^2[a,b]$. Furthermore, we assume that there exists a constant $\delta > 0$ such that

$$p(x) \ge \delta \text{ and } q(x) \ge 0, \quad \text{for all } x \in [a,b].$$

These assumptions are sufficient to ensure that the boundary value problem given in Equations 11.2 and 11.3 have a unique solution.

Let us consider the functional

$$I[u] = \int_a^b F(x, u, u')\, dx \tag{11.4}$$

subject to the boundary conditions

$$u(a) = \alpha \text{ and } u(b) = \beta$$

We assume that F is differentiable sufficient number of times. We wish to find the extremum of the functional (Equation 11.2) subject to the conditions (Equation 11.3). The necessary condition for the existence of an extremum of the functional (Equation 11.2) is that its variation must vanish. From the principal of the calculus of variation, we know that a necessary condition for $I[u]$ to have an extremum is that $y(x)$ must satisfy the following equation.

$$\frac{\partial F}{\partial u} - \frac{d}{dx}\left(\frac{\partial F}{\partial u'}\right) = 0 \tag{11.5}$$

This equation is called the Euler–Lagrange equation. If the solution of the Euler equation is unique, then this solution gives the solution of the variational problem. Conversely, the solution of variational problem provides the solution of the boundary value problem.

Now, we can verify that the variational form of the differential equation (11.3) is given by

$$I[u] = \frac{1}{2}\int_a^b [p(x)[u'(x)]^2 + q(x)[u(x)]^2 - 2f(x)u(x)]dx \tag{11.6}$$

Let

$$F(x,u,u') = \frac{1}{2}[p(x)[u'(x)]^2 + q(x)[u(x)]^2 - 2f(x)u(x)] \tag{11.7}$$

Then, the Euler–Lagrange equation becomes

$$\frac{\partial F}{\partial u} - \frac{d}{dx}\left(\frac{\partial F}{\partial u'}\right) = \left[\frac{\partial}{\partial u} - \frac{d}{dx}\left(\frac{\partial}{\partial u'}\right)\right]\left[\frac{1}{2}[p(x)[u'(x)]^2 + q(x)[u(x)]^2 - 2f(x)u(x)]\right]$$

$$= q(x)u(x) - f(x) - \frac{d}{dx}\big(p(x)u'(x)\big) = 0$$

which yields the Equation 11.2.

The Rayleigh–Ritz method determines the approximate solution $y(x)$ by minimizing the functional, not over all functions in $C^2[0,1]$, but over a smaller set of functions consisting of linear combinations of certain basis functions $\phi_1(x)$, $\phi_2(x)$, …, $\phi_n(x)$. These basis functions are linearly independent and satisfy

$$\phi_i(a) = 0 \text{ and } \phi_i(b) = 0, \quad \text{for each } i = 1,2,\ldots,n$$

Now, let

$$u(x) = \sum_{i=1}^{n} c_i\phi_i(x) \tag{11.8}$$

be an approximation to the solution $y(x)$ of Equation 11.2, which can be determined by finding the unknown coefficients to minimize the functional $I[u]$.

From Equation 11.6, we get

$$I[u] = \frac{1}{2}\int_a^b [p(x)\left[\sum_{i=1}^{n} c_i\phi_i'(x)\right]^2 + q(x)\left[\sum_{i=1}^{n} c_i\phi_i(x)\right]^2 - 2f(x)\sum_{i=1}^{n} c_i\phi_i(x)]dx \tag{11.9}$$

Now, considering I as function of $c_1, c_2, \ldots, c_n$, for minima, we have

$$\frac{\partial I}{\partial c_j} = 0, \quad \text{for } j = 1,2,\ldots,n \tag{11.10}$$

Differentiating Equation 11.9 yields

$$\frac{\partial I}{\partial c_j} = \int_a^b [p(x)\left[\sum_{i=1}^{n} c_i\phi_i'(x)\phi_j'(x)\right] + q(x)\left[\sum_{i=1}^{n} c_i\phi_i(x)\phi_j(x)\right] - f(x)\phi_j(x)]dx \tag{11.11}$$

Next, substituting into Equation 11.10, we obtain

$$\sum_{i=1}^{n}\left[\int_a^b [p(x)\phi_i'(x)\phi_j'(x) + q(x)\phi_i(x)\phi_j(x)]dx\right]c_i - \int_a^b f(x)\phi_j(x)dx = 0, \quad \text{for } j = 1,2,\ldots,n \tag{11.12}$$

The normal equations in Equation 11.12 generate an $n \times n$ linear system

$$\boldsymbol{Tc} = \boldsymbol{b} \tag{11.13}$$

in n number of unknown coefficients $c_1, c_2, \ldots, c_n$, where $\boldsymbol{c} = (c_1, c_2, \ldots, c_n)^T$ is a column vector, the symmetric matrix $\boldsymbol{T}$ has elements

$$t_{ij} = \int_a^b [p(x)\phi_i'(x)\phi_j'(x) + q(x)\phi_i(x)\phi_j(x)]dx = \int_a^b p(x)\phi_i'(x)\phi_j'(x)dx + \int_a^b q(x)\phi_i(x)\phi_j(x)dx,$$

$$\text{for } 1 \le i, j \le n$$

and $\boldsymbol{b}$ is a constant vector defined by

$$b_i = \int_a^b f(x)\phi_i(x)dx, \quad i = 1,2,\ldots,n$$

The matrix elements t_{ij} of the matrix $\boldsymbol{T}$ has been written as the sum of two terms, since the matrix $\boldsymbol{T}$ is often written in this way as the sum of two matrices which are sometimes known as the stiffness matrix and the mass matrix, respectively.

Solving Equation 11.13 for unknown coefficients $c_1, c_2, \ldots, c_n$, we can obtain the approximate solution $u(x)$ in Equation 11.8.

Corollary:

For the differential equation (11.2) with mixed boundary conditions

$$\alpha_0 y(a) + \alpha_1 y'(a) = \gamma_1$$

$$\beta_0 y(b) + \beta_1 y'(b) = \gamma_2$$

the functional can be defined as

$$I[u]=\frac{1}{2}\int_a^b [p(x)(u'(x))^2+q(x)(u(x))^2-2f(x)u(x)]dx+\frac{1}{2}(W_a+W_b)$$

where,

$$W_a=\frac{p(a)}{\alpha_1}[2\gamma_1 y(a)-\alpha_0 y^2(a)]$$

$$W_b=\frac{p(b)}{\beta_1}[\beta_0 y^2(b)-2\gamma_2 y(b)]$$

Note: If $\alpha_1=0$, then set $W_a=0$ and if $\beta_1=0$, then set $W_b=0$.

11.3.1 Algorithm of Rayleigh–Ritz Method

Input: Read the differential equation and read $p(x)$, $q(x)$, $f(x)$ and n.
Output: Print the value of c_i, $i=1,2,\ldots,n$ and print $u(x)$.

Step 1: Define the node points as $a=x_0<x_1<\cdots<x_n=b$.
Step 2: Calculate $h_i=x_i-x_{i-1},\quad i=1,2,\ldots,n$.
Step 3: Define $\phi_i(x)$, $\phi_i'(x)$, $i=1,2,\ldots,n$ as

$$\phi_i(x)=\begin{cases}0, & \text{if } 0\le x\le x_{i-1},\\ \frac{1}{h}(x-x_{i-1}), & \text{if } x_{i-1}<x\le x_i,\\ \frac{1}{h}(x_{i+1}-x), & \text{if } x_i<x\le x_{i+1},\\ 0, & \text{if } x_{i+1}<x\le 1,\end{cases}\quad,\quad \phi_i'(x)=\begin{cases}0, & \text{if } 0<x<x_{i-1},\\ \frac{1}{h}, & \text{if } x_{i-1}<x<x_i,\\ -\frac{1}{h}, & \text{if } x_i<x<x_{i+1},\\ 0, & \text{if } x_{i+1}<x<1,\end{cases}$$

Step 4: for $i=1(1)n$ do
for $j=1(1)n$ do

$$t_{i,j}=\int_a^b\big(p(x)\phi_i'(x)\phi_j'(x)+q(x)\phi_i(x)\phi_j(x)\big)dx;$$

end
end.
Step 5: for $j=1(1)n$ do

$$b_j=\int_a^b f(x)\phi_j(x)dx;$$

end.
Step 6: (Solve the tridiagonal system by Thomas algorithm)

$$\alpha_1=-\frac{t_{1,2}}{t_{1,1}};$$

$$\beta_1=\frac{b_1}{t_{1,1}};$$

Step 7: for $j = 2(1)\overline{n-1}$ do

$$\alpha_j = \frac{t_{j,j+1}}{t_{j,j} + \alpha_{j-1} t_{j,j-1}};$$

end.

Step 8: for $j = 2(1)n$ do

$$\beta_j = \frac{b_j - t_{j,j-1}\beta_{j-1}}{t_{j,j} + t_{j,j-1}\alpha_{j-1}};$$

end.

Step 9: $c_n = \beta_n$;

for $j = \overline{n-1}(1)1$ do

$$c_j = \beta_j + \alpha_j c_{j+1};$$

end.

Step 10: Print $c_1, c_2, \ldots, c_n$.

Step 11: sum = 0;

for $i = 1(1)n$ do

sum = sum + $c_i \phi_i(x)$;

end.

Step 12: Set $u(x)$ = sum;

Step 13: Print $u(x)$.

Step 14: Stop. ■

MATHEMATICA® Program for Rayleigh–Ritz Method for Solving BVP (Chapter 11, Example 11.1)

```
n=4;
h=1/(n+1);
For[i=0,i<=n+1,i++,
  x[i]=i*h];
For[i=1,i<=n,i++,
  phi[i][x]=Piecewise[{{1/h*(x-x[i-1]),x[i-1]<x<=x[i]},
     {1/h*(x[i+1]-x),x[i]<x<=x[i+1]}}];
  dphi[i][x]=Piecewise[{{1/h,x[i-1]<x<x[i]},{-1/h,x[i]<x<x[i+1]}}]];
        n
u[x]= ∑(c[i]*phi[i][x]);
       i=1
p[x]=x;
q[x]=4;
f[x]=4*x²-8*x+1;
For[i=1,i<=n,i++,
  For[j=1,j<=n,j++,
   t[i,j]=Integrate
     [(p[x]*dphi[i][x]*dphi[j][x]+ q[x]* phi[i][x]* phi[j][x]),{x,0,1}]]];
T=Table[t[i,j],{i,1,n},{j,1,n}];
Print[MatrixForm[T]]
For[i=1,i<=n,i++,
  b[i]=Integrate[f[x]*phi[i][x],{x,0,1}]];
a={Table[b[i],{i,1,n}]};
Print[MatrixForm[a]]
c=N[Inverse[T].Transpose[a]];
Print[MatrixForm[c]]
For[i=1,i<=n,i++,
```

```
c1[i]=c[[i]][[1]]]
        n
u[x]= ∑ (c1[i]*phi[i][x]);
      i=1
For[i=0,i<=1,i=i+0.1,
  y[i]=N[u[x]/.x->i];
  yexact[i]=N[i²-i];
  e[i]=Abs[y[i]-yexact[i]];
  Print[i,"     ",y[i],"     ",yexact[i],"     ",e[i]]];
```

Output:

$$\begin{pmatrix} \frac{38}{15} & -\frac{41}{30} & 0 & 0 \\ -\frac{41}{30} & \frac{68}{15} & -\frac{71}{30} & 0 \\ 0 & -\frac{71}{30} & \frac{98}{15} & -\frac{101}{30} \\ 0 & 0 & -\frac{101}{30} & \frac{128}{15} \end{pmatrix}$$

$$\begin{pmatrix} -\frac{31}{375} & -\frac{23}{75} & -\frac{7}{15} & -\frac{211}{375} \end{pmatrix}$$

$$\begin{pmatrix} -0.164404 \\ -0.244262 \\ -0.243367 \\ -0.161953 \end{pmatrix}$$

0	0.	0.	0.
0.1	-0.0822022	-0.09	0.00779778
0.2	-0.164404	-0.16	0.00440445
0.3	-0.204333	-0.21	0.00566682
0.4	-0.244262	-0.24	0.00426191
0.5	-0.243814	-0.25	0.00618569
0.6	-0.243367	-0.24	0.00336671
0.7	-0.20266	-0.21	0.00734001
0.8	-0.161953	-0.16	0.00195327
0.9	-0.0809766	-0.09	0.00902336
1.	$-8.99021*10^{-17}$	$-1.11022*10^{-16}$	$2.11202*10^{-17}$

Example: 11.1

Use Rayleigh–Ritz method to approximate the solution of the boundary value problem

$$-\frac{d}{dx}(xy') + 4y = 4x^2 - 8x + 1, \quad 0 \le x \le 1, y(0) = y(1) = 0$$

with $n = 4$. Compare your results to the actual solution $y(x) = x^2 - x$

Solution:

Here $n = 4$ and $h = 1/(n+1) = 0.2$ with $0 = x_0 < x_1 < x_2 < \cdots < x_n < x_{n+1} = 1$.

Now, $x_0 = 0,\ x_1 = 0.2,\ x_2 = 0.4,\ x_3 = 0.6,\ x_4 = 0.8,\ x_5 = 1$.

According to Rayleigh–Ritz method, the solution can be approximated as

$$u(x) = \sum_{i=1}^{n} c_i \phi_i(x), \tag{11.14}$$

where $\phi_i(x),\ i = 1,2,\ldots,n$ are the basis functions.

We take, here, the basis functions $\phi_i(x),\ i = 1,2,\ldots,n$ such that it satisfies the boundary conditions as

$$\phi_i(x) = \begin{cases} 0, & \text{if } 0 \le x \le x_{i-1} \\ \dfrac{1}{h}(x - x_{i-1}), & \text{if } x_{i-1} \le x \le x_i \\ \dfrac{1}{h}(x_{i+1} - x), & \text{if } x_i \le x \le x_{i+1} \\ 0, & \text{if } x_{i+1} \le x \le 1 \end{cases}$$

for each $i = 1,2,\ldots,n$, and $\phi_i(0) = \phi_i(1) = 0$.

The derivative $\phi_i'(x)$ can be defined as

$$\phi_i'(x) = \begin{cases} 0, & \text{if } 0 < x < x_{i-1} \\ \dfrac{1}{h}, & \text{if } x_{i-1} < x < x_i \\ -\dfrac{1}{h}, & \text{if } x_i < x < x_{i+1} \\ 0, & \text{if } x_{i+1} < x < 1 \end{cases}$$

for each $i = 1,2,\ldots,n$

The unknown functions c_i can be determined by minimizing the functional

$$I[u] = \frac{1}{2}\int_0^1 \left[x\left[\sum_{i=1}^{n} c_i \phi_i'(x)\right]^2 + 4\left[\sum_{i=1}^{n} c_i \phi_i(x)\right]^2 - 2(4x^2 - 8x + 1)\left[\sum_{i=1}^{n} c_i \phi_i(x)\right]\right] dx$$

For minimization of the functional I, we have

$$\frac{\partial I}{\partial c_j} = 0, \quad j = 1,2,\ldots,n$$

or

$$\frac{\partial I}{\partial c_j} = \int_0^1 \left[x\left[\sum_{i=1}^{n} c_i \phi_i'(x)\phi_j'(x)\right] + 4\left[\sum_{i=1}^{n} c_i \phi_i(x)\phi_j(x)\right] - (4x^2 - 8x + 1)\phi_j(x)\right] dx = 0$$

or

$$\sum_{i=1}^{n}\left[\int_0^1 [x\phi_i'(x)\phi_j'(x) + 4\phi_i(x)\phi_j(x)]dx\right]c_i - \int_0^1 (4x^2 - 8x + 1)\phi_j(x)dx = 0, \quad j = 1,2,\ldots,n \quad (11.15)$$

The normal equations (11.15) generates an 4×4 linear system, that is,

$$\boldsymbol{Tc} = \boldsymbol{b}, \quad (11.16)$$

where $\boldsymbol{c} = [c_1,c_2,c_3,c_4]^T$ and

$$T = \begin{bmatrix} \frac{38}{15} & -\frac{41}{30} & 0 & 0 \\ -\frac{41}{30} & \frac{68}{15} & -\frac{71}{30} & 0 \\ 0 & -\frac{71}{30} & \frac{98}{15} & -\frac{101}{30} \\ 0 & 0 & -\frac{101}{30} & \frac{128}{15} \end{bmatrix}, \quad \boldsymbol{b} = \begin{bmatrix} -\frac{31}{375} \\ -\frac{23}{75} \\ -\frac{7}{15} \\ -\frac{211}{375} \end{bmatrix}$$

Solving Equation 11.16, we obtain the values of c_i, $i = 1,2,3,4$ as

$$c_1 = -0.664404,\ c_2 = -0.244262,\ c_3 = -0.243367,\ c_4 = -0.161953$$

Table 11.1 presents the comparison between approximate and exact values of $y(x)$ for different values of x.

Example 11.2

Obtain the Ritz finite element solution of the boundary value problem

$$u'' - u = 2x, 0 < x < 1$$

$$u(0) + u'(0) = 1,\ u(1) = 0$$

using the linear shape functions and two finite elements.

TABLE 11.1

Comparison of Approximate and Exact Results for Example 11.1

x	$u(x)$	$y(x)$	$\|y(x)-u(x)\|$
0	0	0	0
0.1	−0.0822022	−0.09	0.007797788
0.2	−0.164404	−0.16	0.00440445
0.3	−0.204333	−0.21	0.00566682
0.4	−0.244262	−0.24	0.00426191
0.5	−0.243814	−0.25	0.00618569
0.6	−0.243367	−0.24	0.00336671
0.7	−0.20266	−0.21	0.00734001
0.8	−0.161953	−0.16	0.00195327
0.9	−0.0809766	−0.09	0.00902336
1.0	0	0	0

Solution:

Here, $p(x)=-1$, $q(x)=-1$, $f(x)=2x$.

Choose the approximate function $w(x)=(1-x)(a_1+a_2x)$ such that it satisfy the boundary condition $w(1)=0$.

For mixed boundary condition, we have

$$w_a = u^2(0) - 2u(0) \text{ and } w_b = 0$$

The functional for this problem can be written as

$$I[u] = \frac{1}{2}\int_0^1 \left[-(u'(x))^2 - (u(x))^2 - 4xu(x)\right]dx + \frac{1}{2}\left(u^2(0) - 2u(0)\right)$$

We have

$$w(x) = (1-x)a_1 + x(1-x)a_2$$

$$w^2(x) = (1-x)^2 a_1^2 + x^2(1-x)^2 a_2^2 + 2x(1-x)^2 a_1 a_2$$

$$w'(x) = -a_1 + (1-2x)a_2$$

$$w'^2(x) = a_1^2 + (1-2x)^2 a_2^2 - 2a_1a_2(1-2x)$$

$$u(0) \approx w(0) = a_1$$

Now

$$I[w] = \frac{1}{2}\int_0^1 \left[-(w'(x))^2 - (w(x))^2 - 4xw(x)\right]dx + \frac{1}{2}\left(w^2(0) - 2w(0)\right)$$

$$= \frac{1}{2}\left(-\frac{2a_1}{3} - \frac{4a_1^2}{3} - \frac{a_2}{3} - \frac{a_1a_2}{6} - \frac{11a_2^2}{30}\right) + \frac{1}{2}\left(a_1^2 - 2a_1\right)$$

For minimization of the functional, put

$$\frac{\partial I[w]}{\partial a_1} = 0 \Rightarrow 4a_1 + a_2 = -16$$

$$\frac{\partial I[w]}{\partial a_2} = 0 \Rightarrow 5a_1 + 22a_2 = -10$$

Solving these two equations, we have

$$a_1 = -\frac{342}{83}, \quad a_2 = \frac{40}{83}$$

Hence, the approximate solution of this problem is given by

$$w(x) = (1-x)\left(-\frac{342}{83}\right) + x(1-x)\left(\frac{40}{83}\right) = -\frac{2}{83}(20x^2 - 191x + 171)$$

The exact solution of this problem is given by $u(x) = \frac{3}{2}e^x + 2e^{1-x} - \frac{3}{2}e^{2-x} - 2x$ (Table 11.2)

TABLE 11.2
Comparison of Approximate and Exact Results of Example 11.2

x	$u(x)$	$w(x)$	$\lvert u(x)-w(x) \rvert$
0.0	–4.14702	–4.12048	0.0265386
0.1	–3.65188	–3.66506	0.0131812
0.2	–3.19129	–3.21928	0.0279919
0.3	–2.75863	–2.78313	0.0245051
0.4	–2.34757	–2.35663	0.00905252
0.5	–1.95201	–1.93976	0.0122501
0.6	–1.56597	–1.53253	0.0334422
0.7	–1.1836	–1.13494	0.0486586
0.8	–0.799058	–0.746988	0.0520705
0.9	–0.406503	–0.368675	0.0378278
1.0	0	0	0

11.4 THE GALERKIN METHOD

The Rayleigh–Ritz method is a variational technique that involves minimization of functional for the solution of boundary value problems. Therefore, it has a disadvantage regarding existence of functional, which is not always possible to obtain. In order to overcome the above-mentioned difficulty, Galerkin methods may be used. In numerical analysis, Galerkin methods are a class of methods for converting a differential equation to a discrete problem. This method belongs to wider classes of methods, called the *weighted residual methods*, and uses the trial functions (approximating functions) that satisfy the boundary conditions of the problem. The trail function is substituted in the given differential equation, results in the residual. The integral of the weighted residual, taken over a domain, is then set to zero. For further details about this method, see Section 7.6.4. The common thread between all three approaches, that is, Rayleigh–Ritz, Galerkin, and collocation, is that the solution is approximated by a linear combination of trial functions, and the coefficients are obtained by solving a system of equations.

FURTHER READING

Most of the material discussed here can be extended for application to partial differential equations. For further details regarding the mathematical theory and implementation of the FEM, the reader may be referred to the following relevant literature:

Bathe, K. J. and Wilson, E. L., *Numerical Methods in Finite Element Analysis*, Prentice Hall, New Jersey, 1976.

Braess, D., *Finite Elements*, Cambridge University Press, Cambridge, 2001.

Brenner, S. and Scott, L. R., *The Mathematical Theory of Finite Element Methods*, Springer, New York, 2002.

Courant, R., *Variational Method for the Solution of Problems in Equilibrium and Vibrations. Bull. Amer., Math. Soc.*, 49, 1–23, 1943.

Johnson, C., *Numerical Solution of Partial Differential Equations by the Finite Element Method*, Cambridge University Press, Cambridge, 1996.

Mitchell, A. R. and Wait, R., *The Finite Element Method in Partial Differential Equations*, John Wiley & Sons, London, 1977.

Reddy, J. N., *An Introduction to the Finite Element Method*, McGraw-Hill, Singapore, 1993.

EXERCISES

1. Using Rayleigh–Ritz method approximate the solution to the boundary value problem

$$y'' + \frac{\pi^2}{4} y = \frac{\pi^2}{16} \cos \frac{\pi}{4} x, \quad 0 \le x \le 1, \ y(0) = y(1) = 0$$

using $x_0 = 0$, $x_1 = 0.3$, $x_2 = 0.7$, $x_3 = 1$. Compare your results to the actual solution $y(x) = -(1/3)\cos(\pi/2)x - (\sqrt{2}/6)\sin(\pi/2)x + (1/3)\cos(\pi/4)x$.

2. Solve the following boundary value problems by Rayleigh–Ritz method
   a. $(d^2y/dx^2) + y = x^2, y(0) = y(1) = 0$.
   b. $(d^2y/dx^2) + 2x = 0, y(0) = y(1) = 0$.
   c. $(d^2y/dx^2) - 64y + 10 = 0, y(0) = y(1) = 0$.
3. Obtain the Ritz finite element solution of the boundary value problem

$$u'' - u = 2x, 0 < x < 1$$

$$u(0) + u'(0) = 1, u(1) = 0$$

using linear shape functions and two finite elements.

4. Consider the boundary value problem

$$u'' + 2u = x, 0 < x < 1$$

$$u(0) = 0, u(1) = 0$$

Determine the coefficients of the approximate solution function

$$w(x) = x(1 - x)(a_1 + a_2 x)$$

by the Ritz method.

5. Obtain functionals for the following boundary value problems
   a. $(d^2y/dx^2) + ky = x^3, y(a) = 0, \left[\frac{dy}{dx}\right]_{x=b} = 1$.
   b. $x^2(d^2y/dx^2) + 2x(dy/dx) = g(x), y(0) = y(1) = 0$.
   c. $(d^2y/dx^2) + p(x)y + q(x) = 0, y(a) = y(b) = 0$.
   d. $(d^4y/dx^4) + ky = f(x), 0 < x < 1, y = (d^2y/dx^2) = 0$ at $x = 0$ and $x = 1$.
6. Use the Rayleigh–Ritz method to approximate the solution of the boundary value problem

$$-\frac{d}{dx}(xy') + 4y = 4x^2 - 8x + 1, \quad 0 \le x \le 1, \ y(0) = y(1) = 0$$

using $x_0 = 0$, $x_1 = 0.4$, $x_2 = 0.8$, $x_3 = 1$. Compare your results to the actual solution $y(x) = x^2 - x$.

7. Consider the boundary value problem

$$u'' - u = 1, \ 0 < x < 1$$

$$u(0) = 0, \ u(1) = e - 1$$

Determine the coefficients of the approximate solution function

$$w(x) = (e-1)x + x(1-x)(a_1 + a_2 x)$$

by the Ritz method.

8. Use Rayleigh–Ritz method with $n = 9$ to approximate the solution to the boundary value problem

$$-y'' + y = x,\ 0 \le x \le 1,\ y(0) = 1,\ y(1) = 1 + e^{-1}$$

9. Use the Ritz FEM to obtain the difference scheme for the boundary value problem

$$\left[\left(1+x^2\right)u'\right]' - u = 1 + x^2, u(\pm 1) = 0$$

with the linear shape functions and element length h.

10. Determine the Ritz finite element solution of the boundary value problem

$$u'' = u, u'(0) = 0, u(1) = 1$$

using two finite elements and linear shape functions.

11. Show that the boundary value problem

$$-\frac{d}{dx}(p(x)y') + q(x)y = f(x), \quad 0 \le x \le 1,\ y(0) = \alpha,\ y(1) = \beta$$

can be transformed by the change of variable

$$z = y - \beta x - (1-x)\alpha$$

into the form

$$-\frac{d}{dx}(p(x)z') + q(x)z = F(x), \quad 0 \le x \le 1,\ z(0) = 0,\ z(1) = 0$$

12. Consider the boundary value problem

$$u'' + \left(1+x^2\right)u + 1 = 0,\ u(\pm 1) = 0$$

Use the Ritz method to determine the coefficients of the approximate solution

$$w(x) = \left(1-x^2\right)\left(1-4x^2\right)a_0 + \frac{16}{3}x^2\left(1-x^2\right)a_1$$

13. Using the Rayleigh–Ritz method approximate the solution to the following boundary value problems and compare the results to the actual solution:
   a. $-x^2y'' - 2xy' + 2y = -4x^2,\ 0 \le x \le 1,\ y(0) = y(1) = 0$; use $h = 0.1$; actual solution $y(x) = x^2 - x$.
   b. $-\frac{d}{dx}(e^x y') + e^x y = x + (2-x)e^x, 0 \le x \le 1,\ y(0) = y(1) = 0$; use $h = 0.1$; actual solution $y(x) = (x-1)(e^{-x} - 1)$.

c. $-\frac{d}{dx}(e^{-x}y') + e^{-x}y = (x-1) - (x+1)e^{-(x-1)}, 0 \le x \le 1, y(0) = y(1) = 0$; use $h = 0.05$; actual solution $y(x) = x(e^x - e)$.

d. $-(x+1)y'' - y' + (x+2)y = [2-(x+1)^2]e\ln 2 - 2e^x, 0 \le x \le 1, y(0) = y(1) = 0$; use $h = 0.05$; actual solution $y(x) = e^x \ln(x+1) - (e\ln 2)x$.

14. Application of the Ritz FEM, using linear shape functions, to the boundary value problem

$$-u'' = x, u(0) = 0, u(1) = 0$$

leads to the system of equations $\boldsymbol{Tc} = \boldsymbol{b}$. Determine the matrix $\boldsymbol{T}$ and the column vector $\boldsymbol{b}$ for two and four elements of equal length.

15. Apply Galerkin's method to solve the boundary value problems

a. $(d^2y/dx^2) + y = x^2, \quad y(0) = y(1) = 0.$

b. $(d^2y/dx^2) - 64y + 10 = 0, \quad y(0) = y(1) = 0.$

16. Consider the boundary value problem

$$-u'' + u = x, 0 < x < 1$$

$$u'(0) = 1, u'(1) = 2$$

Apply the Ritz method using linear shape functions to compute the finite element approximations for two and four elements.

Answers

CHAPTER 1

1. 2.472, 0.004, 42.308, 9.773
3. a. 46.24
   b. 0.7268
   c. 12.01
   d. 0.02584
5. 4.963×10^{-2}
7. a. (i) $(87)_{10}$, (ii) $(117)_{10}$, (iii) $(93)_{10}$
   b. (i) $(127)_8$, (ii) $(165)_8$, (iii) $(135)_8$
   c. (i) $(57)_{16}$, (ii) $(75)_{16}$, (iii) $(5D)_{16}$
9. b. $(147.550)_8$, $(67.B4)_{16}$
11. 0.048, 0.006
13. a. 4.96×10^{-2}
    b. 9.88×10^{-4}
15. 3%
17. 0.006
19. 0.423
21. 0.004

CHAPTER 2

1. a. 0.5885
   b. 0.607
   c. 1.303
3. a. 2.741
   b. 0.550
5. a. 1.48982
   b. 0.876728
   c. 1.55715
   d. 1.72310
7. Successive values by the bisection method are 2.5, 2.75, 2.625, 2.6875, 2.7187, 2.7344, and 2.7422; the values by false position method are 2.7210, 2.7402, 2.7407, 2.7406, and 2.7406
9. a. −2.105
   b. 1.796
   c. 1.0499
   d. 2.796
11. a. 3.7892
    b. 1.1538
    c. 1.7231
    d. 1.0881
    e. 1.3024
13. a. 0.6190
    b. 0.4656

c. 1.466
d. 1.497
17. a. 1.35106
b. 1.993
c. 2.7589
d. 1.8988
19. Six iteration steps by linear iteration; three iteration steps by Δ^2 process.
21. a. 0.1001
b. 2.140
c. 1.677
d. 1.088
e. 1.314
23. a. 4.493409458
b. 102.092
27. $\alpha = 2.1322677, [a, b] = [1, 1+\pi/2]$
29. a. 3, −2, 1
b. $1.915 \pm 1.907i,\ 0.585 \pm 2.805i$
c. −9.527, 4.092, −0.66
d. 3,2,1
e. $\frac{1}{2}, \frac{-1}{4}, \frac{-1}{4}$
f. $2, \frac{1}{2}\left(1 \pm i\sqrt{15}\right)$
g. 1, 2, 4, 5
h. $1.325,\ -0.6624 \pm 0.5623i$
31. $\begin{pmatrix} x_1 \\ y_1 \end{pmatrix} = \begin{pmatrix} 0.9013 \\ 2.0409 \end{pmatrix}, \begin{pmatrix} x_2 \\ y_2 \end{pmatrix} = \begin{pmatrix} 0.9771 \\ 2.0079 \end{pmatrix}$
33. $x = 2.412249,\ y = -0.643856$
35. Use the Newton–Raphson method with $x_0 = -0.5$. We get $x_1 = -0.504397$, $x_2 = -0.508143$, and $x_3 = -0.508146$. The root is −0.5081.

CHAPTER 3

3. $e^{1.15} = 3.1582$
5. a. $\frac{4}{(x+2)(x+3)(x+4)} + \frac{6}{(x+3)(x+4)(x+5)}$
b. $24 \times 10!$
c. $\frac{2^x(1-x)}{(x+1)!}$
13. 17; 551
15. a. 31
b. 3; 24
c. 22.58; 26.29
d. 246; 180.8
17. a. $2x^{(3)} + 3x^{(2)} + 2x^{(1)} + c$

b. $\frac{1}{4}x^{(4)} + 2x^{(3)} + \frac{9}{2}x^{(2)} + 12x^{(1)} + c$

c. $\frac{1}{2}x^{(4)} + \frac{3}{2}x^{(2)} + 10x^{(1)} + c$

19. Error in the tabulated value $x = 4.0$. Corrected value = 5.75.
21. $x^4 + x^3 + x^2 + x + 1$, 1555
23. 64.8528, 106.8368
25. 5.477
27. a. 4.5904
    b. 1.620536
29. 16.9216
31. 0.6198
33. $1 - \frac{9}{2}x + 3x^2 - \frac{1}{2}x^3$.
35. $y(0.2)|_{\text{Everett}} = 1.14965$, $y(0.2)|_{\text{Newton}} = 1.14964$, $y(0.2)|_{\text{Exact}} = 1.14870$.
39. $\frac{x^4}{16} + \frac{x^3}{4} + 5x + 100$
41. 3.347
43. 1.3757
45. 0.9763
47. 1454
49. 2.174
51. a. 2.528274
    b. 2.528264
53. 0.9856
55. 311
57. 0.453
59. $3x^3 + 5x^3 - 6x - 3$; $f(2.5) \approx 60.125$
61. $0.0068x^5 + 0.002x^4 - 0.1671x^3 - 0.0002x^2 + x$; $f(0.75) \approx 0.6816$; $|E| \le (h^6/4860) = 0.32 \times 10^{-5}$; actual error $= 0.39 \times 10^{-4}$ (Note that data is given only up to four digits).
63. a. Quadratic spline ($f(x), f'(x)$ are continuous at $x = 2$).
    b. Cubic spline ($f(x), f'(x), f''(x)$ are continuous at $x = 0$).
    c. Not a spline ($f''(x)$ is not continuous at $x = 0$).
65. $M_1 = 34/5$, $M_2 = -76/5$. The cubic spline approximations are obtained as

$$P_{31}(x) = (17x^3 - 51x^2 + 94x - 45)/15, \quad 1 \le x \le 2$$

$$P_{32}(x) = (-55x^3 + 381x^2 - 770x + 531)/15, \quad 2 \le x \le 3$$

$$P_{33}(x) = (38x^3 - 456x^2 + 1741x - 1980)/15, \quad 3 \le x \le 4$$

$$f(1.5) \approx P_{31}(1.5) = 2.575, \quad f(2.5) \approx P_{32}(2.5) = 8.525$$

67. 26.625, 29.0
69. 2877.73
71. $2x^4 - x^2 + x + 1$, 10.375
73. $M_1 = 62/5$, $M_2 = 112/5$; $P_3(x) = (-56x^3 + 672x^2 - 2092x + 2175)/15$

CHAPTER 4

1. $y'(1.5) = 5.26094$, $y'(5.8) = 107.073$; $y''(1.5) = 0.4375$, $y''(5.8) = 77.2713$
3. −0.16
5. −4.39085; −4.0067
7. 29.34; 71.33
9. 0.55
11. 3.73 rad/s; 4.48 rad/s^2
13. 3.975
15. Lagrange's interpolation formula: 16.750, 15.000, 6.000
17. 108; 108
19. Min. at $x = 1.7887$; value = 39.8027

CHAPTER 5

1. 1.1084, 1.1108, 1.1107
3. 1.641 s
5. 0.23108, -3.1×10^{-5}
7. 5.16 miles
9. 0.782794; 0.784747; Simpson 0.785398; Rhomberg 0.785398
11. 0.5236
13. $I(0.1) = 0.34351$, $I(0.2) = 0.38994$, $E_r \sim 0.015$, Rhomberg 0.32804
15. 3.1832
17. 0.091111
19. 106.6667
21. Three- and four-point Gauss Legendre: 1.08228 and 1.08979, respectively; three-point Gauss–Chebyshev: 1.08979
23. −0.8948314
25. 1.8521
27. 0.1332
29. 0.5236
31. a. 0.0049988
    b. 0.085116
33. 1.51012, 1.51012, 5-D
35. 3.9975
37. 0.014063
39. −4.21667
41. 3.1507
43. Four-point: 0.927039, six-point: 0.927038, and 5D correct
45. $C = (-9h^5)/10$, $E = (C/4!)h^5 f^{(4)}(\xi)$, $x_0 < \xi < x_3$, $I(h = 1/3) = 0.78462$, and exact = 0.78540.
47. Make the formula exact with $f(x) = 1, x, \ldots, x^5$. We get $A_1 = A_4 = 1/6$, $A_2 = A_3 = 5/6$, and $x_2 = -1/\sqrt{5}$, $x_3 = 1/\sqrt{5}$. Setting $f(x) = x^6$, we get the error term as $R = -(2/23625) f^{(6)}(\xi)$, $-1 < \xi < 1$.

CHAPTER 6

1. a. $x = 3$, $y = 1$, and $z = 2$
   b. $x = 1$, $y = -2$, and $z = 3$

c. $x = 3,\ y = 1,$ and $z = 1$
d. $x = 6.13975,\ y = 2.83596,$ and $z = -2.52874$
e. $x = 1, y = 2, z = 2,$ and $w = 2$
f. $x = 1,\ y = 3,$ and $z = 5$
g. $x_1 = 1,\ x_2 = x_3 = x_4 = 2$

3. a. $x = 3,\ y = 2,$ and $z = 1$
b. $x = y = z = w = 1$
c. $x = 0.8072,\ y = 0.2372,\ z = -0.1046,$ and $w = -0.3581$
d. $x = 0.98202,\ y = 1.00507,$ and $z = 1.56407$

5. a. (3, −1, −2)
b. (3170/151, 4834/151, 130/151, −3005/151)
c. (5655/223, −4374/223, −6446/223, −1740/223)
d. (1.04907, −1.80034, 1.70871)
e. (−2.15343, 1.27419, −1.62357)

7. a. (−0.3797, 1.7197, 1.4232, 2.579)
b. (1.0229, 1.5183, 0.6284)

9. a. (−1.3910, 2.5827, 0.1692)
b. (0.6385, 0.4548, 1.2532)

11. a. $$\boldsymbol{A}^{-1} = \begin{bmatrix} -4/3 & 2 & 7/3 \\ 5/3 & -2 & -8/3 \\ 7/3 & -3 & -10/3 \end{bmatrix}$$

b. $$\boldsymbol{A}^{-1} = \begin{bmatrix} 1.2 & -0.4 & 0.2 \\ -0.2 & -0.1 & 0.3 \\ -0.4 & 0.3 & 0.1 \end{bmatrix}$$

c. $$\boldsymbol{A}^{-1} = \begin{bmatrix} 25 & -41 & 16 & -6 \\ -16 & 27 & -11 & 4 \\ 16 & -27 & 13 & -5 \\ -6 & 10 & -5 & 2 \end{bmatrix}$$

13. a. $$\begin{bmatrix} 2 & -1 & 1 \\ -3 & 2 & -2 \\ 1 & -1 & 2 \end{bmatrix}$$

b. $$\begin{bmatrix} 13.5 & -6 & 2 & -1.5 \\ -6 & 3 & -2 & -1 \\ 2 & -2 & 10 & -3 \\ -1.5 & 1 & -3 & 1 \end{bmatrix}$$

c. $$\begin{bmatrix} 1 & 0 & -2 & 0 \\ -5 & 1 & 11 & -1 \\ 287 & -67 & -630 & 65 \\ -416 & 97 & 913 & -94 \end{bmatrix}$$

15. $$\boldsymbol{L} = \begin{bmatrix} 2 & 0 & 0 \\ -1/2 & \sqrt{15}/2 & 0 \\ 0 & -2/\sqrt{15} & \sqrt{56/15} \end{bmatrix}; \qquad \boldsymbol{x} = [15/56,\ 1/14,\ 1/56]^T$$

17. a. $L = \begin{bmatrix} p & 0 & 0 & 0 & 0 & 0 \\ 0 & p & 0 & 0 & 0 & 0 \\ 0 & 0 & 2.5 & 0 & 0 & 0 \\ 0 & 0 & 0 & p & 0 & 0 \\ 0 & 0 & 1.5 & 0 & 2 & 0 \\ 7/(2p) & 3/(2p) & 0 & 1/(2p) & 0 & q \end{bmatrix}, p = \sqrt{\frac{11}{2}}, q = \sqrt{\frac{31}{11}}.$

b. From $Lz = b$, we get $z = [1/p, \quad 1/p, \quad 0.4, \quad 1/p, \quad 0.2 \quad 0]^T$
From $L^T x = z$, we get $x = [2/11, \quad 2/11, \quad 0.1, \quad 2/11, \quad 0.2 \quad 0]^T$

19. a. Diagonally dominant after rearranging; the solution is $x = 1$, $y = -1$, and $z = 1$.
b. Diagonally dominant after rearranging; the solution is $x = 2$, $y = 2$, and $z = 2$.

21. a. Not diagonally dominant
b. $x = 1$, $y = 2$, and $z = 3$
c. $x = 5$, $y = 4$, and $z = 1$
d. $x = y = z = 2$
e. $x = 0.9936, y = 1.5070,$ and $z = 1.8485$
f. $x = 0.4429, y = 1.5652, z = 0.3229,$ and $u = 0.7238$
g. $x_1 = 1$, $x_2 = 2$, $x_3 = 3$, and $x_4 = 0$

23. a. $A^{-1} = \frac{1}{2}\begin{bmatrix} 1 & 1 & -4 \\ 1 & -1 & 2 \\ -2 & 0 & 4 \end{bmatrix}, \begin{bmatrix} x_1 \\ x_2 \\ x_3 \end{bmatrix} = \begin{bmatrix} -0/2 \\ 5/2 \\ 5 \end{bmatrix}$

b. $A^{-1} = \frac{1}{78}\begin{bmatrix} 22 & -2 & -14 \\ 31 & -17 & -2 \\ 1 & 70 & 10 \end{bmatrix}, \begin{bmatrix} x_1 \\ x_2 \\ x_3 \end{bmatrix} = \begin{bmatrix} 1 \\ 2 \\ -1 \end{bmatrix}$

25. a. (1, 2, 3)
b. (3/4, −9/4, 19/4)

27. a. (3.306, −2.032)
b. (0.6385, 0.4548, 1.2532)

29. a. (1.0587, 1.3718, 1.9655)
b. (2.4316, 3.5833, 1.9269)
c. (1.1168, 2.2267, 3.3367, 4.4467)

31. a. $F(\lambda) = |H_J - \lambda I| = -\lambda^3 = 0$ gives $\rho(H_j) = 0 < 1$. Iteration converges.
b. $F(\lambda) = |H_{GS} - \lambda I| = -\lambda(2-\lambda)^2 = 0$ gives $\rho(H_{GS}) = 2 > 1$. Iteration diverges.

33. Since the coefficient matrix in both systems is strictly diagonally dominant, the Jacobi iteration scheme will converge. It can be verified that $\rho(H_j) < 1$. We obtain
i. $x^{(1)} = [1.0, 1.4, 1]^T$, $x^{(2)} = [0.15, 0.6, 0.2]^T$, $x^{(3)} = [0.75, 1.27, 0.75]^T$. Exact solution is $x = [1/2, 1, 1/2]^T$.
ii. $x^{(1)} = [1.2, 1.0, 0.3]^T$, $x^{(2)} = [0.86, 1.09, 1.10]^T$, $x^{(3)} = [0.984, 0.976, 0.948]^T$. Exact solution is $x = [1, 1, 1]^T$.

35. i. $F(\lambda) = |H_{GS} - \lambda I| = -\lambda(\lambda^2 - (\lambda/9) - (5/3))^2 = 0$; $\rho(H_{GS}) = (1 + \sqrt{841})/18 \approx 1.35 > 1$. Iteration diverges.
ii. $F(\lambda) = |H_{GS} - \lambda I| = -\lambda(2 - \lambda)^4 = 0$; $\rho(H_{GS}) = 4 > 1$. Iteration diverges.

37. i. The characteristic equation of the iteration matrix H_j is
$F(\lambda) = -\lambda(\lambda^2 - 11/18) = 0$. $\mu = \rho(H_j) = 0.7817$.
$w^* = 1.232$, $\rho(H_{SOR}) = 0.232$. Therefore, $v = 0.6345$.

$$(D+wL)v^{(k)} = wr^{(k)}, \text{ where } D+wL = \begin{bmatrix} 3 & 0 & 0 \\ 2.464 & 3 & 0 \\ 0 & -1.232 & 2 \end{bmatrix}$$

Starting with $x^{(0)} = [0.5, 0.5, 0.5]^T$, we get

$$r^{(0)} = [2.5, 2, 0.5]^T, \quad v^{(0)} = [1.0267, -0.0219, 0.2945]^T$$

$$x^{(1)} = [1.5267, 0.4781, 0.7945]^T, \quad r^{(1)} = [-0.5363, 0.3068, -0.1109]^T$$

$$v^{(1)} = [-0.2202, 0.3069, 0.1207]^T, \quad x^{(2)} = [1.3065, 0.7850, 0.9152]^T$$

$$r^{(2)} = [-0.4895, -0.0528, -0.0454]^T, \quad v^{(2)} = [-0.2010, 0.1434, 0.0604]^T$$

$$x^{(3)} = [1.1055, 0.9284, 0.9756]^T$$

Exact solution is $x = [1, 1, 1]^T$

ii. The characteristic equation of the iteration matrix H_j is $F(\lambda) = -\lambda(\lambda^2 - 1/8) = 0$. $\mu = \rho(\boldsymbol{H}_j) = 0.3536$. $w^* = 1.033$, $\rho(\boldsymbol{H}_{SOR}) = 0.033$. Therefore, $\boldsymbol{v} = 1.48$. The SOR iteration scheme is written as

$$(D+wL)v^{(k)} = wr^{(k)}, \text{ where } D+wL = \begin{bmatrix} 4 & 0 & 0 \\ -1.033 & 4 & 0 \\ 0 & -1.033 & 4 \end{bmatrix}$$

Starting with $x^{(0)} = [0.5, 0.5, 0.5]^T$, we get

$$r^{(0)} = [1.5, 1, 1.5]^T, \quad v^{(0)} = [0.3874, 0.3583, 0.4799]^T$$

$$x^{(1)} = [0.8874, 0.8583, 0.9799]^T, \quad r^{(1)} = [0.3087, 0.4341, -0.0613]^T$$

$$v^{(1)} = [0.0797, 0.1327, 0.0184]^T, \quad x^{(2)} = [0.9671, 0.9910, 0.9983]^T$$

$$r^{(2)} = [0.1226, 0.0014, -0.0022]^T, \quad v^{(2)} = [0.0317, 0.0085, 0.0016]^T$$

$$x^{(3)} = [0.9988, 0.9995, 0.9999]^T$$

Exact solution is $x = [1, 1, 1]^T$

39. a. (6.15, 4.2, 3.05)
    b. (1, −2, 3)
    c. (4, 1, 2)
    d. (1.1, 2.2, 3.3, 4.4)
    e. (2.031, 2.684, −1.118, 3.112)
41. Ill-conditioned
43. $x_1 = 2,\ x_2 = -1, x_3 = 3$
45. Solution does not exist.
47. The Cholesky method cannot be applied; solution is (0.92, 1.296, 0.112).

CHAPTER 7

1. 3.005, 3.0202
3. $y_2 = 1 + \dfrac{x}{2} + \dfrac{3}{40}x^5$

5. $y = \frac{4}{3}x^{3/2} + \frac{32}{81}x^{9/2} + \frac{512}{3645}x^{15/2} + \cdots; \quad y(0.9) = 0.7859$
7. 1.005012
9. 1.11686, 1.27730, and 1.5023
11. $y(1.1) = 1.1066$ and $y(1.2) = 1.228$
13. 1.1111; 1.2496
15. $\boldsymbol{A} = \begin{bmatrix} 1-(\alpha^2/2) & \alpha+(h^2/2) \\ -[\alpha+(h^2/2)] & 1-(\alpha^2/2) \end{bmatrix}, \boldsymbol{b} = \begin{bmatrix} h \\ -\alpha h/2 \end{bmatrix}$, where $\alpha = ht_j$

    $\begin{bmatrix} y_1 \\ z_1 \end{bmatrix} = \begin{bmatrix} 0.105 \\ 1.0 \end{bmatrix}; \begin{bmatrix} y_2 \\ z_2 \end{bmatrix} = \begin{bmatrix} 0.219995 \\ 0.997875 \end{bmatrix}$

17. a. $x(-0.1) = 3.1912$; $x(0.1) = 0.6302$ and $y(-0.1) = 2.7675$; $y(0.1) = 2.9645$
    b. 0.9185; −0.9137
    c. 0.9048; −0.9055
19. $y(0.1) = 1.0932$
21. 1.6403 and 2.3625
23. 1.0202
25. a. $4xy = x^4 + 3$; (1.061, 1.057; 1.229, 1.222)
    b. $2x = e^y(1+x^2)$; (−0.0139, −0.0165; −0.0513, −0.0556)
    c. $y(1+e^{-x}) = 1$; (0.5498, 0.5498; 0.5985, 0.5987)
27. a. $y(0) = 0$, $y(1.8) = 1.0795$; error at $x = 1.8$ is −0.8926
    b. $y(0) = 0$, $y(1.8) = 0.2008$; error at $x = 1.8$ is −0.0139
29. a. −0.1001
    b. −0.0146
    c. −0.0111
31. 0.51648
33. 1.0859, 1.0659
35. 0.09548, 0.18864, 0.26679, 0.34659, 0.42557, 0.50488, 0.58850, and 0.67469
37. 1.0048; 1.0196
39. 0.9130; 0.8458
41. a.

t	0	0.4	0.8	1.2
x	1	0.8	0.568	0.2042
y	0.5	0.9	1.364	2.0277
x error	0	−0.0132	−0.1025	−0.4171
y error	0	0.0207	0.1036	0.2594

b.

t	0	0.4	0.8	1.2
x	1	0.7869	0.4654	−0.2134
y	0.5	0.9206	1.4675	2.2875
x error (10^{-3})	0	−0.0026	0.1099	0.3956
y error (10^{-3})	0	0.0922	0.0760	−0.4700

43.

h	$O(h)$	$O(h^2)$	$O(h^3)$
0.3	0.95300		
0.2	1.00576	1.11128	
0.15	1.03273	1.11364	1.11600

45. $y(1) = 1.4983$
47. $y_1 = 0.649$, $y_2 = 0.935$, and $y_3 = 0.941$
49. $y(0) = 1.0$ and $y(10) = -7.4912 \times 10^7$; error becomes dominant as x is increased.
51. $y(0.2) = 2.762239$ (exact value = 2.7616)
    $z(0.2) = -2.060566$ (exact value = -2.368)
53. $y(0.2) = 0.9801$
55. Taylor series: $y(0.1) = 0.909333$ and $y(0.2) = 0.835672$
    Adams–Bashforth method: $y(0.3) = 0.776740$, $y(0.4) = 0.732095$, $y(0.5) = 0.701048$, and $y(0.6) = 0.683506$.
57. 2.0583
59. $y(1.4) = 0.949$
61. 0.9008 and 0.8079
63. $y(0.8) = 2.4366$
65. $x(0.1) = 1.1003$ and $y(0.1) = 1.1102$
67. (2.8597, 0.7873, −1.7713); (2.640, 1.576, −0.051)
69. $y(0.2) = -0.02847$, $y(0.4) = 0.05008$, $y(0.6) = 0.05776$, and $y(0.8) = 0.04389$
71. $y(0.25) = 0.0924571$, $y(0.5) = 0.11937$, and $y(7.5) = 0.0849727$
73. a. Eigenvalues of A are $\lambda_1 = -20.055$ and $\lambda_2 = -1.945$. The system is asymptotically stable.
    b. First order. Apply on $y' = \lambda y$, $\lambda < 0$. We get $|\xi| = |(1-(\lambda h/2)/(1-(3\lambda h/2)| < 1$ for $\lambda < 0$. The method is stable and A-stable.
    c. $h = 0.2$: $y(0.2) \approx [0.753, 0.018]^T \cdot h = 0.1$: $y(0.2) = [0.720, 0.026]^T$.
    Extrapolated values: $[0.687, 0.034]^T$
75. $x(0.4) = 0.4907$ and $y(0.4) = 0.6$
77. a. 0.9048 and 2.0046
    b. 1.1102 and 1.1003

CHAPTER 8

1. −2, 0.5969, and 13.403
3. The eigenvalues of $\boldsymbol{A}$ are 1/2, 5/2, and −1. Using the Gerschgorin theorem, we find the eigenvalues $\tilde{\boldsymbol{A}}$ satisfy $|\lambda_1 - \bar{\lambda}_1| \le 5 \times 10^{-2}$, $|\lambda_2 - \bar{\lambda}_2| = 0$, and $|\lambda_3 - \bar{\lambda}_3| \le 5 \times 10^{-2}$.
5. The given matrix is Hermitian. Therefore, its eigenvalues are real. Using the Gerschgorin theorem, we find the eigenvalues lie in the interval (−3.6503, 3.6503). $\|\boldsymbol{A}\|_2 = \sqrt{\sum |a_{ij}|^2} = \sqrt{19}$.
7. 2, −2.37228, 3.37228. Exact: 2, $(1 \pm \sqrt{33})/2$.
9. a. $\dfrac{5}{\sqrt{2}}, \dfrac{1}{\sqrt{2}}, -\dfrac{1}{\sqrt{2}}; \begin{bmatrix} 1 \\ \sqrt{2} \\ 1 \end{bmatrix}, \begin{bmatrix} -1 \\ \sqrt{2} \\ -1 \end{bmatrix}, \begin{bmatrix} -1 \\ 0 \\ 1 \end{bmatrix}$

    b. 0.585, 3.41, 2; $\begin{bmatrix} 0.707 \\ 1 \\ 0.708 \end{bmatrix}, \begin{bmatrix} -0.709 \\ 1 \\ -0.701 \end{bmatrix}, \begin{bmatrix} 1 \\ 0 \\ -1 \end{bmatrix}$

c. 4, −2, 6; $\begin{bmatrix} \frac{1}{\sqrt{2}} \\ 0 \\ -\frac{1}{\sqrt{2}} \end{bmatrix}, \begin{bmatrix} 0 \\ 1 \\ 0 \end{bmatrix}, \begin{bmatrix} \frac{1}{\sqrt{2}} \\ 0 \\ \frac{1}{\sqrt{2}} \end{bmatrix}$

d. 0.386, −1.313, 5.927; $\begin{bmatrix} 0.565 \\ -0.295 \\ -0.824 \end{bmatrix}, \begin{bmatrix} 0.546 \\ -0.736 \\ -0.401 \end{bmatrix}, \begin{bmatrix} -0.618 \\ -0.676 \\ -0.400 \end{bmatrix}$

11. $\lambda = -1, \begin{bmatrix} 1 \\ -0.33 \\ 0 \end{bmatrix}$

13. $\lambda = 2.00$ with vector $\boldsymbol{x} = \begin{bmatrix} 0 \\ 0 \\ 1 \end{bmatrix}$

15. a. 6; $(2 \quad 1 \quad -2 \quad 1)^T$
    b. 11.6628; $(1\ 0.42180\ 0.02495)^T$
    c. 7.1679; $(0.43074\ 0.21893\ 1)^T$
17. a. $5/\sqrt{2},\ 1/\sqrt{2},\ -1/\sqrt{2}$; $(1,\sqrt{2},1)^T$; $(-1,\sqrt{2},-1)^T$; $(-1,0,1)^T$
    b. 4, −2, 6; $(1,0,-1)^T$; $(0,1,0)^T$; $(1,0,1)^T$
    c. 0.6340, 2.2652, 3.1007; $(0.6280, 0.6280, 0.4597)^T$; $(-0.7726, 0.4319, 0.4655)^T$; $(0.0938, -0.6474, 0.7564)^T$
19. After seven iteration, the largest eigenvalues of $(\boldsymbol{A}-5\boldsymbol{I})^{-1}$ is 2.4142226. We find $\lambda = 5+(1/2.4142) = 5.4142$ as required eigenvalue, $\boldsymbol{X} = [0.7071,1,0.7071]^T$. Exact: $\lambda = 4+\sqrt{2} \approx 0.5142$, $X = [0.7071,1,0.7071]^T$.

21. a. 1, −4, 7; $\begin{bmatrix} 1 \\ 0 \\ 0 \end{bmatrix}, \begin{bmatrix} -2 \\ 5 \\ 0 \end{bmatrix}, \begin{bmatrix} 37/65 \\ 2/11 \\ 1 \end{bmatrix}$

    b. 4, −2, 6; $\begin{bmatrix} 1 \\ 0 \\ -1 \end{bmatrix}, \begin{bmatrix} 0 \\ 1 \\ 0 \end{bmatrix}, \begin{bmatrix} 1 \\ 0 \\ 1 \end{bmatrix}$

    c. 2.5365, −0.0062, 1.4697; $\begin{bmatrix} 0.5315 \\ 0.4615 \\ 0.7103 \end{bmatrix}, \begin{bmatrix} -0.7212 \\ 0.6863 \\ 0.0937 \end{bmatrix}, \begin{bmatrix} -0.4443 \\ -0.5621 \\ 0.6976 \end{bmatrix}$

    d. 0.634, 2.265, 3.101; $\begin{bmatrix} 0.628 \\ 0.628 \\ 0.466 \end{bmatrix}, \begin{bmatrix} -0.773 \\ 0.432 \\ 0.466 \end{bmatrix}, \begin{bmatrix} 0.094 \\ -0.647 \\ 0.756 \end{bmatrix}$

23. a. 729, 81, 9
    b. $10, 1 \pm \sqrt{6}i$
    c. 3,−2, 1

25. a. $\begin{bmatrix} 3 & \sqrt{2} & 0 \\ \sqrt{2} & 5 & 0 \\ 0 & 0 & 1 \end{bmatrix}$, $f_0 = 1$, $f_1 = \lambda - 3$, $f_2 = \lambda^2 - 8\lambda + 13$, $f_3 = (\lambda - 1)(\lambda^2 - 8\lambda + 13)$,

eigenvalues are 1, $4 \pm \sqrt{3}$.

b. $\begin{bmatrix} 2 & \sqrt{10} & 0 \\ \sqrt{10} & 31/10 & 13/10 \\ 0 & 13/10 & -1/10 \end{bmatrix}$, $f_0 = 1$, $f_1 = \lambda - 2$, $f_2 = \lambda^2 - (51710)\lambda + (38/10)$,

$f_3 = \lambda^3 - 5\lambda^2 - 6\lambda + 3$, eigenvalues lie in the intervals (–2, –1), (0, 1) and (5, 6).

c. $\begin{bmatrix} 1 & 2\sqrt{2} & 0 \\ 2\sqrt{2} & 0 & 0 \\ 0 & 0 & 2 \end{bmatrix}$, $f_0 = 1$, $f_1 = \lambda - 1$, $f_2 = \lambda^2 - \lambda - 8$, $f_3 = (\lambda - 2)(\lambda^2 - \lambda - 8)$,

eigenvalues are 1, $(1 \pm \sqrt{33})/2$.

d. $\begin{bmatrix} 2 & 2 & 0 \\ 2 & 17/4 & -\sqrt{3}/4 \\ 0 & -\sqrt{3}/4 & 3/4 \end{bmatrix}$, $f_0 = 1$, $f_1 = \lambda - 2$, $f_2 = \lambda^2 - (25/4)\lambda + (9/2)$,

$f_3 = \lambda^3 - 7\lambda^2 + 9\lambda - 33$, eigenvalues are 1, $3 \pm \sqrt{6}$.

27. a. $\begin{bmatrix} 2 & -\sqrt{2} & 0 \\ -\sqrt{2} & 1 & 0 \\ 0 & 0 & 3 \end{bmatrix}$

b. $\begin{bmatrix} 4 & 2\sqrt{2} & 0 \\ 2\sqrt{2} & 13/2 & 1/2 \\ 0 & 1/2 & 9/2 \end{bmatrix}$

c. $\begin{bmatrix} 2 & 3.16 & 0 \\ 3.16 & 4.3 & -1.9 \\ 0 & -1.9 & 3.9 \end{bmatrix}$

d. $\begin{bmatrix} 4 & \sqrt{5} & 0 \\ \sqrt{5} & 8/5 & 11/\sqrt{5} \\ 0 & -11/5 & 12/5 \end{bmatrix}$

CHAPTER 9

1. a. $-0.2698 + 1.2163x$, $0.2344 - 1.944x + 1.8x^2$

b. $1.375 + x$, $1 + x + x^2$

3. The linear least-square approximations are as follows:
   a. $P_1(x) = 1.833333 + 4x$
   b. $P_1(x) = -1.600003 + 3.600003x$
   c. $P_1(x) = 1.140981 - 0.2958375x$
   d. $P_1(x) = 0.1945267 + 3.000001x$
   e. $P_1(x) = 0.6109245 + 0.09167105x$
   f. $P_1(x) = -1.861455 + 1.666667x$
5. a. $y = (0.557)e^{1.06x}$
   b. $y = (0.1)e^{3.0x}$
7. $P_0(x) = 4.5$. $P_1(x) = 7x - 0.003876$
9. $P_1(x) = x - \frac{\sqrt{3}}{9}$, $P_2(x) = \frac{1}{32}(48x^2 - 18x + 1)$
11. $\phi_0(x) = 1$, $\phi_1(x) = x - \frac{1}{2}$, $\phi_2(x) = x^2 - x + \frac{1}{6}$, and $\phi_3(x) = x^3 - \frac{3}{2}x^2 + \frac{3}{5}x - \frac{1}{20}$

    $p_3(x) = 0.27863x^3 + 0.42125x^2 + 1.01830x + 0.99906$
13. a. $(25/8)T_0 + (5/2)T_1 + (3/2)T_2$
    b. $6T_0 - (5/4)T_1 + 3T_2$
15. $y = 0.025x^2 + 4.4786x - 114.285$
17. $a = 18.9414$, $b = -0.0313$, and $c = 0.0014$
19.

x	$B_4(x)$	$y(x)$	$\lvert y(x) - B_4(x) \rvert$
0.2	−0.160969	−0.16	0.000968885
0.4	−0.240172	−0.24	0.000171743
0.6	−0.240062	−0.24	0.0000615773
0.8	−0.160021	−0.16	0.0000213063

CHAPTER 10

1.

i	j	x_i	t_j	$u_{i,j}$	$u(x_i, t_j)$
3	10	0.942477	0.5	0.48648	0.49069
6	10	1.884955	0.5	0.57189	0.57684
9	10	2.827433	0.5	0.18581	0.18742

3. For $h = 0.4$ and $k = 0.05$

i	j	x_i	t_j	$u_{i,j}$	$u(x_i, t_j)$
2	10	0.8	0.5	0	0
3	10	1.2	0.5	0	0
4	10	1.6	0.5	0	0

5.

$i \backslash j$	0	1	2	3	4	5
0	0	4	12	18	16	0
1	0	4	12	18	16	0
2	0	8	10	10	2	0
3	0	6	6	−6	−6	0
4	0	−2	−10	−10	−8	0
5	0	−16	−18	−12	−4	0

The Crank–Nicolson scheme gives the following results:

i	j	x_i	t_j	$u_{i,j}$	$u(x_i, t_j)$
1	1	0.5	0.05	0.628848	0.652037
2	1	1.0	0.05	0.889326	0.883937
3	1	1.5	0.05	0.628848	0.625037
1	2	0.5	0.1	0.559251	0.552493
2	2	1.0	0.1	0.790901	0.781344
3	2	1.5	0.1	0.559252	0.552493

7. The finite difference scheme gives the following results:

i	j	x_i	y_j	$u_{i,j}$	$u(x_i, y_j)$
1	1	0.5	0.5	0.0	0
1	2	0.5	1.0	0.25	0.25
1	3	0.5	1.5	1.0	1

9. a. Jacobi method

Iteration (n)	u_1	u_2	u_3	u_4
1	0.1875	0.1875	0.4375	0.4375
2	0.15625	0.15625	0.40625	0.40625
3	0.14062	0.14062	0.39062	0.39062
4	0.13281	0.13281	0.38281	0.38281
5	0.12891	0.12891	0.37891	0.37891
6	0.12695	0.12695	0.37695	0.37695
7	0.12598	0.12598	0.37598	0.37598

b. Gauss–Seidal method

Iteration (n)	u_1	u_2	u_3	u_4
1	0.25	0.3125	0.5625	0.46875
2	0.21875	0.17187	0.42187	0.39844
3	0.14844	0.13672	0.38672	0.38086
4	0.13086	0.12793	0.37793	0.37646
5	0.12646	0.12573	0.37573	0.37537

c. SOR method

Iteration (n)	u_1	u_2	u_3	u_4
1	0.275	0.35062	0.35062	0.35062
2	0.16534	0.10683	0.38183	0.37432
3	0.11785	0.12181	0.37216	0.37341

11. The finite difference scheme gives the following results:

a. Thirty iterations required:

i	j	x_i	y_j	$u_{i,j}$	$u(x_i, y_j)$
2	2	0.4	0.4	0.1599988	0.16
2	4	0.4	0.8	0.3199988	0.32
4	2	0.8	0.4	0.3199995	0.32
4	4	0.8	0.8	0.6399996	0.64

b. Twenty-nine iterations required:

i	j	x_i	y_j	$u_{i,j}$	$u(x_i, y_j)$
2	1	1.256637	0.3141593	0.2951855	0.2938926
2	3	1.256637	0.9424778	0.1830822	0.1816356
4	1	2.513274	0.3141593	−0.7721948	−0.7694209
4	3	2.513274	0.9424778	−0.4785169	−0.4755283

c. One hundred and twenty-six iterations required:

i	j	x_i	y_j	$u_{i,j}$	$u(x_i, y_j)$
4	3	0.8	0.3	1.2714468	1.2712492
4	7	0.8	0.7	1.7509414	1.7506725
8	3	1.6	0.3	1.6167917	1.6160744
8	7	1.6	0.7	3.0659184	3.0648542

13. a. Gauss–Seidal method

r	$u_{1,1}$	$u_{2,1}$	$u_{1,2}$	$u_{2,2}$
1	−0.00274	−0.01166	−0.01166	−0.04973
2	−0.00857	−0.02555	−0.02555	−0.05667
3	−0.01552	−0.02902	−0.02902	−0.05841
4	−0.01725	−0.02989	−0.02989	−0.05884
5	−0.01769	−0.03011	−0.03011	−0.05895
6	−0.01780	−0.03016	−0.03016	−0.05898
7	−0.01782	−0.03017	−0.03017	−0.05898
8	−0.01783	−0.03018	−0.03018	−0.05898

b. SOR method

r	$u_{1,1}$	$u_{2,1}$	$u_{1,2}$	$u_{2,2}$
1	−0.00302	−0.01299	−0.01299	−0.05538
2	−0.00981	−0.02871	−0.02871	−0.05854
3	−0.01783	−0.03020	−0.03020	−0.05904
4	−0.01785	−0.03020	−0.03020	−0.05899
5	−0.01784	−0.03018	−0.03018	−0.05899
6	−0.01783	−0.03018	−0.03018	−0.05898

15.

$u_{1,1}$	$u_{2,1}$	$u_{3,1}$	$u_{1,2}$	$u_{2,2}$	$u_{3,2}$	$u_{1,3}$	$u_{2,3}$	$u_{3,3}$
48.5938	57.4609	65.1465	36.8359	44.9805	52.8442	24.8340	32.7661	41.4026
48.5742	57.1753	65.0049	37.0972	44.9707	52.8445	24.9658	32.8348	41.4198
48.5681	57.1359	64.9951	37.1262	44.9854	52.8501	24.9902	32.8489	41.4247
48.5655	57.1365	64.9966	37.1353	44.9927	52.8535	24.9960	32.8534	41.4267
48.5679	57.1393	64.9982	37.1392	44.9963	52.8553	24.9981	32.8553	41.4277
48.5696	57.1410	64.9991	37.1410	44.9982	52.8562	24.9991	32.8562	41.4281
48.5705	57.1419	64.9995	37.1419	44.9991	52.8567	24.9995	32.8567	41.4283
48.5710	57.1424	64.9998	37.1424	44.9995	52.8569	24.9998	32.8569	41.4285
48.5712	57.1426	64.9999	37.1426	44.9998	52.8570	24.9999	32.8570	41.4285
48.5713	57.1427	64.9999	37.1427	44.9999	52.8571	24.9999	32.8571	41.4285

17. The Crank–Nicolson scheme gives the following results:

i	j	x_i	t_j	$u_{i,j}$	$u(x_i, t_j)$
1	1	0.5	0.05	0.628848	0.652037
2	1	1.0	0.05	0.889326	0.883937
3	1	1.5	0.05	0.628848	0.625037
1	2	0.5	0.1	0.559251	0.552493
2	2	1.0	0.1	0.790901	0.781344
3	2	1.5	0.1	0.559252	0.552493

19. $u_{11} = 46.1549$, $u_{21} = 84.6195$, $u_{31} = 92.3117$, $u_{41} = 84.6195$, and $u_{51} = 46.1536$

21.

$t\backslash x$	0.2	0.4	0.6	0.8
0.04	0.399	0.646	0.646	0.399
0.08	0.271	0.439	0.439	0.271

23. a.

$j\backslash i$	1	2	3
1	8.25	12.75	12.75
2	7.625	12	12.125
3	7.083	11.292	11.583
4	6.604	10.639	11.104

b.

$j\backslash i$	1	2	3
1	8.3008	12.7113	12.1580
2	7.6858	11.8815	11.0838
3	7.1289	11.0652	10.1989
4	6.6174	10.2893	9.4531

25. $u_1 = u_4 = 11.00$, $u_2 = 21.50$, and $u_3 = 6.50$

27. $u(1,1) = 2.75$, $u(2,1) = 6.75$, $u(1,2) = 3.25$, and $u(2,2) = 8.25$

29.

t\x	0	1	2	3	4	5
0	**0**	**20**	**15**	**10**	**5**	**0**
1	**0**	**7.5**	**15**	**10**	**5**	**0**
2	**0**	−5	2.5	10	5	**0**
3	**0**	−5	−10	−2.5	5	**0**
4	**0**	−5	−10	−1.5	−7.5	**0**
5	**0**	−5	−10	−1.5	−20	**0**

31.

t\x	0	3	6	9	12
0	**0**	**9**	**13.5**	**13.5**	**9**
3	**0**	8.25	12.75	12.75	**9**
6	**0**	7.625	12	12.125	**9**
9	**0**	7.083	11.292	11.583	**9**
12	**0**	6.604	10.639	11.104	**9**

33. Vertical lines bottom to top from left to right values are (0.18, 0.36, 0.53); (0.21, 0.42, 0.62); and (0.25, 0.49, 0.73)
35. Left-right-bottom-top rows: (0.10, 0.13, 0.10; 0.14, 0.17, 0.14; 0.10, 0.13, 0.10)
37. $u_1 = 1.57$, $u_2 = 3.70$, $u_3 = 6.57$, $u_4 = 2.06$, $u_5 = 4.69$, $u_6 = 8.06$, $u_7 = 2.09$, $u_8 = 4.92$, and $u_9 = 9.00$.
39. Exact solution is $u(x,t) = e^{-t}\sin x$; $u_1^1 = 0.659726$ (error = 0.0837) spline solution is $u_1^1 = 0.6415$ (error = 0.0655).
41. The finite difference scheme gives the following results:

i	j	x_i	t_j	$u_{i,j}$	$u(x_i, t_j)$
2	3	0.2	0.3	0.6729902	0.61061587
5	3	0.5	0.3	0	0
8	3	0.8	0.3	−0.6729902	−0.61061587

43. The forward-difference scheme gives the following results:
    a. For $h = 0.4$ and $k = 0.1$:

i	j	x_i	t_j	$u_{i,j}$	$u(x_i, t_j)$
2	5	0.8	0.5	3.035630	0
3	5	1.2	0.5	−3.035630	0
4	5	1.6	0.5	1.876122	0

For $h = 0.4$ and $k = 0.05$:

i	j	x_i	t_j	$u_{i,j}$	$u(x_i, t_j)$
2	10	0.8	0.5	0	0
3	10	1.2	0.5	0	0
4	10	1.6	0.5	0	0

b. For $h = \pi/10$ and $k = 0.05$:

i	j	x_i	t_j	$u_{i,j}$	$u(x_i, t_j)$
3	10	0.94247780	0.5	0.4864832	0.4906936
6	10	1.88495559	0.5	0.5718943	0.5768449
9	10	2.82743339	0.5	0.1858197	0.1874283

45. The wave equation finite difference scheme gives the following results:

i	j	x_i	t_j	$u_{i,j}$	$u(x_i, t_j)$
2	4	0.25	1.0	−0.7071068	−0.7071068
3	4	0.50	1.0	−1.0000000	−1.0000000
4	4	0.75	1.0	−0.7071068	−0.7071068

CHAPTER 11

1. The approximation is $\phi(x) = -0.0771327\phi_1(x) - 0.0744268\phi_2(x)$; The actual values are $y(x_1) = -0.07988545$ and $y(x_2) = -0.07712903$
3. $u(0.0) \approx -4.090452$ and $u(0.0) \approx -1.924623$
5. a. $$I(v) = \int_a^b \left[\left(\frac{dv}{dx}\right)^2 - kv^2 + 2vx^3\right]dx - v(b)$$

   b. $$I(v) = \int_a^b v\left[2g - \frac{d}{dx}\left(x^2\frac{dv}{dx}\right)\right]dx$$

   c. $$I(v) = \int_a^b \left[\left(\frac{dv}{dx}\right)^2 - pv^2 - 2qv\right]dx$$

   d. $$I(v) = \int_0^1 \left[\left(\frac{d^2v}{dx^2}\right)^2 + kv^2 - 2fv\right]dx$$
7. $a_1 = -0.705204$ and $a_2 = -0.279721$
9. $$A^{(e)} = -\frac{1}{3l^{(e)}}\begin{bmatrix} c & -c \\ -c & c \end{bmatrix} - \frac{l^{(e)}}{6}\begin{bmatrix} 2 & 1 \\ 1 & 2 \end{bmatrix}$$

   $c = 3 + (x_{i+1}^2 + x_i x_{i+1} + 3x_i^2)$, $l^{(e)} = x_{i+1} - x_i$

   $$b^{(e)} = -\frac{l^{(e)}}{12}\begin{bmatrix} 6 + x_{i+1}^2 + 2x_i x_{i+1} + x_i^2 \\ 6 + 3x_{i+1}^2 + 2x_i x_{i+1} + x_i^2 \end{bmatrix}, \quad A^{(e)}u^{(e)} - b^{(e)} = 0$$
13. a.

i	x_i	$\phi(x_i)$	$y(x_i)$
3	0.3	−0.212333	−0.21
6	0.6	−0.241333	−0.24
9	0.9	−0.090333	−0.09

b.

i	x_i	$\phi(x_i)$	$y(x_i)$
3	0.3	0.1815138	0.1814273
6	0.6	0.1805502	0.1804753
9	0.9	0.05936468	0.05934303

c.

i	x_i	$\phi(x_i)$	$y(x_i)$
5	0.25	−0.3585989	−0.3585641
10	0.50	−0.5348383	−0.5347803
15	0.75	−0.4510165	−0.4509614

d.

i	x_i	$\phi(x_i)$	$y(x_i)$
5	0.25	−0.1846134	−0.1845204
10	0.50	−0.2737099	−0.2735857
15	0.75	−0.2285169	−0.2284204

15. a. $v(x) = -\frac{10}{23}x(1-x) - \frac{7}{41}x^2(1-x)$

b. $v(x) = \frac{25}{37}x(1-x)$

Bibliography

Allaire, P. W., *Basics of the Finite Element Method.* Brown, Dubuque, IA, 1985.

Atkinson, K. E., *An Introduction to Numerical Analysis.* John Wiley & Sons, New York, 1989.

Atkinson, K. E., Han, W., and Stewart, D., *Numerical Solution of Ordinary Differential Equations.* John Wiley & Sons, Hoboken, NJ, 2009.

Bathe, K. J. and Wilson, E. L., *Numerical Methods in Finite Element Analysis.* Prentice Hall, Englewood Cliffs, NJ, 1976.

Braess, D., *Finite Elements.* Cambridge University Press, Cambridge, 2001.

Brenner, S. and Scott, L. R., *The Mathematical Theory of Finite Element Methods.* Springer, New York, 2002.

Burden, R. L. and Faires, J. D., *Numerical Analysis.* Brooks/Cole Cengage Learning, Boston, MA, 2011.

Chapra, S. C. and Canale, R. P., *Numerical Methods for Engineers.* McGraw-Hill, New York, 2006.

Conte, S. D. and De Boor, C., *Elementary Numerical Analysis: An Algorithmic Approach.* McGraw-Hill, New York, 1972.

Davis, A. J., *The Finite Element Method.* Clarendon Press, Oxford, 1980.

Dukkipati, R. V., *Applied Numerical Methods Using MATLAB*, 1st ed. New-Age International, New Delhi, India, 2013.

Gear, C. W., *Numerical Initial Value Problems in Ordinary Differential Equations.* Prentice Hall, Englewood Cliffs, NJ, 1971.

Gupta, A. and Bose, S. C., *Introduction to Numerical Analysis*, 3rd ed. Academic Publishers, Kolkata, India, 2009.

Hamming, R. W., *Introduction to Applied Numerical Analysis.* McGraw-Hill, New York, 1971.

Hamming, R. W., *Numerical Methods for Scientists and Engineers.* 2nd ed. McGraw-Hill, New York, 1973.

Hildebrand, F. B., *Introduction to Numerical Analysis.* Dover Publications, New York, 1987.

Hoffman, J. D. and Frankel, S., *Numerical Methods for Engineers and Scientists.* CRC Press, Boca Raton, FL, 2001.

Isaacson, E. and Keller, H. B., *An Analysis of Numerical Methods.* Dover Publications, New York, 1994.

Jain, M. K., Iyengar, S. R. K., and Jain, R. K., *Numerical Methods for Scientific and Engineering Computation*, 4th ed. New Age International, New Delhi, India, 2003.

Jain, M. K., *Numerical Solution of Differential Equations*, 3rd ed. New-Age International, New Delhi, India, 2014.

Jain, M. K., *Numerical Solution of Differential Equations.* Wiley Eastern, New Delhi, India, 1979.

Johnson, C., *Numerical Solution of Partial Differential Equations by the Finite Element Method.* Cambridge University Press, Cambridge, 1996.

Keller, H. B., *Numerical Methods for Two-Point Boundary Value Problems.* Dover, New York, 1992.

Knuth, D. E., *The Art of Computer Programming*, vols. 1–3, 3rd ed. Addison-Wesley, Redwood City, CA, 1998.

Kreyszig, E., *Advanced Engineering Mathematics*, 8th ed. John Wiley & Sons, New York, 1999.

Krishnamurthy, E. V. and Sen, S. K., *Numerical Algorithms*, 2nd ed. Affiliated East-West Press, New Delhi, India, 1986.

LeVeque, R. J., *Finite Difference Methods for Ordinary and Partial Differential Equations.* SIAM, Philadelphia, 2007.

Mathews, J. H. and Fink, K. D., *Numerical Methods Using MATLAB*, 4th ed. Pearson Prentice Hall, New Delhi, India, 2004.

Mitchell, A. R. and Griffiths, D. F., *The Finite Difference Method in Partial Differential Equations.* Wiley Interscience, New York, 1980.

Mitchell, A. R. and Wait, R., *The Finite Element Method in Partial Differential Equations.* John Wiley & Sons, London, 1977.

Morton, K. W. and Mayers, D. F., *Numerical Solution of Partial Differential Equations: An Introduction.* Cambridge University Press, 2005.

Niyogi, P., *Numerical Analysis and Algorithms.* Tata McGraw-Hill, New Delhi, India, 2003.

Pinsky, M. A., *Partial Differential Equations and Boundary Value Problems with Applications*, 2nd ed. McGraw-Hill, New York, 1991.

Pizer, S. J., *Numerical Computing and Mathematical Analysis.* Science Research Associates, Chicago, IL, 1975.

Press, W. H., Flannery, B. P., Teudolsky, S. A., and Vetterling, W. T., *Numerical Recipes in C: The Art of Scientific Computing*, 2nd ed. Cambridge University Press, New York, 1992.

Ralston, A., *A First Course in Numerical Analysis.* McGraw-Hill, New York, 1965.

Reddy, J. N., *An Introduction to the Finite Element Method.* McGraw-Hill, Singapore, 1993.

Sastry, S. S., *Introductory Methods of Numerical Analysis*, 5th ed. Prentice Hall, New Delhi, India, 2013.

Scarborough, J. B., *Numerical Mathematical Analysis*, 6th ed. The Johns Hop-kins Press, Baltimore, 1966.

Smith, G. D., *Numerical Solution of Partial Differential Equations*, 3rd ed. Oxford University Press, New York, 1985.

Sneddon, I. N., *Special Functions of Numerical Physics and Chemistry*, 3rd ed. Longman Higher Education, 1980.

Stoer, J. and Burlirsch, R., *Introduction to Numerical Analysis*, 2nd ed. Springer-Verlag, New York, 1993.

Süli, E. and Mayers, D. F., *An Introduction to Numerical Analysis.* Cambridge University Press, Cambridge, 2003.

Veerarajan, T. and Ramachandran, T., *Theory and Problems in Numerical Methods.* Tata McGraw-Hill, New Delhi, India, 2004.

Wheatley, P. O. and Gerald, C. F., *Applied Numerical Analysis.* Pearson Education, London, 2006.

Wolfram, S., *The Mathematica Book*, 5th ed. Wolfram Media, Champaign, IL, 2003.

Index

Note: Page numbers followed by 'f' and 't' denote figures and tables, respectively.

G

H

I

J

K

L

M

N

O

Q

R

S

U

V

W

Z